Deep Drilling Results in the Atlantic Ocean: Ocean Crust

Maurice Ewing Series

1 Island Arcs Deep Sea Trenches and Back-Arc Basins, Manik Talwani and Walter C. Pitman III (editors)

2 Deep Drilling Results in the Atlantic Ocean: Ocean Crust, Manik Talwani, Christopher G. Harrison, and Dennis E. Hayes (editors)

Maurice Ewing Series 2

Deep Drilling Results in the Atlantic Ocean: Ocean Crust

Edited by
Manik Talwani
Christopher G. Harrison
Dennis E. Hayes

American Geophysical Union
Washington, D. C.

Geodynamics Project:
Scientific Report No. 48

Deep Drilling Results in the Atlantic Ocean: Ocean Crust

ISBN: 0-97590-401-7
Library of Congress Catalog Card Number: 79-88753

Printed in the United States of America by
LithoCrafters, Inc.
Chelsea, Michigan

PREFACE

The second Maurice Ewing Symposium was devoted to the implications of deep drilling results in the Atlantic Ocean. This subject was chosen for two reasons. First, Maurice Ewing was one of the leaders of JOIDES (Joint Oceanographic Institutions For Deep Earth Sampling), the association of oceanographic institutions that was formed to organize and sponsor drilling in the deep ocean, and which has continued to provide scientific advice to the Deep Sea Drilling Project. Second, the first phase of International Program of Ocean Drilling in the Atlantic was finished and it seemed a good time to assess the implications of drilling results in the Atlantic that had been obtained over almost a decade.

During the time this volume was being prepared, discussions were taking place about a new initiative in oceanic drilling, in which a drilling vessel with much enhanced capability might be used. The results presented in this volume thus represent the base on which new drilling plans can be built.

The Maurice Ewing series is based on papers presented at the Maurice Ewing symposia. Two volumes resulted from the second symposium, which was held at Arden House, Harriman, New York on March 19-25, 1978. The symposium was co-sponsored by the Lamont-Doherty Geological Observatory, by JOIDES, and by the Inter-Union Commission in Geodynamics. Financial support for the symposium was provided by the G. Unger Vetlesen Foundation, the U. S. Office of Naval Research, and the U. S. National Science Foundation.

Manik Talwani
Christopher G. Harrison
Dennis E. Hayes

CONTENTS

Preface v

Acoustic Stratigraphy and Structure of the Oceanic Crust, *John I. Ewing and Robert Houtz* 1

Tectonic Processes on Slow Spreading Ridges, *A. S. Laughton and R. C. Searle* 15

Constraints from the Famous Area Concerning the Structure of the Oceanic Section, *Tanya Atwater* 33

The Iceland Crust: Evidence from Drillhole Data on Structure and Processes, *G. Pálmason, S. Arnórsson, I. B. Fridleifsson, H. Kristmannsdóttir, K. Saemundsson, V. Stefánsson, B. Steingrímsson, J. Tómasson and L. Kristjánsson* 43

Geological and Geophysical Investigation of the Midcayman Rise Spreading Center: Initial Results and Observations, *R. Ballard, W. Bryan, H. Dick, K. O. Emery, G. Thompson, E. Uchupi, K. E. Davis, J. Boer, S. DeLong, P. Fox, F. Malcolm, R. Spydell, J. Stroup, W. Melson, and R. Wright* 66

Seismic Velocities, Densities, Electrical Resistivities, Porosities and Thermal Conductivities of Core Samples from Boreholes into the Islands of Bermuda and the Azores, *R. D. Hyndman, N. I. Christensen and M. J. Drury* 94

The Physical State of the Upper Levels of Cretaceous Oceanic Crust from the Results of Logging, Laboratory Studies and the Oblique Seismic Experiment at DSDP Sites 417 and 418, *M. H. Salisbury, R. Stephen, N. I. Christensen, J. Francheteau, Y. Hamano, M. Hobart and D. Johnson* 113

Geomagnetic Reversals and Ocean Crust Magnetization, *William Lowrie* 135

Modeling the Oceanic Magnetic Source Layer, *Hans Schouten and Charles R. Denham* 151

On the Likelihood of Mixed Polarity in Oceanic Basement Drill Cores, *Charles R. Denham and Hans Schouten* 160

A Model for the Structural State of the Upper Half Kilometer of North Atlantic Ocean Layer 1, *James M. Hall* 166

Low Temperature Alteration of the Magnetic Minerals in Ocean Floor Basalts, *N. Petersen, P. Eisenach and U. Bleil* 169

Magnetism of the Mid-Atlantic Crest Near 37°N from Famous and DSDP Results: A Review, *M. Prévot, A. Lecaille and R. Hekinian* 210

Metamorphism in the Ocean Crust, *J. R. Cann* 230

Tectonic and Igneous Emplacement of Crust in Oceanic Transform Zones, *E. Bonatti, A. Chermak and J. Honnorez* 239

Basaltic Glass Erupted Along the Mid-Atlantic Ridge Between 0–37°N: Relationships Between Composition and Latitude, *William G. Melson and Timothy O'Hearn* 249

The Evolution of Ocean-Floor Basaltic Magmas, *J. M. Rhodes and M. A. Dungan* 262

Petrologic Character of the Atlantic Crust from DSDP and IPOD Drill Sites, *W. Bryan, G. Thompson and F. Frey* 273

Nature of Mantle Heterogeneity in the North Atlantic: Evidence from Leg 49 Basalts, *J. Tarney, D. A. Wood, J. Varet, A. D. Saunders and J. R. Cann* 285

The Primary Melt of the Oceanic Tholeiite and the Upper Mantle Composition, *L. V. Dmitriev, A. V. Sobolev and N. M. Suschevskaja* 302

Evolution of the 'Famous' Ocean Ridge Segment: Evidenee from Submarine and Deep Sea Drilling Investigations, *Martin Flower and Paul T. Robinson* 314

Controls on the Major and Minor Element Chemistry of Mid-Ocean Ridge Basalts and Glasses, *A. E. Bence, D. M. Baylis, J. F. Bender and T. L. Grove* 331

Isotope Geochemical Studies of North Atlantic Ocean Basalts and Their Implications for Mantle Evolution, *R. K. O'Nions, N. M. Evensen, S. R. Carter and P. J. Hamilton* 342

Alteration, Fractional Crystallization, Partial Melting, Mantle Properties from Trace Elements in Basalts Recovered in the North Atlantic, *H. Bougault, J. Joron and M. Treuil* 352

Very-Low-Temperature Hydrothermal Alteration of the Oceanic Crust and the Problem of Fluxes of Potassium and Magnesium, *T. W. Donnelly, G. Thompson and P. T. Robinson* 369

Ridge Crest-Hydrothermal Metamorphism at the Galapagos Spreading Center and Reverse Weathering, *J. Edmond, J. B. Corliss and L. I. Gordon* 383

Some Implications of Heat/Mass Ratios in Galapagos Rift Hydrothermal Fluids for Models of Sea Water-Rock Interaction and the Formation of Oceanic Crust, *J. Corliss, L. I. Gordon and J. M. Edmond* 391

Chemistry of Ridge-Crest Sediments from the North Atlantic Ocean, *M. R. Scott, P. F. Salter and L. A. Barnard* 403

Ophiolitic Rocks and Evidence for Hydrothermal Convection of Sea Water Within Crust, *Edward T. C. Spooner* 429

ACOUSTIC STRATIGRAPHY AND STRUCTURE OF THE OCEANIC CRUST

John Ewing

Woods Hole Oceanographic Institution, Woods Hole, Massachusetts 02543

Robert Houtz

Lamont-Doherty Geological Observatory of Columbia University, Palisades, New York 10964

Abstract. With the increase in resolution and the number of measurements, the early seismic model of the igneous crust composed of a 5 km/sec Layer 2 and a 6.8 km/sec Layer 3 evolved into one with at least two subdivisions for each layer. Recent work, in which arrival amplitudes have been given higher diagnostics priority than travel times in the data analysis, has suggested a further modification of the early model in which velocity gradients, with no distinct discontinuities, replace the layered model. Although the gradient model appears to be generally acceptable, there remains much to learn about the degree, scale and pattern of lateral variations in the velocity structure. Both seismic and drilling results indicate a substantial amount of inhomogeneity, particularly in the upper crust.

There is substantial evidence that several features of crustal structure are age-dependent, including:

(1) Topographic roughness of the upper crust diminishes with increasing age.

(2) Low velocity zones, possibly corresponding to magma chambers, are apparently confined to young lithosphere.

(3) The thickness of the low-velocity upper crust (Layer 2A) decreases and/or its wave velocity increases with increasing age. This effect may be due to cementation and lithification of rubble zones and to filling of fractures by mineralization, possibly associated with hydrothermal circulation in the crust.

(4) Total crustal thickness increases with increasing age.

(5) Because of greater sediment accumulation on older crust, and associated higher wave velocity in the basal sediments, the impedance discontinuity at the sediment-basement interface tends to decrease with increasing crustal age; although the effect may be partially or totally counterbalanced by higher velocities in the upper part of the older crust.

(6) Reflections from the crust-mantle transition have been observed often in airgun-sonobuoy profiles and in multi-channel reflection profiles in regions of old crust; much less frequently in young regions. It is not certain whether this is also an age-dependent effect or whether it is related to the difference in crustal topography and interference from diffracted arrivals.

Introduction

The model, or models, of the velocity-depth structure and the inferred composition of the ocean crust have had high priority for testing by the crustal drilling program of JOIDES-IPOD. Other papers in this volume deal specifically with relating the drilled samples to the seismic and to the petrologic and geochemical models. This paper will focus on the development of seismic methods and on the precision of the results.

Our efforts to understand the oceanic crust through seismological studies have been going on for almost thirty years. The first crustal refraction measurement made in the deep sea in 1949 by M. Ewing and others (1950) was initially interpreted as indicating a P wave velocity of 7.58 km/sec directly underlying the sedimentary layer. These data were later re-interpreted by Ewing et al (1954) to give 7.05 km/sec overlying a 7.80 km/sec mantle, a model in closer agreement with measurements made later in 1949 and in the early 1950's, eg., Hersey, et al, 1952; Officer, et al., 1952, Raitt, 1956, Gaskell, Hill, and Swallow, 1958. The main feature of the early model was a crust-mantle discontinuity at a depth of 10-12 km, with the oceanic crust being mainly represented by a wave velocity of 6-7 km/sec and the mantle, like that under continents, represented by a velocity in the vicinity of 8 km/sec. Even though this was a crude model, it had great significance because it demonstrated a fundamental 20 to 30 km difference in crustal thickness between continents and oceans.

Following those earliest results we focused

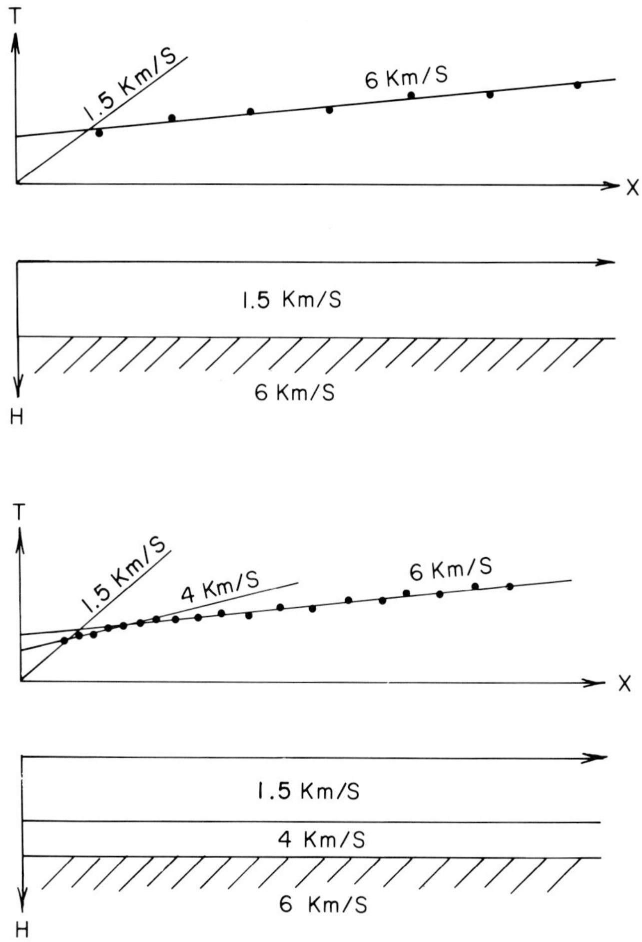

Fig. 1. Upper part shows the refraction data and the derived crustal section based on widely spaced shots. In the lower part, shot spacing has been decreased to about one third, and although the original points are unchanged, the additional data reveals the existence of a 4 km/sec layer.

considerable attention on such question as:

1. Do all oceans have the same structure underneath them?
2. Is the structure under mid-ocean ridges the same as under the basins?
3. What is the structure under aseismic ridges and submarine plateaus?
4. What is the structure under various types of ocean-continent boundaries and in marginal interior basins?

In looking for the answers to these questions, we naturally came up against the need to refine our seismic models, and the refinement has been going on steadily. The refinements have come about by improvements in hardware and techniques, by larger numbers of people working on the problem, and has certainly benefitted from the increase of overall knowledge about the earth, through guidance on where and what to investigate.

Models With a Layered Crust

The first efforts in refinement were largely directed toward increasing the number of measurements, i.e., recording a larger number of shots per kilometer. A simple demonstration of this refinement is shown in Figures 1 and 2. The upper refraction time-distance graph of Figure 1 would probably be interpreted conservatively as indicated, producing the velocity vs. depth model shown underneath. With the larger number of observation points shown in the lower time-distance graph, it is easy to see that an additional layer should be included in the velocity-depth model. The advantage of additional observation points in reflection seismology is obvious. Thus, increasing data density was an important improvement to make. As resources and data handling procedures permitted, we progressed from data points miles apart to data redundancy through the development of single-channel reflection profiling, airgun-sonobuoy

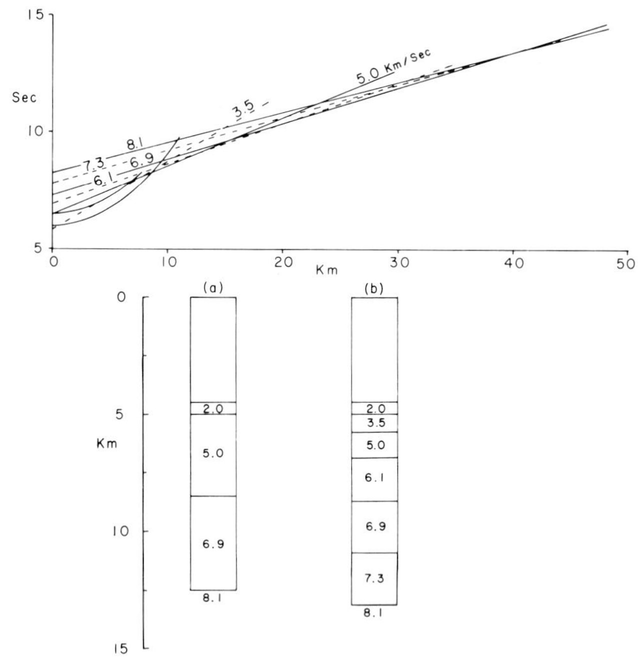

Fig. 2. Solid lines in travel-time graph represent a typical two layer crustal model. Its solution appears under (a). With increased data density additional layers 2A, 2C, and 3B (shown with dashed lines) can be recognized. The solution obtained by incorporating the extra layers is shown in (b). Note that the intercepts of the 5, 6, 7, and 8 km/sec lines remain unchanged in both cases, but that the section thickens as a result of introducing additional layers.

refraction profiling, and multi-channel reflection and expanding spread refraction measurements. These refinements required a substantial amount of time, effort, and hardware.

Significant results of increasing data density were that we learned much more about the configuration of the top of the crust, largely through reflection profiling, and we were able to identify more velocity zones, or layers, in the refraction profiles. The first subdivision in refraction work was to divide the crust into two layers -- layer 2 about 1.5 km thick with an average velocity of about 5 km/sec and Layer 3, about 5 km thick, with an average wave velocity of about 6.7 km/sec (Gaskell, Hill, and Swallow, 1958; Hill, 1952; Raitt, 1956; Ewing and Ewing, 1959). This model seemed to suffice for a number of years, and appeared to be valid for most of the world's ocean basins. It has been published and referred to many times and is shown once more in Figure 2a.

Underlying the sediments is the upper part of the igneous crust (Layer 2) which has been called variously, basement, lithified sediment, basalt layer, serpentinized peridotite, and volcanic layer. This is underlain by Layer 3 which was generally called the gabbroic layer by those petrologically oriented, and the oceanic layer by seismologists. It seemed to deserve a special name among seismologists, because it was almost always observed. This is because Layer 3 refracted arrivals are first events on seismograms over a long range of shot-to-detector separation. With close enough shot spacing, layer 2 was easy to observe, except in deep basins with a thick sediment cover.

The subdivision of the principal layers was most easily accomplished in dealing with Layer 1, the sediments. Thanks to the ease of observation provided by the seismic profiler (Ewing and Tirey, 1961; Hersey, 1963), we could see early-on that the sediment layer is not just one uniform blanket. In some areas it is acoustically homogeneous; in others it is strongly layered. Its thickness varies enormously. Although there has been some question about how many of the apparent interfaces observed in the seismic reflection profiles may be caused by multiple interference, there is no doubt that a large number of them are bona fide geologic features, many of which have been identified by the Deep Sea Drilling Project.

To a lesser degree, high resolution data have permitted the subdivision of layers 2 and 3. Through the use of seismic reflection profiling, coupled with airgun-sonobuoy reflection and refraction profiles, Houtz and Ewing (1976), have proposed that wave velocities in Layer 2 vary both laterally and vertically between about 3.5 and 6 km/sec, higher values corresponding both to increasing depth and to increasing age. With similar data and similar analytical procedures, Sutton et al., (1971) have subdivided Layer 3 into an upper 6.0-6.7 km/sec zone and a lower 7.0-7.6 km/sec zone.

A comparison of the earlier three layer model with a model incorporating these subdivisions is shown in Figure 2b. Notice that as we make more subdivisions, the crust has become thicker even though the intercept times in the original interpretation are unchanged. It is clear that if the subdividing is continued much further, there will be so many layers that a continuous velocity gradient becomes a better approximation to geological reality.

In order to provide a more generalized approach to time-distance analysis of refraction profiles, many experimenters have in recent years used the tau-p approach of Bessonova, et al (1974). This method provides a systematic means of establishing allowable bounding values of velocity vs. depth and has been used to show that marine refraction data sets may be interpreted as convincingly with velocity gradients as with stacks of constant velocity layers, eg. Kennett and Orcutt (1976).

Models with Velocity Gradients

The discussion up to this point has been concerned with deriving models by using only travel time vs. distance for the various arrivals on the seismograms, without considering the amplitudes of the arrivals or their wave forms. The classical analytical procedures are based on the assumption that the observed refracted arrivals are pure head waves, or interface waves, and on the following assumptions about the model:

1. Velocities are higher in successively deeper layers.
2. Velocity is constant within each layer.
3. Layers are thick compared with seismic wave length.
4. The layers are separated by plane boundaries.

Those who developed the layered models recognized that at least some of these assumptions were unlikely to be realized in the real world, particularly when faced with the low energy levels that head wave theory predicted for critically refracted arrivals. However, it was considered that the deduced models would still be good approximations of the real situation and would still be useful in comparing one region with another.

There is no question that a derived model, and its usefulness in comparing one region with another, will be greatly improved if its developed from arrival time vs. distance relationships plus the information contained in arrival amplitudes and wave forms. In essence, this additional information is utilized in an iterative process where a starting model is first developed primarily from time-distance relationships, then it is tested further by computing synthetic seismograms and comparing the synthetics with the observed data. The model can then be modified in a way that seems likely to reduce disagreement between observed and synthetics;

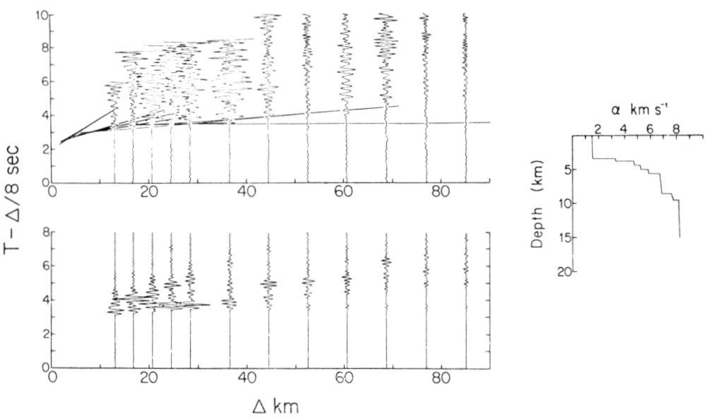

Fig. 3. East Pacific Rise OBS profile on 5 m.y. crust. Observed seismograms, velocity vs. depth model at the right and the synthetic seismograms computed for the model below (From Orcutt, et al., 1976).

new synthetics are computed, and so on, until satisfactory agreement is achieved. The techniques developed for this more complete type of analysis in marine seismology have largely been based on the work of Cagniard, 1939; de Hoop, 1960; Helmberger, 1968; Helmberger and Morris, 1969, 1970; Cerveny and Ravindra, 1971; Fuchs and Muller, 1971; and Kennett and Orcutt, 1976.

In effect, the model to be tested, which may contain gradient zones as well as discontinuities, is approximated by a stack of layers. The reflection response of the stack is computed for an appropriate cone of rays. The response can include all primary and multiple reflection paths and contributions involving mode transformations. These contributions are summed to produce the theoretical response of the model. The model response for an appropriate range is then compared with the actual seismogram recorded at that range. It is also necessary to have taken into account the source characteristics, as well as those of the receiving apparatus, and to account for geometrical spreading losses and attenuation. An excellent review of recent ideas on inversion of seismic travel time data and on testing with models and synthetic seismograms may be found in Kennett (1977).

It is not difficult to imagine that, particularly for models that need many layers in the approximation, this is an expensive procedure, requiring many computations. However, the results seem to justify the effort as demonstrated by the examples in Figures 3, 4, and 5 which are taken from a recent publication by Orcutt, Kennett and Dorman (1976). Note that only the early parts of the arrival trains are shown in the synthetics; the later arriving energy is more complex and more difficult to manage. The refraction arrival amplitudes are related mainly to the steepness of velocity gradient at the appropriate depth for the phase velocities of the arrivals. Reflection amplitudes are primarily a function of the angle of incidence and of the sharpness of impedance contrast between layers. In Figure 3, for example, the refracted arrivals beyond 40 km range were extremely weak, which is an indication of the lack of a positive gradient in the upper mantle. Figure 4 shows slightly stronger mantle arrivals: consequently, a small amount of positive gradient is interpreted to exist. Gradients in the crust account for large amplitudes at ranges of less than 40 km. In both models, substantial discontinuities in the crust-mantle transition zone produce relatively strong moho reflections.

The most dramatic feature in Figure 5 is the offset in travel times and energy partitioning near 20 km range, interpreted as a low velocity zone at about 5 km depth, which Orcutt, Kennett, and Dorman suggest may be indicative of a partial melt region or magma chamber.

Although the comparison between the observed and synthetic seismograms is not exact, the use of new analytical techniques such as these is a refreshing development in the attempt to derive better crustal models. The work is being actively pursued by many research groups, and a substantial number of measurements have been reported, eg., Fowler, 1976; Helmberger, 1977; Lewis and Snydsman, 1977; Reid et al., 1977; Stephen et al., in press. The use of a tight fixed receiving array to better determine phase velocities as a function of range, hence additional opportunity to discriminate between layered vs. gradient models, is reported in Lewis and Snydsman (1977).

Evidence from Reflection Seismology

Perhaps the single most significant result of the recent work has been a consistent indication that Layer 2 is a zone of strong positive velocity gradient and that there is usually no prominent discontinuity between layer 2 and layer 3.

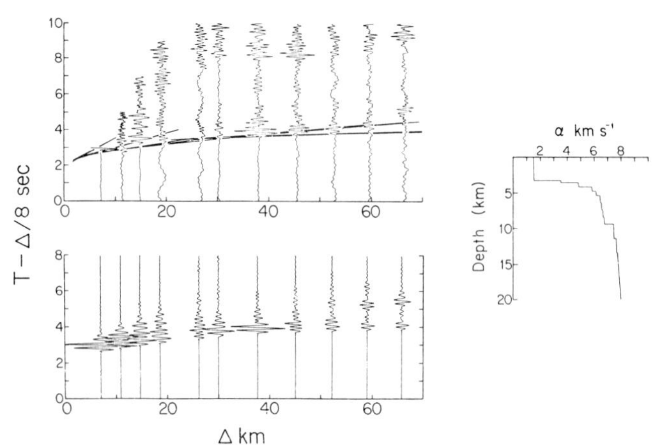

Fig. 4. East Pacific Rise OBS profile on 2.9 m.y. old crust (From Orcutt, et al., 1976).

A strong gradient in the upper crust, changing to a weaker gradient in the lower crust seems to be an appropriate model. The lack of any consistent velocity discontinuity in the crust is in agreement with observations in reflection seismology as well. It should be noted that both the reflection and refraction observations have been made with wave lengths of the order of 1 km. and do not definitely rule out the possibility of discontinuities on a finer scale.

We have had considerable success in observing reflections from the crust-mantle boundary, even at normal incidence, but reflections from within the crust are rare. This is demonstrated in Figure 6 which is from an IPOD-USGS multichannel line southeast of Cape Hatteras reported by Grow and Markl (1977). Mantle reflections have also been observed clearly in CDP profiles over wide areas of the western Pacific basin (Buhl, et al., 1978), and apparently also under the crest of the East Pacific Rise (Paul Stoffa, personal communication). Thus it seems that although velocity vs. depth may vary smoothly in the crust, the crust-mantle transition is in many places a first order discontinuity. In some regions, OBS and surface hydrophone data require a relatively gradual transition at the Moho, (e.g. Helmberger, 1977).

We have been observing for some time what we take to be critical, or near-critical angle reflections from the crust-mantle transition in airgun-sonobuoy profiles. An example is shown in Figure 7. In this profile we not only see very strong arrivals near the critical range, marked CR, but we also observe sub-critical arrivals almost at normal incidence. The subcritical reflections are particularly useful because they permit us to calculate interval velocities and also, are diagnostic of the presence of a first order discontinuity. Houtz (1977) has utilized interval velocity data from a number of such profiles along with measurements of the critical range, to make a case for the presence of a high velocity zone, 7-7.5 km/s in the lower crust.

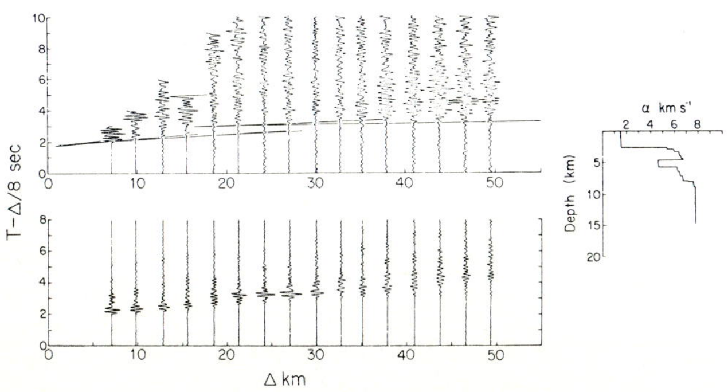

Fig. 5. East Pacific Rise OBS profile at the ridge crest (From Orcutt et al., 1976).

Lateral Variability of Crustal Structures

Given that we have made significant advances in our ability to record seismic data, to invert it, and to refine the model by the use of synthetic seismograms, the question arises about how representative is any model for a substantial portion of a lithospheric plate. What is the degree and scale of lateral inhomogeneity? How does crustal structure vary with age?

The answer to the first question is difficult to find in seismic data alone, because not enough closely-spaced measurements have been made. In the few places where a substantial number of measurements have been made within a few tens of kilometers of each other, the data have been analyzed only from time-distance relationships, but the velocity-depth values determined from slope-intercept analysis vary locally almost as much as they do on a regional scale. For example, groups of 10-20 sonobuoy profiles, all within a 1° square (Naini and Ewing, unpublished) have produced velocity histograms not significantly different from histograms of entire ocean basins. These results, whether interpreted in terms of layers or gradients, suggest lateral variations on the scale of a few kilometers to a few tens of kilometers.

Deep sea drilling results have also given evidence of lateral heterogeneity in the upper crust, even on the scale of hundreds of meters, from closely-spaced drilling sites. Observations from deep submersible investigations on accreting plate margins have suggested models of volcanic and tectonic processes that would be consistent with significant lateral variations in upper crustal structure on the scale of a few kilometers, eg. Ballard and van Andel (1977).

Ocean bottom seismometer (OBS) and other seismic studies of the crustal region of the East Pacific Rise and Mid Atlantic Ridge have not as yet produced a consistent picture of the structure of those features. Some refraction measurements on the axis of the East Pacific Rise have indicated the presence of low velocity zones (magma chambers?) within the crust, eg. Orcutt et al., (1976). Herron et al., (1978) appear to have recorded reflections from the zone at one location. Figure 8 shows a record section from Herron et al., with reflector R4 at the appropriate time (depth) to correspond to the top of the low velocity. Other refraction profiles have not produced evidence for low velocity zones, however, indicating a variability in structure and processes along the ridge axis. Fowler (1976) found no evidence for a low velocity zone in the FAMOUS area of the Mid Atlantic Ridge. Analysis of data on both ridges has given an inconsistent picture with respect to deeper structure. In some cases, strong amplitude arrivals corresponding to phase velocities <8 km/sec suggest appreciable velocity gradients in lower crustal material, while in others the amplitude data suggest strong gradients in velocity >8 km/sec.

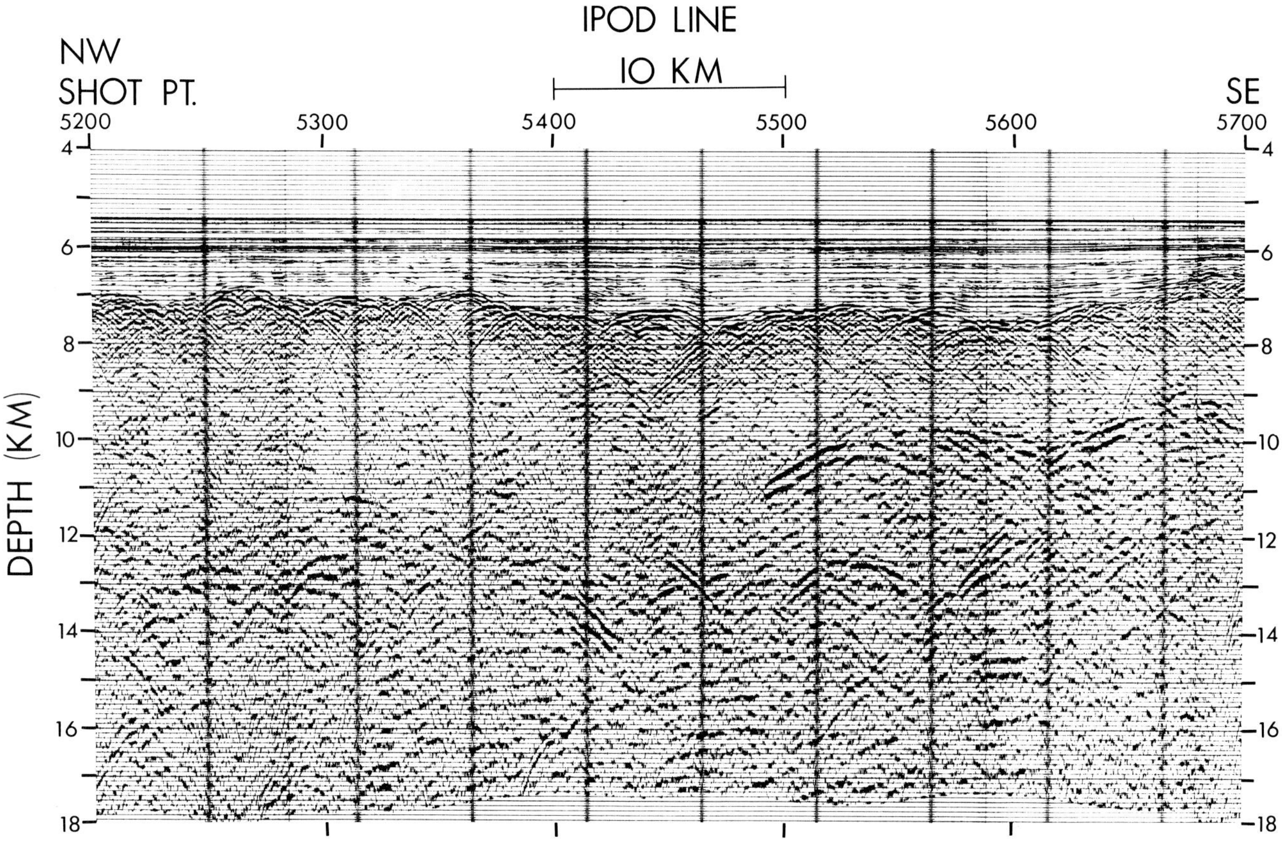

Fig. 6. IPOD multichannel data (from Grow and Markl, 1977) showing reasonably consistent moho reflections in the vicinity of 12 km. depth along a line from Cape Hatteras to the crest of the Mid Atlantic Ridge. The moho reflections were observed only in the older part of the basin; intracrustal reflections at the right are rarely observed.

How much of the apparent variability in axial zone structure remains as the material ages and is moved out into the basins is an important question, and the answer will require a considerable amount of further work. The topography of the crust, which is an important manifestation of lateral variability, appears to be related to age, particularly in the Atlantic, as shown in Figure 9. The upper section is a seismic reflection traverse across the Mid Atlantic ridge. The lower section is partly over the western Bermuda rise and partly over the Hatteras abyssal plain. Each section is about 400 km long and each has approximately 25:1 vertical exaggeration. Some of the apparent topographic smoothing of the old crust can be explained by high velocities in the basal sediments, but the young crust is significantly rougher . A comparison of young and old Pacific crust is shown in Figure 10; the contrast between young and old is much less obvious (vertical exaggeration 25:1). The two types of mid-ocean ridge topography (Pacific vs. Atlantic) probably accounts, at least in part, for the fact that Atlantic seismic data are more scattered than Pacific data. The extremely rough topography of the Mid-Atlantic ridge is usually attributed to the slow spreading rate, resulting in a more brittle (cooler) accreting margin that is more likely to fracture. It is difficult, with the existing data base to make a convincing case for a different seismic structure between the rough and smooth parts. This difficulty comes about partly because seismic measurements made from drifting ships or sonobuoys, particularly in rough terrain, are subject to considerable inaccuracy, and most of the data base were acquired in that way. We simply do not yet have a sufficient amount of tightly-controlled data to make a definitive comparison. To stress this point, it is important to emphasize that the crustal topography, which can be recognized rather easily by the seismic profiler, was not known in detail for many of the earlier refraction profiles, particularly drifting profiles; and even when it is measured, there is no assurance that the topography is mirrored in interfaces or transition zones within the crust.

In an effort to average out the effects of undetected topography and of other lateral vari-

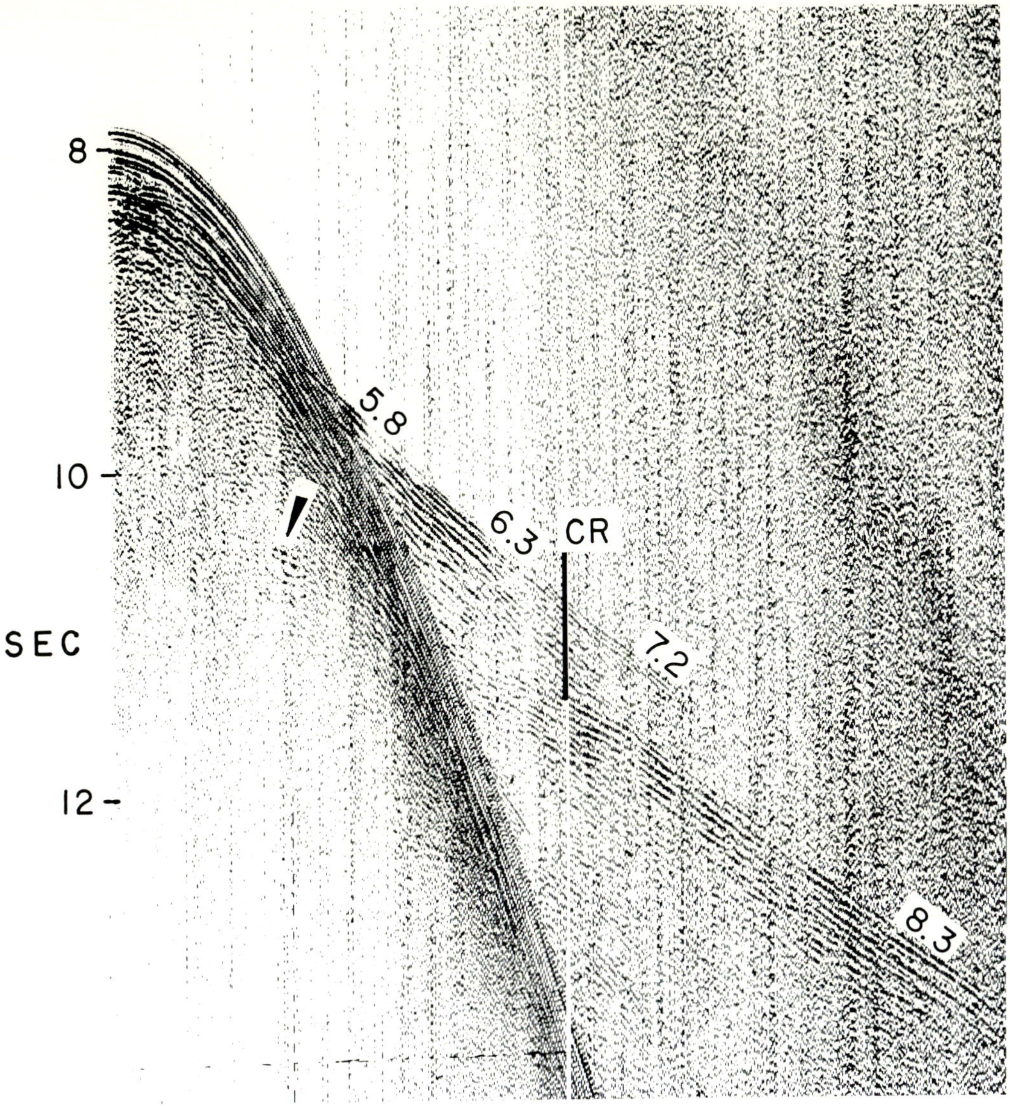

Fig. 7. Airgun-sonobuoy record from the west Pacific. Sub-critical mantle reflections are marked with arrows. The approximate critical reflection point on mantle is marked CR (From Houtz, 1977).

ations, we have systematically examined a large number of airgun-sonobuoy profiles and have arrived at a statistical interpretation that seems to show a definite aging effect in the upper crust (Houtz and Ewing, 1976). The results from that work are summarized in Figures 11, 12, and 13. In Figure 11 upper crustal velocities from all oceans are plotted as a function of crustal age. There were a number of profiles, mainly in young crust, in which refracted arrivals from the uppermost material were not observed, even though the deeper crustal arrivals were observed. Figure 11 does not include such profiles; only recorded values for the top of the crust are plotted. The large amount of scatter in the data is obvious, and it is noteworthy that the values vary about as much in the old, smoother crust as in the young, rough crust. Nevertheless, the mean velocity in the uppermost crust is significantly lower (3-4 km/sec) in young crust. The measured velocities increase gradually with age and stabilize at about 5 km/sec.

We have used only Pacific data in Figure 12, and have divided the crust into layers on the basis of seismic velocity. Velocities below about 4.5 km/sec are put in layer 2A, those between 4.5 and 5.5 are put in 2B and those up to 6 km/sec are called 2C. We have not simply pigeon-holed seismic velocities and put them into layers; we have also considered the observed superposition of layers. The results of these studies have suggested that the low 2A velocities are produced by a badly fractured mixture of pillows, flows, rubble, and sediment.

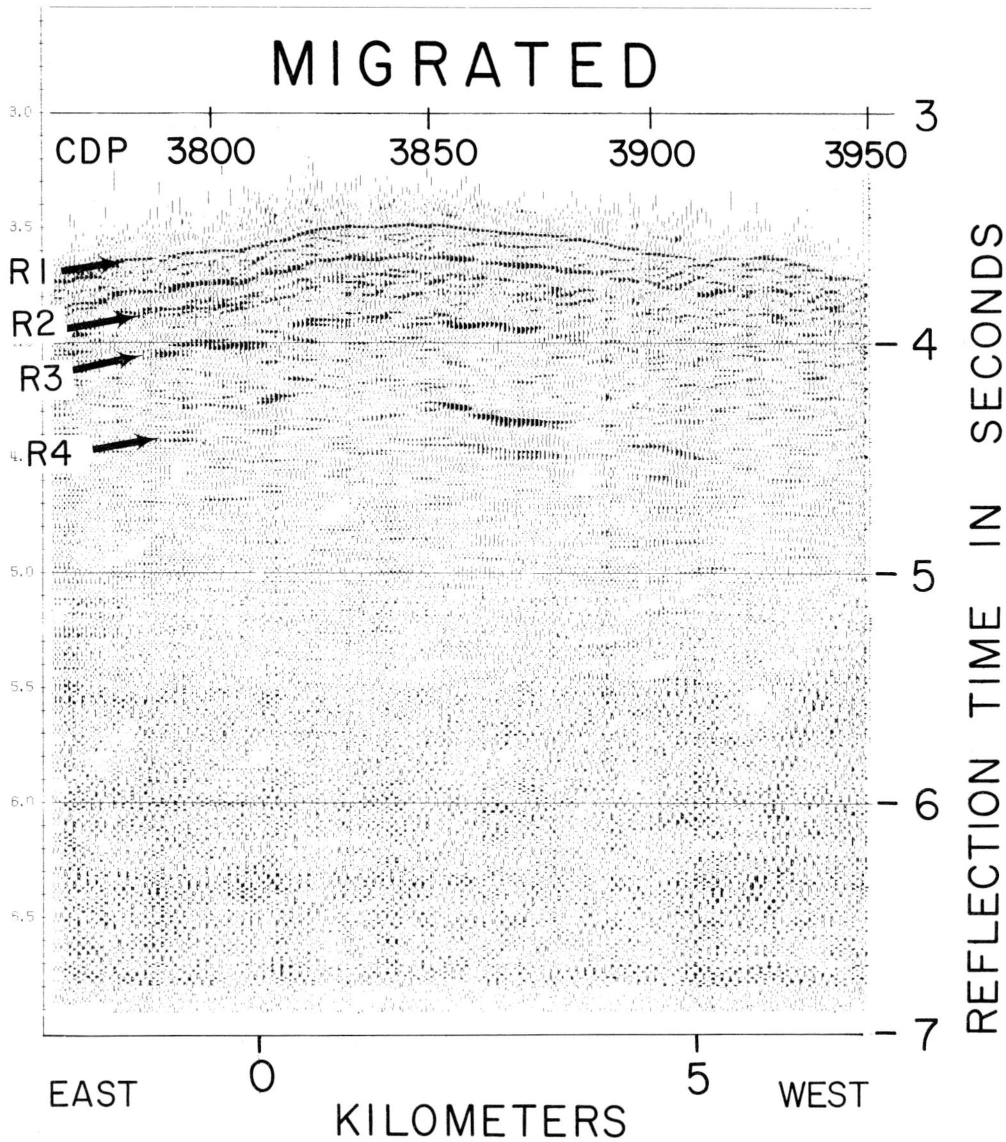

Fig. 8. Migrated section after stacking and deconvolving. This section from the East Pacific Rise shows coherent reflection events within the upper crustal layers. R4 is interpreted as the lid of a low velocity zone (From Herron, et al., 1978).

As this material ages, the cracks and pores are gradually filled by mineralization, possibly associated with hydrothermal circulation, until the whole mixture becomes a competent rock with a wave velocity close to that of its major component, namely basalt. We do not hold strongly to the layered aspect of this model. Instead of pinching out layer 2A on the ridge flank, we would prefer to consider this as analogous to a facies change in seismic velocity, where 3-4 km/sec velocities grade laterally into higher values. Also our data do not require an interface between 2B and 2C and layer 3, as we have derived the models using only arrival time vs. distance. A continuous velocity vs. depth section is not inconsistent with our data. However, we would say that in young crust the velocity vs. depth relationship has different bounding values than it does in old crust.

Figure 13 is a similar demonstration for the North Atlantic. It is dominated by a large amount of layer 2A material in the crestal region, with poor data on the deeper parts of layer 2 and none on layer 3. We consider our data in the older Atlantic to be good but are less satisfied with that in the young part, where the large thickness of layer 2A is indicated. The crestal regions of the Mid Atlantic ridge are

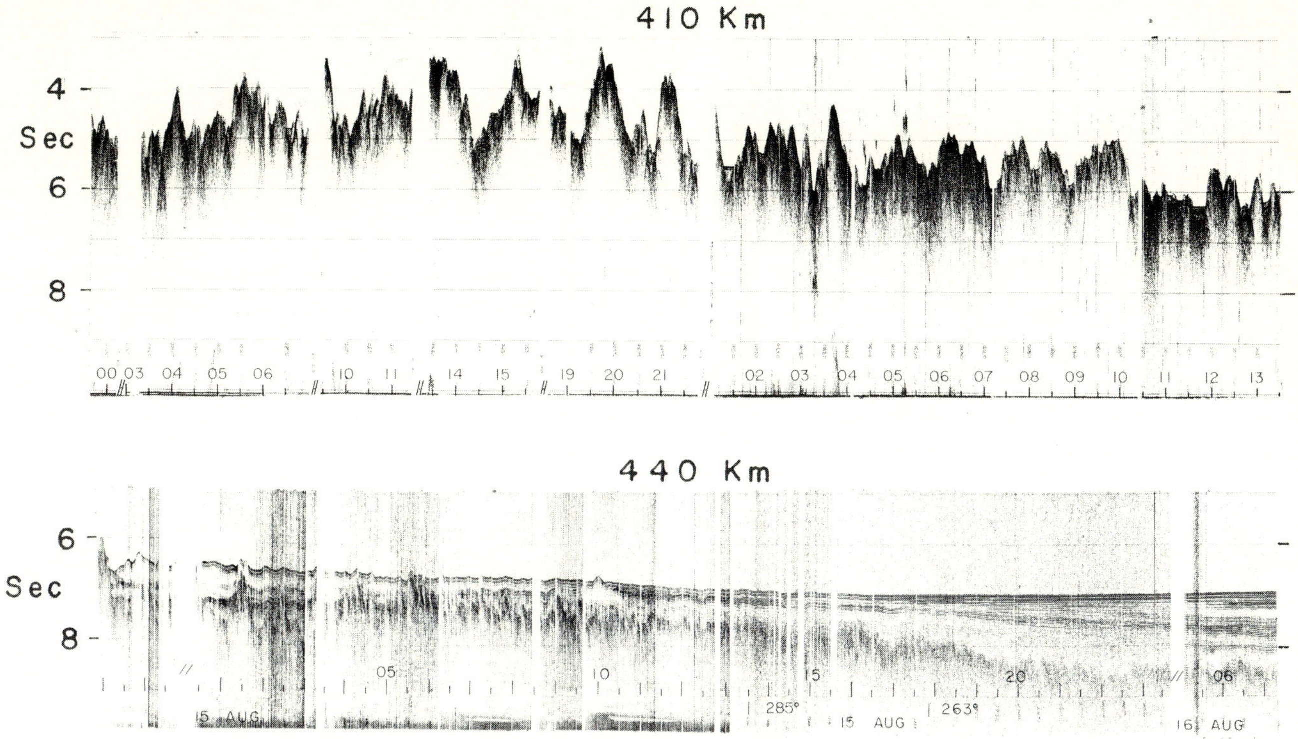

Fig. 9. Representative sections of young Atlantic crust (upper) and old Atlantic crust (lower) suggesting a definite age-dependent change in basement topography.

not well suited for good sonobuoy measurements. It is difficult to find a suitable topographic setting for recording a profile, and the upper crust appears to attenuate seismic energy much more than it does elsewhere. Nevertheless, a similar thickness of material with 2A velocity is measured on the Reykjanes ridge (Talwani et al., 1971), where the topography is not excessively rough.

Refraction data indicating a thickening of the oceanic layer 3 with increasing age were presented by Shor, et al., (1970) and more recently by Lewis and Snydsman (1977). These data are complicated in many instances by the difficulty of identifying mantle arrivals in the crestal regions of the spreading ridges. In these regions, particularly, the crustal structure is extremely difficult to fit into a simple layered model.

Summary

The crustal models, both layered and nonlayered have not yet been thoroughly tested by drilling because of insufficiently deep penetration. The results, from the holes on legs 37, 45 and 46 all in young crust, appear to confirm that the low velocity of layer 2A is primarily due to the physical state of the basalt, a substantial percentage of the whole rock being composed of rubble and fracture zones. (Aumento, F., Melson, W.G. et al., 1977; Rabinowitz, P.A., Melson, W.G. et al., 1976; Dmitriev, L.V., Heirtzler, J.R. et al., 1976) These sites also demonstrate great lateral variability. At Sites 417-418 in crust of Mesozoic age, the drilling results indicate that the upper few hundred meters initially had much the same structure as the crust cored near the ridge axis, but that mineralization had largely sealed up the fractures and voids. (Donnelly, Francheteau, et al., 1977; Bryan, Robinson, et al., 1977; Flower, Salisbury, et al., 1977). Downhole sonic log measurements to depth of about 500 m into basement have given values that are in general agreement with our sonobuoy measurements of 2A velocities near the ridge axis and 2B velocities in old crust (Salisbury, this symposium volume).

A further comparison of marine seismic velocities with sonic velocities measured on laboratory samples from the Bay of Islands ophiolite suite is made possible through the recent work of Salisbury and Christensen (1978). Figure 14 shows their sonic velocity measurements along with the layered crustal model derived from sonobuoy refraction data. Allowing that 2A and possibly most of 2B are apparently missing from the ophiolite suite, the agreement is quite satisfactory. It is clear also that a gradient model would offer good agreement. The sharpness of the transition between the gabbroic and ultramafic rocks is compatible with seismic reflection

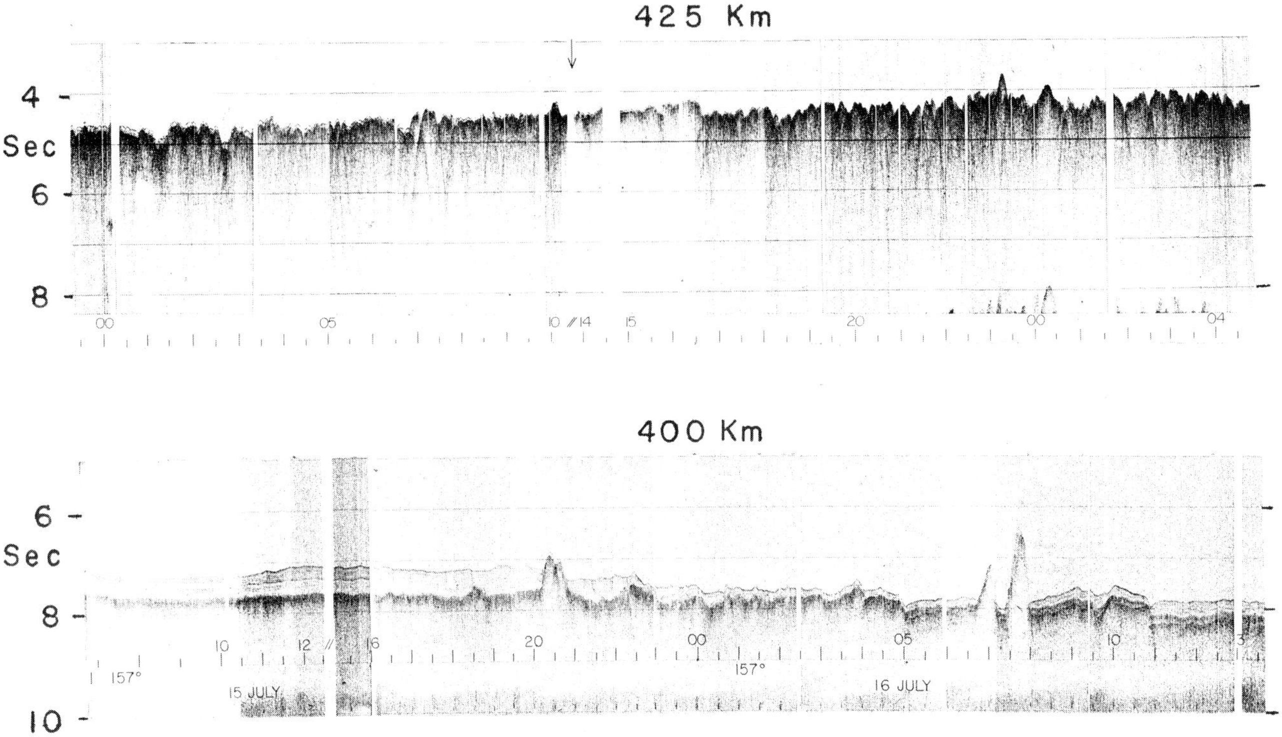

Fig. 10. Representative sections of young Pacific crust (upper) and old Pacific crust (lower). Topographic change with age is less apparent than in the Atlantic.

studies, which indicate that the principal discontinuities associated with the oceanic crust are at its top and its base.

We think that the seismic model of the oceanic crust that has been developed and refined over the past 25 years is a good one in a gross sense. The gradient model is an important modification of the layered model, because it accounts for arrival amplitudes as well as travel times. Further, it is consistent with reflection seismic data, which rarely require discontinuities within the crust. One of the principal shortcomings in our state of knowledge is lack of sufficient data on the degree and scale of lateral heterogeneity at all depths. This is not only an impediment to a more complete analysis of vertical variations, but also of critical importance in understanding the volcanic, tectonic, and geochemical processes involved in the formation of new crust and in its evolution. Fracture zones are obvious candidates for creating heterogeneity in a direction perpendicular to plate motion, and the style and apparent episodicity of crustal accretion, as deduced

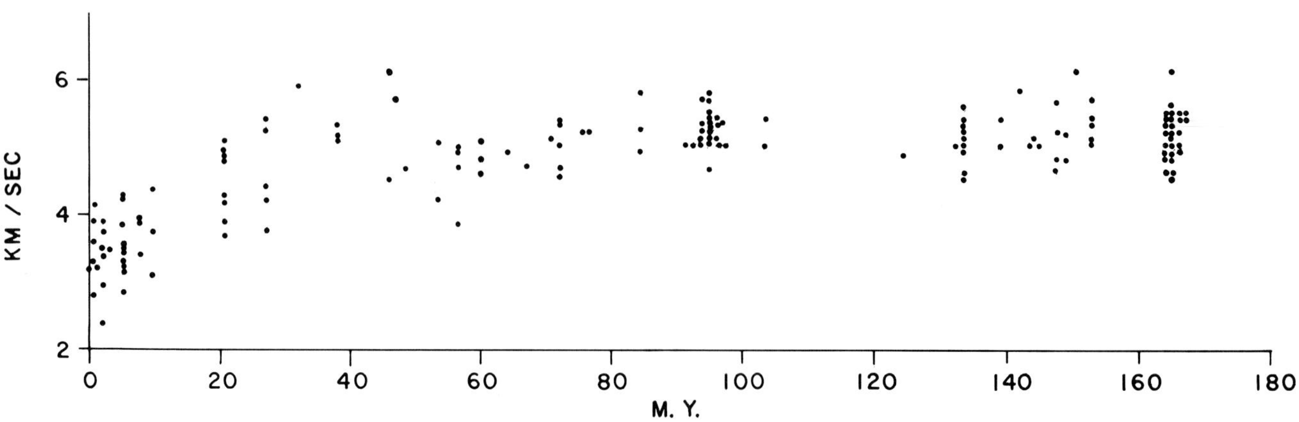

Fig. 11. Refraction velocities of acoustic basement plotted as a function of crustal age.

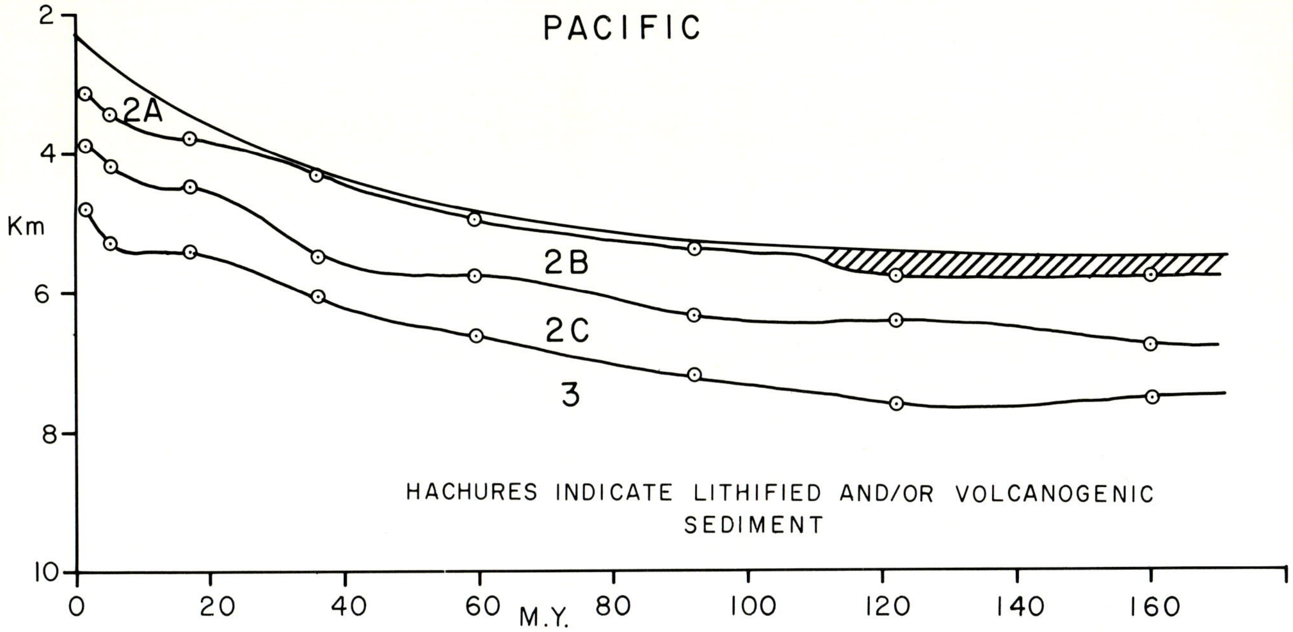

Fig. 12. Pacific crustal structure as a function of age.

from the submarine observations, suggests that heterogeneity is likely to be significant in the direction of plate motion. Work on this problem has begun by several groups using OBS/OBH units in various array configurations, but results are not yet available in sufficient quantity to permit drawing major conclusions. The accurate determination of lower crustal structure is difficult even with very good measurements and interpretative techniques, particularly if we have only refracted compressional wave data to work with. Considerable improve-

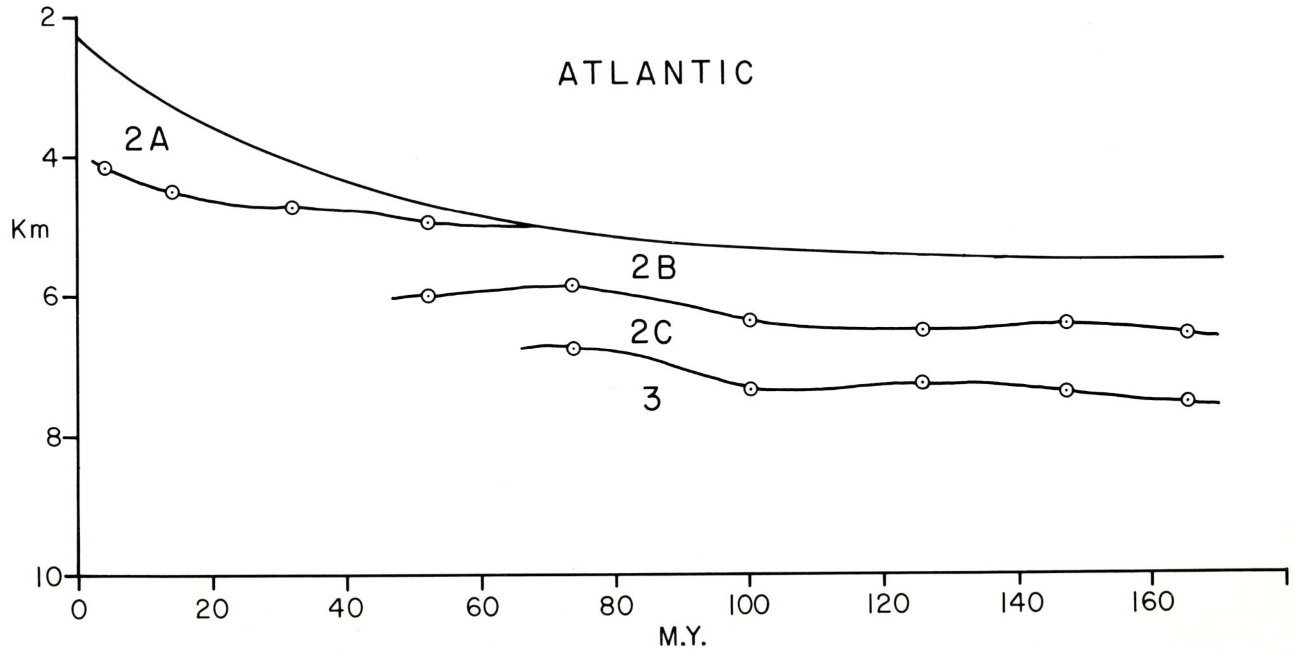

Fig. 13. Atlantic crustal structure as a function of age. Solutions in the younger Atlantic crust are considered to be less reliable than those in the older crust.

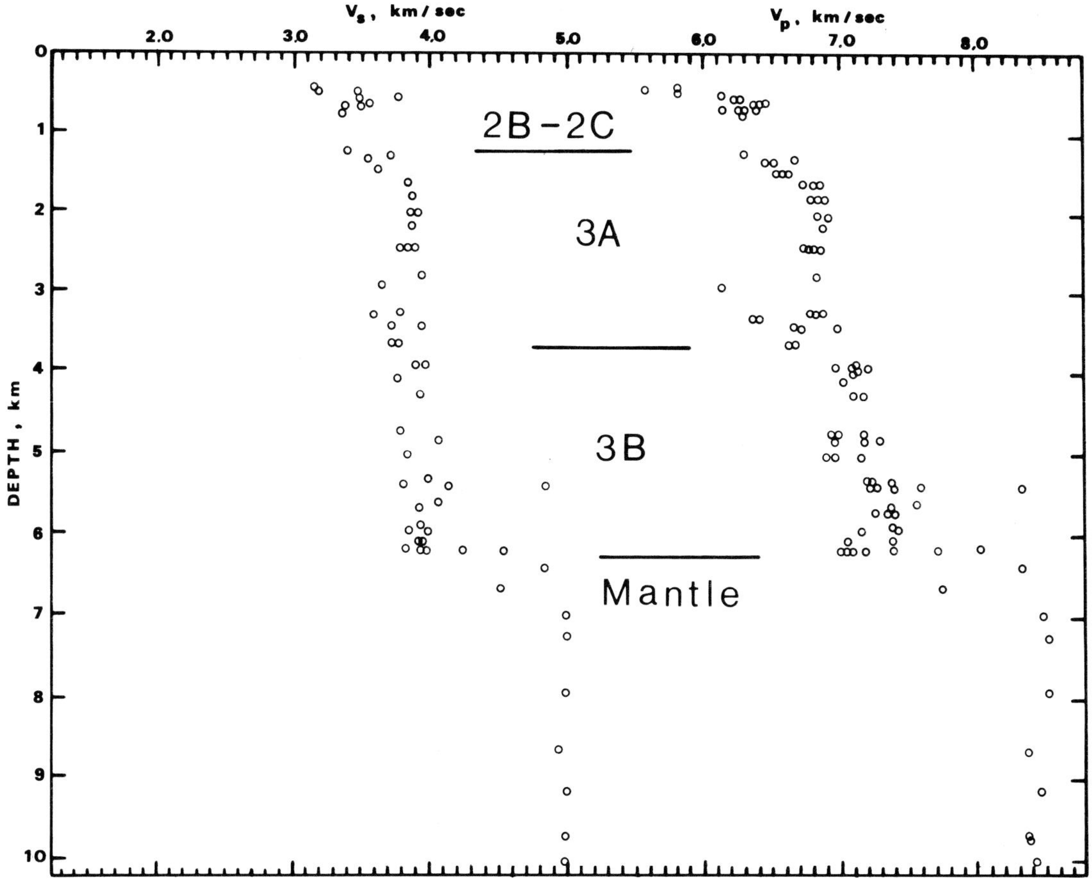

Fig. 14. Sonic-velocity probe measurements from the Bay of Islands ophiolite suite from Salisbury and Christensen (1978). Layer identifications from deep sea refraction measurements are superimposed.

ment may be possible through more experience with, and possibly better designed instruments, which will permit more use to be made of shear waves, both in body wave and surface wave analysis. Steps in this direction are being taken through efforts to better understand coupling of the seismographs to the sea floor and to extending the band width of the measurements.

Acknowledgments. We are pleased to acknowledge the support of the Office of Naval Research and the National Science Foundation over a number of years, permitting us to contribute to the overall understanding of the structure of the earth through marine seismological research. We also owe a large measure of thanks to our colleagues at Lamont-Doherty Geological Observatory and Woods Hole Oceanographic Institution who have collaborated in much of the work. Finally, we are grateful to George Sutton, Brian Lewis, Michael Purdy and Paul Stoffa for very helpful reviews of the manuscript.

References

Aumento, F., W.G. Melson, et al., Initial Reports of the Deep Sea Drilling Project, Volume 37: Washington (U.S. Government Printing Office), 1977.

Ballard, R. and Tj. van Andel, Morphology and tectonics of the inner rift valley at Lat. 36°50'N on the Mid Atlantic Ridge, Bull. Geol. Soc. Amer., 88, 507-530, 1977.

Bessonova, E.N., V.M. Fishman, V.Z. Ryaboyi, and G.A. Sitnikova, The Tau method for the inversion of travel-times - I. Deep seismic sounding data, Geophy. J. Roy. Astron. Soc., 36, 87-108, 1974.

Bryan, W.B., P.T. Robinson, S.M. White, G. Byerly, D.A. Swanson, N. Pertsev, L-E Ricou, S. Levi, G.A. Miles, W.G. Siesser, Y. Hamano, R.A. Stephen, C. Bollinger, and R. Emmermann, Studying oceanic layers, Geotimes, July/August, 1977.

Buhl, P., P.L. Stoffa, T.K. Kan, M. Talwani, and J. Ewing, Observation of reflections from the M-discontinuity in the western Pacific by near vertical incidence multichannel profiling, (abstract) Trans. Am. Geophys. Union, 59, 321, 1978.

Cagniard, L., Reflexion et Refraction des ondes Seismignes Progressives, Gauthier-Villars, Paris, 1939.

Cerveny, V. and R, Ravindra, Theory of Seismic

Head Waves, University of Toronto Press, Toronto, 1971.

Dmitriev, L.V., J.R. Heirtzler, H. Dick, M. Dungan, A. Erickson, F.N. Hodges, J. Honnorez, J. Kirkpatrick, D. Matthews, D. Ohnenstetter, N. Peterson, H. Sato, and H. Schmincke, Drilling into ocean crust, Geotimes, September, 1976.

Donnelly, T.W., J. Francheteau, M. Salisbury, K. Kelts, S.A. Swift, P. Borella, U. Bleil, B.M. Smith, T. Juteau, V. Rusinov, J.M. Sinton, Tadahide Ui, and S. Gartner, Mid-ocean ridge in the Cretaceous, Geotimes, June, 1977.

Ewing, J.I. and G.B. Tirey, Seismic Profiler, J. Geophys. Res., 6, 2917-2927, 1961.

Ewing, J., and M. Ewing, Seismic-refraction measurements in the Atlantic Ocean basins, in the Mediterranean Sea, on the Mid Atlantic Ridge, and in the Norwegian Sea, Geol. Soc. America Bull., 70, 291-317, 1959.

Ewing, M., J.L. Worzel, J.B. Hersey, F. Press, and G.R. Hamilton, Seismic refraction measurements in the Atlantic Ocean basin, Part I, Seismol. Soc. America Bull., 40, 233-242, 1950.

Ewing, M., G.H. Sutton, and C.B. Officer, Jr., Seismic refraction measurements in the Atlantic Ocean, Part VI: Typical deep stations, North American basin, Seismol. Soc. America Bull., 44, 21-38, 1954.

Flower, M.F.J., M.H. Salisbury, D. Bohrer, M. Hobart, D. Johnson, E. Mathez, C. Mevel, N. Pertsev, R.G. Pritchard, H. Puchelt, P.A. Rigotti, and H. Staudigel, Cretaceous crust sought, Geotimes, September 1977.

Fowler, C.M.R., Crustal structure of the Mid-Atlantic Ridge crest at 37°N, Geophys. J. Roy. Astron. Soc., 47, p. 459-491, 1976.

Fuchs, K. and G. Muller, Computation of synthetic seismograms with the reflectivity method and comparison with observations, Geophys. J. Roy. Astron. Soc., 23, 417-433, 1971.

Gaskell, T.F., M.N. Hill and J.C. Swallow, Seismic measurements made by H.M.S. Challenger in the Atlantic, Pacific, and Indian Oceans and in the Mediterranean Sea, 1950-53, Phil. Trans. Roy. Soc. London, A251, 23-83, 1958.

Grow, J.A. and .R.G. Markl, IPOD-USGS multichannel seismic reflection profile from Cape Hatteras to the Mid Atlantic Ridge, Geology, 5, 625-630, 1977.

Helmberger, D.V., The crust-mantle transition in the Bering Sea, Bull. Seism. Soc. America, 58, 179-214, 1968.

Helmberger, D.V. and G.B. Morris, A travel-time and amplitude interpretation of a marine refraction profile: Primary waves, J. Geophys. Res., 74, 483-494, 1969.

Helmberger, D.V. and G.B. Morris, A travel-time and amplitude interpretation of a marine refraction profile: Transformed shear waves, Bull. Seism. Soc. America, 60, 593-600, 1970.

Helmberger, D.V., Fine structure of an Aleutian crustal section, Geophys. J. Roy. Astr. Soc. 48, 81-90, 1977.

Herron, T.J., W.J. Ludwig, P.L. Stoffa, T.K. Kan, and P. Buhl, Structure of the East Pacific Rise crest from multichannel seismic reflection data, J. Geophys. Res., 83, 798-804, 1978.

Hersey, J.B., Continuous reflection profiling, in M.N. Hill (ed.). The Sea, Vol. 3, Interscience, New York, 47-72, 1963.

Hersey, J.B., C.B. Officer, H.R. Johnson and S. Bergstrom, Seismic refraction observations north of the Brownson Deep, Bull. Seis. Soc. Am., 42, 291-306, 1952.

Hill, M.N., Seismic refraction shooting in an area of the eastern Atlantic, Roy. Soc. London Phil. Trans., 244, 561-596, 1952.

Hoop, A. de, A modification of Cagniard method for solving seismic pulse problems, Appl. Sci. Res. B, 8, 349-356, 1960.

Houtz, R., Interval velocities in crustal layers from sonobuoy records of mantle reflections, J. Geophys. Res., 82(33), 5395-5400, 1977.

Houtz, R. and J. Ewing, Upper crustal structure as a function of plate age, J. Geophys. Res., 81, 2490-2498, 1976.

Kennett, B.L.N., Towards a more detailed seismic picture of the oceanic crust and mantle, Marine Geophysical Researches, 3, 7-42, 1977.

Kennett, B.L.N and J.A. Orcutt, A comparison of travel time inversions - marine refraction profiles, J. Geophys, Res., 81, 4061-4070, 1976.

Lewis, B.T.R. and W.E. Snydsman, Evidence for a low velocity layer at the base of the oceanic crust, Nature, 266, 340-344, 1977.

Officer, C.B., M. Ewing, and P.C. Wuenschel, Seismic refraction measurements in the Atlantic Ocean Part IV: Bermuda, Bermuda rise, and Nares basin, Bull. Geol. Soc. America, 63, 777-808, 1952.

Orcutt, J.A., B.L.N. Kennett, and L.M. Dorman, Structure of the East Pacific Rise from an ocean bottom seismometer survey, Geophys. J. Roy. Astron. Soc., 45, 305-320, 1976.

Rabinowitz, P.A., W.G. Melson, H. Bougault, T. Fujii, A.L. Graham, H.P. Johnson, J. Lawrence, J. Natland, E. Prosser, M. Rhodes, and B.P. Zolotarev, Challenger drills on Leg 45, Geotimes, April, 1976.

Raitt, R.W., Seismic refraction studies of the Pacific Ocean basin, Part I, Geol. Soc. America Bull., 67, 1623-1640, 1956.

Reid, I., J.A. Orcutt, and W.A. Prothero, Seismic evidence for a narrow zone of partial melting underlying the East Pacific Rise at 21°N, Bull. Geol. Soc. Am., 88, 678-682, 1977.

Salisbury, M. and N. Christensen, The seismic velocity structure of a traverse through the Bay of Islands Ophiolite complex, Newfoundland, an exposure of oceanic crust and upper mantle, J. Geophys. Res., 83, 805-817, 1978.

Shor, G.G., Jr., H.W. Menard, and R.W. Raitt, Structure of the Pacific basin, in the Sea, V4, Part II, A.E. Maxwell ed., John Wiley and Sons, 664p. 1970.

Stephen, R.A., K.E. Louden, and D.H. Matthews,

The oblique seismic experiment on DSDP Leg 52, submitted for publication in Initial Reports of the Deep Sea Drilling Project, Vol. 52.

Sutton, G.H., G.L. Maynard, and D.M. Hussong, Widespread occurrence of high velocity basal layer in the Pacific crust found with repetitive sources and sonobuoys, Geophys. Monograph, No. 14, American Geophysical Union, 193-209, 1971.

Talwani, M., C. Windisch, and M. Langseth, Reykjanes Ridge crest: a detailed geophysical study, J. Geophys. Res., 76, 473-517, 1971.

TECTONIC PROCESSES ON SLOW SPREADING RIDGES

A.S. Laughton

R.C. Searle

Institute of Oceanographic Sciences,
Wormley, Godalming, Surrey, U.K.

Abstract. The morphology of four areas of the crestal region of the Mid-Atlantic Ridge has been studied by deep ocean side-scan sonar. This has revealed a consistent pattern of closely spaced normal faults parallel to and, for the most part, facing the spreading axis. An analysis of the statistics of the fault orientation, spacing and amplitude indicates that the median valley, typical of slow spreading ridges, is the result of inward-facing normal faulting occurring within a few kilometres of the axis of the ridge and concurrent outward rotation of the fault blocks. A model is proposed, combining these tectonic events with axial volcanism and lithospheric thermal contraction, to explain the observed topographic profile across a slow spreading ridge.

Introduction

It is widely recognised that slow spreading (~1 cm/yr) ridges are characterised by rugged topography and by a median valley, in contrast with the elevated median ridges and comparatively smooth topography of fast spreading ridges (~7 cm/yr) (e.g. Rea, 1975; Lonsdale, 1977).

The typical cross-sectional profile of a slow spreading ridge comprises an inner valley floor with one or more volcanic ridges and small tension fissures, a complex valley wall consisting of scarps and terraces leading up to the crestal mountains some 10-15 km from the axis, and beyond that a rugged topography with ridge crests which progressively deepen with distance from the axis. The majority of slow spreading ridges exhibit this general profile even though the present time is but a snapshot of their evolution. This indicates that the profile is a steady state characteristic acquired by the generation of oceanic crust in the inner valley, modified by tectonism which is spatially related to the axis of spreading.

This profile can be considered as the summation of fault throws and rotations of the fault blocks. (A regional uplift or subsidence is equivalent to the sum of a series of progressively varying positive or negative rotations). Harrison and Stieltjes (1977) have summarised the three possible models in which faulting and rotation can be combined to give the observed type profile of median valley and crestal ridge. Their model 1 comprises normal faulting without rotation near the bottom of the median valley walls, followed by reverse faulting along the same faults in the upper part, thereby cancelling, at the crestal ridges, the staircase progression. This model implies no faults observable beyond the crestal ridge, or at least a significant reduction in inward facing fault amplitude beyond this region. This is contrary to our observations to be presented.

Model 2 maintains the amplitude of the initial inward facing faults but introduces outward-facing normal faults along different fault planes on the upper part of the valley walls, compensating the upwards staircase by a downward staircase. The resulting topography beyond the crestal mountains would then comprise a series of horsts and grabens and the cumulative throw on the outward facing faults would equal that on the inward facing ones. Again our data, contrary to those presented by Macdonald and Luyendyk (1977), suggest that this is not in general the case.

Model 3 allows for inward facing faults with amplitudes increasing with age within the median valley but thereafter constant, combined with a progressively increasing rotation of blocks up to a maximum at the summit of the crestal mountains.

Harrison and Stieltjes reviewed the evidence for and against these models and concluded that all three mechanisms are important and that none is capable alone of explaining the topography.

This paper presents new data on the nature, style, frequency and amplitude of faulting as determined by high resolution morphological studies at a variety of scales. An analysis of these data provides evidence that Model 3 is the main mechanism of median valley formation on the Mid-Atlantic Ridge in the North Atlantic.

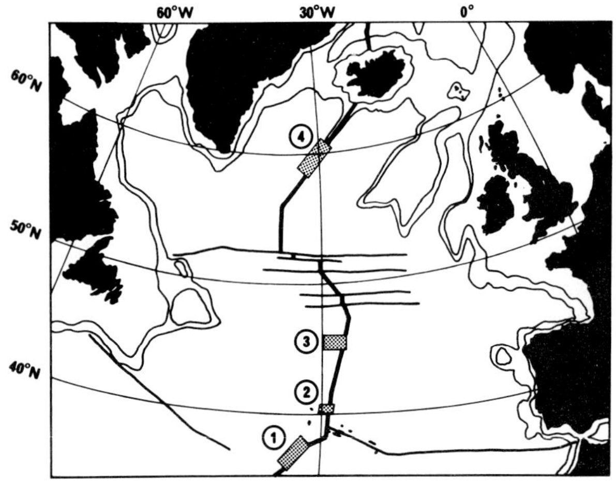

Fig. 1. Location of GLORIA surveys on the Mid-Atlantic Ridge. 1. FAMOUS area; 2. Kurchatov Fracture Zone; 3. 45°N area; 4. Reykjanes Ridge.

New Techniques For Studying Sea Floor Morphology

Until the last five years, ocean floor morphology was studied either by conventional echo-sounding from surface ships, with its inherent limitations of resolution due to the wide beam angle of echosounders and the difficulties of accurate relative navigation between adjacent profiles; or by underwater photography which was limited in range by the poor transmission of light in water and by the difficulties of accurate location of the camera relative to the grosser morphology.

Advances in technology can now add the following techniques:-

(1) Multiple narrow beam echosounding from the surface, only rarely available in non-classified projects (Glenn, 1970, 1976).

(2) Medium range (2.5 km) and long range (30 km) side-scan sonar operated from near the surface (Tucker and Stubbs, 1961; Rusby and Somers, 1977).

(3) Near bottom observations using short range (500 m) side-scan sonar, echosounding, photography and magnetometry, navigated relative to bottom acoustic beacons on vehicles such as "Deep Tow" (Spiess and Tyce, 1973).

(4) Direct visual observations and selective sampling by manned submersibles (e.g. Arcyana, 1975; Ballard and van Andel, 1977a).

The latter two techniques are able to give the sort of data a field geologist on land can obtain by walking over his terrain in misty conditions, while the former bridge the gap between these and the grosser bathymetric charts and give results more analogous to the distant views available to the land geologist from a high vantage point on a clear day. All four techniques were used in the study of the FAMOUS area on the crest of the Mid-Atlantic Ridge, southwest of the Azores (Heirtzler and van Andel, 1977).

The data presented here were obtained from deep ocean side-scan sonars, developed at the Institute of Oceanographic Sciences. They have been used in four separate regions of the Mid-Atlantic Ridge (Fig. 1), largely in areas where the complimentary bathymetric mapping and/or near bottom studies have also been made so that the data can be sensibly integrated and interpreted.

GLORIA (Geological Long Range Inclined Asdic) Mark I (Rusby, 1970) was a long range side-scan sonar (6.4 kHz), used up until 1975, operating from a towed vehicle at 100 m depth. It was able to scan the sea bed out to a range of 27 km to one side of the ship, and could be towed at 7 knots. Mark II (Somers et al., 1978), mounted in a different vehicle and able to be operated in much higher sea conditions, looks both sides simultaneously, covering a swath of sea bed 60 km wide, and has been in operation since summer 1977. GLORIA studies in 1977 on the Mid-Atlantic Ridge were supplemented in depths up to 1500 m by a medium range (2.5 km) side-scan sonar (35 kHz) mounted on a stabilised platform below the hull of the ship. This could also be used with the beam vertical as a narrow beam echosounder (narrow in the fore and aft direction giving high resolution bathymetry when crossing linear features at right angles).

GLORIA Survey Operations and Interpretation

On account of the wide swath coverage and high towing speed, GLORIA can insonify a large area in a short time. An areal coverage of about 30,000 km² can be achieved in about seven days. Regional surveys have been made with GLORIA Mk I using parallel tracks spaced so that there is appreciable overlap between adjacent sonographs, allowing the build up of a mosaic, utilising the far range where slant-range distortion is minimal. In later surveys, using the two-sided Mk II, the tracks were chosen to study individual features and trends, viewing from a variety of directions and resulting in a montage rather than a mosaic.

For a more detailed look at selected features, the effective range of GLORIA can be reduced by a factor of two or four, using an increased pulse repetition rate and achieving a higher resolution. Detailed studies have been made in this way of the axial volcanic zones of all the areas surveyed.

Sonographs are built up by successive scans of the sea floor at right angles to the ship. The echo strengths vary over a wide dynamic range (60 dB) even after time-varied gain has been used to compensate for range. After linear correlation processing the signals are stored on magnetic tape, either with "fixed gain" (independent of signal strength), or AGC (Automatic Gain Control). Whereas the fixed gain channel allows the amplitude contrasts between high and low reflectivity regions to be displayed, it cannot handle the wide dynamic range. The AGC channel tends to suppress differences in signal level over more than 1-2 seconds of

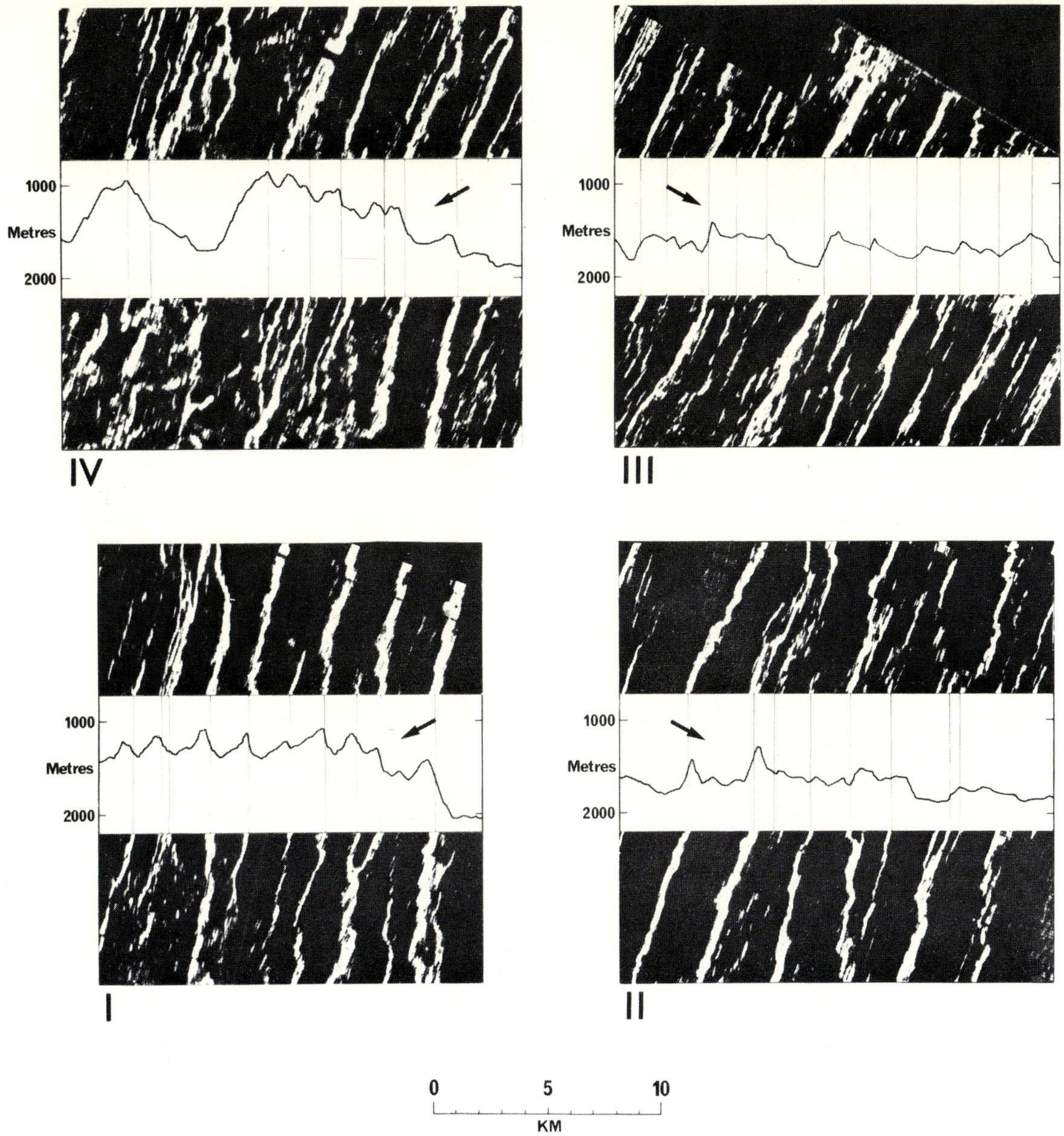

Fig. 2. Sonographs from the Kurchatov Fracture Zone area and associated reduced echosounder profiles. The sonographs have been cut in two halves and separated at the position of the profile. I and IV are west and II and III east of the spreading axis. Arrows show the direction of insonification and vertical lines link echoes with morphological features. Vertical exaggeration of the profiles is 4:1. The steepness of the scarps giving rise to the echoes may be considerably greater than is shown by the profiles.

travel time, but reveals detail throughout the dynamic range. It is, therefore, more suitable for matching adjacent sonographs on a mosaic and has been used in all mosaics presented except on the Reykjanes Ridge. For detailed interpretation of the mosaics, however, it is essential to study both channels.

The GLORIA surveys on the Mid-Atlantic Ridge have highlighted linear targets parallel to the spreading axis, revealing a texture of the sea floor which is superimposed on the grosser morphology indicated by echosounding surveys (Whitmarsh and Laughton, 1976). Detailed comparison between the echoes on the sonographs and reduced (migrated) echosounding profiles (Fig. 2), and an analysis of the acoustic shadows

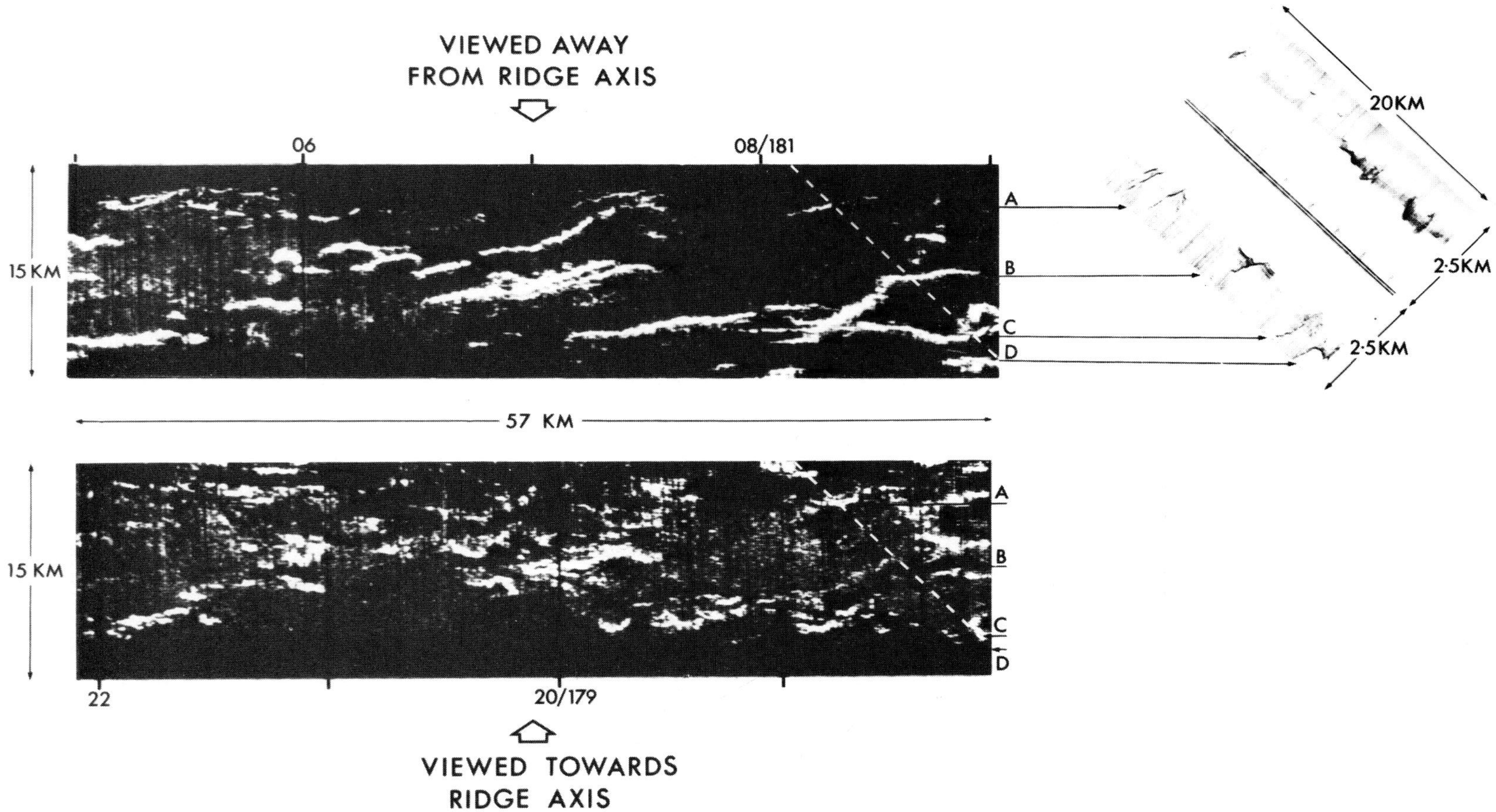

Fig. 3. A comparison of outward and inward viewing sonographs of identical terrain 10 km east of the axis of the Reykjanes Ridge, showing the asymmetry of the ridges. A diagonal medium range sonograph crosses the end of the GLORIA sonograph and shows the transition from side-scan (at maximum range) to vertical profile (at minimum range) and the correlation with GLORIA targets.

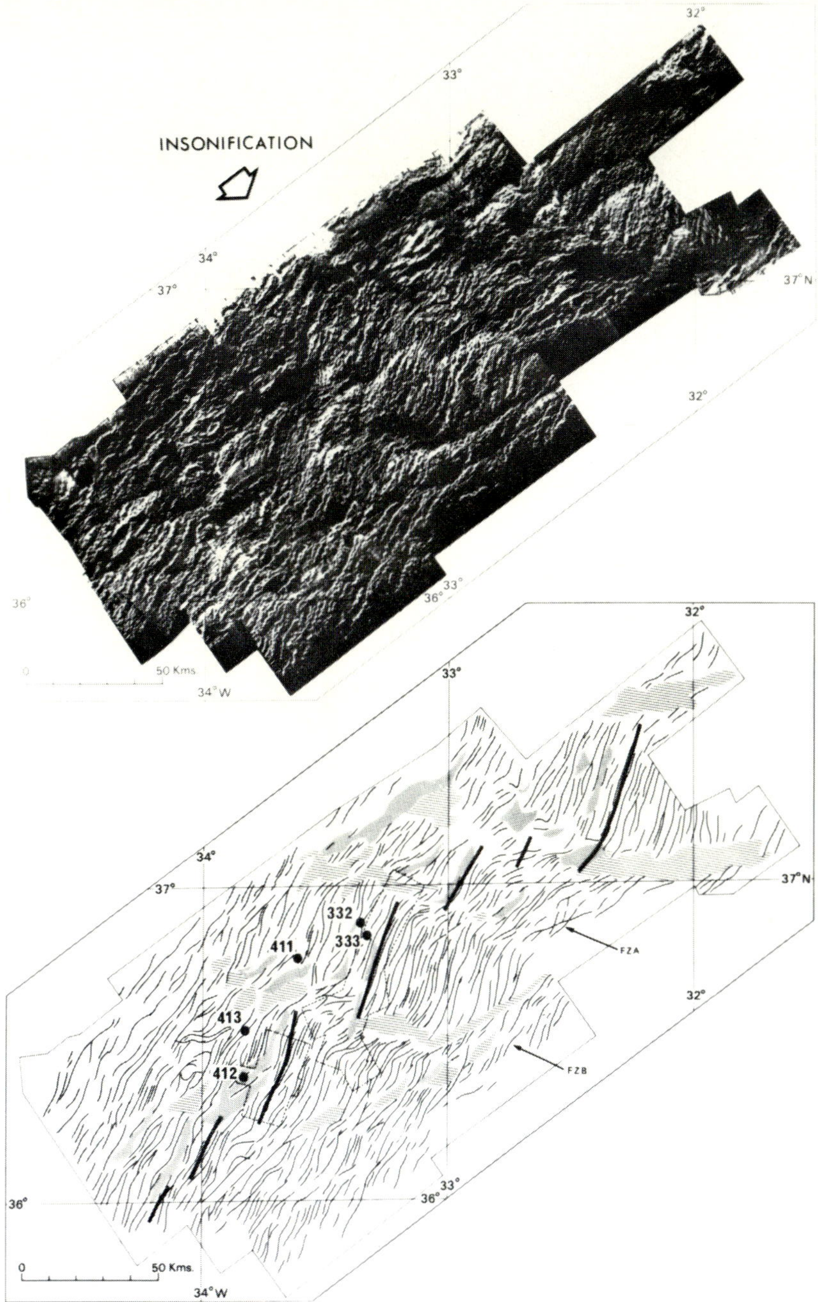

Fig. 4. (a) Sonograph mosaic and, (b) interpretation in terms of faults and fracture zones in the FAMOUS area (from Whitmarsh and Laughton, 1976). Positions of DSDP holes 332, 333, 411-413 are shown. Areas of median valley indicated by heavy lines, sediment ponds by hachure and shadows by stipple.

in relation to the computed angles of incidence of the sound at the bottom (Laughton and Rusby, 1975), have shown that the targets are predominantly steep (40-80°) and linear topographic scarps of height varying between 20 and 800 m. The interpretation of these scarps as normal faults has been confirmed by near bottom observations in the FAMOUS area, both visually by submersible (Ballard and van Andel, 1977b), and by echosounding and side-scan sonar from Deep Tow (Macdonald and Luyendyk, 1977).

The sonographs show other more diffuse and less well-aligned echoes. Within the median valley (or median ridge zone on the northern part of the

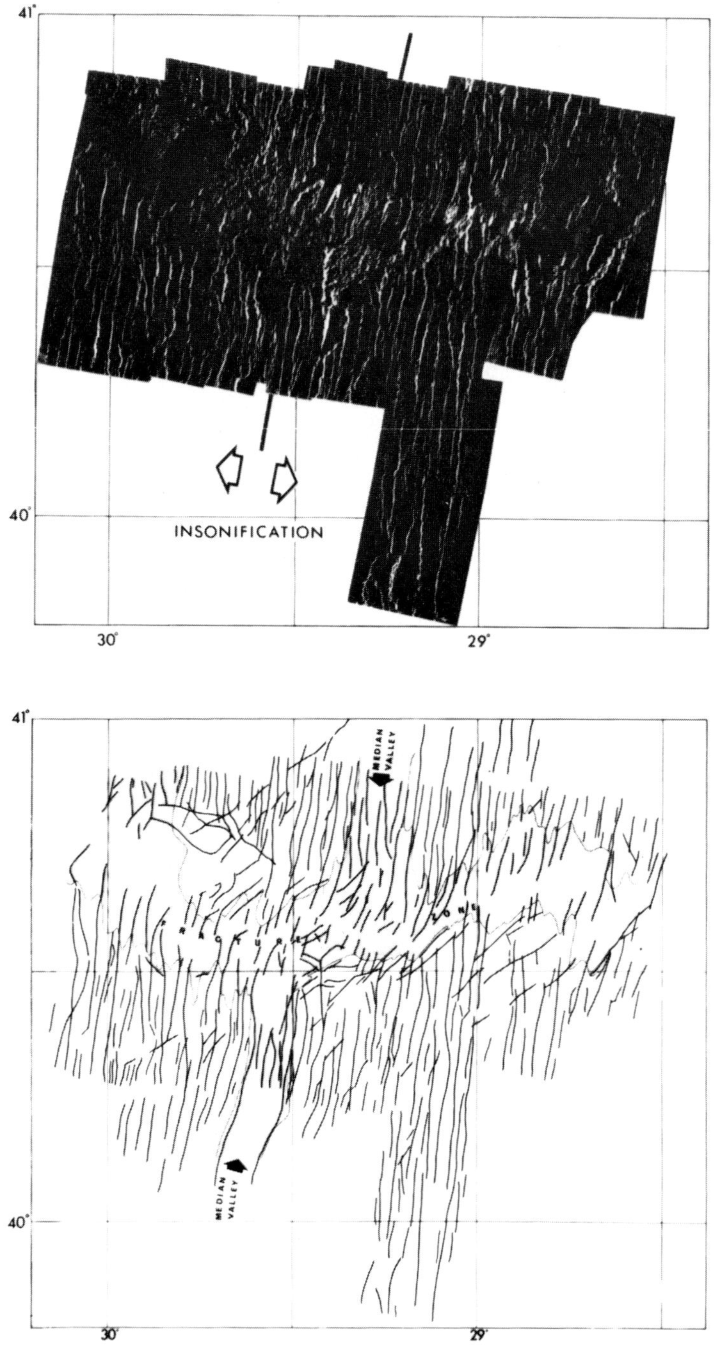

Fig. 5. Sonograph mosaic and interpretation in the Kurchatov Fracture Zone area (from Searle and Laughton, 1977). Depths over 2000 m shown stippled.

Reykjanes Ridge) such echoes have been shown, by submersible observations and by short and medium range sonar, to be extrusive volcanoes aligned approximately normal to the spreading direction. Remnants of these volcanic edifices, which may be partially destroyed by subsequent rifting and by the tectonism associated with the walls of the median valley, are carried away from the axial region on the fault blocks and give rise to irregular echoes between the fault scarps. When the fault blocks are insonified on the dip slope side, the fault scarps are identifiable only as shadows and the dominant features are the irregularities on the fault blocks (Fig. 3). Away from

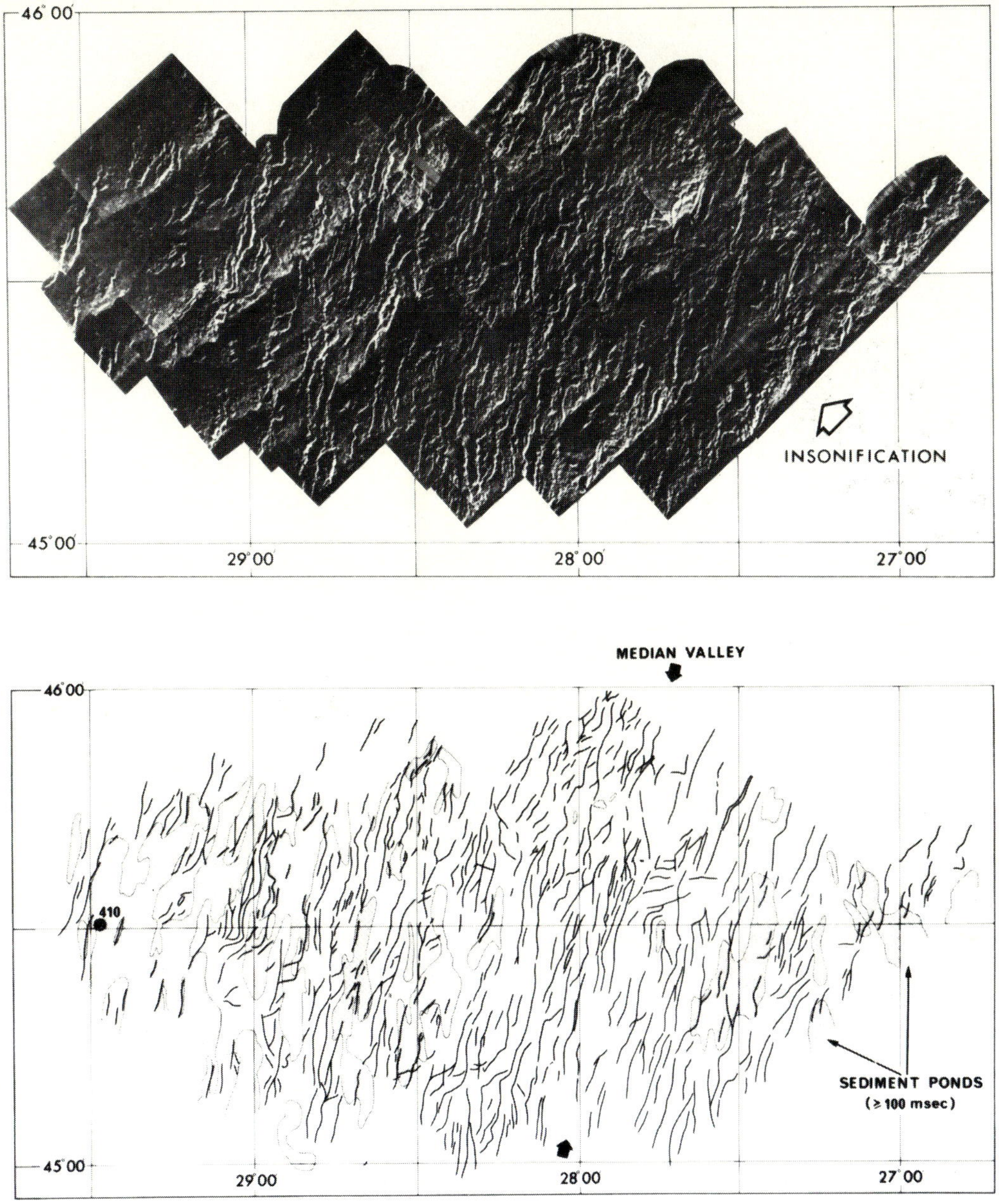

Fig. 6. Sonograph mosaic and interpretation in the 45°N survey area. Position of DSDP hole 410 is shown.

the axial region, sedimentation progressively covers both the volcanic features and the fault scarps.

In the vicinity of fracture zones, which are not discussed in this paper, additional linear targets are seen, both parallel and oblique to the fracture zones.

Mid-Atlantic Ridge Study Areas

Four separate areas of the Mid-Atlantic Ridge, between 36° and 61°N (Fig. 1), have been studied by GLORIA together with other geophysical methods. The most southerly was on the North America-Africa plate boundary, the others on the North America-Eurasia plate boundary. In most cases the survey area included known fracture zones, but these are not discussed here. In some areas the spreading is approximately orthogonal, in others it is oblique. The four surveys are discussed below and in Table I the main faulting characteristics derived from them are tabulated against the spreading parameters.

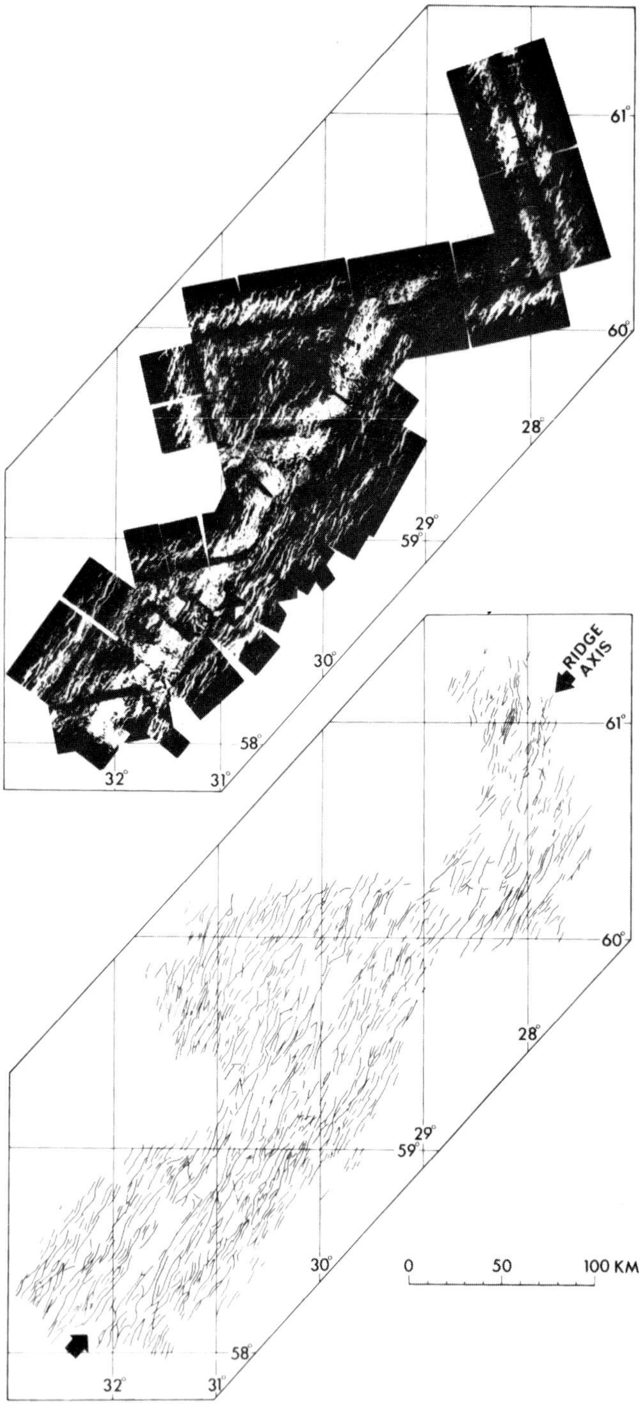

Fig. 7. Sonograph montage and interpretation on the Reykjanes Ridge. The montage differs from the mosaics in Figs. 4, 5 and 6, due to (a) irregular track orientations, (b) simultaneous port and starboard insonification, (c) sonographs are from the fixed gain channel of GLORIA, not the AGC channel; the extended dynamic range clearly shows the axial high-reflectivity zone.

(a) FAMOUS Area (Whitmarsh and Laughton, 1976)

The 30,000 km² survey covered the whole of the greater FAMOUS area with a mosaic (Fig. 4) obtained from tracks oriented 055°, separated by 12 km. Insonification was towards the southeast. East of the spreading axis, strong, narrow echoes were observed, whereas on the west side there were broader and weaker echoes and acoustic shadows often with a sharp linear cut-off on the northwest edge. The sharp echoes on the east side and sharp shadows on the west are interpreted as resulting from linear inward facing fault scarps (the wiggles in the linear echoes are artifacts due to yaw), whereas the diffuse echoes on the west side show the rough volcanic terrain riding on the fault blocks. Some outward facing scarps were identified west of the axis, but they are few compared to the inward facing faults.

Detailed sonar studies (Laughton and Rusby, 1975) of the median valley segments between fracture zones A and B (Rift Valley 2), and between B and C (Rift Valley 3), together with the intensive surveys made by submersible and by Deep Tow (Ballard and van Andel, 1977; Macdonald and Luyendyk, 1977), have shown that steep normal faults occur on the flanks of the median valley forming the inner walls and are responsible for the existence of the valley as a morphological feature. The interpretation of the linear echoes and shadows of the FAMOUS sonar mosaic enabled a fault map to be constructed (Fig. 4). The data to the east of the axis is a more reliable guide to the size and distribution of the inward facing faults since all are adequately insonified. By contrast, on the west side, many such faults may be obscured by the shadows of larger faults, and furthermore, smaller faults may be missed because of the low contrast between the black of a shadow and the sometimes low back-scattering from the dip slope of a fault block.

(b) Kurchatov Fracture Zone (Searle and Laughton, 1977)

The survey (Fig. 5) of 10,000 km² was planned to study the fracture zone but also gave significant data on the crestal region of the Mid-Atlantic Ridge away from the influence of the fracture zone. Knowing the style of faulting to expect, tracks were made parallel to the spreading axis to minimise problems due to vehicle yaw, and the sonar beam was mainly directed outwards from the spreading axis. Equal resolution of inward facing faults, both east and west of the spreading axis, was therefore obtained. There was inadequate coverage of terrain viewed from both directions to establish everywhere the asymmetry of individual ridges from the sonographs alone, but comparison with echo-sounding profiles, reduced to the equivalent of a narrow-beam profile, suggested the presence of asymmetric, tilted fault blocks (Fig. 2). In the few areas viewed inwards towards the axis,

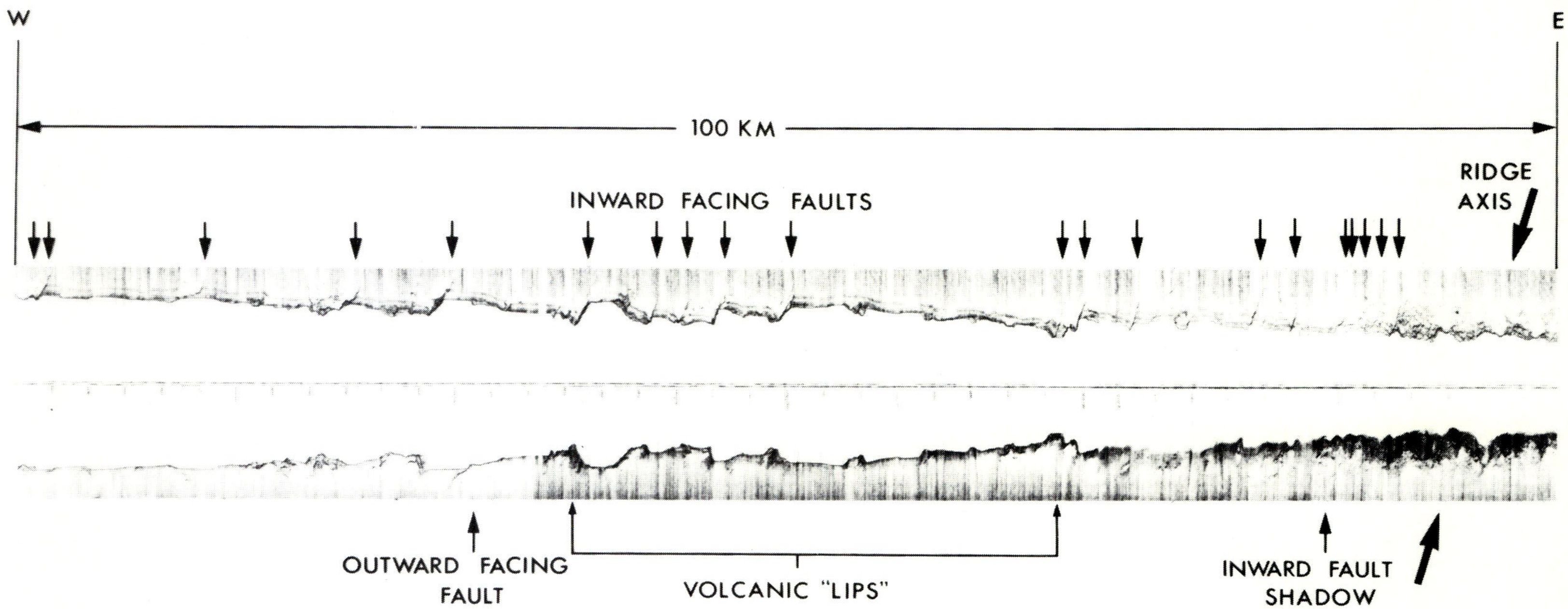

Fig. 8. Part of sonograph from the medium range side-scan sonar extending in direction 261° from the Reykjanes Ridge axis at about 60°N. (GLORIA sonograph taken along the same track is seen in Fig. 7). Note that the linear fault scarps to the north but not to the south of the track are insonified and give echoes indicating the preponderance of inward facing faults. Note also the asymmetry of individual ridges with steep scarps and gentle back tilts, and the occasional perched volcano forming a volcanic "lip" to the fault block.

there was little evidence of outward facing scarps.

(c) 45° North Area

A strip of sea floor across the axis of the Mid-Atlantic Ridge, between 45° and 46°N, was surveyed and sampled intensively, between 1960 and 1969, as part of the Canadian Hudson Geotraverse (Aumento, Loncarevic and Ross, 1971). A GLORIA survey (Fig. 6) of 15,000 km² covered most of the area for which detailed bathymetric charts were available. One objective of the survey was to search for postulated fracture zones, so survey lines were run oblique to the spreading axis and insonification throughout was towards the northwest. Since the median valley lay in the eastern part of the survey area, any inward facing scarps in the western two-thirds of the area would have been insonified, but not those in the eastern third. Linear targets were more prevalent in the western part, although not completely absent in the eastern part, indicating a small, but significant, proportion of outward facing scarps. Beyond 40 km from the axis, sediment accumulations in the valleys progressively obscure some of the basement structure.

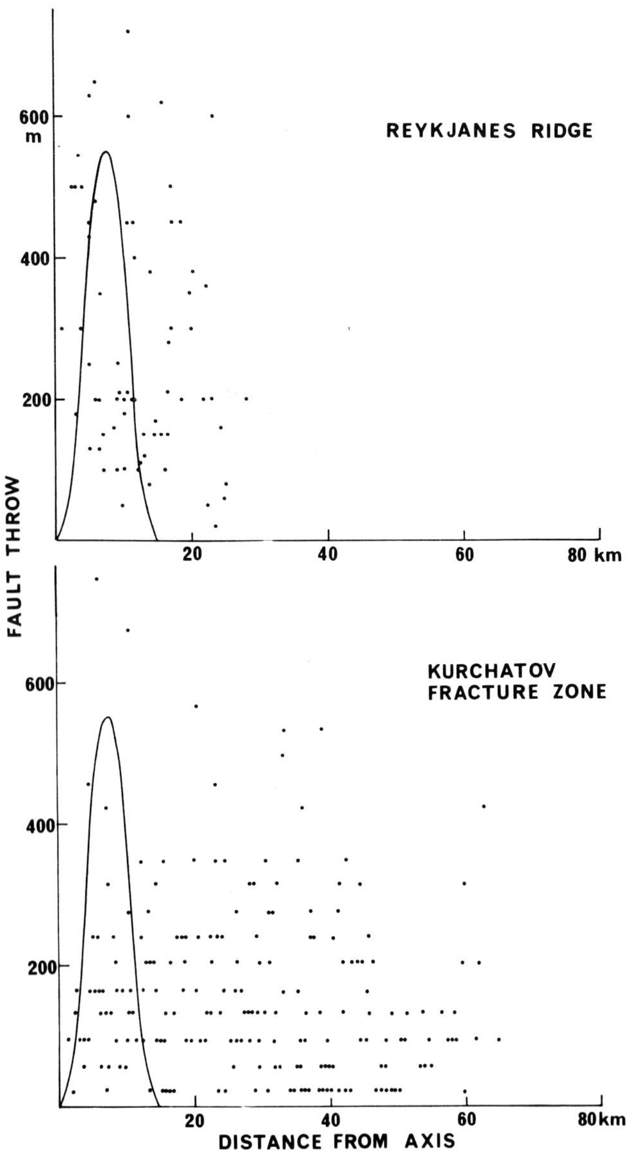

Fig. 9. Fault throws on the Reykjanes Ridge and near the Kurchatov Fracture Zone (Searle and Laughton, 1977), plotted against distance from ridge axis, compared with a theoretical curve appropriate to Model 1 of Harrison and Stieltjes (1976), assuming median valley half width of 15 km.

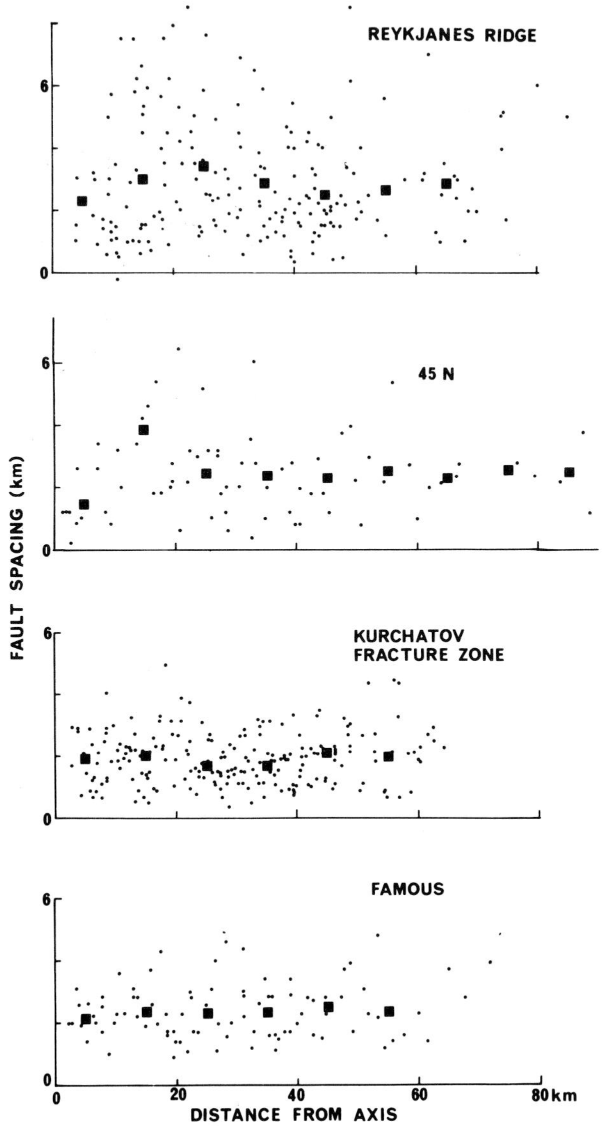

Fig. 10. Fault spacing as a function of distance from the ridge axis for the four areas. Squares indicate means for 10 km intervals.

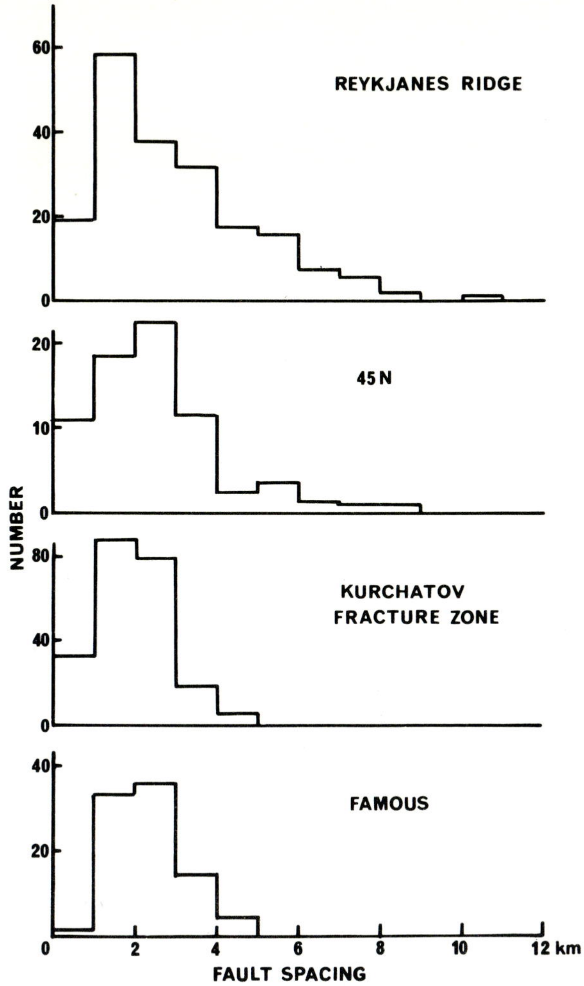

Fig. 11. Histograms of fault spacings measured from typical cross-sections on the mosaic interpretations. Sediment ponds thicker than 75 m, acoustic shadows and areas of poor GLORIA coverage have been excluded.

(d) Reykjanes Ridge

In 1977, GLORIA Mk II, with its ability to look both sides of the ship simultaneously, was used in a survey of the Reykjanes Ridge axis to examine the transition between its rifted and non-rifted axial regions and to obtain data on an oblique spreading ridge (Laughton, Searle and Roberts, in press). The survey montage (Fig. 7) of about 35,000 km² was not obtained from a regular pattern of parallel lines but from a series of reconnaissance lines comprising five oblique crossings of the ridge axis, followed by tracks both parallel and perpendicular to the axial trend. Thus, many parts of the ridge have been viewed from a variety of directions enabling the asymmetry of individual ridges to be assessed (e.g. Fig. 3).

Concurrently with the GLORIA data, medium range side-scan sonographs, out to a slant range of 2.5 km, were obtained both sides of the ship's track, giving a higher resolution of the fault scarps and an effectively narrow-beam profile beneath the ship (Fig. 8). Both sets of sonar data indicate that inward facing faults are heavily preponderant.

The data showed an axial zone, 13 km wide, consisting of a series of en echelon elongated volcanic ridges trending 010°, approximately orthogonal to the spreading direction of 095° and probably reflecting the orientation of the tension cracks and of the feeder dykes. Outside this zone, and in contrast to it, inward facing normal faults trend between 032° and 022°, subparallel to the axial trend of 036°. These trends may, however, comprise a mixture of a fault trend parallel to the ridge axis (i.e. perpendicular to the direction of lithospheric thickening), and relics of the 010° volcanic trend. In the southern part of the survey area the innermost pair of inward facing faults which delineate the median valley are indeed parallel to the axial trend, i.e. 036°.

Analysis Of Faults On The Mid-Atlantic Ridge

(a) Ratio of inward to outward facing faults

The contrast between the inward viewing sonographs on the west side of the FAMOUS area spreading axis and the outward viewing sonographs on the east side, and a study of the geometry of their shadows in relation to the near bottom sound-ray paths led Whitmarsh and Laughton (1976) to conclude that, although there are some outward facing faults, the majority are inward facing. Although correlations have been made with the digitised narrow-beam echosounder profiles obtained by USNS Hayes, it is difficult to determine from these the asymmetry of individual ridges associated with near-vertical fault scarps since the scarps are not easily identified on the echosounder profiles, especially at the rather large vertical exaggerations used.

More detailed profiles have been obtained by Deep Tow which is better able to resolve steep slopes. Macdonald and Luyendyk (1977) and Macdonald and Atwater (1978) concluded that outward facing faults were generated in the upper part of the median valley wall complex, and that beyond 20 km from the axis, they contributed as much to the relief of the crestal mountains as did the inward facing faults, thereby cancelling the shoaling depths of the median valley wall complex. They also concluded that tilting of blocks contributed less than 20% of the contribution by faulting. Thus, these data are in direct conflict with the sonograph data from GLORIA.

A resolution of this conflict may be possible by the direct comparison of the sonographs with individual faults interpreted from Deep Tow data on both sides of the median valley. It is

Table I

Area	Lat.	Plates	Axial bathymetric trend	Spreading				Faulting					
				dir^n.	obliquity	rate mm/y	ref.	trend	obliquity	mean spacing km	max. length km	max. throw m	average throw and s.d.
FAMOUS	36-37°N	AM/AF	017°	090°	17°	23.4	1	020°	20°	2.4±0.9	40	300	-
			023°	095°	18°	23.5	2		15°				
KURCHATOV	40-41°N	AM/EU	008°	098°	0°	21.3	3	010°	2°	1.9±0.8	50	750	147±109
45°N	45-46°N	AM/EU	015°	097°	8°	23.8	4	008°	1°	2.6±1.6	30	-	-
REYKJANES	58-61°N	AM/EU	036°	095°	31°	23.0	5	027°*	22°	2.6±1.6	90	800	274±172

*This is a mean trend. Individual scarps may reflect several trends (see text).

Ref. 1. Macdonald and Luyendyk (1977)
2. Ramberg, Gray and Raynolds (1977)
3. Searle and Laughton (1977)
4. Loncarevic and Parker (1971)
5. Vogt and Avery (1974)

Table II. Spacing of first pair of inward facing faults

Area		Data Source	$2y_1$ km	$\overline{y_1}$ km	$\overline{\frac{\Delta y}{2}}$ km	g km
FAMOUS Rift Valley	1	Ramberg and van Andel (1977)	5?	3.5	1.2	2.3
" " "	2	Macdonald and Luyendyk (1977)	5			
" " "	3	Ramberg and van Andel (1977)	8.5			
" " "	4	Ramberg and van Andel (1977)	10			
Kurchatov FZ,	S	Searle (unpub)	4.2	2.6	0.8	1.8
" "	N	Searle (unpub)	6.2			
45°N survey		Laughton (unpub)	8	4.0	1.3	2.7
Reykjanes Ridge		Laughton (unpub)	9.5	4.7	1.3	3.4

$2y_1$ = separation of inward facing faults
Δy = fault spacing outside median valley
g = distance of fault generation from the axis

possible that GLORIA is more sensitive to near-vertical scarps which will give strong echoes at long range when the ray path is nearly horizontal, whereas the Deep Tow echosounder, still working in a vertical mode, is unable to distinguish between the steep, but not vertical, flow fronts of relic linear volcanoes, and near-vertical faults. Furthermore, it is difficult to determine tilt angles of a rough volcanic terrain since there is no obvious reference surface. If the tilt of large blocks is appreciable, say up to 40° as has been suggested from palaeomagnetic studies of drilled basement rocks (Hall, 1978), this might be confused with outward facing faulting. Submersible observations of the gravity-controlled extrusions from the crests of rotated volcanoes should be able to determine the role of tilting in the crestal mountains and beyond.

There was little data on the ratio of inward to outward facing faults in the Kurchatov Fracture Zone area. However, the high reflectivity of the scarps when viewed outwards and the shortness of the returning echoes (Fig. 2) indicate scarps with a mean slope of 47° on the inward-facing sides of the crestal mountain ridges (Searle and Laughton, 1977), an angle higher than the resolution of the processed echosounder profiles. However, some of these lower-resolution profiles suggest a net asymmetry of the ridges indicating a majority of inward facing scarps. A significant feature found on these profiles, as well as on others on mid-ocean ridges (e.g. in the Gulf of Aden (Laughton, Whitmarsh and Jones, 1970)), and noted by Macdonald and Luyendyk in the FAMOUS area, are the "lips" perched on the uplifted edges of fault blocks (see also Fig. 8). They are interpreted as relic volcanic features and their location indicates that faulting often occurs near the foot of the axial volcanic ridges as shown by Ballard and van Andel (1977b).

In the 45°N survey, no narrow-beam or near bottom profiles were available to examine the asymmetry of the ridges. A comparison of GLORIA data on either side of the median valley indicates, however, a majority of inward facing scarps.

In contrast, in the Reykjanes Ridge survey considerable data were obtained on this issue. In some areas the same piece of sea floor was viewed from both sides and the contrast is evident in Fig. 3. In addition, the medium range side-scan sonar was well able to distinguish between inward and outward facing faults and to relate them to a profile beneath the ship (Fig. 8). Throughout the survey, out to a distance of 75 km from the axis, 192 inward facing and only 6 outward facing faults were crossed. This ratio is supported generally by the comparative views of inward and outward looking GLORIA sonographs.

A preponderance of inward facing faults can also be seen on sonographs collected along passage tracks parallel and close to the Mid-Atlantic Ridge axis between 54° and 37°N. One must conclude that this is, therefore, characteristic of much of the slow spreading ridge system in the North Atlantic.

(b) Fault Throws

Model 1 of Harrison and Stieltjes (1976) (faulting followed by 'unfaulting'), implies that the throw of inward facing faults should be at a maximum part way between the axis and the crestal mountains, and should decrease to zero at the crestal mountains. Sonographs have given data on scarp heights (interpreted as fault throws) in the Kurchatov Fracture Zone area and on the Reykjanes Ridge. Searle and Laughton (1977) correlated target widths from sonograph echoes near the Kurchatov Fracture Zone with scarp heights on selected echosounder profiles and used this relationship to infer scarp heights from the width of the sonar targets over a much wider area. On the Reykjanes Ridge, inferred fault throws were measured from the medium range side-scan sonar profiles directly. When plotted against distance from the axis (Fig. 9), the data points show no correlation with the theoretical distribution required by Model 1. The data on the Reykjanes Ridge are taken from the southerly section of the survey where the median valley is well developed. The gradual decrease in maximum fault throw and in the mean scarp height near the Kurchatov Fracture Zone (Searle and Laughton, 1977, Fig. 11) out to 60 km from the axis, can be attributed to scarp burial by pelagic sedimentation and the growth of talus slopes.

(c) Fault Spacing

The spacing of faults along lines perpendicular to the spreading axis (not necessarily flow lines) was measured for all four areas on several typical sections, avoiding fracture zones, and was plotted against distance from the axis (Fig. 10). The data for the Reykjanes Ridge include both the southerly rifted section and the northerly unrifted section of the survey area. Since there is no significant difference between the fault spacings in the two sections, the two data sets have been combined. Spaces that included sediment basins exceeding 100 msec (approximately 75 m) in thickness or where there is poor sonar coverage due to extreme range or shadow, were excluded from the data. The spaces between the first inward facing pair of faults were also excluded. The plots show no significant change in the fault spacing beyond 10 km from the axis, implying that no new inward facing faults were generated beyond this range, and that all faults were, therefore, formed within the median valley (or median ridge) zone.

The complete data sets can, therefore, be used to give histograms of fault spacings for each area (Fig. 11), and to compute mean values (Table I). There is no significant difference between the mean fault spacings in the four areas. The mean value for all four areas, 2.4 km, agrees closely with that (2.4 km) observed by Macdonald and Luyendyk (1977) in the terrace regions of the FAMOUS area. However, each of the major scarps may in fact consist of a cluster of closely spaced faults producing slivers of crust some tens or hundreds of metres wide as observed on the inner walls of FAMOUS by Deep Tow. We think it likely that each cluster belongs to a single faulting episode.

(d) Length of Faults

A striking feature of the faults is their considerable length compared with their spacing. In the FAMOUS area, some faults can be traced continuously for as much as 35 km, ending at fracture zones. South of the Kurchatov Fracture Zone, fault scarps with only minor offsets were traced for as much as 60 km, although more commonly a single scarp length is 10-20 km. It is difficult to assess an average length since small apparent discontinuities may arise from problems of vehicle stability and quite commonly faults coalesce or branch at small angles.

Long closely spaced faults result in fault blocks that are long and thin. Assuming that a fault is initiated at one place and propagates as a tear, the close spacing of consecutive faults implies a remarkably uniform stress field or a well defined zone of weakness.

(e) Position of Fault Generation

The analysis of the statistics of fault spacing excluded the spacing of the two central inward facing faults either side of the median valley or ridge. It is widely accepted that the morphology of the inner floor of the median valley on mid-ocean ridges is predominantly due to extrusive volcanism and that the floor is bounded by inner walls of inward facing normal faults. These are the first major faults dislocating the newly formed and recently spread lithosphere and the position at which they occur relative to the spreading axis must be where the shear strength of the lithospheric wedge is exceeded by the vertical shear stress resulting from the accretion process.

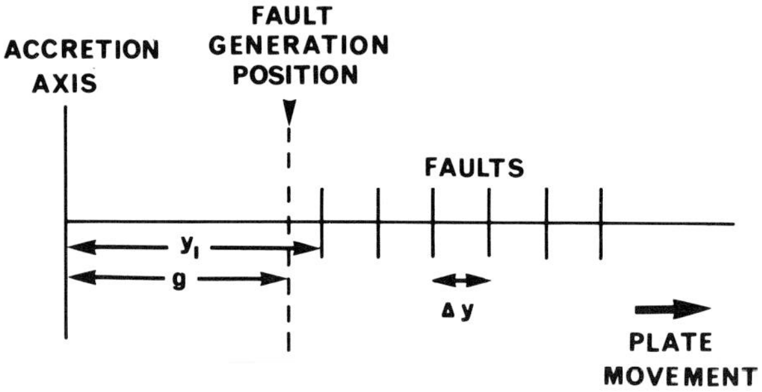

Fig. 12. Definition of measurements in calculation of fault generation position.

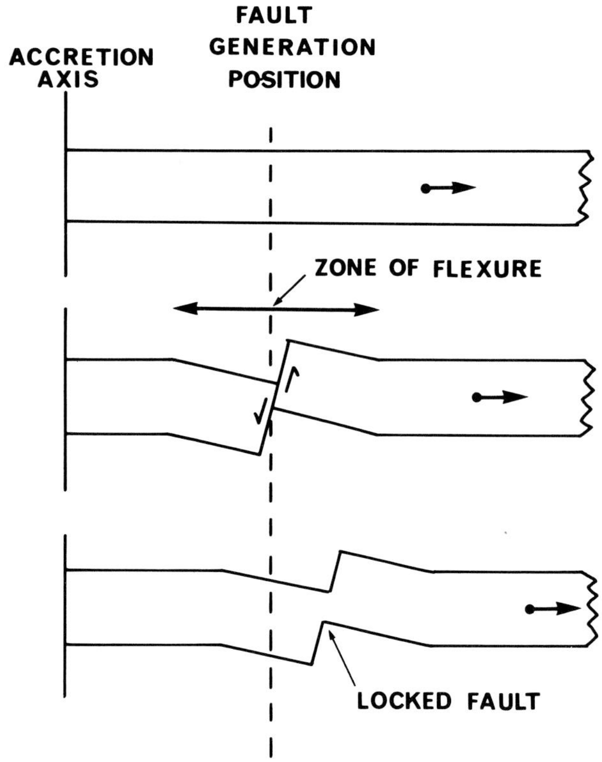

Fig. 13. Three stages, before, during, and after the formation of a fault in laterally moving uniform thickness lithosphere.

Since the major fault blocks have a finite width, the present position of the inner walls will on average be further from the spreading axis than the position of their origin. If g represents the distance of the fault generation position from the axis, y_1 the observed distance from the axis to the first inward facing fault and $\overline{\Delta y}$ the mean fault spacing, then

$$g = \overline{y}_1 - \frac{\overline{\Delta y}}{2} \text{ (Fig. 12)}.$$

Table II shows observations of the width of the spacing of the inner walls (or first pair of major inward facing faults) for the different survey areas, and the calculated values of g, using the appropriate fault spacing. The wide variation of y_1 between segments of the same survey area, separated by fracture zones, reflects the relative ages of the first fault scarp.

Geometry of Fault Generation
on Moving Lithosphere

The observed fault spacing implies an average time interval between the initiation of clusters of new faults of 200,000 years. We believe this characteristic period depends both on the rate of thickening and strengthening of the litho-

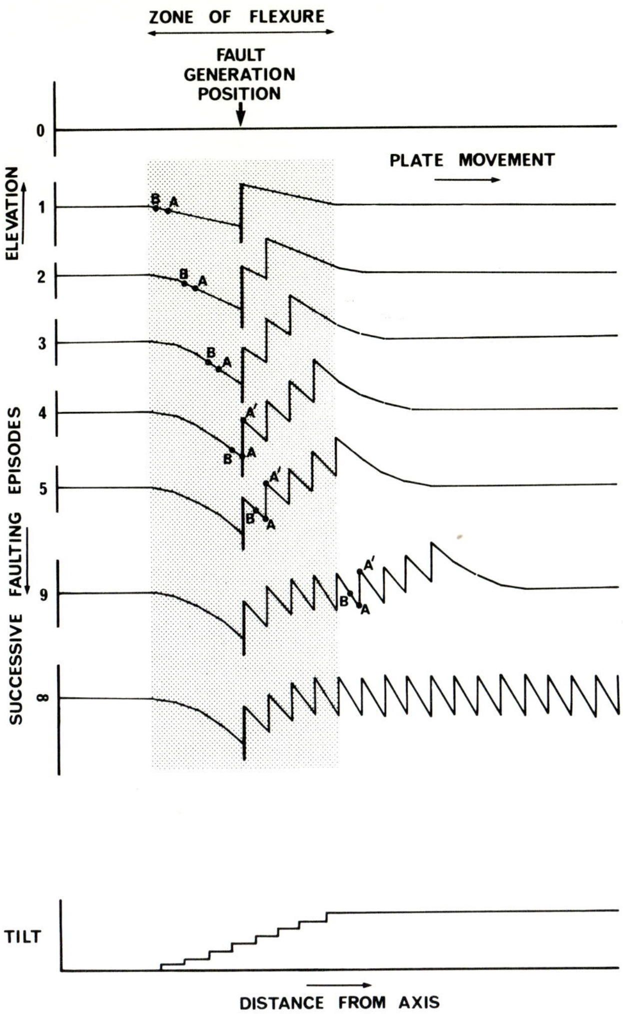

Fig. 14. Geometrical construction of profile obtained by repeated faulting and flexure of uniformly moving lithosphere. It is assumed for illustration purposes that the width of a fault block equals one-eighth of the width of the zone of flexure. The angle of tilt of the fault blocks is shown at the bottom. The flow lines of point A, A' and B are shown in Fig. 15.

sphere, and on the rate of horizontal variation of vertical shear due to viscous flow beneath the ridge crest (Lachenbruch, 1976).

The average inferred height of the innermost fault scarps near the Kurchatov Fracture Zone is not significantly different from the average scarp height; this suggests that fault scarps grow to their maximum height in a time much less than 200,000 years, and this is compatible with measured faulting rates (Francis, 1974; Faure, 1975). Sea-floor spreading must be non-continuous on a time-scale related to intrusive events - about 5 to 250 years for individual dykes 0.1 to 5 m wide (Coleman, 1977), or 10,000 years for the initiation of new extrusive volcanoes which must themselves be fed by intrusive dykes (Ballard and van Andel, 1977). The intervals between faulting episodes are, therefore, an order of magnitude greater than between volcanic episodes and hence, to a first approximation, sea-floor spreading is continuous between the faulting episodes.

Thus our data suggest that major faulting occurs at a small, but finite, distance from the spreading axis, episodically on a relatively uniformly moving lithospheric plate. As new crust moves past the faulting position it is faulted by forces related to the distance from the axis. As it moves away the faults become locked into the crust and subsequently do not change their throw significantly.

In order to determine the flow line of a point on the new crust it is necessary to model the effect of repeated faulting. Let us assume, as a first simplification, that the lithosphere retains a constant thickness after creation and that associated with the faulting is a zone of flexure either side of the fault (Fig. 13). (Some such assumption is needed since the whole lithosphere cannot be uplifted at each fault step). The next fault dislocates new crust, uplifting the now older frozen fault if it still lies in the zone of flexure, and increasing the flexure on the younger block. An equilibrium profile can be constructed by successive superpositions of fault and flexure until one point has moved right through the flexure zone (Fig. 14). One direct consequence of such a simplified geometrical construction is the persistence of a relative depression at the faulting position. (The corresponding relative elevation resulting from the first faulting is carried off with the plate). Throughout the

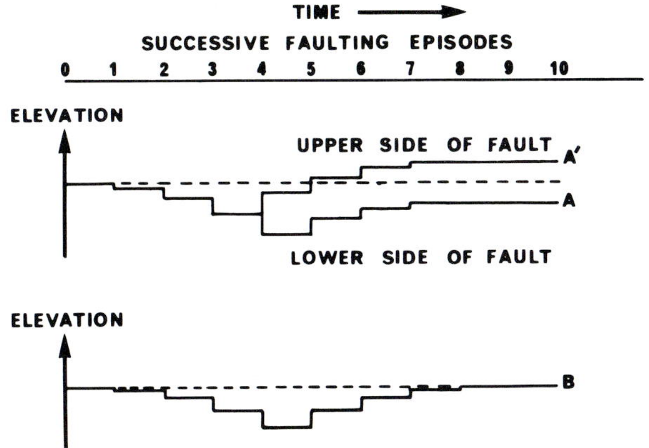

Fig. 15. The flow lines of points A and B (Fig. 14) plotted as elevations against time through ten faulting episodes. Point A has been chosen so that it lies at the position of a future fault and hence is divided by the fault plane into A and A'.

zone of flexure the fault blocks are progressively tilted to a maximum at the outer edge of the flexure zone. For a throw of 200 m and fault spacing of 2 km, the maximum tilt is 6°.

Outward tilting of fault blocks of an average 2° at the inner wall and between 3° and 8° at the outer walls in the FAMOUS area have been observed by Macdonald and Luyendyk (1977). On the Reykjanes Ridge, our medium range side-scan sonar has shown outward tilting as high as 20° to 30°, although more commonly about 10°.

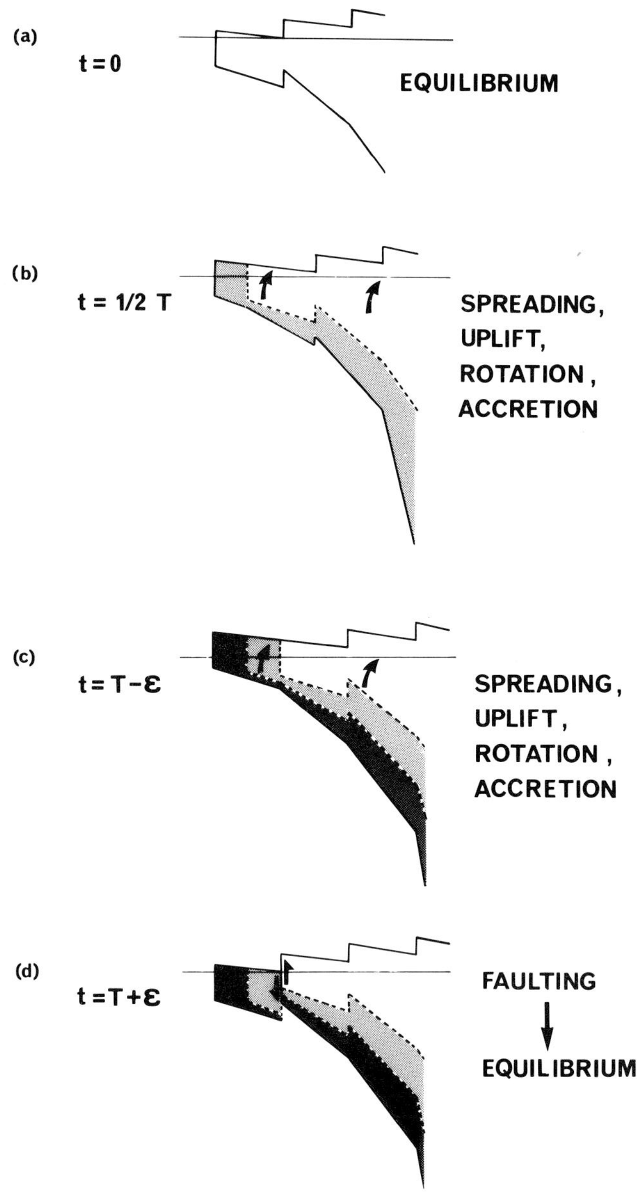

Fig. 16. Faulting and flexure in a more realistic lithospheric profile near the ridge axis assuming faulting with a periodicity of T. Stipple indicates the new lithosphere accreted at each stage.

The flow line of a typical point is dependent on whether it is in the centre of a block or immediately adjacent to a fault. Selected flow lines are shown in Fig. 15, derived from the labelled points in Fig. 14: a point (A) lying at a potential fault site will be depressed until faulting splits it into two points (A and A') on the down and up-faulted sides of the fault respectively. A point in the centre of a block (B), however, will be depressed, then elevated, but never rise above the mean level. These flow lines take no account, however, of adjustments in height that may arise from non-equilibrium forces from below.

If lithospheric cooling and shrinking is superposed (Sclater and Francheteau, 1970) on the synthetic topographic profile in Fig. 14, the ridged profile far from the axis will gradually increase in depth.

This model says nothing about the origin of the stress that creates the fault and is clearly oversimplified by the assumption of uniform lithospheric thickness. A more likely model is one in which a relatively thin rigid lithosphere (i.e. the inner floor of the median valley) overlies the axial region above the mantle conduit or magma chamber, beyond which the lithosphere increases in thickness as material near the conduit or magma chamber walls cools and adds to the plate. At a time immediately following the relief of stress by the generation of an inner wall fault, the lithospheric cross-section may approximate to (a) in Fig. 16. The stresses on the mantle conduit walls, arising from gravity and viscous forces exerted by the rising material (Lachenbruch, 1976) will uplift both the lithospheric wedge and, due to the finite shear strength of the younger lithosphere, the inner floor region (Fig. 16b, c). The latter will be out of equilibrium with the hydrostatic pressure head, reduced by viscous drag, in the rising mantle column, and will eventually collapse giving rise to a normal fault (Fig. 16d). The flexure associated with the uplift will, however, be preserved by the continual underpinning of the lithospheric wedge by mantle accretion on its walls.

Conclusions

The paper has set out to present new evidence on the faulting associated with slow spreading mid-ocean ridges and to provide constraints and perhaps new insights into modelling the tectonic processes at and near the accretion axis.

The following conclusions have been reached:-

(1) There is a pattern of normal faults aligned parallel to the median valley or ridge axis, but not necessarily normal to the spreading direction.

(2) The majority of these faults are facing inwards towards the axis.

(3) There is no significant change in fault throw beyond the crestal mountains, implying that

reverse faulting along pre-existing faults does not take place.

(4) The spacing between faults does not decrease with distance from the axis, implying that no new faults are generated outside the median valley.

(5) Typical fault spacing is 2 to 2.5 km, equivalent to initiation of a new fault, or cluster of faults, about every 200,000 years.

(6) Faults are generated at a more or less constant distance (2-4 km) from the ridge axis as the lithosphere moves past this position.

(7) Fault lengths vary from 5 to 50 km giving rise to long, thin fault blocks, sometimes continuous between fracture zones. This implies that the stress field is extremely uniform in the direction parallel to the axis.

From these data it is concluded that the profile of median valley, valley wall complex, crestal mountains, and ridge flank can be explained qualitatively in terms of a lithosphere continuously generated at the axis being transported laterally, through a region where the shear stresses associated with the edge of the asthenosphere conduit generate normal faults and associated flexure. As repeated flexure is added to the already faulted lithosphere, the fault blocks are progressively tilted to an angle of between 5° and 10° until they reach the crestal mountains. This region (from inner wall to crestal mountains) denotes the horizontal extent of the flexure zone, whose width may be related to the rate of increase of lithospheric thickness away from the axis, and the rate of change of vertical stress. Beyond the crestal mountains the thermal shrinking of the lithosphere accounts for the increasing depth of the sea floor along the ridge flanks.

Acknowledgements. We wish to thank the Masters, Officers, crews, and shipboard scientists of R.R.S. Discovery who all contributed to producing the data used here. We are grateful to Tanya Atwater, Ken Macdonald and Norman Sleep for constructive comments on the manuscript.

References

ARCYANA, Transform Fault and Rift Valley from Bathyscaph and Diving Saucer, Science, 190, 108-116, 1975.

Aumento, F., B.D. Loncarevic and D.I. Ross, Hudson Geotraverse: geology of the Mid-Atlantic Ridge at 45°N, Phil. Trans. Roy. Soc. Lond., A,268, 623-650, 1971.

Ballard, R.D. and Tj.H. van Andel, Project FAMOUS: Operational techniques and American submersible operations, Geol. Soc. Amer. Bull., 88, 495-506, 1977a.

Ballard, R.D. and Tj.H. van Andel, Morphology and tectonics of the inner rift valley at lat $36^{\circ}50$'N on the Mid-Atlantic Ridge, Geol. Soc. Amer. Bull., 88, 507-530, 1977b.

Coleman, R.G., Ophiolites, 229 pp., Springer-Verlag, Berlin, Heidelberg, New York, 1977.

Faure, H., Neotectonics in the Afar (Ethiopia, T.F.A.I.), in Quaternary Studies, pp. 121-126, edited by R.P. Suggate and M.M. Cresswell, Royal Society of New Zealand, 1975.

Francis, T.J.G., A new interpretation of the 1968 Fernandina Caldera collapse and its implications for the Mid-Oceanic Ridges, Geophys. J. R. astr. Soc., 39, 301-318, 1974.

Glenn, M.F., Introducing an operational multibeam array sonar, Int. Hydrogr. Rev., 47, 35-39, 1970.

Glenn, M.F., Multi-narrow beam sonar systems, in Oceans '76, 2nd Annual Combined Conference MTS/IEEE, pp. 8D-1 to 8D-2, Inst. Electrical and Electronics Engineers, Inc., New York, 1976.

Hall, J.M., 2nd Ewing Symposium (this volume), 1978.

Harrison, C.G.A. and L. Stieltjes, Faulting within the median valley, in Present State of Plate Tectonics, pp. 137-144, edited by J. Bonnin and R.S. Dietz, Tectonophysics, 38, (1-2), 1977.

Heirtzler, J.R. and Tj.H. van Andel, Project FAMOUS: Its origin, programs, and setting, Geol. Soc. Amer. Bull., 88, 481-487, 1977.

Lachenbruch, A.H., Dynamics of a Passive Spreading Center, J. Geophys. Res., 81, 1883-1902, 1976.

Laughton, A.S. and J.S.M. Rusby, Long range sonar and photographic studies of the median valley in the FAMOUS area of the Mid-Atlantic Ridge near 37°N, Deep-Sea Research, 22, 279-298, 1975.

Laughton, A.S., R.C. Searle and D.G. Roberts, The Reykjanes Ridge crest and the transition between its rifted and non-rifted regions, Tectonophysics, in press.

Laughton, A.S., R.B. Whitmarsh and M.T. Jones, The evolution of the Gulf of Aden, Phil. Trans. Roy. Soc., A,267, 227-266, 1970.

Loncarevic, B.D. and R.L. Parker, The Mid-Atlantic Ridge near 45°N. XV. Magnetic anomalies and sea-floor spreading, Can. J. Earth Sci., 8, 883-898, 1971.

Lonsdale, P., Structural geomorphology of a fast-spreading rise crest: the East Pacific Rise near $3^{\circ}25$'S, Marine Geophys. Res., 3, 251-293, 1977.

Macdonald, K.C. and T.M. Atwater, Evolution of rifted ocean ridges, Earth Planet. Sci. Lett., 39, 319-327, 1978.

Macdonald, K.C. and B.P. Luyendyk, Deep-Tow studies of the structure of the Mid-Atlantic Ridge crest near 37°N (FAMOUS), Geol. Soc. Amer. Bull., 88, 621-636, 1977.

Ramberg, I.B., D.F. Gray and R.G.H. Raynolds, Tectonic evolution of the FAMOUS area of the Mid-Atlantic Ridge, lat $35^{\circ}50$' to $37^{\circ}20$'N, Geol. Soc. Amer. Bull., 88, 609-620, 1977.

Rea, D.K., Model for the Formation of Topographic Features of the East Pacific Rise Crest, Geology, 3, 77-80, 1975.

Rusby, S., A long range side-scan sonar for use in the deep sea, Int. Hydrogr. Rev., 47, 25-39, 1970.

Rusby, J.S.M. and M.L. Somers, The development of the 'GLORIA' sonar system from 1970-75, in Voyage of Discovery, pp. 611-625, edited by M.V. Angel, Supplement to Deep-Sea Research, 24, 1977.

Sclater, J.G. and J. Francheteau, The implications of terrestrial heat-flow observations on current tectonic and geochemical models of the crust and upper mantle of the Earth, Geophys. J. R. astr. Soc., 20, 509-542, 1970.

Searle, R.C. and A.S. Laughton, Sonar Studies of the Mid-Atlantic Ridge and Kurchatov Fracture Zone, J. Geophys. Res., 82, 5313-5328, 1977.

Somers, M.L., R.M. Carson, J.A. Revie, R.H. Edge, B.J. Barrow and A.G. Andrews, GLORIA II - An improved long range side-scan sonar, Oceanology International 1978, IEEE/IERE Sub-Conference on offshore instrumentation and communications, Tech. Session J, 16-24, BPS Publications Ltd., London, 1978.

Spiess, F.N. and R.C. Tyce, Marine Physical Laboratory Deep-Tow Instrumentation System, Scripps Institution of Oceanography Reference 73-4, 1973.

Tucker, T. and A.R. Stubbs, Narrow-beam echo-ranger for fishery and geological investigations, British J. Applied Physics, 12, 103-110, 1961.

Vogt, P.R. and O.E. Avery, Detailed Magnetic Surveys in the Northeast Atlantic and Labrador Sea, J. Geophys. Res., 79, 363-389, 1974.

Whitmarsh, R.B. and A.S. Laughton, A long range sonar study of the Mid-Atlantic Ridge crest near 37°N (FAMOUS area) and its tectonic implications, Deep-Sea Research, 23, 1005-1023, 1976.

CONSTRAINTS FROM THE FAMOUS AREA CONCERNING THE STRUCTURE OF THE OCEANIC SECTION

Tanya Atwater

Department of Earth and Planetary Sciences, Massachusetts Institute of Technology
Cambridge, Massachusetts 02139

Abstract. Structural data from the Famous Area, near 37°N on the Mid Atlantic ridge, are examined for possible constraints they may place upon the interpretation of materials collected in deep basement drill holes in the Atlantic. Volcanic, petrologic, magnetic and crude age data all suggest that the zone of emplacement of new lavas is 5 to 10 km wide, with possible occasional new additions as much as 15 km from the center. Implications for dike-flow relationships in the section are discussed. The most likely model appears to be one with a narrow, centered focus of volcanism which is then covered with a thinner covering layer by off-axis, late stage volcanism. The dimensions of volcanoes in the Famous Area, combined with magnetic layer thickness studies suggest a volcanic layer about three to six volcanoes thick. Each volcano consists of a pile of about 25 flows of highly variable thickness which erupt over a relatively short time span compared to the 5,000 to 10,000 years between major eruptions.

Rubble in the section is likely to be generated in two ways: (1) as horizontal breccia sheets which broke off of steep flow fronts and then were overrun as the fronts advanced and (2) as fault zone breccias. The spacing and geometry of faults suggest that there is a 30% probability that a random 500m drill hole will encounter at least one major fault zone.

Two possible origins for large tectonic tilts are considered. Individual volcanic units may be rotated during burial, then larger coherent blocks may suffer additional tectonic rotations as they pass the rift mountain tops. The latter tilts probably affect less than 10% of the crust.

In the transform faulted region of the Famous area (age 5 m.y. and younger), about 3/4 of the crust appears to be "typical" oceanic material vs that which is highly tectonized by transform faults or which was originally emplaced there.

Introduction

In this paper I shall attempt to outline what we can tell about the structure of Atlantic Ocean crust from the point of view of its original formation. I shall draw most heavily upon data collected at the Mid Atlantic ridge near 36-37°N during the FAMOUS Project, adding a few preliminary remarks based upon a return expedition to that area, AMAR78, reported by Atwater et al. (1978), Macdonald and Atwater (1978), Atwater and Macdonald (1978), Crane and Ballard (1978) and Ballard et al. (1978). Comparisons to other parts of the ridge will also be made. Structurally, at least, the Famous section of the ridge crest appears to be quite typical, having a central rift valley with typical relative depth and width, fault block topography, mountainous flank relief, and rough but just discernable Vine-Matthews Anomalies. I shall summarize some of the volcanic and structural results from this area and discuss their implications for materials which might be encountered in a deep sea drilling basement hole. This is not an easy task, since most of the Famous data concern the horizontal upper surface of the ocean floor while the basement holes examine the vertical section to depth. Still, the former do include some implications for the vertical section. In particular, they put constraints upon the internal composition of the volcanic layer, the origin of rubble within the section, and the tilting of rocks during and after their emplacement. In addition, they allow one to estimate the amount of sea floor area which is "typical" oceanic crust versus that which was generated or was seriously deformed in fracture zones.

Structure and History of the Volcanic Layer

Width of the Zone of Extrusion. Before the FAMOUS project, the most common type of model for the internal structure of the volcanic layer was that illustrated in Figure 1a. It assumes that the zone of addition of new lavas is narrow and well centered. Thus, each new volcano splits the previous one in two, building a new edifice overlapping the older ones in a quite orderly fashion.

Split volcano models such as the one in Figure 1a imply that only one major volcanic vent, a.

b.

c.

d.

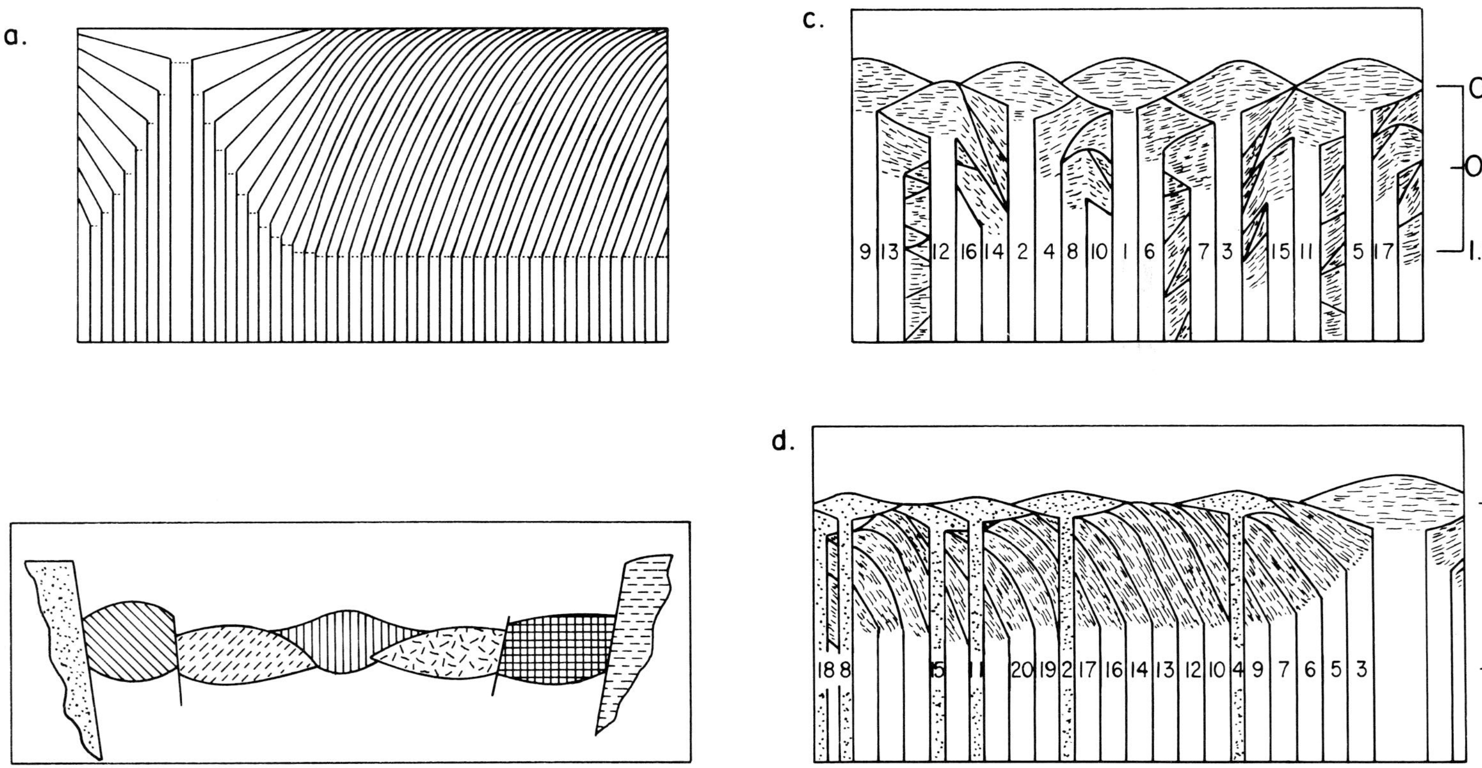

Fig. 1. Models proposed for the internal composition of the oceanic volcanic layer. a. Centered, split volcano model, from Cann, 1974. One half of each volcano is carried to each side. Surface consists entirely of flow toes beyond the central volcano. b. Whole volcano model, from Ballard and van Andel (1977). Each new volcano is emplaced at one foot or the other of the previously active center. Surface consists of whole, intact volcanoes, as observed in Famous area. c. Buried whole volcano model, about five volcanoes thick, built up over wide extrusion zone. Surface consists of whole volcanoes, as observed. Section at depth shows pervasive mixing of flows and dikes, unlike ophiolite sections. d. Split volcano model with additional off center late stage volcanism. Surface consists of whole volcanoes while section at depth is relatively dike free.

the central one, should be present on the sea floor. All older volcanoes would be represented on the surface only by their tilted flow toes. This is not what was found in the Famous area. Rather, the rift valley floor consisted of a number of hills flanking the central one and all the hills which were studied seemed to be intact volcanoes. Their flow directions were still oriented downhill and their summits were decorated with cones and "haystacks" (Ballard and van Andel, 1977). From this observation Ballard and van Andel postulated a somewhat different family of models for volcanic emplacement, illustrated in Figure 1b. They suggested that each new volcanic episode begins at the foot of an older edifice so that in general, whole volcanoes are carried away to one side or the other. Unfortunately, dating techniques for such young oceanic rocks are very poor so that it is not possible to distinguish the sequence of emplacement of the hills. Ballard and van Andel develop a model in which active volcanic centers are only slightly off center, each new one originating at one foot or the other of the previously active volcano, but they note that the data are equally compatible with a model in which activity occurs anywhere over some broader area and in a more random manner. The nearly centered model was motivated by the fact that the most recent volcanoes in the American submersible area are centered. On the other hand, one very recent volcano to the north, Mt. Mercury, is actually perched upon the first step of the western wall (Arcyana, 1975; Macdonald and Luyendyk, 1977) suggesting a more random emplacement of volcanoes.

Crustal emplacement models constructed to match magnetic effects also favor a wider zone of volcanic additions. In particular, Schouten (this volume) constructed a simplified model assuming that new discrete volcanoes are distributed about the center in a Gaussian way. When he tried to model various observed magnetic characteristics (polarity transition widths, magnetic anomaly variability and the presence of mixed polarities in deep crustal holes) he found that they required a standard deviation of about 2.5 km or a little less. In other words, most new volcanoes are added over a zone about 10 km wide and about 70% of them are spread over the central 5 km (in or near the rift valley floor).

Another study of the emplacement width was made by Maddonald (1977) when he estimated the bulk magnetic intensity of individual ocean floor peaks and scarps. Magnetic intensity decays very rapidly with age, from 20-30 $A.m^{-1}$ for features less than a few hundred thousand years old to about 7 $A.m^{-1}$ at 1 million years. Thus, off-axis late stage volcanoes should have higher magnetic intensities than their surroundings. While most of Macdonald's intensities cluster about a smooth decay curve, a small number were clearly too intense for their distances from the center (figure 2). A possible explanation for these features is that they were late additions upon the older

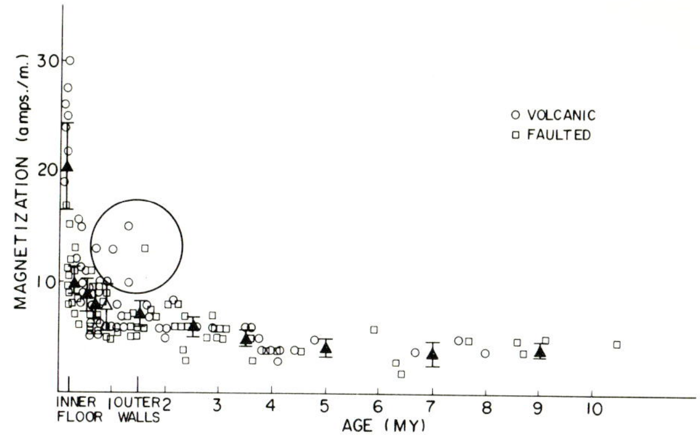

Fig. 2. Bulk magnetic intensity of individual topographic features as a function of expected age, after Macdonald (1977). Triangles show average values and bars are 4 standard errors in length. Circle shows features with intensity greater than that typical for their expected ages, perhaps indicating that they were formed by younger, late stage volcanism.

crust away from the center. In the region where this effect is easily detectable (.5 - 2 m.y. expected age), about 10% of the surface features seem to be .5 to 1.3 m.y. younger than their surroundings, i.e., they were emplaced 5 to 15 km outside the center of emplacement. This conclusion should be approached with caution, however, since the AMAR expedition found very little young volcanism associated with Mt. Saturn which had a high intensity deep tow magnetic anomaly (Atwater, et al, 1978). The source of that anomaly is presently being investigated.

Petrologic, geochemical and aging data from the rocks collected in the Famous Area also suggest a wider zone of emplacement, though not so wide as just discussed. Bryan and Moore (1977) found a strong zonation in rock type, from primitive in the center to differentiated at the edges of the valley floor. This they interpreted to be the result of an evolving, continually leaking magma chamber, so that all the surface lavas overlying the chamber are quite young. Crude age dating measures (manganese thickness and palagonitization (Bryan and Moore, 1977) and rock magnetic intensity (Johnson and Atwater, 1977) likewise suggest that the surface rocks all the way to the edge of the valley floor are younger than would be expected from a narrow, centered emplacement. Bryan and Moore supposed that the late stage volcanism is confined to the rift valley floor (a region 3-5 km wide). It is often supposed that the bounding faults mark the edge of the underlying magma chamber. However, very little is known about the ages of the rocks on the walls and beyond, so that this is not a compelling constraint at this time.

In conclusion, most data from the Famous rift valley suggest that the zone of extrusion is at least as wide as the valley floor, with possible occasional volcanic additions as much as 15 km from the center. This implies that the difference in age between the bottom and top flows in a given section will commonly be 200,000 years and occasionally will be 1 million years or more. The implications of such a wide region for volcano emplacement are very strong for models of the structure of the deeper volcanic section. Namely, the relationship between flows and feeder dikes and the ratio of one to the other at given depths would be entirely different from those in the countered split volcano model. Any whole volcano model requires that a significant number of dikes should be present high up in the volcanic section, and their number should increase rapidly with depth. Compare figures 1a with 1c, for instance. Any increase in the width of the emplacement zone increases this intermingling of dikes and flows. Occasional clear crossings of dikes have been reported in the deep crustal holes (eg. Bryan et al., 1977), and Flower et al., 1977) and others could easily go undetected as such. On the other hand, ophiolite sequences seem to show relatively little intermixing of dikes and flows (Cann, personal communication).

Volcanoes in the Vertical Section. Ballard and van Andel considered only the surface section in their emplacement models. The three deep crustal holes which have penetrated beyond 500 m in the basement were still encountering extrusives when they stopped drilling (Melson et al., 1974; Rabinowitz et al., 1976 and Flower et al., 1977). Thus, it seems clear that the oceanic layer 2A is generally more than one volcano thick. One simple extension of the intact volcano model is to assume that the volcanic layer consists of buried whole volcanoes. Figure 1c shows such a possibility. Given this assumption, we can use the dimensions of various volcanoes studied in the Famous Area to estimate the number of whole volcanoes which might be encountered in a given crustal hole.

Individual volcanoes in the Famous Area are about 500 to 700 m wide and 2 km long. The height of the edifice built during a given volcanic episode is not known, but if it is assumed that the present central volcanoes, Mts. Venus and Pluto, grew to fill intervolcano depressions such as the one which presently lies between them, then they are 200-300 m thick at their centers. This, however, may be an underestimate, since faulting surrounding them suggests that they sink into the crust after and probably during their construction (Ballard and van Andel, 1977).

The thickness of the volcanic layer is not known. Macdonald (1977) found an equivalent thickness of about 700 m by assuming that the entire layer has the same magnetization as near surface features. This is probably an underestimate since more and more instrusives and altered sections would be included with increasing depth and these would have lower magnetic intensities than the young lavas. Also, as noted above, direct samples from the bottoms of deep crustal holes included pillow basalts to nearly 600 m depth into the basement. For the sake of discussion, then, we might suppose that the layer which includes significant amounts of quenched lavas extends to one km depth. If such a layer is built of whole volcanoes similar to Mts. Venus and Pluto, as described above, it would be three to six volcanoes thick.

In any given surface crossing of the valley floor, parallel to the spreading direction, the volcanoes identified by Ballard and van Andel (1977) occur at spacings of about 600 m. At the spreading rate here, this is equivalent in time to a spacing of 30,000 years. If, indeed, the volcanic layer contains two to five buried volcanoes for every one at the surface, then the average time between new volcanic starts would be 5,000 to 10,000 years.

It is not known how long a given volcanic episode lasts: whether there is a more or less regular outpouring of lava so that a given volcano takes 5,000 years to be formed, or whether the volcanoes are built very quickly, with long quiet periods in between. A number of lines of evidence suggest the latter hypothesis. The divers reported that all the lavas explored on each of the two youngest volcanoes, Venus and Pluto, seemed to be of similar age, in contrast to the older rocks surrounding them. The presence of thick units in the deep crustal holes within which geochemical, petrologic and magnetic properties are nearly constant, also argues for a very rapid building of thick piles.

Another promising model for the structure of the volcanic layer deserves consideration. It is essentially a hybrid, with a centered, split volcano structure for the lower portion of the construction plus an addition of superficial volcanoes from smaller amounts of late stage, off-axis volcanism. This is illustrated in figure 1d. This model satisfies the observations of a wide extrusion zone and intact volcanoes at the surface while minimizing and localizing the number of dikes in the volcanic section at depth. It constructs a section which is similar to that reported from Troodos, figure 3, and is similar to the model proposed by Bryan and Moore to explain their geochemical gradients in the Famous Area. It also shares many features with the structure found on the fast spreading East Pacific Rise at 3.5°S (Lonsdale, 1977), although the various components are somewhat different and are differently developed. On the East Pacific Rise the central volcano is by far the dominant volcanic feature. It is linear, continuous and quite uniform, quite different from the alternating central highs and lows of the Famous Area. Off center volcanism at the East Pacific Rise appears to be represented only by small knobs scattered sparsely upon the terrain of the dominant central volcano. Thus, at Famous we must postulate an

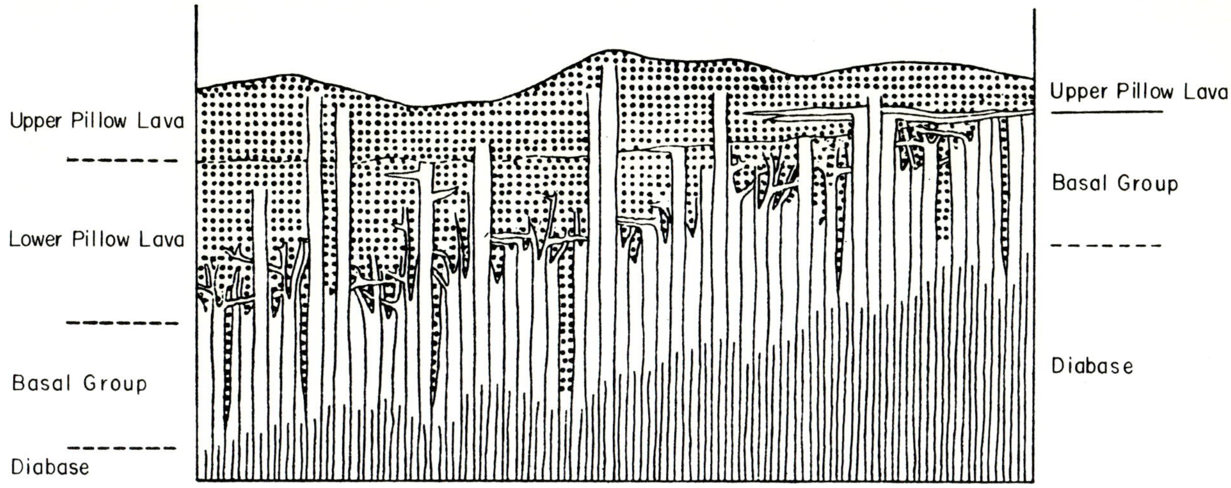

Fig. 3. Diagrammatic sketch showing the relationships in cross-section between dikes, flows and diabase in Troodos, Cyprus, after Wilson (1959) as published in Coleman (1977), Figure 68. Dike/flow relationships suggest that the Lower Pillow Lava group may have been formed by a centered split volcano emplacement while the Upper Pillow Lava group represents off-axis, late stage volcanism, as illustrated in Figure 1d.

intermittant central edifice and a much more active late phase, one which conveniently hides most of the evidence for the earlier stage.

Flow Structure within Volcanoes. The Famous data may also be of use in understanding the lava structure encountered within a given volcano. Many flow fronts were examined during the Famous dives. Figure 4 shows a cross-sectional sketch of an advancing flow front as surmised by Ballard and Moore (1978) from these observations. The flow can be viewed as a sloping lake, budding on top and advancing by overflowing its dam and building new foresets of elongate pillows down the front and then moving over them. Van Andel and Bryan (personal communications) estimate that these volcanic fronts are typically 10 to 20 m high, although a great variability was seen. If we take this as a rough estimate for the thickness of flows and assume that a given volcano is 300 m thick, then a given section through the edifice might consist of about 25 flow units. Given the many uncertainties in such an estimate, however, it must be considered only a very rough working number. A given vertical section downward through such a flow would consist of top bulbous pillows (sometimes hollow and collapsed), a more massive core, lower elongate foreset pillows, and then a rubble sheet put down by materials spalling off of the advancing front and subsequently being run over.

Sheet Flows in the Section. An additional complication in the volcanic section will be the presence of sheet flows. These have been described in the Galapagos spreading center (Lonsdale, 1977, and Ballard et al., in press) but were not previously reported in the Atlantic. During the AMAR78 expedition we encountered these thin platey lavas forming lakes in many of the lows (Atwater, 1978, Ballard et al., 1978). In the various fault scarps examined we estimate that sheet flows constituted 10 to 15% of the vertical sections. This percentage probably increases with depth, since the sheets occur in the lowest spots and would be preferentially buried.

Rubble and Talus Within the Section

Volcanic rubble is one of the most common rock types found in deep crustal holes. What is the origin of this rubble?

Surface Debris. One of the more dramatic aspects of the ocean floor is the great amount of talus found there. However, it is probably a mistake to suppose that equal amounts have been buried. Most of the large, impressive talus piles are found at the bases of the major fault scarps and it appears that much of the large offset faulting occurs as the crust is uplifted away from the valley floor, after the completion of volcanism in a given section. Thus, the really large talus ramps are only surface phenomena and will not, in general, be found buried within the volcanic section.

The central valley floor, i.e., the part most likely to be buried by lavas in the future, is relatively barren of talus. Small amounts of rubble are seen where hollow pillows and tubes have collapsed and at the feet of small scarps and flow fronts. Still, the total area covered by talus reported by Ballard and van Andel in their detailed dive track descriptions, their figure 2, averages only 9% while deep tow photo

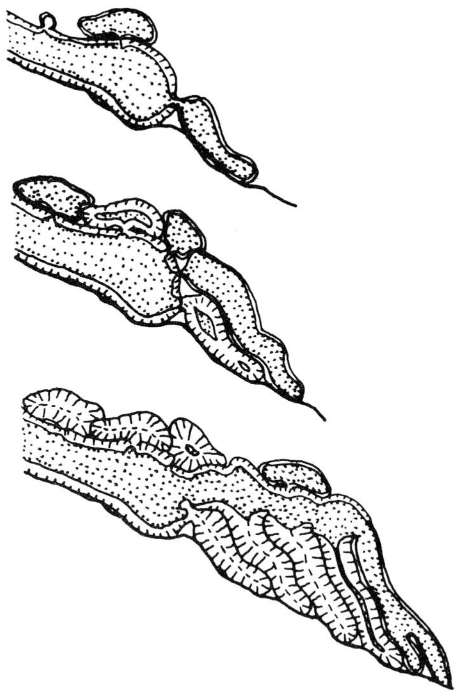

Fig. 4. Structure of an advancing pillow flow front as deduced by Ballard and Moore (1977) figure 21, from Famous Area observations. Dotted pattern indicates molten lava.

analysis indicates only 15-20% (Luyendyk and Macdonald, 1977).

Another probable source of buried rubble is implied by the observations of flow fronts by Ballard and Moore (1978). As a front advances, the steep elongate foreset pillows tend to break away, forming a rubble heap at the foot of the front. Thus every flow should rest upon a sheet of its own rubble which broke off of the oversteepened flow front as it advanced. The rubble formed by this process is different from that found in the fault scarp talus ramps in that it consists of whole or half pillow fragments while the latter generally consists of wedge shaped fragments (Atwater et al., 1978). It is not clear that such a whole pillow talus would be distinguishable in a drill core from the surrounding intact pillow flows. Some magnetic oddities such as discrepancies between magnetic and segregation vesicle orientations (Gruver, 1978) could arise from this ambiguity.

Fault Shearing and Gouge. Another way in which to form broken rock is by faulting. It appears that every section of crust formed in the Famous rift undergoes three episodes of deformation before it is finally welded on to the passive oceanic plate. First, it is pervasively fissured and faulted in the valley floor, then it is block faulted to form the rift valley walls. Finally it is tilted or faulted yet again to form the undulating rift mountains.

One of the most impressive features of the rift valley floor is its pervasive parallel system of small faults and fissures. These features are remarkably linear and parallel to the valley trend and have nearly vertical dips. The fissures presumably overlie dikes emplaced at depth which did not reach the surface. Their upper, empty parts become depositories for sediments and rubble. The ubiquity of these fine scale fractures can be seen in the table presented by Ballard and van Andel (1977) which lists tectonic features observed from the submersible Alvin. These data cover 7.5 (projected) km of east-west tracks within the valley floor during which they observed nearly 100 fractures or fracture groups, with an average spacing of about 80 m and offsets of 0.2 to 25 m. It is likely that many additional fractures of this type have been subsequently buried by lavas so that their frequency probably increases rapidly with depth in the section.

When oceanic crust moves outward to the edge of the valley floor it reaches a point where it begins to ascend the valley walls. This is accomplished by faulting the crust into long, narrow blocks and uplifting the separate blocks to form a staircase of sorts. These faults tend to have much greater throw (100 m average) and wider spacing (1.5 km average) than the valley floor fractures and are quite consistently inward dipping. Many of these "master faults" when examined in detail are represented at the surface by fault zones 100 m to 1 km wide consisting of a number of faults and slivers (e.g. see Figures 4 and 5 of Ballard and van Andel, 1977).

The actual values of the dips of these faults are difficult to ascertain. Macdonald and Luyendyk (1977) measured the slopes of the upper, talus free parts of 115 fault scarps crossed by deep tow and found an average slope of 50°. This is probably an underestimate of the actual fault dips since both erosion and the presence of undetected fault slivers and gouge would tend to produce a scarp slope which is less than the fault dip. The largest, clearest, best studied scarp, the inner west wall, has dips exceeding 75°. It is possible that this gives a better estimate for all fault dips and that other faults, with better resolution data, would be as steep. On the other hand, the inner west wall is clearly anomalous in its large throw and may be in dip as well.

The depth to which these faults penetrate as brittle features can only be surmised. As a rule of thumb, materials lose their ability to store stress at about one half of their melting temperatures (in degrees K), i.e., at 700-800°C for basalt. In thermal models, this temperature is reached at about 1-1.5 km depth. On the other hand, these master faults may well become conduits of easy access for cold water penetration, cooling a halo around them and allowing the cracks to propagate deeper, cool deeper, propagate, cool, etc. (Lister, 1974). This could in fact be the mechanism for solidification of the underlying magma chambers at this point in space, as is often assumed to occur (e.g., Bryan and Moore, 1977).

It is interesting to consider what the chances are of encountering a fault or fault zone in a randomly placed drill hole. Given the above parameters the probability of hitting at least one large offset normal fault in a 500 m vertical hole is quite large, about 30%.

The nature of the third, mountain top deformation episode was investigated during the AMAR78 expedition and is presently being studied. Large scale flexures seem to characterize this zone (Atwater et al., 1978, Macdonald and Atwater, 1978). Their manifestation at depth is problematical.

Tectonic Tilt

Basalts recovered from deep sea drilling holes often have magnetic inclincations which are radically different from those predicted (e.g., see Hall and Ryall, 1977, or Lowrie, this volume). Furthermore, these "anomalous" inclinations can be constant over thick portions of the section. These variations are larger than would be reasonably predicted from secular variations of the field. The most popular explanation for them is that some sort of tectonic tilt occurred after they were emplaced. Large tilts of 30 degrees or more are often required.

During the AMAR78 Expedition, we collected evidence near the rift mountain tops which suggests tectonic rotations of substantial blocks of crust by as much as 30° outward (Atwater et al., 1978, Macdonald and Atwater, 1978). This is in opposition to previous interpretations of morphologic data which appeared to preclude post formational tilts of more than a few degrees (Macdonald and Luyendyk, 1977, Macdonald and Atwater, 1978). In the ridge flank profiles presented in figure 10 of Macdonald and Luyendyk (1977), about 10% of the surface has a substantial outward slope. Some of these slopes are surely fault scarps. We may conclude that less than 10% of the area of the crust has been involved in these large mountain top flexures.

The tilts just described were observed at the surface and thus represent only post formational events. Tilts may also be acquired during formation. Indeed the models in Figures 1a and 1d require quite large consistent and predictable tilts. In the whole volcano model of 1c, large tilts are also quite possible for the buried volcanoes, since we have very little idea of how they sink out of sight. In deep holes, the observation that inclination changes often occur at lithologic breaks may also support a model in which each volcano in the vertical section experiences a somewhat different tilting history during burial. These ideas are more fully developed by Hall (this volume).

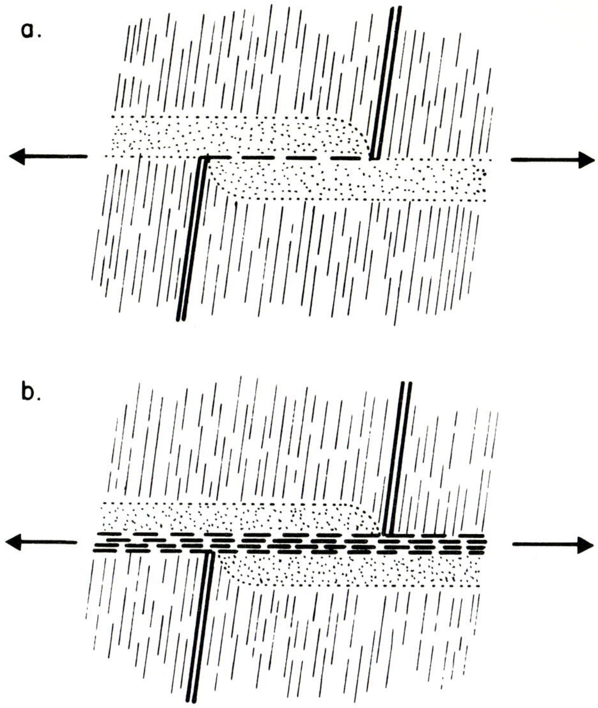

Fig. 5. Aerial patterns of fracture zone regime crust: (a) when all crust is formed at the ridge crest and then some is deformed during transform faulting (dotted area) and (b) when some additional area is added within the transform fault itself (dashed area). Double lines show centers of spreading. Arrows show direction of motions of plates. Assuming east-west spreading, the pattern in the Famous Area is like that in (b) with three quarters of the fracture zone area appearing to be formed by ordinary spreading processes and thereafter deformed.

Typical Crust Versus Fracture Zone Crust

When we speak about a "typical" oceanic section we usually think of crust which was formed at a spreading center, faulted, and then carried away, pretty much undisturbed. In fact, a certain, not insignificant portion of the sea floor is either formed in fracture zones or highly deformed by them. It would be interesting to know what percentage of the ocean floor surface belongs to each category, and thus to know the chances of hitting fracture zone terrain when drilling in terrain which is less than perfectly mapped.

Simple topographic maps suggest that a very large part of the Atlantic sea floor belongs to the fracture zone domain. While the rift mountains are generally envisioned as linear features, in fact they appear as nearly circular features on the charts both in the Famous area and in many other well mapped, rifted sections of the Mid Atlantic ridge (e.g., Tag, Reykjanes). However, more detailed studies show that bathymetry is a very poor tool for delimiting the ridge and fracture zone domains. As discussed above, "typical" oceanic crust has a very strong, consistent lineation in both the small and large scale tectonic features. In deep tow and multibeam data, individual step faults in the valley walls can be traced from one fracture zone to the next. However, these features, and indeed the valley floor itself, are not level but rather are bowed, being lower at their ends and higher at the center as shown by Macdonald and Luyendyk (1977) in their figure 16. This bowing causes the circular bathymetric appearance of the rift mountains on charts.

Thus it seems that the best indicator of "typical" crust is probably its lineated character. We might define "typical" oceanic crust as that which shows a clear tectonic grain parallel to the ridge crest. The Gloria side scan information is a very good detector of this type of fabric. The lineation map produced by Whitmarsh and Laughton (1976, their figure 6) was taken and sections where the ridge crest grain is disrupted or non existent were shaded. The shaded area accounts for about 25% of the area less than 5 m. y. old. (The transform faults in this region are believed to have been formed during a spreading direction change starting about 5 m. y. ago, Atwater and Macdonald, 1977.)

The multibeam bathymetric maps published of the area by Phillips and Fleming (1978) allow a fine scale interpretation to be made of the fracture zone - ridge crest junctions. In their maps of the AB Rift Valley - Fracture Zone B intersection, the interaction of the rift and fracture zone tectonics are especially clearly shown. On the passive, eastern side of the rift floor, the linear rift valley scarps are continuous, undisrupted southward well into the fracture zone depression. On the western side, the side which faces the active transform fault, these same linear features are present at the edge of the valley floor but are broken up by east-west faults farther up the wall. Some square pieces of ocean floor, bordered by these two fault sets can be detected here. Farther to the west, these features disappear. Presumably they have been disrupted beyond bathymetric recognition as transform tectonism continues. These observations suggest that we must consider two types of material in the fracture zone domain: that which was formed as ordinary ocean crust and later tectonized, and that which was formed in the fault itself. If all material in the fracture zone were of the former type, fracture zone domains might be distributed as shown in figure 5a. In fact, the Gloria and multibeam maps show a pattern of disruption like that in figure 5b, in which a region in the center of the transform fault may never have been part of the "ordinary" crust. Assuming an east-west spreading direction in the Famous Area, this type of area appears to constitute about one quarter of the total fracture zone domain in the Famous Area.

Summary and Conclusions

In conclusion, we find that using the structural and volcanic observations from the FAMOUS Project and other mid ocean ridge work we can make some comments concerning what is being encountered in deep basement drill holes in the Atlantic.

Concerning the structure of the volcanic section, morphologic, magnetic, and petrologic data suggest that the zone of volcanic additions is probably as wide as the rift valley floor with possible occasional additions as much as 15 km from the center. This suggests that individual vertical sections should commonly include lavas with ages spanning a few hundred thousand years and may occasionally include as much as a million years or more. The evidence for a wide extrusion zone also leads one to models in which the number of dikes included in the volcanic section increases rapidly with depth. This is in opposition to the situation reported from ophiolites. If we wish to reconcile these two approaches, we can call upon a quite special model, one in which the bulk of the volcanic layer is emplaced by a very localized, centered process, but then a small amount of lava is extruded randomly on top, covering the surface with late stage, whole volcanoes. On the other hand, it may be a mistake to try to reconcile ophiolite and mid Atlantic models. There is a large variation among spreading centers, especially with respect to spreading rate. The ophiolite structure seems much more compatible with fast spreading ridges, where a continuous, centered site of extrusions is present and flank volcanism is relatively minor.

Famous Area observations suggest that a given section might commonly include three to six major volcanic episodes, that each episode might consist of about 25 flows and that flows might average 10 to 20m thick, though all these parameters show large variations. Flows may commonly be separated by layers of flow front rubble, consisting of whole or half pillow pieces. A vertical section downward through a pillow flow is expected to consist of bulbous pillows, a massive layer, elongate pillows, flow front rubble, and then (perhaps) some sediments. Sections of thin sheet flows are also expected. They might comprise 10 to 15% of the section near the top and greater amounts at depth.

Three sources of broken rock in the section are clear. Flows should be separated by flow front talus, described above; fault scarps may create large talus ramps, mostly on or near the top of the section; and faults may create zones of sheared rock at depth. Given the parameters associated with large normal faults in the Famous Area, 30% of deep crustal holes might expect to encounter fault zones at depth. These three sources generate quite different types of broken rock which should be distinguishable in drill cores. The first sheds whole or large pieces of pillows; the second gives wedge shaped pieces, small sections of pillows; and the third generates lath shaped, sheared pieces.

If the transform faulted area of the Famous Area is typical, a randomly placed Atlantic drill site has a 75% chance of hitting material which was created at a mid ocean ridge and accreted to the crust without major disruption. The rest of the area was either formed within the transform fault itself (less than 10%) or was created at the ridge and then severely disrupted thereafter.

Acknowledgements

Tj. H. van Andel and W.B. Bryan were very generous with their time and ideas during the preparation of this summary. Discussions during the Ewing Symposium sessions were helpful; the comments and suggestions of R. Cann and W.B. Bryan being especially useful. Tj. H. van Andel and C.G.A. Harrison kindly reviewed the manuscript. This work was done under the partial support of the National Science Foundation.

References

Atwater, Tanya, Robert Ballard, Kathleen Crane, Nancy Gruver, Clifford Hopson, H. Paul Johnson, Bruce Luyendyk, Kenneth Macdonald, John Peirce, John Shih, Loren Shure, Debra Stakes, Nicolas Walker, and Tjeerd van Andel, Expedition cruise report, AMAR78 - A coordinated submersible, photographic mapping and dredging program in the Mid Atlantic Rift valleys near 36.5°N, M.I.T. Internal Report, 203 pp., 1978.

Atwater, T.M., and K.C. Macdonald, Are Spreading centers perpendicular to their transform faults? Nature, 270, 715-719, 1978.

Atwater, Tanya, and K.D. Macdonald, AMAR78 Preliminary results II: Gradiometer results in the Narrowgate magnetic anomaly region (abstract), Eos Trans AGU, 59, 1198, 1978.

Ballard, R.D., T. Atwater, D. Stakes, K. Crane, and C. Hopson, AMAR78 Preliminary results IV: Amar rift valley - Evidence for cycles in the evolution of rift valleys (abstract), Eos Trans AGU, 59, 1198.

Ballard, R.D., R.T. Holcomb, and Tj. H. van Andel, The Galapagos Rift at 86°W, 3. Sheet flows, collapse pits and lava lakes of the rift valley, J. Geophys. Res., in press.

Ballard, R.D., and J.G. Moore, Photographic Atlas of the Mid-Atlantic Ridge Rift Valley, Springer-Verlag, N.Y., 114 pp., 1977.

Ballard, R.D., and Tj. H. van Andel, Morphology and tectonics of the inner rift valley at lat 36°50'N on the Mid-Atlantic Ridge, Geol. Soc. Amer. Bull., 88, 507-530, 1977.

Bryan, W.B. and J.G. Moore, Compositional variations of young basalts in the Mid Atlantic Ridge rift valley near lat 36°49'N, Geol. Soc. Amer. Bull., 88, 556-570, 1977.

Bryan, W.B., P.J. Robinson, S.M. White, G. Byerly, D.A. Swanson, N. Pertsev, L. E. Ricou, S. Levi, G.A. Miles, W.G. Siesser, Y. Hamano, R.A. Stephen, C. Bollinger, R. Emmermann, Studying oceanic layers, Geotimes, 22, 22-26, 1977.

Cann, J.R., A model for oceanic crustal structure developed, Geophys. J.R. astr. Soc., 39, 169-187, 1974.

Crane, K. and R.D. Ballard, AMAR78 Preliminary results: Narrowgate tectonics and structure (Angus results) (abstract), Eos Trans AGU, 59, 1198, 1978.

Coleman, R.G., Ophiolites, Springer-Verlag, Pub., 229 pp., 1977.

Flower, M., M.H. Salisbury, D. Bohrer, M. Hobart, D. Johnson, E. Mathez, C. Mevel, N. Pertsev, R.G. Pritchard, H. Puchelt, P.A. Rigotti, H. Staudigel, Cretaceous crust sought at the Bermuda Rise, Geotimes, 22, 20-22, 1977.

Gruver, N.A., The significance of segregation vesicles to the interpretation of the tectonics of oceanic crust (abstract), Eos Trans AGU, 59, 1056, 1978.

Hall, J.M., and P.J.C. Ryall, Paleomagnetism of basement rocks, Leg 37, Initial Reports of the Deep Sea Drilling Project, 37, p. 425-445, 1977.

Lister, C.R.B., On the penetration of water into hot rock, Geophysical J. R. astr. Soc., 39, 465-509, 1974.

Lonsdale, Peter, 1977, Abyssal pahoehoe with lava coils at the Galapagos rift, Geology, 5, 147-152.

Luyendyk, B.P., and K.C. Macdonald, Physiography and structure of the inner floor of the Famous rift valley: Observations with a deep-towed instrument package, Geol. Soc. Amer. Bull., 88, 648-663, 1977.

Macdonald, K.C., Near bottom magnetic anomalies, asymmetric spreading, oblique spreading, and tectonics of the Mid-Atlantic Ridge near lat 37°N, Geol. Soc. Amer. Bull., 88, 541-555, 1977.

Macdonald, K.C., and T.M. Atwater, Evolution of rifted ocean ridges, Earth Plan. Sci. Lett., 39, 319-327, 1978.

Macdonald, K.C., and T.M. Atwater, AMAR78 Preliminary results: I Evolution of the median rift (abstract), Eos Trans AGU, 59, 1198, 1978.

Macdonald, K.C., and B.P. Luyendyk, Deep-tow studies of the structure of the Mid-Atlantic Ridge crest near lat 37°N, Geol. Soc. Amer. Bull., 88, 621-636, 1977.

Melson, W.G., F. Aumento, J. Ade-Hall, H. Bougault, L. Dmitriev, J.F. Fischer, M. Flower, R. C. Howe, R. Miles, P.J. Robinson, T.L. Wright, Deep Sea Drilling Project, Leg 37 - The Volcanic layer, Geotimes, 19, 16-18, 1974.

Phillips, J.D., and H.S. Fleming, Multibeam sonar study of the Mid-Atlantic Ridge rift valley, 36°-37°, Maps MC-19, Geol. Soc. Amer., pub., chart III, 1978.

Rabinowitz, P.A., W.G. Melson, H. Bougault, F. Fujii, A.L. Graham, H.P. Johnson, J. Lawrence,

J. Natland, E. Prosser, M. Rhodes, and B.P. Zolotarev, Challenger drills on leg 45 - along the Mid-Atlantic Ridge, Geotimes, 21, 20-23, 1976.

Whitmarsh, R.B., and A.S. Laughton, A long range sonar study of the Mid-Atlantic Ridge crest near 37°N (Famous Area) and its tectonic implications, Deep Sea Research, 23, 1005-1023, 1976.

Wilson, R.A.M., The geology of the Xeros-Troodos area, Cyprus Geol. Survey Dept. Mem. I, 184 pp., 1959.

THE ICELAND CRUST: EVIDENCE FROM DRILLHOLE DATA ON STRUCTURE AND PROCESSES

Gudmundur Pålmason, Stefån Arnórsson, Ingvar Birgir Fridleifsson,

Hrefna Kristmannsdóttir, Kristjån Saemundsson, Valgardur Stefånsson,

Benedikt Steingrímsson, and Jens Tómasson.

Orkustofnun, Reykjavík.

Leó Kristjånsson.

Science Institute, University of Iceland, Reykjavík.

Abstract. Over 120 drillholes deeper than 1000 meters have been drilled in Iceland, the maximum depth being 3085 m. Most of the holes are in geothermal areas where crustal temperatures are controlled by hydrothermal convection. Drillhole data are presented from type localities in structurally different areas such as active rift zones and central volcanoes as well as in dissected flood basalt and central volcanic sequences. The drillholes penetrate strata ranging in age from about 16-0 m.y.

The chlorine content of waters issued from the deep wells is highly variable (<20-19000 ppm) which is largely due to different proportions of meteoric water and sea-water in the rock. The relative mobility of all the other major elements in these deep waters seems to be governed solely by various temperature dependent mineral/solute equilibria.

Rock alteration ranges from the zeolite to the greenschist metamorphic facies. Four alteration zones have been correlated with distinct rock temperature intervals. Fossil alteration zones and retrograde alteration of secondary minerals frequently demonstrate previous thermal conditions.

Magnetic susceptibility of drill chips has been measured in 15 holes and shows no systematic decrease with depth.

A regional survey of heat flow, based mostly on shallow drillholes located outside known hydrothermal areas, is summarized. The heat flow pattern is closely related to the history and development of the active zones of rifting and volcanism.

Introduction

A fairly large number of holes have been drilled into the Icelandic crust, mainly in the last 20 years, for the purpose of producing geothermal water and steam for house heating, electric power production, and various industrial uses. The holes are of varying depth, mostly from about 500 m to 2000 m. The deepest one at present is 3085 m. Most of the holes are located in geothermal areas within the axial rift zone (high-temperature areas) or in the flood basalt areas flanking the axial rift zone (low-temperature areas).

The purpose of this paper is to review drilling carried out in Iceland, with particular regard to results that may be of significance for understanding the oceanic crust, its structure and the processes at work within it. Among the topics discussed are: geological structure, stratigraphic logs, proportion of intrusives, resistivity logs, temperature profiles, hydrothermal convection, water chemistry, rock alteration, magnetic properties, regional heat flow, and crustal stresses. Typical drillhole data are presented from both high-temperature and low-temperature areas, followed by a general discussion based on all available data.

Rotary drilling rigs have mostly been used for the deeper holes. Continuous cores are not taken, but cuttings are regularly sampled at 2 m intervals from the circulating fluid, which is usually water. The uncertainty in sampling depth appears to be of the order of 5-10 m, depending on depth and penetration rate.

Location of drillholes with regard to geology

The geological history of the land area of Iceland goes back about 15-16 My (Moorbath et al., 1968). Conventionally, three stratigraphic units are identified (Fig. 1.) on the basis of

Table I. The chemical composition of water from deep drillholes in Iceland. Concentrations in ppm. Location numbers refer to Fig. 1.

Location	Sample no.[a]	Temp. °C	Downhole sample m	pH/°C	SiO_2	Na	K	Ca	Mg	CO_2[e]	SO_4	H_2S[f]	Cl	F	Diss. Solids
01 Vík í Mýrdal, 1	77-0153	21[b]		9.69/23	52	473	9.4	24.4	0.59	28.4	211	<0.1	648	4.01	1502
02 Vestmannaeyjar,	73-0011	35[b]		6.55/20	148	6000	134	660	575	645	923	<0.1	9840	0.65	18994
03 Stórólfshvoll, 1	77-0003	43[b]	500	7.35/20	48	1082	27.8	406	2.61	21.1	293	<0.1	2161	1.19	4545
04 Laugaland in Holt, 3	77-0155	69[b]	212	9.78/21	80	106	1.64	6.3	0.07	21.1	102	<0.1	76.3	0.93	393
06 Reykjaból, 1	74-0028	152[c]		7.23/169[d]	233	87.4	6.6	1.0	0.02	128	72.6	2.9	29.2	1.80	518
07 Reykholt, 1	74-0031	132[c]		7.67/140[d]	167	105	4.8	2.2	0.06	51.4	74.4	1.7	73.4	2.49	480
08 Selfoss, 8	75-0082	82[b]		8.44/20	74	172	5.1	31.4	0.10	15.0	56.4	0.2	268	0.80	667
09 Árbær, 1	75-0083	96[b]		9.67/20	95	68.2	1.4	1.9	0.05	29.7	42.9	0.7	35.2	0.80	311
10 Eyrarbakki, 1	72-0169	70[c]	750	7.95/20	38	1360	11.0	1240	0.90	15.0	398	<0.1	3880	0.15	7617
11 Nesjavellir, 5	72-0193	268[d]		7.43/255[d]	555	133	24.3	1.1	0.04	278	31.8	163	13.4	1.74	950
12 Hveragerdi, 4	73-0082	184[c]		6.99/199[d]	270	151	11.9	1.5	0.16	169	63.2	26.4	112	1.80	681
14 Thorlákshöfn (Bakki) 1	77-0023	126[c]		6.84/128[d]	131	391	21.1	74.9	0.03	13.7	124	0.4	596	0.39	1515
15 Reykjanes, 8	77-0147	282[d]		5.58/246[d]	501	9050	1271	1736	3.36	1234	68.5	20.6	19966	0.16	31572
16 Svartsengi, 4	74-0063	242[c]		5.47/237[d]	446	6444	974	987	5.16	436	34.0	1.5	12647	0.11	28777
17 Krísuvík, 5	71-0132	151[c]	800	8.85/20	164	233	16.7	16.5	0.51	63.0	325	1.3	52.0	0.60	896
18 Trölladyngja, 6	71-0106	218[c]	800	7.30/20	304	596	64.0	40.0	0.44	59.5	40.1	1.7	914	0.30	2020
20 Seltjarnarnes, 3	77-2052	102[b]		8.44/28	116	368	10.8	144	0.17	5.0	205	0.1	685.0	0.73	1631
21 Reykjavík, G-17	77-2008	132[b]		9.64/18	150	58.5	2.9	2.5	0.02	20.7	25.2	0.1	37.0	0.80	323
22 Reykjavík, G-23	77-2011	93[b]		9.70/20	94	41.8	1.3	2.4	0.01	17.2	19.2	0.1	17.4	0.42	219
23 Mosfellssveit, MG-16	77-2044	101[b]		9.65/23	110	50.8	1.5	2.3	0.01	20.5	28.6	1.4	24.2	0.95	230
24 Akranes, 4	67-	110[c]	810	7.20/20	100	1120	17.0	540	0.18	8.3	53.0	<0.1	2485	0.10	5600
25 Leirá, 4	76-0017	128[b]		6.29/20	219	213	20.6	30.9	0.37	279	56.5	0.3	240	2.34	894
26 Hvalstöd, 4	77-0045	12[b]		9.20/26	25	76.8	1.00	8.5	0.48	17.8	122	<0.1	50.4	3.20	290
27 Laugarholt, 1	77-0133	91[b]		9.16/22	116	108	3.25	14.5	0.02	12.9	74.7	0.5	125	1.99	438
28 Kolvidarnes, 4	77-0091	60[b]		7.97/28	77	129	2.5	24.9	0.25	40.2	52.0	<0.1	187	0.66	481
29 Reykhólar, 5	75-0073	108[b]		9.61/20	140	58.3	1.9	2.7	0.03	15.3	43.4	0.7	33.7	0.50	302
30 Ísafjördur, 2	76-0023	27[b]		9.90/19	59	90.3	0.6	4.2	0.01	9.6	53.4	<0.1	77.4	1.92	327
32 Súgandafjördur, 2	76-0111	64[b]		9.80/17	59	72.1	1.0	6.1	0.02	9.5	71.2	0.1	63.4	0.30	296
33 Reykir in Midfjördur, 4	69-0151	96[b]		9.20/23	97	141	2.3	22.7	0.03	14.0	143	0.2	140	2.00	613
34 Reykir at Reykjar, 5	76-0191	69[b]		9.64/21	116	54.8	1.8	2.9	0.01	24.0	59.0	1.2	13.3	5.84	306
35 Saudárkrókur, 10	69-0145	70[b]		9.92/23	74	51.0	1.1	3.0	0.03	16.2	37.7	0.5	19.7	1.40	232
36 Siglufjördur, 8	77-0125	68[b]		10.08/25	99	38.3	1.0	2.1	0.01	15.4	14.5	<0.1	9.5	0.38	226
37 Ólafsfjördur, 8	69-0125	56[b]		10.20/24	58	39.0	0.8	1.8	0.06	11.5	5.3	<0.1	2.9	0.10	150
38 Dalvík, 10	77-0148	64[b]		10.29/22	95	46.7	0.5	2.2	0.01	13.5	15.5	<0.1	10.1	0.50	220
39 Laugaland in Hörg., 2	69-0121	88[b]		9.78/24	130	55.5	1.6	2.0	0.02	24.0	31.3	0.2	13.7	0.60	288
40 Laugaland in Eyjafj., 8	77-0173	92[b]	2300	9.99/23	116	72.1	2.6	3.9	0.09	27.7	63.9	<0.1	13.8	0.53	304
41 Stórutjarnir, 2	69-0119	65[b]		9.69/24	117	51.5	1.5	2.8	0.06	20.0	30.9	0.9	18.2	0.80	253
42 Laugar, 2	75-0091	62[b]		10.58/20	90	45.5	0.5	2.0	<0.01	7.5	25.9	<0.1	4.7	0.76	198
43 Árnes, 1	75-0162	97[c]	1160	9.65/21	122	51.2	1.1	2.8	0.10	11.3	37.4	<0.1	11.9	0.69	282
44 Húsavík, 1	66-	94[b]	500	7.40/20	80	840	18.7	17.6	10.6	5.0	82.0	-	1633	0.20	3600
45 Krafla, 9	77-1178	300[c]		7.12/303[d]	681	143	20.2	1.1	0.02	10013	73.1	92.4	11.8	0.46	1014
46 Námafjall, 10	76-0156	300[c]		7.31/272[d]	607	118	22.8	0.5	0.02	231.5	26.6	232.0	82.3	0.75	1036
47 Urridavatn, 3	76-0025	43[b]		9.95/20	56	63.4	0.9	5.4	0.03	8.5	47.6	<0.1	41.9	0.60	253

[a] year of sampling-sample no. [b] measured temperature of well discharge. [c] measured temperature at aquifer inflow. [d] computed pH of the reservoir water. [e] total carbonate ($H_2CO_3 + HCO_3^- + CO_3^{-2}$). [f] total suphide ($H_2S + HS^- + S^{-2}$).

the geomagnetic time scale and paleoclimatic evidence. They are (1) Tertiary rocks formed up to 3.1 My. ago, composed mainly of subaerial lava and only minor sediment and tuff, (2) Plio-Pleistocene rocks, formed between 3.1 and 0.7 My. ago, containing a substantial amount of sediment and subglacial volcanic rocks intercalated between subaerial lava flows, (3) Late Quaternary rocks of the active volcanic zones, formed during the Brunhes epoch, i.e. since 0.7 My. ago.

The zones of recent volcanism comprise the axial rift zones, which erupt tholeiitic rocks, and are the present locus of active plate growth, and the flank zones or lateral rift zones (Jakobsson, 1972; Saemundsson, 1978), which are characterized by alkaline or transitional rocks. The axial rift zones have in some places shifted position during the evolution of Iceland. In northwestern Iceland there is evidence of an extinct axial zone, marked by anomalous heat

Table II. Data on drillholes used to determine conductive heat flow in the Icelandic crust.

Borehole	Latitude N	Longitude W	Elevation m	Depth m	Gradient °C/km	Heat flow* mW/m²
Arnarholt	64°14.'8	21°51.'9	20	240	165	281
Ferstikla	64 24.3	21 36.2	(5)	100	145	247
Akranes I	64 19.4	22 04.6	(5)	1400	129	219
Akranes II (Innrihólmur)	64 18.3	21 56.2	(5)	100	153	260
Akranes III (Hvítanes)	64 22.0	21 57.9	(5)	100	150	255
Borgarnes (Borg)	64 33.7	21 55.6	(5)	100	103	175
Stadastadur	64 48.4	23 01.0	(5)	100	66	112
Ólafsvík (Hólmkelsá)	64 52.1	23 49.8	10	163	86	146
Ólafsvík (Innri Bugur)	64 53.1	23 37.5	10	102	95	162
Grundarfjördur (Kirkjufell)	64 55.2	23 17.7	15	53	116	197
Grundarfjördur (Hamrar)	64 56.1	23 12.7	5	77	88	150
Stykkishólmur	65 04.5	22 44.3	15	82	110	187
Stykkishólmur (Svelgsá)	64 59.5	22 40.0	15	100	100	170
Búdardalur	65 06.8	21 46.5	(5)	102	75	128
Tindar	65 18.8	22 13.6	15	85	111	189
Flatey	65 22.3	23 55.9	5	102	124	211
Thingeyri	65 52.5	23 29.1	(5)	102	53	90
Flateyri	66 02.9	23 29.4	20	156	47	80
Súdavík (Svarfhóll)	65 58.3	23 04.8	20	105	86	146
Hólmavík	65 41.4	21 41.8	(5)	102	65	111
Hvammstangi	65 23.9	20 56.8	(10)	103	70	119
Skagaströnd	65 50.1	20 19.1	(5)	101	70	119
Hólar	65 43.9	19 07.6	140	103	58	99
Akureyri	65 41.1	18 06.1	40	100	64	109
Háls (Fnjóskadal)	65 44.3	17 51.5	120	129	62	105
Árnes (Adaldal)	65 53.3	17 24.5	20	1250	80	136
Brúar (Adaldal, LM2)	65 48.9	17 19.0	110	46	98	167
Thorshöfn (Gunnarsstadir)	66 09.2	15 26.1	10	100	55	94
Vopnafjördur	65 44.8	14 52.8	20	100	56	95
Eidar	65 22.9	14 20.3	40	100	37	63
Egilsstadir (Ullartangi)	65 16.8	14 26.1	15	44	77	131
Vallholt (Fljótsdal)	65 03.5	14 52.2	20	46	67	114
Seydisfjördur	65 16.2	13 59.4	5	100	70	119
Eskifjördur	65 04.0	14 02.3	10	102	84	143
Reydarfjördur	65 02.1	14 14.4	10	102	82	139
Fáskrúdsfjördur	64 57.1	14 07.6	80	102	80	136
Stödvarfjördur	64 50.0	13 51.0	20	102	56	95
Breiddalsvík	64 47.5	14 01.1	10	100	63	107
Lón (Skyndidalsá)	64 25.8	14 58.4	60	130	82	139
Bjarnanes (Hornafirdi)	64 18.7	15 14.6	20	101	50	85
Prestbakki	63 49.4	18 02.6	40	100	47	80
Vík	63 25.4	19 00.6	40	551	53	90
Vestmannaeyjar	63 26.7	20 17.2	20	1565	63	107
Thykkvibaer	63 44.7	20 37.5	10	90	86	146
Hella	63 50.4	20 24.2	30	397	72	122
Kópsvatn	64 09.9	20 18.2	90	620	150	255
Eyrarbakki	63 51.9	21 09.2	(5)	752	88	150
Árbaer	63 56.7	21 02.8	20	937	142	241

*) Thermal conductivity assumed 1.7 W/m°C

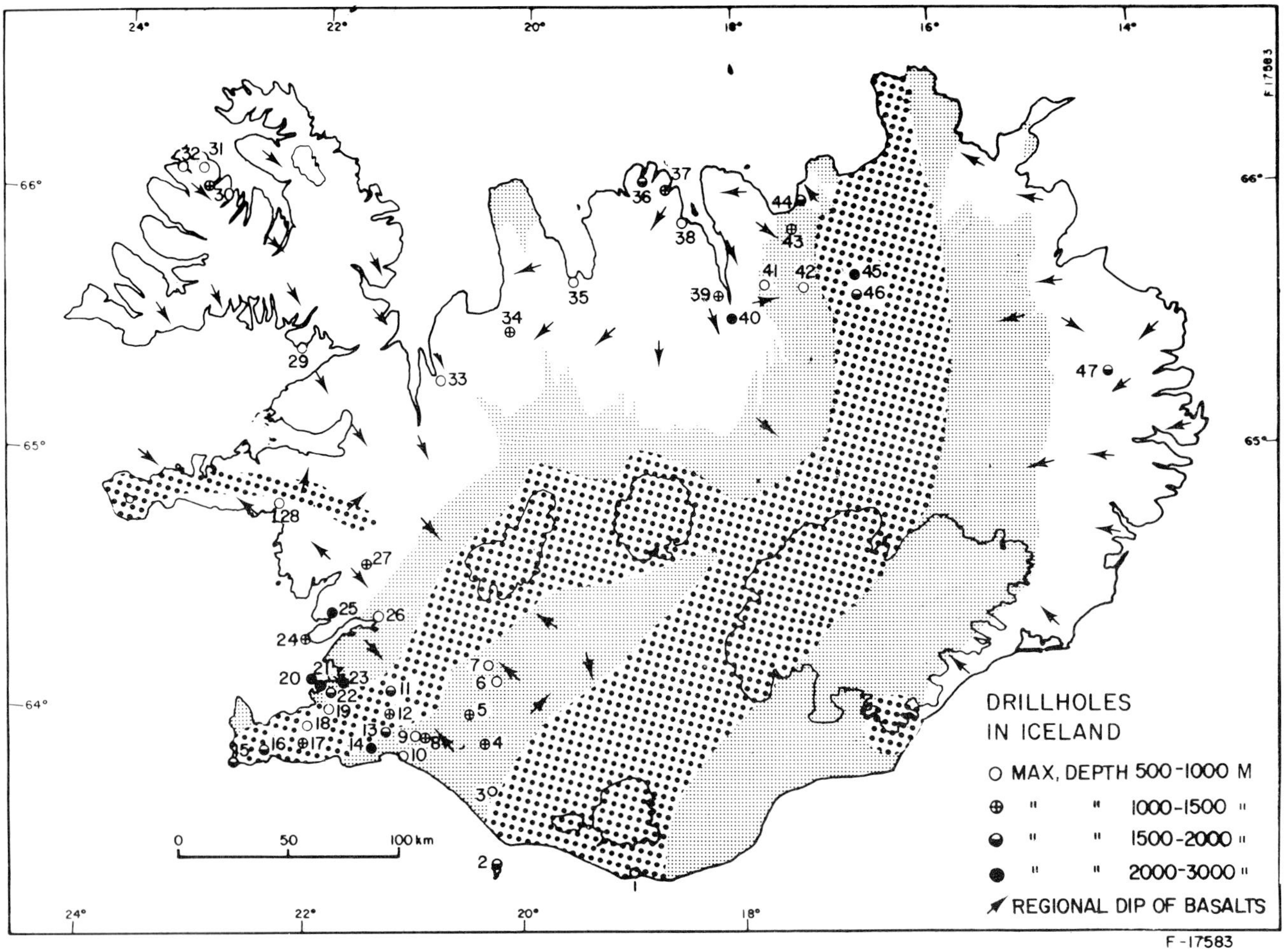

Fig. 1. Location of drillholes over 500 m. The drilling areas are numbered for reference in the text. Coarsely dotted areas: Late Pleistocene to Holocene volcanics (younger than 0.7 m.y.). Finely dotted area: Plio-Pleistocene volcanics (0.7 - 3.1 m.y.). Unshaded areas: Miocene - Pliocene volcanics (older than 3.1 m.y.)

flow (cf.Fig. 17) and a synclinal structure of the basalts.

Deep holes have been drilled in all three stratigraphic units. The location of the drilling areas is shown in Fig. 1. In some of the areas there is only a single hole, while in others there are up to several tens of holes. The holes are in different structural environments, and penetrate rocks of the entire age spectrum of the volcanic succession. In the axial rift zones all the holes except one, Kaldårsel (loc. 19 in Fig. 1), have been drilled in high-temperature geothermal areas, which are often but not always associated with central volcanoes.

It should be noticed that the depth of the holes does not always correspond to the depth from the original surface of the crust. This is especially true in the coastal areas and in valleys of the Tertiary flood basalts in eastern, northern and northwestern Iceland, where glacial erosion has removed in places as much as 1500 m or more from the original surface. For example, the second deepest hole in Iceland at present (2820 m) is located in a valley in northern Iceland (Laugaland, loc. 40), where about 1500 m have been removed by erosion. The hole, together with the exposures alongside the valley, reaches about 4300 m into the crust as originally formed. In the axial rift zone on the other hand, drilling starts from a present-day surface that has not yet suffered erosion.

In the next section four examples will be given of data collected from drillholes in four different areas, two from high-temperature areas in the axial rift zone, and two from low-temperature areas in Tertiary and Plio-Pleistocene flood basalts outside the axial zone.

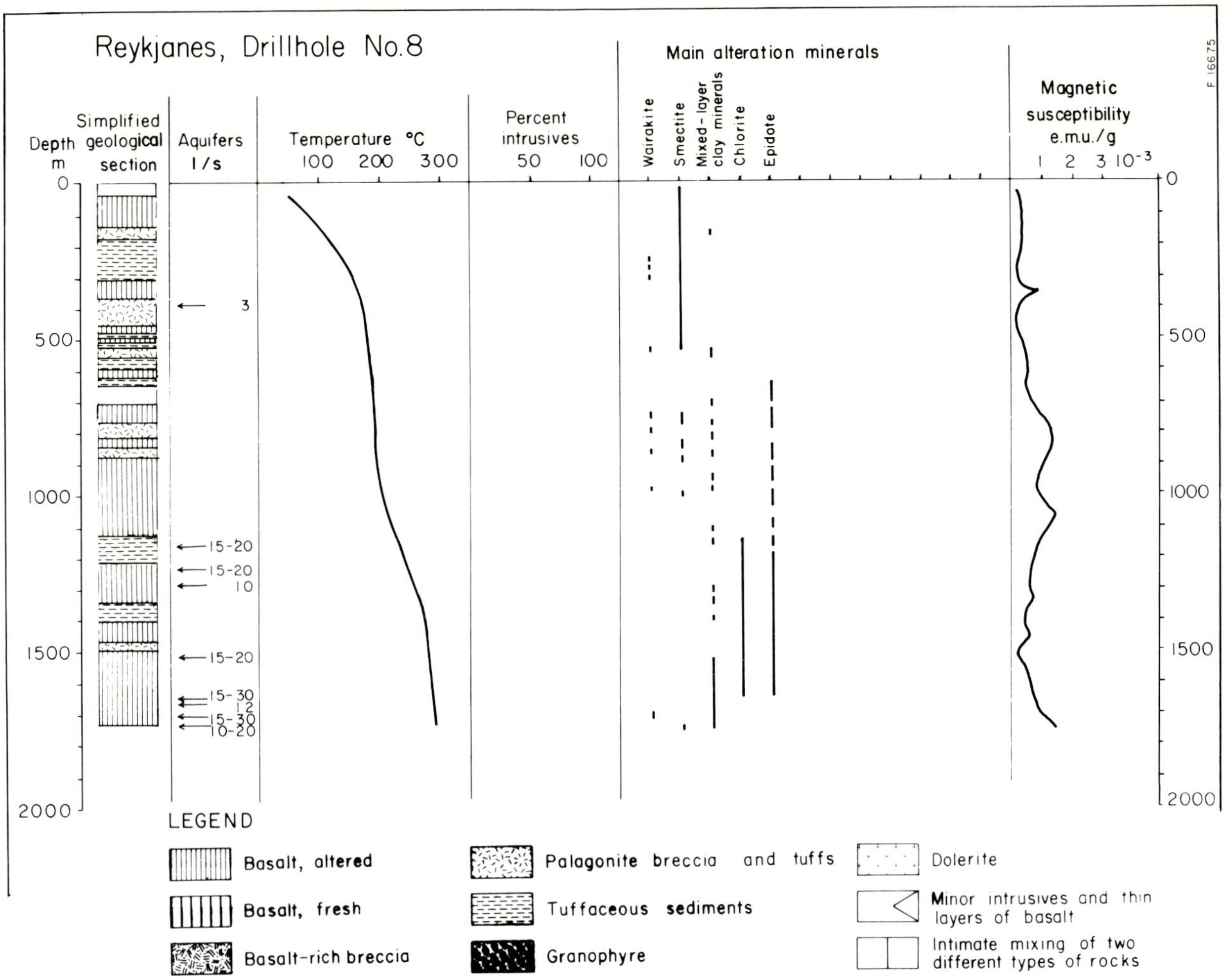

Fig. 2. Selected data from drillhole No. 8, Reykjanes (loc. 15) showing: a simplified geological section[1)], location of the main aquifers, temperature measurements[1)], percent intrusives, distribution of the main alteration minerals and the magnetic susceptibility of drill cuttings. The aquifers are located from recordings during drilling of loss or gain of circulation water and from temperature measurements. The numbers refer to l/s of lost or gained drilling fluid. The temperature measurements are selected in such a way that they represent nearly undisturbed conditions. The percentage of intrusives is calculated for every 100 m of penetration. As the samples are cuttings some finegrained dykes may not have been identified. As dykes are cut under small angles the thickness will be highly exaggerated in the profile. Of the alteration minerals the most significant were selected and others omitted. The magnetic susceptibility represents smoothed values of measurements on samples from every ten meters of penetration.
1) Revised from Tómasson and Kristmannsdóttir, 1972.

Examples of drillhole data

Reykjanes high-temperature area (loc. 15)

This thermal area is located at the tip of the Reykjanes peninsula, where the axis of the Reykjanes Ridge comes ashore. Fig. 2 gives data on stratigraphy, aquifers, temperature, alteration and magnetic susceptibility of the rocks penetrated in drillhole No. 8 which is the deepest hole in the area, 1754 m. The hole is close to sea level elevation, and located right at the axis as defined by earthquake epicenters (Klein et al., 1973) and recent volcanism. Some results of exploration and drilling in the Reykjanes area have been given by Björnsson et al., (1970, 1972).

The upper part of the stratigraphic section to a depth of about 900 m consists of mainly hyaloclastite and about one third of lava. In the lower part of the section basalt lavas predominate and are about 80% of the section. No dykes have been positively identified in this hole. The transition from the upper part to the lower

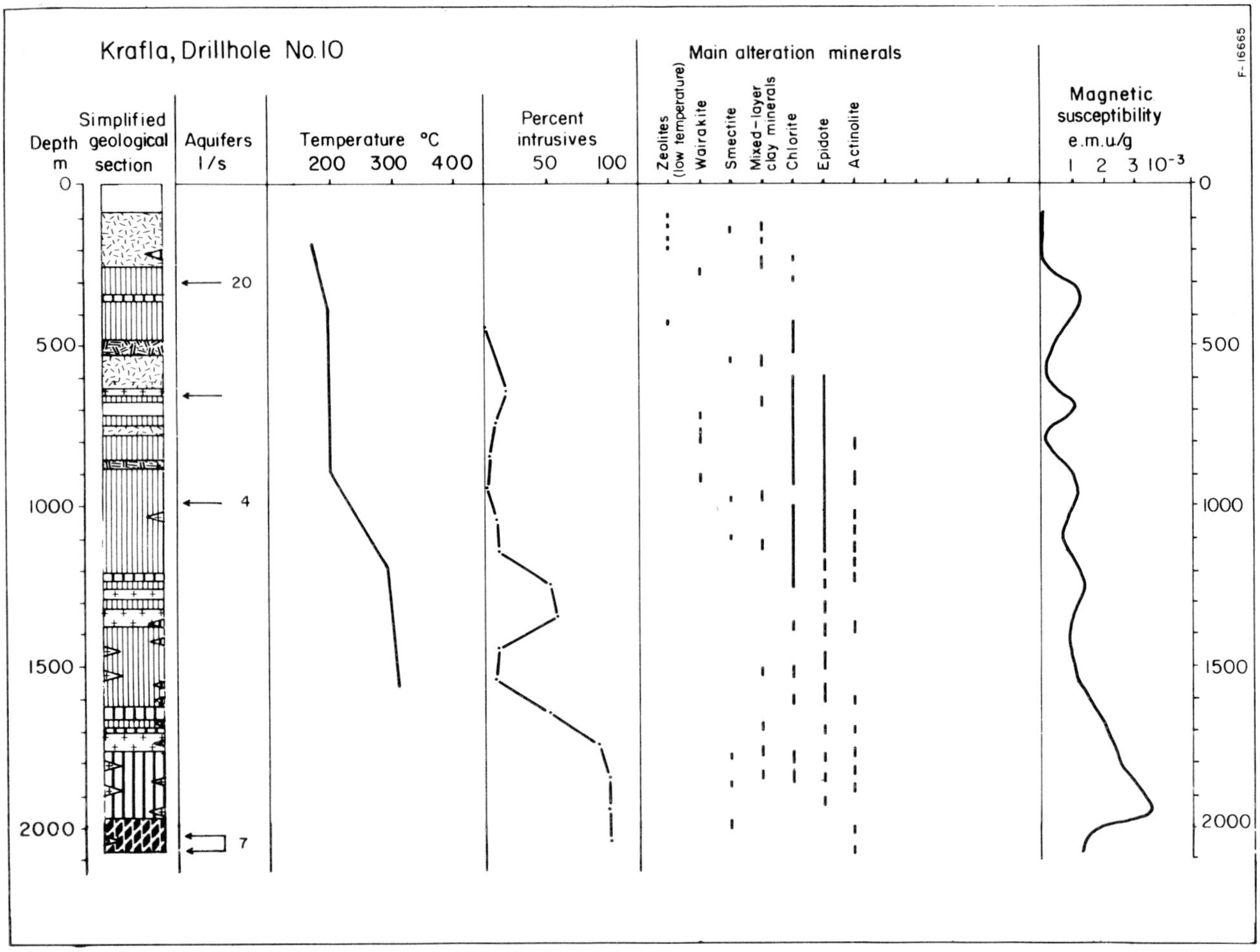

Fig. 3. Selected data from drillhole No. 10, Krafla (loc. 45). Legend as in Fig. 2.

part appears to coincide with a change in seismic P-wave velocity from 3 km/s to about 4 km/s, as inferred from seismic refraction profiles (Pålmason, 1971).

The lavas and the character of the sediments in the stratigraphic section indicate accumulation on dry land and in shallow water (Björnsson et al., 1972). This observation has a bearing on the mechanism of crustal accretion in the axial rift zone, and shows that crustal subsidence accompanies the deposition of volcanics at the surface. This process was already suggested by Walker (1960) to explain the regional structure of the lava pile in eastern Iceland.

The hydrothermal areas on the Reykjanes peninsula are somewhat unique among the Icelandic high-temperature areas in that the convecting fluid is largely sea-water, in which the chemical composition has been modified by interaction with the crustal rocks (Björnsson et al., 1972). The water is enriched in K, Ca and SiO_2 but depleted in Mg and SO_4 relative to sea-water.

The main aquifers in this hole are found in the lower part, where the basalt lavas predominate. The alteration pattern reflects the relatively low temperature in the uppermost 2-300 meters. Chlorite and continuous occurrence of epidote are first obtained below about 1100 m.

Krafla high-temperature area.

The Krafla area is located in the axial rift zone of northeastern Iceland (loc. 45). Twelve holes have been drilled there in the last 4 years for a 60 MW geothermal power station which is under construction. The thermal area is located within a caldera structure, which is traversed by a volcanic fissure system, which has been tectonically and volcanically active for the past 3 years (Björnsson et al., 1977).

Fig. 3 shows sample data from one of the holes, KG-10. The uppermost 800 meters are mostly hyaloclastite, split into two parts by a 250 m thick suite of basalt lavas. Basalt lavas are predominant below 800 m depth, and hyaloclastites are rare. Dykes are common below 400 m, and from 1700 m the rocks are entirely intrusive and mostly fresh to the bottom of the hole at 2080 m.

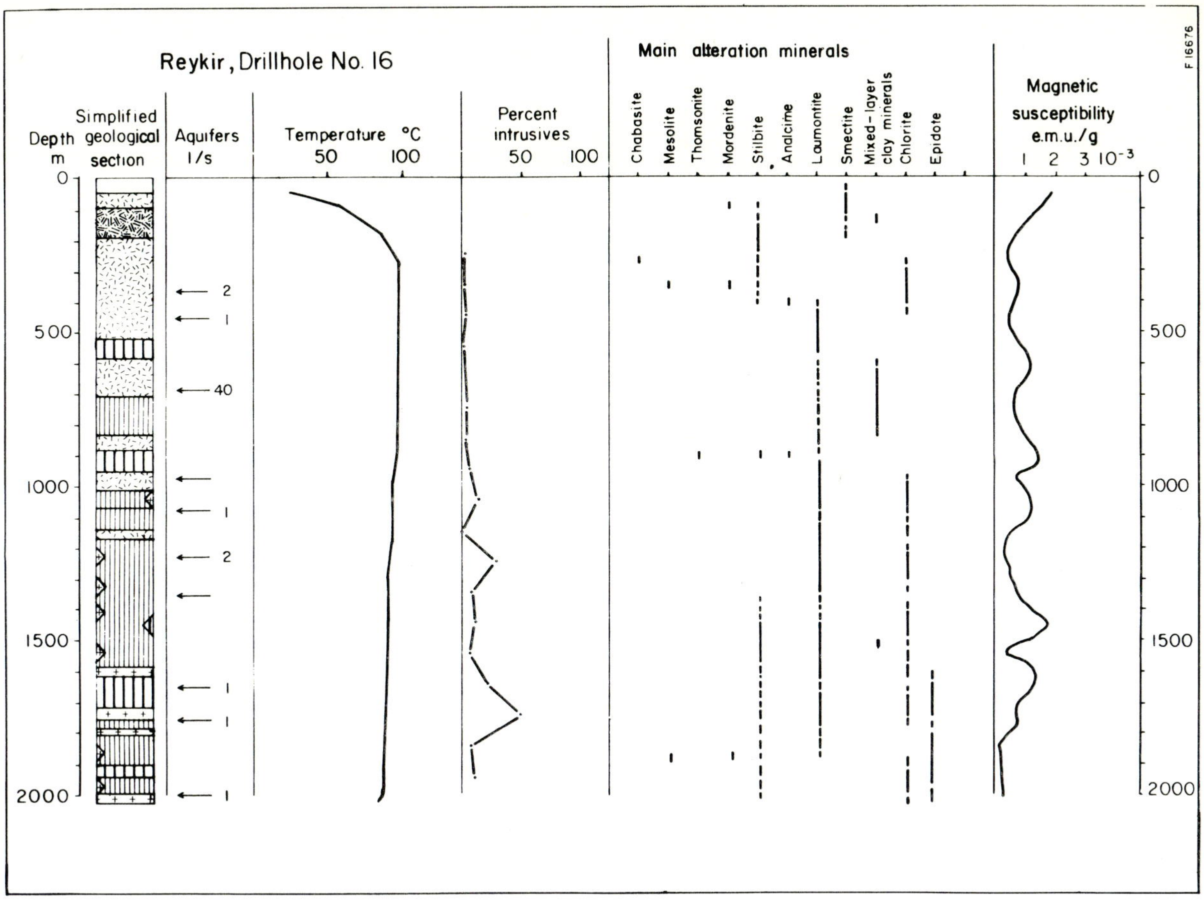

Fig. 4. Selected data from drillhole No. 16, Reykir (loc. 23). Legend as in Fig. 2.

The temperature values measured in the Krafla drillholes include the highest yet found in Iceland, about 340°C. The alteration pattern shows a fairly regular zoning, with some indication of a retrograde transformation of the sheet-silicates at depths less than 1200 m, where some cooling appears to have taken place (Kristmannsdóttir, 1977a). The magnetic susceptibility tends to increase with depth in this hole. One might perhaps be tempted to correlate this with the increase in the intrusives fraction with depth.

Reykir low-temperature area

This low-temperature area (Tómasson et al., 1975) is located close to a Plio-Pleistocene central volcano (Fridleifsson, 1973) on the western flank of the axial zone in southwestern Iceland (loc. 23). It is one of the main sources of hot-water for the Reykjavík district heating system, which serves about 120.000 inhabitants. Between 30 and 40 holes about 2000 m deep have been drilled in the area.

Fig. 4 shows data from one of the holes, MG-16, which is typical for the area. Hyaloclastites are dominant to about 1000 m depth. Below, there are mostly lavas, but hyaloclastite horizons are found to the bottom of the hole. Dykes are rare in the uppermost 1000 m, but their number tends to increase with depth. The temperature profile shows a maximum at a depth of 300-600 m. The bottom hole temperature is very low compared to the expected temperature based on thermal conduction models; this illustrates the strong disturbance of the crustal temperature field by water convection in this highly permeable part of the crust. The hydrothermal system is characterized by horizontal aquifers. The best producing aquifers are found at the contacts of hyaloclastites and lavas. Fig. 5 shows a thermal cross section through a part of the Reykir thermal area, based on data from a number of holes. It shows two main tongues of ascending thermal waters, but otherwise the section is characterized by a relatively constant temperature, as one commonly finds in hydrothermal convective systems.

The alteration pattern in the Reykir area is very complex. Low-temperature alteration is superimposed on former high-temperature alteration associated with the extinct central volcano (Tómasson and Kristmannsdóttir, 1974).

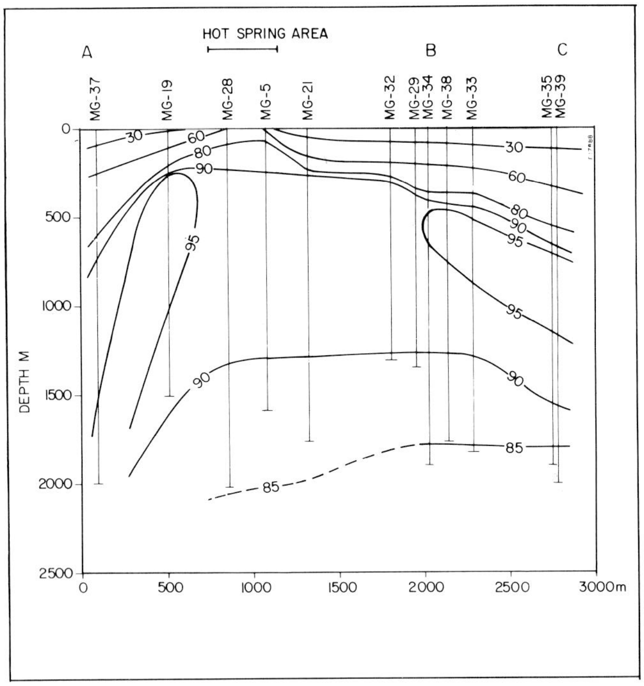

Fig. 5. A thermal cross section through a part of the Reykir low-temperature area. The area of original hot spring activity is also shown.

A retrograde transformation of chlorites has been demonstrated.

Laugaland low-temperature area

Fig. 6 shows data from the Laugaland low-temperature area, located in the Tertiary flood basalt pile of northern Iceland (loc. 40). This is the second deepest hole yet drilled in Iceland, 2820 m. Basalt lavas with thin intercalated sediments are the dominant rock type. Quartz-tholeiitic basalt is most common, but olivine-tholeiites make up approximately 20 percent of the pile. Dykes are very common, especially in the deeper parts below about 800 m (Kristmannsdóttir and Gudmundsson, in prep).

The temperature profile shows a fairly constant temperature of about 100°C. At the bottom of the hole this is very low compared to the expected temperature based on regional thermal gradients in central northern Iceland (cf. section on heat flow). A strong hydrothermal convective cooling of the crust at 2-3 km depth is thus inferred. The aquifers are mainly connected with dykes, in particular where they cut scoriaceous and sedimentary interlayers between lava flows.

The alteration is mostly a fossil one, and effects of the present geothermal system appear to be minimal. This may be due to the relatively poor permeability in the old lavas, and the rather narrow aquifers in this type of geothermal system. The magnetic susceptibility is fairly constant, and shows no trend of variation with depth.

As mentioned previously, this drillhole is located in a valley where approximately 1500 m have been eroded from the original top of the lava pile, as indicated by the zeolite zones. The hole thus reaches to well over 4 km into the original crust of northern Iceland.

Geophysical logging

Besides temperature measurements, which are made routinely both during and after the drilling, other geophysical methods are being adopted on an increasing scale. Resistivity (E-log) and diameter logging (caliper log) have been used in many holes for the last 3 years, and equipment has recently been obtained for making neutron-neutron and gamma-gamma logs, although no results from these last methods are available yet for presentation.

The resistivity and caliper logs have proved to be useful as complementary methods to studies of drill cuttings in the basaltic crust of Iceland. Due to mixing in the hole, rock contacts obtained from the cuttings are not precise. Frequent changes in rock types without significant changes in the penetration rate are especially easy to overlook. A combined interpretation using the resistivity and caliper logs as well as the drill cuttings and the penetration rate, gives a much more detailed picture. Location of contacts is more exact, allowing thin layers and strata of contrasting porosity in the same rock unit to be distinguished.

In the following, an example will be given to illustrate the usefulness of the resistivity and caliper logs.

Fig. 7 shows a stratigraphic section, resistivity and caliper logs from a section of the drillhole LJ-8 at Laugaland in northern Iceland (loc. 40). The drillhole passes through a dolerite dyke in the section shown. It is not possible to locate exactly the boundaries of the dyke from the cuttings because of mixing with altered basalt lava cuttings above and below the dyke. This is indicated by two rock types shown in parallel in the stratigraphic section. The dolerite is predominant in the cuttings from about 1770 to 1940 m depth, and the penetration rate is also somewhat lower than in the surrounding lavas.

The caliper log has a rugged character at the upper margin, then shows a smooth surface to about 1940 m, below which the walls of the hole are very rugged, the diameter expanding to more than 30 cm in several places. This log locates the boundaries of the dyke at approximately 1780 and 1940 m, assuming the smooth part of the log to be associated with the dyke.

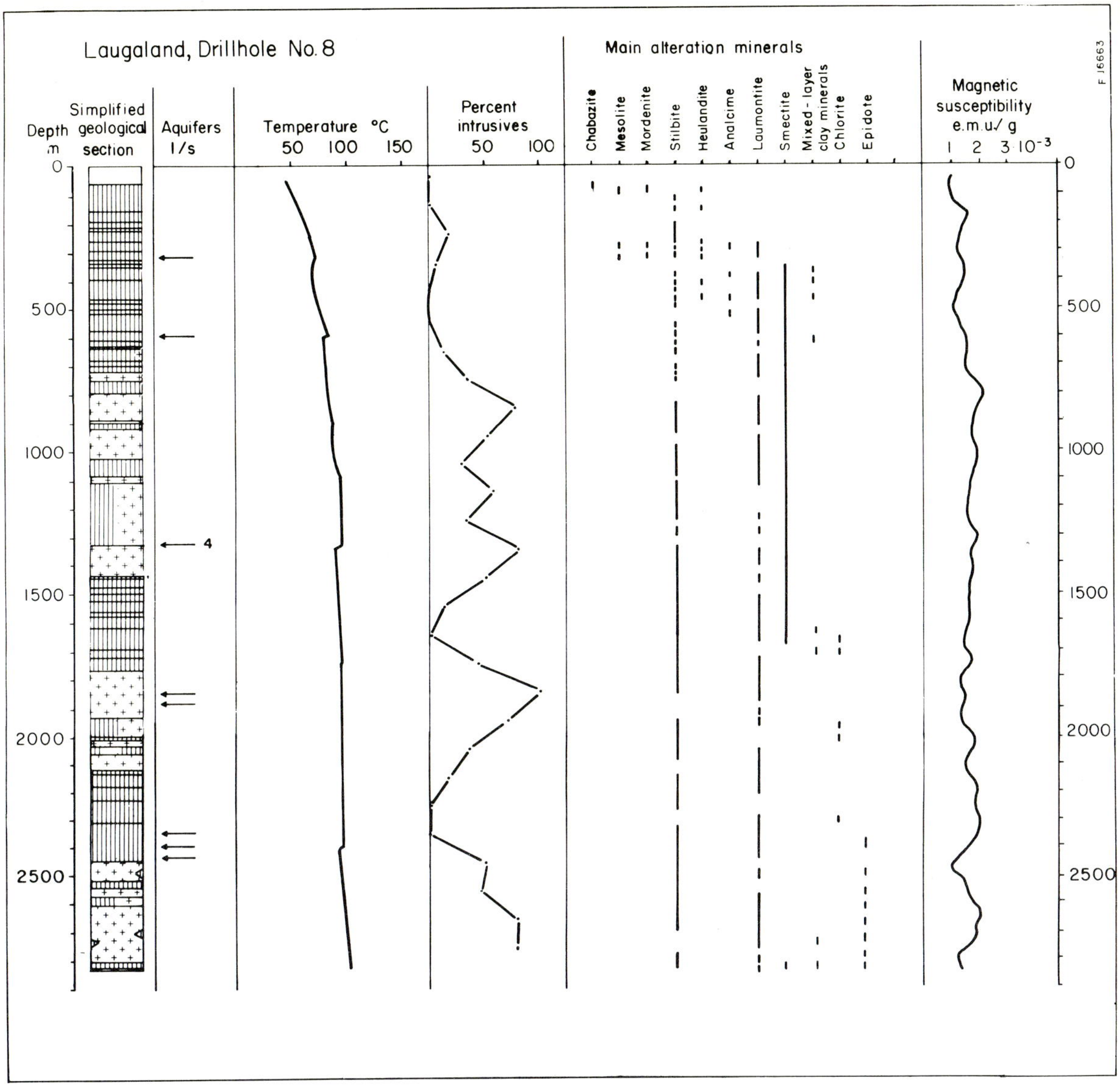

Fig. 6. Selected data from drillhole No. 8, Laugaland (loc. 40). Legend as in Fig. 2.

The E-log shows a conspicuous high resistivity associated with the dyke. The upper margin is not clearly indicated on this log, and the hole may well be following the dyke margin for some length before entering the dyke. At the lower margin the cuttings were a mixture of dolerite and lava fragments from 1940 to 2000 m depth. The E-logs in this depth range on the other hand are typical for a sequence of lava layers, with low resistivity porous boundaries between the less porous inner part of the lavas. The E-logs indicate a relatively sharp transition from the dyke to the underlying lava section. The caliper log is in good agreement with this interpretation.

The porosity is the main parameter affecting the resistivity measurements. On the basis of the cuttings and the caliper log the upper boundary of the dyke is located at about 1770 m. The resistivity logs then show that considerable variations in porosity exist within the dyke. The upper part in the section is highly porous, perhaps due to fractures, while the lower part has a very low porosity.

Water Chemistry

Chlorine variations.

The content of dissolved solids in waters from the deep wells is highly variable (150-29000 ppm). The waters in young formations near the coast tend tend to be highest in dissolved solids (Fig. 8); the increased salt content observed has probably

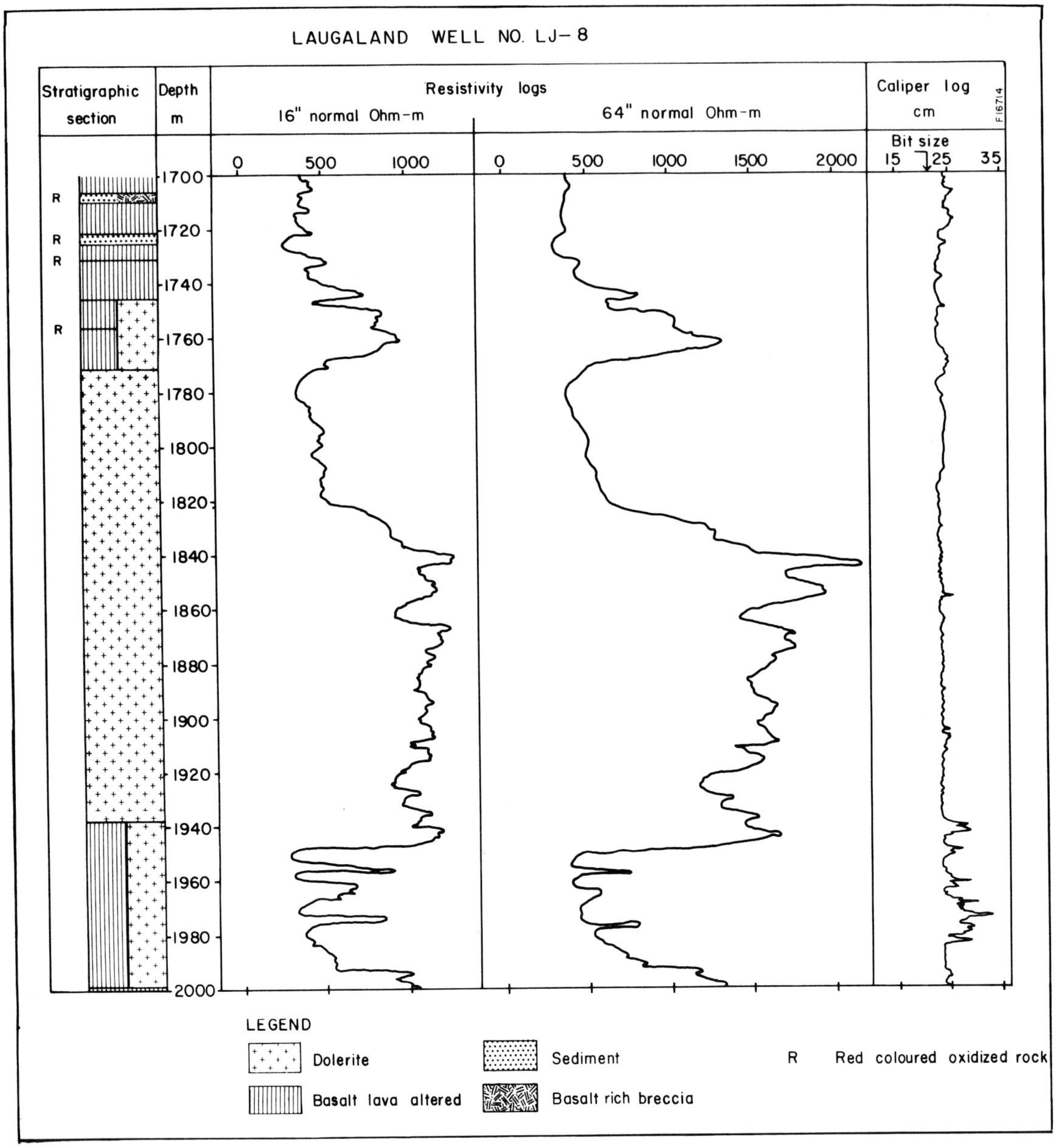

Fig. 7. A stratigraphic section and resistivity and caliper logs from a section of drillhole No. 8, Laugaland (loc. 40).

been derived from sea-water. Two different mechanisms can explain the transfer of salt from sea-water into the geothermal systems. They are: (1) leaching of salts from sediments deposited in marine environments (Tómasson, 1967), (2) percolation of sea-water into the geothermal systems and mixing with waters of meteoric origin either during the present sea level or during a higher sea level at the end of the Pleistocene when large parts of the lowlands were submerged (see Arnórsson, 1978a). It is thought that process (1) is of little significance in affecting the salt content of the geothermal waters in Iceland.

Thermal waters, which do not seem to have been affected by sea-water mixing, tend to be low in dissolved solids as reflected in their low chlorine content, which lies within about 5-50 ppm. It is thought that this low chlorine content is due to the low chlorine content of the Icelandic basalts (see Sigvaldason and Óskarsson, 1976).

Mineral/solute equilibria.

In the following discussion the calculated activities of the various dissolved chemical species are always fitted to equilibrium cons-

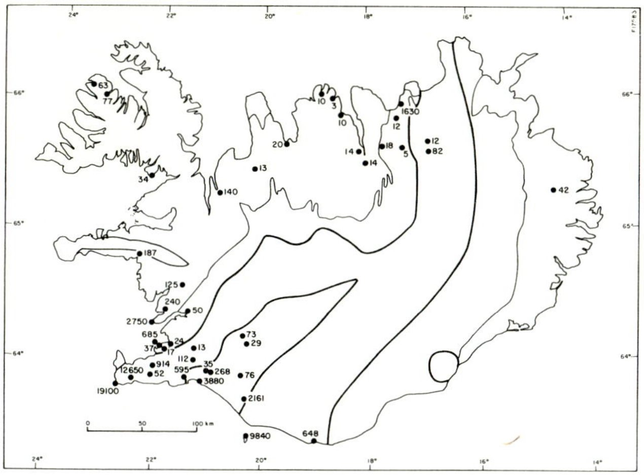

Fig. 8. The chlorine content (in ppm) of thermal waters from deep drillholes in Iceland.

tants for mineral solubilities or ionic exchange reaction. The method of calculation has been described by Arnórsson et al. (1978).

The thermal waters tend to be saturated with chalcedony when the water temperatures are below about 180°C but at higher temperatures equilibrium with quartz is attained (Fig. 9A). It is generally found that other dissolved components fit better mineral equilibria at the chalcedony and quartz equilibrium temperatures than at the measured temperatures. This indicates that mineral/solute equilibria do not respond faster to cooling than quartz/chalcedony and dissolved silica do. The major exception is re-equilibration with calcite.

In general the waters are just saturated with calcite at the chalcedony/quartz equilibrium temperature (Fig. 9B). Only 5 samples show significant supersaturation. Three of these (1, 2, and 25) are significantly chalcedony supersaturated as can be inferred from Fig. 9A and Table 1.

Calculations show, however, that they are just saturated at the measured water temperature. It is thus concluded that these waters have reequilibrated with calcite upon cooling although this has not been so for chalcedony.

Samples 11 and 15 are from wet-steam wells. Their supersaturation with respect to calcite is explained by flashing in the feeding aquifers and some loss of steam and gas at the same time. This steam loss produces a gas deficient and calcite supersaturated well discharge (Arnórsson, 1978b).

Anhydrite is common in altered rocks associated with saline geothermal waters. Calculations show anhydrite saturation for the more calcium rich saline waters and the hottest water (e.g. loc. 45), but other waters are undersaturated with respect to this mineral (Fig. 9C). There is an overall tendency for undersaturation to increase with decreasing salinity. It may be that sulphate in the more dilute waters is controlled by the following redox equilibrium:

$$H_2S + 4H_2O = SO_4^{-2} + 10H^+ + 8e^-$$

where H_2S and the redox potential would in turn be governed by pyrite/pyrrhotite saturation and water pH.

Except for a few of the cooler drillholes all the waters are distinctly fluorite undersaturated (Fig. 9D). Waters from drillholes which are located in the vicinity of outcrops of acid rock tend to be higher in fluorine (25, 26, and 34) than waters that only have been in contact with basaltic rocks. It seems likely that thermal waters flowing through acid rocks may equilibrate with fluorite due to sufficient supply of fluorine from these rocks as has indeed been observed in the western U.S.A. (Nordstrom and Jenne, 1977). In basaltic rocks an ionic exchange equilibrium between fluoride and hydroxide ions is thought to govern the concentration of fluoride in solution (Fig. 11A). It is not known what OH-bearing silicates may be involved in this reaction but the plot in Fig. 11A suggests that probably two different minerals are involved. One of these minerals would be stable above about 100°C (1000/T less than 2.7) given a log F^-/OH^- ratio for the associated water of a little more than 0 and no variation with temperature. The other mineral is stable below about 100°C, giving log F^-/OH^- ratios between some -0.5 and -1.5 . Variations with temperature are not clear. Samples 11, 45, and 46 plot at a low log F^-/OH^-. This is considered to be the result of high pH calculated for these waters, giving a high figure for OH^- as discussed below.

From Fig. 10 it can be concluded that the relative activities of the major cations (Na, K, Ca, and Mg) and protons are such that knowledge of the activity of one cation and the water temperature allows estimation of the activities of the other cations.

Samples 11, 45, and 46, plotted in Fig. 10, are from wet-steam wells and yield high metal/proton ratios for all the cations. Flashing occurs in the aquifers feeding these wells. Some steam (and gas) loss accompanies this flashing giving a gas deficient well discharge. This results in a high calculated pH value relative to the unflashed reservoir water, which in this case is estimated to amount to about 1 pH unit.

The same samples tend on the whole to show the largest discrepancy from the best fit lines for individual metal/proton ratios, suggesting that the deviation may be mostly due to erroneous pH.

The distribution of Na^+, K^+ and H^+ fits with equilibrium with alkali feldspar and K-mica over the whole temperature range (Figs. 11B, and 11C). The low-temperature waters seem to be wollastonite saturated (Fig. 11D). Equilibrium with

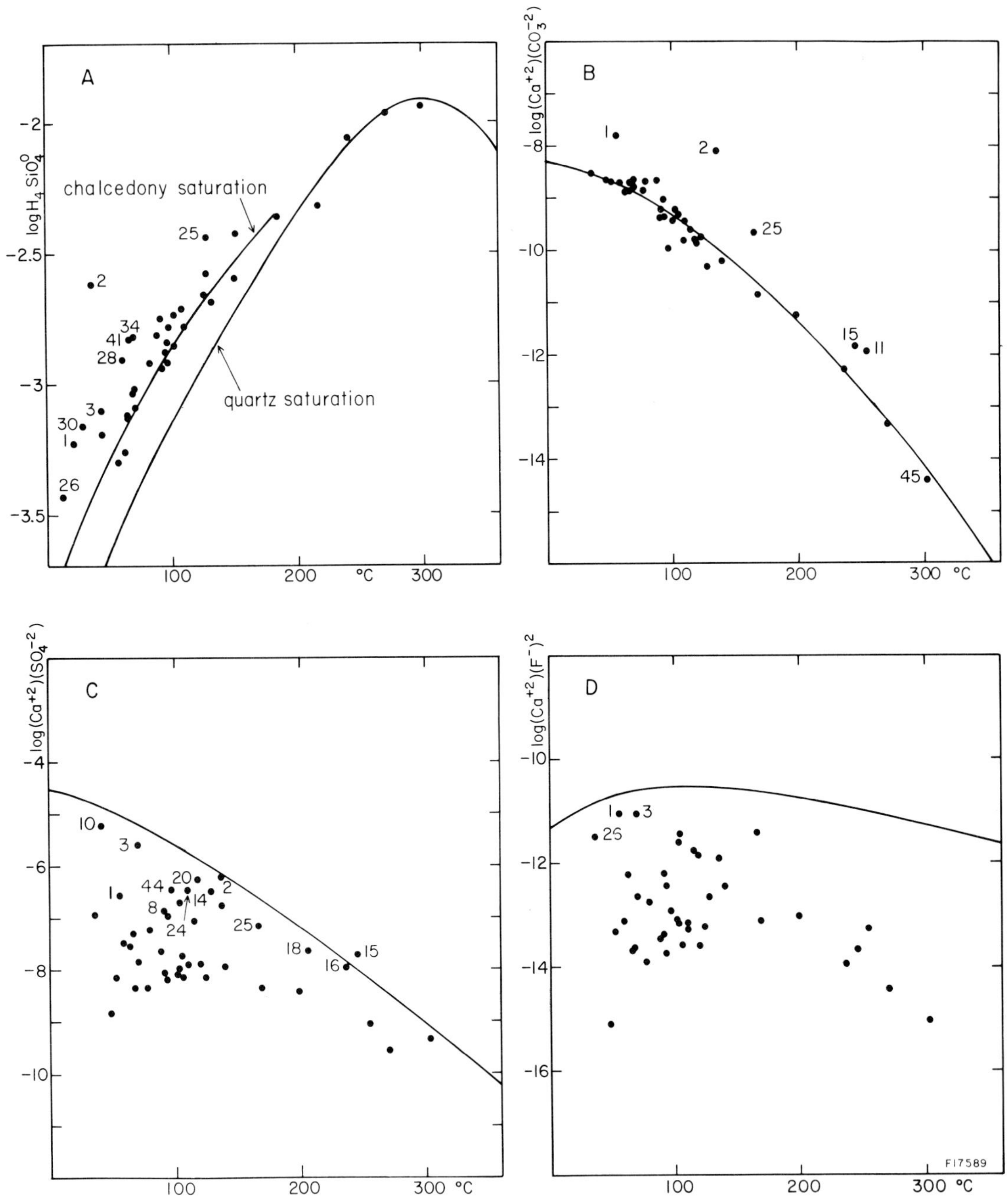

Fig. 9. The state of A) chalcedony/quartz, B) calcite, C) anhydrite, and D) fluorite saturation in deep well waters in Iceland. The low flow rate of water from wells labelled 1, 2, 25, 26 and 30 in part A indicates that conductive cooling is the cause of chalcedony supersaturation. Numbers refer to localities in Fig. 1 and Table 1.

wollastonite and chalcedony (or quartz) would fix the ratio $\sqrt{Ca^{+2}}/H^+$ in the thermal waters. The more saline waters and waters above 100°C tend to be wollastonite undersaturated. It is not known what minerals control the $\sqrt{Ca^{+2}}/H^+$ ratio in these waters. The Ca^{+2}/Mg^{+2} ratio fits with equilibrium with montmorillonite as deduced from thermodynamic data of Helgeson (1969).

Analytical data for iron are not available for the waters from the deep wells. Preliminary studies of some other geothermal waters suggest, however, pyrite/pyrrhotite saturation for dilute water and pyrite saturation for geothermal sea - water (Mottl, 1976; Gunnlaugsson, 1977) and a

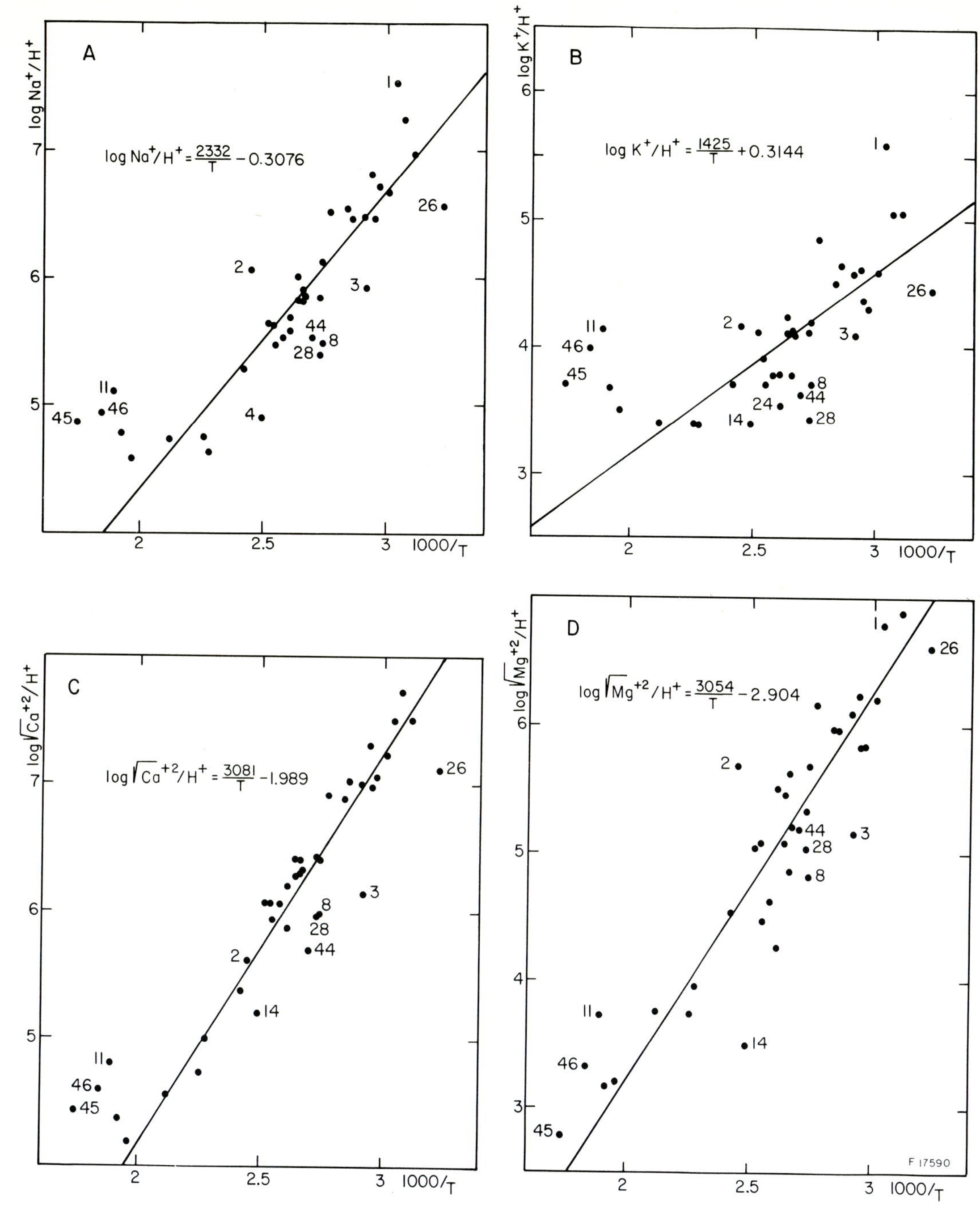

Fig. 10. The distribution of major cations and protons in water from deep wells in Iceland. The formulas indicate the best fit lines through the plotted points.

$\sqrt{Fe^{+2}}/H^{+}$ temperature relation similar to that for the other cations (see Fig. 10).

Rock alteration

During growth and burial of the volcanic pile in the active zones of volcanism the rocks become altered to a varying degree. The degree of alteration increases with depth. High-temperature geothermal areas will locally create cupolas of high temperature alteration reaching to shallow levels.

The effects of a low grade burial metamorphism can be studied in the Tertiary lava pile in the deeply cut fjord landscape of eastern, northern and western Iceland. A flat lying sequence

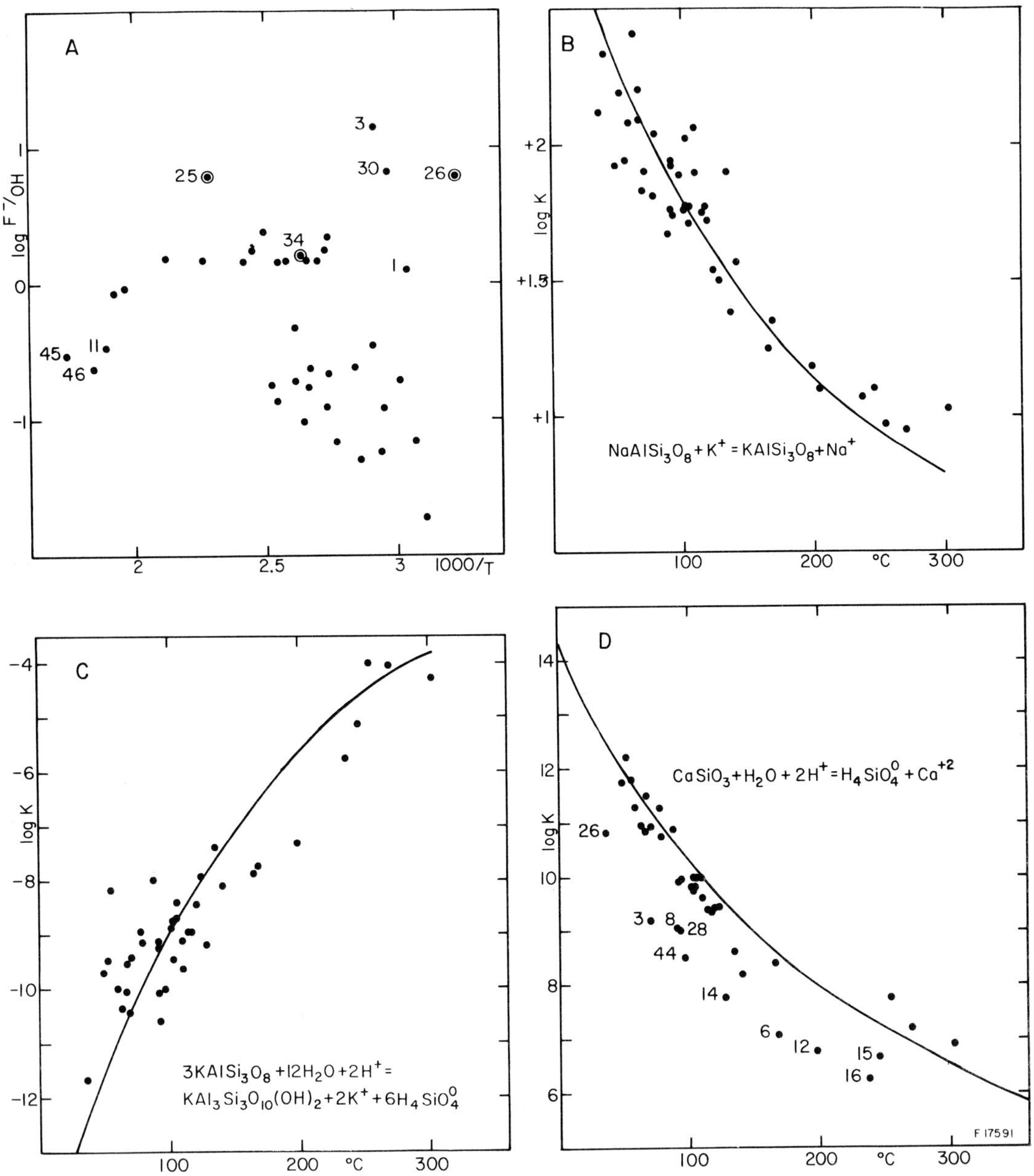

Fig. 11. A: F^-/OH^- activity relations in the deep well waters. Parts B, C, and D demonstrate that the waters tend to be in equilibrium with alkalifeldspar, K-mica, and wollastonite.

of mineral zones was first demonstrated by Walker (1960) and has since been found to apply fairly well to other flood basalt areas studied in Iceland. The boundaries between secondary mineral zones represent fossil isotherm surfaces. Aureoles of a higher degree of alteration are exposed in the eroded roots of central volcanoes. Those are the depth equivalents of former high-temperature geothermal areas.

Two cross sections are presented here which show the extrapolation of secondary mineral zones downwards based on the examination of cuttings from drillholes. One of the sections (Fig. 12) lies along the west side of Eyjafjordur in central northern Iceland, where erosion at sea level amounts to some 1200-1500 m. No central volcanoes occur along the section, so that the results are probably representative of conditions in the Tertiary flood basalt pile in general. The dyke intensity varies along the section. It is between 5 and 10% in the northernmost part and in the extreme south. In between it is sligh-

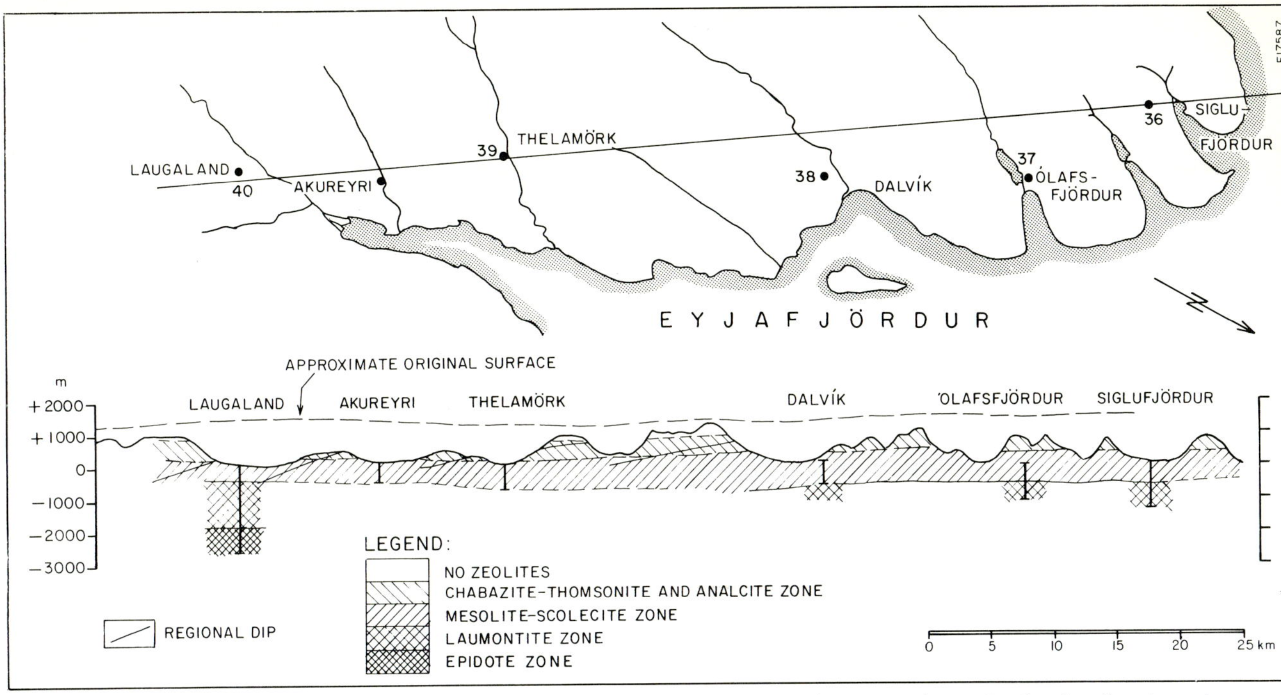

Fig. 12. A cross section along the western coast of Eyjafjördur, northern Iceland, showing the subsurface extension of the alteration zones. The original surface of the lava pile is indicated in the figure.

tly less. The regional dip of the flood basalts cuts across the more or less horizontal mineral zoning.

The drillhole results show that the laumontite zone is extensively developed at depths generally between about 500 and 2200 m below sea level. The top of this secondary mineral zone is in other parts of Iceland visible only at the deepest level of exposure where dykes and intrusions exceed 10%. The laumontite zone is underlain by a zone with abundant epidote. This alteration zone is only exposed in the most extensively altered core of central volcanoes where intrusions form a significant proportion of the rock.

The second cross section (Fig. 13) lies across the axial rift zone in southwestern Iceland and the adjacent Plio-Pleistocene and Tertiary rocks. The regional lava dips are synclinal about the axis. The section passes through or near several central volcanoes with a high proportion of intrusions (Saemundsson, 1967; Fridleifsson, 1973, 1977). The secondary mineral zoning is disturbed around the intrusive complexes where hydrothermal convective systems have produced a zone of propylitic alteration with extensive development of chlorite, calcite, pyrite and epidote.

Correlation of rock alteration with temperature in active geothermal systems

Drilling in an active geothermal area gives an opportunity to observe metamorphism in progress. Most parameters affecting the process can be measured or calculated. The factors controlling the alteration pattern include temperature, pressure, texture, structure (including porosity and permeability) and composition of rocks and also the original fluid composition.

At depths of 0-2 km the effect of pressure on the alteration pattern can be considered rather small. In rocks of similar composition, with geothermal fluids of similar origin, the main factors controlling the alteration pattern are temperature and permeability. The rocks in Icelandic geothermal areas investigated by deep drilling are largely basaltic in composition. The geothermal fluid is mostly meteoric water (see Årnason, 1976). There are differences in the lithology of the rocks, which affect more the kinetics than the products of the alteration process. The same main groups of alteration minerals are found to be formed in the holocrystalline basalts as in the basaltic hyaloclastites.

In the active high-temperature geothermal areas one comes closest to an equilibrium situation between mineral assemblages and rock temperature. In the low-temperature areas on the other hand one must be cautious of the relics from former alteration processes under different temperature conditions than those prevailing today. The appearance or disappearance of certain groups of alteration minerals and transformation of one clay mineral group to another has been used to define successive alteration

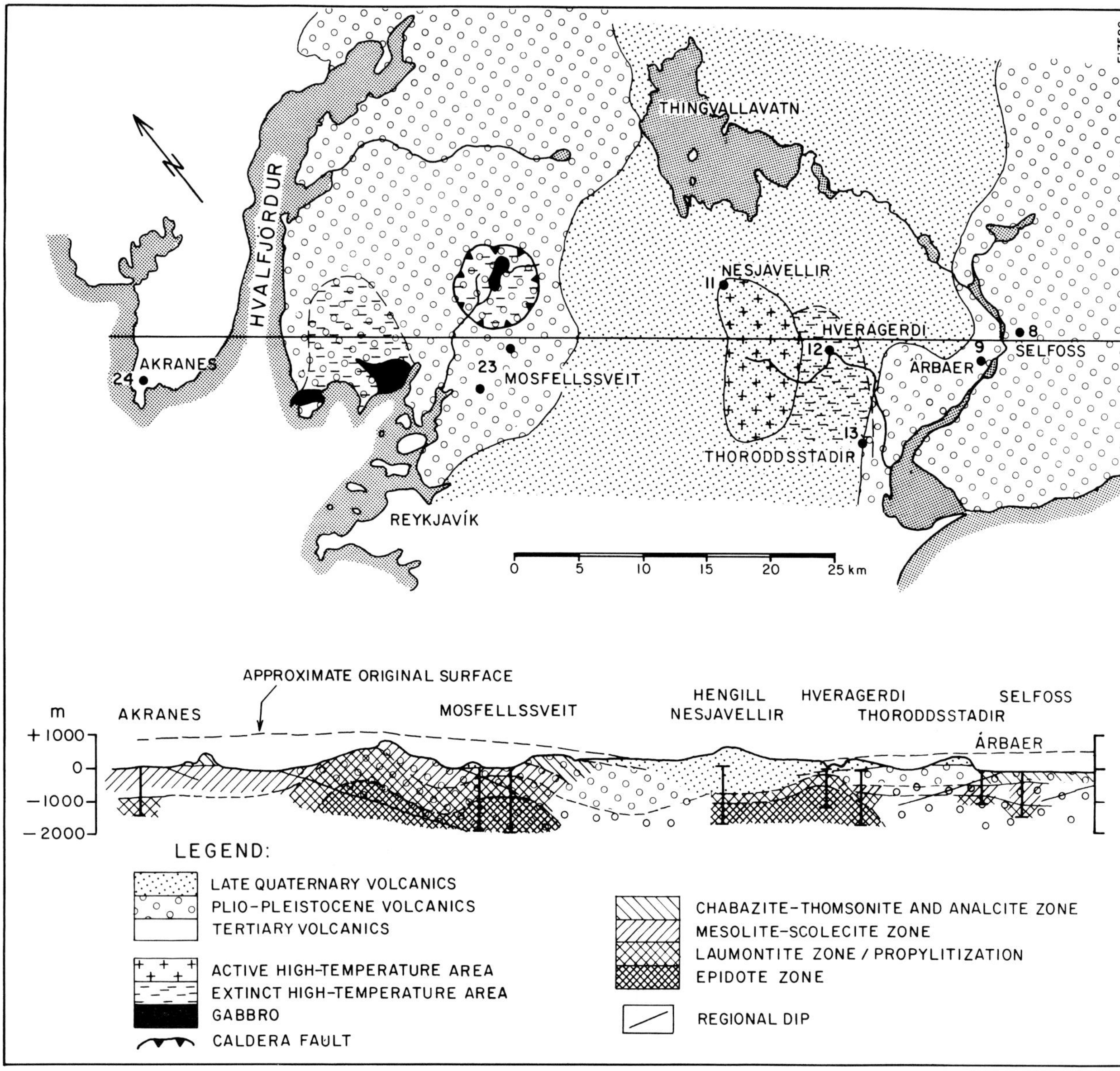

Fig. 13. A cross section through the axial rift zone in southwestern Iceland and the surrounding Plio-Pleistocene and Tertiary rocks. The section shows the subsurface extension of alteration zones obtained from drillhole data. The location of the section is shown on a very simplified geological map of the area. The cupolas shown by the epidote zone coincide with the cores of central volcanoes.

zones (Kristmannsdóttir; 1975, 1977b, 1978) in the basaltic rocks of the Icelandic geothermal areas (Fig. 14). The zones can be correlated with certain metamorphic facies. Rock temperature is estimated from temperature measurements in the drillholes. Distribution of alteration minerals from all investigated areas is compared, and results from areas showing signs of relict alteration and disequilibrium are discarded. From such compilation and critical study of the data the temperature interval for the formation of a certain mineral assemblage is fixed. The estimated transformation temperatures for the clay minerals are assumed to be quite reliable. The range of stability for laumontite is assumed fairly reliable and this is also valid for wairakite and epidote. The stability ranges for other zeolites is considered more uncertain and the same is true for actinolite. For the calcium silicates, prehnite and a few other alteration minerals in the Icelandic crust no stability temperature interval

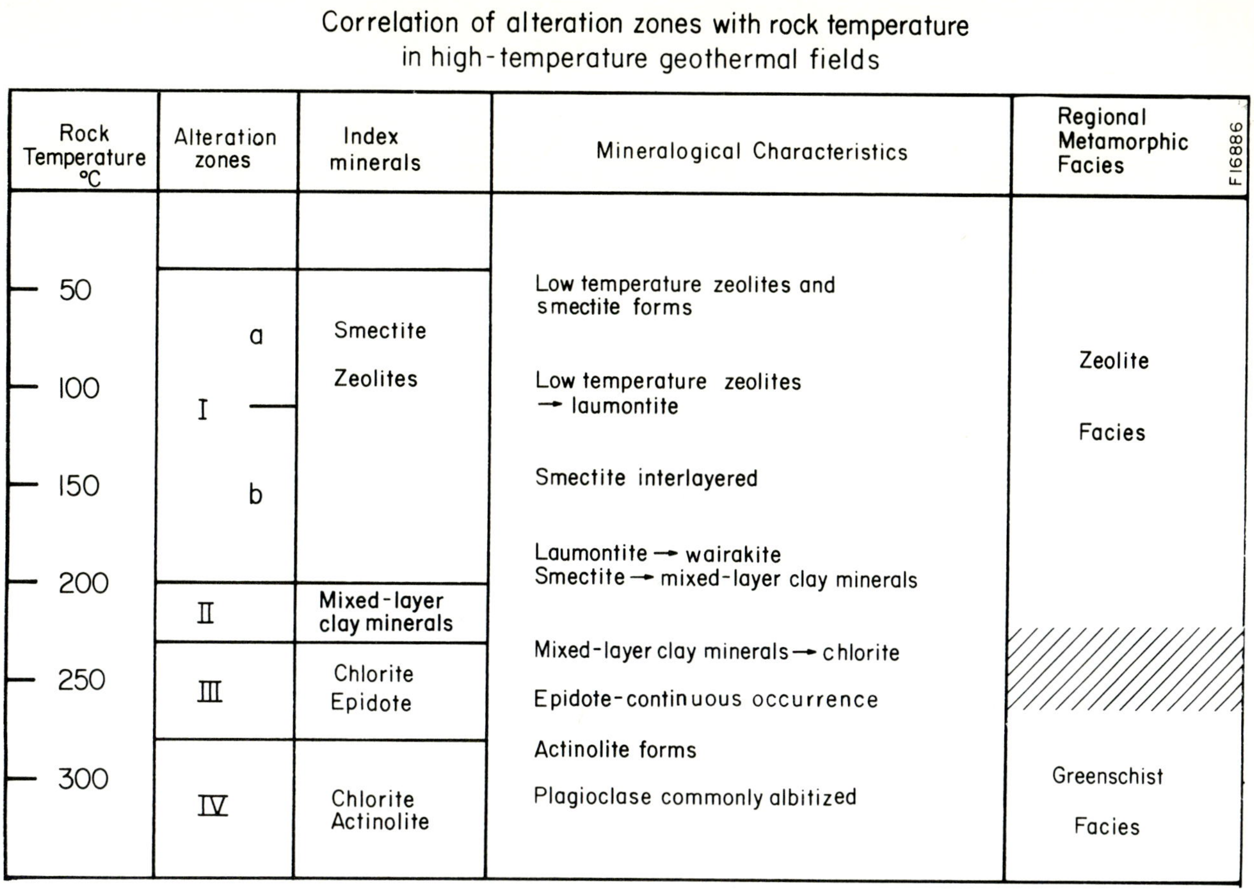

Fig. 14. Successive alteration zones observed with increasing depth and temperature in active high-temperature areas in Iceland. A correlation is made with rock temperature and regional metamorphic facies. Also shown are the index minerals and mineralogical changes characterizing each of the zones.

has been determined due to too few and in some cases contradictory data.

Fig. 14 is based mainly on data from high-temperature geothermal areas. The "low-temperature zeolites" are grouped together since a distinction can not be made between individual zeolite zones in the high-temperature geothermal areas due to their steep thermal gradient. A zeolite zonation comparable to that observed in the Tertiary basalt pile is observed in some low-temperature areas on the flanks of the axial zone (Kristmannsdóttir and Tómasson, 1976).

Due to the complex thermal history of the low-temperature geothermal areas it is more difficult to relate the low-temperature zeolites to certain temperature intervals than is the case for the high-temperature alteration minerals.

Magnetic properties

Remanence

From various paleomagnetic surveys we infer that the polarity of primary magnetic remanence in deep drillholes outside the volcanic zones of Iceland is likely to change sign on average some eight times per km of depth, including short events. However, in the almost complete absence of core material from Icelandic deep holes, it has not yet been possible to utilize the remanence polarity for mapping or stratigraphic correlation in these. Little is also known about remanence intensities in subsurface formations (Kristjánsson and Watkins, 1977). Logging devices for downhole magnetic fields are difficult to construct because of the severe environment met with, and interpretation is subject to interference from lateral inhomogeneities, secondary and induced magnetization components, and variations in hole dimensions.

A proton magnetometer, developed by T. Sigurgeirsson and B. Jónsson, has been successfully tested to 470 m depth in Reykjavík (Jónsson, 1976). In basalts it shows rapid and irregular field intensity variations, commonly ± 10 kγ (10 μT) in a few meters, while in thick tuff beds the field variations are of order of ± 0.1 kγ.

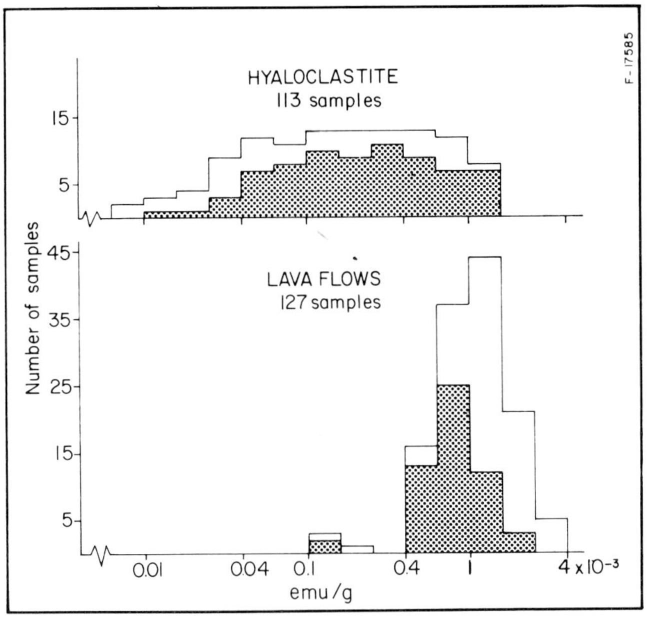

Fig. 15. Histograms of magnetic susceptibility values in samples of drill cuttings at 10 m intervals in Reykjanes hole No. 8 (shaded) and Krafla hole No. 4, grouped by rock type. Data from thin units are excluded.

Magnetic minerals

The magnetic properties of basalts from deep drillholes in Iceland appear to be mostly due to cation-deficient magnetite. Strong field thermomagnetic curves obtained on 25 samples of drill cuttings from 14 deep holes (Kristjånsson and Watkins, 1977, and subsequent work) generally yield a single high Curie point in the range 540°-610°C. The major exception is two samples from below 1400 m in Heimaey island, both having a main Curie point near 350°C.

The magnetite probably exsolves from titanomagnetites on oxidation, in some cases during initial cooling of the igneous bodies, in others by the action of hot ground water during later burial. Incompletely exsolved titanomagnetites or maghemites are likely to occur at depth only in rapidly cooled rock units in areas where the thermal gradient has been low, as in Heimaey.

Susceptibility

Room temperature magnetic susceptiblility χ of drill cuttings has been measured at 10 m intervals for holes reaching to about 1000 m or more in the following 14 localities (2, 11, 14, 15, 17, 19, 21, 23, 25, 30, 40, 45, 46, 47).

Manually smoothed graphs of χ for some of these holes are included in Fig. 2, 3, 4 and 6, and for a few others in Kristjånsson and Watkins (1977).

The susceptibility in high Curie point samples is a good indicator of the amount of magnetite present, that is broadly speaking the original magnetite content of the rock less what has been converted to non-magnetic minerals by secondary processes. The frequency distribution of observed χ values depends on the rock type. As an example we have plotted histograms of χ values from two high temperature drillholes (Fig. 15) where two main rock types occur. In contrast to hyaloclastite, susceptibility values of lava are high and tightly clustered. This shows that in basalt units the conversion of magnetite is a slow process, even at above 250°C, if they are massive enough to limit water circulation.

In the volcanic zone, the χ values from each drillhole, after smoothing, reflect to some extent the lithology (Fig. 2,3,4 and 6). A remarkable feature is the occurrence of relatively fresh and highly magnetic basalts at 300-500 m and below 1000 m in Krafla (45) (Fig. 3), at 1600-1700 m at Nåmafjall (46), and below 1500 m in Reykjanes (15) (Fig. 2). Some of these basalts may be quite recent intrusions, cf. section on Krafla above. In some holes in the volcanic zone

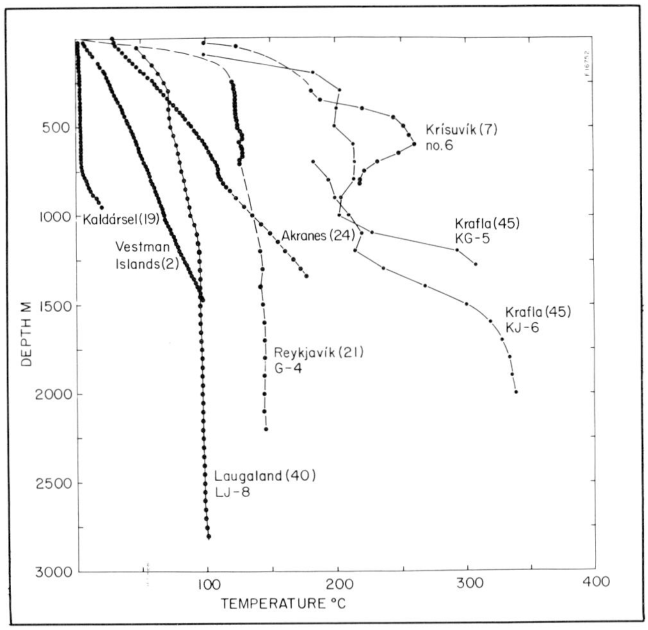

Fig. 16. Some typical temperature profiles from deep drillholes showing disturbances of the crustal temperature by water convection. Numbers refer to localities in Fig. 1. No. 7 and 45 are high-temperature areas within the active volcanic zone, 21 and 40 are low-temperature areas on the flanks of the active zone, 2 and 24 are areas of relatively undisturbed thermal state of the crust, and 19 is an area within the active zone, where strong cooling by water convection has taken place. The thermal disturbance in the upper part of the Akranes hole is due to a small upward flow of water within the hole.

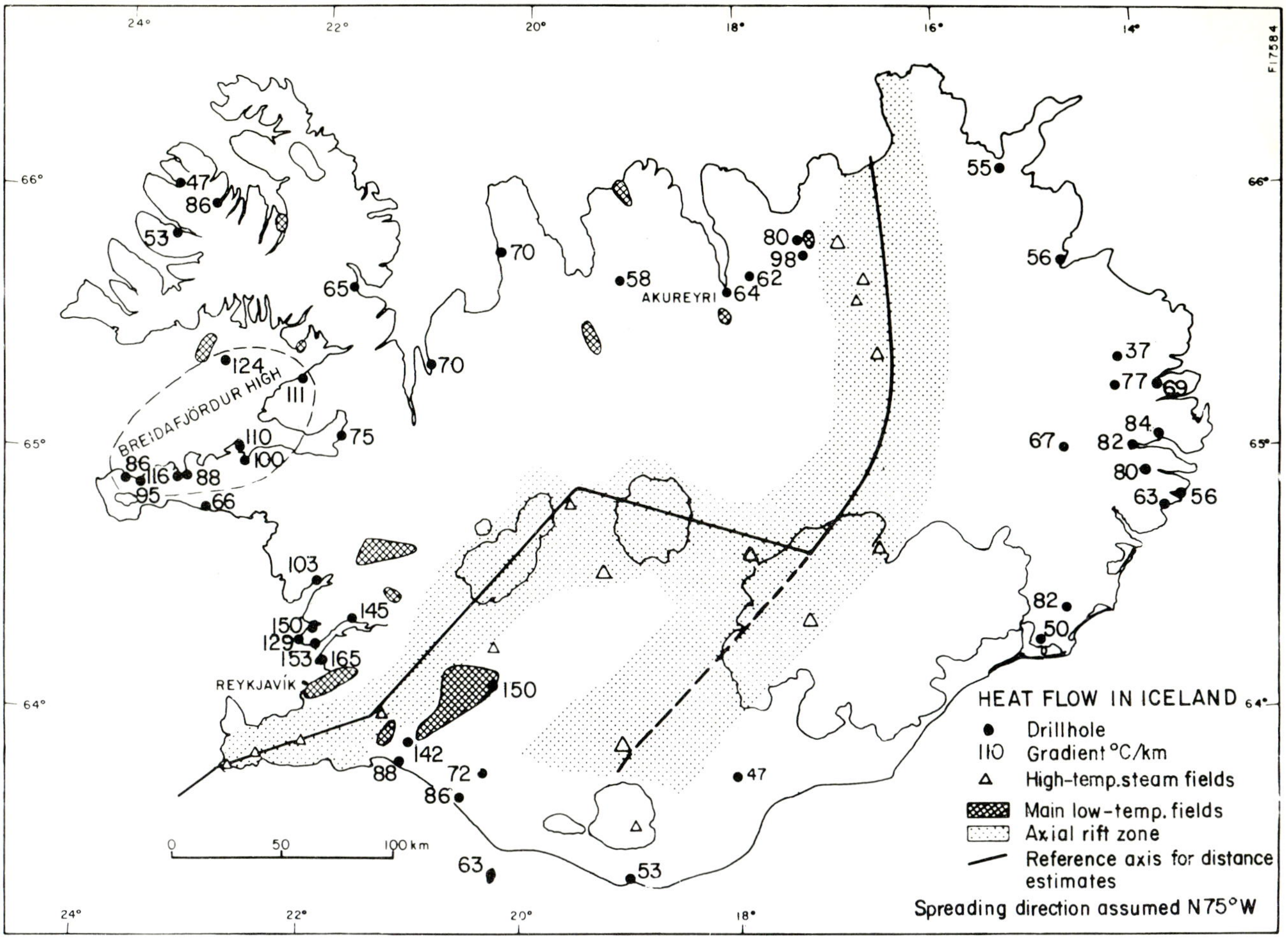

Fig. 17. The location of drillholes used to determine thermal gradients in Iceland. The main centers of geothermal activity are shown on the map.

the susceptibility is uniformly low (Heimaey (2), Kaldårsel (19)), which may reflect the petrochemistry of the rock, or even almost zero due to alteration (Nåmafjall (46)). However, on the whole a trend towards low χ values with increasing depth is not observed so far.

Outside the volcanic zone the susceptibility lies mostly between 0.5 and 2 times 10^{-3} emu/g. It is very rare to find χ values lower than 0.5 10^{-3} units occurring consistently over a depth range of more than one hundred m (Fig. 4 and 6). No general trend towards low susceptibilities with increasing depth in a drillhole is observed, but there is a distinct increase in hole-average χ with increasing age of the area drilled. It has not been attempted to correlate these results with local magnetic anomalies in detail, but Kristjånsson and Watkins (1977) have discussed the general implication of such observations to the "magnetic layer" problem in oceanic areas.

Heat flow

Heat is transported to the surface in Iceland by the three processes of heat conduction, water convection and volcanism. The following discussion will be confined mainly to the conductive heat flow. The convective part, i.e. the heat output of the geothermal areas, is in most cases poorly known. It has been roughly estimated to be 10^9 cal/s or about 4000 MW (Bodvarsson, 1961), of which the major part is from the high-temperature geothermal areas within the axial zone. The heat output of extrusive volcanism may be derived approximately from the rate of production of volcanics which is on the average about 0.045 km^3/y. (see Jakobsson,1972), giving a heat output of about 6000 MW (Pålmason, 1973).

Determining the conductive heat flow from shallow drillholes is complicated by the disturbing effects of convective hydrothermal systems, primarily in the neighbourhood of surface geothermal manifestations. Within the youngest axial zone, where the surface rocks are highly permeable, the movement of cold ground water by the effect of gravity also severely disturbs the shallow crustal temperature field. A few examples of temperature profiles disturbed by the effect of water convection are shown in Fig. 16.

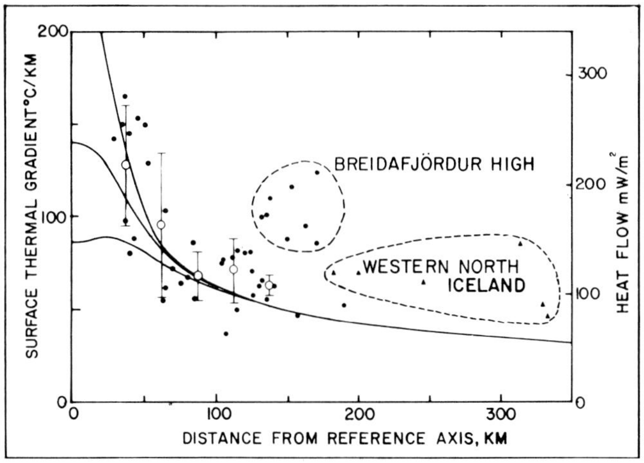

Fig. 18. The relation of the measured thermal gradients in the Icelandic crust to the distance from a reference spreading axis (Fig. 17). The error bars denote the sample standard deviations for grouped values within each 25 km distance interval. Theoretical curves for the thermal gradient versus distance from the spreading axis are shown for reference.

The location of the drillholes is shown in Fig. 1.

An attempt has been made to map the conductive heat flow pattern, by a careful selection of data to avoid the thermal effects of convecting water as much as possible. The main criteria in selecting drillholes for this purpose are that the temperature profiles be linear and that the holes be located outside known hydrothermal systems. There is of course always the possibility that hydrothermal systems may occur at depths below the bottom of the holes, but such systems would probably show as irregularities in the overall heat flow pattern of the area, if they existed.

The thermal conductivity of the rocks is not known for each individual hole, but since the lithology is similar in most of the holes, the gradient values will give a fairly good idea about the variation in heat flow values from one hole to another. Some 150 samples have been taken for conductivity measurements (unpublished measurements by B. Polyak, V.Kononov and A Jessop), and they indicate that a value of about 1.7 W/m°C is representative for a typical drillhole in the Icelandic crust. This value has been used to estimate the heat flow from the gradients.

Fig. 17 shows the thermal gradients from the holes selected. The holes are of varying depth, from about 100 meters to over 1500 meters. Data on some of the holes have been published by Pålmason (1973), others are more recent. A summary of the available data is given in Table 2.

We have tried to relate the gradient values to distances from an axis of spreading, using a reference axis as shown in Fig. 17. The reference axis follows the Reykjanes-Langjökull volcanic zone in southwestern Iceland and then shifts over to the eastern zone in northeastern Iceland. The heat flow values in central southern Iceland which are rather low near the eastern zone, support the assumption that the eastern zone in southern Iceland is relatively young, and that the Reykjanes-Langjökull zone has been the main zone of the rifting and volcanism during most of the geological history of Iceland (Saemundsson, 1974). In northern Iceland the situation is less clear. The first heat flow measurements from eastern Iceland indicated relatively low values suggesting that the main axial zone might have been a direct northward continuation of the Reykjanes-Langjökull zone (Pålmason, 1973). When a larger number of new gradient values were obtained in eastern Iceland, they were generally higher, suggesting that eastern Iceland crust might have been generated near the present axial zone in northeastern Iceland. This is in agreement with recent radiometric dating studies from eastern Iceland (Ross and Mussett, 1976; McDougall et al.,1976). We have therefore chosen here to locate the reference axis in northern Iceland along the presently active zone there.

The gradient values are plotted in Fig. 18 versus distance from the reference axis. The direction of drift is assumed to be N75°W,corresponding to the pole of rotation deduced by Pitman and Talwani (1972), and Talwani and Eldholm (1977). For comparison theoretical curves are shown which are based on model calculation of crustal accretion for the Icelandic segment of the Mid-Atlantic Ridge (Pålmason, unpublished). For distance greater than about 100 km the theoretical curve is identical with the result of Parker and Oldenburg (1973), while near the axis different curves are obtained for different assumptions regarding the mechanism of crustal accretion.

The data from southwestern, southern and eastern Iceland agree reasonably well with the theoretical curves. On the other hand the northwestern Iceland gradients are noticeably high. A conspicuous gradient high occurs on the northern part of the Snaefellsnes peninsula extending out into Breidafjördur bay. These values are shown encircled in the diagram, termed the Breidafjördur high. Other values from western North-Iceland are also high as shown separately in the diagram. It seems likely that an explanation for these high values in northwestern Iceland is to be sought in an extinct axis, possibly located along the Breidafjördur high. This possibility is also supported by the synclinal structure of the Tertiary basalt pile on both sides of this high. The continuation of such an extinct axis to the northeast of the Breidafjördur high is not clear, however.

Stress measurements in drillholes

The first direct measurements of the state of stress at shallow levels in the Icelandic crust outside the axial rift zone were made by N. Hast (Hast, 1969). His measurements, which were made by an overcoring technique, indicated compressive horizontal stress with average magnitudes greater than the overburden pressure at the depth of measurement. The depths at which Hast's measurements were made, were in most cases less than 30 meters.

In 1976 the first stress measurements by hydrofracturing techniques were made in two drillholes in Reykjavík (Haimson and Voight, 1977). The measurements were made at depths between 180-375 meters. Horizontal compressive stresses were also found here of a similar magnitude as in the measurements of Hast. An important new observation was that a change in the stress environment with depth was indicated, with associated changes in principal stress orientations.

The results of all the drillhole stress measurements made so far indicate that the maximum horizontal compressive stress at shallow levels on the flanks of the axial rift zone is oriented approximately perpendicular to the zone.

Earthquake focal mechanism solutions in the axial rift zone indicate on the other hand a minimum horizontal compression in a direction perpendicular to the main trend of the axial zone (Björnsson and Einarsson, 1974). Thus a significant change in the crustal stress appears to take place from the axial rift zone to the adjacent flanks. A possible explanation for this, which has been discussed in some detail by Haimson and Voight (1977), is that the cooling and thickening of the lithosphere as it moves away from the axial zone may cause horizontal compressive stresses, with the stress orientations changing with depth, as indicated by the Reykjavík hydrofracturing results. This possibility needs to be tested further by model calculations, but it appears in a qualitative way capable of reconciling the earthquake focal mechanism results and the result of static stress measurements in drillholes by overcoring and hydrofracturing methods.

Conclusions

On the basis of data from drillholes into the Iceland crust several conclusions may be drawn regarding the structure of the crust and the processes at work within it. Some of the more significant points are enumerated below.

1. Subaerial lavas are the main rock type to the bottom of all the holes, the deepest of which reaches to more than 4000 meters below the original uneroded surface of the crust.
2. In general, the fraction of intrusives increases with depth.
3. Aquifers occur in the drillholes to a depth of at least 2-3 km. Disturbance of the thermal state of the crust indicates that hydrothermal circulation reaches well below 3 km in some areas.
4. The magnetic susceptibility of the basaltic rocks generally shows no systematic variation with depth.
5. None of the holes yet drilled is deep enough to reach the high-velocity ($v_p \simeq 6.5$ km/s) lower crust indicated by seismic refraction surveys. Some shallower seismic boundaries have been penetrated by drillholes.
6. The chlorine content of the geothermal waters is mostly low, but high values are found near the coast, especially within the active volcanic zone, and are caused by mixing with sea-water.
7. Other chemical species in the geothermal waters appear to be governed by temperature dependent mineral/solute equilibria.
8. The alteration pattern observed in surface outcrops due to low grade burial metamorphism is also found in the drillholes, and there extended farther downwards. The deeper holes reach into the epidote zone, which in surface exposures is only found in the intensively altered cores of central volcanoes.
9. The rock alteration in active high-temperature hydrothermal systems has been correlated with rock temperature. Greenschist facies metamorphism occurs at temperatures above about 260°C.
10. The conductive heat flow pattern, based on data from 48 holes, shows certain regularities with respect to the active zone of rifting and volcanism, with decreasing heat flow values with distance from the zone. In certain areas, however, irregularities exist in the overall pattern.
11. Stress measurements by overcoring and hydrofracturing methods in relatively shallow holes show horizontal compressive stresses on the flanks of the active zone. The axis of maximum compression is approximately perpendicular to the strike of the zone of rifting and volcanism.

References

Árnason, B., Groundwater Systems in Iceland, Soc. Sci. Islandica, Publ. 42, 236 pp., 1976.

Arnórsson, S., Major element chemistry of the geothermal sea-water at Reykjanes and Svartsengi, Iceland, Min. Mag. 167-180, 1978 a.

Arnórsson, S., Precipitation of calcite from

flashed geothermal waters in Iceland, Contr. Mineral. Petrol. 25-28, 1978 b.

Arnórsson, S., Grönvold, K., and Sigurdsson, S., Aquifer chemistry of four high-temperature geothermal systems in Iceland, Geochim. Cosmochim. Acta, 523-536, 1978.

Björnsson, A., Saemundsson, K., Einarsson, P., Tryggvason, E. and Grönvold, K., Current rifting episode in north Iceland, Nature, 318-323, 1977.

Björnsson, S., Arnórsson, S., Tómasson, J., Exploration of the Reykjanes brine area, Geothermics, Spec. iss. 2, 1640-1650, 1970.

Björnsson, S., Arnórsson, S. and Tómasson, J., Economic evaluation of Reykjanes thermal brine area, Iceland, Am. Ass. Petr. Geol. Bull., 56 (12), 2380-2391, 1972.

Björnsson, S. and P. Einarsson, Seismicity of Iceland. In: Geodynamics of Iceland and the North Atlantic Area (ed. L. Kristjánsson), D. Reidel Publ. Co., 225-239, 1974.

Bodvarsson, G., Physical characteristics of natural heat resources in Iceland, Jökull (Reykjavík), 11, 29-38, 1961.

Fridleifsson, I.B., Petrology and Structure of the Esja Quaternary Volcanic Region, Southwest Iceland. (D. Phil. thesis), Oxford University, 208 pp, 1973.

Fridleifsson, I.B., Distribution of large basaltic intrusions in the Icelandic crust and the nature of the layer 2 - layer 3 boundary. Geol. Soc. Amer. Bull., 88, 1689-1693, 1977.

Gunnlaugsson, E., The origin and distribution of sulphur in fresh and geothermally altered rocks in Iceland, Unpublished Ph.D. thesis, Univ. of Leeds, 192 p., 1977.

Haimson, B.C. and B. Voight, Crustal stress in Iceland, Pure and Appl. Geophys., 115 (1), 153-190, 1977.

Hast, N., The state of stress in the upper part of the earth's crust, Tectonophysics, 8, 169-211, 1969.

Helgeson, H.C., Thermodynamics of hydrothermal systems at elevated temperature and pressure, Am. Jour. Sci., 729-804, 1969.

Jakobsson, S.P., Chemistry and distribution pattern of Recent basaltic rocks in Iceland, Lithos, 5, 365-386, 1972.

Jónsson, B.B., A proton precession magnetometer for drillhole measurements. Science Inst. University of Iceland, Report no. 17, 19 pp, 1976.

Klein, F.W., P. Einarsson and M. Wyss, Microearthquakes of the Mid-Atlantic plate boundary on the Reykjanes peninsula in Iceland, J. Geophys. Research, 78, 5084-5099, 1973.

Kristjánsson, L., and N.D. Watkins, Magnetic studies of basalt fragments recovered by deep drilling in Iceland, and the "magnetic layer" concept. Earth Planet. Sci. Lett., 34, 365-374, 1977.

Kristmannsdóttir, H., Hydrothermal alteration of basaltic rocks in Icelandic geothermal areas. Proceedings of the Second U.N. Symp. of Development and Use of Geothermal Resources, San Francisco 1975, 441-445, 1975.

Kristmannsdóttir, H., Interaction between hydrothermal alteration and contact metamorphism - Krafla high-temperature geothermal area, Iceland (abstract). IASPEI/IAVCEI Joint General Assemblies, Durham, p. 245, 1977a.

Kristmannsdóttir, H., Hydrothermal formation of clay minerals in Icelandic geothermal fields, Volume of summaries. Third Meeting of the European Clay Groups, Oslo, 87-90, 1977b.

Kristmannsdóttir, H., Alteration of basaltic rock by hydrothermal activity at 100-300°C. Proceedings of the Sixth International Clay Conf., Oxford, 1978, in press.

Kristmannsdóttir, H. and Tómasson, J., Zeolite zones in Geothermal areas in Iceland. Natural zeolites. Occurrence, Properties, Use (ed. Sand, L.B. and Mumpton F.A.). Pergamon Press, Oxford and N.Y., 277-284, 1978.

Mc Dougall, I., N.D. Watkins and L. Kristjánsson, Geochronology and paleomagnetism of a Miocene-Pliocene lava sequence at Bessastadaá, Eastern Iceland, Am. J. Sci., 276, 1078-1095, 1976.

Moorbath, S., Sigurdsson, H. and Goodwin, R., K-Ar ages of the oldest exposed rocks in Iceland, Earth Planet. Sci. Lett., 4, 197-205, 1968.

Mottl, M., Chemical exchange between sea water and basalt during hydrothermal alteration of the oceanic crust. Unpubl. Ph. D. Thesis, Stanford Univ., 138 pp, 1976.

Nordstrom, D.K., and Jenne, E.A., Fluorite solubility equilibria in selected geothermal waters, Geochim, Cosmochim, Acta, 175-188, 1977.

Pálmason, G., Crustal Structure of Iceland from

Explosion Seismology, Soc. Sci. Islandica, Publ. 40, 187 pp., 1971.

Pålmason, G., Kinematics and heat flow in a volcanic rift zone, with application to Iceland, Geophys. J.R. astr. Soc., 33, 451-481, 1973.

Parker, R.L. and D.W. Oldenburg, Thermal model of ocean ridges, Nature Phys. Sci., 242, 137-139, 1973.

Pitman, W.C., III, and M. Talwani, Sea-floor spreading in the North Atlantic, Geol. Soc. Am. Bull, 83, 619-646, 1972.

Ross, J.G., and A.E. Mussett, $^{40}Ar/^{39}Ar$ dates for spreading rates in eastern Iceland. Nature, 259, 36-38, 1976.

Saemundsson, K., Vulkanismus und Tektonik des Hengill - Gebietes. Acta, Nat. Isl. II, 7, 101 pp., 1967.

Saemundsson, K., Evolution of the axial rifting zone in northern Iceland and the Tjörnes fracture zone, Bull. Geol. Soc. Am., 85, 495-504, 1974.

Saemundsson, K., Fissure swarms and central volcanoes of the neovolcanic zones of Iceland, Geol. Journal, Spec. Issue 10, 415-432, 1978.

Sigvaldason, G.E., and Óskarsson, N., Chlorine in basalts from Iceland, Geochim. Cosmochim. Acta, 40, 777-789, 1976.

Talwani, M. and O. Eldholm, Evolution of the Norwegian - Greenland Sea, Geol. Soc. Am. Bull., 88, 969-999, 1977.

Tómasson, J., On the origin of sedimentary water beneath Westmann Islands, Jökull (Reykjavík), 17, 300-311, 1967.

Tómasson, J. and Kristmannsdóttir, H., High temperature alteration minerals and thermal brines, Reykjanes, Iceland. Contr. Mineral. and Petrol., 36, 123-134, 1972.

Tómasson, J. and Kristmannsdóttir, H., Reykir - Reykjavík. Investigation of three low-temperature areas in Reykjavík and its neighborhood. Proceedings Int. Symp. on Water-Rock interaction, Praha, 243-249, 1974.

Tómasson, J., Fridleifsson, I.B. and Stefánsson, V. A hydrological model for the flow of thermal water in southwestern Iceland with special reference to the Reykir and Reykjavík thermal areas. Proceedings of the Second U.N. Symp. on Development and Use of Geothermal Resources, San Francisco 1975, 643-648, 1975.

Walker, G.P.L., Zeolite zones and dyke distribution in relation to the structure of the basalts in eastern Iceland. J. Geol., 68, 515-528, 1960.

Preface

In early 1976 we embarked on a multi-institutional field program to investigate the Mid-Cayman Rise Spreading Center, located along the North American-Caribbean plate boundary (Figure 1). Our study was designed to elucidate the tectonic behavior and petrologic processes associated with a short (110 km long) accreting plate boundary that is truncated by transform faults. This paper represents a general summary of our field program that spanned two field seasons and that involved traditional surface ship marine geophysical and geological techniques, a deep-towed camera system (ANGUS), a submersible (DSRV ALVIN), a bathyscaph (DSRV TRIESTE II), and the United States Navy multi-narrow beam sonar mapping system (SEABEAM). These diverse, intensive and sophisticated tools provided investigators with a plethora of data that is still undergoing analysis. Consequently, our presentation is not meant to be a complete synthesis of the program but rather an opportunity to identify the types of data that we recovered, to summarize the results of our work to date and to outline the aims of our ongoing investigations. This general and integrated presentation is not meant to supplant the specific and in-depth papers, authored by a subset of the authors of this paper, that will follow during the coming year.

GEOLOGICAL AND GEOPHYSICAL INVESTIGATION OF THE MIDCAYMAN RISE SPREADING CENTER: INITIAL RESULTS AND OBSERVATIONS

*CAYTROUGH

Introduction

Regional Morpho-Tectonic Setting

The 1600 km long, east-west trending Cayman Trough (Figure 1) extends eastward from the Gulf of Honduras and terminates at the Windward Passage to the southeast of Cuba. The north side of the Trough, bounded by the Cayman Ridge, is characterized by precipitous escarpments. The Nicaraguan Plateau, rising roughly 5000 m above the floor of the Trough, flanks it to the south. Bathymetric and seismic reflection surveys made by Holcombe, et al. (1973) clearly indicate that the central portion of the Cayman Trough is comprised of rugged, north-south trending elongated ridges and troughs that have relief of 200 to 2000 m. Holcombe and coworkers found these lineated morpho-tectonic elements to be bilaterally symmetric about a central, deeply incised (to ~6000 m), north-south trending V-shaped valley. Seismic reflection data (Ewing, et al., 1960; Edgar, et al., 1971; Erickson, et al., 1972; Holcombe, et al., 1973) indicate that the sediment blanket thins toward the axis of this central valley. The observations of Holcombe, et al., that the basement morphology and the distribution of sediment within the Cayman Trough are virtually indistinguishable from the distinctive topographic fabric and sediment patterns associated with some slowly accreting plate boundaries, led them to suggest that the central valley is the axis of a slowly accreting plate boundary. They called this plate boundary the Mid-Cayman Rise and suggested that the Oriente Fracture Zone, which terminates the axial valley to the north, and the Swan Fracture Zone terminating the axial valley to the south, are transform faults. In support of their interpretation, Holcombe and coworkers

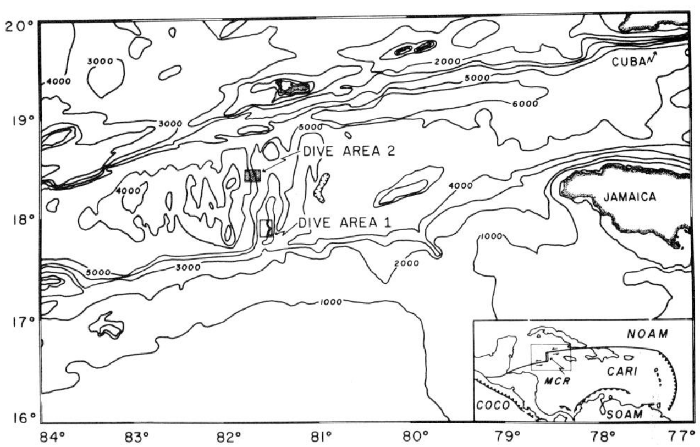

Fig. 1. Bathymetric map of the Cayman Trough showing location of the study area and two dive areas. The inset (after Jordan, 1975) shows the relation of the map area to the plate boundaries of the Caribbean. NOAM = North American Plate; COCO = Cocos Plate; SOAM = South American Plate; CARI = Caribbean Plate; MCR = Mid-Cayman Rise.

*Robert Ballard, Wilfred Bryan, Henry Dick, K.O. Emery, Geoffrey Thompson and Elazar Uchupi of Woods Hole Oceanographic Institution, Woods Hole, MA 02543; Karleen E. Davis, Department of Earth and Planetary Sciences, Massachusetts Institute of Technology, Cambridge, MA 02139; Jelle de Boer, Department of Geology, Wesleyan University, Wesleyan, CT 06457; Stephen E. DeLong, Paul J. Fox, Frieda L. Malcolm, Randall Spydell and Janet Stroup, Department of Geological Sciences, State University of New York at Albany, Albany, NY 12222; William G. Melson, Smithsonian Institution, Washington, D.C. 20560; Raymond Wright, Department of Mines and Geological Survey, Hope Gardens, Kingston 6 Jamaica.

__________: participated in field program.

noted that earthquake activity is confined to the axial valley and to the Oriente and Swan transforms; in addition, first motion studies along the Oriente and Swan transforms yield fault plane solutions consistent with left-lateral strike-slip motion parallel to them (Sykes and Ewing, 1965; Molnar and Sykes, 1969; Jordan, 1975).

Other marine geophysical and geological data are consistent with the interpretation of Holcombe and coworkers. Seismic refraction experiments (Ewing, et al., 1960) indicate that the crust of the Cayman Trough has a velocity structure which is similar to that of oceanic velocity sections, although the crustal velocity structure in the Cayman area appears to be much more variable than normal oceanic crust. Analysis of free air gravity anomaly patterns and Bouguer anomaly values suggests that, like the Mid-Atlantic Ridge, the Mid-Cayman Rise is compensated and that it is underlain by a low density root (Bowin, 1968). Erickson, et al. (1972) have shown that high heat flow values (2.0 HFU) are associated with the central portion of the Mid-Cayman Rise and the transform boundaries. The distinctive magnetic anomaly pattern recognized in ocean basins characterizes the Cayman Trough region and appears to be bilaterally symmetric about the central valley (Matthews, 1974; Macdonald and Holcombe, 1978). In addition, Macdonald and Holcombe suggest that the Mid-Cayman Rise has been a symmetrically-spreading accreting plate boundary for at least the past 6 m.y., at a total opening rate ranging from 2.0 cm/yr (0 to 2.4 m.y.B.P.) to 4.0 cm/yr (2.4 to 6.0 m.y. B.P.). Eggler, et al. (1973) and Perfit and Heezen (1978) report the recovery of rocks identical to those recovered from other ocean basins; Perfit (1977) and Perfit and Heezen (1978) present chemical data demonstrating that the basaltic rocks recovered from the Mid-Cayman Rise are similar to abyssal tholeiitic basalts recovered from many parts of the mid-oceanic ridge system.

The geological and geophysical properties of the Cayman Trough are consistent with the suggestion that the Trough is underlain by oceanic crust and that presently oceanic crust is being accreted along a short (110 km long), north-south trending spreading center. To the north and south the Mid-Cayman Rise is truncated by two left-lateral transform faults. The bilateral symmetry of topography and sediment distribution about the Rise axis, the gentle decay of sea floor topography away from the flanking rift valley mountains, and the correlatable and bilaterally symmetric magnetic anomaly pattern all lend support to the suggestion that the locus of sea floor spreading has been positioned beneath the central valley of the Mid-Cayman Rise for at least 6 m.y.

The accreting plate boundaries of the Earth are continuously evolving as the geometry of the plate mosaic adjusts and the dynamics of the driving mechanism change. Accreting plate boundaries exhibit a wide range in morpho-tectonic character: rates of crustal accretion range from a few cm/yr to as high as 18 cm/yr; some accreting plate boundaries are characterized by frequent ridge-transform-ridge intersections, others are unbroken for hundreds of kilometers. If we are to understand fully the salient geologic processes operative during plate accretion and how these processes vary in time and space, then we must study representative segments of the accretionary continuum. The Mid-Cayman Rise spreading center represents an end-member example of this continuum in that its rate of spreading is at the low end of crustal accretion rates (≈ 2 cm/yr), and it is an isolated short accreting plate boundary truncated by two long transform faults that juxtapose cold, thick lithosphere against the spreading center. Our intention during the Cayman Trough project was to document the structural and petrologic character of this accreting segment and ask the question "how do these geologic properties compare with the rest of the plate accretion continuum?"

The Cayman Trough Project

The Cayman Trough Project took place between January 1976 and July 1977. During this period four separate field programs were conducted. The first effort in January and February 1976 involved the use of the submersible ALVIN and the deep-towed ANGUS photographic system, with the ALVIN conducting a detailed sampling and observational program on the steep fault scarps that border the rift valley of the Mid-Cayman Rise spreading center and the ANGUS conducting a comprehensive photography program in the rift valley itself. ALVIN, aided by its advanced acoustical transponder navigation system (ALVIN), carried out a series of 15 dives at two sites on either side of the rift valley (Figure 2). While these dive traverses were being conducted, scientists aboard the R/V KNORR used a similar tracking system (ACNAV) to conduct 46 dredge stations (Figure 2) and 21 camera lowerings (Figure 3) in various portions of the rift valley and flanking terraces. The ANGUS system used in our investigation consists of a 35 mm color-camera system mounted in a heavy-duty steel frame.

The initial field program was followed in April 1976 by a detailed bathymetric survey of the region using the USNS BOWDITCH. This U.S. Naval Oceanographic Office ship is outfitted with an advanced multi-narrow beam sonar mapping system capable of surveying large areas of the sea floor very rapidly (Glenn, 1970). The multi-narrow beam sonar system (SEABEAM) aboard the USNS BOWDITCH consists of a hull-mounted sonar array and a sophisticated inertial navigation system. The hull-mounted sonar is a T-shaped unit having two separate arrays. The fore and aft array generates a narrow signal which sweeps out an arc perpendicular to the direction the ship is traveling. The reflected acoustical signals are received by the thwart ship array, which divides them electrically into a large number of narrow beams. This acoustical information is combined with satellite and inertial navigational data to produce a swath of bathymetric information printed

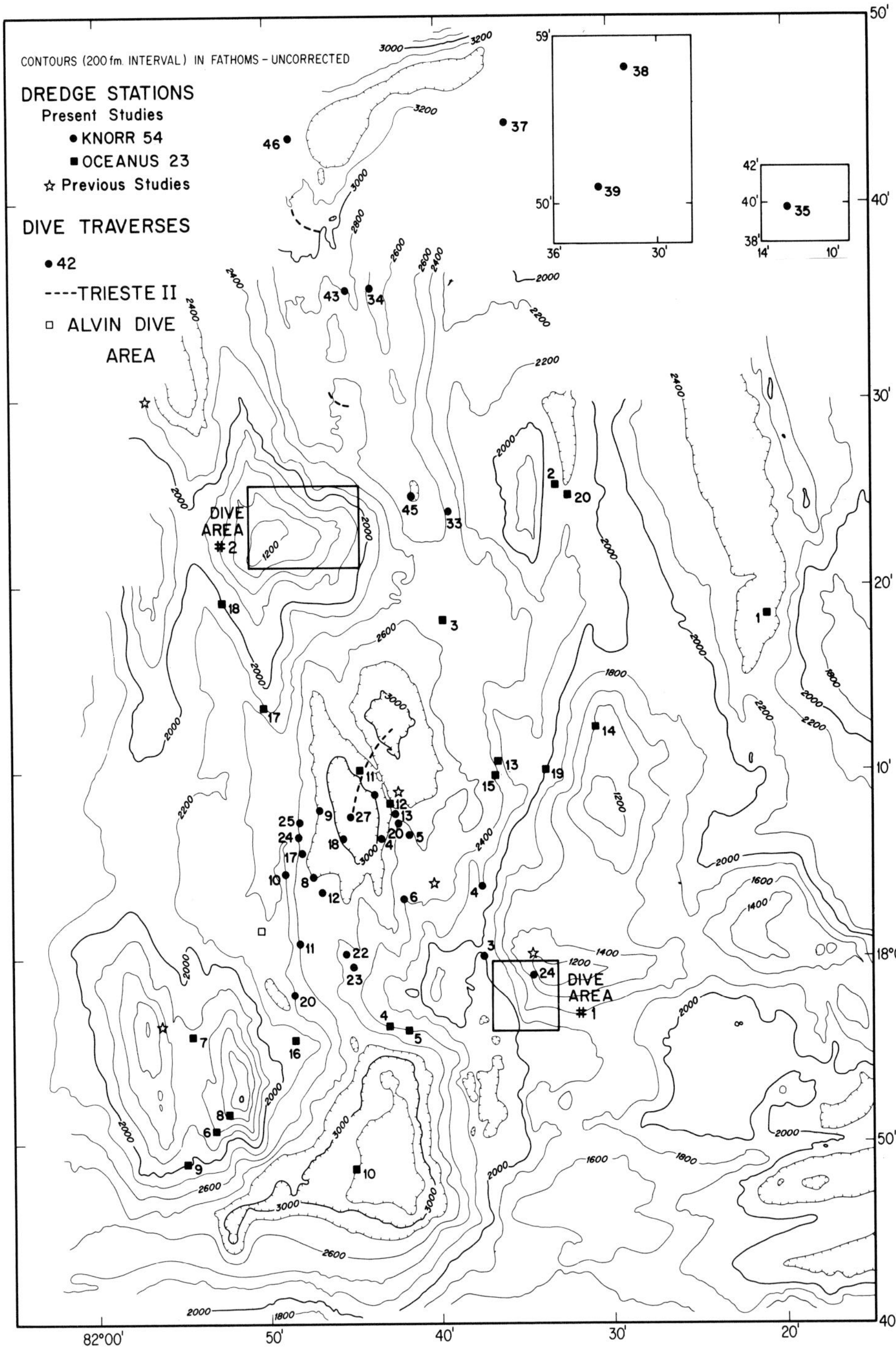

Fig. 2. Map of the study area on the Mid-Cayman Rise showing location of dive areas for dives conducted by the submersible ALVIN and the bathyscaph TRIESTE II and the location of dredge stations conducted by R/V KNORR and R/V OCEANUS and by previous investigations (Perfit and Heezen, 1978).

in real-time at a preselected scale and contour interval. The width of the swath varies with depth, but by running the ship back and forth across the survey area the desired coverage can be attained. During a one-week period, the ship surveyed a one degree square centered on the Mid-Cayman Rise spreading center extending from its intersection with the Oriente transform fault in

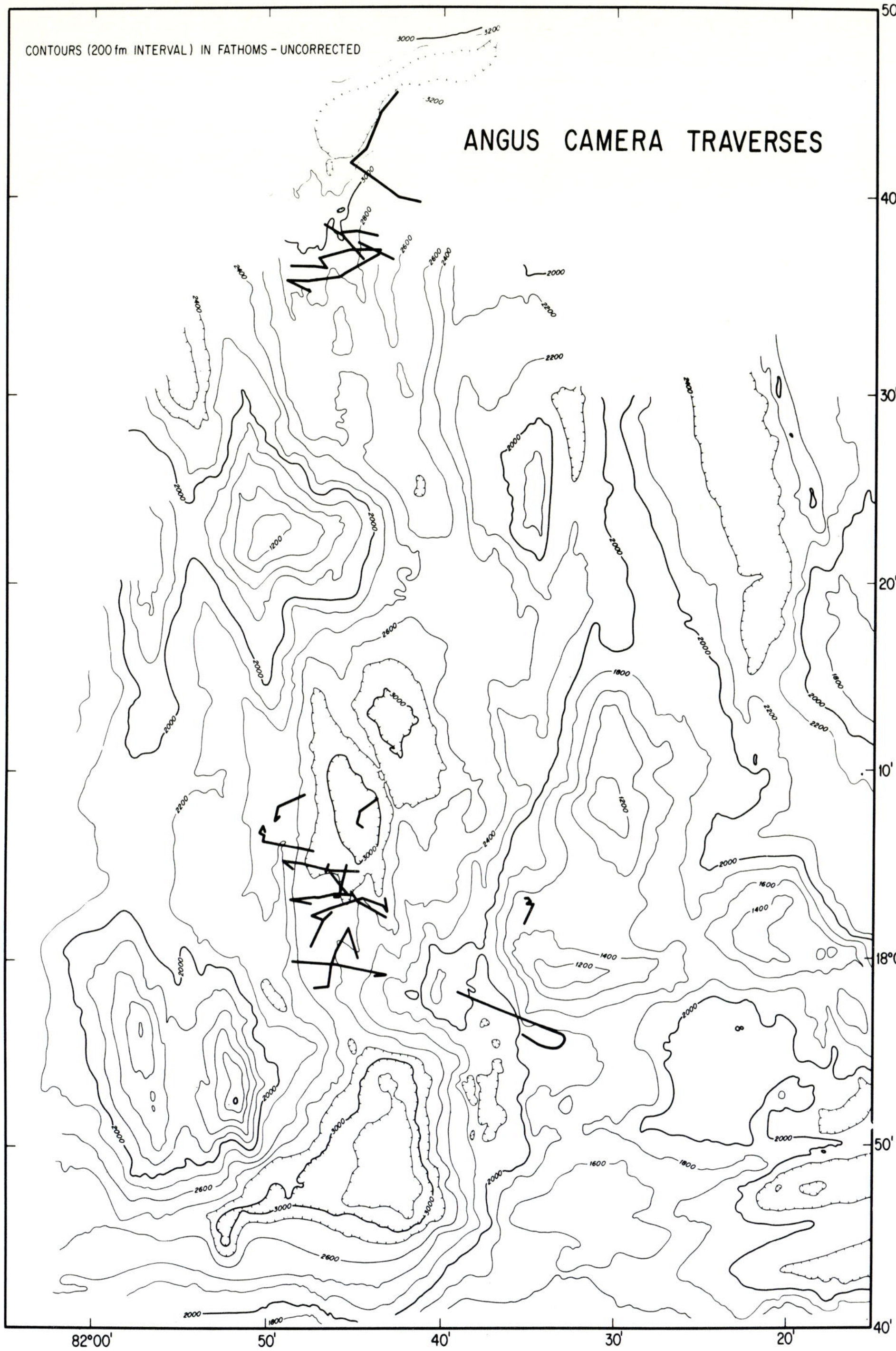

Fig. 3. Location of the camera lowerings made by the deep-tow system ANGUS.

the north to the intersection with the Swan transform fault in the south. The rift valley received 100% insonification, while the flanking scarp regions averaged 80%. The initial map to be produced from this survey was at a scale of 1:136,000 (Figure 4) and a contour interval of 50 fathoms (uncorrected). These topographic data, which are recorded in a digital format, can subsequently be displayed at smaller scales.

Based upon an analysis of the data collected on the first two field programs, a return visit by the submersible ALVIN and the research vessel OCEANUS was conducted in March of 1977. This field program had two primary purposes; the first was to use ALVIN to make a series of eight additional dive traverses up the steep faulted terrain on the eastern side of the rift valley, where outcrops of Layer 3 were found the previous year. The purpose of the OCEANUS program, on the other hand, was to conduct another dredging program throughout the rift valley and flanking terrace regions to delineate further the central zone of volcanic activity, as well as to obtain samples of ultramafic material exposed in the nearby fault scarps that are below ALVIN's diving limit of 3600 meters.

The final phase of the Cayman Trough Project was carried out in July of 1977 using the U.S. Navy's bathyscaph TRIESTE II. The KNORR and OCEANUS programs using ANGUS and precisely navigated dredge stations had succeeded in delineating the recent zone of volcanic activity. Using this information and the detailed multi-narrow beam topographic maps, a series of three dive traverses were made by the TRIESTE II in the northern, central and southern portions of the rift valley (Figure 2) during which observations, bottom photographs and basalt samples were collected.

General Topography of the Mid-Cayman Rise Spreading Center

The Inner Floor

The bathymetric map compiled from the SEABEAM data clearly defines the first-order morphotectonic character of the 110 km long axial valley. The inner floor is outlined by the 2500 fm (4575 m) contour and has a general north-south trend as defined by the strike of the flanking rift valley walls. The volcanic-tectonic grain of the rift valley floor is more complicated, however, than the reconnaissance surface-ship-bathymetric surveys (Holcombe, et al., 1973) had originally indicated. The SEABEAM results (Figure 4) show that the inner floor is not continuous along its 110 km length, but appears to be disrupted and offset in a right lateral sense by a short, several kilometer long, northeast-southwest trending discordant zone (nascent transform fault?) positioned at a point midway between the truncating Oriente and Swan transform faults.

The rift valley segment north of the discordant zone is characterized by a closed-contour, 3500 fm (6402 m) deep depression at $18^{o}45'$W where it intersects the Oriente transform (Figure 4). The inner floor proximal to the transform intersection is 15 km wide, and the volcanic/tectonic terrain comprising the floor is linear, strikes north-south, and has relief on the order of a few hundred meters. Away from the intersection towards the south, the floor of the rift valley progressively shallows to 2800 fm (5120 m) and continuously narrows to a width of a few kilometers; in addition, the axis of the inner floor is offset to the east by several kilometers (to its most easterly point at $18^{o}20'$W), and the lineated elements of the inner floor are progressively rotated from a trend of 180° at the northern end to 160° in the south at the intersection of the floor with the discordant zone ($18^{o}20'$W). This intersection is characterized by a well-defined, conical volcanic hillock.

As in the north, the intersection of the southern inner floor with the Swan transform at $18^{o}43'$W is characterized by a closed-contour, 3600 fm (6585 fm) depression. The inner floor is approximately 15 km wide at the rift valley-transform intersection, but over a distance of 20 km the floor narrows to a width of several kilometers and shoals to a depth of 2700 fm (4940 m). Northwards from this narrow gate, approaching the intersection with the discordant zone, the inner floor becomes progressively wider (reaching a maximum of ~18 km) and deeper (characterized by a 3000 fm (5487 m) deep closed-contour depression). The linear volcanic/tectonic grain of the southern inner floor strikes approximately 180° near the terminus of the inner floor with the Swan transform, but towards the north, adjacent to the discordant zone, the elongated closed-contour depressions and an intervening 500 fm (915 m) high ridge strike 160°.

The inner valley floor of the Mid-Cayman Rise spreading center exhibits many of the morphotectonic characteristics that have been shown to be typical of slowly accreting plate boundaries. The deep, closed-contour depressions defined at the northern and southern terminus of the inner floor are a characteristic feature of slowly accreting ridge/transform fault intersections (Fox, et al., 1969; Sleep and Biehler, 1970). Like the rift valley segments of the FAMOUS region (Ballard and van Andel, 1977; Macdonald and Luyendyk, 1977) the width of the inner floor is variable, ranging along strike from a few kilometers to many kilometers, and the morphologic elements of the inner floor are characterized by a series of linear ridges and troughs striking roughly parallel to the overall strike of the rift valley and with relief on the order of several tens of meters.

The ANGUS photographic data and TRIESTE observations indicate that the ridges are characterized by linear volcanic edifices comprised of elongated pillow flows stacked in a chaotic manner (Figure 5a, d). The flanks of the constructional piles are dominated by steeply dipping flow fronts draped by elongated pillow forms pointing downslope. In general, the troughs between the parallel volcanic ridges are constructional depressions formed by interlocking pillow fronts emanating from the flanking ridges. In the northern valley segment, however, the zone of most recent volcanism is also characterized by an

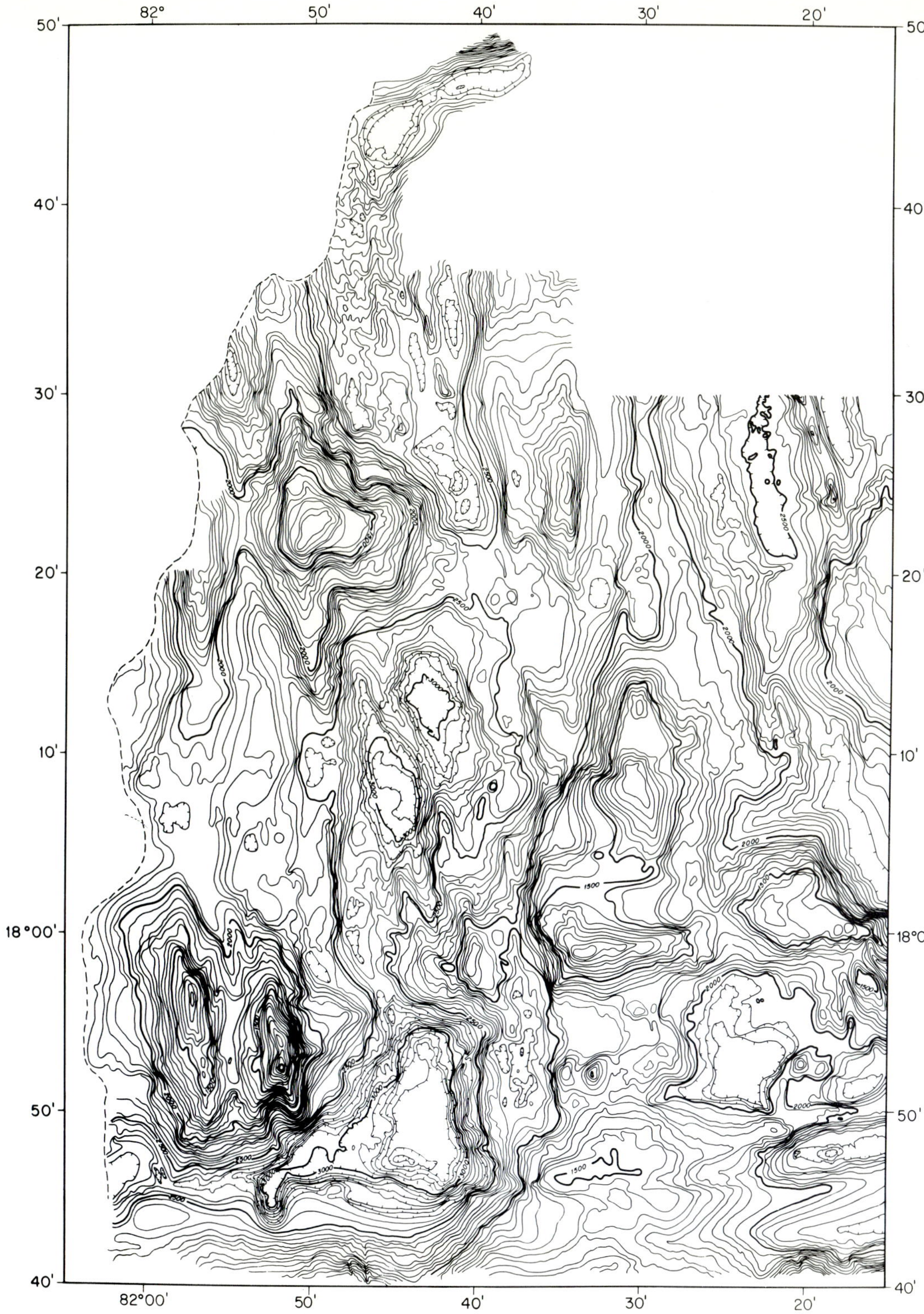

Fig. 4. Topographic map of the Mid-Cayman Rise spreading center compiled by the U.S. Naval Oceanographic Office using multi-narrow beam data. Contour interval of 50 fathoms (uncorrected).

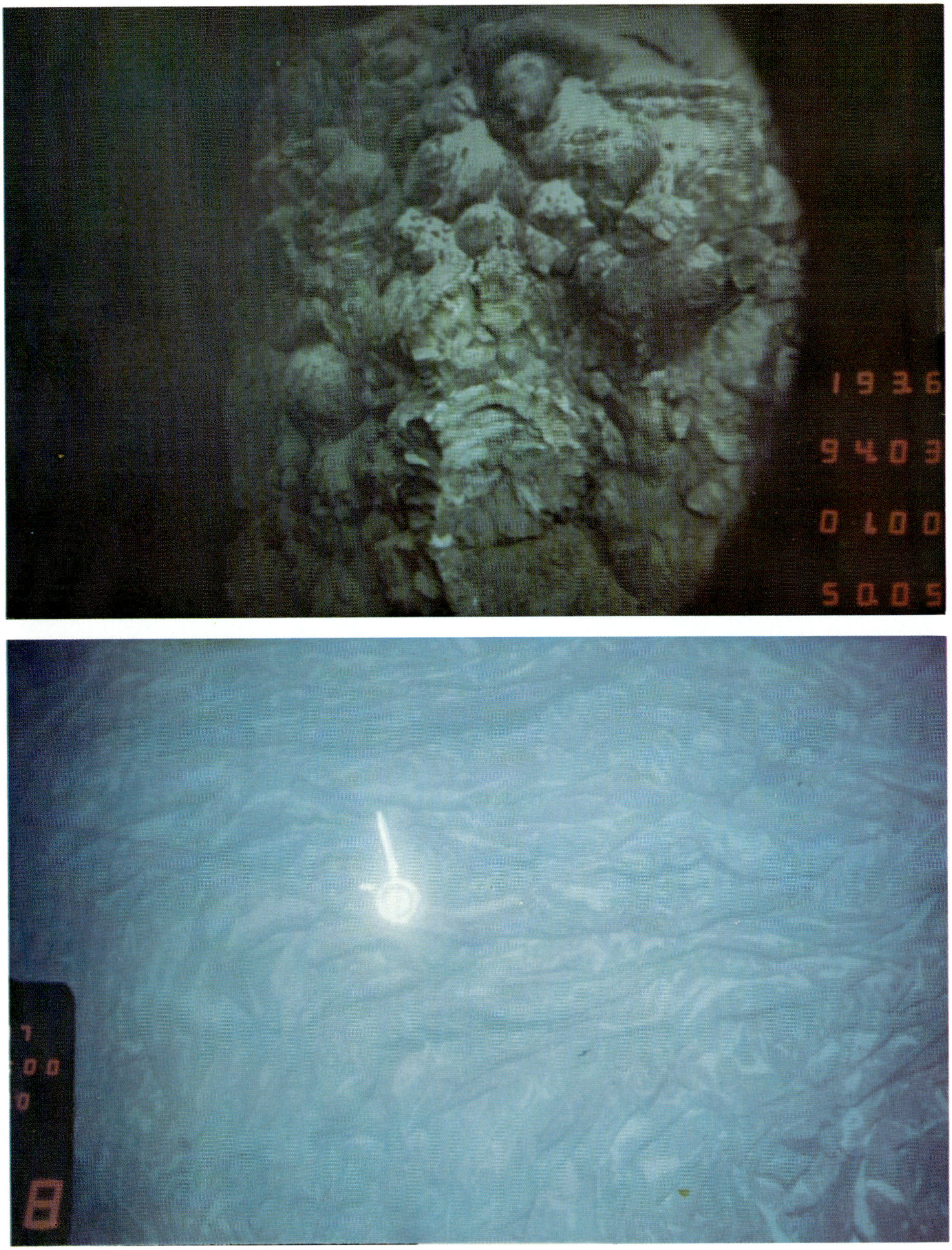

Fig. 5. (a) Photograph taken by ANGUS camera system of typical pillow lava terrain. Pillow form in lower right-hand corner has been broken, exposing its hollow interior. (b) ANGUS photograph of a sheet flow.

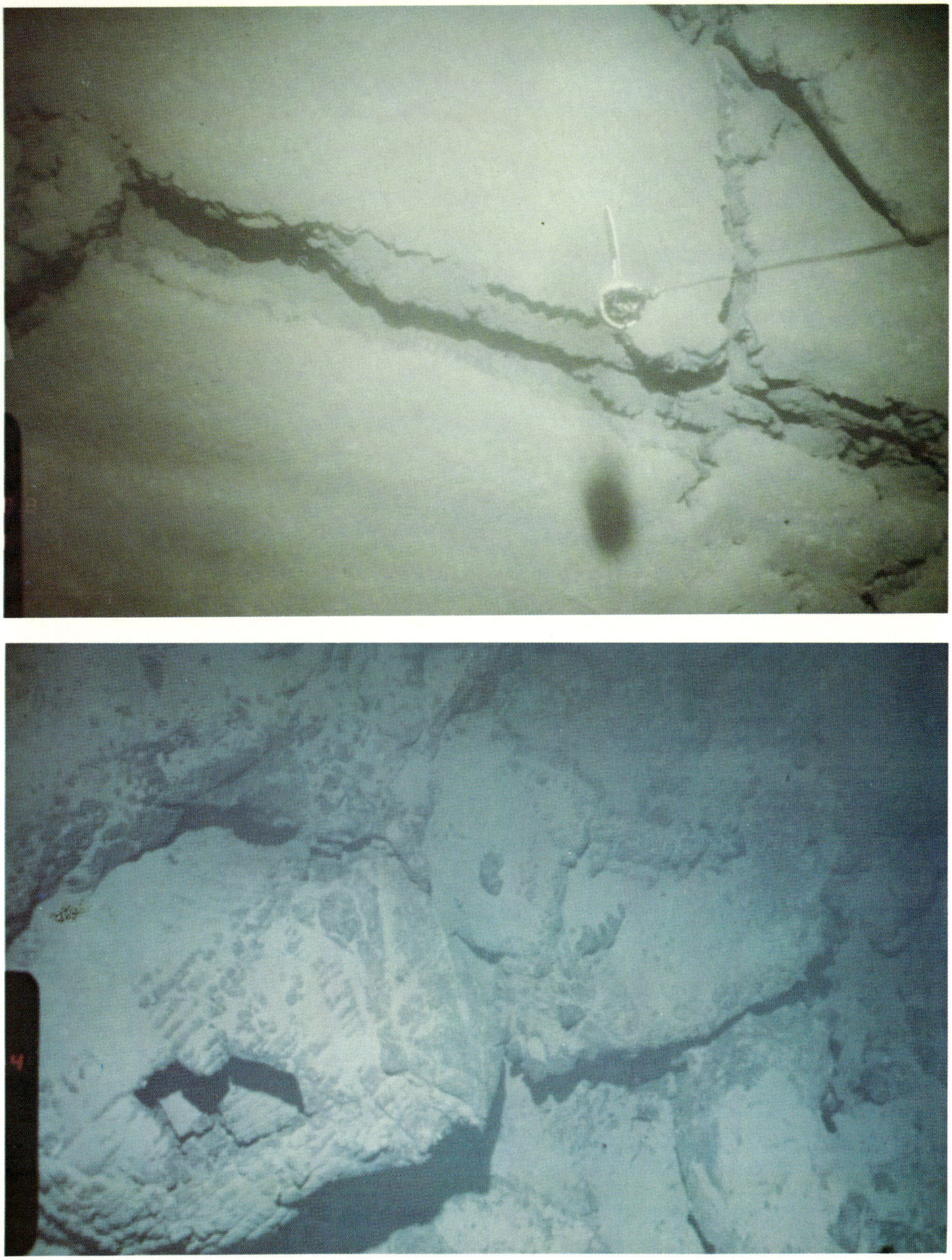

(c) ANGUS photograph of open fissures cutting across the older volcanic terrain which flank the central zone of volcanism. (d) Picture taken by the bathyscaph, TRIESTE II, near the summit of a young central pillow volcano in the northern rift valley segment. The vertical face shown in the picture is a small fault scarp exposing the truncated cross-section of pillow forms.

extensive sheet flow (Figure 5b) which has filled a linear depression adjacent to a central pillowed ridge. Massive sheet flows have been observed along the axis of accreting plate boundaries exhibiting a wide range in spreading velocities (Ballard, et al., in press), and they appear to represent an important phase in the volcanic cycle that creates the extrusive carapace of the oceanic crust. In general, where constrained by ANGUS and TRIESTE data, the zone of most recent volcanism within the inner floor is 2 to 3 km wide and is characterized by sediment-free, elongated volcanic constructional piles similar to those observed in the FAMOUS region (Ballard and van Andel, 1977; Ballard and Moore, 1977). The one exception to this observation is obtained by the TRIESTE and ANGUS traverses across the elongated depressions and ridge located at the northern terminus of the southern inner valley floor, revealing that this area has not experienced recent volcanic activity, since the terrain is blanketed by a thin veneer of sediments and only rarely is the pillowed oceanic basement exposed. ANGUS and TRIESTE results indicate that the areas flanking the zone of most recent volcanic activity are characterized by older sedimented volcanic terrain that is undergoing distension, as exhibited by open fissures (Figure 5c) and small fault scarps flanked by talus ramps.

The Rift Valley Walls

The morpho-tectonic boundary between the inner floor of the rift valley and the bounding inward-facing walls is abrupt and defined by north-south striking escarpments which rise in a series of steps several hundred to over a thousand meters above the inner floor (Figure 4). These near-vertical elements, undoubtably comprised of a family of small-throw faults, are linked to one another by broad terraces that dip gently toward the inner floor. This morphologic fabric, known to be characteristic of the topography developed along the rift valley walls of slowly accreting plate boundaries, integrates to create over 2000 fm (3658 m) of relief from the inner floor to the crest of the bounding mountain tops. The well-defined north-south tectonic grain of the rift valley walls is interrupted in places by a second order east-west grain that, when combined with the north-south grain, outlines a number of blocks that flank the inner floor. Two of these blocks were chosen to be investigated during the Cayman Trough Project by the submersible ALVIN. The first area of investigation, referred to as Dive Area 1, was visited in 1976 (Dives 611-616) and 1977 (Dives 737-742), and is located on the eastern wall of the rift valley near the southern terminus of the Mid-Cayman Rise spreading center (Figure 2). Dive Area 2 was visited in 1976 (Dives 620-625) and is located on the western wall of the rift valley on a block that is midway between the Oriente transform to the north and the Swan transform to the south (Figure 2).

The 18 dives made by ALVIN in the two dive areas provide a perspective of the morphologic elements of the sea floor that can be attained only by this type of in situ observation. Even at the relatively small scale (50 to 100 meter contour intervals) depicted on the base maps for each dive area (Figures 6 and 7), observed variability in topography cannot be accurately represented. As can be seen on the dive cross-sections (Figures 8 and 9), the regional slope of the rift valley walls is actually created by a series of N-S striking, steep (slopes range from 60° to vertical), inward-facing rock scarps with relief on the order of a few meters to several tens of meters. These near-vertical elements are linked by inward-facing carbonate/talus ramps characterized by widths ranging from a few meters to several hundred meters and by slopes ranging from 10° to 45°. The steeper carbonate slopes contain numerous talus fragments and rock bolsters that protrude through and are incorporated in the sediment blanket. ALVIN survey data indicate that the character of a given topographic element can change markedly along strike. Over a distance of a few tens to a few hundred meters a carbonate/talus ramp may vary considerably in slope and/or width, or a well-defined escarpment may lose definition and disappear.

Superimposed upon the N-S striking inward-facing escarpments and carbonate/talus ramps is a family of generally E-W striking valleys and promontories which modify the dominant N-S grain and produce a saw-tooth morphology. The vertical extent of these E-W features varies from a few tens to several tens of meters, and reconnaissance mapping on the bottom indicates that the E-W trending valleys and promontories are discontinuous over distances of several tens to a few hundred meters. In general, the valleys are characterized by moderately dipping carbonate slopes (10°-30°) covered by talus and bolsters, but occasionally rock escarpments are observed with relief of a few tens to several tens of meters suggesting that these E-W trending elements are fault controlled.

The faces of individual scarps are cut by a family of intersecting joint sets (N-S and E-W, and a third set which appears to be roughly horizontal) and by anastamozing cracks which interrupt the structural continuity of the massive, joint-bounded blocks. The intensity of jointing and cracking varies from relatively unfractured faces which are continuous over several tens of meters (Figure 10a) to outcrops at which the rocks are fragmented into cobble-sized blocks. The faces of a few escarpments that are not badly disrupted by jointing and fracturing exhibit the polished and grooved surfaces typical of slickensides (Figure 10b). We suggest that all of the steep rock escarpments represent relief elements created by faulting, but that gravitational collapse of joint-bounded blocks has resulted in removal of the original faulted surface.

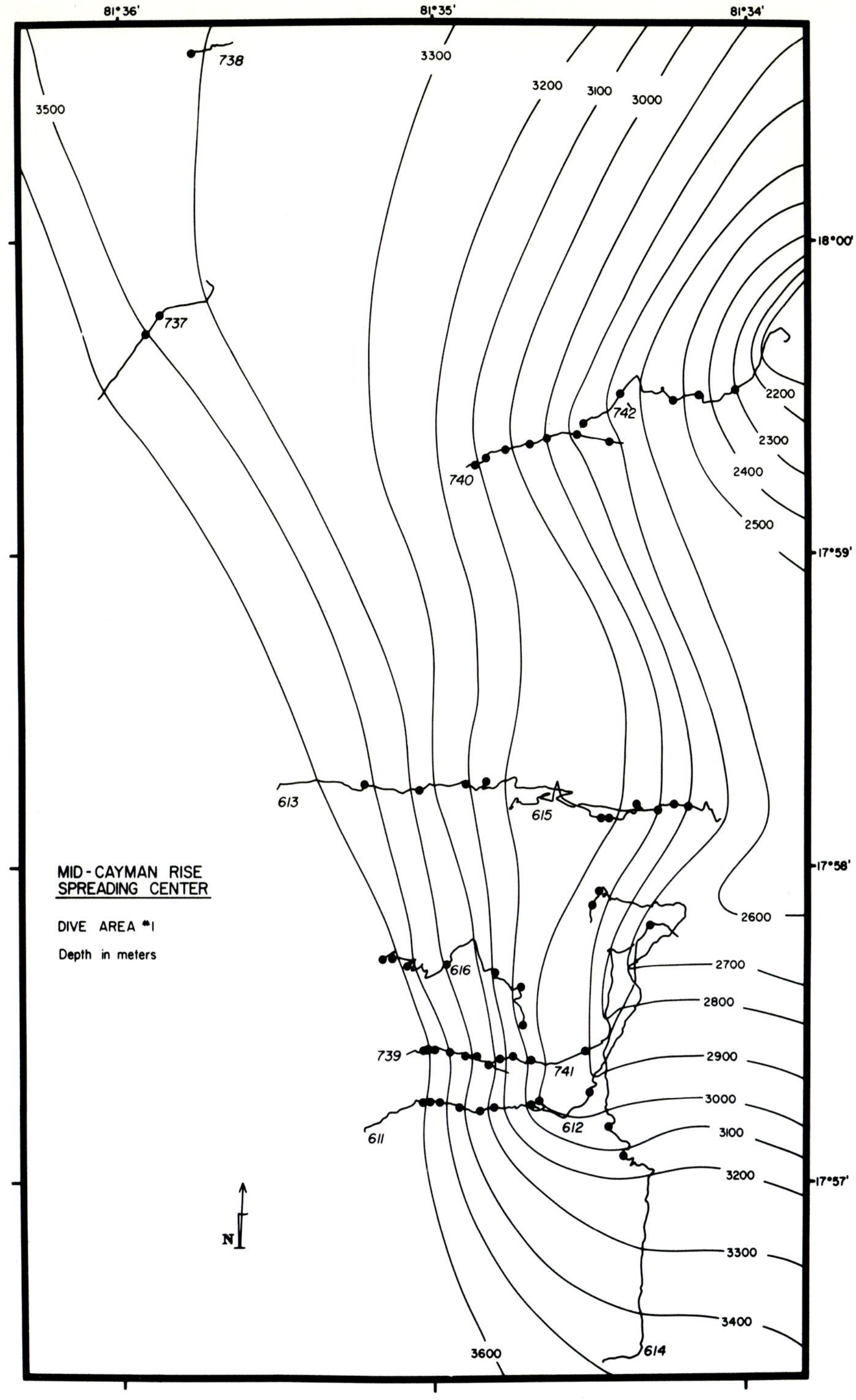

Fig. 6. Generalized bathymetric map of Dive Area #1, showing locations of dive tracks and sampling stations (black dots). All dives were conducted moving upslope (see Fig. 8). Black dots represent sampling stations. Contour interval of 100 m (corrected).

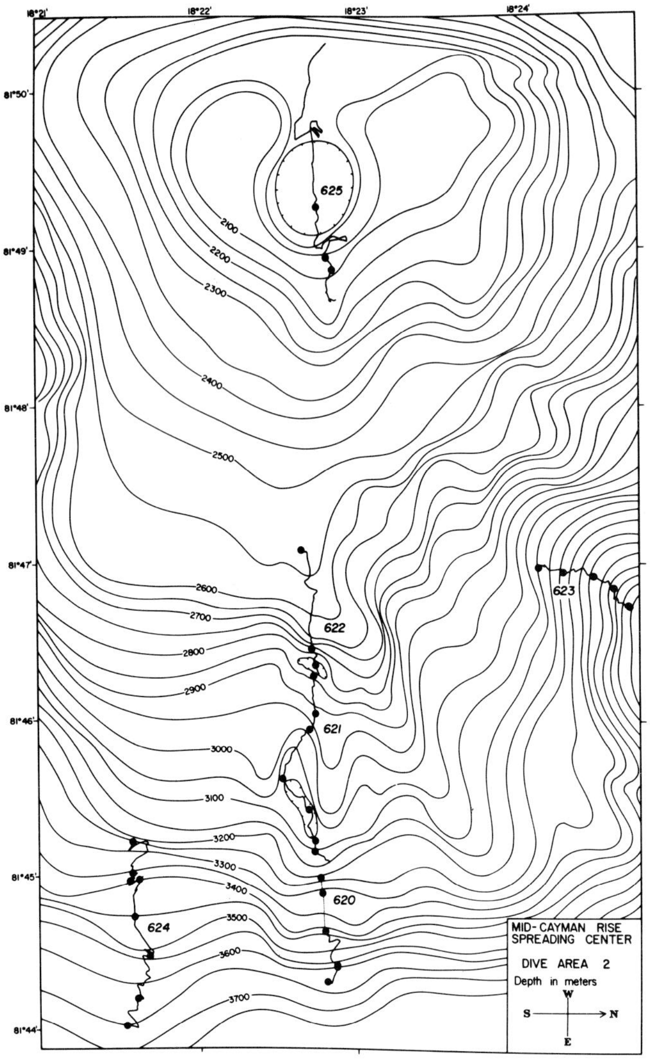

Fig. 7. Generalized bathymetric map of Dive Area #2, showing locations of dive tracks and sampling stations. All dives were conducted moving upslope (roughly, east to west; see Fig. 9). Dots on Dive 620 represent loss of signal for dive track. Contour interval of 50 m (corrected).

Rocks From the Mid-Cayman Rise Spreading Center

Our bottom sampling program during the Cayman Trough project had two thrusts. The first of these was an extensive dredging program carried out during two cruises: R/V KNORR, 1976 and R/V OCEANUS, 1977. Numerous dredge stations (Figure 2, Tables 1 and 2) were positioned along the length of the rift valley inner floor, and these results confirmed that a central lineated zone along the inner floor is characterized by recent volcanism as evidenced by glass-encrusted pillow lavas. Fewer dredge stations were located along the flanks of the rift valley (Figure 2, Tables 1 and 2); these dredges recovered weathered basalts, metabasalts, gabbroids and ultramafic rocks. Our sampling program on the rift valley walls focused on two blocks that flank the inner floor and was carried out by the submersible ALVIN (Figure 2).

During ALVIN's 18 traverses across the bottom an effort was made to sample the outcrops as frequently as possible. Sampling was limited by time and by the difficulty of recovering rocks from massive outcrops with a mechanical arm. Of necessity some samples were recovered from the flanks of bolsters or rarely from isolated talus blocks. Most frequently samples were obtained directly from rock escarpments or from loose fragments lying on an escarpment or at its base.

Ninety-four sampling stations were occupied during our dive program, at which 142 samples were collected. A diverse range of rock types, all typical of those recovered from other areas of the ocean floor, are represented in our collection. Gabbroic rocks are the most abundant. In situ sampling along some outcrop faces and closely spaced sampling up an escarpment indicates that the predominantly gabbro terrain is occasionally cut both by foliated amphibolites (Figure 10c), and by ultramafic screens that alter the country rock, often resulting in the production of rodingitized gabbro. At other localities, intensely fractured zones of altered and deformed gabbro lie adjacent to gabbro that is fresh and undeformed. No basalts were collected in situ during our investigation of Dive Area 1, and no outcrops were observed there which exhibited macroscopic features suggestive of extrusive basalt. During one dive (742, station 5 at 2800 m depth), gabbros were collected to within 250 meters of the crest of the rift valley wall, and on this same traverse, 13 meters above station 5, cross-cutting structures suggestive of dikes were observed (Figure 10d). In Dive Area 2 basalt samples were collected in situ from pillow lava terrain during two dives (624 and 625).

All of the exposed rock surfaces are covered by a veneer of manganese, and therefore only the macroscopic texture of the rock outcrop can be unequivocally characterized. The outcrops sampled during the submersible program exhibited four distinct textural expressions. Most frequently we observed massive blocks characterized by a rough surface. If an outcrop of this nature was cut by only widely spaced intersecting joint sets, then the rock type recovered was relatively unaltered troctolite or gabbro (Figure 10a). In some instances at these outcrops we observed a subtle subhorizontal foliation suggestive of layering. When the continuity of the rock, although still coarsely textured, was disrupted by anastamozing fractures, we recovered altered and deformed gabbro. The second outcrop type exhibits a moderately well-defined foliation and has a rubbly appearance (Figure 10f). Samples recovered

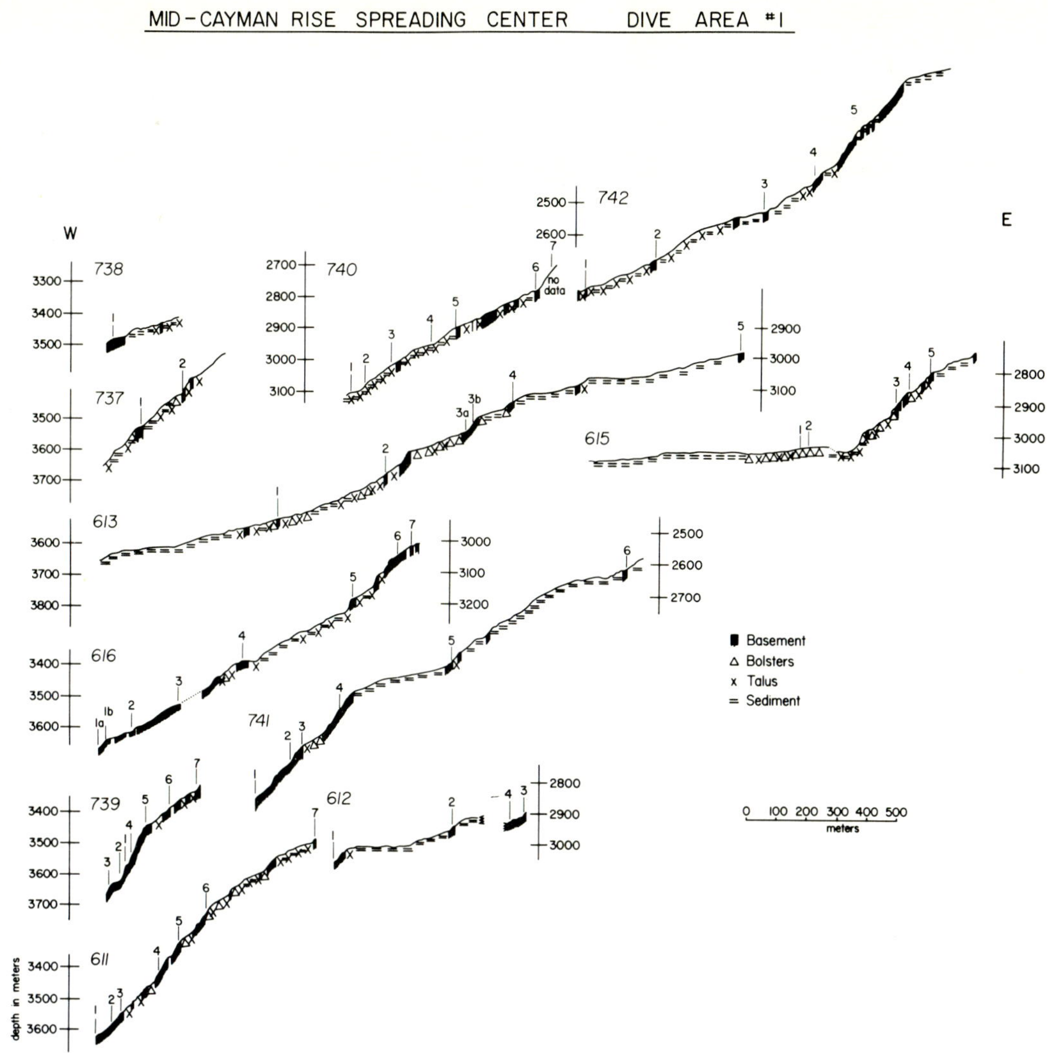

Fig. 8. Individual dive cross-sections for Dive Area #1. Numbers refer to sampling stations (see correlation with Fig. 6).

from these outcrops are foliated amphibolites. The third outcrop type is heterogeneous and chaotic in character. In some cases the outcrop has the appearance of a fruitcake with large ovoid fragments of rock randomly set in a finer rubbly matrix (Figure 10e). Rocks recovered from exposures with this texture are serpentinized ultramafics. The fourth outcrop type was observed only in Dive Area 2 and is characterized by very gently dipping slopes covered by pillow lava.

Petrology of Basalts from the Inner Floor

There is some difference in the external character of the basalts recovered from different areas within the spreading center. Basalts collected from the volcanic constructional highs which define the axis of the rift valley floor are fresh and glassy with little or no manganese surface coating; these samples exhibit no evidence of low temperature alteration. Basalts that were sampled from the sides of the rift valley floor and from the crest of the flanking rift valley mountains are encrusted by a thin veneer of manganese and do exhibit the effects of low temperature alteration.

The majority of the basalts recovered by dredging are aphyric or sparsely porphyritic. When phenocrysts are present they occur as olivine and plagioclase; in a few samples olivine is the sole phenocryst phase. Clinopyroxene phenocrysts have not been observed. Microphenocrysts of spinel are rarely present, and large corroded xenocrysts of olivine and plagioclase are present in a small number of samples. The range in phenocryst abundance is 0 to 6% of the total volume. Most samples are microscopically

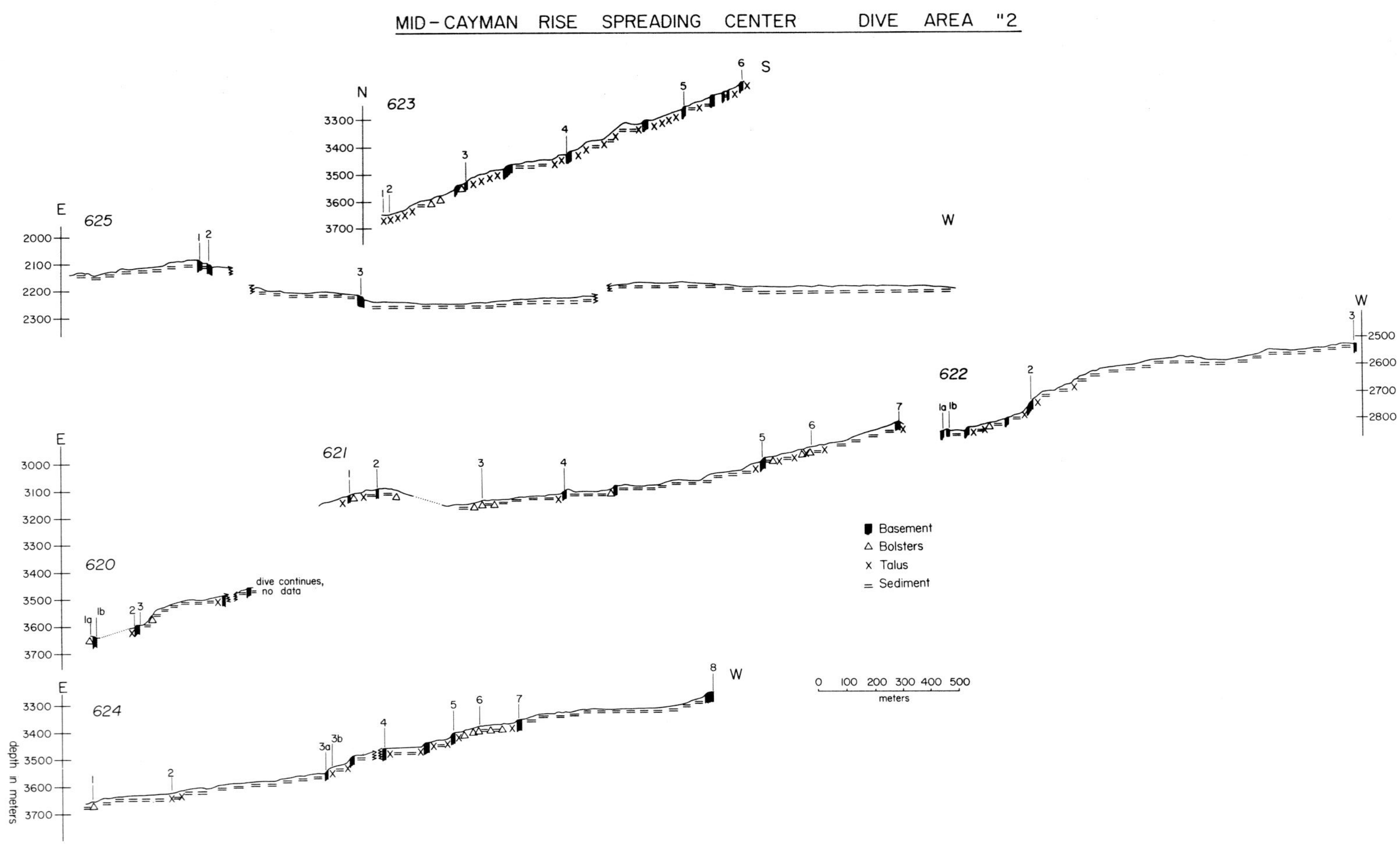

Fig. 9. Individual dive cross-sections for Dive Area #2 showing location of sampling stations (see Fig. 7).

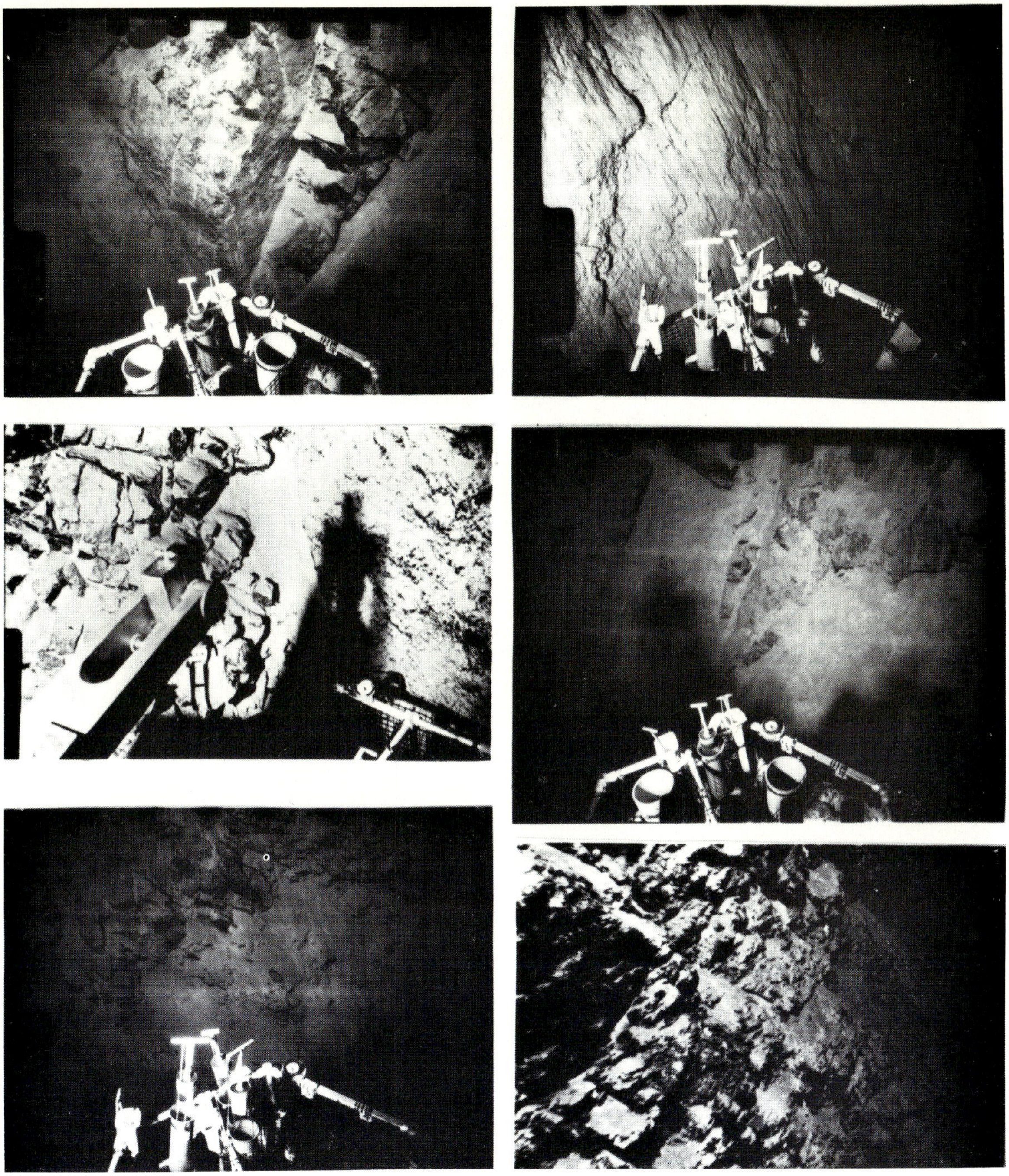

Fig. 10. ALVIN bottom photographs representative of the following features: (a) massive outcrop exhibiting some jointing; (b) slickensides on scarp face highlighted by veneer of pelagic sediment; (c) visible contact between gabbroic country rock (left) and amphibolite screen; (d) possible dykes observed at the top of dive traverse 742, Dive Area #1; (e) "fruitcake"-like appearance of ultramafic outcrop; (f) rubbly surface exhibited by amphibolite outcrop.

TABLE 1. Successful Dredge Results from R/V KNORR Cruise 54.

Station No.	Depth (m)	Latitude (N)	Longitude (W)	Rock Type
1*	4892	18°04.2'	81°44.7'	Fresh, glassy pillow basalt
2*	4795	18°02.6'	81°43.1'	Fresh, glassy basalt
11	2855	18°00.9'	81°36.8'	Metagabbros and pyroxenite
12	3484	18°03.6'	81°37.3'	Metagabbro
13	4514	18°06.0'	81°42.1'	Metabasalt
14	4324	18°03.4'	81°41.9'	Metabasalt and metagabbro
23	4892	18°04.4'	81°47.9'	Fresh and weathered basalt
24	5011	18°07.1'	81°46.9'	Slightly weathered basalt
26	4001	18°04.5'	81°48.8'	Fresh basalt
27	4413	18°01.2'	81°48.6'	Fresh basalt
29	4956	18°02.7'	81°46.9'	Fresh basalt
30	5411	18°05.4'	31°44.9'	Fresh and weathered basalts, metabasalts, gabbros, serpentinite
33	4874	18°05.8'	81°42.6'	Fresh and slightly weathered basalt
37	5157	18°05.7'	81°46.7'	Weathered basalt
39	5502	18°06.9'	81°45.4'	Fresh, glassy basalt
40	4828	18°01.6'	81°46.3'	Fresh, glassy basalt and slightly weathered basalt
42	4360	17°58.1'	81°48.7'	Weathered basalt
44	4610	18°01.6'	81°44.9'	Fresh and slightly weathered basalt
46	4623	17°59.9'	81°45.6'	Fresh, glassy basalt and weathered basalt
47	5075	18°06.0'	81°46.7'	Weathered basalt
48	5212	18°07.0'	81°46.6'	Metabasalt
51	5684	18°07.4'	81°43.1'	Fresh, glassy basalt
52	5393	18°07.3'	81°46.0'	Fresh, glassy basalt
54	2047	17°59.1'	81°34.5'	Metabasalt, metagabbros and serpentinite
58	5329	18°07.7'	81°43.9'	Weathered basalt
62	4665	18°25.0'	81°40.8'	Weathered basalt, gabbro
63	5123	18°35.7'	81°41.3'	Fresh, glassy basalt
67	4118	18°38.7'	81°12.6'	Weathered basalt, gabbro
73	4701	18°56.7'	81°31.4'	Weathered basalt and hyaloclastites
74	5657	18°49.0'	81°32.9'	Serpentinite
90*	4755	18°37.0'	81°44.0'	Weathered basalt
91	4683	18°31.1'	82°08.1'	Weathered basalt and hyaloclastities
94	5084	18°35.2'	80°46.2'	Slightly weathered basalt
98	5084	18°26.0'	80°41.2'	Weathered basalt
102	6142	18°44.7'	81°48.5'	Slightly weathered basalt
103*	5176	18°37.6'	81°44.8'	Fresh, glassy basalt

*Recovered on camera frame. All other recoveries by conventional rock dredges.

TABLE 2. Successful Dredge Results from R/V OCEANUS Cruise 23.

Station No.	Depth (m)	Latitude (N)	Longitude (W)	Rock Type
1	4250	18°19.3'	81°23.0'	Weathered basalt
4	4820	17°56.5'	81°45.0'	Weathered basalt
5	5330	17°54.7'	81°41.0'	Weathered basalt
8	3180	17°51'	81°52'	Weathered basalt
9	4850	17°48.2'	81°54.2'	Weathered basalt
11	5380	18°10.7'	81°46'	Slightly weathered
12	5400	18°06'	81°43'	Weathered basalt
13	4400	18°10.3'	81°37'	Spinel harzburgite tectonite
14	3280	18°12.5'	81°32.5'	Gabbro, plagioclase spinel dunite, troctolite and serpentinite
16	3800	17°53'	81°50'	Weathered basalt
17	3870	18°06.7'	81°51.5'	Brecciated serpentinite

vesicular, containing 0.10% to 1.5% of vesicles by volume.

To date our petrologic studies of basaltic rocks have concentrated on the fresh glassy pillows recovered from the axis of recent volcanism centered along the rift valley floor. Major element chemistry of 33 glass specimens from pillow lavas were analyzed by electron microprobe at the Department of Mineral Sciences, Smithsonian Institution. Carefully separated glass fragments were hand picked from the same samples and cleaned of alteration products or phenocrysts and analyzed for their trace element content. B, Ba, Co, Cr, Cu, Li, Ni, Sr, V, Y and Zr were determined by direct-reading emission spectrometry.

The 33 major and trace element analyses indicate (Tables 3 and 4) that when compared with typical mid-ocean ridge basalts (MORB), the Cayman basalts are higher in TiO_2, Na_2O, K_2O, P_2O_5, Zr, Y, Sr and Ba contents and generally lower in CaO content for a given FeO/MgO ratio. Transition metals and large ion lithophile (LIL) element covariances are similar, but not identical to MORB. Overall, the Cayman glasses seem transitional between MORB and 'plume' or 'hot spot' related oceanic basalts, but the Cayman samples clearly have some unique features of their own (e.g. high TiO_2, Na_2O, Zr, Sr and Y contents).

The range in composition and tight correlation between LIL-elements (Figs. 11 and 12) suggest that the liquids producing our suite of samples are derived from a common source, and that the variable composition of these liquids is controlled by fractional crystallization. Preliminary modelling of the observed compositional variations is compatible with a sequence of crystallization involving plagioclase, olivine and clinopyroxene. The lack of clinopyroxene phenocrysts in the basalts indicates that clinopyroxene is not an equilibrium phase in the low pressure environment of a magma chamber at shallow depth. Aluminum-chrome diopside is present in the plutonic samples collected by ALVIN from the rift valley walls and may represent an early fractionating (high pressure) phase. Observed

TABLE 3. Analysis of Fresh Basalt Glass from the Mid-Cayman Rise Spreading Center.

Major Elements[1]	Range of Composition (33 samples)		
SiO_2	50.51	-	52.23
TiO_2	1.57	-	2.49
Al_2O_3	14.53	-	16.97
FeO*	8.62	-	11.08
MgO	5.54	-	7.80
CaO	9.02	-	10.39
Na_2O	3.38	-	4.18
K_2O	0.18	-	0.39
P_2O_5	0.17	-	0.38
FeO*/MgO	1.13	-	1.90
Trace Elements[2]			
Cr	125	-	250
Co	30	-	34
Ni	75	-	122
V	215	-	305
Y	35	-	60
Zr	170	-	265
Sr	133	-	176
Ba	17	-	38

[1]Major elements by electron microprobe.

[2]Trace elements by direct reading emission spectrometry.

TABLE 4. Average Composition of Cayman Glasses Compared with other Oceanic Basalts--all with Similar FeO*/MgO (1.1 to 1.2).

%Wt.	Cayman Glass	MORB	FAMOUS (Mid-Atlantic Ridge) Axis, 36°N	Mid-Atlantic Ridge Fracture Zone (36°N)
TiO_2	1.7	1.2	1.3	1.3
Na_2O	3.5	2.5	2.5	2.5
K_2O	0.25	0.12	0.20	0.30
ppm				
Ba	23	5	55	75
Zr	170	70	70	80
Sr	160	100	110	110
Y	36	25	25	22

*Total iron reported as FeO.

variations in the Cayman basalt chemistry can be explained by a two-stage fractionation model: selective removal of an aluminum-chrome diopside having the chemistry of those observed in the plutonic rocks, followed by shallow-level fractionation of plagioclase and olivine. If, as our results suggest, the gabbros and basalts are genetically related, then there is long-term consistency of the magmatic processes that produce both the basalts dredged from the rift valley floor and the gabbros sampled from the flanking rift valley walls.

Petrology of Gabbros from the Rift Valley Walls

The principal rock types sampled from the flanking walls of the rift valley (in both dive areas 1 and 2) include "normal" gabbro, olivine gabbro, orthopyroxene gabbro, troctolite, norite, serpentinized ultramafic, amphibolite, diabase (one sample) and crystalline basalt. The samples were recovered from the faces of individual scarps, were sampled as loose fragments associated with a given rock outcrop, or (rarely) were recovered from talus piles in otherwise heavily sedimented terrain. Almost all specimens were encrusted with a thin veneer of manganese. The samples exhibit a wide range in degree of alteration and deformation.

Our freshest samples vary from medium to extremely coarse-grained, with individual minerals ranging in size from 0.2 to 4 cm, and are equigranular to slightly porphyritic. Primary phases include plagioclase, clinopyroxene, olivine and orthopyroxene, and a minor amount of hornblende and primary opaque minerals. The principal accessory phases are apatite, zircon and sphene. Ophitic to subophitic intergrowths of plagioclase

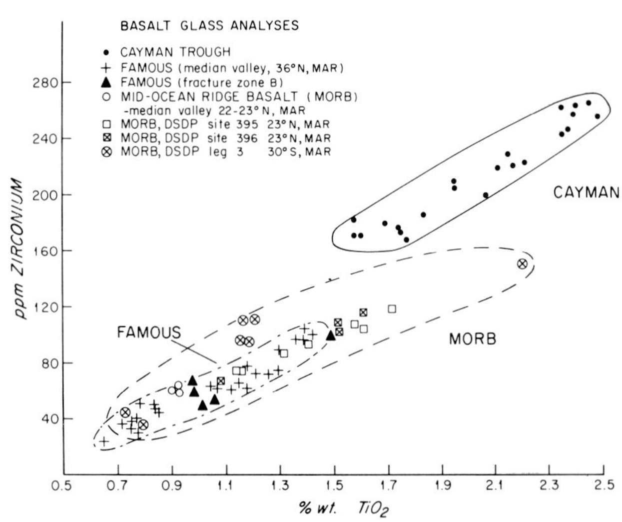

Fig. 11. TiO_2 versus Na_2O contents of Cayman glasses.

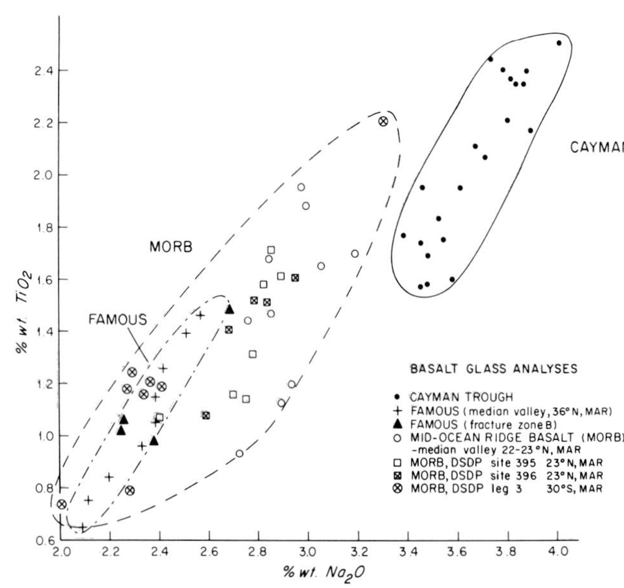

Fig. 12. TiO_2 versus Zr contents of Cayman glasses.

and clinopyroxene and poikilitic intergrowths of olivine and plagioclase or olivine and clinopyroxene are sometimes present. Although mineral layering or banding is not observed in our collection, other textural evidence (such as an apparent volumetric excess of plagioclase) suggests a cumulate origin for some samples.

Primary plagioclases are subhedral to euhedral and many show simple twinning, but only very rarely are optically zoned. Polysynthetic twins are common but in most cases are clearly deformation-related. Clinopyroxene exhibits a variety of exsolution textures, and minerals present as exsolution lamellae are brown hornblende, opaques and orthopyroxene. The lamellae may be wide and stubby, very thin and closely spaced, or, in rare instances, wormy, similar to myrmekitic texture. Olivines within our suite of samples are anhedral to subhedral and within even the freshest rocks are cross-cut by fractures filled by alteration phases. Orthopyroxene is not a common phase.

Brown hornblende is a primary phase in many of the fresh or only slightly altered gabbroic rocks. Chestnut-brown pleochroic hornblende commonly occurs in rims around mafic minerals in pyroxene-rich rocks and very rarely in troctolites. Brown strongly pleochroic hornblende appears in small, irregular and isolated patches in a few samples.

Two styles of alteration characterize the Cayman Trough gabbros. The first style is a relatively high temperature recrystallization of primary igneous phases (typically plagioclase, clinopyroxene and hornblende) involving changes in phases, grain size and composition. The second type of alteration which is associated with brittle fracturing of the samples and which is characterized by static replacement of a pre-existing phase by a new phase or phases, produces reaction rims, pseudomorphs and veins of secondary minerals. Most of the gabbroic samples exhibit some evidence of the low temperature, static alteration. Incipient alteration is characterized by the filling of brittle fractures with secondary minerals (e.g. chlorite, actinolite, etc.). More extensive alteration is characterized by the replacement of mafic phases by a host of secondary minerals; pervasive alteration is indicated by the complete replacement of mafic phases and the alteration of plagioclase.

At least two phases of deformation also appear to have affected the plutonic rocks collected from the rift valley walls of the Mid-Cayman Rise spreading center. An early ductile phase is indicated by bending, twinning, kinking, subgrain formation and recrystallization of primary phases. A later brittle phase is indicated by penetrating fractures and broken grains. The degree of intensity of these features varies considerably within the suite. Some specimens preserve original ig-

TABLE 5. General Rock Types Collected by ALVIN from the Rift Valley Walls of the Mid-Cayman Rise Spreading Center.

Dive Number	Rock Types Collected (number of samples)
	I Dive Area 1 1976
611	gabbroid (6); serpentinized ultramafic (1); amphibolite (1)
612	gabbroid (4); spinel harzburgite (2)
613	gabbroid (2); spinel harzburgite (4); spinel lherzolite (1); basalt (1)
614	gabbroid (1); amphibolite (1)
615	gabbroid (6); amphibolite (2)
616	gabbroid (10); spinel harzburgite (2)
	II Dive Area 2 1976
620	gabbroid (4); plagioclase dunite (3); troctolite dunite (1); basalt (1)
621	gabbroid (9); serpentinized ultramafic (1)
622	gabbroid (7)
623	gabbroid (9); spinel harzburgite (1)
624	gabbroid (4); spinel harzburgite (3); basalt (3)
625	basalt (3)
	III Dive Area 1 1977
737	gabbroid (2); amphibolite (1)
738	gabbroid (2)
739	gabbroid (10)
740	gabbroid (7)
741	gabbroid (6); amphibolite (2)
742	gabbroid (8); basalt (1)

TABLE 6. Major Element Chemical Compositions of Selected Rocks from the Mid-Cayman Rise Spreading Center.

Rock Type	Crystalline Basalt	Olivine Gabbro	Orthopyroxene Gabbro	Gabbro	Amphibolite
Sample Number	624 7-1	611 1-1	611 4-1A	613 3-1	615 2-2
SiO_2	51.09	49.28	53.16	46.23	44.97
TiO_2	0.18	0.30	1.38	5.76	5.75
Al_2O_3	17.43	20.82	16.79	13.12	12.35
FeO*	5.07	5.64	7.11	16.16	12.18
MnO	0.09	0.04	0.14	0.22	0.16
MgO	10.67	10.38	6.77	5.43	7.55
CaO	13.36	9.60	10.69	9.26	10.56
Na_2O	2.34	2.99	4.00	3.59	3.93
K_2O	0.04	0.11	0.28	0.27	0.23
P_2O_5	0.10	0.02	0.38	0.12	1.94
Total	100.37	99.18	100.70	100.16	99.62
FeO*/MgO	0.475	0.543	1.050	2.976	1.613

(FeO* means total iron as FeO; i.e., FeO + 0.9 Fe_2O_3).

neous textures whereas others have been almost entirely recrystallized and/or are greatly fractured. Even on the thin-section scale, narrow zones of intense shearing and recrystallization cut through domains much less affected by the deformation.

Whole-rock major element chemistry of selected representatives of the principal rock types sampled from the walls of the Mid-Cayman Rise spreading center is presented in Table 6. One hundred and thirteen plutonic rocks and eight crystalline basalts were analyzed on the electron microprobe after preparation of glass beads by rapid fusion at 1700°C in a molybdenum strip furnace under argon atmosphere. Detailed major element data on these samples and trace element data on a subset of sixty samples will be presented elsewhere.

Unlike the chemistry of the basalts dredged from the Cayman area, the bulk chemistry of the gabbro samples appears to be identical to that reported for other oceanic gabbros and abyssal tholeiitic basalts. All but one of the entire suite are olivine-normative and all but twenty are hypersthene-normative. The percentage of K_2O is uniformly low, ranging from 0 to 0.43% and averaging 0.18%; this is comparable to the range defined by previous determinations on fresh abyssal tholeiitic basalts collected from several oceanic regions, which have uniformly low potash contents varying from 0.03 to 0.40% (Shibata and Fox, 1975), and slightly lower than the range defined by Shibata and Fox for fresh abyssal tholeiitic basalts from the Oceanographer Fracture Zone, North Atlantic (0.16 to 0.75%). The range of K_2O for Cayman gabbros is slightly larger than the range of values for gabbros collected from the Indian and Atlantic Oceans (0 to 0.30%; Fox and Stroup, in prep.). SiO_2 for individual analyses of the Cayman rocks varies between 35 and 55 percent and averages roughly 49 percent. Al_2O_3 content is moderately high, averaging 18%, which is only 1-2 percent higher than the average Al_2O_3 given by other workers (i.e., Miyashiro, et al., 1970, for oceanic gabbros and Shibata and Fox, 1975 for abyssal tholeiitic basalts). The range in FeO*/MgO ratio (FeO* is total iron calculated as FeO) in the Cayman gabbro samples is 0.27 to 3.24, as compared with a range of 0.32 to 2.84 presented by Miyashiro, et al. (1970) for other oceanic gabbros. It is interesting to note that the variability in chemistry observed within the gabbroic samples from this single, geographically well-constrained area of the ocean floor is as large as that observed in gabbroic samples collected from all of the world's oceans.

The effects of two distinct processes, primary igneous crystallization differentiation and secondary alteration must be evaluated in a preliminary analysis of major element variations. Our results might appear to suggest that alteration has not significantly modified the primary igneous chemistry of the Cayman gabbroic samples (reflected in the composition of the least altered aphyric basalt 624 7-1, Table 6). When the population of Cayman plutonic rocks is plotted relative to gabbros from elsewhere in the world's oceans, as shown in Figure 13, the Cayman samples follow the tholeiitic differentiation trend defined by Miyashiro, et al. (1970) for gabbros dredged from the Mid-Atlantic Ridge. In addition, many of our samples are relatively less fractionated than other oceanic gabbros and appear to be the relatively unfractionated extension of (or cumulate compositions complementary to) the basalt glasses dredged from the axial valley of the Mid-Cayman Rise spreading center.

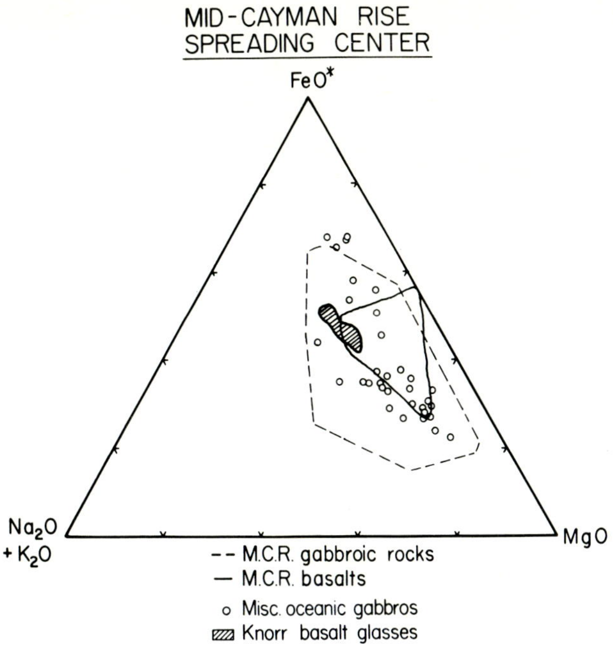

Fig. 13. FMA ternary diagram showing fields of Cayman basalts and gabbros (sampled by ALVIN) and basalt glasses (dredged by the R/V KNORR), and plotted compositions of gabbros dredged from other oceanic regions (analysis made by Miyashiro, et al.,1970; Melson and Thompson, 1970; Bonatti, et al.,1971; Thompson, 1973; Engel and Fisher, 1975).

Positive correlation of FeO*/MgO with TiO_2, as shown in Figure 14, would appear to confirm that variations in total iron and magnesium abundances in the Cayman plutonic rocks are a function of differentiation processes, since TiO_2 has been suggested to be stable under a wide range of hydrothermal alteration conditions (Pearce and Cann, 1973). However, more recent evidence suggests that seawater leaching of late-stage magmatic fluids enriched in alkalis, Fe, Ti and the light rare earth elements from some parts of the oceanic crust could result in the enrichment of these elements in other parts of the crust (Stakes, 1978). These leaching reactions would produce chemical variations mimicing those produced by fractional crystallization. The suite of gabbros collected from the Mid-Cayman Rise provides an opportunity to examine in detail the effects of and hopefully to determine the contribution to chemical variations made by these two processes; this examination is currently in progress.

An initial attempt to model fractionation of the major phases to produce the range of chemistries of the plutonic rocks recovered from the Mid-Cayman Rise spreading center is shown in Figure 15. Variability of Al_2O_3 versus FeO*/MgO in some of the samples appears to reflect accumulation or removal of plagioclase from an initial liquid having the composition of basalt 624 7-1.

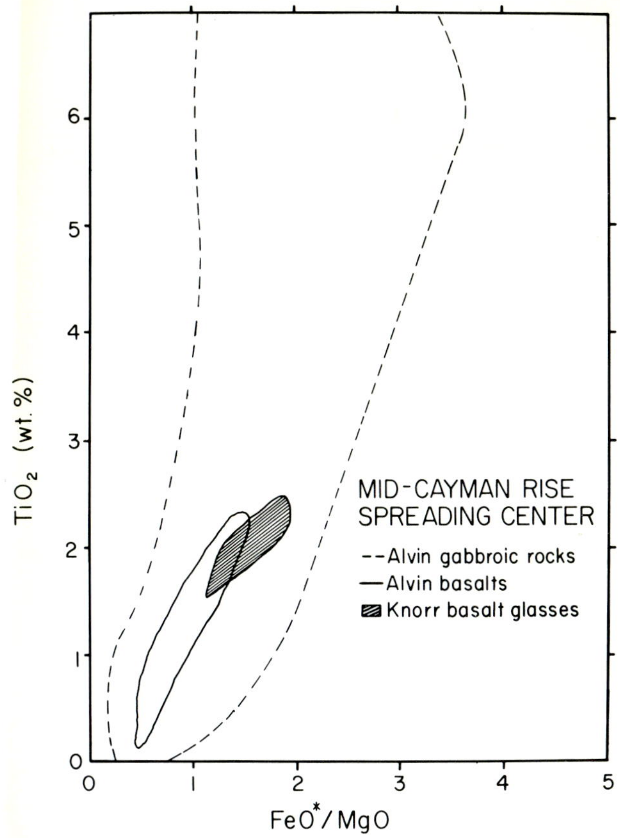

Fig. 14. Diagram showing the variation of TiO_2 relative to FeO*/MgO ratio in the ALVIN-sampled plutonic rocks and dredged basalt glasses. The highest TiO_2 value is 8.9%.

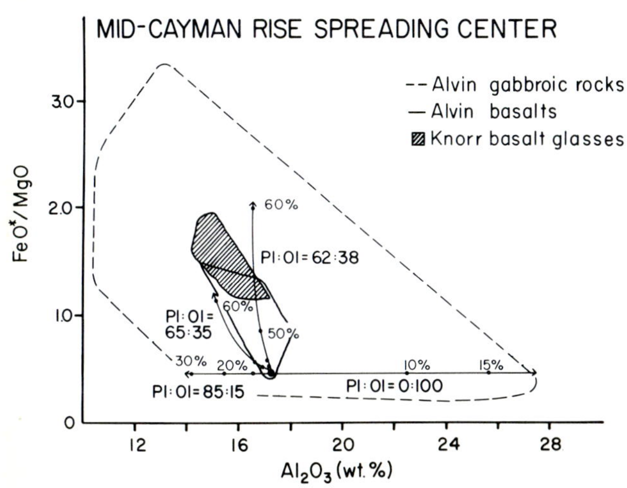

Fig. 15. Diagram of Al_2O_3 versus FeO*/MgO ratio, showing fractionation curves for different proportions of the crystallizing phases plagioclase and olivine and the percent crystallization that has occurred for any given proportion.

The extreme degree of crystallization, however, necessary to produce other gabbros and the KNORR basalt glasses with fractionation involving only plagioclase and olivine suggests either that (1) derivation of these gabbro and basalt compositions may also involve fractionation of clinopyroxene, or (2) crystallization differentiation is not the sole mechanism producing the observed chemical variations. The former suggestion is compatible with the basalt glass data indicating that high-pressure removal of an aluminum-chrome diopside present in the Cayman gabbros could produce a derivative liquid having the chemistry of the KNORR basalts (see discussion on basalt chemistry above). Further work is essential, however, before either of these two suggestions for producing the range observed in gabbro chemistry can be validated.

Thirty-eight representative samples were selected from the gabbroic suite for detailed microprobe analysis of the primary minerals plagioclase, olivine, clinopyroxene, orthopyroxene and amphibole. Secondary plagioclase, clinopyroxene and amphibole phases can often be distinguished by granoblastic texture, irregular grain shape, multi-crystal aggregates, reduced grain size and irregular optical zoning. In some cases the chemical analyses clearly support the secondary origin of particular grains.

The total compositional range of olivine in twelve samples is Fo_{70-88}, but within individual samples variation is limited to 3 mole percent Fo or less. The olivines show no primary zoning; alteration along cracks and grain boundaries has apparently not affected the chemistry of the relict olivine. Figure 16 plots these data against Fe/Fe + Mg (molar) of each host bulk-rock composition. Most data points are compatible with a distribution coefficient ($K_D = (MgO/FeO)_{liquid}/(MgO/FeO)_{olivine}$ somewhat higher than the value of 0.30 determined by Roeder and Emslie (1970). This may reflect the less fractionated nature of these rocks if they represent liquid compositions; alternatively, if the rocks are cumulates, the curve for $K_D = 0.30$ could be used to estimate Fe/Fe+Mg in the original liquids from which these olivines crystallized. Values of Fe/Fe+Mg of the liquid that are determined in this way are not substantially different from the range of values actually observed in the gabbros.

The majority of all analyzed pyroxene grains are clinopyroxenes that plot in a tight cluster near the diopside corner of the quadrilateral (coincident with the three largest enclosed areas in the upper left of Figure 17). The overall range of Mg/Mg + Fe (~ 0.9 to 0.5) is relatively large in the whole suite, however, and is notably greater in pyroxenes than olivines. Coexisting clino- and orthopyroxene compositions are shown in Figure 17 and are broadly similar to trends found in DSDP Site 334 samples by Hodges and Papike (1976), although we have not yet explored the exsolution effects that they discuss.

Plagioclase compositions range from An_6 to An_{84}. Primary plagioclases (about An_{50} or more calcic) show no significant optical or chemical zoning. Secondary plagioclases may be considerably less calcic (up to 30% An) than immediately adjoining relict primary plagioclase; this is reflected in the somewhat greater range of An contents in the more altered gabbros as compared with the range observed in relatively fresh gabbros. Compositional differences within a single specimen are greatest where the extent of deformation, recrystallization and/or alteration is most variable; intra-sample differences are least in unaltered samples and in severely altered or deformed samples (which have presumably been "homogenized" by these later processes).

The amphiboles show a large range in Mg/Mg + Fe, but have a relatively constant proportion of Ca (Figure 18). Brown Ti-rich hornblendes rim and are intergrown with pyroxenes and occur as

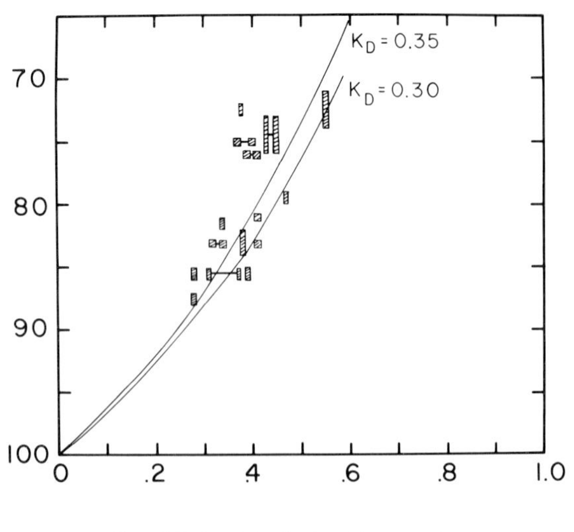

Fig. 16. Mole % Fo in olivine vs. molar Fe/Fe+Mg ratio in host rock for Cayman gabbroic rocks. $K_D = (MgO/FeO)_{liq}/(MgO\ FeO)_{oliv.}$

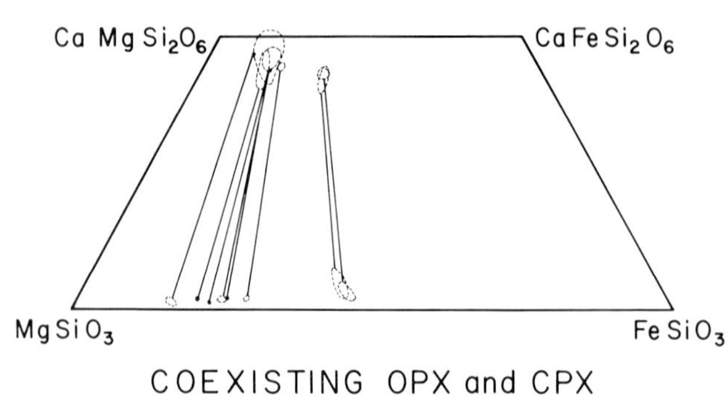

Fig. 17. Pyroxenes plotted in Ca-Mg-Fe. Tie-lines connect clinopyroxene compositions (upper) and coexisting orthopyroxene compositions (lower).

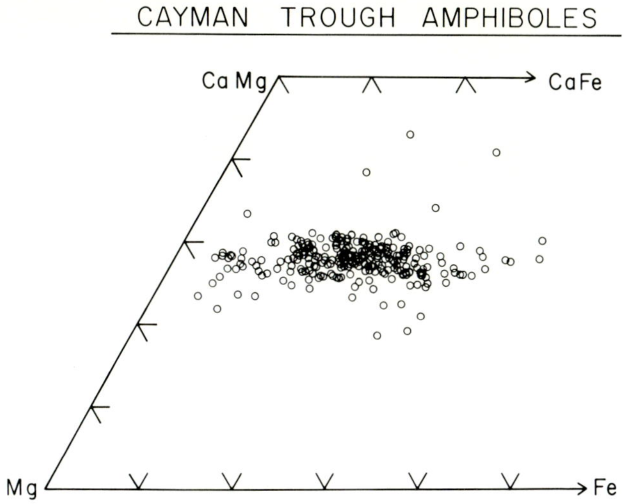

Fig. 18. Amphiboles plotted in the Ca-Mg-Fe quadrilateral.

isolated crystals, suggesting a primary or deuteric origin; these brown titaniferous hornblendes are also found in recrystallized zones with plagioclase and opaque minerals and without pyroxene. Actinolitic amphiboles are primarily Mg-rich and commonly form as alteration products after olivine and orthopyroxene rather than clinopyroxene.

Petrology of Ultramafic Rocks from the Rift Valley Walls

Partially serpentinized peridotites were recovered during six Alvin dives: three in Dive Area #1 and three in Dive Area #2. Peridotites in varying degrees of serpentinization were also recovered from six dredge sites along the eastern and western rift valley walls. At the dive sites where outcrop was sampled directly, the peridotites come from relatively narrow zones which cut gabbro and metagabbro country rocks. These peridotites are often highly deformed, and we suggest that the heavily serpentinized peridotites were emplaced along major zones of weakness from deeper sections of the oceanic crust and upper mantle. On the other hand, two dredges (Oceanus Dredge 13, 14) taken from the upper and lower face of a promontory on the eastern wall of the rift valley at 18°10'N and 81°35'W both contained over 90% relatively unaltered, though heavily weathered, peridotite suggesting that the promontory may be a relatively large (15-20 km^2) fault block of peridotite.

A variety of lithologies are included in the peridotite suite from the escarpments of the Mid-Cayman Rise spreading center (see Tables 1, 2 and 5). The three principal lithologies are spinel harzburgite, plagioclase-spinel harzburgite, and spinel troctolite or plagioclase-spinel dunite. In addition, dunite, olivine websterite and gabbro are all found in small patches or layers within the harzburgite samples.

The spinel harzburgites are tectonites with well-developed pseudoporphyritic textures in which round and abraded, or stretched and greatly elongated, enstatites are enclosed in a matrix of recrystallized olivine. Clinopyroxene, present in amounts varying between one and seven percent, is characteristically finer grained than coexisting enstatite. Spinel is present as a ubiquitous accessory phase in amounts up to one percent. The spinel typically is anhedral, occurring in irregular patches, often intergrown with pyroxene. Olivine has a composition ranging from $Fo_{90.0}$ to $Fo_{91.1}$ (5 analyses). Enstatite contains 3.7 to 5.3 weight percent Al_2O_3, 0.30 to 0.61 weight percent Cr_2O_3, and has Mg/(Mg+Fe) ratios from 0.904 to 0.909 (5 analyses). One diopside analysis gives an Mg/(Mg+Fe) ratio of 0.909 and 6.24 weight percent Al_2O_3 and 0.94 weight percent Cr_2O_3. Chromian spinel is generally aluminous with Cr/(Cr+Al) ratios between 0.17 to 0.48 and $Mg/(Mg+Fe^{2+})$ ratios from 0.72 to 0.79.

The plagioclase-spinel harzburgites (70-85% olivine, 10-20% orthopyroxene, 1-5% diopside, 1-5% plagioclase, < 1% spinel) are less tectonized than the spinel harzburgites, noticeably lacking the strongly deformed enstatites of the latter. Undulatory extinction and kinking of enstatite are present in many thin sections, but not common. Spinel occurs as relatively rare euhedral and subhedral grains and in highly irregular patches and stringers intergrown with olivine and pyroxene. Plagioclase is generally altered to nearly isotropic fine-grained calc-silicates. It occurs as angular lenses that are clearly interstitial to olivine and pyroxene. Similar textures have been attributed to the crystallization of trapped melt within peridotite at the Lanzo Massif by Boudier and Nicolas (1977). There appears to be no reaction-relation between spinel and plagioclase and the two appear to be in equilibrium. Olivine from one sample has a composition of $Fo_{90.1}$. Enstatite from the same sample has 2.6 weight percent Al_2O_3, 0.31 weight percent Cr_2O_3 and Mg/(Mg+Fe) of 0.895. Diopside from one sample has 4.19 weight percent Al_2O_3, 1.04 Cr_2O_3 and Mg/(Mg+Fe) of 0.918. Chromian spinel from one sample has a restricted composition with an intermediate chrome content (Cr/(Cr+Al) = 0.452-0.473).

Plagioclase-spinel dunites and troctolites were recovered during ALVIN dive 620 (Dive Area 2) and in one dredge haul (Oceanus 23-14). Olivine in these rocks is weathered and serpentinized, while plagioclase is completely replaced by white calc-silicates. Spinel is generally fresh though in some sections it is pseudomorphed by ferrichromite. Plagioclase and diopside clearly crystallized late and form angular lenses along olivine grain boundaries and triple junctions. Where these rocks are least deformed the olivine grains appear to have been equant to elongated anhedral or subhedral crystals. Spinel grains in the relatively undeformed dunites are euhedral and may be occluded within olivine grains or, more frequently, lie in the plagioclase-filled interstices between them. Layering can be seen in one sample, defined largely by variations in the olivine to plagioclase ratio.

Dunite, olivine websterite and gabbro all are found as minor segregations in the different major lithologies. Dunite occurs as a small patch in one spinel harzburgite and as a layer in a feldspathic dunite. Olivine websterites, similar to those described by Dick (1977), are found as 1-2 cm thick layers in the spinel harzburgites. Gabbroic segregations found in the plagioclase peridotites are similar to those in peridotites of the Lanzo massif described by Boudier and Nicolas (1977). These authors suggest that such segregations represent basaltic liquid generated in the peridotite during anatexis.

The spinel and plagioclase-spinel harzburgites have all undergone some degree of hydrothermal metamorphism, lower temperature serpentinization and weathering. Much of the deformation of these rocks appears to have occurred prior to serpentinization. The degree of hydrothermal alteration is highly variable and its affects are often slight, expressed chiefly by the replacement of some primary olivine and pyroxene by talc, amphibole and secondary diopside. A significant, often large, variation of spinel composition was seen in many thin sections with increases in Cr_2O_3 and FeO contents attributed to hydrothermal alteration of primary spinel. In many rocks serpentine has completely replaced the olivine and most of the pyroxene. The serpentinization appears to have occurred late and, at least in part, post-deformation. In addition, many of the peridotites are criss-crossed by late aragonite veins.

It is clear from the textural data alone that the feldspathic dunites are olivine-spinel cumulates. The spinel harzburgite has all the features typical of presumed mantle harzburgites from alpine peridotites: a tectonite fabric, a similar restricted range in silicate compositions, and a relatively wide range of spinel compositions. The plagioclase-spinel peridotites, however, have features that seem to make them intermediate between the feldspathic dunites and the spinel harzburgites. Plagioclase-spinel harzburgites and feldspathic dunites occur together at locations where they were collected.

The intrusive, cross cutting relationships of the partially serpentinized peridotites collected during ALVIN traverses indicate that, at least locally, partially serpentinized peridotite derived from upper mantle sources can intrude the plutonic foundation of the oceanic crust along fractures which cut deeply into the crust. The peridotites recovered in two carefully navigated dredges located on the same tectonic block but at different levels suggests that at least in this locality upper mantle rocks are exposed. The deepest dredge (Oceanus 13; 4400 m) recovered predominantly tectonized spinel harzburgite whereas the shallower dredge (Oceanus 14; 3280 m) recovered plagioclase bearing peridotites and troctolite. This apparent stratigraphic order as defined by two dredges is similar to that described from the transition zone in many ophiolite suites (e.g. England and Davies, 1973; Menzies and Allen, 1974) and is thought to represent the transition across the "petrologic moho" from mantle tectonite to cumulate peridotites at the base of the lower crust. To our knowledge, this represents the first time that this sequence has been described in oceanic rocks collected essentially in situ.

Magnetic Properties of Gabbros from the Rift Valley Walls

The large number of plutonic samples collected during the Cayman Trough Project has permitted the characterization of the magnetic properties of the plutonic foundation of the oceanic crust in the Cayman area. We have measured 1) the intensity of the natural remnant magnetization (NRM); 2) the stability of this intensity (mean destructive field and $NRM_{time\ 1}/NRM_{time\ 2}$); 3) the susceptibility of these rocks to the earth's field; 4) the relative importance of the NRM and susceptibility (the Konigsberger ratio, Q); and 5) the directional stability of the NRM (maximum angular change after exposure to a 250 oersted alternating field). Curie temperatures were measured for a variety of lithologies to determine the dominant magnetic mineralogy. A total of 58 plutonic rock samples was cored for magnetic measurements. These were cut into one inch lengths, producing 138 specimens.

The NRM of the Cayman Trough plutonic suite (all rocks excluding basalts) ranges from 10^{-7} to 10^{-2} Gauss (G), and has an arithmetic mean of 6.5×10^{-4}G. (Although the NRM and susceptibility values are approximately log normally distributed, total magnetization is the sum of all the magnetic vectors and therefore the arithmetic mean is considered a better indicator of total magnetization than the geometric mean (Lowrie, 1977).) The susceptibility of the entire plutonic suite ranges from 10^{-5} to 10^{-2}G/emu. The susceptibility of the gabbros alone has an arithmetic mean of 6.5×10^{-4}G/emu. The NRM and susceptibility of all specimens measured are plotted in Figure 19.

The relative importance of the remnant magnetization and the susceptibility is expressed as the Konigsberger ratio, Q = NRM/(susceptibility x geomagnetic field intensity at sample site). Thus, for samples with $Q > 1$, the remnant paleomagnetization dominates and a remnant paleomagnetic vector direction might be retained. A line representing $Q = 1$ is drawn in Figure 19. Two-thirds of all the specimens measured in this study lie above this line. Clearly, in order to contribute to the magnetic anomalies observed from the ocean's surface, a major portion of the

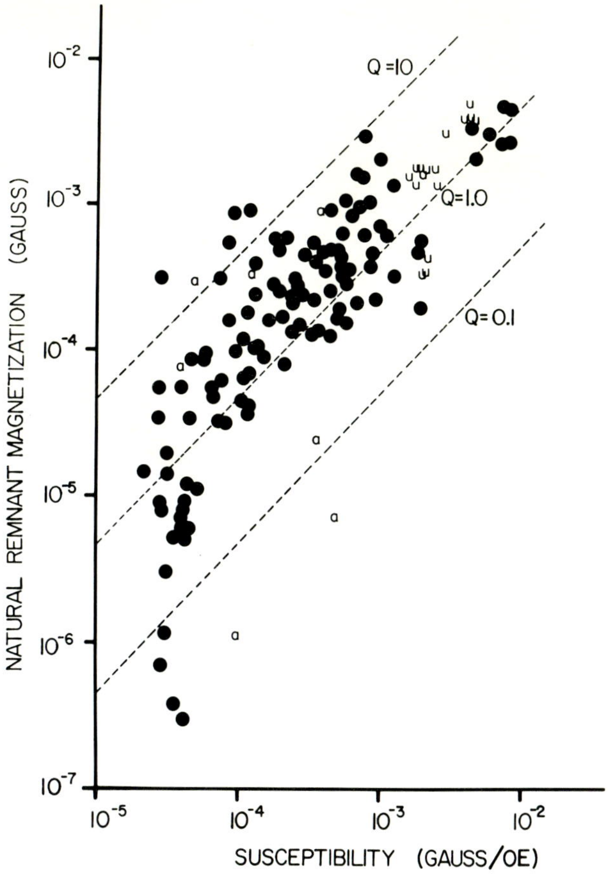

Fig. 19. Natural remnant magnetization and susceptibility of Cayman plutonic rocks (plotted log/log). Solid dots represent gabbroic rocks; a = moderately to severly deformed amphibolite; u = serpentinized ultramafic.

plutonic complex must have a Q > 1. The rocks sampled from the Mid-Cayman Rise spreading center meet this requirement.

Eighty-five specimens representing 54 samples were subjected to alternating field stepwise demagnetization to ascertain the stability of the magnetic vector in these rocks. NRM measurements were repeated immediately before demagnetization because a 3 to 6 month period had lapsed since the original NRM measurements had been made. The ratio of these two measurements is an indicator of the short-term stability of the rocks. Of the 45 gabbro samples subjected to this storage test only seven include specimens which show an NRM variation greater than 10%; all of these had Konigsberger ratios less than one. This storage test also provided a measure of the directional stability of a rock magnetic vector. The angular distance between the initially determined NRM direction and that determined just prior to demagnetization has been measured for 42 gabbro specimens for which Q > 1. Ninety percent of these showed a directional change of 5° or less.

The results of alternating field demagnetization of our gabbro samples are depicted graphically in Figures 20 and 21. Sixty-seven percent of these rocks have median destructive fields of 200 oe or more. By comparison, only 36% of the DSDP basalts have MDF of 200 oe or more (Lowrie, 1977). The resistance of the NRM vector direction to an applied field was also observed during stepwise alternating field demagnetization. In this study a parameter called the "250 oersted diameter" is used to describe the results. The 250 oersted diameter is the apical angle of a cone which circumscribes the vector directions obtained during stepwise demagnetization up to 250 oersteds. In most cases, this data set is comprised of 6 measurements made at 50 oersted intervals. A histogram of 250 oersted

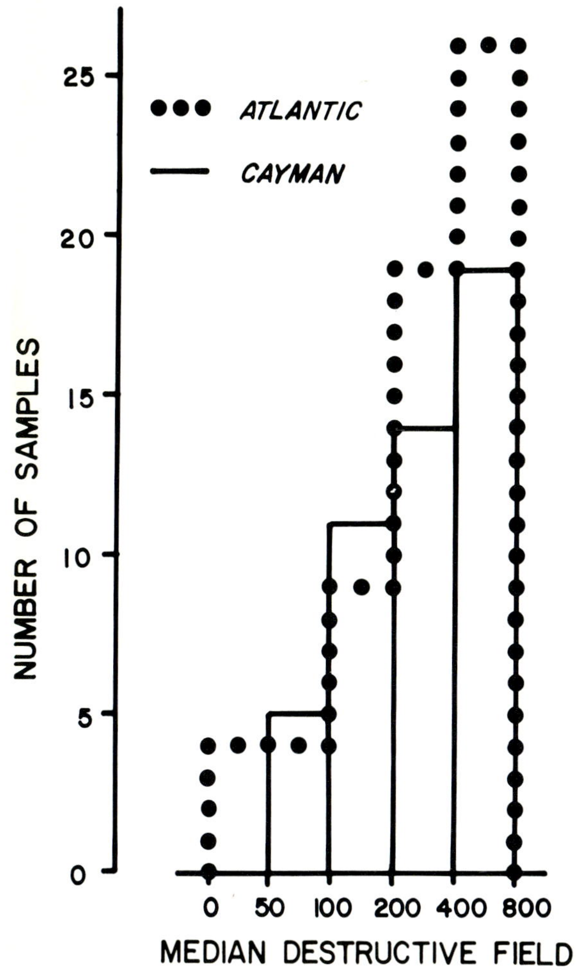

Fig. 20. Histogram showing the median destructive field distribution of gabbros collected from the Mid-Cayman Rise spreading center compared to gabbros collected from localities along the Mid-Atlantic Ridge.

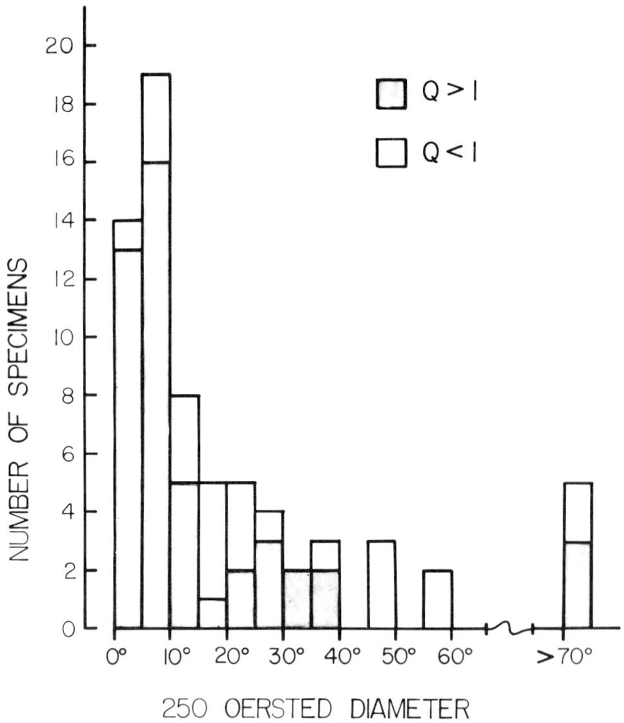

Fig. 21. NRM vector direction stability during alternating field demagnetization for Cayman Trough gabbros and troctolites. 250 oersted diameter is defined as the apical angle of a cone which circumscribes the NRM vector directions obtained during stepwise demagnetization up to 250 oersteds.

diameters for 70 specimens of gabbro is presented in Figure 21.

Curie temperatures obtained for 9 samples chosen to be representative of the entire collection in terms of both mineralogy and magnetic character are tabulated in Table 7. The values range from 535°C to 580°C. These are interpreted as evidence that magnetite (Curie temperature 570°) with low Ti content is the dominant magnetic mineral in all rock types.

From their study of ocean floor gabbros from the Atlantic, Kent and others, (1978), concluded that the intensity and stability of the magnetization in such rocks is a result of deuteric alteration or originally homogeneous titanomagnetite. During this process, ilmenite lamellae are exsolved, isolating large numbers of single domain magnetites and armoring them against further oxidation. The high Curie temperatures reported here and their association with a wide variety of Ti concentration (see Table 7) supports this idea.

The magnetic properties of ocean floor plutonics dredged in the Atlantic Ocean have been determined by Opdyke and Hekinian (1967), Fox and Opdyke (1973) and most recently by Kent and others (1978). The magnetic properties of the plutonic rocks cored at DSDP site 334 are reported in Aumento and others (1977). Data from these studies compare favorably with ours for the Cayman plutonics. The arithmetic mean NRM of the Atlantic dredge samples is 8.8×10^{-4}G (Kent and others, 1978). For the Cayman gabbros this value is about a factor of two lower, 4.8×10^{-4}G.

The Konigsberger ratio averages slightly higher for the Atlantic plutonics and the fraction of samples that have $Q > 1$ is high in both data sets: .75 in the Atlantic and .66 in the Cayman. The stability of these rocks to alternating field demagnetization is also comparable. For samples with $Q > 1$, 70% of the Cayman and 81% of the Atlantic samples have mean destructive fields of 200 oersteds or more.

Detailed magnetic studies of basalt from DSDP cores have cast some doubt on the assumption that the magnetic character of 0.5 km of

TABLE 7. Curie Temperature Analyses

Sample	Rock Type	Curie Temperature 1)	Whole Rock Ti Concentration
616 1-1 Y A	ultramafic	569°C, 580°C	n.d.
616 4-1 A	gabbro	565°C, 563°C	0.30%
616 7-1 D	gabbro	559°C	n.d.
620 3-1 B	gabbro	535°C	8.90%
623 6-1 A	gabbro	567°C	0.00%
737 1-1 B	gabbro	542°C	1.26%
739 2-2 B	gabbro	550°C	0.60%
739 7-1 C	gabbro	562°C	0.13%
742 1-1 B	gabbro	568°C	0.09%

1) Determined graphically by the method of Gromme, et al. (1969).

extrusive basalt and shallow intrusive basalt which comprises seismic layer 2a is sufficient to explain the amplitude of magnetic anomalies observed in the oceans (e.g. Ryall, et al., 1977; Lowrie, 1977). These authors have found that the basalt layer is far more magnetically heterogeneous than is accounted for in models of magnetic anomalies, and the primary result of this inhomogeneity is to decrease the total bulk magnetization.

Layer 3 is an increasingly appealing additional source for magnetic anomalies. The NRM of the plutonics is surprisingly stable, apparently due to the deuteric isolation of single domain magnetites. The remnant magnetization is more important than susceptibility in the large majority of plutonic samples. Perhaps most importantly, the great thickness (3-5 km) of layer 3 compensates for the low average NRM (about 1/10 the average for ocean floor basalts). Kent, et al., (1978) have calculated that for block widths greater than about 20 km, a deep, 5 km thick layer of NRM 8×10^{-4}G will actually contribute more intensity to the observed anomaly than a shallow 0.5 km thick layer of NRM 40×10^{-4}G. Although the relative importance of its contribution to the marine magnetic anomalies is difficult to evaluate, the plutonic foundation of the oceanic crust must make a contribution and should not be neglected when considering the source region for magnetic anomalies.

Summary and Conclusions

The intensive survey of the Mid-Cayman Rise spreading center made possible by the use of sophisticated marine geological and geophysical tools has provided a wealth of data enabling us to define the tectonic and petrologic character of a short (≈2 cm/yr full rate) plate boundary that is truncated at either end by two long transform faults.

Our high resolution bathymetric data set, deep-towed photographic results, bathyscaphe observations and bottom sampling all confirm the original proposition by Holcombe, et al. (1973) that the north-south trending axial valley of the Mid-Cayman Rise is the locus of recent basalt volcanism and represents a zone of plate accretion. Like other rift valley segments, the inner floor is characterized by a narrow zone (2-3 km) of recent basaltic volcanism arranged in elongated ridges composed of a chaotic assemblage of pillow lavas. Although this type of volcanism appears presently to dominate the extrusive mode operating along the inner floor, at one locality near the northern end of the inner floor a massive sheet flow was observed filling a trough between two constructional ridges. The major and trace element chemistry of basalt glasses recovered from the recent volcanic lineaments show that the extrusive rocks have first order chemical properties similar to ocean floor tholeiites, although they exhibit some differences in chemistry from typical mid-ocean ridge basalts. That portion of the inner floor which flanks the narrow zone of recent volcanism is characterized by a blanket of sediment that thickens away from the volcanic axis, and is experiencing tectonic modification as evidenced by the formation of fissures. The inner floor is bounded by steep inward facing fault scarps and terraces which parallel the inner floor and which create morphologic elements typical of rift valley terrain.

When one considers the tectonic and petrologic details associated with the Mid-Cayman Rise, however, it is clear that this short accreting plate boundary is distinctive. The rift valley floor resides at depths from 2000 to 3000 m deeper than the floors of other rift valleys. The width of the inner floor varies along strike from a few kilometers to 18 kilometers and the trend of the volcanic/tectonic lineaments changes down the length of the inner floor. Furthermore, the continuity of the rift valley floor is interrupted and offset by a short, several kilometer long NE-SW trending discordant zone positioned midway along the Mid-Cayman Rise. Large tectonic blocks comprise segments of the bounding rift valley walls and seem to be the product of both N-S and E-W tectonic lineaments. We suggest that these tectonic subtleties, played against the processes operative and typical of slowly accreting plate boundaries, are the consequences of the North American-Caribbean plate geometry in this region. The Mid-Cayman Rise represents a short accreting plate boundary truncated by two, long (> 1000 km) transform faults. We propose that although the strike slip tectonics associated with the transform domains of the Oriente and Swan transform faults are concentrated along narrow zones to the north and south of the Mid-Cayman Rise, the newly created lithosphere along the Rise axis is not completely independent of the effects of strike slip motion of the two transforms. Although the Mid-Cayman Rise is principally an extensional regime, a left lateral shear-couple, a function of the juxtaposition of two long left lateral transform faults against a short accretionary plate boundary, perturbs and disrupts the morpho-tectonic grain of this slowly accreting ridge segment.

The in situ sampling by the submersible ALVIN and our dredging results document another peculiar characteristic of the Mid-Cayman Rise spreading center. The well navigated, intensive in situ sampling by ALVIN of escarpments in two areas representative of the rift valley walls indicates that gabbroic rocks outcrop in abundance. In fact, the sampling results argue persuasively that the volcanic carapace (extrusive basalt and shallow intrusives) must be as thin as 250 m. The plutonic component of the oceanic crust may be thin as well (≈1000 m), because dredge results suggest that ultramafic rocks, which exhibit structural and petrologic characteristics considered to be representative of the petrologic Moho, are exposed in the escarpments at the base of the rift valley wall. The apparent thin nature of the oceanic crustal components along the Mid-Cayman Rise can be explained in several ways. Firstly, they could be tectonically thinned by faulting out portions of the crustal column. In order to thin the oceanic crust to the extremes recognized in the Cayman Trough, normal or reverse faults with throws of several hundreds of meters would have to be envoked. Our detailed bathymetric studies and bottom observations provide no evidence which documents the existence of faults with these offset magnitudes. A second explanation is that the processes which control the production of basaltic magmas beneath rift valley floor are perturbed and less magma is produced thereby creating a thinner oceanic crust. It has been suggested that the Mid-Cayman Rise may be relatively cold because of the proximal, cold truncating transform boundaries and this environment results in the production of less basaltic melt in the rising asthenospheric wedge beneath the rift valley axis (Fox, 1978, and in prep.). A third possibility is that the gabbroic and ultramafic rocks represent intrusive bodies which have melted their way up into overlying oceanic crust of normal thickness. Perhaps these intrusive bodies follow penetrative zones of weakness caused by the tectonic setting characteristic of the Mid-Cayman Rise spreading center.

Our work on the CAYTROUGH data set is ongoing and in time the ideas briefly outlined above will be developed more rigorously. The preliminary results, however, indicate clearly that although new oceanic lithosphere is being accreted along the Mid-Cayman Rise spreading center in a way that develops a morpho-tectonic fabric typical of slowly accreting plate boundaries, important structural and petrologic differences are developed which must reflect this particular plate tectonic setting. Indeed, although the Mid-Cayman Rise spreading center represents an extreme example of an accreting plate margin dominated by transform tectonics, we suggest that processes operative along this plate boundary may be operative, albeit more subtly developed, along other ridge-transform-ridge boundaries. For example, the Mid-Atlantic Ridge south of the Azores-Gibralter triple junction is characterized by frequent ridge-transform-ridge plate boundaries with transform faults of varying offset lengths, occurring every 50 to 100 km. If our initial results from the Cayman Trough Project stand the test of continued investigation, then a major conclusion will be that the crust formed proximal to and sandwiched between transform boundaries may be structurally and petrologically much more complex than previously suspected.

Acknowledgments. We would like to thank the Captains, officers and crews of the USNS BOWDITCH, R/V KNORR and R/V OCEANUS and the pilots, engineers and technicians who work with the submersible ALVIN and bathyscaphe TRIESTE for their dedicated support throughout the field program. The mineral chemical data reported in this paper was done while one of us (F.L.M.) was a Visiting Scientist at the Lunar and Planetary Institute which is operated by the Universities Space Research Association under contract no. NSR-09-051-001 with the National Aeronautics and Space Administration. The Field programs and the other laboratory investigations were supported by the Submarine Geology Branch, National Science Foundation through grants OCE-75-18533 (WHOI) and OCE-76-21882 (SUNYA and Wesleyan), and Office of Naval Research grant #N00014-74-C-0262 (R.B.). Woods Hole Oceanographic Institution Contribution No. 4278

Lunar and Planetary Institute Contribution No. 354.

References

Aumento, F., Melson, W.G., et.al., 1977, Initial reports of the Deep Sea Drilling Project, 37: Washington (U.S. Government Printing Office), 1008 p.

Ballard, R.D. and Moore, J.G., Photographic Atlas of the Mid-Atlantic Ridge: Springer-Verlag, NY, 114 pp., 1977.

Ballard, R.D. and van Andel, Tj., Morphology and tectonics of the inner rift valley at Lat. 36° 50'N on the Mid-Atlantic Ridge: Geol. Soc. Amer. Bull., v. 88, p. 507-530, 1977.

Ballard, R.D., Holcombe, R. and van Andel, Tj., The Galapagos Rift at 86°W, 3: sheet flows, collapse pits and lava lakes of the rift valley: submitted to Jour. Geophys. Res., in press.

Bonatti, E., Honnorez, J. and Ferrara, G., Peridotite-gabbro-basalt complex from the equatorial Mid-Atlantic Ridge: Phil. Trans. Roy. Soc. London, Ser. A, 268, 385-402, 1971.

Boudier, F. and Nicolas, A., Structural controls on partial melting in lanzo peridotites, in Magma Genesis, Bull. 96, Proc. Am. Geophys. Union, H.J.B. Dick, ed., 63-78, 1977.

Bowen, C.O., Geophysical study of the Cayman Trough: Jour. Geophys. Res. 73, 5159-5173, 1968.

Dick, H., Partial melting of the Josephine peridotite, I: the effect of mineral composition and its consequence for geothermometry and geobarometry: Amer. J. Sci., 277, 801-832, 1977.

Edgar, N.T., Ewing, J.I. and Hennion, J., Seismic refraction and reflection in the Caribbean Sea: Am. Assoc. Petroleum Geologists Bull., v. 55, p. 833-870, 1971.

Eggler, D.H., Fahlquist, D.A., Pequequat, W.E. and Herndon, J.M., Ultrabasic rocks from the Cayman Trough, Caribbean Sea: Geol. Soc. Amer. Bull., 84 (6), 2133-2138, 1973.

Engel, C.G. and Fisher, R.L., Granitic to ultramafic rock complexes of the Indian Ocean ridge system, western Indian Ocean: Geol. Soc. Amer. Bull., 86, 1553-1578, 1975.

England, R.N. and H.L. Davies, Mineralogy of

ultramafic cumulates and tectonites from eastern Papua: Earth Planet. Sci. Lett., 17, 416-425, 1973.

Erickson, A.J., Helsley, C.E. and Simmons, G., Heat flow and continuous seismic profiles in the Cayman Trough and Yucatan Basin: Geol. Soc. Amer. Bull., v. 83, p. 1241-1260, 1972.

Ewing, J., Antoine, J. and Ewing, M., Geophysical measurements in the western Caribbean Sea and in the Gulf of Mexico: Jour. Geophys. Res., v. 65, p. 4087-7126, 1960.

Fox, P.J., 1978, The effect of transform faults on the character of the oceanic crust, abs.: Geol. Soc. Am. Abstracts with Programs, 10, (7), p. 403.

Fox, P.J. and N.D. Opdyke, Geology of the oceanic crust: magnetic properties of oceanic rocks: Jour. Geophys. Res., v. 78, p. 5139-5154, 1973.

Fox, P.J., A. Lowrie, Jr. and B.C. Heezen, Oceanographer fracture zone, Deep Sea Research, v. 16, p. 59-66, 1969.

Fox, P.J. and Stroup, J.B., in press, The geology of the plutonic foundation of the oceanic crust, in: Emiliain, C., Ed., The oceanic lithosphere, vol. VII of The Sea, John Wiley and Sons, NY.

Glenn, M., Introducing an operational multibeam-array sonar: Intern. Hydrogr. Rev., 47, 35-40, 1970.

Hodges, F.N. and J.J. Papike, DSDP site 334: magmatic cumulates from oceanic layer 3: J. Geophys. Res., 81, 4135-4151, 1976.

Holcombe, T.L., Vogt, P.R., Matthews, J.E. and Murchison, R.R., Evidence for sea-floor spreading in the Cayman Trough: Earth and Planet. Sci. Lett., v. 20, p. 357-371, 1973.

Jordan, T.H., The present-day motions of the Caribbean plate: Jour. Geophys. Res., v. 80, p. 4433-4440, 1975.

Kent, D.V., Honnorez, B.M., Opdyke, N.D. and Fox, P.J., Magnetic properties of dredged oceanic gabbros and the source of marine magnetic anomalies: submitted to Jour. Geophys. Res., in press.

Lowrie, W., Intensity and direction of magnetization in oceanic basalts: J. Geol. Soc., v. 133, p. 61-82, 1977.

Matthews, J.E., The geomagnetic field of the Caribbean Sea: Geol. Soc. Amer. Abs. with Programs, v. 6, no. 7, p. 859, 1974.

Macdonald, K.C. and Holcombe, T.L., Inversion of magnetic anomalies and sea-floor spreading in the Cayman Trough, submitted to Earth Planet. Sci. Lett., in press.

Macdonald, K.C. and Luyendyk, B.P., 1977, Deep-tow studies of the structure of the Mid-Atlantic Ridge crest near lat. 37°N: Geol. Soc. Amer. Bull., 88, 621-636.

Melson, W.G. and Thompson, G., Layered basic complex in oceanic crust, Romanche Fracture, equatorial Atlantic Ocean: Science, 168, 817-820, 1970.

Menzies, M. and Allen, C., Plagioclase lherzolite-residual mantle relationships within two eastern Mediterranean ophiolites: Contrib. Min. Pet., 45, 1974.

Miyashiro, A., Shido, F. and Ewing, M., Crystallization and differentiation in abyssal tholeiites and gabbros from mid-ocean ridges: Earth Planet. Sci. Lett., 7, 361-365, 1970.

Molnar, P. and Sykes, L.R., Tectonics of the Caribbean and Middle America regions from focal mechanisms and seismicity: Geol. Soc. Amer. Bull., v. 80, p. 1639-1664, 1969.

Opdyke, N.D. and Hekinian, R., Magnetic properties of some igneous rocks from the Mid-Atlantic Ridge: Jour. Geophys. Res., v. 72, p. 2257-2260, 1967.

Pearce, J.A. and Cann, J., Tectonic setting of basic volcanic rocks using trace element analyses: Earth Planet. Sci. Lett., 19, 290-300, 1973.

Perfit, M.R., Petrology and geochemistry of mafic rocks from the Cayman Trench: evidence for spreading: Geology, 5, 105-110, 1977.

Perfit, M.R. and Heezen, B.C., The geology and evolution of the Cayman Trench: submitted to Geol. Soc. Amer. Bull., 1978.

Roeder, P.L. and Emslie, R.F., Olivine-liquid equilibrium: Contrib.Mineral. Petrol., 29, 275-289, 1970.

Ryall, P.J.C., Hall, J.M., Clark, J. and Milligan, T., Magnetization of oceanic layer 2--results and thoughts after DSDP Leg 37: Can. J. of Earth Sci., v. 14, p. 684-706, 1977.

Shibata, T. and Fox, P.J., Fractionation of abyssal tholeiites: samples from the Oceanographer Fracture Zone (35°N, 35°W): Earth Planet. Sci. Lett., 27, 62-72, 1975.

Sleep, N.H. and Biehler, S., 1970, Topography and tectonics at the intersections of fracture zones with central rifts; Jour. Geophys. Res., 75 (14), 2748-2752.

Stakes, D.S., 1978, Submarine hydrothermal systems: variations in mineralogy, chemistry, temperatures and the alteration of oceanic layer II, thesis, Oregon State University, 189 pp.

Sykes, L.R. and Ewing, M., The seismicity of the Caribbean region: Jour. Geophys. Res., v. 70, p. 5065-5070, 1965.

Thompson, G., Trace element distributions in fractionated oceanic rocks, 2. gabbros and related rocks: Chem. Geol., 12, 99-111, 1973.

SEISMIC VELOCITIES, DENSITIES, ELECTRICAL RESISTIVITIES, POROSITIES AND THERMAL CONDUCTIVITIES OF CORE SAMPLES FROM BOREHOLES INTO THE ISLANDS OF BERMUDA AND THE AZORES

R. D. Hyndman

Pacific Geoscience Centre, Earth Physics Branch, Dept. of Energy, Mines and Resources, Sidney, B.C., V8L 4B2, Canada

N. I. Christensen

Dept. of Geological Sciences, Graduate Program in Geophysics and Dept. of Oceanography, University of Washington, Seattle, Washington

M. J. Drury

Division of Geomagnetism, Earth Physics Branch, Dept. Energy, Mines and Resources, Ottawa, Ontario, K1A 0Y3, Canada

Contribution of the Earth Physics Branch 751

Abstract. A detailed study of physical properties was made on core samples from 1 km deep boreholes into islands of Bermuda and the Azores. The properties measured are seismic velocity, density, electrical resistivity, porosity and thermal conductivity. The properties of subaqueous lava flow samples from Bermuda and the Azores are very similar to those for deep ocean tholeiites. Subaerial flows and pyroclastics from the Azores have quite different properties, reflecting their higher porosity. Lamprophyric intrusive sheets from Bermuda also have significantly different and unusual properties, notably lower velocity and higher density than the flows. It is concluded that aseismic ridges, islands and seamounts may have significantly different physical properties from normal upper oceanic crust if they contain subaerially erupted material as on Sao Miguel, Azores or are produced by intrusion of lamprophyric sheets or dykes into older oceanic crust as inferred for Bermuda.

Introduction

Laboratory data on the physical properties of rocks are an essential element in the interpretation of surface geophysical measurements, and for structural and petrological models. Much work has been done on oceanic crustal samples, particularly those from the Deep Sea Drilling Project, that gives a reasonably clear picture of the upper part of the normal oceanic crust (e.g. Christensen and Salisbury, 1975; Hyndman and Drury, 1976.) However, there are some anomalous oceanic areas of particular significance. These are the aseismic ridges and oceanic islands, which reflect higher than normal vulcanism and are referred to as hot spots or hot spot traces. They may arise as the surface expression of deep mantle plumes (Wilson, 1965; Morgan, 1971) or from propagating lithospheric fractures (e.g. Shaw and Jackson, 1973). In either case, compared to normal oceanic crust they are characterized by greater crustal thickness and by different composition.

This article reports seismic velocities, densities, electrical resistivities, porosities, and thermal conductivities for basaltic samples from approximately 1 km deep boreholes into the islands of Bermuda and São Miguel, Azores (Fig. 1 and 2) and discusses the geophysical significance of these measurements.

An 802 m deep borehole was drilled into the islands of Bermuda at 32°22'N, 64°42'W (Fig. 1) at an elevation of 10 m during the summer of 1972 for Dalhousie University and Lamont-Doherty Geological Observatory. The hole was cored with almost complete recovery for geological and geophysical studies. The islands of Bermuda lie approximately 1100 km east of the North American coast, about one third of the distance to the mid-Atlantic ridge. They have a maximum elevation of 30 m and occupy the southern 7% of the 116 km^2 platform that makes up the top of the Bermuda pedestal. Only limestones are exposed, the volcanic platform lies at an average of about 80 m below sealevel (Gees and Medioli, 1970). Bermuda is the northernmost and largest of three seamounts of the Bermuda Rise. From potassium-argon dating, the Bermuda seamount appears to have been formed 33 m.y. ago by the intrusion of numerous lamprophyric sheets into oceanic crust that was then about 80 m.y. old (Reynolds and Aumento, 1974; Gees, 1969). The surrounding seafloor is now about 110 m.y. old (Larson and Pitman, 1972). This age of the seamount is substantiated by the age of the oldest limestones overlying the volcanics (see

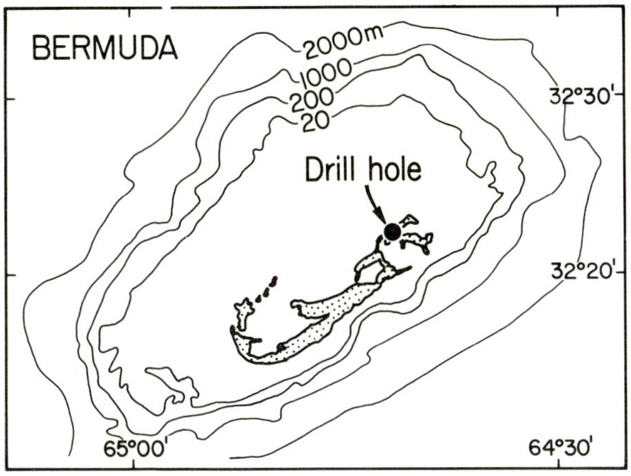

Fig. 1. Location of the Bermuda borehole.

Wilson, 1963) and by the low heat flow and the subsidence history of the platform (Hyndman et al., 1974a). Temperatures in the drilled section are low, reaching only about 38°C at the hole bottom (Fig. 3). In addition to radioactive dating, heat flow and heat production studies, extensive rock magnetic studies have been performed on the core (Ade-Hall et al., 1973) as well as detailed petrology (Aumento and Ade-Hall, 1973) and geochemistry (Aumento et al., 1976; and F. Aumento and B. Gunn, personal communication). A preliminary report of laboratory velocity measurements was presented by Barrett et al. (1973).

The borehole first penetrated 36 m of limestone, then 772 m of volcanic rocks made up of over 1000 igneous units. The whole core shows evidence of hydrothermal alteration both in petrology and in high Curie temperatures. Of the igneous units, 64% were altered subaqueously extruded tholeiite lavas and 36% were thin lamprophyric sheets of unusual composition, e.g. low silica, high Ti, K etc. It appears that any subaerial volcanics that were present have been eroded by a subsequent uplift before final subsidence. Such uplift could have been produced by the intrusion of the sheets.

During the summer of 1973, a 981 m deep borehole was drilled into the island of São Miguel, Azores at 25°31'W, 37°48.9'N (Fig. 2) at a surface elevation of 72 m for Dalhousie University and Lamont-Doherty Geological Observatory. Core recovery was almost complete except in the upper 148 m. The Azores are a group of 9 islands aligned in a NW-SE chain which crosses the mid-Atlantic ridge near 39°N. The region is marked by a change in trend of, and by a broadening of the ridge into a large platform (Krause and Watkins, 1970) and a positive regional gravity anomaly (e.g. Kaula, 1972). The Terceira rift, along which the main islands are located, trends from the ridge toward Gibraltar, probably defining a plate boundary that originated about 45 m.y. ago (Krause and Watkins, 1970). São Miguel is the largest island and lies 400 km east of the mid-Atlantic ridge crest. The drill site was located on the lower southern flank of the volcano Agua de Pau, which has erupted 5 times in the past 4,600 years, the last in 1563. Numerous hot springs and sporadic seismic activity indicate the volcano to be only dormant at present.

Extrusive lavas make up 72% of the Azores core in 140 flows averaging 5 m in thickness (Muecke et al., 1974). The main rock types are alkali basalts, hawaiites and mugearites. Three trachyte flows make up 6% of the total. The upper 763 m consists of 3 distinct subaerial volcanic eruptive sequences. This section is underlain by a 107 m transition sequence of basaltic sandstones and basaltic breccias, then by pillows and massive basaltic flows of subaqueous origin.

Temperatures were measured in the hole at intervals during drilling until the hole was terminated by the eruption of steam. Only bottom hole measurements appeared to represent in-situ temperatures, (Fig. 3) giving gradients up to 250°C/km. Temperatures in the hole nearer the surface were dominated by in-hole convection (see Muecke et al., 1974). The geothermal gradient is dominated by probably near-horizontal water flow so a meaningful geothermal heat flux cannot be ascertained. The temperature at the bottom of the hole was 200°C. Two K-Ar dates of 117,000 ± 24,000 yr at 57 m and 280,000 ± 140,000 yr at 950 m (Muecke et al., 1974) and that all of the core was normally magnetized presumably in the Brunhes polarity epoch that extends to 690,000 yr ago, indicate the youth and rapid formation of this volcanic edifice. It appears that the island has subsided over 900 m in less than 0.5 m.y. probably in isostatic response to the increasing volcanic load. A

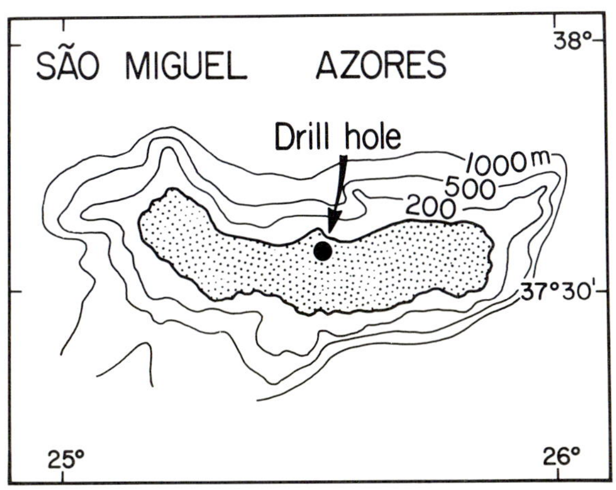

Fig. 2. Location of the Azores borehole.

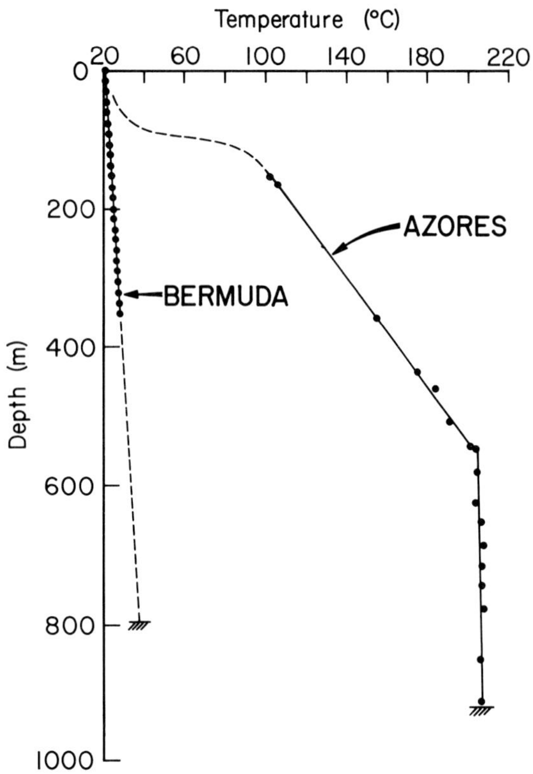

Fig. 3. Temperature-depth profiles in the Bermuda and Azores boreholes.

detailed description of the petrology of the core has been made by McGraw (1976) and extensive rock magnetic studies have been made (J.M. Hall, N.D. Opdyke and W. Lowrie, personal communication).

Density

The bulk densities of water saturated samples (stored in water from time of drilling) were determined by weighing the cores in air and obtaining the volume either by measuring the weight loss with the samples suspended in distilled water or by measurements of sample dimensions (Tables 1 and 2). The estimated accuracy is $\pm$ 0.01 g cm^{-3}. The grain or mineral densities for samples for which porosities are reported were determined simply by subtracting the mass and the volume of the pore fluid and recomputing the densities. Since the porosities tend to be underestimated because of incomplete drying (i.e. effective porosity), the grain density also will be too low. The estimated error limits are $+$ 0.02 to $-$0.01 g cm^{-3}.

The mean bulk density of 43 samples from Bermuda is 2.83 $\pm$ 0.02 g cm^{-3} with no systematic variation with depth (Fig. 4). However, the intrusive lamprophyric sheets have a significantly higher mean of 2.90 $\pm$ 0.03 g cm^{-3} (16 samples) and larger variation compared to the flows, with a mean of 2.79 $\pm$ 0.01 g cm^{-3} (27 samples), the higher mean perhaps reflecting the high density of the pyroxene in the former since the porosities are similar. Taking the mean sheet and flow densities and their relative proportions of 36% and 64% respectively in the drilled section gives a section mean of 2.83 g cm^{-3}, identical to the sample mean indicating that there was a representative sampling of sheets and flows. The mean grain density of 17 Bermuda samples is 2.85 $\pm$ 0.02 g cm^{-3}; the mean for 17 samples of the flows is 2.81 $\pm$ 0.01 and for 8 samples of the sheets is 2.92 $\pm$ 0.05 g cm^{-3}.

TABLE 1. Physical Properties of Samples from Bermuda

Property	Lava Flows		Lamprophyric Sheets		All Samples	
	Number	Mean	Number	Mean	Number	Mean
Compressional velocity (0.4 kbar) km s^{-1} (Vp)	12	5.97 ± 0.08	8	5.89 ± 0.07	20	5.94 ± 0.05
Shear velocity (0.4 kbar) km s^{-1} Vs)	12	3.22 ± 0.06	8	3.07 ± 0.05	20	3.16 ± 0.04
Poisson's ratio (σ)	12	0.295 ± 0.003	8	0.314 ± 0.006	20	0.303 ± 0.004
Bulk density, g cm^{-3}	27	2.79 ± 0.01	16	2.90 ± 0.003	43	2.83 ± 0.02
Grain density, g cm^{-3}	17	2.81 ± 0.01	8	2.92 ± 0.005	25	2.85 ± 0.02
Porosity, % vol.	17	2.6 ± 0.4	8	3.4 ± .05	25	2.9 ± 0.3
Electrical resistivity ohm m	17	1320 ± 400	6	340 ± 120	23	780 ± 210
Thermal conductivity W m^{-1} K^{-1}	32	2.08 ± 0.02	21	2.19 ± 0.07	53	2.12 ± 0.03

TABLE 2. Physical Properties of Samples from the Azores

Property	Subaerial		Subaqueous		All Samples	
	Number	Mean	Number	Mean	Number	Mean
Compressional velocity (0.4 kbar) km s^{-1} (Vp)	15	4.92 ± 0.17	3	5.60 ± 18	19	5.08 ± 0.15
Shear velocity (0.4 kbar) km s^{-1} (Vs)	15	2.49 ± 0.12	3	2.97 ± 0.09	19	2.60 ± 0.11
Poisson's ratio (σ)	15	0.328 ± 0.007	3	0.307 ± 0.003	19	0.323 ± 0.006
Bulk density, g cm^{-3}	41	2.60 ± 0.004	11	2.79 ± 0.04	56	2.64 ± 0.03
Grain density, g cm^{-3}	19	2.88 ± 0.003	3	2.85 ± 0.03	23	2.87 ± 0.02
Porosity	19	10.6 ± 1.7	3	4.4 ± 1.2	23	9.4 ± 1.5
Electrical resistivity ohm m	19	82 ± 38	3	230 ± 170	23	110 ± 50
Thermal conductivity W m^{-1} K^{-1}	31	1.75 ± 0.05	5	1.73 ± 0.04	40	1.75 ± 0.04

The mean bulk density of 56 samples from the Azores borehole is 2.64 ± 0.03 g cm^{-3}. There is some systematic increase in density with depth (Fig. 4). The mean is much lower and the scatter of values is more than double that for the Bermuda samples reflecting the wide variation and generally high porosity of the Azores rock types, particularly the pyroclastics in the subaerial section. The mean for the subaerial section is 2.60 ± 0.04 g cm^{-3} (41 samples). The mean for the igneous-sedimentary transition sequence is 2.71 ± 0.09 g cm^{-3} (4 samples), and the mean for the subaqueous sequence is 2.79 ± 0.04 g cm^{-3} (11 samples). As expected, the mean density of the subaqueous Azores rocks is the same as the mean of 2.79 for the flows from Bermuda, and for deep ocean tholeiitic basalts (e.g. Hyndman and Drury, 1976). The mean grain density of 23 Azores cores is 2.87 ± 0.02 g cm^{-3}, not greatly different from the Bermuda value, indicating that most of the bulk density differences arise from variations in porosity or water content. Bulk density is plotted as a function of porosity in Fig. 5 for both Bermuda and Azores samples. The intercept of about 2.85 gm cm^{-3} is in agreement with the above grain density estimates. This lower grain density and lower porosity, compared to the mean respective values for fresh deep ocean basalts (2.95 g cm^{-3} and 7.8%) suggests that initially free pore water in the Bermuda and Azores rocks is now bound in hydrous minerals, there being little change in total water in the alteration process.

For both the Bermuda and Azores core there is some sampling bias, low density fracture zones and rubble not being recovered, so the insitu density is lower than the means from the core. We believe that the effect is significant only for the upper several hundred meters of the Azores hole where core recovery was poor.

The densities of subaerial volcanic flow samples have been reported for the island of Hawaii, which is in a similar setting to São Miguel, by Manghnani and Woollard (1968) and Kinoshita (1965). They found that most surface and borehole samples had a density of 2.3 g cm^{-3}, in close agreement with the one near surface sample from the Azores borehole but considerably lower than the mean of 2.60 g cm^{-3} for the subaerial section. They found a number of dykes and plugs with much higher densities of 3.0 to 3.2 g cm^{-3}. Woollard (1951) estimated the mean density of the Hawaii island of Oahu to be 2.3 g cm^{-3} from that required to minimize the effect of topography on island gravity values. However, taking a larger area, including marine gravity and a more sophisticated model, Walcott (1970) and Watts and Cochran (1974) estimated a mean density of 2.80 g cm^{-3} for the Hawaiian Islands, close to our Bermuda mean and Azores subaqueous samples suggesting only the near surface rocks have low density (see also Worzel and Harrison, 1963). A similar density has been estimated from gravity measurements for the Great Meteor seamount (Watts et al., 1975). Lower values of about 2.5 g cm^{-3}, however, were estimated for the Caryn seamount by Worzel and Talwani (1959) and of 2.3 g cm^{-3} for the Jasper seamount off California by Harrison and Brisbin (1959). A possible explanation for this low density is that they may have carbonate caps. In a detailed gravity study of the Cook Islands, Robertson (1967) estimated that the island platforms had an average density of 2.32 g cm^{-3} with large central plugs of 2.88 g cm^{-3} (see also data in Malahoff, 1969). From Bermuda gravity measurements Woollard (1954) computes a

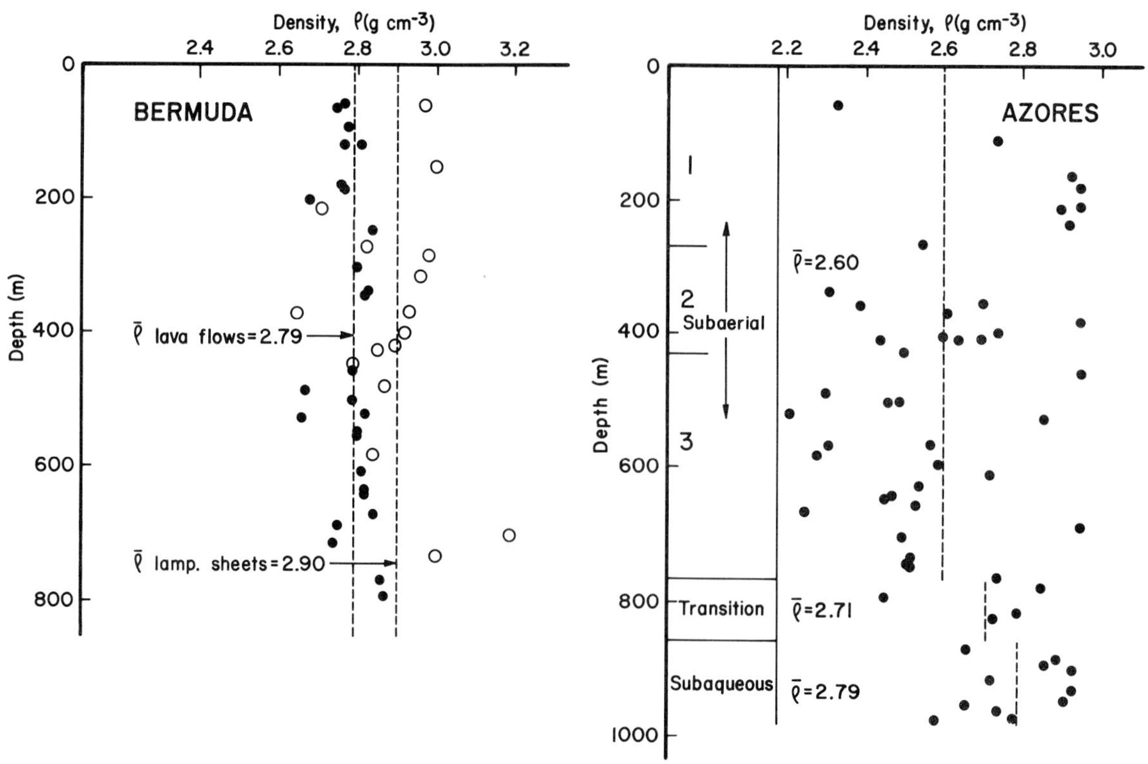

Fig. 4. Density as a function of depth in the Bermuda and Azores boreholes.

mean density of 2.80 for the seamount structure itself with lower density material, 2.50 g cm^{-3} of several km thickness on the adjacent sea floor.

On the basis of our sample densities from Bermuda and the Azores and the above gravity estimates for seamounts and islands we conclude: The mean density for the bulk of oceanic volcanic islands and seamounts is about 2.8 g cm^{-3}, which is appropriate for subaqueous pillow lava flows with a few higher density dykes and sills. Density means as low as 2.3 g cm^{-3} and commonly 2.6 g cm^{-3} exist for the subaerially erupted volcanics which make up the upper part of active volcanic islands. As occurs in São Miguel, subaerially erupted volcanoes may extend to a kilometer or more below sea level in an active volcanic island because of loading subsidence (see Muecke et al., 1974). As in the case for Bermuda, most old seamounts will have the subaerial volcanics removed by erosion, so that only the higher density subaqueous volcanics remain (see discussion in Hyndman et al., 1974a). A thick low density coral cap, of course, exists on the tops of many shallow sea mounts. The fairly low density material apparently on the seafloor surrounding many islands and seamounts, probably consists of pyroclastics and volcanic sediments such as found in our Azores borehole transition sequence. The high density of the intrusive lamprophyric sheets could be a useful diagnostic parameter for seamounts or aseismic ridges formed by later intrusions into older seafloor, compared to those made up primarily by subaqueous lava flows, but the contrast probably is not sufficient to be detected by gravity measurements.

Porosity

Porosities (% pore volume) were determined by weighing the 2.5 cm dia. 5 cm long minicores seawater saturated (preserved in water from time of drilling) and after drying at abut 70°C under vacuum for two weeks. There was significant weight loss for the first 7 days but little thereafter. This treatment should remove most of the pore water except in very low permeability samples and it should produce only minor dehydration of hydrous minerals. However, we note that there probably is a continuum between free pore water and strongly bound water in hydrous minerals, some water being loosely bound. A small correction was made for the residual salt in the porosity estimate, assuming an original salinity of 35 ppt and that all of the salt was left behind on drying. We estimate the accuracy to be $\pm$ 10% to -5% of the value for porosity and water content.

The porosities of 20 samples from the Bermuda borehole range from 1.1 to 8.2% with a mean of 2.9 $\pm$ 0.3%. The mean for the lava flows of

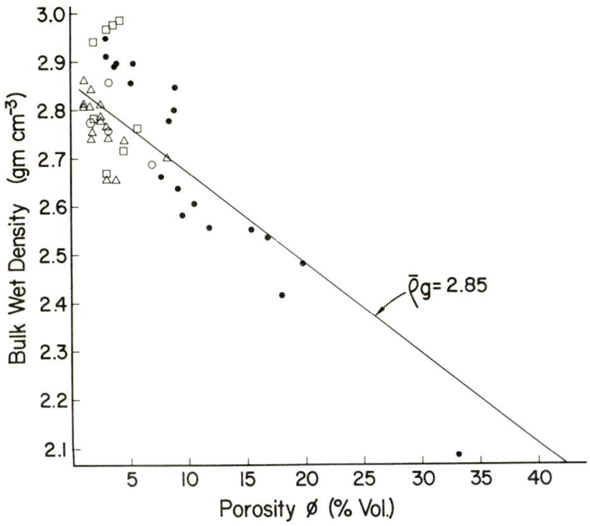

Fig. 5. Density as a function of porosity for Bermuda and Azores samples. (●) Azores subaerial; (◐) Azores transition; (○) Azores subaqueous; (△) Bermuda flows; (□) Bermuda sheets.

2.6 ± 0.4% is lower than the mean of 3.4 ± 0.4% for the sheets. These values are significantly lower than the means for fresh young seafloor basalts, (e.g. 7.8%, Hyndman and Drury, 1976), indicative of some pore space filling by secondary minerals during hydrothermal alteration in the Bermuda rocks.

The porosities of 19 samples from the Azores borehole range from 3.0 to 33.1% with a mean of 9.4 ± 1.5%. The mean of the subaerial section is 10.6 ± 1.7%, while that for 3 subaqueous samples is 4.4 ± 1.2%, slightly higher than for Bermuda lavas and lower than young subaqueously extruded tholeiitic lavas. The porosity is presented as a function of depth for both boreholes in Fig. 6.

The permeabilities of 14 Bermuda samples by Boyle's Law gas expansion in a Hassler Holder were measured by Core Laboratories Ltd., Calgary. The all are less than 0.01 millidarcys the resolution limit of the method, except for one sample cut by a fracture with a value of 7.7. Low values are also suggested by the low drying rates. Of course water flow in the oceanic crust must be primarily in cracks and fractures. The permeability of the Azores subaerial rocks probably is much higher.

Seismic Velocities

Compressional and shear wave velocities were measured on 2.5 cm diameter, approximately 5 cm long water saturated samples to pressures of 6 kbars. The technique is essentially as described by Birch (1960) and Christensen and Shaw (1970). The accuracy estimated is 0.5% for compressional and ± 1.0% for shear wave velocities. The fluid pressure medium was excluded from the samples by a thin copper foil jacket and pore pressures were maintained at values much less than external pressure by placing 100 mesh screens between the samples and jackets.

The usefulness of laboratory measurements depends on how well the insitu conditions are simulated, notably the pressure, temperature, extent of water saturation and sample orientation if the rocks are anisotropic. It is also important to consider how representative the samples are of the section being investigated. In this study we wish to estimate the insitu velocities both in the sections penetrated by the boreholes and in other more general situations where the rocks may be similar such as in the upper few kilometers of an aseismic ridge. The velocities of most rocks increase significantly with increasing pressure, up to about 2.0 kbars (e.g. 7 km depth) primarily from the closing of microcracks and some pores. In addition to the effect of pressure insitu, the drilling process probably causes microcracks. These cracks should close at small confining pressure. We take the 0.4 kbar velocities to be representative of insitu values in the sections penetrated by the boreholes and the 2.0 kbar values as representative of depths greater than a kilometer in the crust. We

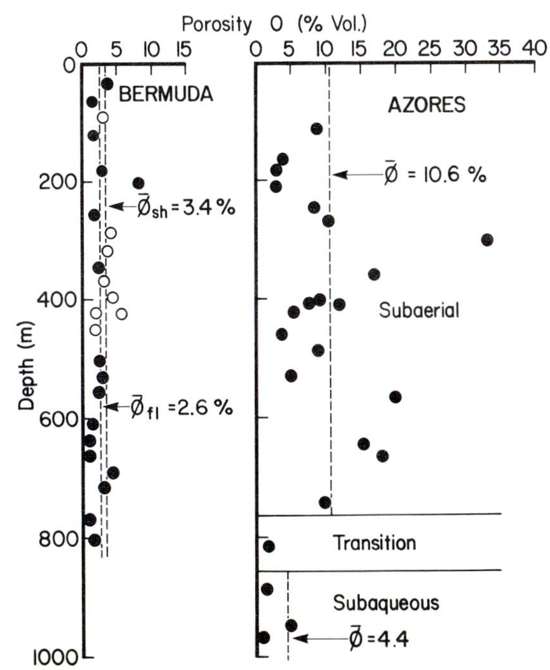

Fig. 6. Porosity as a function of depth in the Bermuda and Azores boreholes. Solid circles are Bermuda flows, open circles are Bermuda sheets.

TABLE 3. Mean Variations of Velocity with Pressure

	Pressure (Kbar) 0.2	0.4	0.6	0.8	1.0	2.0	4.0	6.0
	Bermuda Lava Flows (N = 12)							
Vp	5.95 ±.27	5.97 ±.27	6.00 ±.27	6.02 ±.27	6.03 ±.27	6.09 ±.26	6.17 ±.25	6.23 ±.24
Vs	3.20 ±.20	3.22 ±.20	3.23 ±.19	3.23 ±.19	3.24 ±.19	3.27 ±.18	3.30 ±.18	3.32 ±.17
σ	.296 ±.011	.295 ±.011	.296 ±.010	.297 ±.011	.294 ±.011	.298 ±.010	.299 ±.011	.301 ±.012
	Bermuda Lamprophyric Sheets (N = 8)							
Vp	5.86 ±.22	5.89 ±.21	5.92 ±.20	5.94 ±.20	5.96 ±.20	6.03 ±.20	6.13 ±.20	6.19 ±.20
Vs	3.05 ±.15	3.07 ±.15	3.08 ±.15	3.10 ±.15	3.11 ±.15	3.16 ±.15	3.21 ±.16	3.24 ±.17
σ	.314 ±.018	.314 ±.016	.314 ±.016	.313 ±.018	.313 ±.016	.311 ±.016	.310 ±.016	.311 ±.015
	Azores Subaerial (N = 15)							
Vp	4.88 ±.67	4.92 ±.66	4.95 ±.65	4.98 ±.65	5.00 ±.64	5.07 ±.62	5.21 ±.56	5.33 ±.52
Vs	2.45 ±.48	2.49 ±.46	2.53 ±.44	2.55 ±.43	2.57 ±.42	2.64 ±.38	2.73 ±.34	2.77 ±.31
σ	.332 ±.032	.328 ±.028	.325 ±.024	.324 ±.022	.319 ±.020	.314 ±.015	.311 ±.015	.313 ±.013
	Azores Subaqueous (N = 3)							
Vp	5.58 ±.30	5.60 ±.31	5.62 ±.31	5.63 ±.31	5.64 ±.31	5.70 ±.32	5.77 ±.29	5.83 ±.27
Vs	2.96 ±.17	2.97 ±.16	2.98 ±.16	2.98 ±.16	2.99 ±.15	3.01 ±.15	3.04 ±.15	3.06 ±.15
σ	.303 ±.006	.307 ±.006	.307 ±.006	.303 ±.006	.303 ±.006	.307 ±.006	.310 ±.000	.310 ±.000

stress that laboratory pressure can be related only approximately to equivalent depth in the crust. In our laboratory samples we have attempted to maintain close to zero pore pressure (see Hyndman and Drury, (1976) for a discussion of the importance of pore pressure), so the effective confining pressure is close to the external fluid pressure. The mean effect of pressure on velocities of Bermuda and Azores samples is given in Table 3 and Fig. 7.

Increasing temperature decreases the velocities of most rocks. Only a few measurements have been reported and over very limited temperature ranges. The effect on mafic rocks to 300°C ranges between -0.6×10^{-4} and -15×10^{-4} per °C (Birch, 1958; Hughes and Maurette, 1957; Nafe and Drake, 1968; Kroenke et al., 1976). Recent detailed measurements on basalts from the East Pacific Rise show -5×10^{-4} to be a representative value for oceanic basalts (Christensen, in preparation). Thus, the difference from a laboratory temperature of 25°C and the 200°C in the lower part of the Azores borehole (Fig. 3) requires a correction to the measured values of -0.1 km s^{-1}. The correction is negligible for the Bermuda borehole where the maximum is about 40°C (Hyndman et al., 1974a). The temperature in the crust under aseismic ridges and other old volcanic islands will rarely exceed the 200°C found in the Azores borehole. The temperature may be significantly higher at depth under active volcanic islands, requiring a greater correction.

All of the measured samples have been water saturated before measurement to simulate insitu conditions (Christensen and Salisbury, 1975). Velocities were measured in the vertical direction only since previous measurements have shown negligible anisotropy in extrusive oceanic volcanic rocks (e.g. Christensen and Shaw, 1970; Christensen and Salisbury, 1972).

Relations Between Velocity and Other Parameters

The relations between compressional and shear velocities and bulk density are important because they permit the estimation of density from seismic refraction data. Birch (1961) showed that there is one nearly linear velocity-density relation for all rocks with a particular mean atomic weight. The mean atomic weight of unaltered basaltic rocks varies only slightly so a single relation is expected. The relation might be different for altered rocks since the high water content associated with alteration will lower the mean atomic weight significantly perhaps without a corresponding decrease in bulk density. Christensen and Salisbury (1975) found a better fit with non-linear relations.

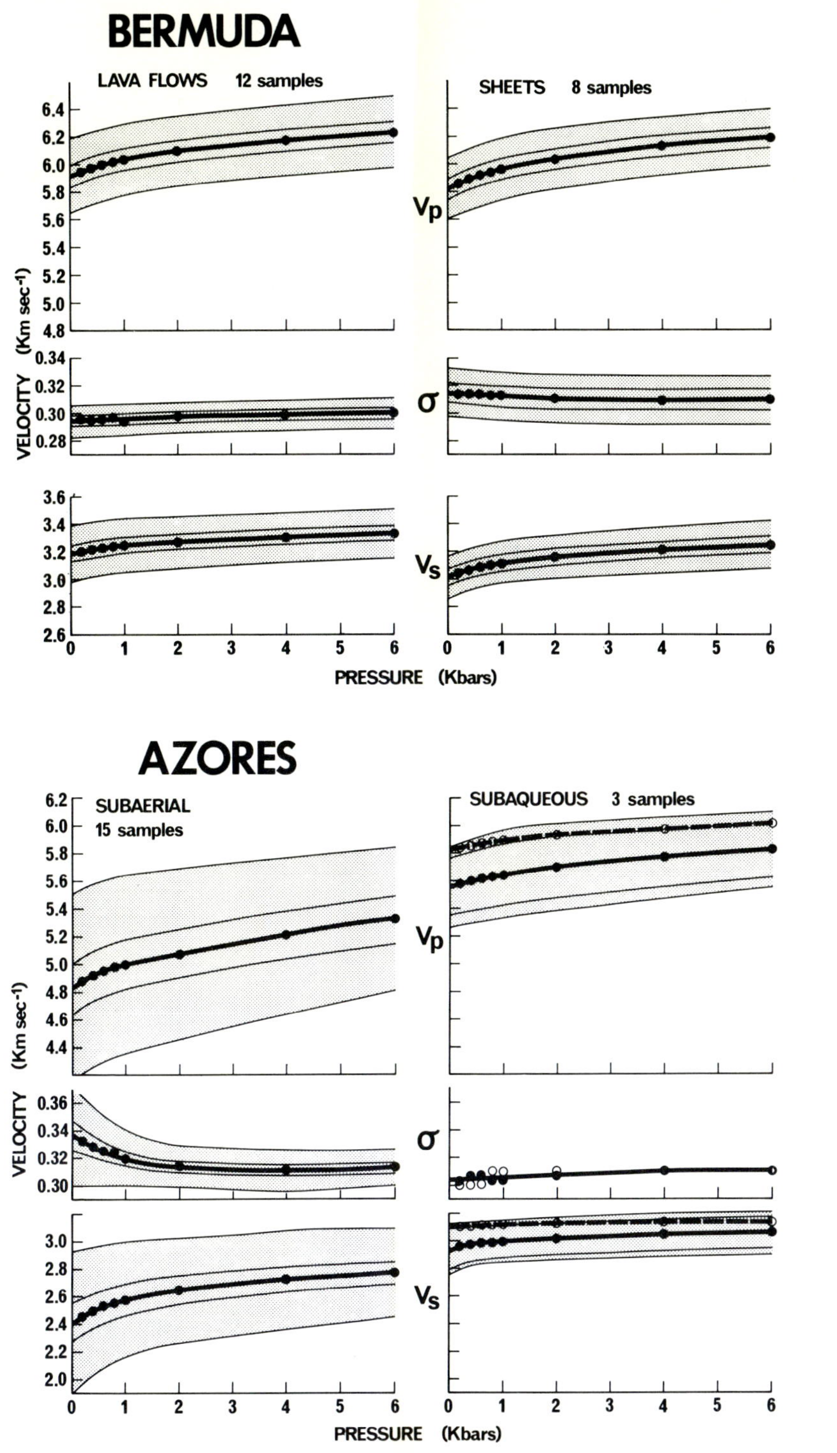

Fig. 7. Compressional and shear wave velocity as a function of pressure in the Bermuda and Azores samples. The inner bounds are the standard error estimate of the mean, the outer bounds are the standard deviation. The open circles are the Azores transition sample.

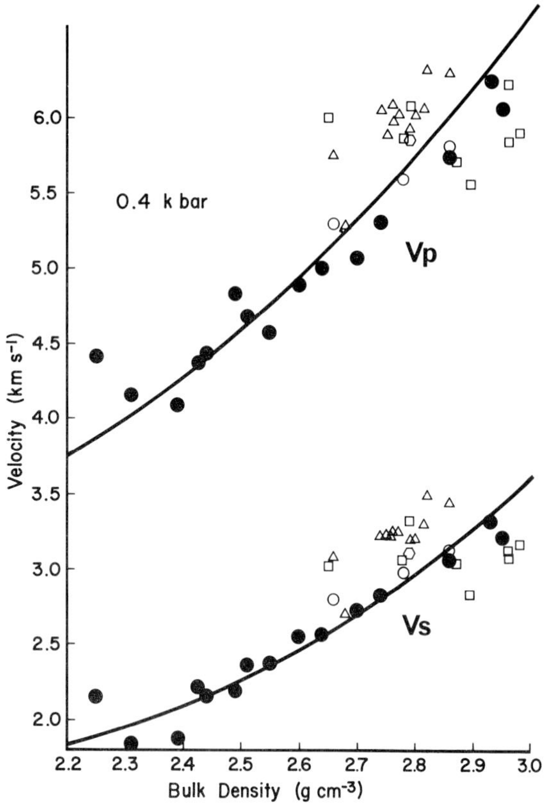

Fig. 8. Compressional and shear wave velocity as a function of bulk density for Bermuda and Azores samples. Symbols as in Fig. 5. The curves are from Christensen and Salisbury (1975).

The relation for the Bermuda and Azores samples is shown in Figure 8 along with the non-linear best fits for 77 sea floor basalts from the Deep Sea Drilling Project given by Christensen and Salisbury. An additional 79 samples from deep holes on the mid-Atlantic ridge fit their relations well although the shear velocities for the latter are slightly higher for a particular density (Hyndman and Drury, 1976). Only the Azores subaerial volcanics have a significant range of density but they define similar relations. The one transition and 3 subaqueous Azores samples also lie close to the lines. The lamprophyric sheet samples from Bermuda generally fit the relation although they scatter widely. In contrast, all of the altered tholeiitic lavas show much less scatter and have both compressional and shear velocities averaging 0.3 km s^{-1} above the relation of Christensen and Salisbury (1975). This difference may reflect a lower mean atomic weight produced by the extensive hydrothermal alteration (see Aumento and Ade-Hall, 1973), and possibly the filling of vesicles with alteration products. The contrast between the Bermuda lavas and fresh sea floor tholeiites is also evident, for example, in the high thermal conductivity (see below). It is important to note that the Azores subaerial samples, which have low velocities primarily because of high porosity in flows and pyroclastics, have the same velocity-density relation as the deep ocean samples reported by Christensen and Salisbury (1975) in which low velocities arise primarily from low temperature weathering. Thus, the relation should be valid for the upper oceanic crust, of any age of either normal ridge origin, or an origin associated with a hot spot.

The relation between Poisson's ratio (a measure of the ratio of compressional to shear velocity) and compressional velocity for Bermuda and Azores samples is shown in Figure 9. Only the Azores subaerial samples have a sufficient range to define a trend. Poisson's ratio increases systematically with decreasing velocity below about 5.0 km s^{-1}, in agreement with the general trend found for many seafloor tholeiitic basalts obtained by dredging or shallow drilling (Christensen, unpublished data; Hyndman, 1979). But the trend is in contrast to that of increasing Poisson's ratio with increasing velocity found by Hyndman and Drury (1976) for young fresh basalts from Deep Sea Drilling Project, 0.5 km deep holes on the mid Atlantic ridge, in the velocity range 5.5 to 6.5 km s^{-1}. The opposite dependences probably reflect the different nature of the porosity structure associated with the velocity variation. The young fresh tholeiites from the mid Atlantic ridge deep holes have porosity primarily in nearly spherical vesicles that are poorly connected, while in the Azores subaerial samples the porosity is primarily along well connected grain boundaries (see discussion in Hyndman, 1979).

On a plot such as Figure 10 (see below) the mid-Atlantic ridge rocks have a mean pore aspect ratio of 1/2 compared to about 1/10 for Azores subaerial volcanics. Two highly vesicular basalts from the Lau ridge with a 5.0 km s^{-1} compressional velocity (Christensen and Salisbury, 1975) also have a low Poisson's ratio of 0.27. In contrast, the low velocity samples from dredging or shallow drilling have porosity primarily from low temperature weathering which occurs along grain boundaries giving well connected spaces. Similarly, the low velocity Azores subaerial volcanics have well interconnected porosity, particularly in the pyroclastic sections and well fractured flows. We thus emphasize that variations in Poisson's ratio of unaltered basaltic rocks with velocity probably arise mainly from the porosity structure rather than from differences in mineralogy. Alteration products such as chlorite have high Poisson's ratios so Poisson's ratios of basalts may increase with very extensive hydrothermal alteration that also decreases the compressional velocity.

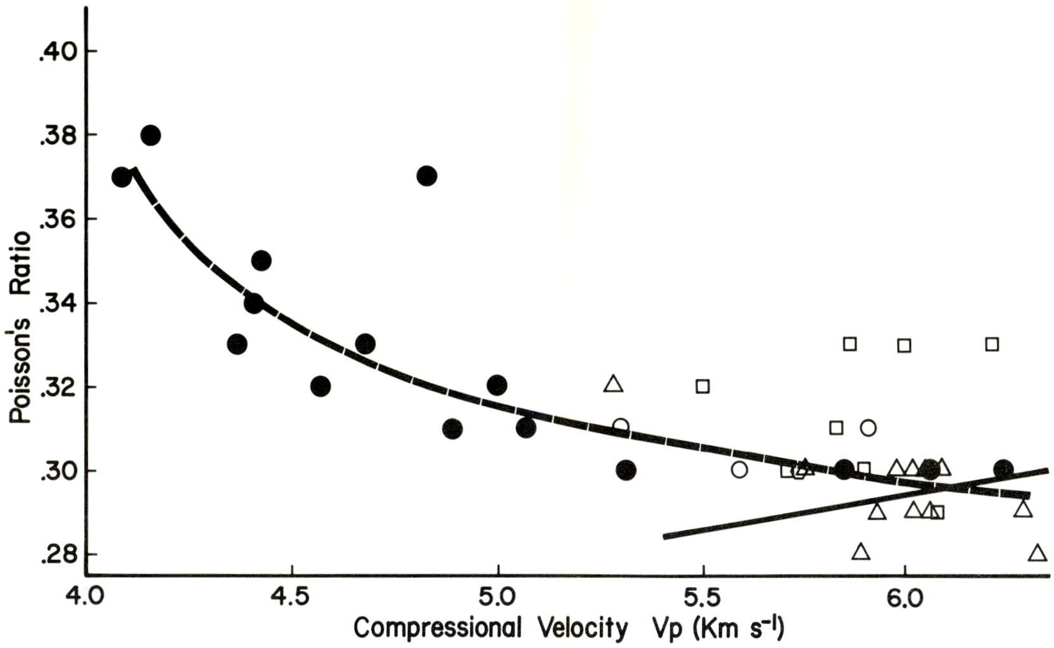

Fig. 9. Poisson's ratio as a function of compressional velocity for Bermuda and Azores samples at 0.4 kbar pressure. Symbols as in Fig. 5. The dashed line illustrates the general trend. The solid line is the relation of Hyndman and Drury (1976) for basalts from deep drilling on the mid-Atlantic ridge.

Figure 10 shows the relation between compressional velocity and porosity for the samples. A knowledge of the porosity and its form in the upper oceanic crust and in islands is important for an understanding of alteration processes, and for rough estimates of the permeability available for hydrothermal circulation processes (e.g. Francis, 1976; Whitmarsh, 1978). The relation may be useful for the estimation of upper crustal porosities from seismic refraction measurements, although much of the upper crustal pore space probably is in large fractures and voids so is not represented by small samples. The relation of Figure 9 is largely based on the data for the Azores subaerial rocks. A good correlation has been expected from the relation between velocity and density (Fig. 8) and the relation between density and porosity (Fig. 5). Figure 10 also shows the predicted velocity-porosity relations for various pore aspect ratios (or elongations) from the non-interaction theory of Toksöz et al. (1976) as applied to the upper oceanic crust by Whitmarsh (1978). A mean aspect ratio of 1/10 gives the best fit in rough agreement with the large scale upper crustal values suggested by Whitmarsh.

Mean Elastic Parameters For Drilled Sections

Compressional and shear wave velocities for Bermuda samples as a function of depth are shown in Figure 11. There is no systematic variation. There are, however significant differences in both velocities and in Poisson's ratio between the lava flows and lamprophyric

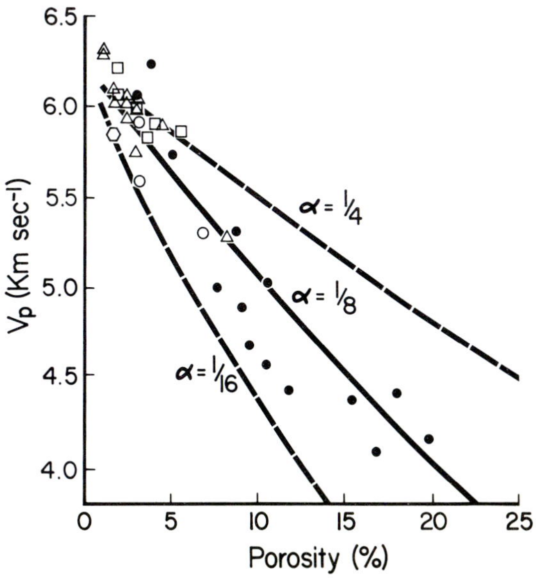

Fig. 10. Compressional velocity at 0.4 kbar pressure as a function of porosity. Symbols as in Fig. 5. The curves are the predicted relations for different pore aspect ratios from Whitmarsh (1978).

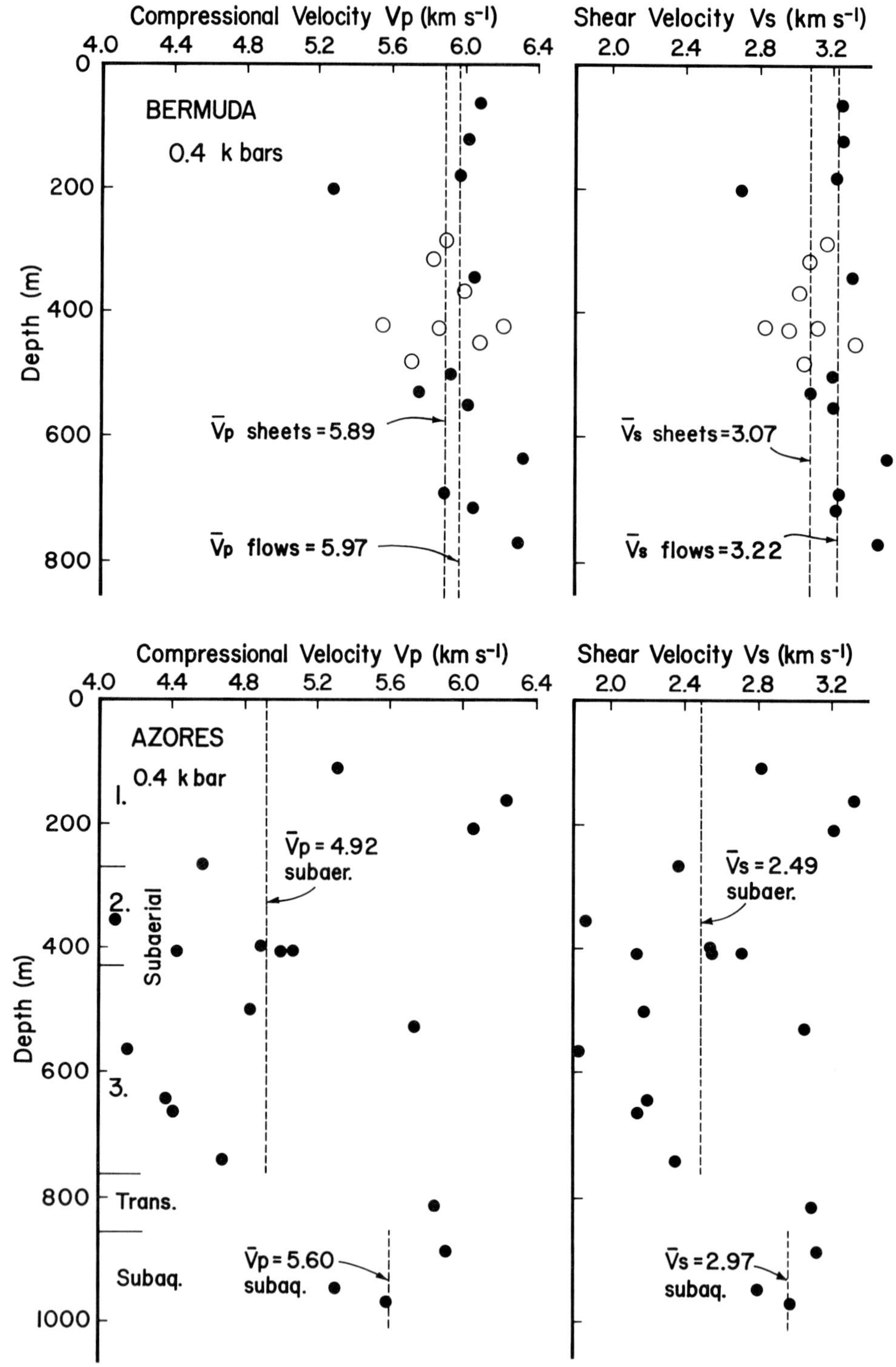

Fig. 11. Compressional and shear wave velocities at 0.4 kbar pressure as a function of depth in the Bermuda and Azores boreholes.

sheets. The mean compressional velocities are 5.97 ± 0.08 and 5.89 ± 0.07, mean shear velocities 3.22 ± 0.06 and 3.07 ± 0.05 km s^{-1} respectively. The mean Poisson's ratios are 0.295 ± 0.003 and 0.314 ± 0.006 (Table 1). The velocities for the flows are higher and Poisson's ratios lower than for the sheets even though the flow densities are lower (Fig. 4). Christensen and Salisbury (1975) show how appropriate differences in mineralogy can affect velocity and Poisson's ratio. The mean velocities for the lavas are higher than for most sea floor samples that have been measured, reflecting the low porosity of these hydrothermally altered rocks. The mean velocities and Poisson's ratio of the flows are almost identical to those for samples from deep drilling on the mid-Atlantic ridge (Hyndman and Drury, 1976). The mean compressional velocity of 5.94 km s^{-1} for the drilled section is somewhat higher than the velocity of 5.1 km s^{-1} estimated for under the Bermuda platform by Woollard (1954).

The compressional and shear wave velocities for the Azores samples as a function of depth are shown in Figure 11. There is no systematic variation with depth in the subaerial section which has a mean compressional velocity of 4.92, mean shear velocity of 2.49 km s^{-1} and mean Poisson's ratio of 0.328. The scatter of all parameters is large. The high Poisson's ratios are associated with low velocity (Fig. 9). The velocities of the 3 subaqueous samples are significantly higher than those for the subaerial samples, the mean compressional velocity being 5.60, mean shear velocity 2.97 km s^{-1} and Poisson's ratio 0.307. The latter is typical of the higher velocity subaerial samples, but is slightly higher than for either fresh sea floor basalts (e.g. Hyndman and Drury, 1976) or the hydrothermally altered Bermuda subaqueous lava flows of 0.295. The mean velocities of the Azores subaqueous samples also is slightly lower than for the latter two rock types. One sample from the transition zone has similar values to the subaqueous rocks.

Electrical Resistivity

The electrical resistivity of the Bermuda and Azores minicores were measured by applying a 0.5V, 10 or 50 Hz signal to the ends which had been painted with electrically conducting epoxy resin (see Hyndman and Drury, 1976). The precision of the measurements is better than 1% but variations in surface water makes the values reproducible only to about ± 10%. All of the data reported are for samples saturated with seawater except for one Bermuda sample measured dry to high temperature. The rocks beneath Bermuda and the Azores are undoubtedly water saturated. Seawater salinity probably is a good approximation for beneath Bermuda and for the subaqueous section penetrated by the Azores borehole. The salinity of the insitu fluid in the Azores subaerial section might be lower than that of seawater so that our measured resistivities may be too low. In order to relate the measured resistivities to variable insitu conditions and to investigate conduction mechanisms, the variation of resistivity with porosity, pressure and temperature were measured on a number of samples. The resistivity change with different degrees of dehydration and with varying signal frequency for some Bermuda and Azores cores have been presented by Drury (1977).

Resistivity as a Fuction of Porosity

The resistivity of rocks for which conduction is primarily through fluid filled pore spaces, is found to correlate closely with the rock porosity. The empirical relation is (Archie, 1942):

$$\rho = A\rho_f \, \phi^{-n}$$

where ρ_f is the resistivity of the fluid, ϕ is the porosity and A and n are constants depending primarily on the extent of interconnection between the pore spaces. A is usually close to 1. If the porosity consists of randomly spaced spherical pores n is about 2. (Brace et al., 1965; Brace and Orange, 1968; Shankland and Waff, 1974). Values of n greater than 2 are possible if the pores are less than randomly interconnected.

The resistivity (at atmospheric pressure and 22°C) of all Bermuda and Azores samples correlates well with porosity (Fig. 12) with an n value of 2.4, and a seawater resistivity of 0.2 ohm-m, in agreement with thin sections which indicate most of the porosity to be in roughly spherical pores perhaps with less than random connection. Deep sea floor basalts also have shown high n values of 2.5 (Hyndman and Drury, 1976) suggesting less than random interconnection. The correlation clearly implies conduction primarily through pore fluid in the Bermuda and Azores rocks.

Resistivity as a Function of Pressure

The effect of pressure on resistivity provides important diagnostic information on rock conduction mechanisms. If pore fluid conduction dominates, resistivity should increase with increasing pressure as microcracks and pores are closed, reducing the available conduction paths. If conduction through mineral grains dominates, resistivity should decrease with increasing pressure, as the conducting grains are brought into more intimate contact (e.g. Brace and Orange, 1968). The effective pressure in either situation is the confining pressure, i.e. the external pressure minus the internal pore pressure.

The problem of relating laboratory sample confining pressure to depth in the crust is discussed by Hyndman and Drury (1976).

The variation of resistivity of seawater saturated samples from Bermuda was measured to 2 kbars pressure at 20 to 22°C. The sample cylinders are placed between two stainless steel electrodes of the same diameter and the assembly encased in heat-shrink plastic tubing to exclude the hydraulic fluid pressure medium. The electrodes have perforations leading to small reservoirs, that hold the fluid squeezed from the sample with increasing pressure. Thus, the pore pressure is close to zero and the confining pressure close to the external pressure.

The resistivity increases with increasing pressure for all of the samples measured indicating that pore fluid conduction is the dominant mechanism. The increase is by about 50% to a pressure of 0.5 kbar and 100% to 2.0 kbar (Fig. 13).

Resistivity as a Function of Temperature

The resistivity of most rocks changes rapidly with temperature. If conduction is through pore fluid the temperature dependence expected is of the form exhibited by the fluid itself. At moderate temperatures seawater resistivity decreases with increasing temperature (e.g. Horne, 1969). Its behaviour at higher temperatures probably is similar to NaCl solutions which have decreasing resistivity to about 300°C because of the increasing ionic mobility from decreasing viscosity. At still higher temperatures resistivity increases because of a decrease in ionic concentration due to density and association effects (e.g. Quist and Marshall, 1968). At very high temperatures or for dry rocks semiconduction through the mineral grains may dominate. The resistivity is then given by

$$\frac{1}{\rho} = \frac{1}{\rho_o} e^{-E/kT}$$

where ρ_o is the limiting resistivity at very high temperature, E is the activation energy (in electron volts) and K is Boltzmann's constant. Previous measurements of sea floor basalts have shown a more rapid decrease in resistivity with temperature than for sea water or saline solutions even though pore fluid conduction in the basalts was indicated by the dependence on porosity and on pressure. The effect might arise through conduction partially through clay minerals (Hyndman and Drury, 1976).

The resistivity of seawater saturated basalts from Bermuda at a pressure of 0.3 kbar was measured up to 250°C (Fig. 14). One sheet

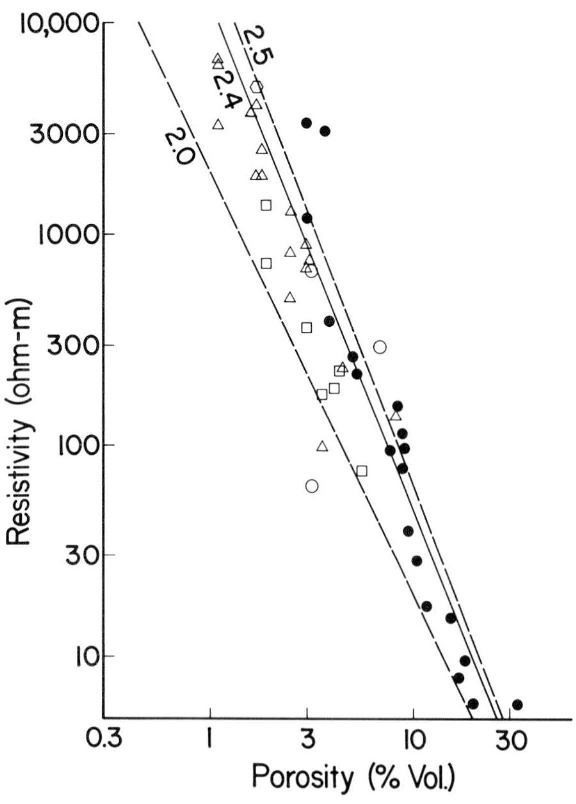

Fig. 12. Electrical resistivity as a function of porosity for Bermuda and Azores samples. Symbols as in Fig. 5. The lines are for different values of the exponent in Archie's Law.

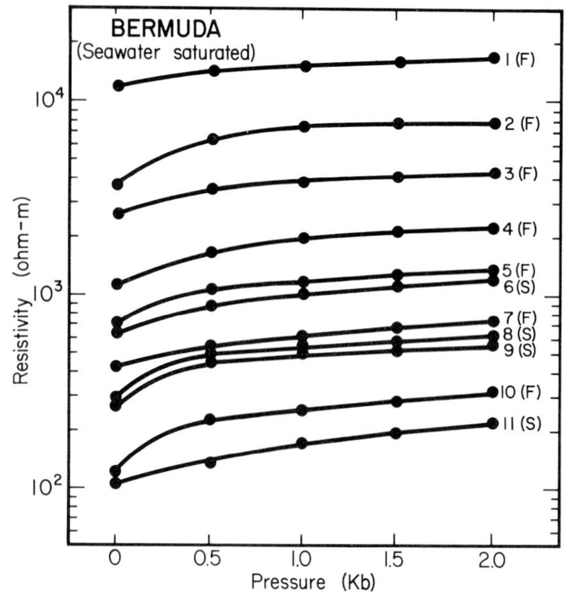

Fig. 13. Electrical resistivity as a function of pressure for Bermuda samples, at 23°C and seawater saturated.

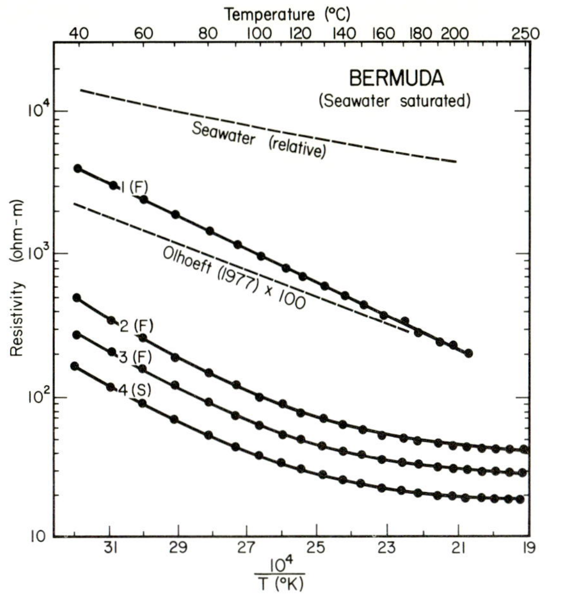

Fig. 14. Electrical resistivity as a function of temperature for Bermuda samples at 0.3 Kbar pressure, seawater saturated. The sample measured by Olhoeft (1977) had 14% porosity and was saturated with 1.7 ohm-m NaCl solution.

and two flow samples show a more rapid decrease in resistivity with increasing temperature than seawater at low temperatures, becoming similar to seawater above about 150°C. We suggest clay mineral conduction may be important in these samples (see Drury, 1977). One lava flow sample exhibits a linear decrease of log resistivity with inverse of temperature which suggests a semiconduction primary mechanism. This is a low porosity, high resistivity sample.

Clay minerals may be important to conduction in some basaltic rocks, producing through ion exclusion and cation exchange mechanisms, the more rapid change in resistivity with temperature than expected for pore fluid conduction. A more complex interaction appears to exist among pore fluid, clay, and mineral grain conduction to produce this behaviour (see the discussion in Drury, 1977). Clay mineral conduction probably occurs but is less pronounced for subaerial basalts than for subaqueous tholeiite pillows lavas. Olhoeft (1977) also found that some process in addition to simple pore fluid conduction was necessary to explain the effect of temperature on the resistivity of a Hawaiian tholeiitic basalt. (See Fig. 14).

One Bermuda sample was measured dry to 400°C (Fig. 15). It exhibits a linear dependence indicating semiconduction in the mineral grains. All of the pore water should have been removed and the clays dehydrated between 100 and 200°C (e.g. Deer et al., 1966). The activation energy is 0.5 eV and $\rho_o = 10^{-2}$ ohm-m, close to the values of 0.34 to 0.99 eV and 10^{-1} to 10^3 ohm-m found for deep ocean tholeiites by Schloessin and Dvorak (1976) to 1000°C. In the ocean crust mineral semiconduction probably dominates above about 300°C, if such temperatures occur. Pore fluid and clay mineral conduction dominate at lower temperatures.

Mean Resistivities for Drilled Sections

The mean resistivity of 23 Bermuda samples at 22°C and atmospheric pressure is 780 ohm-m with no trend with depth (Fig. 16). The mean for 17 lava flows is 1320 ohm-m and for 6 sheets is 340 ohm-m. The mean resistivity for the sheets is slightly lower reflecting their higher mean porosity. The mean resistivity is significantly higher than for deep ocean tholeiites recovered by the Deep Sea Drilling Project which have a mean of about 250 ohm-m (Hyndman and Ade-Hall, 1974; Hyndman and Drury, 1976; Drury, 1977). The higher resistivity of the Bermuda rocks reflects their reduced porosity resulting from hydrothermal alteration. The insitu mean resistivity of the drilled Bermuda section to 800 m depth should be close to the 800 ohm-m measured since the slightly higher average insitu temperature of about 30°C (Hyndman et al., 1974a) 8°C above

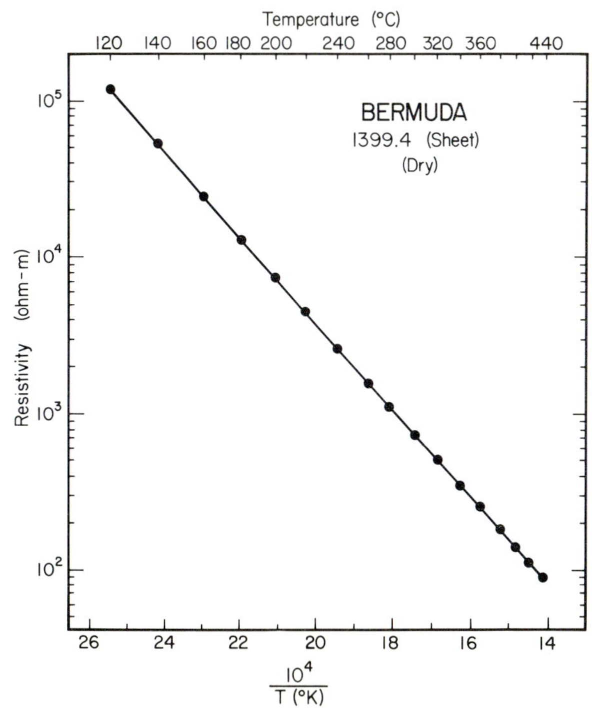

Fig. 15. Electrical resistivity as a function of temperature for one dry Bermuda sample at atmospheric pressure.

the laboratory temperature, will be about offset by the higher insitu pressure. At greater depth the resistivity should be significantly lower, the effect of the increasing temperature, 22°C km^{-1}, being more important than the effect of increasing pressure. At a depth of 5 km similar rocks would have a resistivity of about 200 ohm-m.

The mean resistivity of 23 Azores samples at 22°C and atmosphere pressure is 110 ohm-m. There is a general increase with depth (Fig. 16). The mean for 19 samples of the subaerial sequences is 82 ohm-m, for one transition zone sample it is 4900 ohm-m, and for 3 subaqueous samples it is 230 ohm-m. The very low resistivities of the subaerial rocks simply reflects their very high porosity. The subaqueous pillow lavas have resistivities very close to the mean for deep sea samples. The mean insitu resistivity of the drilled subaerial section should be about 5 ohm-m, significantly lower than the laboratory mean because of the much higher mean insitu temperature of about 140°C (Muecke et al., 1974). The subaqueous rocks under insitu conditions will have a resistivity of about 50 ohm-m. The temperature probably increases only slowly at depths greater than that penetrated by the borehole, unless a recent intrusion is approached, so the resistivity will be only slightly less than the 50 ohm-m,

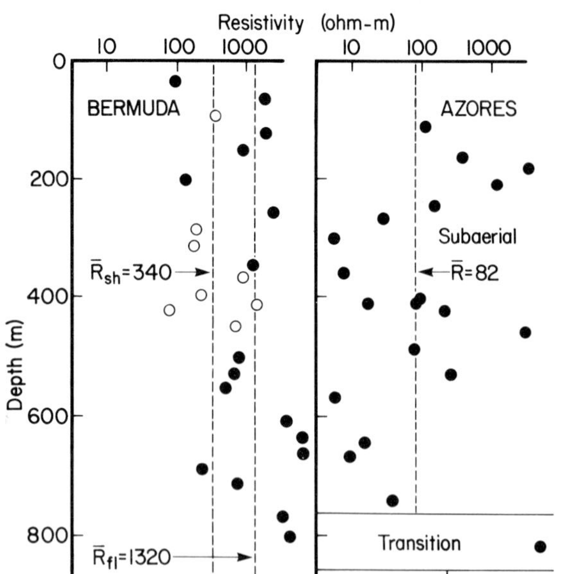

Fig. 16. Electrical resistivity as a function of depth in the Bermuda and Azores boreholes, for samples at 23°C and atmospheric pressure, seawater saturated. The dashed lines are the geometric means.

perhaps 10 ohm-m at 5 km, for similar pillow lavas.

Thermal Conductivity

The thermal conductivity of samples from the Bermuda and Azores boreholes was measured on 32 mm diameter, 10 mm thick disks with a divided bar instrument of the type described by Jessop (1970). The samples were water saturated, at a mean temperature of 25°C and subject to an axial pressure of 25 bars. Corrections were applied for disk-bar contact resistance, lateral heat loss and small variations in disk diameter. Calibration was with crystalline quartz and fused silica using the conductivity values of Ratcliffe (1959). Measurements on individual disks were reproducible to ± 2% and we estimate the accuracy to be better than ± 5%.

The mean thermal conductivity of 53 samples from the Bermuda core (Fig. 17), 32 flows and 21 sheets is 2.12 ± 0.03 W m^{-1} K^{-1} (5.07 ± 0.07 mcal cm^{-1} s^{-1} °C^{-1}). The mean of sheets of 2.19 ± 0.07 (5.23 ± 0.16) is significantly higher than the 2.08 ± 0.02 (4.96 ± 0.05) for the flows. The variation indicated by the standard deviation of 0.31 (0.74) also is much greater than that for the flows of 0.13 (0.31). The conductivity for the sheets appears to reflect primarily the content of pyroxene which has relatively high conductivity (e.g. Clark, 1966). The mean for the flows is also much higher than the 1.7 (4.0) representative of most previously measured fresh oceanic tholeiitic lavas (e.g. Hyndman and Drury, 1976). The higher conductivity in this case appears to be associated with the extensive hydrothermal alteration of the Bermuda lavas that probably occurred at the time of the intrusion of the lamprophyric sheets.

The mean thermal conductivity of 40 samples from the Azores borehole (Fig. 16) is 1.75 ± 0.04 W m^{-1} K^{-1} (4.19 ± 0.10 mcal cm^{-1} s^{-1} °C^{-1}). The mean of 31 subaerial samples is 1.75 ± 0.04 (4.18 ± 0.11) and of 5 subaqueous lavas is 1.73 ± 0.04 (4.13 ± 0.10). The subaerial volcanics have a similar conductivity to that for similar rocks from Hawaii with porosity from 5 to 10% (Robertson and Peck, 1974). The large variation (s.d. = 0.25 (0.59)) reflects primarily large variations in porosity, although the composition also is highly variable, (Muecke et al., 1974). The subaqueous lavas have a much smaller scatter (s.d. = 0.09 (0.22)). The mean is slightly higher than for most fresh tholeiitic flows, reflecting the significant hydrothermal alteration which was actually still in progress at the time of drilling (see temperature profile, Fig. 3). We conclude from both the Bermuda and Azores cores that hydrothermal alteration increases the

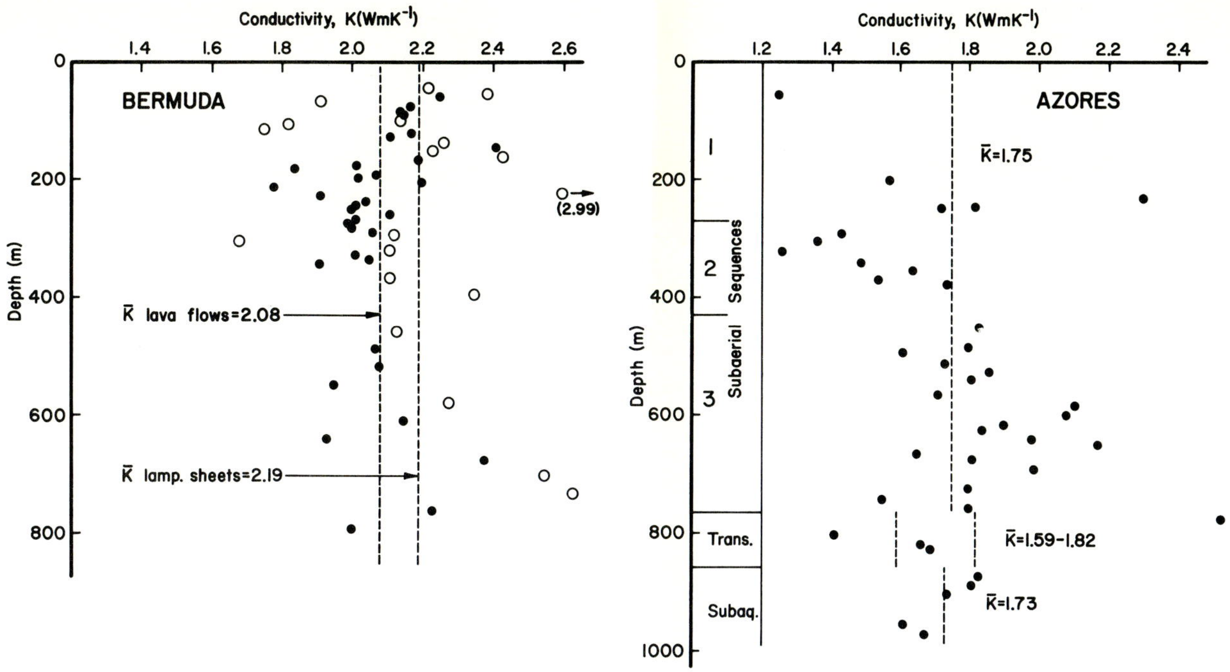

Fig. 17. Thermal conductivity as a function of depth for the Bermuda and Azores boreholes. The dashed lines are the simple arithmetic means.

conductivity of submarine tholeiitic lavas. In contrast, low temperature weathering appears to decrease the conductivity, in one case to 1.55 (3.7) (Hyndman et al., 1974b; Hyndman and Drury, 1976).

The effect of pressure on the thermal conductivity of basalts is quite small and can be neglected at least for upper crustal pressure of 0.5 kbar. The effect of temperature is significant being a decrease of about 0.054 (0.13) per 100°C increase (see review of data in Hyndman and Drury (1976)) implying a mean insitu conductivity at the bottom of the Azores borehole where the temperature is 200°C, of 1.64 (3.92). Such high temperatures and thus low thermal conductivity will occur locally in the upper crust of many young areas. The temperature effect can be neglected in old areas such as Bermuda at least for the upper few kilometers of the crust. The insitu conductivity may also be slightly lower than the sample means because of fractured sections of low conductivity, not represented in the core. In our study, we think these zones are important only for the upper part of the Azores borehole.

Conclusions

The physical properties of basalt samples from boreholes into the islands of Bermuda and the Azores (Tables 1 and 2) are generally similar to those from the deep ocean floor. The most significant differences are for samples from the Azores upper subaerial section. They have high porosity and as a consequence lower density, lower seismic velocities, higher Poisson's ratio, lower electrical resistivity and lower thermal conductivity as compared to deep ocean tholeiitic basalts. Mineralogical differences affect the measured physical properties but much less than differences in porosity. The lamprophyric sheets in the Bermuda section have higher density, lower seismic velocities, higher Poisson's ratio, and higher thermal conductivity than the lava flows from Bermuda or from the deep ocean floor. The hydrothermal alteration of the Bermuda lava flows appears to have reduced their density, particularly the grain density, reduced their porosity, increased their electrical resistivity and increased their thermal conductivity. This is assuming that their original properties before alteration were similar to samples from the deep ocean crust. The hydrothermal alteration apparently has not significantly affected the seismic velocities. The subaqueous Azores samples exhibit a similar but smaller difference compared to deep ocean lavas, probably from less pronounced hydrothermal alteration at the present 200°C insitu temperature.

The only physical properties of those measured, that are different for aseismic ridges and volcanic island seamount chains or 'hot spot traces' compared to normal oceanic crust, that might be detected by surface geophysical measurements are the slightly higher density and higher Poisson's ratio compared to the surrounding oceanic crust, suggested by the Bermuda samples.

Acknowledgements. We wish to acknowledge the efforts of Dr. F. Aumento who was primarily responsible for the drilling effort in Bermuda and Dr. J.M. Hall who organized the drilling in the Azores. Dr. M.J. Keen supported and encouraged both projects. D.L. Barrett undertook some of the early velocity measurements on the Bermuda core and was of considerable assistance in the setting up of some of the measurement equipment. J. Hull assisted in the velocity measurements and analysis of the data. The Bermuda drilling was financed by Dalhousie University, the National Research Council of Canada and by the Lamont-Doherty Geological Observatory. Financial support for the Azores drilling came from the U.S. National Science Foundation (International Decade of Ocean Exploration), the Research Corporation, the National Research Council of Canada, Dalhousie University, and the Empressa Insular de Electricidade (Ponta Delgada) S.A.R.L. Financial support for the measurements was provided by the Office of Naval Research Contract N-00014-75-C-0502 and the National Research Council of Canada. The drilling and some of the measurements were undertaken while R.D. Hyndman and M.J. Drury were at Dalhousie University.

References

Ade-Hall, J.M., W. Lowrie, N.D. Opdyke, and F. Aumento, Deep drill 1972: The paleomagnetism of a long succession of submarine lavas from Bermuda (Abstract), E.O.S., Trans. Am. Geophys. Un., 54, 485, 1973.

Archie, G.E., The electrical resistivity log as an aid in determining some reservoir characteristics, Trans. A.I.M.E., 146, 54-67, 1942.

Aumento, F. and J.M. Ade-Hall, Deep drill 1972: Petrology of the Bermuda drill core (Abstract), E.O.S., Trans. Am. Geophys. Un., 54, 485, 1973.

Aumento, F., W.S. Mitchell, and M. Fratta, Interaction between seawater and oceanic layer 2 as a function of time and depth - 1. Field evidence, Canadian Mineralogist, 14, 269-290, 1976.

Barrett, D.L., F. Aumento, J.M. Ade-Hall, and R.D. Hyndman, Deep drill 1972: Elastic properties of the rocks and the inferred oceanic crustal model (Abstract), E.O.S., Trans. Am. Geophys. Un., 54, 485, 1973.

Birch, F., Interpretation of the seismic structure of the crust in the light of experimental studies of wave velocities in rocks, in Contributions in Geophysics in Honor of Beno Gutenberg, 158-170, Pergamon, New York, 1958.

Birch, F., The velocity of compressional waves in rocks to 10 kbar, 1., J. Geophys. Res., 65, 1083-1102, 1960.

Birch, F., The velocity of compressional waves in rocks to 10 kbar, 2., J. Geophys. Res., 65, 2199-2224, 1961.

Brace, W.F., A.S. Orange, and J.R. Madden, The effect of pressure on the electrical resistivity of water-saturated crystalline rocks, J. Geophys. Res., 70, 5669-5678, 1965.

Brace, W.F. and A.S. Orange, Further studies on the effect of pressure on the electrical resistivity of rocks, J. Geophys. Res., 73, 5407-5420, 1968.

Christensen, N.I. and G.H. Shaw, Elasticity of mafic rocks from the mid-Atlantic ridge, Geophys. J., 20, 271-284, 1970.

Christensen, N.I., and M.H. Salisbury, Sea floor spreading progressive alteration of layer 2 basalts, and associated changes in seismic velocities, Earth Planet. Sci. Lett., 15, 367-375, 1972.

Christensen, N.I. and M.H. Salisbury, Structure and constitution of the lower oceanic crust, Rev. Geophys. Space Phys., 13, 57-86, 1975.

Clark, S.P., Jr., Thermal conductivity, in Handbook of Physical Constants, S.P. Clark, Jr. (Ed.), Geol. Soc. Am. Memoir 97, 459-482, 1966.

Deer, W.A., R.A. Howie, and J. Zussman, An Introduction to Rock Forming Minerals, Longman, London, 264-269, 1966.

Drury, M.J., The electrical properties of oceanic basalts, Unpublished Ph.D. thesis, Dalhousie University, Halifax, Canada, 1977.

Francis, T.J.G., The ratio of compressional to shear velocity and rock porosity on the axis of the mid-Atlantic ridge, J. Geophys. Res., 81, 4361-4364, 1976.

Gees, R.A., The age of the Bermuda seamount, Maritime Sediments, 5, 56-57, 1969.

Gees, R.A. and F. Medioli, A continuous seismic survey of the Bermuda Platform, Maritime Sediments, 6, 21-25, 1970.

Harrison, J.C. and W.C. Brisbin, Gravity anomalies off the west coast of North America, 1: Seamount Jasper, Bull. Geol. Soc. Am., 70, 929-934, 1959.

Horne, R.A., Marine Chemistry, Interscience, New York, 563 p., 1969.

Hughes, D.S. and C. Maurette, Variation of elastic wave velocities in basic igneous rocks with pressure and temperature, Geophysics, 22, 23-31, 1957.

Hyndman, R.D. and J.M. Ade-Hall, Electrical resistivity of basalts from D.S.D.P. leg 26, in Initial Reports of the Deep-Sea Drilling Project, vol. 26, U.S. Government Printing Office, Washington, D.C., 505-508, 1974.

Hyndman, R.D., G.K. Muecke, and F. Aumento, Deep drill 1972: Heat flow and heat production in Bermuda, Can. J. Earth Sci., 11, 809-818, 1974a.

Hyndman, R.D., A.J. Erickson, and R.P. Von Herzen, Geothermal measurements on D.S.D.P. leg 26, in Initial Reports of the Deep-Sea Drilling Project, vol. 26, U.S. Government Printing Office, Washington, D.C., 451-463, 1974b.

Hyndman, R.D. and M.J. Drury, The physical properties of oceanic basement rocks from deep-drilling on the mid-Atlantic ridge, J. Geophys. Res., 81, 4042-4052, 1976.

Hyndman, R.D., Poisson's Ratio in the oceanic crust - a review, Tectonophysics, in press, 1979.

Jessop, A.M., The effect of environment on divided bar measurements, Tectonophysics, 10, 39-49, 1970.

Kaula, W., Global gravity and mantle convection. Tectonophysics, 13, 341-359, 1972.

Kinoshita, W.J., A gravity survey of the Island of Hawaii, Pacific Sci., 19, 339-340, 1965.

Krause, D.C. and N.D. Watkins, North Atlantic crustal genesis in the vicinity of the Azores, Geophys. J., 19, 261-283, 1970.

Kroenke, L.W., M.H. Manghnani, C.S. Rai, P. Fryer, and R. Ramananantoandro, Elastic properties of selected ophiolitic rocks from Papua, New Guinea: Nature and composition of oceanic lower crust and upper mantle, in The Geophysics of the Pacific Ocean Basin and its Margin, S.H. Sutton, M.H. Manghnani and R. Moberly eds., Am. Geophys. Un. Monogr. 19, A.G.U., Washington, D.C., 407-421, 1976.

Larson, R.L. and W.C. Pitman, World wide correlation of Mesozoic magnetic anomalies, and its implications, Bull. Geol. Soc. Am., 83, 3645-3662, 1972.

Malahoff, A., Gravity anomalies over volcanic regions, in The Earth's Crust and Upper Mantle, P.J. Hart, ed., Am. Geophys. Un. Monograph 13, A.G.U., Washington, D.C., 364-379, 1969.

Manghnani, M.H. and G. Wollard, Elastic wave velocities in Hawaiian rocks at pressures to ten kilobars, in The Crust and Upper Mantle of the Pacific Area, L. Knopoff, C.L. Drake and P.J. Hart, eds., Am. Geophys. Un. Monogr. 12, A.G.U. Washington, D.C., 501-516, 1968.

McGraw, P.A., A petrological/geochemical study of rocks from the São Miguel, Azores. Unpubl. M.Sc. thesis, Dalhousie University, Halifax, Canada, 1976.

Morgan, W.J., Convection plumes in the lower mantle, Nature, 230, 42-43, 1971.

Muecke, G.K., J.M. Ade-Hall, F. Aumento, A. MacDonald, P.H. Reynolds, R.D. Hyndman, J. Quintino, N. Opdyke, and W. Lowrie, Deep drilling in an active geothermal area in the Azores, Nature, 252, 281-285, 1974.

Nafe, J.E. and C.L. Drake, Physical properties of rocks of basaltic composition, in Basalts, the Poldervaart Treatise on Rocks of Basaltic Composition, H. Hess, ed., Interscience, New York, 483-502, 1968.

Olhoeft, G.R., Electrical properties of water saturated basalt, preliminary results to 506K (233°C), U.S. Geol. Surv. Open File Report D-77-785, 7p, 1977.

Quist, A.S. and W.L. Marshall, Electrical conductances of aqueous sodium chloride solutions from 0 to 800°C and at pressures to 4000 bars, J. Phys. Chem., 72, 684-703, 1968.

Ratcliffe, E.H., Thermal conductivities of fused and crystalline quartz, Brit. J. Appl. Phys., 10, 22-25, 1959.

Reynolds, P. and F. Aumento, Deep drill 1972. Potassium-argon dating of the Bermuda drill core, Can. J. Earth Sci., 11, 1269-1273, 1974.

Robertson, E.I., Gravity survey in the Cook Islands, New Zealand J. Geol. Geophys., 10, 1484-1498, 1967.

Robertson, E.C. and D.L. Peck, Thermal conductivity of vesicular basalt from Hawaii, J. Geophys. Res., 79, 4875-4888, 1974.

Schloessin, H.H. and Z.D. Dvorak, Physical properties of samples from JOIDES, leg 37 Deep Sea Drilling Project in Initial Reports of the Deep Sea Drilling Project, Vol. 37, U.S. Government Printing Office, Washington, D.C., 403-416, 1977.

Shankland, T.J. and H.S. Waff, Conductivity in fluid-bearing rocks, J. Geophys. Res., 79, 4863-4868, 1975.

Shaw, H.R. and E.D. Jackson, Linear island chains in the Pacific: Result of thermal plumes or gravitational anchors, J. Geophys. Res., 78, 8634-8652, 1973.

Toksöz, N., C.H. Cheng, A. Timur, Velocities of seismic waves in porous rocks, Geophysics, 41, 621-645, 1976.

Walcott, R.I., Flexure of the lithosphere at Hawaii, Tectonophysics, 9, 435-446, 1970.

Watts, A.B. and J.R. Cochran, Gravity anomalies and flexure of the lithosphere along the Hawaiian-Emperor seamount chain, Geophys. J., 38, 119-141, 1974.

Watts, A.B., J.R. Cochran, and G. Selzer, Gravity anomalies and flexure of the lithosphere: a three-dimensional study of the Great Meteor seamount, northeast Atlantic, J. Geophys. Res., 80, 1391-1398, 1975.

Whitmarsh, R.G., Seismic refraction studies of the upper igneous crust in the North Atlantic and porosity estimates for layer 2, Earth Planet Sci. Lett., 37, 451-464, 1978.

Wilson, J.T., Evidence from islands on the spreading of ocean floors, Nature, 197, 536-538, 1963.

Wilson, J.T., Evidence from oceanic islands suggesting movement in the earth, in A Symposium on Continental Drift, P.M.S. Blackett, E.C. Bullard and S.K. Runcorn, eds, Phil. Trans. Roy. Soc. Lond. A, 258, 145-167, 1965.

Woollard, G.P., A gravity reconnaissance of the island of Oahu, Trans. Am. Geophys. Un., 32, 358-368, 1951.

Woollard, G.P., The crustal structure beneath oceanic islands, Proc. Roy. Soc. Lond. A., 222, 361-387, 1954.

Woollard, G.P., Crust and mantle relations in the Hawaiian area, in Continental Margins and Island Arcs, Can. Geol. Surv. Paper 66-15.

Worzel, J.L. and J.C. Harrison, Gravity at sea: in The Sea, Vol. 3, M.N. Hill, ed., Interscience, New York, 134-174, 1963.

Worzel, J.L. and M. Talwani, Gravity anomalies on seamounts, Bull. Geol. Soc. Am., 70, 1702-1703, 1959.

THE PHYSICAL STATE OF THE UPPER LEVELS OF CRETACEOUS OCEANIC CRUST FROM THE RESULTS OF LOGGING, LABORATORY STUDIES AND THE OBLIQUE SEISMIC EXPERIMENT AT DSDP SITES 417 AND 418

Deep Sea Drilling Project and Geological Research Division, Scripps Institution of Oceanography, University of California, San Diego, La Jolla, California 92093

Matthew H. Salisbury and Ralph Stephen[1]

Department of Geodesy and Geophysics, University of Cambridge, Cambridge, England

Nikolas I. Christensen

Department of Geological Sciences and Graduate Program in Geophysics, University of Washington, Seattle, Washington 98195

Jean Francheteau

Centre Oceanologique de Bretagne, B.P. 337, 29273 Brest Cedex, France

Yozo Hamano

Geophysical Institute, University of Tokyo, Tokyo 113, Japan

Michael Hobart

Lamont-Doherty Geological Observatory, Columbia University, New York, New York 10027

Douglas Johnson[2]

Program in Geosciences, University of Texas at Dallas, Richardson, Texas

Abstract. The combined results of logging, physical properties studies and the oblique seismic experiment conducted during DSDP Legs 51-53 at Sites 417 and 418 in Cretaceous crust at the southern end of the Bermuda Rise allow the first detailed evaluation of the physical state of the upper levels of old oceanic crust.

From first arrival and synthetic seismogram analysis of the results of the oblique seismic experiment, the P-wave velocity increases linearly from 4.8 ±0.2 km/sec at the top of layer 2 to 6.4 ±0.2 km/sec at a sub-basement depth of 1.3 km. The P-wave velocity of layer 3 at approximately 1.5 km is 6.7 ±0.2 km/sec. The S-wave velocity is 2.6 ±0.1 km/sec at the top of layer 2 and 3.7 ±0.1 km/sec at the top of layer 3.

The average value of V_p (4.8 km/sec) measured by logging in the uppermost basement in Hole 417D is in excellent agreement with the value obtained from the oblique seismic experiment, but is lower than the formation velocity (5.3 to 5.6 km/sec) reconstructed for the site from laboratory measurements of velocity through recovered core material. This requires that the formation contain cracks on a scale finer than the resolution of the logging and oblique seismic experiments but greater than that of laboratory samples. From these results and petrologic constraints imposed by the core, the upper crust in Hole 417D consists of 90% basalt with an average grain boundary porosity of 8%, less than 1% interpillow limestone, 5% smectite consisting of about 50% water, and 5% open cracks filled with standing water. The formation porosity thus resides in two domains, grain boundaries and open cracks, and totals 13-14%. This value is confirmed by electrical resistivity logs which indicate, in addition, that the cracks are interconnected, giving the formation an average permeability in the thousands of darcies, with lower values in the less fractured massive basalts.

A comparison of these results with logging data obtained in young crust in Hole 396B on the Mid-Atlantic Ridge indicates that although the porosity and permeability of the upper levels of the crust at Site 417 are much lower than at the ridge crest, the formation is still not entirely sealed. Although water circulation is thus still possible in old crust, it may be limited by the presence of massive basalts and the absence of shallow sources of heat.

---

[1] Now with the Department of Geology and Geophysics, Woods Hole Oceanographic Institution, Woods Hole, Massachusetts 02543

[2] Now with the Graduate Program in Geophysics, University of Washington, Seattle, Washington 98195

Introduction

During the past several decades, our perception of the structure of the oceanic crust (Table 1) has evolved from simple layered models derived from surface refraction data (e.g., Raitt, 1963), through multiple layered models based on OBS and sonobuoy data (e.g., Hussong, 1972; Peterson et al., 1974; Houtz and Ewing, 1976) to detailed velocity gradient models derived from velocity and amplitude analysis of reflection and refraction data (e.g., Helmberger and Morris, 1970; Orcutt et al., 1976; Spudich et al., 1978). Although the velocity structure of the crust can now be determined in considerable detail, the composition of the crust has remained largely conjectural beyond the limits of direct sampling because of the difficulties of interpreting velocities in terms of petrology in the absence of *in situ* data on the distribution of cracks and voids in the basement and their effects upon *in situ* pressure, water saturation and formation velocity.

During the past five years, several attempts have been made to drill deep into the oceanic basement in order to determine its petrology and geophysical behavior directly. Although the composition of the crust is known from such efforts to depths of more than 0.5 km at several sites in the Atlantic (Table 2), low core recovery has prevented reconstruction of its geophysical behavior.

Recently, two attempts have been made to measure the *in situ* physical properties of the crust by means of downhole geophysical experiments. During the first such experiment (Kirkpatrick, in press), a 200 meter section of basement was logged in DSDP Hole 396B in 10 m.y. old crust at 23°N on the Mid-Atlantic Ridge. Although the operating conditions were not ideal, the experiment nonetheless clearly demonstrated that *in situ* compressional wave velocities in young crust are consistent with those of Layer 2A (Houtz and Ewing, 1976) and that the porosity and fluid permeability of the uppermost levels of the basement are extremely high due to the presence of numerous water-filled cracks and voids.

A second such attempt, consisting of both downhole logging and the first successful application of the oblique seismic experiment (Stephen et al., in press), was recently completed in DSDP Hole 417D in conjunction with drilling conducted in 108 m.y. old crust at the southern end of the Bermuda Rise during DSDP Legs 51-53 (Figure 1). As can be seen in Figure 2 and Table 3, four holes were drilled during this program for a cumulative basement penetration of 1129 meters and a maximum penetration in one hole (418A) of 544 meters. The results of laboratory studies of the rocks obtained from these holes (Site 417 and 418 Reports in Bryan et al., in press; Christensen et al., in press; Hamano, in press; Johnson, in press *a*, in press *b*), together with those of the downhole logging and oblique seismic experiments in Hole 417D (Salisbury et al., in press; Stephen et al., in press), represent the most comprehensive body of geophysical data in existence on the *in situ* properties of the crust. The purpose of the present paper is to determine the physical

TABLE 1. MODELS OF OCEANIC SEISMIC STRUCTURE

3 Layer Model[1]			Multiple Layer Model[2,3]		
Layer	Velocity V_p, km/sec	Thickness, km	Layer	Velocity V_p, km/sec	Thickness, km
1	∿2.0	∿0.5	1	1.7 - 2.0	0.5
2	5.07±0.63	1.71±0.75	2A	2.5 - 3.8	0.5 - 1.5
			2B	4.0 - 6.0	0.5 - 1.5
3	6.69±0.26	4.86±1.42	3A	6.5 - 6.8	2.0 - 3.0
			3B	7.0 - 7.7	2.0 - 5.0
Mantle	8.13±0.24	--	Mantle	8.1	--

1) After Raitt, 1963.

2) After Peterson et al., 1974.

3) Houtz and Ewing (1976) have proposed dividing layer 2 into layers 2A, B and C having average velocities of 3.64, 5.19 and 6.09 km/sec, respectively.

TABLE 2. DEEP BASEMENT PENETRATION SITES IN THE ATLANTIC

Leg	Hole	Location	Basement Penetration (m)	Recovery in Basement (%)	Age m.y.	Downhole Experiments
37	332A[1)]	36°52.72'N, 33°38.46'W	230	10	3.5	
37	332B[1)]	36°52.72'N, 33°38.46'W	589	21	3.5	
37	333A[2)]	36°50.45'N, 33°40.05'W	300	8	3.5	
45	395A[3)]	22°45.35'N, 46°04.90'W	580	18	7	
46	396B[4)]	22°59.14'N, 43°30.90'W	256	23	10	Logging[5)]
49	409[6)]	62°36.98'N, 25°57.17'W	240	24	2.3	
51-53	417A[7)]	25°06.63'N, 68°02.48'W	209	61	108	
"	417D[7)]	25°06.69'N, 68°02.82'W	366	72	108	Logging[8)] Oblique Seismic Exp.[9)]
"	418A[10)]	25°02.10'N, 68°03.44'W	544	72	108	

1) Site 332 Report in Aumento, Melson et al., 1977.

2) Site 333 Report in Aumento, Melson et al., 1977.

3) Site 395 Report in Melson, Rabinowitz et al., in press.

4) Site 396 Report in Dmitriev, Heirtzler et al., in press.

5) Kirkpatrick, in press.

6) Site 409 Report in Cann, Luyendyk et al., in press.

7) Site 417 Report in Bryan et al., in press.

8) Salisbury et al., in press.

9) Stephen et al., in press.

10) Site 418 Report in Bryan et al., in press.

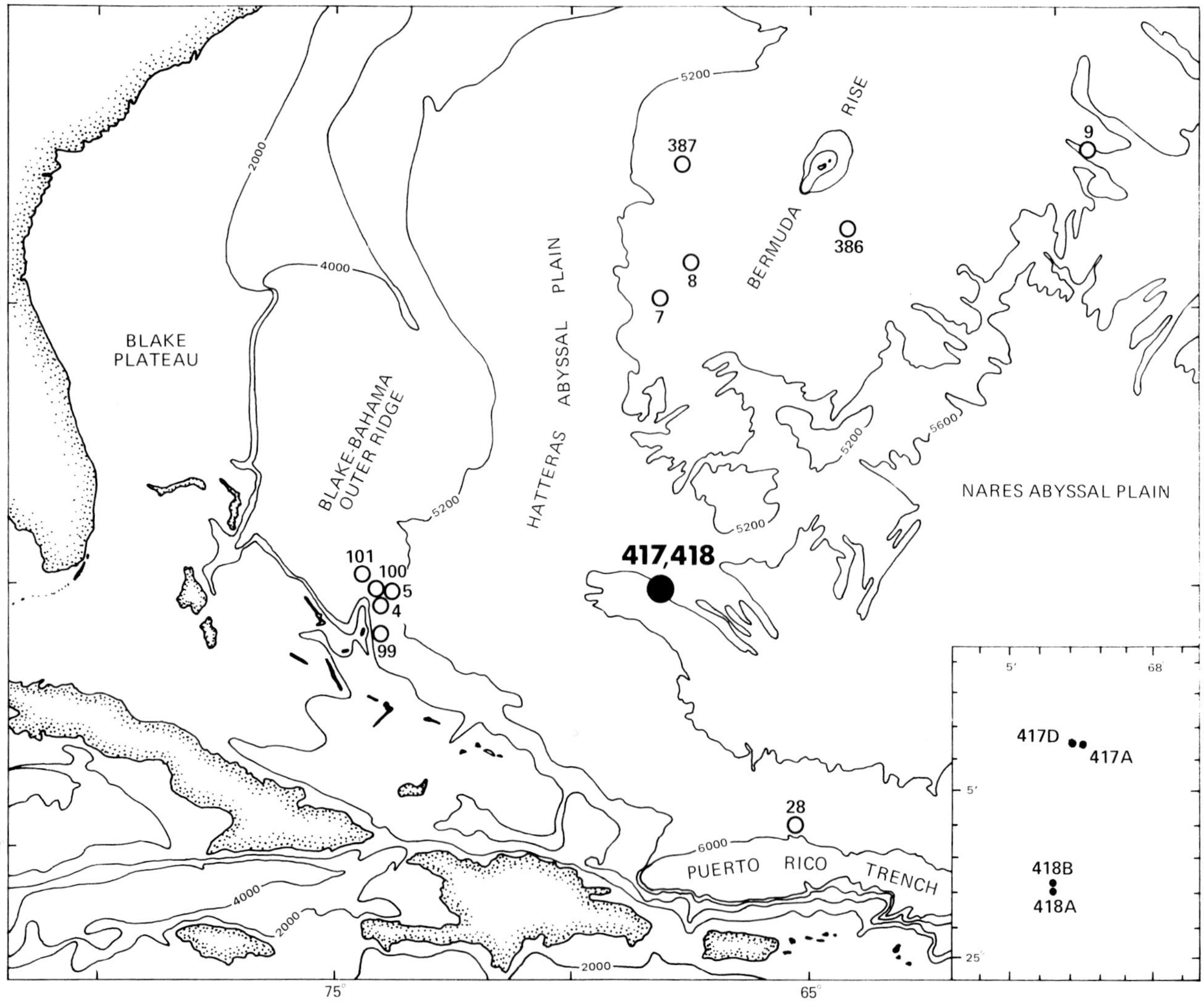

Figure 1 Location of DSDP 417A, 417D, 418A and 418B. Oblique seismic and logging experiments were conducted in Hole 417D.

state of Layer 2 from a review of this data and a comparison of results determined at different scales of investigation in the same hole.

Basement Lithology and Physical Properties

The section drilled at Sites 417 and 418 consists of approximately 300 meters of unconsolidated to semiconsolidated pelagic clay, claystone and chalk of Recent to early Aptian age underlain by pillow basalt, breccia and massive basalt, the latter cut by dikes in the bottom of the deepest holes (417D and 418A). The basalts are plagioclase-phyric with subordinate phenocrysts of olivine ± clinopyroxene and tend to be fairly uniform in composition. Despite their age, they are quite fresh with the exception of those recovered from the upper levels of Hole 417A, which are among the most profoundly altered basalts ever recovered from the sea floor (Donnelly et al., in press). Since the basal Aptian to Cenomanian section is missing in Hole 417A (Figure 2) and the hole was drilled into a small basement high, it has been suggested that the basement section in the hole was exposed to sea water for millions of years before its eventual burial (Site 417 Report in Bryan et al., in press).

As can be seen in Table 3, the basement material recovered at Sites 417 and 418 consists of 71% pillow basalt, 22% massive basalt and 7% breccia. Of this material, approximately 6% consists of smectite and calcite fillings in cracks and between pillows. Although studies of the distribution of cracks in the core (Johnson, in press b) suggest that the basement is strongly fractured in situ, the basement recovery at Sites 417 and 418 was unusually high (70%) due to partial healing of the crust by such material. This suggests

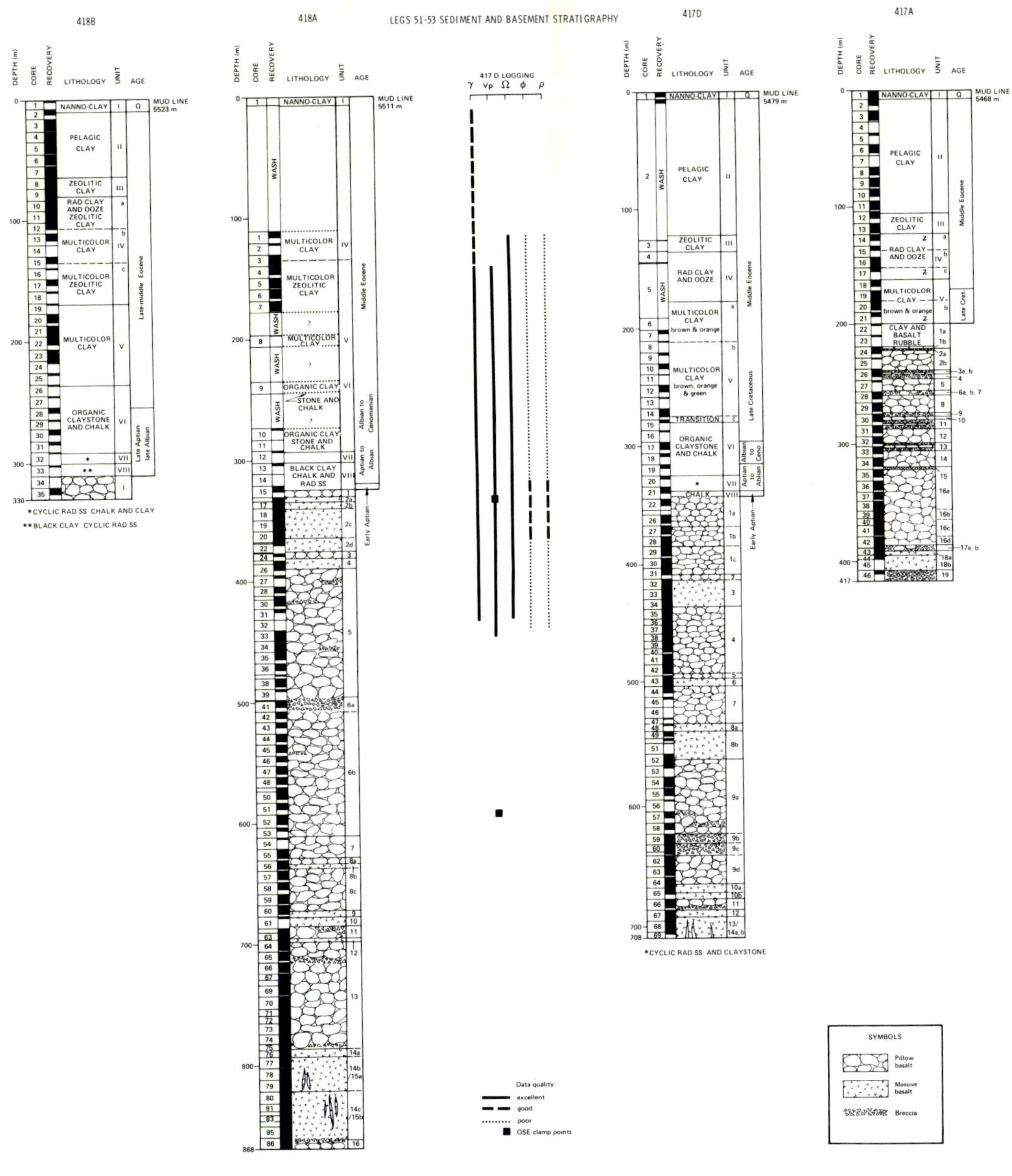

Figure 2 Lithology vs. depth in DSDP Holes 417A, 417D, 418A and 418B. Also shown are estimates of the quality of the logging data and the positions of the geophone during the oblique seismic experiment in Hole 417D. γ, V_p, Ω, ϕ and ρ stand for the natural gamma ray, velocity, resistivity, porosity and density logs, respectively.

that the figures cited in Table 3 are representative of the drilled section, and thus of the upper levels of layer 2 in old crust.

Detailed studies of the physical properties of the material recovered from these holes have been completed by several investigators and are reported elsewhere (Site 417 and 418 Reports in Bryan et al., in press; Christensen et al., in press; Hamano, in press; Johnson, in press a, in press b). These include measurements at room temperature and pressure of wet bulk density, porosity, compressional and shear wave velocity, permeability, thermal conductivity, and electrical resistivity as well as measurements of compressional wave velocity as a function of hydrostatic confining pressure. The values of wet bulk density, porosity and compressional wave velocity obtained at room temperature and pressure from these studies, together with calculated values of grain density and acoustic impedance, are shown as a

TABLE 3. BASEMENT LITHOLOGY AT DSDP SITES 417 AND 418[1)]

Hole	Drilled (m)	Recovery (%)	Recovered Lithology: Pillow Basalt (%)	Massive Basalt (%)	Breccia (%)
417A	209.0	61	75	9	16
417D	365.5	72	71	24	5
418A	544.0	72	69	27	4
418B	10.1	74	100	0	0
TOTAL	1128.6	70	71	22	7

[1)] Site 417 and 418 Reports in Bryan et al., in press.

function of depth for Hole 417A, Hole 417D and Holes 418A and 418B in Figures 3, 4 and 5 respectively. Also shown for each hole are synthetic physical property profiles based on the values shown and section-by-section estimates of the relative abundance of basalt, smectite and calcite. The density profile was reconstructed from the relation,

$$\rho_s = x_b \rho_b + x_{sm}\rho_{sm} + x_{ls} \rho_{ls} \quad (1)$$

where ρ_s is the mean density of the depth interval or section under consideration, x_b, x_{sm} and x_{ls} are the respective volumetric fractions of basalt, smectite and limestone in the interval, ρ_b is the nearest measured density value in basalt, ρ_{sm} is the mean density of smectite in the hole (2.20, 2.40 and 2.45 g/cm^3 in Holes 417A, 417D and both Holes 418A and B, respectively) and ρ_{ls} is the density of limestone (2.7 g/cm^3).

Similarly, the velocity, impedance and porosity profiles were determined as a function of depth from the relations,

$$V_{p_s} = \left[\frac{x_b}{V_{p_b}} + \frac{x_{sm}}{V_{p_{sm}}} + \frac{x_{ls}}{V_{p_{ls}}}\right]^{-1} \quad (2)$$

$$Z_s = (V_{p_s})\,(\rho_s) \quad (3)$$

and

$$\phi_s = x_b\phi_b + x_{sm}\phi_{sm} + x_{ls}\phi_{ls} \quad (4)$$

where V_{p_s}, V_{p_b}, $V_{p_{sm}}$ and $V_{p_{ls}}$ are respectively the compressional wave velocity of the section in question, the nearest measured velocity value in basalt, the mean smectite velocity in the hole (2.9, 3.8 and 4.0 km/sec in Holes 417A, 417D and both Holes 418A and B, respectively) and the velocity of limestone (5.9 km/sec); where Z_s is the acoustic impedance of the interval; and where ϕ_s, ϕ_b, ϕ_{sm} and ϕ_{ls} are respectively the porosity of the section, the nearest measured porosity value in basalt, the mean porosity of smectite in the hole (54% in Holes 417A and D and 20% in Holes 418A and B), and the porosity of limestone (equated to that of the nearest basalt). The profiles shown were calculated on a section-by-section basis, that is, in 1.5 m long increments roughly equivalent to the sensor spacing in down-hole logging tools. Since for the purposes of calculation, the formation was considered to be crack-free, the synthetic velocities, densities and impedances shown will be maximum values and the porosities will be minimum values.

A comparison between these profiles and the laboratory measurements upon which they are based indicates that the density and velocity measurements do not represent formation properties but rather maximum density and velocity envelopes for each hole, while the porosity values represent minimum envelopes. The discrepancy is largely due to sampling bias: although care was taken to sample each component (basalt, smectite or limestone) according to its overall abundance in the core, this was often impractical within specific lithologic units, with the principal result that the physical properties of the smectites are often inadequately represented in the laboratory data. The resulting discrepancy between laboratory values and the synthetic profiles can be substantial, particularly in zones of strong alteration. For example, the synthetic velocities of lithologic Units 4, 9 and 13 in Hole 417A are considerably lower than the

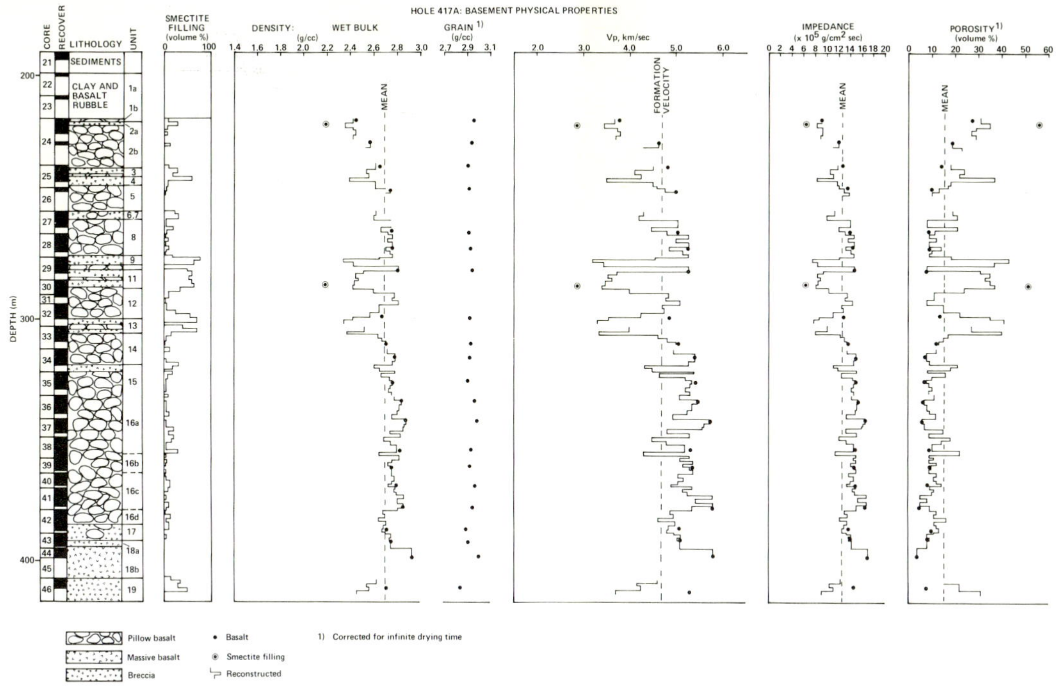

Figure 3 Lithology, interstitial fillings and physical properties vs. depth in DSDP Hole 417A. Synthetic profiles based on laboratory measurements and estimates of the relative abundance of basalt, smectitie and calcite in each section (see text).

laboratory velocities of the relatively unaltered samples which bracket these units above and below. Since the synthetic profiles are based on laboratory data but reduce the effects of sampling bias, they are considered to be a more realistic presentation of the data.

It is clear in Figures 3, 4 and 5 that many (but not all) of the units distinguished on lithologic grounds can also be delineated on the basis of physical properties. For example, in Hole 417A, Units 1-2 can be clearly distinguished from Units 3-4, 5-8, 9-11, 12, 13, 14-18 and 19, while in Hole 417D, Units 1-2 can be distinguished from Units 3, 4, 5-9b, 9c, 9d-11 and 12-14 and in Hole 418A, Units 1-4 can be distinguished from Units 5-6a and 6b-16. Among these subdivisions, the breccias are notable for their high smectite content (locally greater than 50% by volume) and consequent high porosity, low density and low velocity, while the massive basalts display low porosities, high densities and high velocities beneath a strongly altered, low velocity cap formed by prolonged interaction with sea water at elevated temperatures. Not unexpectedly, the physical properties of the pillow basalts tend to range irregularly between these two extremes. Although the units described above can be distinguished in terms of physical properties, it is unlikely that many can be detected geophysically. The only likely exceptions are the breccias which may be detectable as sub-basement reflectors because of their low acoustic impedances.

From the physical property data summarized in Table 4, it is clear that there is a pronounced difference between basalts recovered from holes drilled in topographic depressions (Holes 417D and 418A) and basalts from topographic highs (417A). The compressional wave velocities and wet bulk densities of samples from Holes 417D and 418A, for example, average 5.5 km/sec and 2.8 g/cm^3, respectively, while the porosities average 8.4%. These values are in marked distinction to those observed in Hole 417A, which average 5.0 km/sec, 2.7 g/cm^3 and 13.7%, respectively. Since the age and lithology of the holes are virtually identical, the differences in physical properties are attributed to differences in alteration due to varying exposure to sea water, a suggestion consistent with the presence of sharp physical property gradients in the upper levels of Hole 417A. If this conclusion is correct, it is clear from the more subtle, but no less striking gradients in Hole 418A (Figure 5), that the effects of alteration are felt to considerable depths in the crust; a projection of the laboratory data, for example, suggests that the porosity approaches zero and that the wet bulk density approaches the grain density (i.e., the physical properties approach intrinsic values) at about 700 meters sub-basement.

Although of considerable interest, it must be born in mind that the physical properties discussed to this point were measured on small samples which were free of large cracks. Since the velocities of fresh basalts measured in the laboratory tend to be considerably higher than the

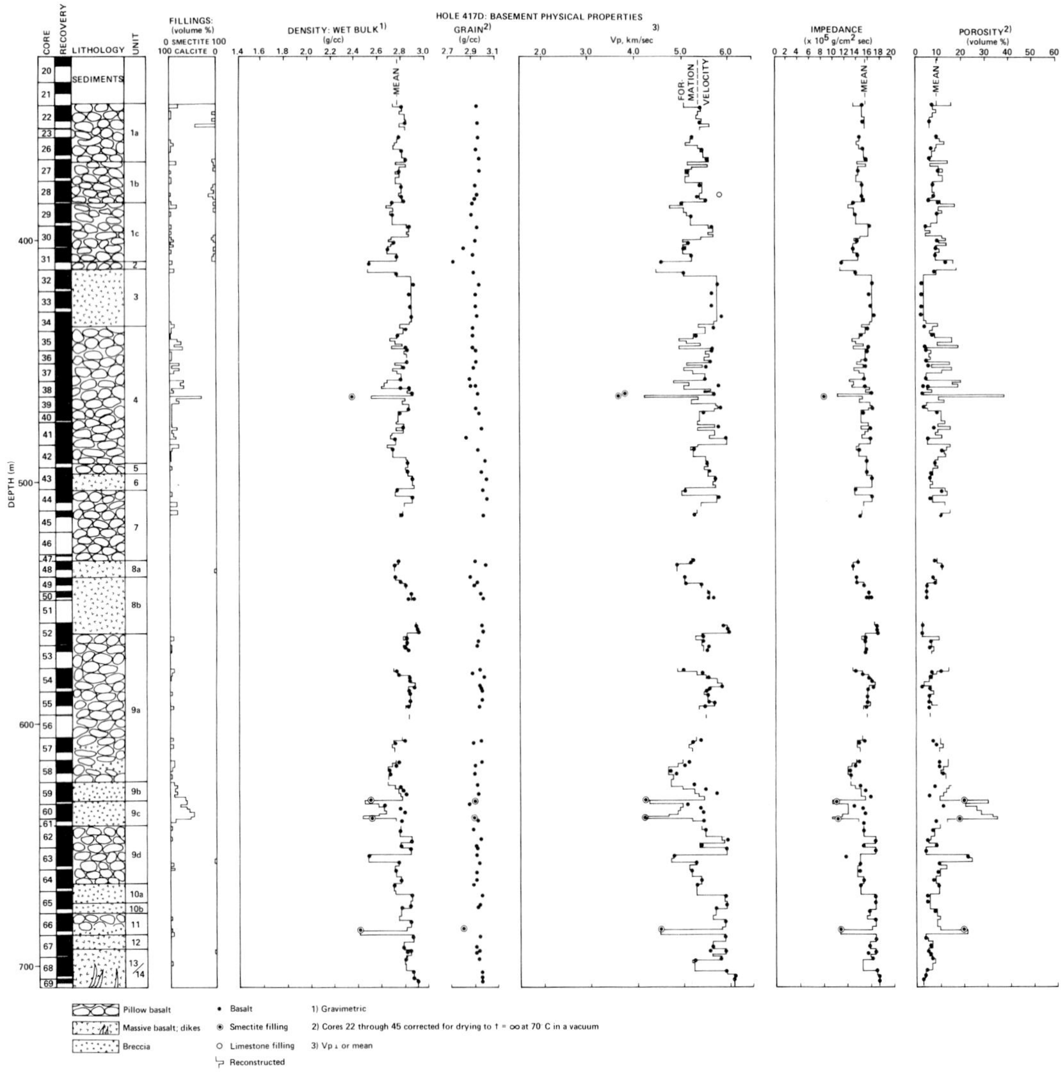

Figure 4 Lithology, interstitial fillings and physical properties vs. depth in DSDP Hole 417D. Synthetic profiles based on laboratory measurements and estimates of the relative abundance of basalt, smectite and calcite in each section (see text).

velocities observed for the upper levels of layer 2, it is clear that at least the upper levels of the crust are strongly fractured. To understand the significance of such cracks, we must examine data obtained at a larger scale of investigation.

Downhole Logging in Hole 417D

As can be seen in Figure 2, an extensive program of downhole geophysical logging was conducted in Hole 417D in an effort to obtain in situ data on the physical properties of both the sediment column and the underlying basement. Although the uppermost levels of the sediments could not be logged directly (the hole was cased by drill pipe to a depth of 114 meters and later, to 144 meters, to prevent caving), the open hole below the pipe was logged in four stages to an obstruction in the basement at 445 meters sub-bottom using Schlumberger high resolution temperature, natural gamma ray, density, porosity, velocity, resistivity and caliper tools in various combinations (Table 5). In addition, the uppermost sediments were logged through the pipe using the natural gamma ray and temperature tools. The results of this experiment, together with a discussion of the tools and their limitations are presented in detail elsewhere (Salisbury et al., in press). Only the results from the basement section logged between 343 and 445 meters sub-bottom will be discussed here.

Since many of the properties measured by logging are sensitive to temperature, pressure, the

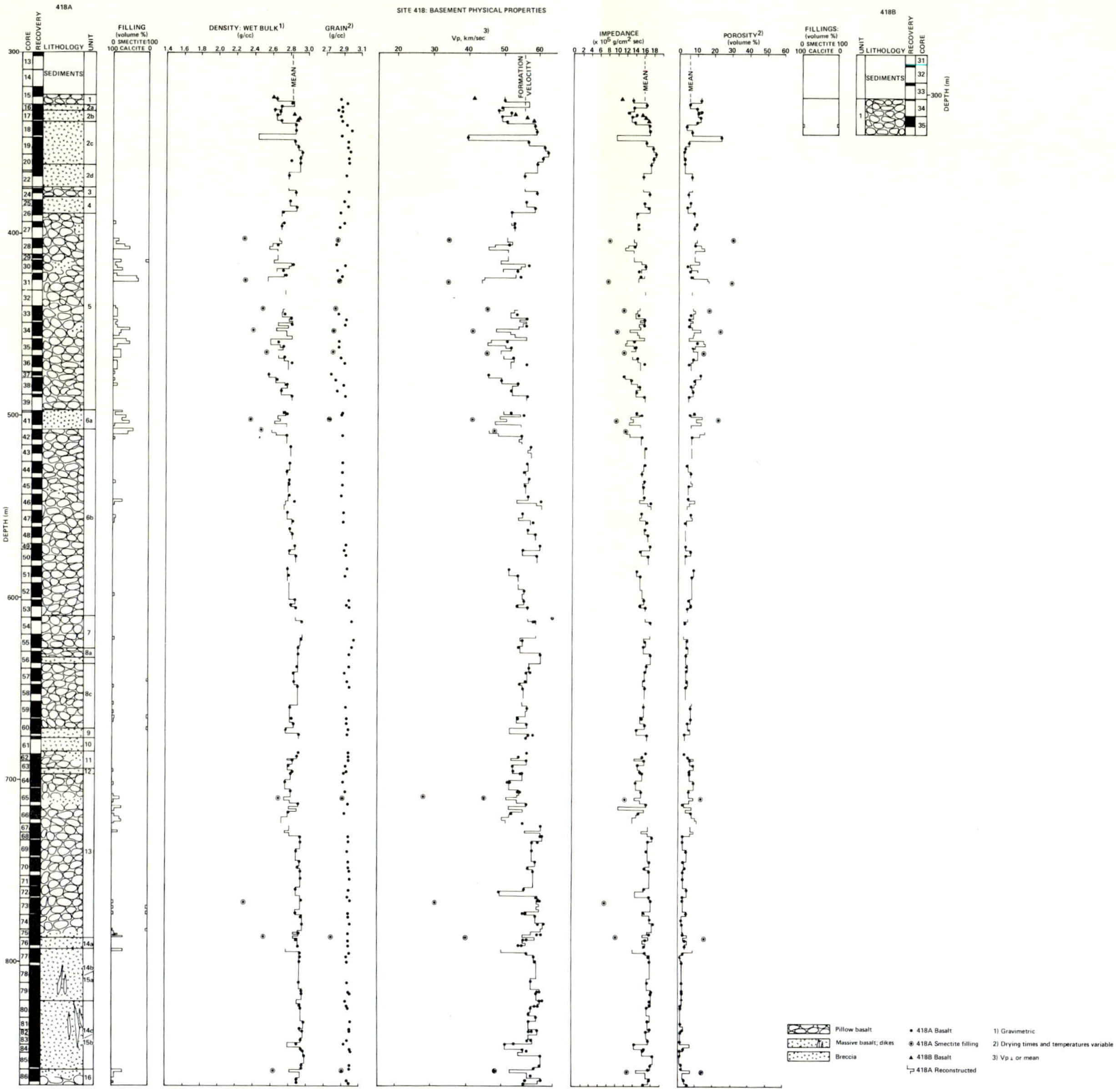

Figure 5 Lithology, interstitial fillings and physical properties vs. depth in DSDP Holes 418A and 418B. Synthetic profiles based on laboratory measurements and estimates of the relative abundance of basalt, smectite and calcite in each section (see text).

hole diameter and the properties of the fluid in the hole, the data obtained was corrected to in situ conditions assuming a 10" hole filled with sea water, an average hydrostatic confining pressure of 0.6 kbar, a temperature at the mudline of 2.4°C (from the temperature log), a temperature gradient of 5.4 x 10^{-2}C°/m in the sediments (also from the temperature log) and thus, a gradient of 2.2 x 10^{-2}C°/m in the basalts and a temperature of about 21°C at the top of layer 2 (as required by uniform heat flow). The corrected logging data for the basement, together with the lithology of the drilled section and visual estimates of the relative abundance of smectite and calcite are presented in Figure 6.

Although the basalts in Hole 417D all have essentially the same composition, it is clear from Figure 6 that most of the units and subdivisions delineated on the basis of lithology can also be distinguished on the basis of logging. The massive basalts of Unit 3, for example, have distinctly higher and more uniform velocities than the pillow basalts of Units 1, 2 and 4 and individual flows within the pillow basalts can often be dis-

TABLE 4. SITES 417 AND 418 AVERAGE PHYSICAL PROPERTIES[1)]

	Laboratory[2,3)]		Synthetic[2)]			
	Hole 417A	Holes 417D and 418A	Hole 417A	Hole 417D	Hole 418A	Hole 418B
Penetration (m)	209	910	209	366	544	10
Interstitial Fillings (Volume %)	13.2	4.2	13.2[4)]	5.7[5)]	3.6[6)]	1.5[7)]
Wet Bulk Density, ρ (g/cc)	2.70±0.19 (N = 23)	2.79±0.13 (N = 185)	2.69	2.77	2.83	2.83
Grain Density (g/cc)	2.94±0.03 (N = 21)	2.91±0.04 (N = 104)	--	--	--	--
Porosity (Volume %)	13.7±13.6 (N = 23)	8.4±5.5 (N = 104)	15.4	9.6	5.4	--
Velocity, V_p[8)] (km/sec)	4.99±0.79 (N = 23)	5.48±0.48 (N = 189)	4.68	5.35	5.58	5.31
Velocity, V_s[8)] (km/sec)	--	3.10±0.21 (N = 65)	--	--	--	--
Impedance, $V_p \times \rho$ ($\times 10^5$ g/cm^2-sec)	13.6±2.8 (N = 23)	15.3±2.1 (N = 185)	12.6	14.8	15.8	15.0
Poisson's Ratio	--	0.28±0.01 (N = 65)	--	--	--	--
Electrical Resistivity[8)] (ohm-m)	--	120 (N = 48)	--	--	--	--
Air Permeability (cm^2)	--	1.1×10^{-16} (N = 16)	--	--	--	--
Thermal Conductivity (mcal/cm-sec °C)	--	4.31±0.17 (N - 64)	--	--	--	--

1) Assuming no large scale cracks

2) From Site 417 and 418 Reports in Bryan et al., in press

3) From Hamano, in press

4) Smectite

5) 5.0% smectite, 0.7% limestone

6) 3.5% smectite, 0.1% limestone

7) 0.8% smectite, 0.7% limestone

8) Samples water saturated, at room temperature and pressure

tinguished by a sharp decrease in density, resistivity and velocity and a marked increase in porosity and natural gamma ray activity associated with brecciation, alteration and potassium enrichment along their contacts (Donnelly et al., in press).

For the present purpose, however, the most important application of the basement logs is the determination of the *in situ* physical properties of the crust for comparison with laboratory data and with geophysical data obtained from the surface. Since there appears to be no formation

TABLE 5. 417D GEOPHYSICAL LOGGING RUNS

Run	Tools	Scale of Investigation (m)	Depth Interval (m sub-bottom Sediments/Basement	Data Quality	Remarks
1	High Resolution Temperature (HRT)	--	0 - 44	Fair	Through pipe
			44 - 46	Fair	Open hole; terminated by caving
2	Borehole Compensated Velocity (BHC)	0.6	144 - 343 - 445	Excellent	Open hole
	Natural Gamma Ray	0.15	0 - 144	Fair	Through pipe
			144 - 343 - 445	Excellent	Open hole
	Caliper	--	--	Poor	Signal lead broken, spot readings only
3	Gamma Ray Density	0.15	320 - 343 - 369	Fair	Open hole
			144 - 343 - 443	Poor	Open hole; excentralizer broken; data not shown
	Neutron Porosity	0.15	320 - 343 - 369	Fair	Open hole
			114 - 343 - 443	Poor	Open hole; excentralizer broken
	Natural Gamma Ray	0.15	114 - 343 - 443	Fair	Open hole
4	Electrical Resistivity				
	Induction Log Medium (ILM)	1.5	117 - 343 - 434	Fair	Open hole; no centralizer
	Induction Log Deep (ILD)	2.5	117 - 343 - 434	Excellent	Open hole; no centralizer
	Laterolog 8 (LL8)	1.0	117 - 343 - 434	Excellent	Open hole; no centralizer
	Natural Gamma Ray	0.15	117 - 343 - 434	Fair	Open hole

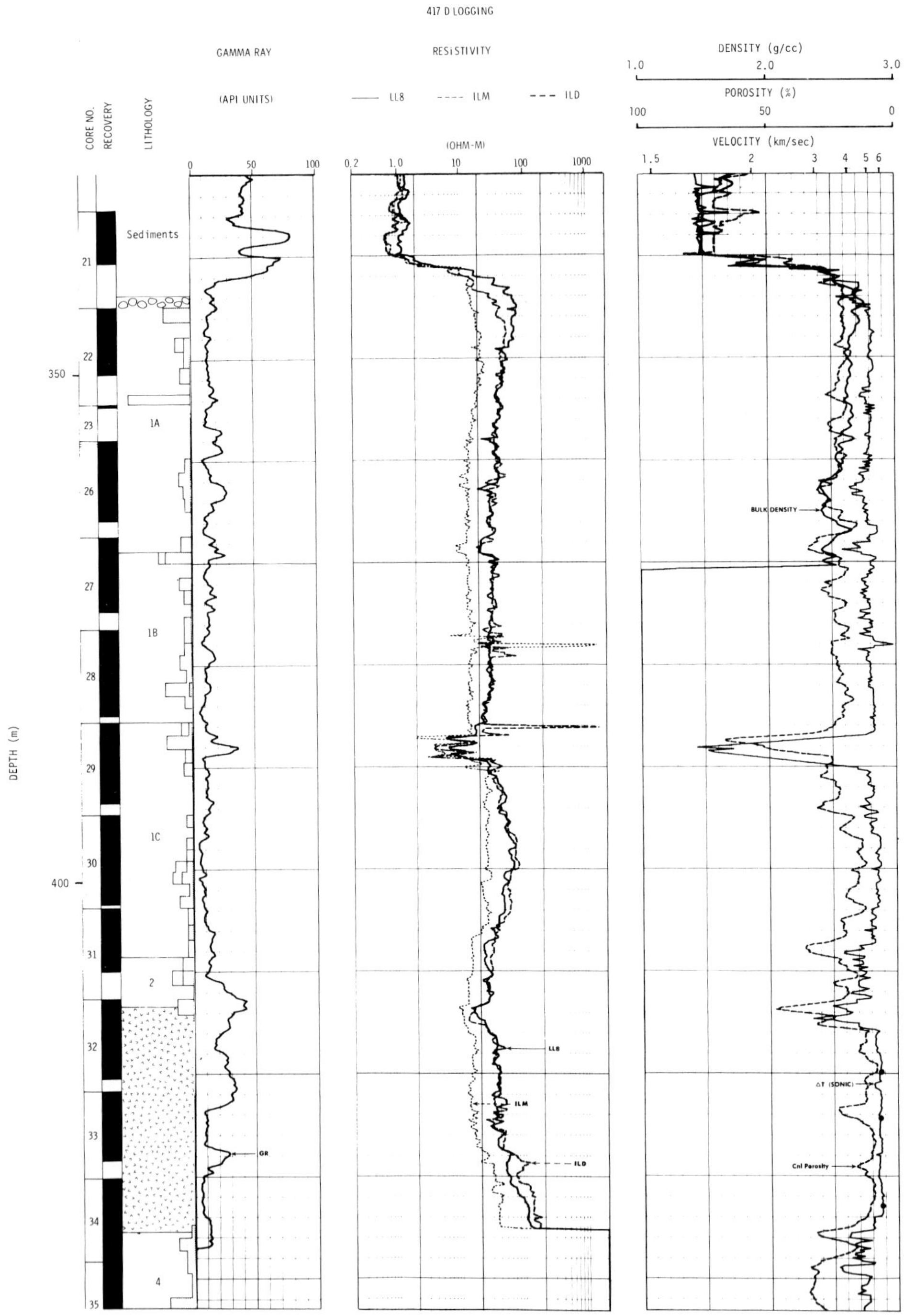

Figure 6 Basement logging data and lithology vs. depth in DSDP Hole 417D. Units 1, 2, and 4 are composed of pillow basalt while Unit 3 (hachured) is composed of massive basalt; insets in lithology column represent abundance of limestone (plain) and smectite (light gray) on a scale of 0-50%; circles (•) on velocity curve in Unit 3 represent laboratory velocities (V_p) at an effective confining pressure of 0.1 kbar.

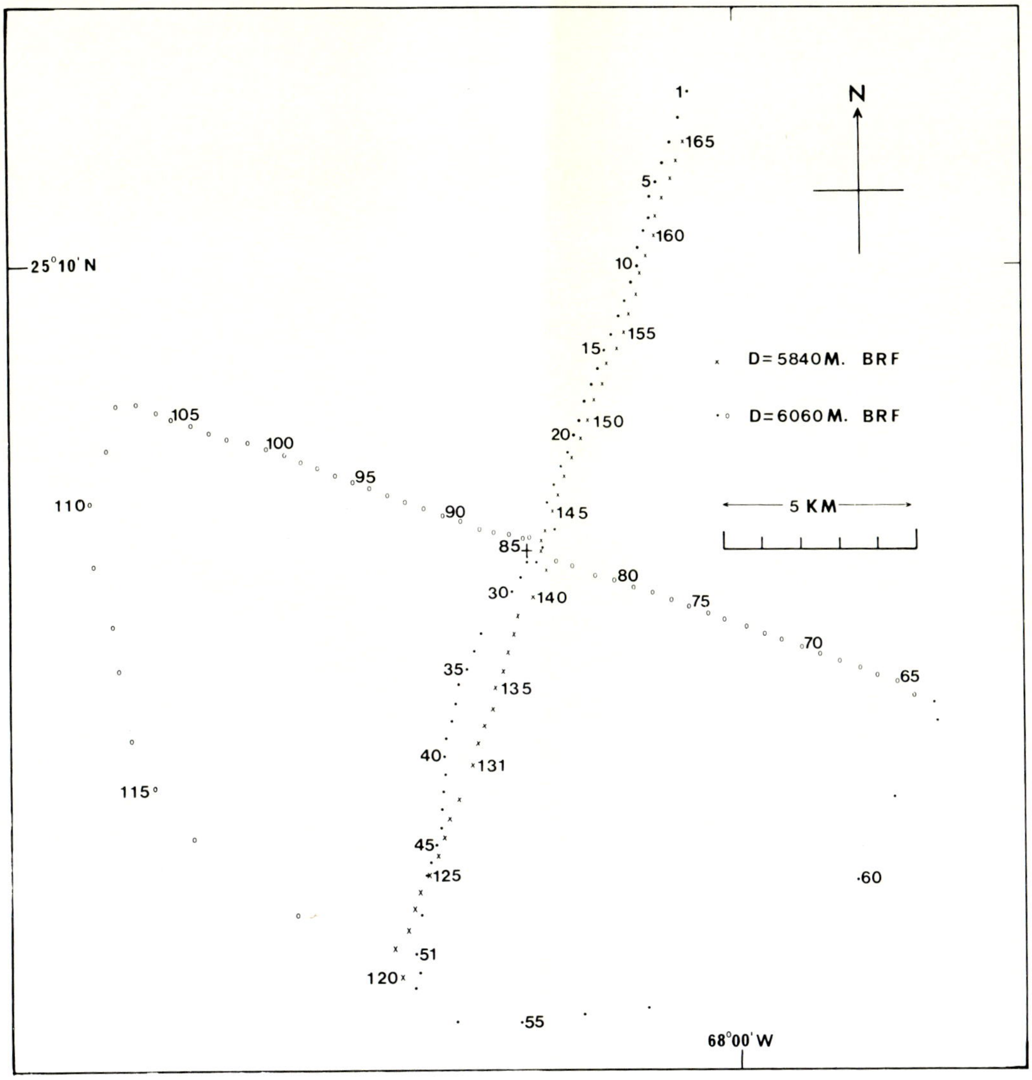

Figure 7 Plan view of the oblique seismic experiment shooting pattern in the vicinity of Hole 417D (from Stephen et al., in press). Plus (+) represents the position of the GLOMAR CHALLENGER; the crosses (x) and circles (•o) represent shots to the geophone in the 8 and 228 m sub-basement positions, respectively. Magnetic lineations at Site 417 strike parallel to the SSW-NNE shooting line.

damage (the Laterolog 8 and ILD data coincide throughout most of the basement), this can be accomplished provided the logging data can be used quantitatively.

Of the four active tools employed in Hole 417D, the resistivity and velocity tools functioned properly while the density and porosity tools malfunctioned intermittently due to a broken ex-centralizer. Thus the resistivity and velocity data are considered to be of high quality throughout the logged section of the hole, while the density and porosity data are considered marginal between 343 and 369 meters sub-bottom and useless below 369 meters with the possible exception of the low porosity values observed in Unit 3 between 425 and 433 meters sub-bottom when the tool swung intermittently against the side of the hole.

As can be seen in Figure 6, the electrical resistivity of the uppermost levels of the basement in Hole 417D ranges for the most part between 30 and 80 ohm-meters, with lower values (3 ohm-meters) observed in breccias and higher values (up to 200 ohm-meters) observed in massive basalts. Since these <u>in situ</u> values are lower than the average laboratory value of electrical resistivity for water-saturated basalts from this interval (97 ohm-meters; see Table 4) and markedly lower than the resistivities (10^3 to 10^6 ohm-meters) reported by Hyndman and Hall (1974) for dry basalts at room temperature, it is clear that the basement in Hole 417D is cracked and that the cracks must be filled with sea water (0.25 ohm-meters).

Turning to the velocity data, it is apparent in Figure 6 that the compressional wave velocities in the hole range for the most part between 4.7 and 5.3 km/sec with lower velocities displayed in breccias and higher velocities (up to 5.85 km/sec) displayed in massive basalts, while the formation velocity derived from integrated transit time data for the entire logged basement interval is 4.8 km/sec (Salisbury et al., in press). Although there may be, under certain circumstances, a small systematic error in data obtained using the Schlumberger BHC velocity tool (Wang and Simmons, in press; S. Raymer, personal communication), we see no compelling reason to question the numerical accuracy of the present data. In fact, the logging velocities measured in relatively unfractured sections of the hole (for example, the massive basalts between 425 and 433 meters sub-bottom) are indistinguishable from the laboratory velocities of samples from this interval measured at the effective confining pressure appropriate to *in situ* conditions (0.1 kbar).

Interpretation of the porosity data, however, is on less certain ground. Taken at face value, the porosity logs run between 343 and 369 meters before the excentralizer broke suggest that the porosity in the upper part of the hole ranges between 15 and 35% with an average of 21 to 22%. The minimum value recorded as the tool swung against the side of the hole in the 425 to 433 meter section noted above, however, was 10%. Since the basalts in this interval have a measured porosity of 3%, it is apparent that the logging porosity must be corrected by approximately minus 7%. As demonstrated by Salisbury et al. (in press), this correction may be derived independently by comparing the velocity vs. density data obtained by logging to velocity-density data obtained in the laboratory at 0.1 kbar for rock samples from the same interval. A density correction of plus 0.1 g/cm^3 is necessary to superimpose the two sets of data. The porosity correction is then the decrease in porosity necessary to restore the porosity vs. (corrected) density curve obtained by logging to the porosity-grain density solution observed in the laboratory. The porosity correction obtained by this means (minus 8%) agrees quite well with that determined above (minus 7%) and suggests that Unit 1A has an average porosity of 13-14% while Unit 3 has a porosity of about 3%. By the same token, if the density data shown in Figure 6 is corrected by plus 0.1 g/cm^3, it is apparent that Unit 1A has an average density of about 2.7 g/cm^3.

Since the velocity and density values obtained in the laboratory are higher than those obtained by logging, while the laboratory porosities are lower, it is obvious that a comparison of the data may be used to estimate the crack porosity and perhaps the permeability of the upper levels of the crust. Before attempting such a comparison, however, the results of the oblique seismic experiment will be reviewed to determine the behavior of the crust in the vicinity of Sites 417 and 418 at a scale of investigation typical of marine geophysical surveys.

The Oblique Seismic Experiment

As noted above and reported in detail by Stephen et al., (in press), an oblique seismic experiment was conducted in Hole 417D in order to determine the velocity structure of the crust in the vicinity of the borehole, to analyze the role of fissures and large cracks in determining the velocity of the crust, to look for anisotropy caused by cracks with a preferred orientation and to measure crustal attenuation.

The experiment consisted of shooting explosive charges to a three-component clamped borehole geophone lowered into the hole from the GLOMAR CHALLENGER (Figure 2). A cross pattern (Figure 7) with 12 km long arms was fired with the geophone clamped 228 meters into basement (571 meters into the hole or 6060 meters below the rig floor). The cross was oriented with lines parallel and perpendicular to the spreading axis in the area. A single line 24 km long was then fired parallel to the spreading axis (SSW-NNE) with the geophone clamped 8 meters into basement. The shooting was carried out by the VIRGINIA KEY, a research vessel from the Atlantic Oceanographic and Meteorological Laboratories of the National Oceanic and Atmospheric Administration in Miami.

After processing to remove ship noise, the travel times and amplitudes were examimed for evidence of anisotropy, attenuation and topographic effects. The amplitude analysis (Figure 8) was carried out using the reflectivity synthetic seismogram method (Fuchs and Müller, 1971; Stephen, 1977). From the results of such analyses (e.g., Figure 9), it is clear that the velocity model which best reproduces the oblique seismic experiment data for layer 2 at Site 417 is one with a uniform increase in compressional wave velocity from 4.8 km/sec at the top of the basement to 6.4 km/sec at a depth of 1.3 km and a similar increase in shear wave velocity from 2.6 to 3.6 km/sec over the same interval.

Although fissures and large normal faults oriented parallel to the ridge have been observed in the axial zone of the Mid-Atlantic Ridge, no undisputed evidence for preferred crack orientation was found in layer 2 at Site 417 from either velocity or amplitude studies. As can be seen in Table 6 from Stephen et al. (in press), the mean north-south shear wave velocity (2.56 ± 0.04 km/sec) is not significantly different from the mean east-west velocity (2.66 ± 0.14 km/sec). In general, shear waves are well suited to the study of seismic anisotropy since a) the direct shear wave arrival plots as a straight line for about 3 km whereas the direct compressional wave arrival falls on a curve, b) shear wave amplitudes increase with range in contrast to compressional

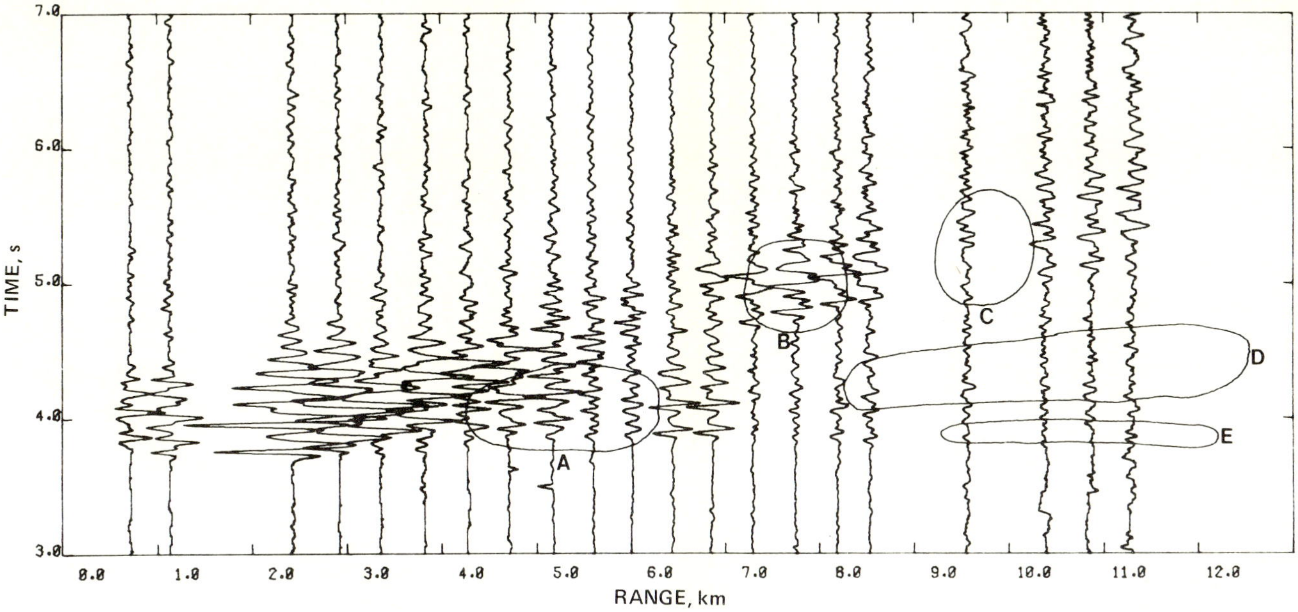

Figure 8 Observed N-S horizontal component data from South line of oblique seismic experiment at Site 417; geophone 228 meters into basement (6060 meters below rig floor); amplitudes weighted by $(\text{range}/7.0)^{**2.9}$ for ranges greater than 7.0 km; time reduced to 6.0 km/sec. Note signal character in regions A-E. Figure from Stephen et al. (in press).

wave amplitudes which decrease (Ergin, 1952) and c) shear waves are theoretically more sensitive to the presence of water-filled cracks than compressional waves (Anderson et al., 1974). Since the present experiment was well suited to the detection of azimuthal anisotropy, it appears that large, oriented fissures are not present at Site 417.

The situation in layer 3, however, is less clear. The average compressional wave velocities for the north-south and east-west lines are 6.71 ± 0.30 and 6.61 ± 0.41 km/sec, respectively, whereas the average shear wave velocities for the same lines are 3.77 ± 0.10 and 3.60 ± 0.09 km/sec, respectively. These values are not convincing evidence of anisotropy in layer 3, but if we assume that most of the spread in the errors is caused by small-scale topography and accept the mean values, then shear wave anisotropy on the order of 0.2 km/sec may exist.

Data Comparison

The scale of investigation of the three sets of data presented above ranges from approximately 10^{-2}m for the laboratory data through 10^{0}m for the logging data to about 10^{4}m for the data collected by means of the oblique seismic experiment. Since all three types of data were obtained for Hole 417D, it should be possible from a comparison of the data, to determine the volume fraction and scale of open cracks and fractures in the upper levels of the crust at Site 417. To this end a summary of the porosity, density, velocity and electrical resistivity data is presented in Table 7 for the interval in Hole 417D examined by all three techniques (Units 1-4). Also shown are summaries of the data for the pillow basalts of Unit 1A (for which all of the logging tools were working) and the relatively unfractured massive basalts at the base of Unit 3 which served as a calibration unit for the logging tools.

Several features of this table are quite striking. First, the average compressional wave velocity obtained by logging for the uppermost 100m of the crust (4.8 km/sec for Units 1-4) is virtually identical to that obtained by the oblique seismic experiment but both are considerably lower than the velocity derived for the same interval from laboratory data (5.3 km/sec). This indicates that the crust is fractured on a scale larger than that of laboratory samples but generally less than the spacing between the transmitter and the receiver in the BHC velocity tool (0.6 m). Such spacing would be consistent with the cracks between individual pillows and the radial cracks within them.

Secondly, the average porosity of Unit 1A as determined by logging (13%) is about 5% greater than the porosity determined from laboratory measurements (8.3%). This indicates that approximately 5% of the formation by volume consists of open cracks, a value consistent with the 0.1 g/cm^3 discrepancy in density observed between laboratory and logging data in the same interval. Since the CNL porosity tool measures porosity by determining the total amount of hydrogen in the formation, these cracks, like those between indi-

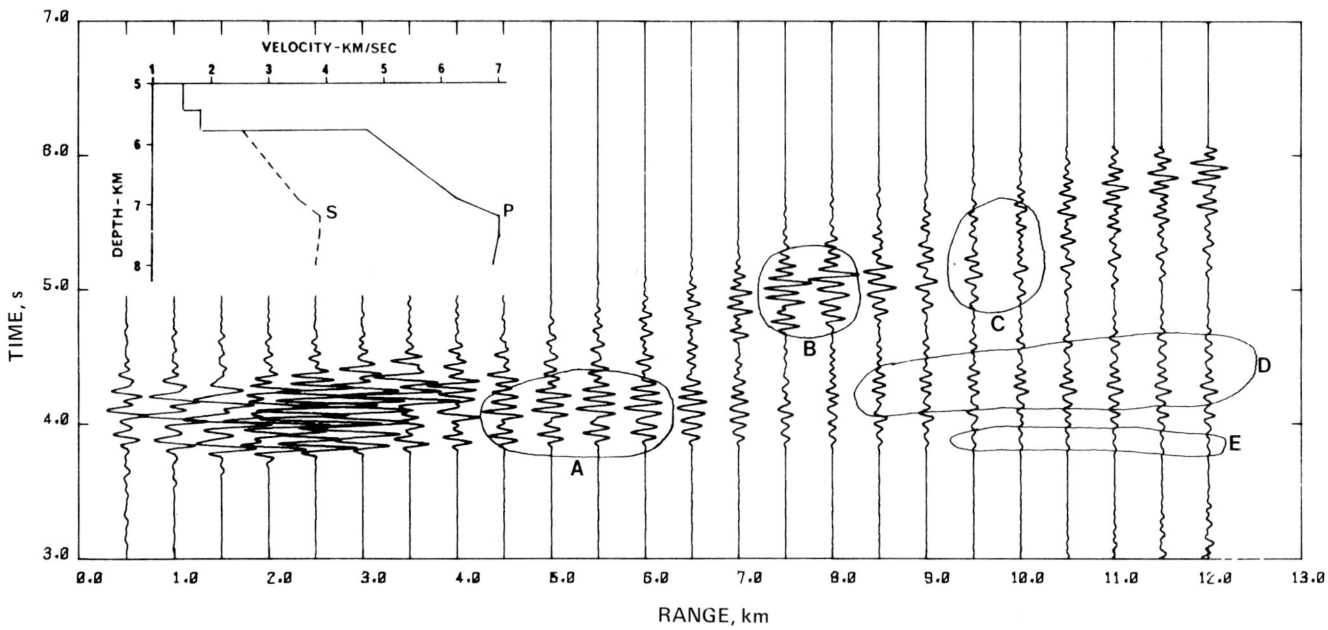

Figure 9 Velocity model and resultant horizontal component synthetic seismogram representing best fit to salient features (regions A-E) of Figure 8. Velocity model represents compressional and shear wave velocity structure of crust at Site 417. Figure of Stephen et al. (in press).

vidual grains, are apparently filled with water.

Finally, it should be noted in Table 7 that the formation velocity determined by logging for Units 1-4 is identical to that of Unit 1A. We may thus be justified in extrapolating the conclusions drawn above for Unit 1A to the entire logged interval.

In order to understand the distribution of porosity within the logged interval, however, it is necessary to juxtapose the velocities determined by the three techniques as in Figure 10 in which the velocity profiles reconstructed from zero-pressure laboratory data, integrated logging transit times and the results of the oblique seismic experiment are shown as a function of depth. Since the discrepancy between the logging and laboratory data is a measure of the volume fraction of cracks, Figure 10 suggests that the formation porosity associated with open cracks resides largely within the pillow basalts while the massive basalts contribute relatively little to the porosity of the formation beyond their measured grain boundary porosity of about 3%. Figure 10 also suggests that thin zones of very high crack porosity (up to 20% at the top of Unit 1C) exist locally in the crust in association with breccias between flows. That the low velocities associated with the case noted here are not due to caving but to an aquifer of some antiquity is suggested by the high natural gamma ray count seen at this level in Figure 6. It should be noted, incidentally, that in a rigorous comparison, laboratory velocities appropriate to in situ conditions would be used instead of the zero-pressure data shown in Figure 10. This procedure was not taken in the present case because the velocity correction is variable (from 0.0 - 0.25 km/sec, depending on density) and because the error in the estimate of porosity (2%) due to a maximum velocity correction of 0.25 km/sec is almost certainly less than those inherent in the Wyllie relation on which the estimate is based.

Conclusions

Since the beginning of the basement drilling program in 1974, approximately 1200 m of basalt have been recovered in the Atlantic, of which 790 m, or more than half, were recovered at DSDP Sites 417 and 418. If the sections drilled at these sites are representative of Cretaceous oceanic basement, it is clear from the presentation above on the nature and physical properties of the core and the results of the logging and oblique seismic experiments in Hole 417D, that a number of important conclusions can be drawn concerning the composition and physical state of the upper levels of layer 2 in old oceanic crust:

1) Composition. The upper levels of the crust down to at least 0.5 km consist of approximately 90% basalt, 5% smectite, less than 1% interpillow limestone and 5% open cracks filled with standing water. Of the basalts, 71% consist of fractured pillow basalt, 22% are relatively unfractured massive basalts and 7% consist of partially cemented basalt rubble. Dikes begin to appear at depths of 350 to 500 meters in the basement but are not volumetrically significant.

TABLE 6. SEISMIC ANISOTROPY AT SITE 417[1)]

Shooting Line	Geophone Position (m below rig floor)	Layer 2 Shear Wave Velocity[2)] (km/sec)	Layer 3 Compressional Wave[3)] Velocity (km/sec)	Layer 3 Shear Wave Velocity[3)] (km/sec)
North	6060	2.49 ±0.02 (N = 5)	6.97 ±0.44 (N = 6)	3.75 ±0.12 (N = 11)
North	5840	2.48 ±0.04 (N = 6)	6.66 ±0.20 (N = 7)	3.88 ±0.02 (N = 6)
South	6060	2.61 ±0.05 (N = 4)	6.70 ±0.30 (N = 11)	3.63 ±0.08 (N = 8)
South	5840	2.67 ±0.04 (N = 3)	6.50 ±0.27 (N = 7)	3.81 ±0.19 (N = 5)
East	6060	2.72 ±0.21 (N = 4)	6.75 ±0.44 (N = 8)	3.68 ±0.11 (N = 6)
West	6060	2.60 ±0.01 (N = 4)	6.46 ±0.37 (N = 8)	3.51 ±0.06 (N = 7)
Mean		2.60 ±0.06	6.67 ±0.34	3.71 ±0.10

1) From Stephen et al., in press

2) Direct

3) Refracted

TABLE 7. COMPARISON OF LABORATORY, LOGGING AND OBLIQUE SEISMIC DATA IN HOLE 417D

Parameter	UNIT 1A (pillow basalt)			BASE OF UNIT 3 (unfractured massive basalt)			UNITS 1-4 (entire logged interval)		
	Laboratory[1] Results	Downhole Logging	Oblique Seismic Experiment	Laboratory[2] Results	Downhole Logging	Oblique Seismic Experiment	Laboratory[1] Results	Downhole Logging	Oblique Seismic Experiment
Density (g/cm^3)	2.79	2.7	--	2.88	--	--	2.78	--	--
Porosity (volume %)	8.3	13	--	3.2	3	--	8.2	--	--
Velocity, V_p (km/sec)	5.36 (5.6)	4.8	--	5.37 (5.87)	5.8	--	5.34 (5.6)	4.8	4.8
Resisitivity (ohm-m)	74	40	--	--	120	--	97	35	--

1) Density, porosity and velocity data from synthetic profiles of Figure 4; values in parentheses are estimated velocities at 0.1 kbar effective confining pressure; resistivity data from Hamano (in press).

2) Measured values in Core 34-Section 2; velocity in parentheses measured at 0.1 kbar confining pressure.

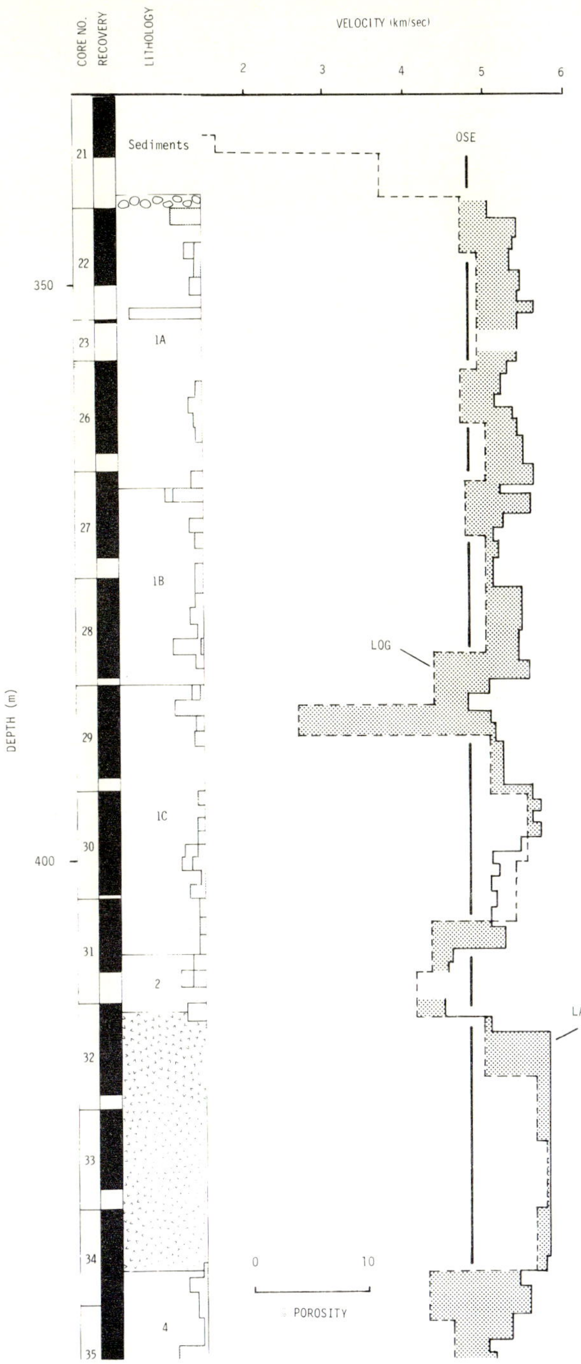

Figure 10 Comparison of compressional wave velocities obtained from logging, physical property studies and the oblique seismic experiment in Hole 417D (see text). Discrepancy between laboratory and logging data (stippled area) is a measure of formation porosity due to open cracks. Units 1, 2 and 4 are composed of pillow basalt while Unit 3 (hachured) is composed of massive basalt; insets in lithology column represent abundance of limestone (plain) and smectite (light gray) on a scale of 0-50%.

2) Internal Structure. Although the crust displays marked downhole variations in physical properties, it is unlikely that such variations will be detectable using surface geophysical techniques. It is possible, however, that the low velocity zones associated with breccias within the basement might be detected by deep tow seismic reflection techniques.

3) Resistivity. The *in situ* electrical resistivity of the basement ranges from 3 Ω-meters in the breccias to about 200 Ω-meters in the massive basalts, with most values falling between 30 and 80 Ω-meters. Since the average laboratory resistivity of water-saturated basalts from the section is 97 Ω-meters and the resistivity of dry basalts at room temperature ranges from 10^3 to 10^6 Ω-meters, it is clear that the basement is cracked and saturated with sea water having a low *in situ* resistivity (0.25 Ω-meters at 21°C).

4) Velocity. Compressional wave velocities in the uppermost levels of the basement range for the most part between 4.7 and 5.3 km/sec with lower velocities displayed in breccias and velocities as high as 5.8 km/sec displayed in the massive basalts. The average formation velocity derived from both the oblique seismic experiment and from integrated logging transit times is 4.8 km/sec, a value consistent with those reported for layer 2B by Houtz and Ewing (1976). This value is considerably higher than the layer 2A formation velocity of 3.6 km/sec reported by Kirkpatrick (in press) for the basement section drilled in Hole 396B in young crust along the same flow line and is consistent with the observation that layer 2A does not commonly persist in the Atlantic beyond the 60 million year isochron. Examination of the core material from Hole 417D clearly indicates that the increase in formation velocity in the upper levels of layer 2 with age is due to the infilling of cracks and interpillow voids by products of low temperature alteration such as calcite, smectite, and zeolites.

5) Velocity vs. Depth. From the results of the oblique seismic experiment it is clear that the compressional and shear wave velocities at Site 417 increase respectively with depth from 4.8 and 2.6 km/sec at the top of layer 2 to 6.4 and 3.6 km/sec near its base. This could be due, in varying combinations, to decreasing alteration, to a decrease in cracks or to changing mineralogy with depth. The laboratory velocities shown in Figure 11 suggest that the velocity gradient in layer 2 can be explained in terms of the first two phenomena alone: the laboratory velocities increase with depth in response to decreasing alteration, reaching the velocity of fresh basalt (6.4 km/sec) at a projected depth of about 700 m, which would appear to be the maximum depth of alteration. The discrepancy between the laboratory and oblique seismic experiment data confirms

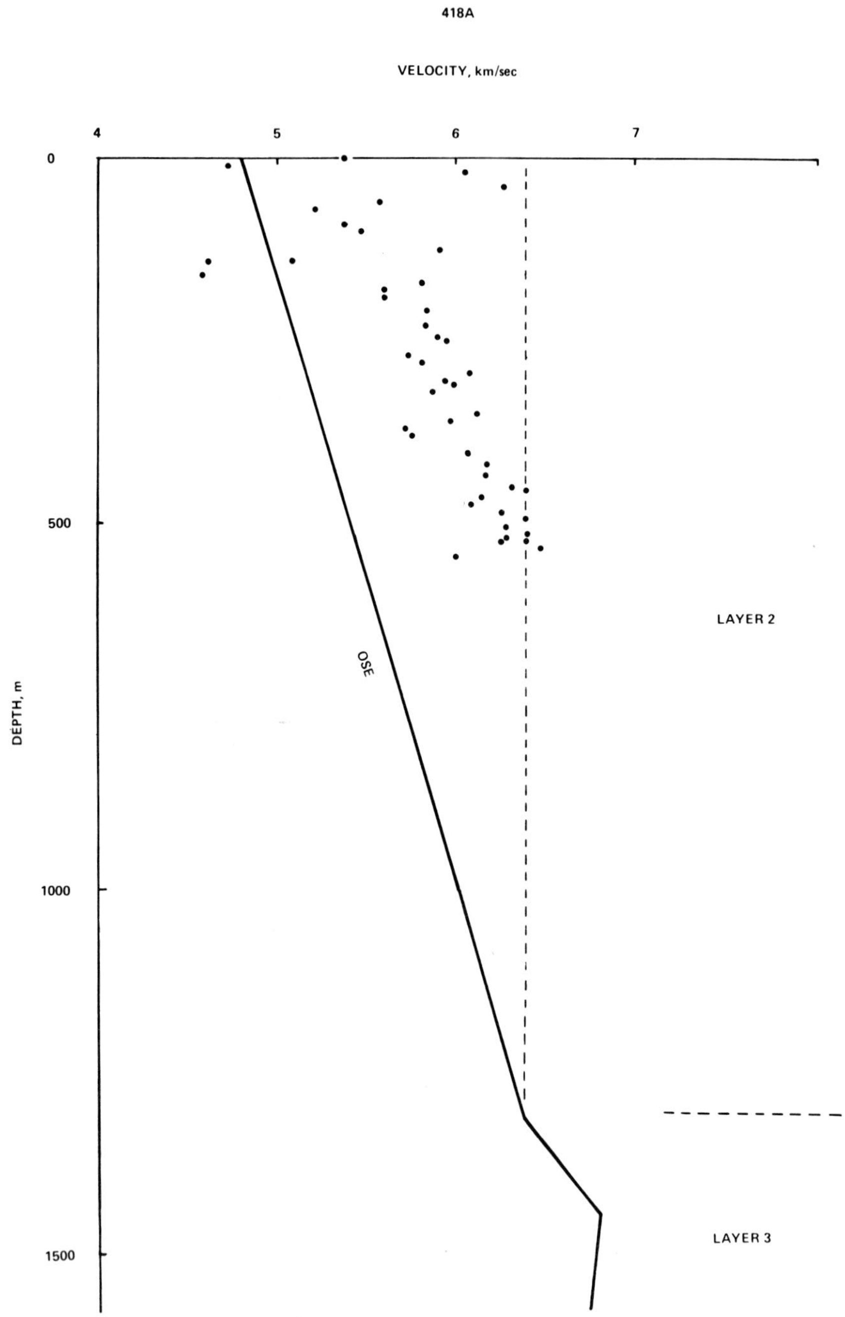

Figure 11 Comparison of laboratory velocities of samples from Hole 418A (Christensen et al., in press) and the results of the oblique seismic experiment in Hole 417D (Stephen et al., in press). Laboratory velocities measured at an effective confining pressure of 0.1 kbar. Vertical dashed line represents velocity of fresh basalt at 1.0 kbar.

that the crust is cracked, but the increase in the formation velocity with depth suggests that the cracks decrease downward until the formation velocity reaches the velocity of a fresh, crack-free basalt at the base of layer 2. Since this velocity is still lower than the velocity of layer 3, the layer 2/layer 3 transition must be due to a change in mineralogy.

6) Anisotropy. No convincing evidence of seismic anisotropy was found in layer 2 from either velocity or amplitude analysis of the oblique seismic experiment data. Thus the large, open

fissures observed parallel to the Mid-Atlantic Ridge in the floor of the median valley have either been filled with rubble and sediments or closed as they evolved into normal faults in the walls of the median valley. There is still weak evidence, however, of seismic anisotropy in layer 3 with V_s slow by about 0.1 to 0.2 km/sec parallel to the direction of spreading. Since open cracks cannot be invoked at this depth, this anisotropy may be due to the grain of sheeted dikes.

7) Porosity, density and permeability. Although the crust at Sites 417 and 418 has experienced considerable infilling, the basement is strongly fractured in situ. The total porosity of the logged basement interval in Hole 417D ranges from 3% in the massive basalts to as much as 20% in the breccias, while the average for the pillow basalts and the logged interval as a whole is 13%, of which about 8% is due to grain boundary porosity in basalt and 5% is due to open cracks filled with standing water. These cracks decrease the formation density by about 0.05 g/cm^3 to an average of 2.70 g/cm^3.

The most important influence of such cracks, however, is upon formation permeability. This can be estimated from the relation,

$$K = \frac{h^2 R_w}{4\pi S R} \quad (5)$$

where K is the permeability of the formation in darcies, h is the average crack width in centimeters, R_w is the resistivity of the interstitial water in Ω-meters, R is the average formation resistivity, and S is a shape factor close to unity (Sheidegger, 1960). Using the formation resistivity determined by logging (50 Ω-meters), a sea water resistivity of 0.25 Ω-meters and an average crack width of 0.25 to 0.50 cm (estimated from the crack density and porosity of the formation), the permeability of the upper levels of the crust is found to range between 2500 and 10,000 darcies, in excellent agreement with the results of Johnson (in press, b) from visual estimates of the density and distribution of cracks in the core. Such values imply that convective heat transfer in old oceanic crust is limited not by the permeability of the crust but by the availability of heat and the presence of overlying sediments and that the alteration of submarine basalts is only terminated by subduction.

Acknowledgments. We wish to thank the Captain and crew of the GLOMAR CHALLENGER and the VIRGINIA KEY for their assistance during the drilling, logging and oblique seismic operations, Loïc Beurdeley of Schlumberger, Ltd. for running the logging tools, R. Jones, J. Vohs and S. Bearman for their assistance in processing the logging data, S. Blair, R. H. Wilkens, J. Hull, G. Bussod and R. Prior for their assistance in the laboratory velocity studies and J. Kirkpatrick, K. E. Louden and D. H. Matthews for their many helpful suggestions during the course of this study. The laboratory velocity studies we supported by Office of Naval Research contract N-00014-75-C-0502.

References

Anderson, D. L., B. Minster and D. Cole, 1974. The effect of oriented cracks on seismic velocities, J. Geophys. Res., 79, p. 4011-4016.

Aumento, F., Melson, W. G., et al., 1977, Site 332, In Aumento, F., Melson, W. G., et al., Initial Reports of the Deep Sea Drilling Project, v. 37: Washington (U.S Government Printing Office), p. 15-200.

Aumento, F., Melson, W. G., et al., 1977. Site 333, In Aumento, F., Melson, W. G., et al., Initial Reports of the Deep Sea Drilling Project, v. 37: Washington (U.S. Government Printing Office), p. 201-238.

Bryan, W. et al., in press. Site 417, In Bryan, W. et al., Initial Reports of the Deep Sea Drilling Project, v. 51-53: Washington (U.S. Government Printing Office).

Bryan, W. et al., in press. Site 418, In Bryan, W. et al., Initial Reports of the Deep Sea Drilling Project, v. 51-53: Washington (U.S. Government Printing Office).

Cann, J. R., Luyendyk, B., et al., in press. Site 409, In Cann, J. R., Luyendyk, B., et al., Initial Reports of the Deep Sea Drilling Project, v. 49: Washington (U.S. Government Printing Office).

Christensen, N. I., S. C. Blair, R. H. Wilkens and M. H. Salisbury, in press. Compressional wave velocities, densities and porosities of basalts from Holes 417A, 417D and 418A, DSDP Legs 51-53, In Bryan, W. et al., Initial Reports of the Deep Sea Drilling Project, v. 51-53: Washington (U.S. Government Printing Office).

Donnelly, T. W., G. Thompson and M. H. Salisbury, in press. The chemistry of altered basalts at Site 417, DSDP Leg 51, In Bryan, W., et al., Initial Reports of the Deep Sea Drilling Project, v. 51-53: Washington (U.S. Government Printing Office).

Ergin, K., 1952. Energy ratio of the seismic waves reflected and refracted at a rock-water boundary, Bull. Seism. Soc. Am., 42, p. 342-372.

Fuchs, K. and G. Muller, 1971. Computation of synthetic seismograms with the reflectivity method and comparison with observations, Geophys. J. Roy. Astron. Soc., 23, p. 417-433.

Hamano, Y., in press. Physical properties of basalts from Holes 417D and 418A, In Bryan, W. et al., Initial Reports of the Deep Sea Drilling Project, v. 51-53: Washington (U.S. Government Printing Office).

Helmberger, D. V. and G. B. Morris, 1970. A travel time and amplitude interpretation of a marine refraction profile: transformed shear waves, Bull. Seism. Soc. Am., 60, p. 593-600.

Houtz, R. and J. Ewing, 1976. Upper crustal

structure as a function of plate age, J. Geophys. Res., 81, p. 2490-2498.

Hussong, D. M., 1972, Detailed structural interpretation of the Pacific oceanic crust using ASPER and ocean bottom seismometer methods, Ph.D. thesis, 165 pp., Univ. of Hawaii, Honolulu.

Hyndman, R. D. and J. M. Ade-Hall, 1974. Electrical resistivity of basalts from DSDP Leg 26, In Davies, T. A., Luyendyk, B. P. et al., Initial Reports of the Deep Sea Drilling Project, v. 26: Washington (U.S. Government Printing Office), p. 505-508.

Johnson, D. M., in press a. Fluid permeability of oceanic basalts, In Bryan, W. et al., Initial Reports of the Deep Sea Drilling Project, v. 51-53: Washington (U.S. Government Printing Office).

Johnson, D. M., in press b. Crack distribution in the upper oceanic crust and its effect upon seismic velocity, seismic structure, formation permeability and fluid circulation, In Bryan, W. et al., Initial Reports of the Deep Sea Drilling Project, v. 51-53: Washington (U.S. Government Printing Office).

Kirkpatrick, R. J., in press. Results of downhole geophysical logging Hole 396B, DSDP Leg 46. In Dmitriev, L., Heirtzler, J., et al., Initial Reports of the Deep Sea Drilling Project, v. 46: Washington (U.S. Government Printing Office).

Melson, W. G., Rabinowitz, P. D., et al., in press. Site 395, In Melson, W. G., Rabinowitz, P.D. et al., Initial Reports of the Deep Sea Drilling Project, v. 45: Washington (U.S. Government Printin Office).

Orcutt, J. A., B. L. N. Kennett and K. M. Dorman, 1976. Structure of the East Pacific Rise from an ocean bottom seismometer survey, Geophys. J. Roy. Astron. Soc., 45, p. 305-320.

Peterson, J. J., P. J. Fox and E. Schreiber, 1974. Newfoundland ophiolites and the geology of the oceanic layer, Nature 247, p. 194-196.

Raitt, R. W., 1963. The crustal rocks, In Hill, M.N., Ed., The Sea, v. 3, John Wiley, New York, p. 85-102.

Robinson, P. T., J. M. Hall, F. Aumento, W. G. Melson, H. Bougault, L. Dmitriev, J. F. Fischer, M. Flower, R. C. Howe, R. D. Hyndman, G. A. Miles and T. L. Wright, 1977. Leg 37 cruise synthesis: The lithology, structure, petrology, and magnetic history of Layer 2. In Aumento, F., Melson, W. G., et al., Initial Reports of the Deep Sea Drilling Project, v. 37: Washington (U.S. Government Printing Office).

Salisbury, M. H., T. W. Donnelly and J. Francheteau, in press. Geophysical Logging in DSDP Hole 417D, In Bryan, W., et al., Initial Reports of the Deep Sea Drilling Project, v. 51-53: Washington (U.S. Government Printing Office).

Sheidegger, A. E., 1960. The Physics of Flow Through Porous Media, University of Toronto Press, Toronto.

Stephen, R. A. 1977. Synthetic seismograms for the case of the receiver within the reflectivity zone, Geophys. J. Roy. Astron. Soc., 51, p. 169-181.

Stephen, R. A., K. E. Louden and D. H. Matthews, in press. The oblique seismic experiment, In Bryan, W. et al., Initial Reports of the Deep Sea Drilling Project, v. 51-53: Washington (U.S. Government Printing Office).

Spudich, P. K. P., M. H. Salisbury and J. A. Orcutt, 1978. Ophiolites found in oceanic crust?, Geophys. Res. Lett., 5, p. 341-344.

Wang, H. F. and G. Simmons, in press. Microcracks in crystalline rocks from 5.2 km depth in the Michigan Basin.

GEOMAGNETIC REVERSALS AND OCEAN CRUST MAGNETIZATION

William Lowrie

Institut für Geophysik, ETH Hönggerberg,
CH-8093 Zürich, Switzerland

Abstract. The Cenozoic and Mesozoic history of geomagnetic field reversals is derived mostly from interpretation of oceanic magnetic anomalies. Our knowledge of the magnetization of the oceanic crust is still somewhat incomplete, in particular concerning the sources of the magnetic anomalies. A review is given here of what is known about the polarity, direction and intensity of magnetization of the different layers of the oceanic crust.

Large parts of the oceanic magnetic polarity sequence remain unverified by independent continental paleomagnetic studies. Only the Late Cenozoic and Cretaceous magnetic reversal sequences have been confirmed. Several hypotheses have been offered to explain the origins of oceanic magnetic smooth zones.

The source of oceanic magnetic anomalies has most commonly been identified with basaltic Layer 2. Magnetometer measurements from surface ships or deep-towed vessels have suggested that the entire signal could derive from the uppermost 500 m. of the basalt layer. Accumulating evidence, including power spectral analysis of anomaly amplitudes and direct measurement of dredged and drilled (DSDP) basalt samples with a broad geographic provenance, now indicates that a much thicker source layer is needed. Deep penetration of Layer 2 in recent DSDP holes provides evidence that Layer 2 has a complex structure and history; large variations of magnetization intensity and inclination are found and there are occasional polarity alternations. In none of the deep holes was the bottom of the strongly magnetized layer reached. It now appears likely that the gabbroic Layer 3, although more weakly magnetized than Layer 2, may contribute significantly to the oceanic magnetic anomaly pattern.

Introduction

Large areas of the oceanic crust are characterized by lineated magnetic anomaly patterns [Mason and Raff, 1961; Raff and Mason, 1961], which are interpreted in terms of alternating geomagnetic field polarity during sea-floor spreading [Vine and Matthews, 1963]. World-wide correlation of oceanic anomalies yielded a geomagnetic polarity scale which, by comparison with radiometrically dated continental lava sequences, allowed ages to be associated with different parts of the ocean basins since the late Cretaceous [Heirtzler et al., 1968]. By matching anomalies on both sides of the spreading center at the mid-Atlantic ridge the relative motions of the Atlantic-bordering continents have been reconstructed quasi-continuously during the opening of the Atlantic ocean [Pitman and Talwani, 1972]. From the point of view of geomagnetic theory, frequency analysis of the reversal time scale leads to constraints upon models of the origin of the geomagnetic field [Cox, 1968].

The magnetization of the oceanic crust is therefore of considerable geophysical interest. The magnetization polarity sequence has been derived in many marine geophysical investigations. Although mutually consistent reversal sequences are obtained in different ocean basins there are difficulties in verifying the sequence independently. There is not even unanimous agreement as to where and how the anomalies originate. Much effort has been expended to understand the ocean crust magnetization by recording the anomalies in surveys at the ocean surface or from deep-towed instrument packages, and by studying the magnetic properties of rocks recovered from the ocean bottom by dredging, by sampling from submersibles in the FAMOUS project, by shallow drilling and by deeper penetration of the ocean crust in the Deep Sea Drilling Project (DSDP) and its continuation, the International Program of Ocean Drilling (IPOD).

In this paper a review is given of investigations to confirm the oceanic reversal sequence and to establish the origin of the corresponding oceanic magnetic anomaly source layer.

Geomagnetic Polarity

The magnetization of the oceanic crust is our prime source of knowledge of the history of reversals of the geomagnetic field since the

Early Jurassic, that is, in the last 180 M.y. Four distinct episodes may be identified. A continuous sequence of well-defined reversals characterizes the time interval from the present to the Late Cretaceous, the Cenozoic sequence [Heirtzler et al., 1968]. In oceanic crust of Middle and Early Cretaceous age the anomalies have generally reduced amplitudes and are uncorrelatable; this is referred to as the Cretaceous magnetic smooth zone. In the Early Cretaceous and Late Jurassic are again found well-defined, correlatable anomalies, the Mesozoic M-sequence [Larson and Pitman, 1972, 1975; Larson and Hilde, 1975]. In crust older than the M-sequence is found the Jurassic magnetic smooth zone. The dating of the two sequences of magnetic anomalies has been periodically revised, and the reasons for the two magnetic smooth zones have been frequently speculated upon.

The Cenozoic Reversal Sequence

The refinement of the K/Ar method made it possible to date lavas radiometrically with an accuracy of about 2 to 5%. By the middle 1960's a well-defined geomagnetic polarity sequence containing long epochs and shorter events of constant polarity had been developed in Pleistocene and Pliocene lavas for the last 3.7 M.y. [Cox et al., 1966]. The sequence was extended to about 4.5 M.y. with the aid of deep sea sediments data. Beyond this age a 2% error exceeds the duration of some events and derivation of a polarity-time scale from the combination of K/Ar dating and polarity measurements on stratigraphically unrelated samples becomes limited by the lack of resolution of the radiometric dating method [Cox, 1969].

The polarity-time scale has been extended further by studying the polarity of K/Ar-dated, stratigraphically related samples from thick lava sequences in Iceland, which have been laid down almost continuously. The errors of individual dates were partially compensated by linear regression analysis and optimum ages for the magnetic reversals were derived by interpolation. Using this method McDougall et al. [1977] extended the polarity-time scale to 7 M.y. and demonstrated good agreement with the oceanic anomalies polarity sequence of Talwani et al. [1971]. In lavas with ages from 8.7 M.y. to 12 M.y. [McDougall et al., 1976] the age dates were more scattered but the polarity sequence agreed quite well with the oceanic sequence derived for the interval near anomaly 5 by Blakely [1974].

The reversal record in deep sea sediments unequivocally confirms the sequence derived from lavas back to epoch 5, or about 5 M.y. [Opdyke, 1972]. It has been extended to epoch 11 (Late Miocene) in a continuous long core [Foster and Opdyke, 1970]. Further extension of the sequence through the Miocene has also been attempted in overlapping cores [Opdyke et al., 1974; Theyer and Hammond, 1974]. The polarity sequence from a single core is less reliable than a sequence confirmed in several cores [Opdyke, 1972] and the correlation of the replicate sections of overlapping cores is not always unambiguous. Nevertheless geomagnetic polarity intervals from deep sea sediments and from oceanic magnetic anomalies have been used to calibrate the Neogene biostratigraphic zonation [Ryan et al., 1974].

The polarity-time scale derived from oceanic anomalies by Heirtzler et al. [1968] has undergone subsequent refinement and revision, but its correctness is remarkable. Reversals were dated by correlation with radiometrically dated lava sequences for the last 3.5 M.y. at its younger end, and by linear extrapolation to older ages equivalent to 20 times this base.

The oceanic magnetic polarity sequence has not yet been confirmed in land sections for most of the Paleogene. Reversal time scales have been pieced together from a number of partial sections for the Paleocene and Late Cretaceous [Watkins, 1975; Keating et al., 1975]. The first continuous magnetic stratigraphy at a single locality for this period was obtained from paleomagnetic studies in a biostratigraphically complete pelagic limestone section near Gubbio, Italy, where the record of geomagnetic reversals was found to match closely the sequence inferred from marine magnetic anomalies in the Early Paleocene and Late Cretaceous [Roggenthen and Napoleone, 1977; Lowrie and Alvarez, 1977a, 1977b]. Zonation of planktonic foraminifera [Premoli Silva, 1977] provided the biostratigraphic framework for the magnetic reversal sequence in the Gubbio section which has been proposed as a magnetostratigraphic type section [Alvarez et al., 1977]. The Late Cretaceous magnetic stratigraphy has been verified subsequently in a check section at Moria [Alvarez and Lowrie, 1978].

The principal conclusions to be drawn from the Gubbio and Moria results [Fig. 1] are: (1) the Cretaceous-Tertiary boundary falls near the top of the reversed polarity interval between anomalies 29 and 30 (as proposed by Larson [1974] and Sclater and Fisher [1974]), (2) the existence of a controversial reversed interval between anomalies 33 and 34 is established, and (3) a long period of normal polarity ending near the Campanian-Santonian boundary and beginning prior to the Cenomanian is correlated with the Cretaceous magnetic smooth zone.

In order to compare the limestone magnetic stratigraphy with the oceanic magnetic anomaly reversal sequence both must be plotted against a common time scale. Radiometric ages are not yet securely established for most major stage boundaries. The numerical ages given by Van Hinte [1976] have been used to prepare Fig. 1, but these may be subsequent to revision.

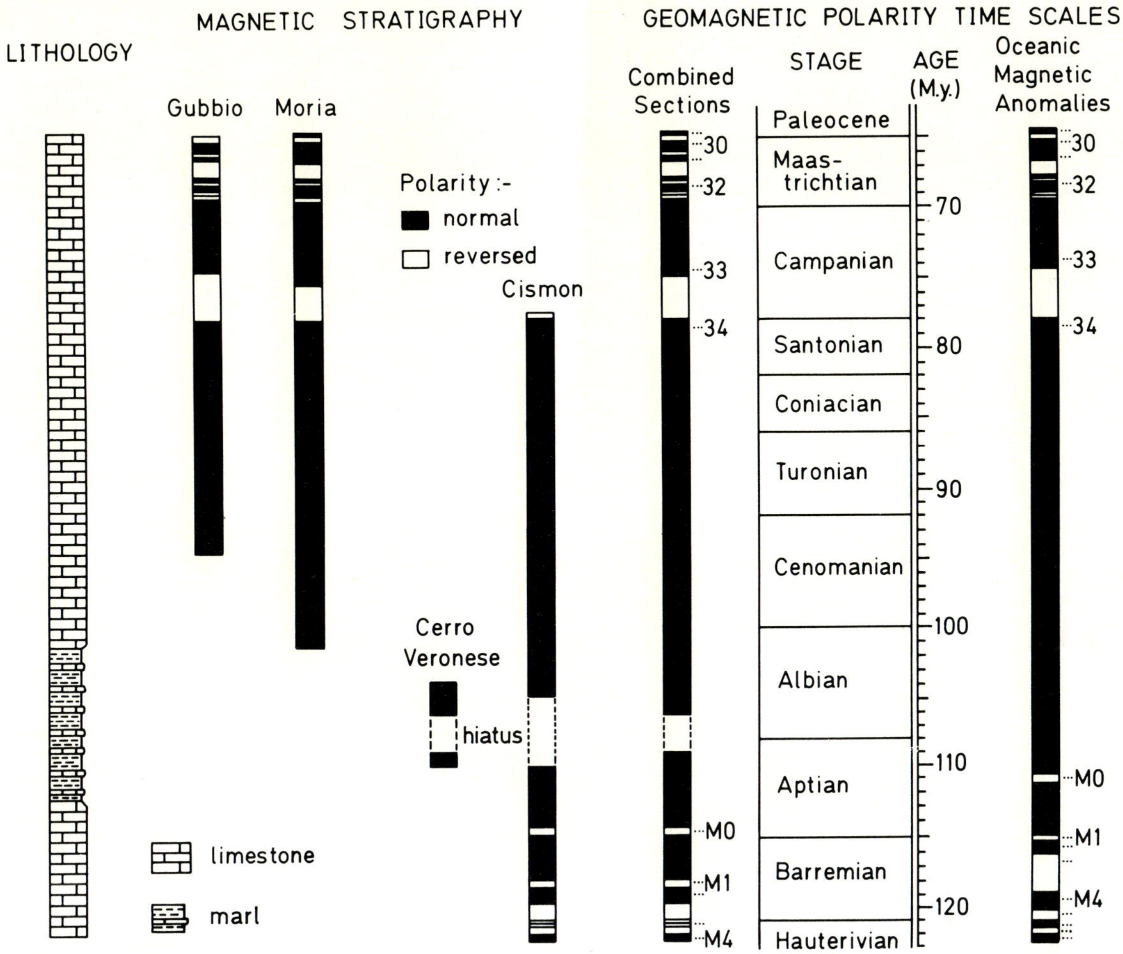

Fig. 1. Geomagnetic polarity sequence during the Middle and Late Cretaceous. The magnetic stratigraphies of pelagic limestone sections at Gubbio (Lowrie and Alvarez, 1977b) and Moria (Alvarez and Lowrie, 1978) in Umbria, and at Cismon and Cerro Veronese in the Southern Alps (Channell et al., 1979) are combined to form a composite limestone magnetic polarity sequence which is compared to that derived from oceanic magnetic anomalies (Larson and Hilde, 1975; LaBrecque et. al., 1977). Both are plotted against the time scale of Van Hinte (1976).

For example, LaBrecque et al. [1977] assumed two key radiometric ages of 65 M.y. for the Tertiary-Cretaceous boundary and 3.32 M.y. for anomaly 2A and produced a revised magnetic polarity time scale for Late Cretaceous and Cenozoic time. The ages of Cenozoic anomalies 0-29 were obtained by interpolation and those of Late Cretaceous anomalies 30-34 by extrapolation. The age obtained for anomaly 34 which ends the Cretaceous quiet interval is almost exactly 80 M.y., slightly older than the value of 78 M.y. given by Van Hinte [1976] and used in this paper [Fig. 1].

The revision of the Cenozoic polarity time scale by LaBrecque et al. [1977] results in a 7% decrease of the age of anomalies dated by the Heirtzler et al. [1968] time scale. Recent recommended revision of the K/Ar decay constants [Steiger and Jaeger, 1977] would cause a partially compensating increase in the radiometric ages of about 2.6%, which is not incorporated in the new LaBrecque et al. [1977] time scale.

The Mesozoic Reversal Sequence

Few attempts have yet been made to verify the Mesozoic magnetic polarity sequence of the Late Jurassic and Early Cretaceous. The dating of these marine magnetic anomalies (M0 to M25) is rather imprecise, and indirect. It is made by determining optimum paleontological ages for the sediment immediately overlying basement at 3 DSDP drillholes in the Early Cretaceous and 2 DSDP drillholes in the Late Jurassic [Larson and Hilde, 1975]. This dating is complicated by the possibility of hiatuses between the formation of igneous basement and deposition of the overlying sediments. Absolute ages have been ascribed to the paleontological dates by comparison with the Geological Society of London [1964] time scale. The approximate nature of the assignment of numeric ages to the stage boundaries in this time scale naturally compounds the errors of determining absolute ages for the Mesozoic magnetic anomalies.

Recently paleomagnetic stratigraphic studies have been extended backward in time to cover all of the Cretaceous with the exception of the Berriasian. The first successful data have been realized in pelagic limestones from the Cismon section in the southern Alps [Channell et al., 1979]. Although very weak, the remanences possess stable directions and yield a clearly defined magnetic stratigraphy. Ammonites are not present in the section, thus it is extremely difficult to locate the Valanginian-Hauterivian and Hauterivian-Barremian stage boundaries accurately. The polarity sequence in this interval consequently may be stretched or compressed; it does not yet correlate unambiguously with the corresponding M-anomalies. Nevertheless, the distinctive reversal pattern which gives the key anomalies M0 to M4 is clearly present [Fig. 1]. The paleontological dating of the section places the youngest of the correlated M-anomalies (M0) at the base of the Aptian, instead of in the Late Aptian as it was dated by Larson and Hilde [1975].

The Cretaceous Smooth Zone extends from there to anomaly 34 at the base of the Campanian. However, there is a depositional hiatus in the Lower Albian-Upper Aptian part of the Cismon section. Preliminary results from other Lower Cretaceous sections in Umbria show an additional short reversed zone in the Late Aptian. If this unconfirmed reversal proves to be real it might represent an as yet unidentified but correlatable anomaly younger than the rest of the M-sequence. This is a distinct possibility; anomaly M0 was not identified in the first world-wide correlation of the Mesozoic anomalies [Larson and Pitman, 1972], and cursory examination of the youngest parts of the profiles correlated by Larson and Hilde reveals several fairly large amplitude anomalies within the Cretaceous smooth zone.

The Magnetic Smooth Zones

At least three sets of magnetic smooth zones may be identified. These are (1) the Cretaceous, (2) the Jurassic and (3) the marginal smooth zones. It is neither necessary nor possible that a single explanation can suffice for all.

The Cretaceous magnetic smooth zones are the best understood. In the Pacific ocean the anomalies are smooth; in the Atlantic ocean more pronounced anomalies exist but they do not correlate to form an extensive lineation pattern, and may only represent bathymetric trends [Larson and Pitman, 1972]. As first suggested by Helsley and Steiner [1968] and demonstrated in the Gubbio, Moria and Cismon sections [Fig. 1] independent paleomagnetic evidence associates the Cretaceous magnetic smooth zone with a long period of normal geomagnetic polarity lasting from the Early Aptian until the Early Campanian.

The Jurassic magnetic smooth zone can not yet be explained unambiguously. It is probably a geomagnetic phenomenon, but diversely conflicting interpretations as to its nature have been suggested. The amplitudes of anomalies M21 to M25 taper uniformly towards the smooth zone. This tapering amplitude envelope was interpreted by Larson and Hilde [1975] in terms of geomagnetic field intensity increasing away from an anomalously low intensity in the Jurassic smooth zone. Cande et al. [1977] have subsequently proposed that the Jurassic smooth zone represents a period of weak field intensity coupled with rapidly alternating polarity.

Both of these explanations rely on direct interpretation of marine magnetic anomalies and assume that the oceanic crust acts as a perfect recorder of the ancient geomagnetic field. Not all oceanic basalts possess the requisite magnetic stability, however, and many display instability of the remanent magnetization over long time intervals which manifests itself in the acquisition of viscous remanent magnetization [Lowrie and Kent, 1978]. VRM is not a rare phenomenon in oceanic basalts but has been reported at 40% of all DSDP sites (up to Leg 41) for which basalt magnetic properties have been published. Within a few weeks of laboratory observation, VRM in some cases exceeded NRM and in 1/4 of the samples studied exceeded 40% of NRM. These VRM results are probably not representative of all DSDP basalts. However, if it is a substantial and widespread magnetization component in the oceanic crust it may not always be appropriate to interpret the existence, absence or amplitudes of marine magnetic anomalies in terms of geomagnetic field behavior.

Paleomagnetic data from continental rocks do not definitely resolve the origin of the Jurassic smooth zone. Summaries of Jurassic paleomagnetic pole positions [McElhinny and Burek, 1971; Irving and Pullaiah, 1976] have indicated that the polarity of the geomagnetic field was dominantly, but not exclusively, normal during this period. Recently Steiner [1978] presented an updated summary of Jurassic paleomagnetic studies. Disregarding data whose age uncertainty exceeds two paleontological stages, the Jurassic polarities cited by Steiner are found to represent six unequal time intervals in each of which the number of reports of normal or reversed polarity can be counted [Fig. 2]. The Tithonian-Kimmeridgian mixed polarities correspond to the M-sequence oceanic anomalies. The Toarcian and older polarities are equally divided between normal and reversed but are too old to be observed in the marine record. Probably the mid-Bathonian to Toarcian

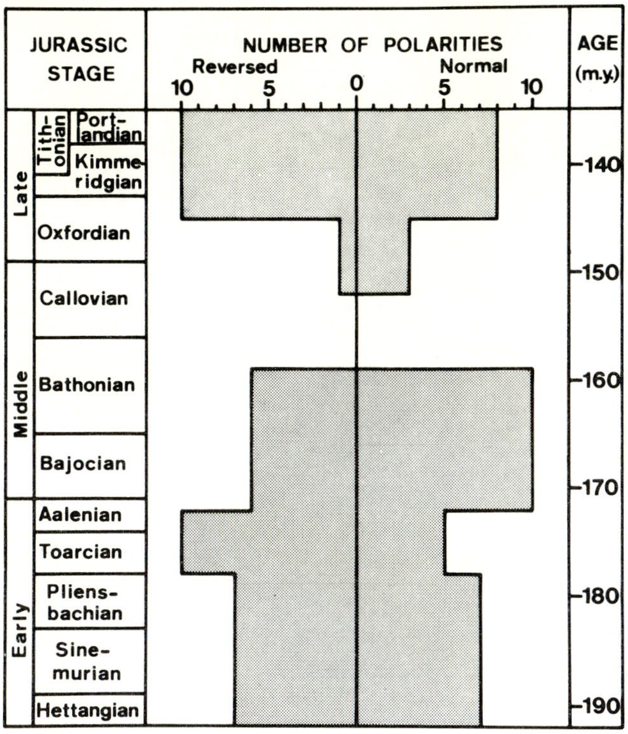

Fig. 2. The number of paleomagnetic normal and reversed polarities for different time intervals in the Jurassic [data after Steiner, 1978].

reversals would be lost in the marginal smooth zone adjacent to the Atlantic continental margins. Only the youngest part of the Jurassic magnetic smooth zone corresponding to the Late Bathonian, Callovian and Oxfordian stages (160-145 M.y.) lends itself to direct comparison with the paleomagnetic record. Unfortunately the data are so scarce for this interval that no conclusions regarding the predominant polarity can be made. The land-based paleomagnetic data are consistent with the interpretation of the Jurassic smooth zone as corresponding to a period of constant normal geomagnetic polarity, but are insufficient to require it [Irving and Pullaiah, 1976]. Once again data from continuous sequences offer the best means of resolving the issue.

Paleomagnetic investigations of the Middle Jurassic Summerville and Curtis formations indicate mainly negative polarity [Steiner, 1978]. Unfortunately these rocks are unfossiliferous, flat-lying red beds which were found to be strongly influenced by secondary components. No fold test or rock magnetic evidence concerning the age of the magnetization was presented. Thus the ages of these rocks and of their magnetizations are uncertain. In studies of shallow marine limestones from Franconia in southern Germany Heller [1978, 1977] has eliminated remagnetized sections in the Malm δ (Late Kimmeridgian) and shown that the Malm α, β and γ -- equivalent to Oxfordian and Early Kimmeridgian-- are normally magnetized. This result is in harmony with associating the Jurassic smooth zone with a long period of normal geomagnetic polarity, but would place the onset of the M-anomalies in the Middle Kimmeridgian at the earliest, a full stage later than inferred by Larson and Pitman [1972].

The magnetic smooth zones found adjacent to the margins of certain continents, such as southern Australia and Antarctica are not isochrons. Likewise, the landward Jurassic smooth zone boundary appears to cut across magnetic lineations in the eastern North Atlantic [Hayes and Rabinowitz, 1975]. Several different explanations have been offered for these marginal smooth zones. Because their landward boundaries are not isochrons they are probably not of geomagnetic origin but are due to a local geological cause related to the time of initial rifting. It has been proposed that these are areas of foundered continental crust rather than true oceanic crust [Talwani and Eldholm, 1972]. Even if true oceanic crust is present the submarine basalts form under conditions that are different from later steady state sea floor spreading. Irving [1970] assumed that conditions at initial rifting would favor formation of coarse-grained basalts with low NRM intensities and stabilities. Such basalts typically develop high VRM and it has been suggested that viscous remagnetization may be an important process in the production of the marginal smooth zones [Lowrie and Kent, 1978].

Cretaceous Spreading Rates

Larson and Pitman [1972] described the Cretaceous magnetic quiet interval as lasting from 112 M.y. (anomaly M1) to 85 M.y. (anomaly 34, which they did not number). They found that the computed average spreading rates for this interval were anomalously high at all the spreading centers in both the Atlantic and Pacific oceans. Subsequent discovery of anomaly M0 does not affect this conclusion which is based on the dating of anomalies 34 and M1.

Anomaly 34 has been identified in the Gubbio section close to the Santonian-Campanian boundary, whose age according to Van Hinte [1976] is 78 M.y., much younger than the age used by Larson and Pitman [1972]. Identification of anomaly M0 at the base of the Aptian [Channell et al., 1979], on the other hand, makes the youngest of the M-anomalies older than they were estimated by Larson and Pitman [1972] or Larson and Hilde [1975]; the age of M1 becomes approximately 118 M.y. (Early Barremian rather than Late Barremian). The Cretaceous quiet interval therefore lasted for 40 M.y., and Larson and Pitman's spreading rates for this interval are about one third too high.

The spreading rates during the formation of

TABLE 1. Average Spreading Rates (cm/yr) Recomputed From Larson and Pitman [1972].

Period	Spreading Center: Central Atlantic	South Atlantic	Phoenix-Pacific	Farallon-Pacific
Early Tertiary	1.6	2.1	1.8	4.0
Late Cretaceous	2.6	2.1	4.3	4.7
Cretaceous Quiet Zone	1.8	3.2	11.3	N4.4 S6.3
Early Cretaceous	1.2	no data	5.9	3.2

the M-anomalies were determined by Larson and Pitman [1972] with the aid of only two calibration points--estimated numeric ages of 155 M.y. for DSDP site 105 and 120 M.y. for DSDP site 166. The time interval covering the anomalies M1 to M22 lasted about 36 M.y. After revision of the age of M1 as above the spreading rates during formation of the Mesozoic anomalies must increase by about 15%. The net effect of the above considerations is to reduce the difference between Cretaceous quiet interval spreading rates and the spreading rates on either side of it. Recomputed spreading rates are given in Table 1.

In the central Atlantic the spreading rate in the Cretaceous quiet interval is essentially the same as in the Late Cretaceous lineated sequence, but both are still about double the rate in the Early Cretaceous preceding the quiet interval. In the South Atlantic estimates of quiet interval spreading rate are inexact for other reasons. The Cretaceous smooth zone extends there as far as the continental margins (where it may become a marginal type smooth zone) and the onset of spreading can not be timed from magnetic anomalies. Larson and Pitman [1972] assumed initiation of spreading in the South Atlantic at 110 M.y.B.P., but drew attention to the existence of lineated anomalies south of the Walvis Ridge, near the African margin, which could represent the last two or three anomalies of the M-sequence. After revision of the ages of the youngest M-sequence anomalies, the onset of spreading appears to be nearer to 120 M.y.B.P., and the time base for calculating the quiet interval spreading rate becomes 40 M.y. (120-80 M.y.B.P.) instead of 25 M.y. (110-85 M.y.B.P.). However, in spite of these adjustments the quiet interval spreading rate is about 50% higher than in the following Late Cretaceous and Cenozoic period of lineated anomalies.

In the Pacific ocean the re-computed quiet interval spreading rates are higher than in the preceding and following intervals at both the Phoenix-Pacific and Farallon-Pacific spreading centers. Larson and Pitman's [1972] conclusion that spreading rates increased sharply at all Pacific and Atlantic spreading centers during the Cretaceous magnetic quiet interval stands up to this re-examination. However, at the Central Atlantic and Farallon-Pacific spreading centers the higher spreading rate was not confined to the quiet interval but persisted for the remainder of the Cretaceous.

Direction of Magnetization of Oceanic Basalts

The source of oceanic magnetic anomalies has long been assigned to the strongly magnetized basaltic Layer 2, which has been sampled in a variety of ways. In basalt samples that were collected by dredging or from a submersible the inclination of remanence was determined relative to orientable volcanic features. DeBoer *et al.* [1969] found normal polarities in dredged

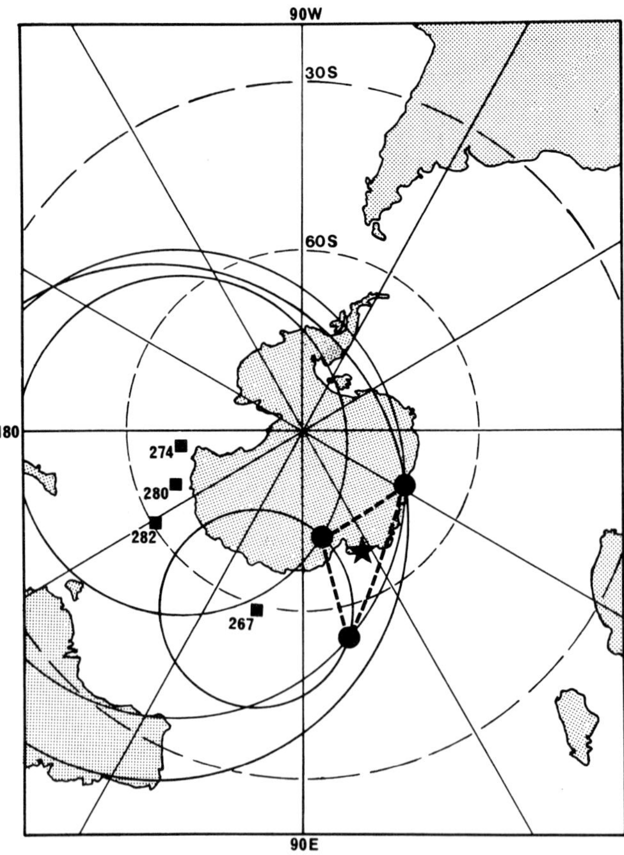

Fig. 3. Paleolatitude circles defining possible VGP loci for Atlantic DSDP sites. The optimum Early Tertiary VGP position is indicated by the star in the intersection triangle [from Lowrie, 1977].

basalts from the axial zone of the Reykjanes Ridge. Prévot et al. [1976] reported exact agreement with local axial dipole inclinations in a small collection of 3 dredged basalts and 6 basalts collected from the submersible Archimède. On the other hand mixed polarities were reported in samples obtained by shallow drilling within the MAR axial zone using an ingenious pressure-powered drill [Ade-Hall et al., 1973].

Direct sampling of Layer 2 in the Deep Sea Drilling Project (DSDP) provided a large supply of samples from depths up to nearly 600 m. in which the inclinations of the magnetizations of Layer 2 could be studied. The DSDP samples were not oriented in azimuth, but the inclination could be used to determine the paleolatitude at which the basalt was magnetized, after applying a correction factor to compensate for the absence of declination data [Briden and Ward, 1966]. The paleolatitude defines a circle as the locus of possible virtual geomagnetic poles (VGP) for a given DSDP site. At a few places anomalies of sufficiently similar age have been drilled so that the VGP position could be further located by the intersections of the paleolatitude circles. For example, an Early Tertiary VGP position was found for the Antarctic plate, lying some 15° from the present geographic pole [Fig. 3] and close enough to it to support the concept that the Antarctic plate has occupied its present high latitude throughout the Tertiary [Lowrie and Hayes, 1975; Lowrie, 1977].

At most DSDP sites, however, large discrepancies are found between the paleolatitudes derived from basalt magnetization inclinations and those expected, even after correction for plate motion [Lowrie, 1974, 1977]. There was no overall tendency for shallow or steeper inclinations at 55 geographically distributed DSDP sites [Fig. 4(a)]. At most sites only the top few meters or few tens of meters of basalt were sampled; the paleolatitude discrepencies might not be physically representative of Layer 2 but may result from incomplete averaging out of secular effects.

In recent years, however, some very deep holes have penetrated the basalt layer to depths of half a kilometer, and have yielded very curious results. When occasional samples in earlier shallower holes were found to have different polarities from their neighbors, this was usually attributed to misorientation of the sample. However, in the data from site 332A of Leg 37, zones of both positive and negative polarities are found, and at any chosen depth there is a very large scatter in the measured inclination [Hall, 1976]. Even after averaging over units or subunits to reduce this (as yet unexplained) large inclination scatter, there are found to be very large differences down the hole [Ryall et al., 1977]. Most strikingly, the inclinations are too shallow for the latitude at which the magnetizations were acquired, which could not be appreciably different from the present site latitude.

The inclination distribution that might be expected at a given latitude due to secular variation dispersion is similar to the Fisherian kind. Harrison and Watkins [1977] calculated the inclination expectation distribution for a latitude of 37°N and compared it with the observed inclinations in the deep holes of Leg 37 [Fig. 4(b)]. The observed low inclinations in Leg 37 were attributed by Harrison and Watkins to tectonic rotation about a horizontal axis, possibly as a result of collapse of the material above the magma chamber. The effect of such a rotation depends on the orientation of the rotation axis to the spreading center and on the sense of the rotation. Rotation about an axis parallel to an East-West trending ridge would produce a shallowing of the inclination on one side of the ridge and a steepening on the other side, for example.

Shallow inclinations have been reported at a few other deep holes, as also have some anomalously steep inclinations. Indeed it has not yet been established that shallow inclinations are typical for the oceanic basalt layer. The DSDP data from many shallow holes [Fig. 4(a)] indicate no preference for shallow values over steep values. The argument has been advanced that most of these holes are too shallow to be physically representative of Layer 2. Similarly, however, it is premature to conclude on the basis of the few deep holes that have been studied that they constitute a statistically representative sampling of Layer 2.

As puzzling as the anomalous inclinations in DSDP holes is the discovery of distinct zones with opposite polarities in the same hole. The mixed polarities and anomalous inclinations at hole 332 of Leg 37 appear to result from this hole having been drilled in a transition zone instead of in a negative zone as originally planned [MacDonald, 1976]. However, anomalous inclinations and polarities have also been reported at several other holes for which this explanation does not serve. Harrison [1976] has pointed out that one polarity change in a vertical section would not be surprising if the zone of injection of the dikes feeding the extrusive lavas were narrow; successive lava flows would not extend as far as their predecessors and would lead to an inclined boundary. However, the DSDP data require a more complex mode of flow emplacement, possibly involving episodic vulcanism at the ridge crest.

The remarkable symmetry of lineated magnetic anomaly patterns on spreading centers--such as the Reykjanes or Juan de Fuca ridges--constitutes one of their most striking features. However, on close examination the individual "stripes" are seen to be irregular in outline and sometimes discontinuous. The width of a given anomaly changes from profile to profile normal to the ridge axis; on certain profiles

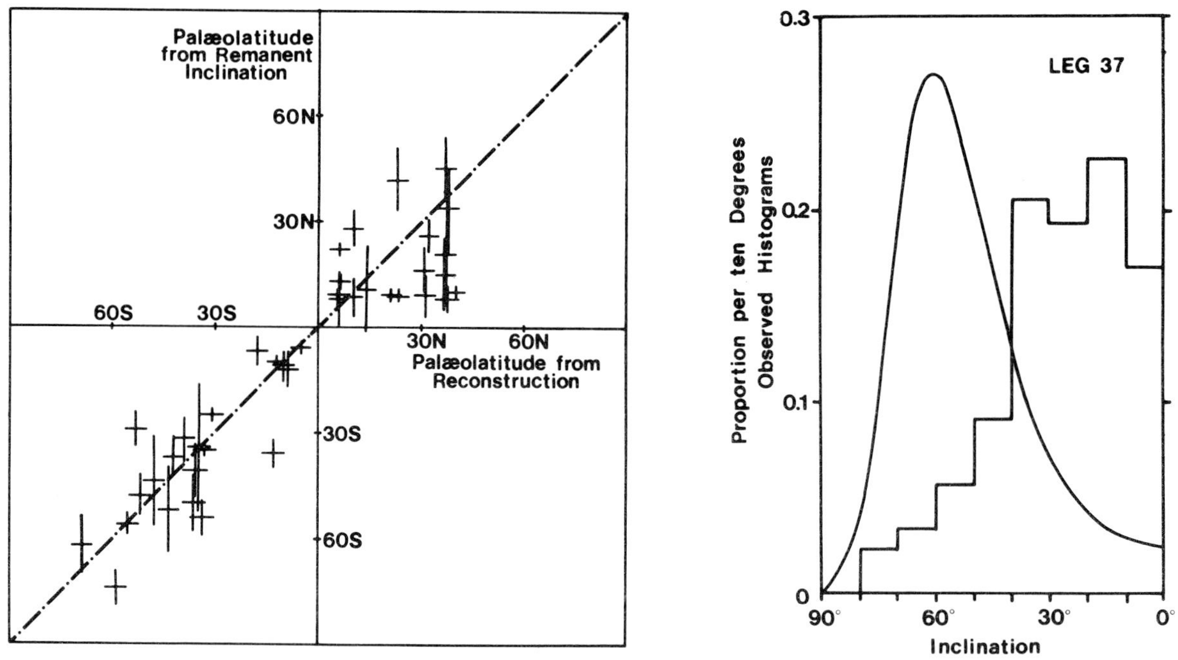

(a) (b)

Fig. 4. (a) Paleolatitudes derived from basalt magnetization inclinations at 55 DSDP sites are compared to the paleolatitudes expected from plate tectonic reconstructions (from Lowrie, 1977). Although scattered, the DSDP values show no preference to be higher or lower than expected. (b) The histogram of inclinations measured in basalt samples from DSDP Leg 37 sites is compared to a theoretical distribution curve (after Harrison and Watkins, 1977). The observed inclinations are predominantly shallower than expected.

the anomalies corresponding to the Jaramillo and Olduvai events are missing. This could reflect in part the degree of magnetic resolution in the survey or interference from topographic effects. However, the irregularity and discontinuity of the stripes may also be evidence that accretion of new crust is episodic along the ridge on a time scale of 10^4-10^5 years.

Rock magnetic studies on basalts dredged from the axial zone of the Reykjanes ridge showed systematic variations of titanomagnetite content and magnetic properties with distance along the ridge [deBoer, 1975]. Similarly, variations of remanent magnetic properties were observed in basalts sampled from within the mid-Atlantic ridge axial zone during the FAMOUS project [Prévot and Lecaille, 1976]. Although Curie temperatures--used as an indication of composition--did not vary significantly within the axial zone, local and regional variations in remanent intensity were attributed to corresponding fluctuations in concentration and grain size of the magnetic minerals. Apparently not only the accretion but also the magnetic mineralogy of the basaltic lavas may fluctuate with time.

The Magnetic Anomaly Source Layer

Oceanic Basalt Magnetic Mineralogy

Oceanic basalts generally correspond to deuteric oxidation class 1, although occasionally higher oxidation classes have been reported [Ade-Hall, 1974]. The primary magnetic mineral is commonly a skeletally shaped titanomagnetite

$$(x\ Fe_2TiO_4 \cdot (1-x)Fe_3O_4)$$

with ulvöspinel content (x) in the range 0.4-0.7. The grain size may be as large as 100μ [e.g. Ade-Hall, 1974], but more commonly is finer than 10μ and frequently is micron to submicron size. The grains progressively become converted to titanomaghemite at ambient ocean floor temperatures, which is visible on polished sections as a characteristic whitening along cracks and on the rims of coarser grains. Fine grains may be entirely maghemitized. Alteration also causes volume changes, which manifest themselves as cracks in the grain due to shrinkage. These cracks can be filled secondarily with sulfides [see Ryall *et al.*, 1977 for examples].

The alteration leads to the phenomenon of granulation in older, altered basalts [Ade-Hall et al., 1971]. Accompanying alteration there is a reduction in spontaneous magnetization and in the remanence intensity.

Intensity of Magnetization

Estimates of the intensity of magnetization, and identification of the magnetic source layer have been made by a variety of methods. Talwani et al. [1971] measured magnetic anomaly profiles parallel to the axis of the Reykjanes Ridge on each of three lineations and estimated the magnetization by correlating the longitudinal anomalies with topography. They deduced magnetizations of 30 A/m for the central anomalies, 12 A/m for the negative anomaly younger than anomaly 4 at distance 75 km, and 7 A/m for anomaly 5 at distance 100 km. Using a value of 12 A/m for the uniform magnetization of a block model they accounted for the anomaly amplitudes on profiles perpendicular to the ridge by assuming an anomaly source layer thickness of 400 m. This implied that the anomalies derived entirely from the thin uppermost, pillow basalt Layer 2A, in contrast to earlier models which, using lower magnetizations, had associated the source of the anomalies with the entire thickness of Layer 2 [e.g. Heirtzler et al., 1968] or even the entire thickness of the oceanic crust [e.g. Vine and Matthews, 1963; Vine and Wilson, 1965].

In subsequent investigations of Pacific ocean spreading centers using deep-towed instruments, the thin 0.5 km source layer was usually adopted. These results demonstrated a marked lateral decrease in magnetization intensity with increasing distance from the spreading center. Anomalous magnetizations of 30-50 A/m are found near the axis, and decrease to values around 5-15 A/m on the flanks of spreading systems.

A semi-logarithmic plot of some deep-tow results [Fig. 5(a)] shows that two stages of relaxation are involved, characterized by quite different relaxation times, τ. Near the ridge axis the data are significantly correlated and yield relaxation times of around 1 M.y. or less. Over oceanic crust older than about a million years relaxation times of 6 to 20 M.y. are found, but only the Pacific-Antarctic Ridge data are significantly correlated.

In a systematic, detailed investigation of the magnetic properties of submarine basalts dredged from the North Mid-Atlantic Ridge high NRM intensities up to 100 A/m were measured in axial zone samples [Irving et al., 1970]. Within the axial zone the intensities were quite variable, probably indicating compositional and grain size variation of the magnetic minerals in the fresh basalts. The intensity fell off rapidly to around 5 A/m on the MAR flanks [Fig. 5(b)]. Again two processes are involved with relaxation times of around 1 M.y. near the ridge axis and 20 M.y. further away. However, again the data defining the second process are not significantly correlated.

The similarity of these results from different data sets is striking. The deep tow data represent the total magnetic signal derived from a uniformly magnetized layer assumed to be 0.5 km thick. The NRM intensity data reflect the properties of a surficial rock collection. Evidently the basalt magnetic behavior is fairly representative of the magnetic source layer, and the causal process of the intensity decay affects the entire thickness of the magnetized layer.

The initial rapid decrease of magnetization intensity has been attributed to low-temperature oxidation of the original titanomagnetite as a result of ocean-floor weathering [Irving et al, 1970; Irving, 1970; Marshall and Cox, 1972]. The oxidation is probably assisted by mild hydrothermal activity in the axial zone, and the initial decay may be enhanced by isothermal relaxation of the original thermoremanence [Irving et al., 1970], a process that could continue further away from the ridge. The data of Fig. 5 are generally not significantly correlated beyond an age of 1 M.y. and do not provide experimental support for the supposition that further decrease actually occurs.

The intensities of DSDP basalts are lower than those of dredged basalts. This is because most of the DSDP basalts were drilled in areas away from ridge crests where sediment cover is already thick enough to permit successful drilling, while most dredged basalts have come from axial zones. The DSDP basalt remanences and susceptibilities display wide variations, reflecting considerable variation in magnetic mineralogy and grain size. Because of the scatter of inclinations at each site, the effective mean remanent intensity must be substantially smaller than the arithmetic mean [Ryall et al., 1977]. The between-site variations in magnetic properties are larger than the within-site variations, which suggests that there must be correspondingly large horizontal variations in magnetization of the magnetic source layer [Lowrie, 1977].

When the DSDP remanences are averaged for the various ocean basins, a remarkably consistent value of around 4 A/m results, except for the Antarctic and South Atlantic oceans where the intensities are much weaker [Table 2]. This average value is close to the 5 A/m observed by Irving et al. [1970] in basalts dredged from the flanks of the MAR, but is much lower than the corresponding value of 5-15 A/m inferred from deep-tow studies.

The Thickness of the Magnetic Anomaly Source Layer

In a study of DSDP basalts from the South Atlantic ocean Lowrie et al. [1973] showed that

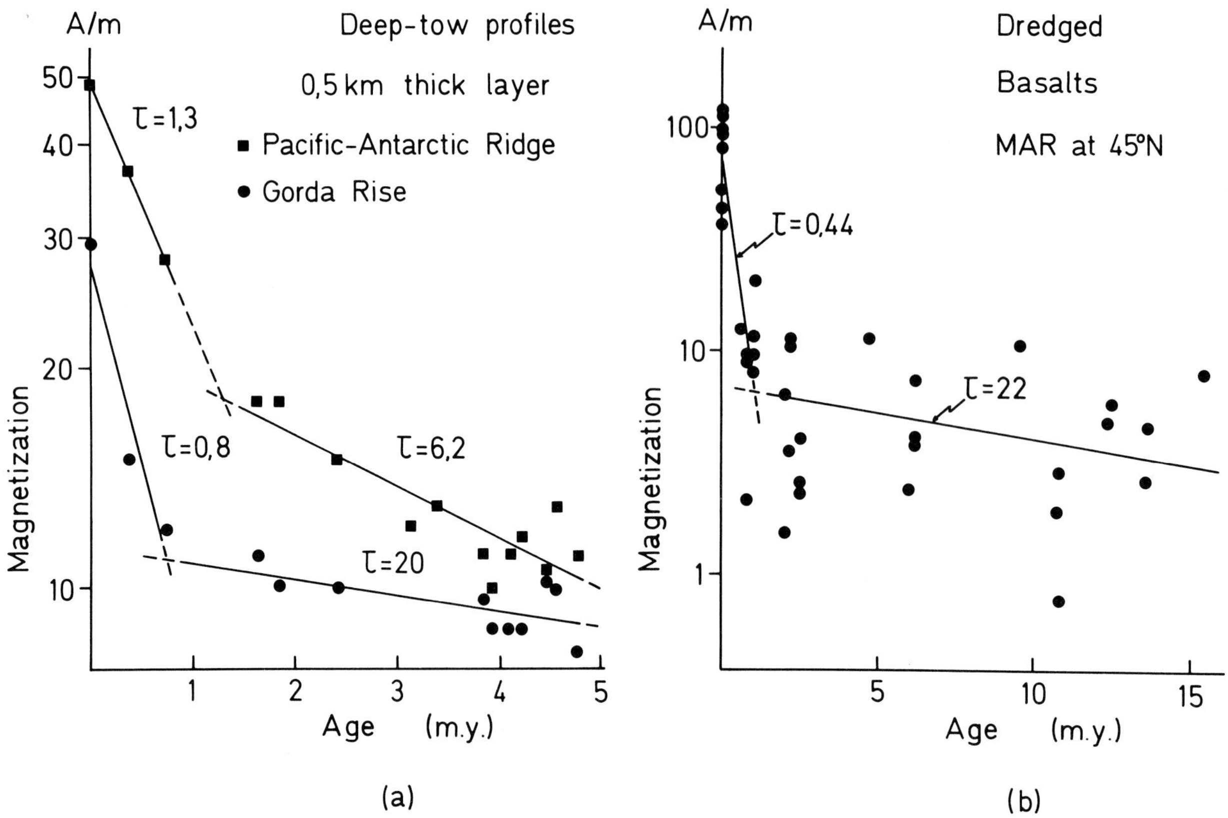

Fig. 5. The magnetization of the ocean crust decreases with distance from the spreading axis more rapidly than a simple exponential function. Two separate mechanisms may be involved, corresponding to different magnetization relaxation processes (straight line segments) at different distances from the ridge. (a) Data from deep-towed magnetometer surveys in the Pacific ocean (after Klitgord et al., 1974). (b) Data from dredged basalt samples from the mid-Atlantic ridge (after Irving et al., 1970).

the remanent intensities were too low to account for the amplitudes of oceanic anomalies on a nearby profile with a magnetized layer thickness of 0.5 km, and suggested a thickness of about 2 km, corresponding to the entire thickness of Layer 2.

Harrison [1976] compared the layer magnetizations and thicknesses used by many authors in modelling the source layer and demonstrated their inverse relationship. He normalized the magnetizations to equatorial values which are about 40% lower than the observed magnetizations at latitudes 20°-60° where most DSDP holes have been drilled. Thus the ridge-flank magnetization deduced from deep-tow studies becomes an equatorial value of about 9 A/m in a 0.5 km thick layer; this is equally satisfied by the observed DSDP basalts magnetization (equatorial value of 3 A/m) in a layer 1.5 km thick.

The depth to the source of magnetic anomalies may be determined from the slope at high wavenumbers of a plot of the natural log of power against wavenumber [Spector and Grant, 1970]. The portion of the power spectrum of an aeromagnetic profile over oceanic crust [Blakely et al., 1973] which was due to lineated magnetic anomalies was limited to wavenumbers less than 0.3 rad/km by noise at higher wavenumbers. Harrison [1976] derived theoretical power spectra for various spreading rates, source depths and source layer thicknesses, and compared their slopes at wavenumbers 0.1 to 0.3 rad/km with the slope measured between these

TABLE 2. Arithmetic Mean NRM Intensities in DSDP Basalts. N is Number of Samples or Sites.

	Samples		Sites	
Ocean	N	NRM (A/m)	N	NRM (A/m)
Indian	113	4.44	16	4.54
Pacific	59	4.17	12	3.67
Caribbean	23	4.18	5	4.54
N. Atlantic	64	4.69	10	4.54
S. Atlantic	10	1.21	4	1.00
Antarctic	25	2.45	7	2.22

numbers in the power spectrum of Blakely et al. [1973]. To satisfy the observed slope Harrison deduced that (a) the reversal rate had to be much less than that measured, or (b) the depth of the magnetic source layer was much greater than the surface of the igneous basement, or (c) the source layer was much thicker than 0.5 km. Harrison rejected the first two possibilities as less likely than the third.

Further support for a thicker source layer comes from the deep holes of Leg 37 [Ryall et al., 1977]. It was found that: (1) the remanent intensity showed no significant correlation with depth (at the 95% confidence level) throughout the 589 m of basement penetrated in hole 332B, or in the other deep holes, and (2) the measured remanent intensities are too low to account for the magnetic anomaly amplitudes in the vicinity of the Leg 37 holes.

Seismic evidence appears to limit the thickness of the strongly magnetized pillow basalt Layer 2A to about 0.5 km, whereas the available evidence from magnetic anomaly modelling or inversion and from rock magnetic studies indicate that at least part of the anomaly sources are more deeply located within the oceanic crust. A realistic model of ocean crust magnetization may need to include deeper layers consisting of other rock types. At present the magnetic properties of these rocks are known only from ophiolite studies and measurements on dredged samples.

Ophiolites

A complete ophiolite suite consists of an ordered rock sequence made up of pelagic sediments, pillow basalt, a sheeted dike complex, gabbro, peridotite and serpentinite. Frequently one or more members may be missing. Ophiolites occur on the continents in many parts of the world. Because of the resemblance of their layered structure and physical properties to those of oceanic crust ophiolites are regarded in light of plate tectonic theory as former oceanic crust that has been emplaced tectonically on the continents [see, for example, Moores and Vine, 1971]. The mode of emplacement is obscure, but may involve obduction of oceanic crust at a consuming plate boundary. Whether the ophiolite suite is typical of oceanic crust formed at an accreting plate boundary or in the marginal basin behind a consuming plate boundary is currently a matter of contention. In principle, paleomagnetic investigation of these accessible rock sequences ought to give a vertical profile of the magnetization of the oceanic crust, which would elucidate in particular the contribution of the deeper layers to oceanic magnetic anomalies.

Unfortunately, it is often found that the members of an ophiolite suite have undergone extensive alteration and metamorphism which have modified their magnetic properties to the extent that in many cases they no longer are representative of original oceanic crust. For example, in the Smartville ophiolite complex the remanent intensities of altered pillow lavas averaged only 0.05 A/m [Levi and Banerjee, 1977] which is two orders of magnitude less than the average values reported in DSDP or dredged basalts [Lowrie, 1974, 1977].

Rock magnetic studies have suggested that two processes may be active in altering the ophiolite magnetic properties. These are (a) ocean floor weathering and (b) metamorphism. For example, Butler et al. [1976] described thermomagnetic evidence for maghemitization of the original magnetic mineralogy in unmetamorphosed pillow basalts, with complete conversion to magnetite in metamorphosed pillows. Levi and Banerjee [1977] found a correspondence of remanent magnetic properties with degree of lateration. Sheeted dike samples had reversible thermomagnetic curves with magnetite Curie points, and their remanent intensities were higher than in altered pillow basalts. Provided the high Curie points and remanences are not the result of emplacement metamorphism the sheeted dikes complex may make a strong contribution to oceanic magnetic anomalies [Levi and Banerjee, 1977; Butler et al., 1976]. This conclusion is opposite to that of Kent et al. [1978] who concluded from studies of dredged samples that the sheeted dike complex consists of meta-basalts and probably makes no appreciable contribution to oceanic magnetic anomalies, although this point is not crucial to the model they proposed.

The metamorphism in ophiolites could result from hydrothermal processes near an accreting plate boundary, from burial effects, or from the tectonic emplacement of the ophiolites on to a continent. Evaluation and separation of the effects of these different processes requires detailed petrographic and structural geological evaluation of the ophiolite complex. In a study of ophiolite complexes in southern Chile, de Wit and Stern [1976] distinguished two metamorphic episodes. The second, later episode was a low-grade effect related to regional deformation of the rocks surrounding the ophiolites. The first, earlier episode represented ocean-floor metamorphism influenced by hydrothermal circulation near a spreading center and produced a metamorphic overprint that determined the physical properties of the respective layers of the oceanic crust. In particular the model of de Wit and Stern [1976] accounts for the seismic layering of the oceanic crust and correlates magnetic properties with each layer. The sheeted dikes and uppermost gabbros are affected by the hydrothermal circulation, while the pillow basalts are little affected and the deep-seated gabbros are unaffected. Kent et al. [1978], using a larger data base of dredged samples, invoked the same relationships in a model of the oceanic crust magnetization in which the contribution of the gabbro layer to

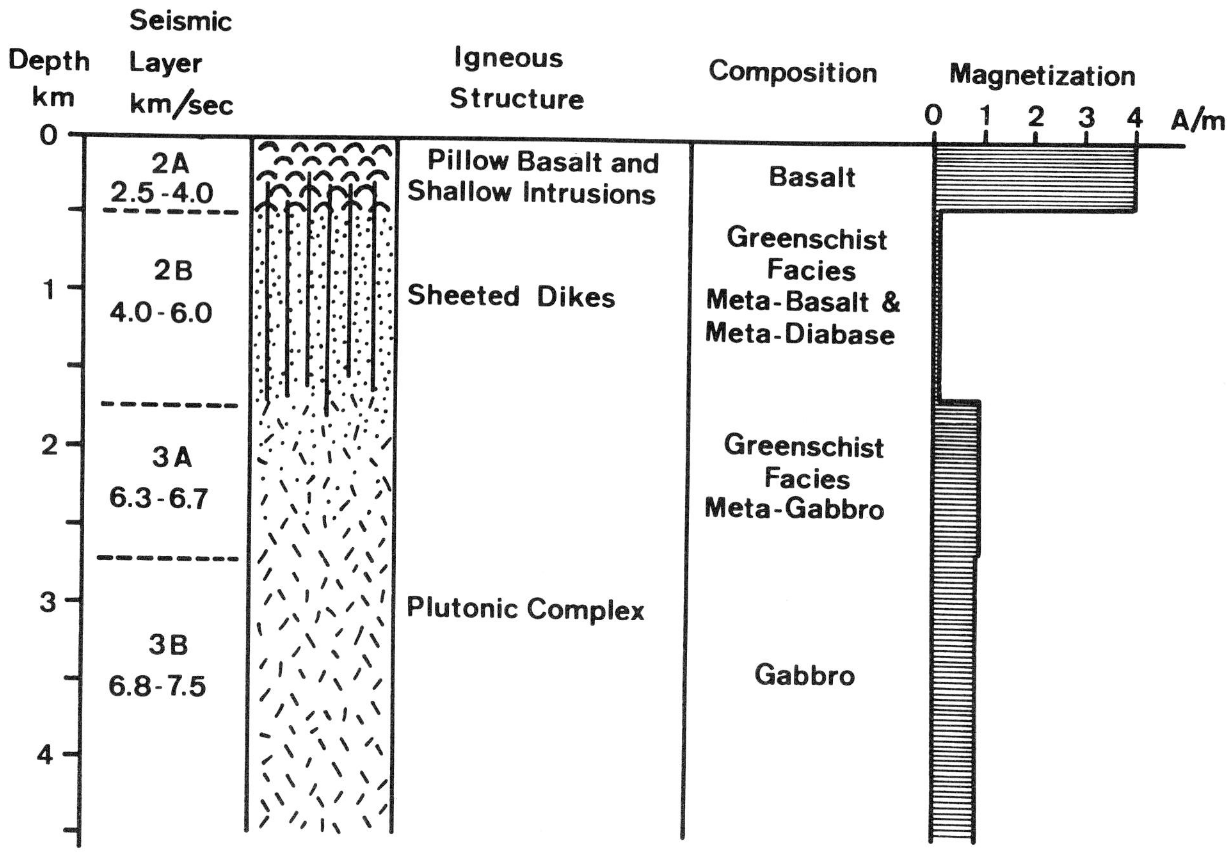

Fig. 6. Model of the magnetization structure of the oceanic crust [After Kent et al., 1978].

oceanic magnetic anomalies is appreciable.

It is apparent from the above discussion that the degree to which ophiolite suites can be used to determine the vertical stratification of magnetization in the oceanic crust hinges largely on the ability to separate samples on the basis of their degree of alteration and its cause.

Multi-Layer Models of Magnetic Anomaly Sources

Rock types typical of deeper layers of the oceanic crust have not yet been studied as intensively as pillow basalts. Meta-basalts probably represent the sheeted dikes of Layer 2B. Dredged samples have been found to be only weakly magnetized [Fox and Opdyke, 1973] and it is unlikely that they could make an appreciable contribution to magnetic anomalies. Serpentinites are quite strongly magnetized, but there is some conflict of opinion as to whether they occur as a uniform layer or are dispersed throughout Layer 2 or 3. Possibly they are tectonically emplaced along the fracture zones from which dredged samples have been predominantly recovered.

Recently, two-layer models of the sources of magnetic anomalies have been developed independently by different workers [Blakely, 1976; Cande and Kent, 1976]. A common feature of the models is that the pillow basalts cool by rapid chilling, while the deeper layer cools more slowly by conduction and passes through the Curie point at some distance from the spreading center. The interface between adjacent magnetized blocks in the deeper layer therefore slopes away from the spreading center.

Blakely [1976] observed that the width of the transition zone between adjacent oppositely magnetized blocks of ocean crust appeared to increase with crustal age. He invoked a two-layer model for the anomaly sources, corresponding to the pillow basalt and sheeted dike layers, as in the model of Butler et al. [1976]. A narrow transition zone in the upper layer and a wider transition zone in the lower layer (due to slower cooling) resulted in sloping interfaces between adjacent blocks. The apparent increase in transition width with age was ascribed to progressive deterioration of the magnetization of the upper layer.

Seismic data indicate that Layer 3 is composed of gabbro, and is much thicker than the basaltic Layer 2. Oceanic gabbros have also been dredged mainly from fracture zones and as a result their magnetic properties may not be typical of in situ material. Kent et al. [1978]

found that the average intensities of magnetization of meta-gabbros and unaltered gabbros were approximately 0.9 A/m and 0.8 A/m, respectively, or about one fifth that of the basalt layer. The Curie point of the gabbros was around 550°C, they were in a higher state of deuteric oxidation than the basalts, and their remanences were stable against demagnetization. Kent et al. [1978] integrated these observations in a new model of the magnetization of the oceanic crust [Fig. 6]. Most of the observed magnetic signal derives from the basalts in Layer 2A; the meta-basalts in Layer 2B make negligible contribution; the meta-gabbros in Layer 3A and the unaltered gabbros in Layer 3B participate jointly as a gabbroic source layer.

Cande and Kent [1976] showed that the signal contributed by the deeper gabbroic source layer was shifted in phase relative to the signal from the shallower basaltic layer. This results in an anomalous skewness of the magnetic anomalies after they have been transformed to their paleomagnetic pole. By varying the relative amplitude contributions of the upper and lower source layers they best approximated the observed profile EL33 with a ratio 3 to 1. If the upper layer consists of 0.5 km of pillow basalts with average magnetization 4 A/m, and the lower layer of gabbro with magnetization 0.8 A/m, the optimum modelling of the amplitudes and skewness of the magnetic anomalies on profile EL33 would be achieved with a gabbro layer thickness of about 1 km.

Discussion and Conclusions

The geomagnetic polarity sequence inferred from oceanic magnetic anomalies is of fundamental importance for determining sea-floor spreading rates and for dating relative plate motions and their tectonic consequences. Yet major parts of the reversal sequences in the Tertiary and in the Mesozoic have not been confirmed independently of the anomalies themselves. It is not even known what polarity state the geomagnetic field was in during formation of the Jurassic smooth zones. Paleomagnetic investigations of continental rock sequences are essential to verify and date more precisely the entire reversal sequence of the last 180 M.y. and to clear up the uncertainties associated with it.

The distribution of magnetization in the magnetic source layer is still somewhat uncertain. For example, slow spreading centers appear to have unusually large magnetic anomalies associated with their axial zones, while fast spreading centers do not. At most spreading centers a rapid decrease of crustal magnetization with distance from the spreading center has been inferred. This intensity reduction appears to take place within about a million years accompanying low temperature oxidation of the magnetic minerals. There is no strong evidence that the oceanic crust magnetization continues to decrease in intensity as has occasionally been claimed and in fact, oceanic basalt intensities are as high in some Cretaceous and Jurassic DSDP holes as in much younger Tertiary holes. The gentle decreases over spreading center flanks [Fig. 5] mostly are not statistically significant. Also, the suggestion that old crust is a less faithful recorder of short polarity events [Blakely, 1976] is not well documented. For example, the Cretaceous anomalies 30 and 31 are separated by a very short negative polarity zone [Fig. 1] lasting about 80,000 yrs [LaBrecque et al., 1977], but this reversed zone was found on all oceanic magnetic profiles to which the Gubbio magnetic stratigraphy was correlated [Lowrie and Alvarez, 1977a, 1977b]. In spite of the large amount that is known, our accumulated knowledge of oceanic crustal magnetization is still fragmentary. Generalizations are both inapt and deceptive.

Scattered inclinations (mostly shallower than expected, but also steeper) and occasional polarity reversals have been found in several deep DSDP holes. Although too few deep holes have yet been drilled to establish how prevalent these characteristics are, they indicate a complex magnetic structure in the basaltic layer. Remanent intensities of oceanic basalts within individual DSDP holes and at a broad geographic distribution of holes vary over several orders of magnitude but average around 4 A/m in several oceanic basins. The effective magnetization is reduced further by the complexity of the basaltic layer magnetizations. Deep-tow studies near to the ocean bottom and modelling experiments on anomalies measured at the ocean surface are able to resolve quite short magnetic reversals and imply that at least part of the oceanic magnetic signature derives from comparatively shallow sources, most probably from the basalt layer. A growing body of evidence now supports the existence of a second, deeper source layer whose effect is to broaden polarity transition zones, and produce phase shifts in the surface magnetic anomalies. It has not been positively established whether the second source layer corresponds to the sheeted dikes or the deeper gabbroic Layer 3, but rock magnetic studies indicate that the gabbros may contribute to oceanic magnetic anomalies [Kent et al., 1978]. Perhaps the most promising means of quickly and inexpensively advancing our knowledge of the magnetization of the oceanic crust is by rock magnetic studies on ophiolites. These will only be meaningful if accompanied by detailed petrographic study of the effects of ocean-floor and emplacement metamorphism on the members of the ophiolite suite.

Acknowledgements

I am grateful to Dr. D. V. Kent for giving me a preprint of the article on oceanic gabbros,

Drs. J. E. T. Channell, C. G. A. Harrison, F. Heller, E. Irving, D. V. Kent and F. J. Vine for their constructive suggestions as to how the manuscript could be improved, and to M. S. Lowrie for typing the original and final camera-ready copies. Institut für Geophysik, ETH-Zürich, Contribution Nr. 228.

References

Ade-Hall, J.M., The opaque mineralogy of basalts from DSDP Leg 26, In: T.A. Davies and B.P. Luyendyk et al., Initial Reports of the Deep Sea Drilling Project, Vol. 26 (U.S. Govt. Printing Office), 533-539, 1974.

Ade-Hall, J.M., H.C. Palmer, and T.P. Hubbard, The magnetic and opaque petrological response of basalts to regional hydrothermal alteration, Geophys. J. R. astr. Soc., 24, 137-174, 1971.

Ade-Hall, J.M., F. Aumento, P.J.C. Ryall, R.E. Gerstein, J. Brooke, and D.L. McKeown, The Mid-Atlantic ridge near 45°N. XXI. Magnetic results from basalt drill cores from the Median Valley, Can. J. Earth Sci., 10, 679-696, 1973.

Alvarez, W., and W. Lowrie, Upper Cretaceous paleomagnetic stratigraphy at Moria (Umbrian Apennines, Italy): confirmation of the Gubbio section, Geophys. J. R. astr. Soc., 55, 1-18, 1978.

Alvarez, W., M.A. Arthur, A.G. Fischer, W. Lowrie, G. Napoleone, I. Premoli Silva, and W.M. Roggenthen, Upper Cretaceous-Paleocene magnetic stratigraphy at Gubbio, Italy. V. Type section for the Late Cretaceous-Paleocene geomagnetic reversal time scale, Geol. Soc. Amer. Bull., 88, 383-389, 1977.

Blakely, R.J., Geomagnetic reversals and crustal spreading rates during the Miocene, J. Geophys. Res., 79, 2979-2985, 1974.

Blakely, R.J., An age-dependent, two-layer model for marine magnetic anomalies, The Geophysics of the Pacific Ocean Basin and Its Margin, AGU Geophysical Monograph 19, 227-235, 1976.

Blakely, R.J., A. Cox, and E.J. Eufer, Vector magnetic data for detecting short polarity intervals in marine magnetic profiles, J. Geophys. Res., 78, 6977-6983, 1973.

Briden, J.C., and M.A. Ward, Analysis of magnetic inclination in borecores, Pure Appl. Geophys., 63, 133-152, 1966.

Butler, R.F., Banerjee, S.K., and J.H. Stout, Magnetic properties of oceanic pillow basalts: evidence from Macquarie Island, Geophys. J. R. astr. Soc., 47, 179-196, 1976.

Cande, S.C., and D.V. Kent, Constraints imposed by the shape of marine magnetic anomalies on the magnetic source, J. Geophys. Res., 81, 4157-4162, 1976.

Cande, S.C., J.L. LaBrecque, and R.L. Larson, Marine magnetic anomaly data from the Jurassic and Cretaceous quiet zones: Implications for long period intensity variations of the paleomagnetic field (abs.), Eos Trans. AGU, 58, 740, 1977.

Channell, J.E.T., W. Lowrie, and F. Medizza, Middle and early Cretaceous magnetic stratigraphy from the Cismon section, northern Italy, Earth Planet. Sci. Lett., in press, 1979.

Cox, A., Lengths of geomagnetic polarity intervals, J. Geophys. Res., 73, 3247-3260, 1968.

Cox, A., Geomagnetic reversals, Science, 163, 237-245, 1969.

Cox, A., D.M. Hopkins, and G.B. Dalrymple, Geomagnetic polarity epochs: Pribiloff Islands, Alaska, Bull. Geol. Soc. Amer., 77, 883-910, 1966.

deBoer, J., Variations of the magnetic properties of postglacial pillow lavas along the Reykjanes Ridge, J. Geophys. Res., 80, 3769-3776, 1975.

deBoer, J., J.G. Schilling, and D.C. Krause, Magnetic polarity of pillow basalts from Reykjanes Ridge, Science, 166, 996-998, 1969.

de Wit, M.J., and C.R. Stern, A model for ocean-floor metamorphism, seismic layering and magnetism, Nature, 264, 615-619, 1976.

Foster, J.H., and N.D. Opdyke, Upper Miocene to Recent magnetic stratigraphy in deep-sea sediments, J. Geophys. Res., 75, 4465-4473, 1970.

Fox, P.J., and N.D. Opdyke, Geology of the oceanic crust: Magnetic properties of oceanic rocks, J. Geophys. Res., 78, 5139-5154, 1973.

Geological Society, Phanerozoic time-scale, Quart. J. Geol. Soc. Lond., 120S, 260-262, 1964.

Hall, J.M., Major problems regarding the magnetization of oceanic crustal layer 2, J. Geophys. Res., 81, 4223-4230, 1976.

Harrison, C.G.A., Magnetization of the oceanic crust, Geophys. J. R. astr. Soc., 47, 257-284, 1976.

Harrison, C.G.A., and N.D. Watkins, Shallow inclinations of remanent magnetism in Deep-Sea Drilling Project igneous cores: Geomagnetic field behavior or postemplacement effects?, J. Geophys. Res., 82, 4869-4877, 1977.

Hayes, D.E., and P.D. Rabinowitz, Mesozoic magnetic lineations and the magnetic quiet zone off northwest Africa, Earth Planet. Sci. Lett., 28, 105-115, 1975.

Heirtzler, J.R., G.O. Dickson, E.M. Herron, W.C. Pitman III, and X. le Pichon, Marine magnetic anomalies, geomagnetic field reversals, and motion of the ocean floor and continents, J. Geophys. Res., 73, 2119-2136, 1968.

Heller, F., Rock magnetic studies of Upper Jurassic limestones from southern Germany, J. Geophys., 44, 525-543, 1978.

Heller, F., Palaeomagnetism of Upper Jurassic limestones from southern Germany, J. Geophys., 42, 475-488, 1977.

Helsley, C.F., and M.B. Steiner, Evidence for long periods of normal magnetic polarity in the Cretaceous period, Geol. Soc. Amer. Spec. Paper 121, 133, 1968.

Irving, E., the mid-Atlantic ridge at 45°N: XIV. Oxidation and magnetic properties of basalt; review and discussion, Can. J. Earth Sci., 7, 1528-1538, 1970.

Irving, E., and G. Pullaiah, Reversals of the geomagnetic field, magnetostratigraphy, and relative magnitude of paleosecular variation in the Phanerozoic, Earth Sci. Rev., 12, 35-64, 1976.

Irving, E., W.A. Robertson, and F. Aumento, The mid-Atlantic ridge near 45°N: VI. Remanent intensity, susceptibility and iron content, Can. J. Earth Sci., 7, 226-238, 1970.

Keating, B., C.E. Helsley, and E.A. Pessagno, Jr., Late Cretaceous reversal sequence, Geology, 3, 73-76, 1975.

Kent, D.V., B.M. Honnorez, N.D. Opdyke, and P.J. Fox, Magnetic properties of dredged oceanic gabbros and the source of marine magnetic anomalies, Geophys. J. R. astr. Soc., 55, 513-537, 1978.

Klitgord, K.D., S. P. Huestis, J.D. Mudie, and R.L. Parker, An analysis of near-bottom magnetic anomalies: Sea-floor spreading and the magnetized layer, Geophys. J. R. astr. Soc., 43, 387-424, 1975.

LaBrecque, J.L., D.V. Kent, and S.C. Cande, Revised magnetic polarity time scale for the Late Cretaceous and Cenozoic, Geology, 5, 330-335, 1977.

Larson, R.L., An updated time scale of magnetic reversals for the late Mesozoic (abs.), Eos Trans. AGU, 55, 236, 1974.

Larson, R.L., and T.W.C. Hilde, A revised time scale of magnetic reversals for the early Cretaceous and late Jurassic, J. Geophys. Res. 80, 2586-2594, 1975.

Larson, R.L., and W.C. Pitman III, World-wide correlations of Mesozoic magnetic anomalies, and its implications, Geol. Soc. Amer. Bull., 83, 3645-3662, 1972.

Larson, R.L., and W.C. Pitman III, World-wide correlation of Mesozoic magnetic anomalies: Reply, Geol. Soc. Amer. Bull., 86, 270-272, 1975.

Levi, S., and S.K. Banerjee, The effects of alteration on the natural remanent magnetization of three ophiolite complexes: Possible implications for the oceanic crust, J. Geomag. Geoelectr., 29, 421-439, 1977.

Lowrie, W., Oceanic basalt magnetic properties and the Vine and Matthews hypothesis, J. Geophys. Res., 40, 513-536, 1974.

Lowrie, W., Intensity and direction of magnetization in oceanic basalts, J. Geol. Soc., 133, 61-82, 1977.

Lowrie, W., and W. Alvarez, Upper Cretaceous-Paleocene magnetic stratigraphy at Gubbio, Italy. III. Upper Cretaceous magnetic stratigraphy, Geol. Soc. Amer. Bull., 88, 374-377, 1977a.

Lowrie, W., and W. Alvarez, Late Cretaceous geomagnetic polarity sequence: detailed rock- and palaeomagnetic studies of the scaglia rossa limestone at Gubbio, Italy, Geophys. J. R. astr. Soc., 51, 561-581, 1977b.

Lowrie, W., and D.E. Hayes, Magnetic properties of oceanic basalt samples, In: D.E. Hayes and L.A. Frakes et al., Initial Reports of the Deep Sea Drilling Project, Vol. 28, (U.S. Govt. Printing Office), 869-878, 1975.

Lowrie, W., and D.V. Kent, Characteristics of VRM in oceanic basalts, J. Geophys., 44, 297-315, 1978.

Lowrie, W., R. Løvlie, and N.D. Opdyke, The magnetic properties of Deep-Sea Drilling Project basalts from the Atlantic Ocean, Earth Planet. Sci. Lett., 17, 338-349, 1973.

MacDonald, K.C., Geomagnetic reversals and the deep drill hole at DSDP site 332, J. Geophys. Res., 81, 4163-4165, 1976.

Marshall, M., and A. Cox, Magnetic changes in pillow basalt due to sea-floor weathering, J. Geophys. Res., 77, 6459-6469, 1972.

Mason, R.G., and A.D. Raff, A magnetic survey off the west coast of North America 32°N to 42°N, Geol. Soc. Amer. Bull., 72, 1259-1265, 1961.

McDougall, I., N.D. Watkins, G.L. Walker, and L. Kristjansson, Potassium-argon and paleomagnetic analysis of Icelandic lava flows: limits on the age of anomaly 5, J. Geophys., Res., 81, 1505-1512, 1976.

McDougall, I., K. Saemundsson, H. Johannesson, N.D. Watkins, and L. Kristjansson, Extension of the geomagnetic polarity time scale to 6.5 M.y.: K-Ar dating, geological, and paleomagnetic study of a 3500 m lava section in western Iceland, Geol. Soc. Amer. Bull., 88, 1-15, 1977.

McElhinny, M.W., and P.J. Burek, Mesozoic paleomagnetic stratigraphy, Nature, 232, 98-102, 1971.

Moores, E.M., and F.J. Vine, The Troodos Massif, Cyprus, and other ophiolites as oceanic crust: evaluation and implications, Phil. Trans. R. Soc. London, Ser. A., 268, 443-466, 1971.

Opdyke, N.D., Paleomagnetism of deep-sea cores, Rev. Geophys. Space Phys., 10, 213-249, 1972.

Opdyke, N.D., L.H. Burckle, and A. Todd, The extension of the magnetic time scale in sediments of the central Pacific Ocean, Earth Planet. Sci. Lett., 22, 300-306, 1974.

Pitman, W.C., III, and M. Talwani, Sea floor spreading in the North Atlantic, Geol. Soc. Amer. Bull., 83, 619-646, 1972.

Premoli Silva, I., Upper Cretaceous-Paleocene magnetic stratigraphy at Gubbio, Italy. II. Biostratigraphy, Geol. Soc. Amer. Bull., 88, 371-374, 1977.

Prévot, M., and A. Lecaille, Comments on paper by K.D. Klitgord, "Sea-floor spreading: The central anomaly magnetization high", Earth Planet. Sci. Lett., 33, 164-168, 1976.

Prévot, M., A. Lecaille, J. Francheteau, and R. Hekinian, Magnetic inclination of basaltic lavas from the mid-Atlantic ridge near 37°N, Nature, 259, 649-653, 1976.

Raff, A.D., and R.G. Mason, Magnetic survey off the west coast of North America, 40°N-52°N latitude, Geol. Soc. Amer. Bull., 72, 1267-1270, 1961.

Roggenthen, W.R., and G. Napoleone, Upper Cretaceous-Paleocene magnetic stratigraphy at Gubbio, Italy. IV. Upper Maastrichtian-Paleocene magnetic stratigraphy, Geol. Soc. Amer. Bull., 88, 378-382, 1977.

Ryan, W.B., M.B. Cita, M.D. Rawson, L.H. Burckle, and T. Saito, A paleomagnetic assignment of Neogene stage boundaries and the development of isochronous datum planes between the Mediterranean, the Pacific and Indian Oceans in order to investigate the response of the World Ocean to the Mediterranean "salinity crisis", Riv. Italiana Paleontologia, 80, 631-688, 1974.

Ryall, P.J.C., J.M. Hall, J. Clark, and T. Milligan, Magnetization of oceanic crustal layer 2--results and thoughts after DSDP Leg 37, Can. J. Earth Sci., 14, 684-706, 1977.

Sclater, J.G., and R.L. Fisher, Evolution of the east central Indian Ocean, with emphasis on the tectonic setting of the Ninetyeast Ridge, Geol. Soc. Amer. Bull., 85, 683-702, 1974.

Spector, A., and F.S. Grant, Statistical models for interpreting aeromagnetic data, Geophys., 35, 293-302, 1970.

Steiger, R.H., and E. Jäger, Subcommission on geochronology: Convention on the use of decay constants in geo- and cosmochronology, Earth Planet. Sci. Lett., 36, 359-362, 1977.

Steiner, M.B., Magnetic polarity during the middle Jurassic as recorded in the Summerville and Curtis formations, Earth Planet. Sci. Lett., 38, 331-345, 1978.

Talwani, M., and O. Eldholm, Continental margin off Norway: A geophysical study, Geol. Soc. Amer. Bull., 83, 3575-3606, 1972.

Talwani, M., C. Windisch, and M.G. Langseth, Jr., Reykjanes ridge crest: a detailed geophysical study, J. Geophys. Res., 76, 473-517, 1971.

Theyer, F., and S.R. Hammond, Cenozoic magnetic time scale in deep-sea cores: Completion of the Neogene, Geology, 2, 487-492, 1974.

Van Hinte, J.E., A Cretaceous time scale, Amer. Assoc. Petrol. Geol. Bull., 60, 498-516, 1976.

Vine, F.J., and D.H. Matthews, Magnetic anomalies over oceanic ridges, Nature, 199, 947-949, 1963.

Vine, F.J., and J.T. Wilson, Magnetic anomalies over a young oceanic ridge off Vancouver Island, Science, 150, 485-489, 1965.

Watkins, N.D., Correlating stratigraphic zones and magnetic polarities, Geotimes, 20, 26-27, 1975.

MODELING THE OCEANIC MAGNETIC SOURCE LAYER

Hans Schouten and Charles R. Denham

Woods Hole Oceanographic Institution, Woods Hole, Massachusetts 02543

Abstract. The texture of an extrusive magnetic source layer and the variability of marine magnetic anomalies are studied as a function of two statistical parameters, λ and σ, that describe the temporal and spatial behavior of overlapping extrusive units that accumulated in an active spreading center, and subsequently, were transported outside. We use numerical simulations of an extrusive magnetic source layer and associated magnetic anomalies to combine the variability observed in seafloor spreading anomalies and Deep Sea Drilling Project cores with the FAMOUS observations. A simple two-parameter process of statisitcally controlled temporal (λ = 5-10 units/km spreading) and spatial (σ <2.5 km) emplacement of major extrusive units, adequately models most of the marine magnetic anomaly and in situ observations.

Introduction

The uppermost igneous part of the ocean crust is predominantly of extrusive origin. The estimated thickness (500 - 1000 m) and the magnetization intensity of recovered pillow basalts and massive flows are sufficient to account for the amplitude of the observed seafloor spreading anomalies.

The marine magnetic anomalies are caused by lateral magnetization contrasts in this source layer. These contrasts can be: changes in magnetized volume (i.e. layer thickness); changes in magnetization intensity or susceptibility (i.e. magnetic properties) of the material that builds the magnetic source layer; and changes in amplitude and direction of the remanent magnetization induced by the field at the time of the emplacement of the magnetic source layer. The most dramatic changes in direction of the remanent magnetization are, of course, induced by the polarity reversals of the Earth's paleofield.

Seafloor spreading and the emplacement of the magnetic source layer make the marine magnetic anomalies the most continuous record of the Earth's paleo-magnetic field behavior over the past 200 m.y. (Heirtzler et al., 1968; La Brecque et al., 1977; Larson and Hilde, 1975). Analyses of 0-10 mybp subareal paleomagnetic material indicate an essentially geocentric axial dipolar paleo-field (Wilson et al., 1972) changing polarity 2-3 times per m.y., with a polarity transition period of probably less than 10,000 years (Cox & Dalrymple, 1967; Opdyke et al., 1973).

Interpretations of marine magnetic anomalies generally justify the extrapolation of the "squarewave" paleofield behavior beyond 10 mybp. However, the magnetic source layer does not contain an ideal recording of this simple signal. When converted into time, the marine magnetic anomalies indicate polarity transitions longer than 10,000 years (spatial equivalent); also they demonstrate considerable variability between adjacent measurement profiles and between spreading regimes.

Prevalent Models

Matthews and Bath (1967) and Harrison (1968) modeled the crustal variability and polarity transitions by a process of dike injection randomly distributed over a zone of finite width, the rift valley. Atwater and Mudie (1973) used a more realistic model of a combination of injected feeder dikes and extruded lava flows.

The inner floor of the rift valley contains the bulk of the most recent volcanic activity (Macdonald, 1977). It is bounded by inward-facing scarps that mark the transition from a regime of predominant volcanic emplacement to one of tectonic deformation (the scarps will effectively constrict the extent of the lava flows). Hence, it is commonly assumed that the width of the original emplacement zone is less than or equal to the width of the inner valley floor.

In the FAMOUS area, the floor of rift valley #3 averages 11 km in width, or five times the average width of the inner floor in the well-studied rift valley #2 (Macdonald and Luyendyk, 1977). Considerable floor width variation in space and time thus can be expected. This allows for a wide range of widths of original emplacement of the magnetic source layer, and, consequently, a variable recording of the simple Earth field reversal signal.

Another factor which significantly influences the recording is the episodicity of emplacement. FAMOUS observations indicate that the extrusive volcanism which builds the volcanic layer is not continuous; estimates of the frequency of major volcanic activity range from 100 units/my (Ballard and van Andel, 1977) to 30 units/my (Atwater, 1979).

The Deep Sea Drilling cores in the central North Atlantic (Hall et al., 1975; Hall, 1976) have shown that the volcanic layer consists of ≥5 volcanic units emplaced on top of one another. The paleomagnetism of these cores also indicate a number of separate units. The remanent magnetization inclinations are constant within each unit, but they may differ from one unit to the other. Reversals have been found in many of the holes. Although the scatter of the observed inclinations cannot all be attributed to the paleosecular variation, the consistency of inclinations within each unit suggests that the units acquired their remanent magnetization in a relatively short time.

Statistical Emplacement Model of the Magnetic Source Layer

At the active spreading center, steady state is only achieved statistically, since the accretion process is a series of discrete successive events (Francheteau and Tapponnier, 1978). Discrete random emplacement (in time and space) of the magnetic source material implies discontinuous recording of the simple reversal signal. Discontinuous recording adds noise to the recorded signal. The noise will increase with decreasing frequency of emplacement.

The random emplacement of extrusives within the limits given by the width of emplacement zone can be described adequately by a Gaussian probability distribution of distances from the mean spreading axis. This Gaussian distribution gives the probability of <u>where</u> extrusives will be emplaced with respect to the mean spreading axis. The flow width will determine the amount of overlap which occurs. The frequency of emplacement can be described by a Poisson probability distribution of intervals between extrusive emplacement. The distribution gives the probability of <u>when</u> extrusive volcanism will take place with respect to the last volcanic event. Kono (1973) has shown that recent subaerial eruptions follow a simple Poisson model; we will adopt it for the oceanic situation.

The spatial distribution of eruptions will be assumed to be Gaussian, centered on the ridge axis. There is a 95% probability that emplacement occurs within the central four standard deviations (σ) of the Gaussian distribution. The predominance of youngest extrusives in the center of the inner floor, together with rare evidence for the off-axis volcanism beyond the first inward facing scarps, make a Gaussian distribution a fair approximation of the spatial emplacement probability.

The Poisson probability distribution $\lambda te^{-\lambda t}$ (λ is the frequency of emplacement averaged over an infinite period of time; t is the time interval between consecutive emplacements) is the most general statistical description of an event which occurs "every now and then", the length of the next time interval being statistically independent of the preceding interval.

The Gaussian and Poisson distributions are the simplest statistical generalizations of the facts known about the process of emplacement and as such are a first-order approximation to a broad band of possible processes and actual distributions.

<u>Final Distribution vs. Original Distribution, Figure 1</u>

Figure 1 shows the evolution of an original Gaussian distribution of extrusives emplaced at T_0, from the time of its emplacement T_0 to the final emplacement distributions at T_3 well outside the active rifting zone. At T_1, the original distribution has been stretched by the process of dike injection, open fissures, and normal faulting which account for the actual spreading of the volcanic layer in the active zone. The most pronounced stretching (spreading) takes place in the center. Stretching tapers off to the sides, leaving the flanks of the original distribution relatively unaffected. The outer flanks progressively move outside the active zone of stretching, while the inner flanks continue to be stretched. The small arrows indicate the position of the center (median) of the original distribution, and their separation indicates the total amount of spreading since T_0.

The two final distributions at T_3 are bell-shaped, for which another Gaussian distribution is an adequate first order approximation (Denham and Schouten, 1979); their widths are approximately equal to the width of the original distribution, they are only slightly skewed, tapering toward the active rifting zone and centered about the arrows which indicate the center (median) of the original distribution. The T_0-T_3 distributions are probability distributions. They indicate the probability (as function of time and distance to the spreading center) of finding extrusives that were emplaced during a finite time interval at time T_0.

The actual distribution of extrusives in the source layer depends on the frequency of emplacement and the size of the individual units. This is illustrated in the lower half of Figure 1. All three realizations conform to the same probability distribution T_3. They overlap source material emplaced earlier, and they are covered almost completely by later extrusives. The least-frequent emplacement distribution (bottom) displays the highest variability (i.e. deviation from the T_3 probability distribution). The shape

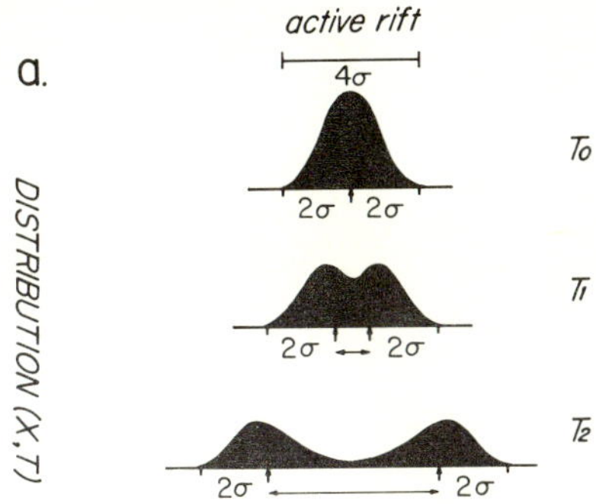

Fig. 1a. Evolution of a Gaussian distribution of extrusives originally emplaced in the active rift at time T_0, to two final emplacement distributions at T_3 when all the T_0 material has been transported outside the active rifting zone by seafloor spreading. In this example the standard deviation of Gaussian stretching (spreading) active within the rift zone is equal to the standard deviation of the T_0 distribution of extrusives ($\sigma_s = \sigma_e$). The arrows indicate the center of the original T_0 distribution, their separation indicates the amount of seafloor spreading since T_0. The bell shape of the final distributions can be described adequately by another Gaussian distribution.

of the final probability distribution T_3 will vary, depending on the standard deviation of the original emplacement (σ_e) and the standard deviation of stretching (σ_s). In Figure 1, $\sigma_e = \sigma_s$.

Width of the Final Magnetization Distribution

Atwater and Mudie (1973) explored the effect of the emplacement mechanism on the shape and width of the final magnetization transition in an extrusive source layer (the final magnetization transition is the integral of the final magnetization distribution). In their model, the stretching (spreading) is provided only by the dike injection that feeds the extrusives ($\sigma_e \geq \sigma_s$). They show that the final width of the magnetization transition (read also "final magnetization distribution") generated by this model ranges from roughly half the width of the original emplacement zone for a pure lava flow model (flow width >> zone of injection) to roughly equal the width of the original emplacement zone for a pure dike model. Stretching of the magnetic source layer by normal faults (Macdonald and Luyendyk, 1977) and by open fissures (Ballard and van Andel, 1977) decouples the stretching from the emplacement of extrusives. The assumption of significant tectonic stretching outside the original emplacement zone extends the range of possible emplacement models to $\sigma_e < \sigma_s$. Denham and Schouten (1979) show that for $\sigma_e < \sigma_s$, the standard deviation of the final distribution roughly equals σ_s (i.e. the width of the final magnetization distribution roughly equals the width of the stretching (spreading) zone).

If significant tectonic stretching also occurs outside the zone of extrusive emplacement of the magnetic source layer, it will be essentially impossible to detect the width of the median valley, or the width of the original distribution of extrusive activity in the median valley, by analysis of the observed magnetic anomalies. However, the addition of tectonic stretching to the source layer emplacement model can explain wide magnetic anomaly transition zones originating from narrow volcanic emplacement zones; also, it can explain why, contrary to Kidd's (1977a) predictions, relatively few feeder dikes have been encountered in the DSDP extrusive layer drill holes, since in this model normal faults and open fissures partly replace injected feeder dikes as the spreading agent of the extrusive layer.

Computer Simulation of the Magnetic Source Layer and Anomalies

For sake of simplicity in the numerical modeling, we did not simulate the actual emplacement and stretching process in the active spreading center, but rather we made use of the final emplacement distributions (Denham and Schouten, 1979) that evolve on either side of the zone of spreading from the one distribution originally emplaced at the spreading center. Our computer program was designed to sample the reversal time scale at Poisson distributed intervals and to emplace a magnetized flow, representing each sampling event, in a two-dimensional source layer array, conforming to a Gaussian probability distribution of distances between the center of the flow and the sampled time event-times-spreading rate position. The final distributions resulting from most processes of emplacement and stretching in the active rift can be approximated adequately by Gaussian distributions. The consecutive individual flows will be widely scattered or emplaced almost on top of one another.

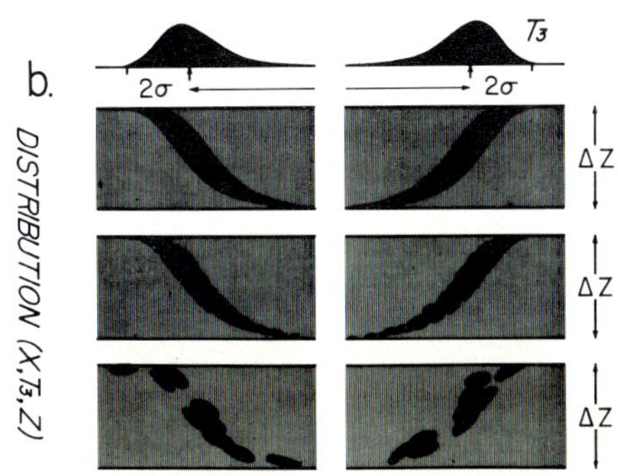

Fig. 1b. Cartoon of the final distribution of extrusives in a source layer of thickness ΔZ, for continuous, frequent and less frequent extrusive emplacement in the active rift.

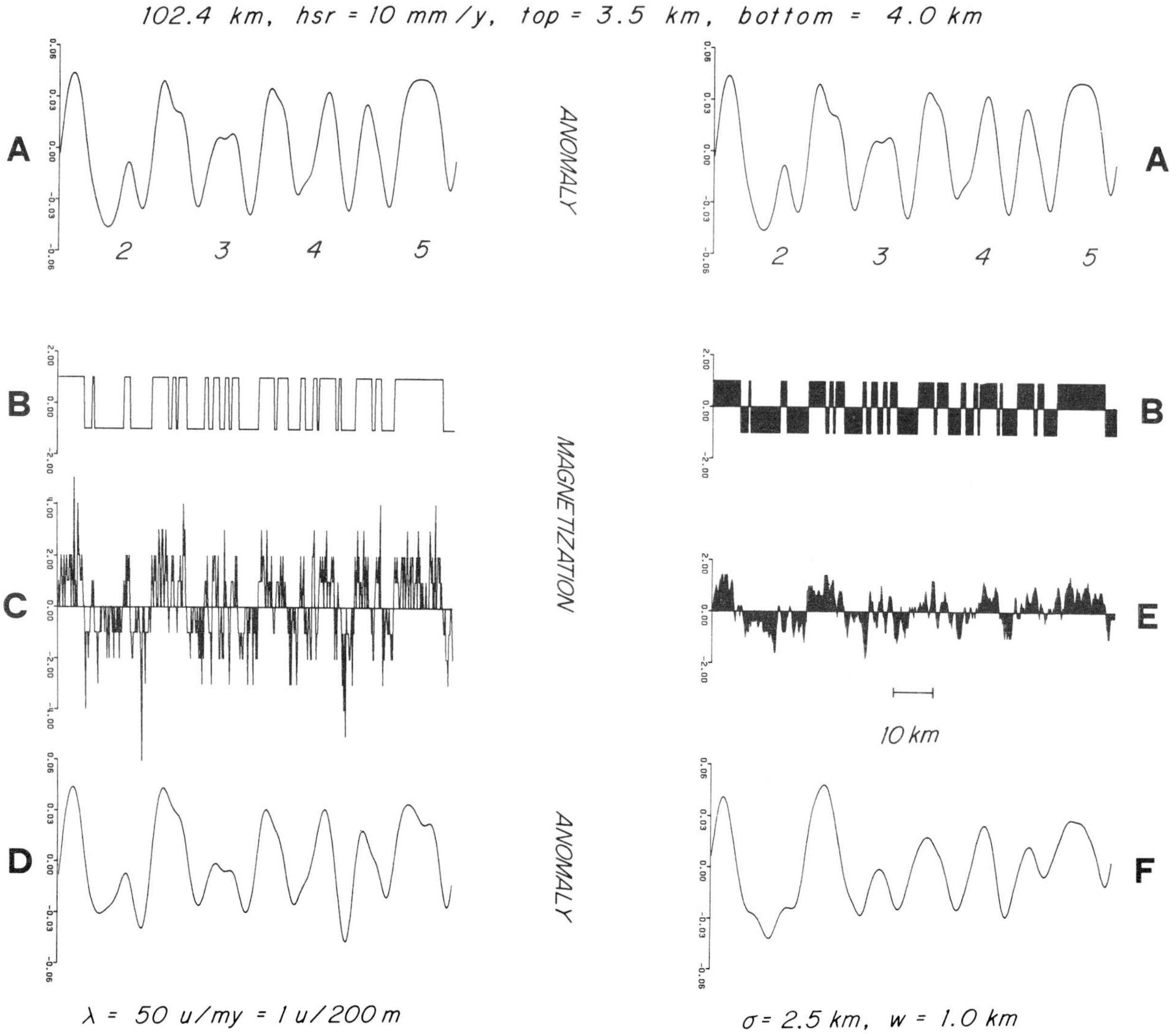

Fig. 2. Example of the effect of the statistical emplacement parameters λ (average frequency of emplacement) and σ (standard deviation of the final distribution of magnetic source material) on the magnetic anomalies. Curve A is the anomaly over the ideal reversal block model B, 0-10 mybp. In C the reversal time scale is sampled at an average frequency of 50 units/my (this corresponds to 100 unit/my emplacement activity at the spreading center). At this slow spreading (half spreading rate hsr - 10 mm/yr), the Poisson sampling by itself has very little effect on the magnetic anomalies D. The distance of the anomalies to the source (top 3.5 km - bottom 4.0 km) effectively suppresses the short wavelength fluctuations in C. The Gaussian distribution (σ = 2.5 km) of the Poisson sampled extrusives increases the scatter in the resulting magnetization distribution E, while the 1.0 km flow width acts as a smoothing function. The magnetization distribution E is a poor reflection of the uncontaminated magnetization distribution B, as is its magnetic anomaly F.

An Example, Figure 2

A sample of the procedure and of the effect of the two emplacement parameters λ and σ on the magnetic anomalies is shown in Figure 2. The Poisson sampling (left side of Figure 2) in itself has very little effect on the magnetic anomalies since the distance to the source effectively suppresses the short wavelengths due to the irregular sampling, while the large number of major polarity reversals are dominant at this slow spreading rate. In contrast, the standard deviation of 2.5 km (right side of Figure 2), scatters the individual flows over 10 km, which is wider than the average separation between the major polarity reversals, resulting in a poor recording of the reversal signal.

It is interesting to note that the apparent transition widths in the magnetization distribution E of Figure 2 suggest a much smaller value than the actual 10 km (4σ) that controlled the simulated process of emplacement. The noise caused by infrequent and widely scattered emplacement inhibits the recognition of the true 10 km wide transition zone.

Simulated Slow Spreading Anomalies, Figure 3

In Figure 3 we compare the effect of σ = 0.0, 0.5 and 2.5 km on the simulated magnetic anomalies (average half-frequency of emplacement of 50 units/my). The anomalies for σ = 0.0 and 0.5 km are not significantly different from the ideal reversal block model and hence display a low variability. However, the three examples of anomalies for σ = 2.5 km (zone of emplacement and tectonic stretching is 10 km) are poorly developed and show considerable variability between themselves. These three sets of magnetic anomalies refer to the same reversal time scale and statistical emplacement parameters λ and σ, while their variability is caused by the random character of the Poisson and Gaussian distributions. They can be viewed as three observed magnetic anomaly profiles over oceanic crust of the same age in a seafloor spreading regime that is determined by the same statistical parameters: average spreading rate, depth, and volume, average frequency λ and standard deviation σ of final emplacement.

The variability of these simulated anomalies is quite similar to the variability that characterizes a large part of the North Atlantic seafloor spreading magnetic anomalies, first noted by Matthews and Bath (1967). Faster spreading rates and higher frequency of emplacement tend to reduce the anomaly variability; the former reduces the σ (scatter) relative to the width of the polarity intervals, the latter provides a more continuous (i.e. less variable) recording of the polarity reversal signal.

Simulated Magnetic Source Layer, Figure 4

The variability of the magnetic source of these simulated slow spreading anomalies (Figure 3) is shown in Figure 4, confirming the obvious: variable magnetic anomalies have variable sources. In the source section for σ = 0.5 km, the material representing each polarity interval appears as one magnetic body overlapping the previous one. Each unit consists of an integer number of one km-wide flows. The considerable variation in source layer thickness has no significant effect on the associated near-perfect magnetic anomalies seen in Figure 3. The variable thickness of the source layer, which is a natural consequence of a random temporal emplacement

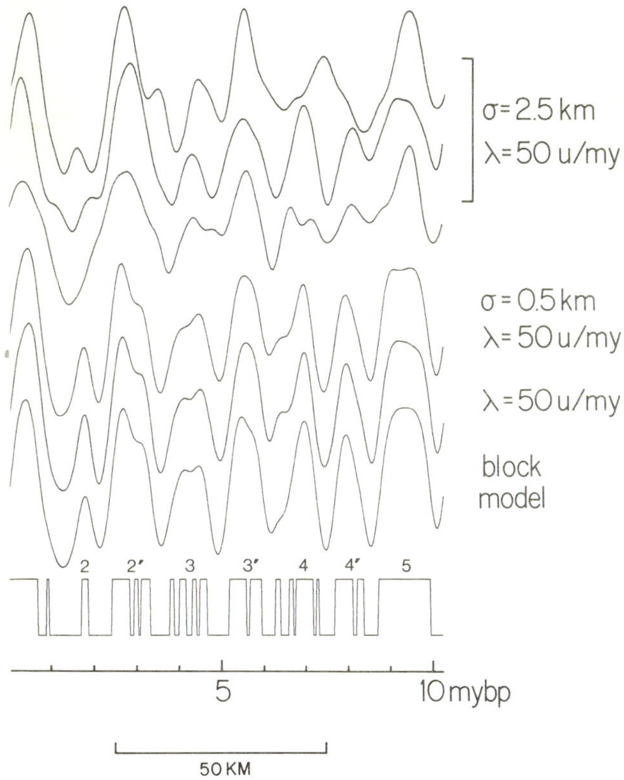

Fig. 3. The variability of simulated magnetic anomalies as a function of the standard deviation σ of the final emplacement distribution and the average frequency of emplacement λ. The low sampling frequency of 50 units/my (this corresponds to 100 units/my emplacement activity at the spreading center) combined with the large standard deviation of the scatter of the individual emplacement units, create the variability between the top three sets of magnetic anomalies that were calculated from three computer simulated extrusive source layers using the same statistical parameters. The smaller standard deviation generates near perfect (low variability) magnetic anomalies. The high variability between the σ = 2.5 km anomalies is qualitatively similar to the variability that characterizes a large part of the North Atlantic seafloor spreading magnetic anomalies.

process, also indicates the high probability for tilting of extrusives after their original emplacement. Because younger flows only overlap older ones, there will be predominant tilt toward the younger part of the crust, but many cases of opposite tilt also can be observed as a result of the scatter of the infrequent emplacement process. The variability of the source layer thickness is identical for σ = 2.5 km and σ = 0.5 km in Figure 3 (the thickness distribution is the Poisson distribution convolved with the

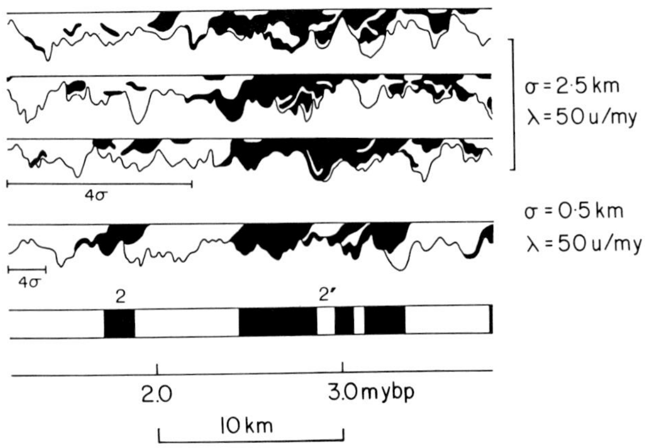

Fig. 4. The anomaly 2 and 2' section of the computer simulated source layers that were used to calculate the anomalies in Figure 3. The width of the individual flows is 1 km, the average thickness of the source layer is 5 times the average flow thickness, 4σ indicates the scatter of the individual flows and, consequently, the theoretical transition width. The upper 3 layers illustrate the variable texture of the extrusive source layer and the high probability of finding vertical sections of mixed magnetic polarity. The variable texture also indicates the high probability for tilting of the individual flows after their original (horizontal) emplacement. The variability of the source layer thickness depends on λ and the width of the flows only, which are identical for the upper 4 layer simulations. The vertical exaggeration is 2 for an average extrusive layer thickness of 500 m.

width of the individual flows and is independent of the standard deviation of emplacement). The large σ relative to the 1 km width of the flows has scattered the flows emplaced during the same polarity event and there exists a high probability of finding vertical sections of mixed magnetic polarity (Denham and Schouten, 1979).

Simulated Fast Spreading Anomalies, Figure 5

In Figure 5, we show the effect of spatial emplacement frequency on 50 mm/yr simulated magnetic anomalies. The variability of these anomalies is primarily caused by the variability of the source layer thickness (the anomalies are normalized to the maximum amplitude and clearly illustrate the increasing variability with decreasing frequency of emplacement).

In order to simulate the less variable seafloor spreading anomalies observed in the faster spreading Pacific, the higher half-frequency of emplacement of 500 units/my is required. This represents a spatial emplacement frequency of 10 units/km spreading, which is roughly equal to the Atlantic observations in FAMOUS.

Simulations and Observations

This simple emplacement model generates an extrusive source layer having the complex nature which has become apparent from the Deep Sea Drilling Project holes into the North Atlantic basement (Hall, 1976). The variability is a direct consequence of the fact that the emplacement of the source layer is not continuous, but rather it is derived from a discrete number of major extrusive events at irregular time intervals and randomly distributed over an emplacement zone of finite-width. Intrusive dikes and sills, faults, and off-axis volcanism will further contaminate the detailed nature of the source layer, but the magnetic anomalies will suffer only a small amount, due to the smoothing effect of distance to the source.

In the FAMOUS area the estimated frequency of the emplacement of major volcanic units is 100 units/my (Ballard & van Andel, 1977). At a total spreading rate of 20 mm/yr, this means an emplacement every 200 m of spreading on the average (5 units/km); for an average cross section of the units of 100,000 m^2, the average thickness of the extrusive source layer will be

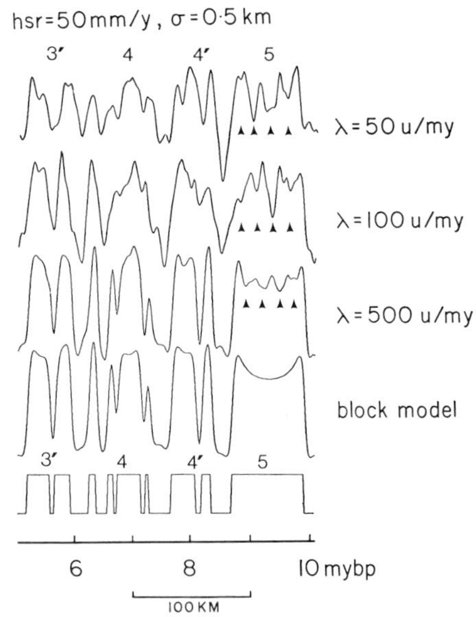

Fig. 5. The variability of magnetic anomalies simulated for a "Pacific" spreading rate of 50 mm/yr as a function of the average frequency of emplacement λ and a "Pacific" standard deviation of 0.5 km (Atwater and Mudie, 1973). The λ = 500 units/my simulates magnetic anomalies qualitatively similar to the low variability anomalies observed in most of the fast spreading Pacific. The spatial emplacement frequency is 10 units/km spreading, which is roughly equal to the 5 units/km estimated for the North Atlantic. The arrows indicate the apparent short wavelength information present in the anomaly 5 simulations which is caused by the variable source layer thickness.

500 m; for an average width of the volcanic units of 1000 m, the extrusive layer will be on average 5 units thick. These general statistics based on observations in the presently active FAMOUS rift zone agree quite well with the results of Deep Sea Drilling in the older central North Atlantic crust (DSDP Leg 52 Staff, 1977) which suggest a 500 - 1000 m thick extrusive layer consisting of 5-10 major volcanic units. Estimates of the magnetic source layer thickness based on observed magnetic anomalies (Talwani et al., 1971; Huestis and Parker, 1977) indicate a similar range.

Our anomaly simulations suggest that the variability of slow and fast spreading anomalies can be explained by a spatial emplacement frequency of 5-10 major extrusive units/km spreading. For the slow spreading anomalies, an additional standard deviation of final emplacement of ~ 2.5km is needed which, together with the low frequency of emplacement, generates the variable quality of the Atlantic observed anomalies. For the faster spreading anomalies, the polarity transitions are more widely spaced and the standard deviation of the final emplacement (<2.5 km) only modifies the individual transitions. The spatial emplacement frequency of 5-10 units/km spreading generates a variability in between the polarity transitions which is qualitatively similar to the short wavelength "wiggles" observed in the faster spreading anomalies (Blakely, 1974; Cande and LaBrecque, 1974).

Short Wavelength Information in Fast Spreading Anomalies, Figures 5 and 6

Considerable short wavelength information is present in observed anomalies from widely separated regions of equivalent age (Blakely et al., 1975). Cande and LaBrecque (1974) and Blakely (1974) correlated short wavelength anomalies in the Pacific, which they attribute, respectively, to fluctuations in magnetic intensity and to short polarity reversals of the paleomagnetic field.

In Figure 5, the arrows indicate the apparent short wavelength information present in the anomaly 5 simulations which is caused by the fluctuations in the source layer thickness. The average wavelength of the short wavelength anomalies is 12.5 km for all λ's. This apparent uniform wavelength can easily be understood. The Poisson emplacement of extrusive units generates a random source layer thickness variation. The spectrum of the magnetization distribution representing a source consisting of magnetic units emplaced at Poisson distributed space intervals is white (i.e. equal for all wavelengths) except for the mean value which represents the average layer thickness-times-magnetization (Lee, 1967, p. 244). The magnetic anomaly observed at a distance to the source layer is a band-passed representation of the source signal (Schouten, 1971; Schouten and McCamy, 1972). Depending on the depth and thickness of the source layer (the parameters which determine the shape of the bandpass), the magnetic anomalies will be dominated by a narrow band of wavelengths that suffer minimum attenuation. At first glance, this composite of a narrow wavelength band appears to be a one-wavelength sinusoidal anomaly which is very correlatable to another sinusoidal anomaly of the same apparent wavelength. (Note: Although

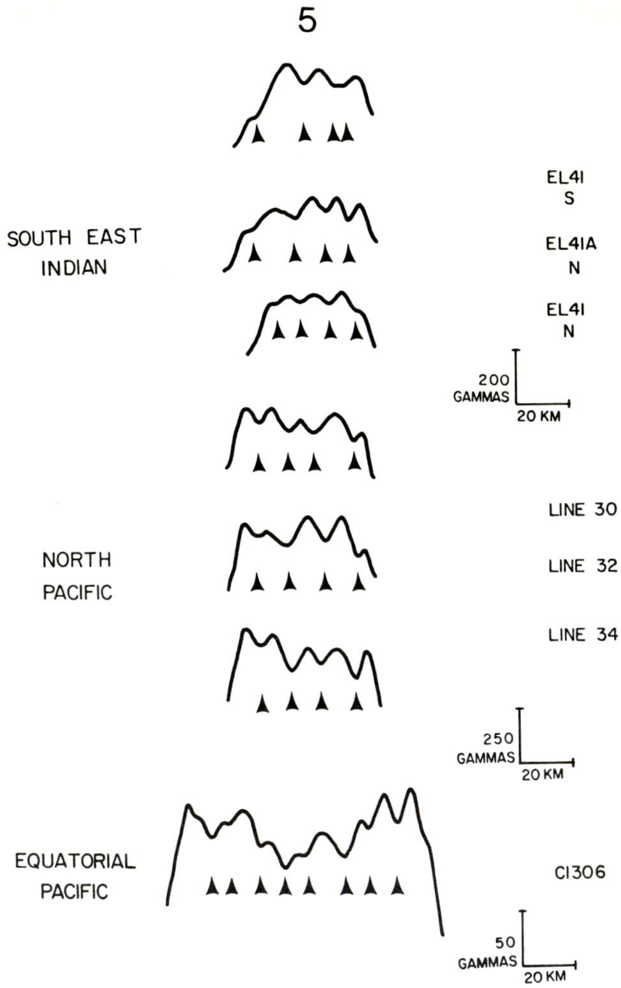

Fig. 6. Short wavelength anomalies observed in Indian Ocean and Pacific anomalies 5 (from: Cande and LaBrecque, 1974). The arrows indicate a dominant wavelength of 10-15 km, independent of the spreading rate at which the anomalies were formed. This apparent uniform wavelength is similar to the wavelength that dominates magnetic anomalies simulated over a white noise source at anomaly 5 basement depth. The amplitude of these short wavelength anomalies relative to the polarity reversal anomalies can be modeled by an average emplacement frequency of 5-10 units/km seafloor spreading.

wavelengths are the same, the phases are not coherent. Consequently, the apparent correlation will break down when submitted to rigorous cross-correlation.)

The short wavelength anomalies observed in Pacific and Indian Ocean anomalies 5, shown by Cande and LaBrecque (1974) and reproduced in Figure 6, display a dominant wavelength of 10-15 km, including anomaly 5 in the C1306 profile in the Equatorial Pacific which contains twice the number of peaks and troughs at twice the spreading rate. The uniform wavelength of the observed short wavelength in anomalies 5 (Figure 6) appears to be independent of spreading rate and is similar to the wavelength which dominates magnetic anomalies simulated at the same distance (3.5 km) to a white noise source (Figure 5).

Although the contribution of magnetic intensity fluctuations or short polarity reversals cannot be ruled out completely, it is highly plausible that most of the short wavelength information present in observed anomalies from widely separated regions of equivalent age is caused by the variable source layer thickness resulting from discontinuous emplacement.

Discussion and Conclusion

We are dealing only with qualitative estimates in our comparison of computer simulations and actually observed magnetic anomalies. In our model, the temporal distribution of emplacement of an extrusive source layer is treated in the same statistical manner (i.e. emplacement after Poisson distributed time intervals) as the spatial distribution of emplacement (i.e. Gaussian distributed emplacement and stretching at the accreting boundary). Stretching refers to the actual process of spreading of the brittle magnetic source layer in the active rift, and is assumed to be normally distributed about the mean spreading center. Stretching occurs by injection of dikes which feed extrusive magnetic source material, injection of dikes which do not feed extrusive volcanism, normal faulting and open fissuring. The latter three stretching activities can also occur outside the zone of initial volcanic emplacement; in that case, the width of the final distribution of magnetic source material will roughly equal the width of the zone of volcanic and tectonic stretching. Consequently, it will be essentially impossible to detect the width of the zone of initial extrusive emplacement (i.e. inner floor) by analysis of the observed magnetic anomaly transition zones.

In the slow spreading Atlantic, the quality of the marine magnetic anomalies is quite sensitive to both the frequency of extrusive emplacement λ, and the standard deviation of final emplacement σ. In the fast spreading Pacific, the frequency of extrusive emplacement is the parameter to which the quality of the magnetic anomalies are most sensitive. An emplacement rate of 5-10 units/km seafloor spreading, and a standard deviation < 2.5 km is appropriate for both the poor quality of the Atlantic anomalies and the easily recognized Pacific anomalies.

The discontinuous emplacement model simulates an extrusive source layer having the complex nature which has become apparent from the Deep Sea Drilling Project holes into the North Atlantic basement (Hall, 1976). A standard deviation σ = 2.5 km scatters the flows emplaced during the same polarity event and there exists a high probability of finding vertical sections of mixed polarity in the extrusive layer. The variable thickness of the simulated source layer, which is a natural consequence of the random temporal emplacement process (and independent of the standard deviation), also indicates the high probability for tilting of the extrusives after their original emplacement (i.e. finding variable magnetic inclinations in basement holes). The variable thickness of the magnetic source layer causes noise which either drowns the subtleties of the reversal time scale signal or masquerades as relevant short-wavelength information on the paleofield behaviour in fast spreading magnetic anomalies.

The apparent wavelength of the "little wiggles" observed in Pacific anomalies 5, is found to be independent of spreading rate and similar to the wavelength that dominates magnetic anomalies simulated over a white noise source (i.e. a random variable thickness magnetic source layer) placed at anomaly 5 basement depth.

Many other processes may effectively influence the recording in the magnetic source layer of the Earth's reversal signal: for example, heterogeneity of the magnetic source material, dike injection, tectonism, off-axis volcanism, hydrothermal alteration of the magnetic source material. However, our model simulations suggest that unless the spatial frequency of major extrusive emplacement is significantly higher than the estimated 5-10 units/km spreading, most of the magnetic anomaly and in situ observations can be modeled by a simple two-parameter process of statistically controlled temporal and spatial empalcement of major extrusive units.

The total source of the seafloor spreading anomalies includes a second layer of relative low magnetization consisting of sheeted dikes and layered gabbro (Kidd, 1977b). Blakely (1976) presented this two-layer model to explain wide transitions estimated from surface anomalies by progressive weathering of the highly magnetized extrusive layer which would reduce its contribution to the observed anomalies. However, the mixed polarities of the extrusives found in the Deep Sea Drilling holes indicate that also wide scatter of final emplacement occurs and that the extrusive layer remains a likely source for the observed magnetic anomalies. We conclude that the discontinuous nature of extrusive emplacement is the most likely contributor to the variability of the seafloor spreading anomalies. Intrusive emplacement of the second layer will be of a more continuous nature making a smooth and considerably less variable contribution to the observed anomalies.

Acknowledgements. We very much appreciate the contribution which Robert Detrick and Christopher Tapscott made toward this work during early discussions. We thank Kim Klitgord for his critique and suggestions for improvement of the initial manuscript. And we thank Janet Bear for her outstanding typing of the initial and final versions.

The research was fully supported by the Ocean Science and Technology Division of the Office of Naval Research, Contract N00014-74-C0262; NR083-004. Woods Hole Oceanographic Institution Contribution Number 4233.

References

Atwater, T., Constraints from FAMOUS area concerning the structure of the oceanic section, In Implications of Deep Drilling Results in the Atlantic Ocean: Ocean Crust, Proc. Second Ewing Mem. Symp., this volume, 1979.

Atwater, T. and J.D. Mudie, Detailed near-bottom geophysical study of the Gorda rise, J. Geophys. Res., 78, 8665-8686, 1973.

Ballard, R.D. and Tj. van Andel, Morphology and tectonics of the inner rift valley at lat 36°50'N on the Mid-Atlantic Ridge, Geol. Soc. Amer. Bull., 88, 507-530, 1977.

Blakely, R.J., Geomagnetic reversals and crustal spreading rates during the Miocene, J. Geophys. Res., 79, 2979-2985, 1974.

Blakely, R.J., An age-dependent, two-layer model for marine magnetic anomalies, In The Geophysics of the Pacific Ocean basin and its margin, Amer. Geophys. Union Geophysical Monograph, 19, 227-235, 1976.

Blakely, R.J., K.D. Klitgord, and J.D. Mudie, Analysis of marine magnetic data, Rev. Geophys. Space Phys., 13, 182-185, 1975.

Cande, S.C. and J.L. LaBrecque, Behavior of the earth's paleomagnetic field from small scale marine magnetic anomalies, Nature, 247, 26-28, 1974.

Cox, A. and G.B. Dalrymple, Statistical analysis of geomagnetic reversal data and the precision of potassium-argon dating, J. Geophys. Res., 72, 2603-2614, 1967.

Denham, C.R. and H. Schouten, On the likelihood of mixed polarity in oceanic basement drill cores, In Implications of Deep Drilling Results in the Atlantic Ocean: Ocean Crust, Proc. Second Ewing Mem. Symp., this volume, 1979.

DSDP Leg 52 Staff, Studying oceanic layers, Geotimes, July/August, (22-26), 1977.

Francheteau, J., and P. Tapponnier, Necking of the lithosphere and the mechanics of slowly accreting plate boundaries, J. Geophys. Res., 83, 3955-3970, 1978.

Hall, J.M., Major problems regarding the magnetization of oceanic crustal layer 2, J. Geophys. Res., 81, 4223-4230, 1976.

Hall, J.M. and 11 others, Sources of magnetic anomalies on the mid-Atlantic Ridge, Nature, 255, 389-390, 1975.

Harrison, C.G.A., Formation of magnetic anomaly patterns by dyke injection, J. Geophys. Res., 73, 2137-2142, 1968.

Heirtzler, J.R., G.O. Dickson, E.M. Herron, W.C. Pitman III, and X. LePichon, Magnetic anomalies, geomagnetic field reversals, and motions of the ocean floor and continents, J. Geophys. Res., 73, 2119-2136, 1968.

Huestis, S.P. and R.L. Parker, Bounding the thickness of the oceanic magnetized layer, J. Geophys. Res., 82, 5293-5303, 1977.

Kidd, R.G.W., A model for the process of formation of the upper oceanic crust, Geophys. J. Roy. astr. Soc., 50, 149-183, 1977a.

Kidd, R.G.W., The nature and shape of the sources of marine magnetic anomalies, Earth Planet. Sci. Lett., 33, 310-320, 1977b.

Kono, M., Geomagnetic polarity changes and the duration of volcanism in successive lava flows, J. Geophys. Res., 78, 5972-5982, 1973.

LaBrecque, J.L., D.V. Kent, and S.C. Cande, Revised magnetic polarity timescale for late Cretaceous and Cenozoic time, Geology, 5, 330-335, 1977.

Larson, R.L., and T.W.C. Hilde, A revised time scale of magnetic reversals for the Early Cretaceous and Late Jurassic, J. Geophys. Res., 80, 2586-2594, 1975.

Lee, Y.W., Statistical Theory of Communication, pp. 509, Wiley (New York), 1967.

Macdonald, K.C., and B.P. Luyendyk, Deep-tow studies of the structure of the Mid-Atlantic Ridge crest near lat 37°N, Geol. Soc. Am. Bull., 88, 621-636, 1977.

Matthews, D.H. and J. Bath, Formation of magnetic pattern of Mid-Atlantic Ridge, Geophys. J. Roy. astr. Soc., 13, 349-357, 1967.

Opdyke, N.D., D.V. Kent, and W. Lowrie, Details of magnetic polarity transitions recorded in a high deposition rate deep-sea cores, Earth Planet. Sci. Lett., 20, 315-324, 1973.

Schouten, J.A., A fundamental analysis of magnetic anomalies over oceanic ridges, Mar. Geophys. Res., 1, 111-114, 1971.

Schouten, H. and K. McCamy, Filtering marine magnetic anomalies, J. Geophys. Res., 77, 7089-7099, 1972.

Talwani, M., C.C. Windisch, and M.A. Langseth, Jr., Reykjanes Ridge crest: a detailed geophysical study, J. Geophys. Res., 76, 473-517, 1971.

Wilson, R.L., P. Dagley, and A.G. McCormack, Paleomagnetic evidence about the source of the geomagnetic field, Geophys. J. Roy. astr. Soc., 28, 213-224, 1972.

ON THE LIKELIHOOD OF MIXED POLARITY IN OCEANIC BASEMENT DRILL CORES

Charles R. Denham and Hans Schouten

Woods Hole Oceanographic Institution, Woods Hole, Massachusetts 02543

Abstract. The probability of striking mixed magnetic polarities in a deep oceanic crustal drill hole depends most heavily on the spreading rate, the width of the median valley where lavas accumulate and the width of the active spreading (stretching) zone. The temporal rate of extrusive activity has a relatively small effect. Numerical simulations and statistical estimates show that in slow spreading rate crust, mixed polarity is very likely to occur both near to and far away from the polarity boundaries which are inferred from the seafloor spreading magnetic anomalies.

Introduction

The variability of oceanic crustal emplacement is vividly demonstrated by the common occurrence of mixed magnetic polarity in basaltic cores raised by the Deep Sea Drilling Project (eg. Hall, 1976). Although some of the mixed polarity observations may be due to sample misorientation, it is now abundantly clear that genuine magnetic polarity reversals have been recorded within the basaltic column. Since geomagnetic polarity reversals probably require several thousands of years to occur (Cox and Dalrymple, 1967; Opdyke et al., 1973), it is reasonable to infer that the emplacement process itself also requires at least a similar length of time to be completed at a particular site.

Schouten and Denham (1979) modeled the oceanic crustal accretion as the combined realization of several statistical processes, to study the variability of marine magnetic anomalies. Here, we will adopt the same model of overlapping lava flows from episodic volcanic activity to investigate the likelihood that two or more magnetic polarity zones would be observed in a vertical basement drill core. The assessment will be based on the ranges of emplacement parameters which are inferred from the magnetic anomalies and from actual geological observations.

Statistical Model

We will invoke a Gaussian spatial behavior and a Poisson temporal behavior for the accumulation of overlapping extrusive units of finite width. The half-spreading-rate is the parameter which links the various spatial and temporal aspects of the model, including the magnetic polarity timescale. We will calculate the probability of mixed polarity as a function of age for several reasonable values of the key statistical parameters, and will show that mixed polarity is nearly inevitable over substantial portions of the slowly-spreading Atlantic system, but rare in the faster spreading Pacific system. The most important parameter here is the characteristic width of the final lava distribution; the least important is the frequency of extrusive activity.

The final spatial distribution of extrusive material depends upon a combination of three factors: 1) the width of the zone of extrusion, which we characterize by the standard deviation σ_e of a Gaussian distribution; 2) the widths of the flows, which we assume are a constant w, centered on their vents; and 3) the width of the stretching zone, characterized by Gaussian σ_s, where the actual spreading takes place by dike injection, open fissuring, and normal faulting. The extrusion and stretching zones are centered on the ridge axis, and the majority of both activities occurs in its near vicinity (95% within ± 2 standard deviations). Their combined result is itself roughly Gaussian, so we have characterized it also by a standard deviation, σ. The correct probability density curves are derived and illustrated in Appendix I.

A. Gaussian Spatial Distribution

In order to simplify the calculations, we assumed at the outset that the final probability density distribution of lavas is Gaussian, with

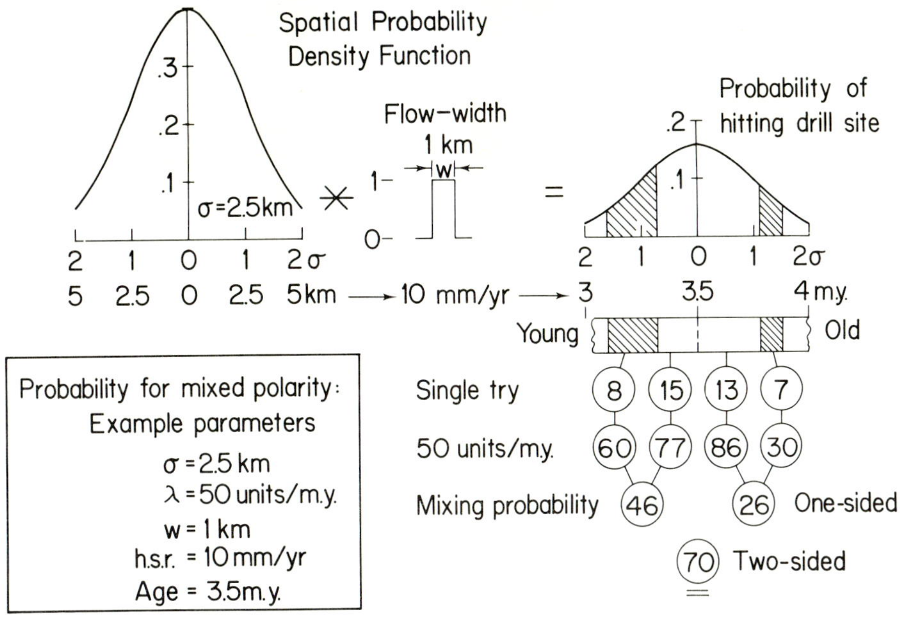

Fig. 1. Schematic of the procedure for calculating the probability of striking mixed magnetic polarity in a deep basement drill hole at a particular site. The spatial probability density function (top left) is convolved with the flow width (top middle) to convert from probability density to probability (top right). The polarity sequence, converted from time to distance via the spreading rate, is superimposed on the probability curve, and the joint probabilities for mixed polarity (right side of figure) are computed by conventional means. (Numbers are percentages.)

standard deviation σ, rather than the slightly skewed forms shown in Appendix I.

As illustrated at the top of Figure 1, the probability density function is converted to a probability function by convolving it with a unit-height rectangular waveform of width w, representing the flow width. That is, the probability $P_o(x)$ at distance x is:

$$P_0(x) = \frac{1}{\sigma\sqrt{2\pi}} \int_{x-w/2}^{x+w/2} \exp(u^2/2\sigma)du \qquad (1)$$

The geomagnetic polarity timescale (converted to distance) is then superimposed on the new curve, whose median is positioned at the drillsite. The result shows the probability that a single flow of a certain polarity from a particular age would affect the site.

B. Poisson Temporal Distribution

By specifying the rate at which successive lavas issue forth under the Gaussian spatial behavior, we can investigate whether a particular site would be affected by any material from a particular geomagnetic polarity (time) interval. The combination of probabilities from all the polarity intervals then leads to the probability for mixed polarity at that site.

As shown in Appendix II, the only values needed for merging a particular polarity interval with the Poisson distributed (temporal) extrusive activity are the average single-flow probability within the interval, and the number of flows likely to issue forth during the interval. The extrusion rate is characterized by λ, the average number of flows per million years.

If the i-th polarity interval is defined between distances x_1 and x_2, then the probability of at least one flow from that interval touching the site is:

$$P_i\ (x_1, x_2) = 1-e^{-\beta(x_2 - x_1)} \qquad (2)$$

where β is a combination of parameters:

$$\beta = \frac{\lambda}{(hsr)} \frac{1}{x_2-x_1} \int_{x_1}^{x_2} P_o(x)\ dx \qquad (3)$$

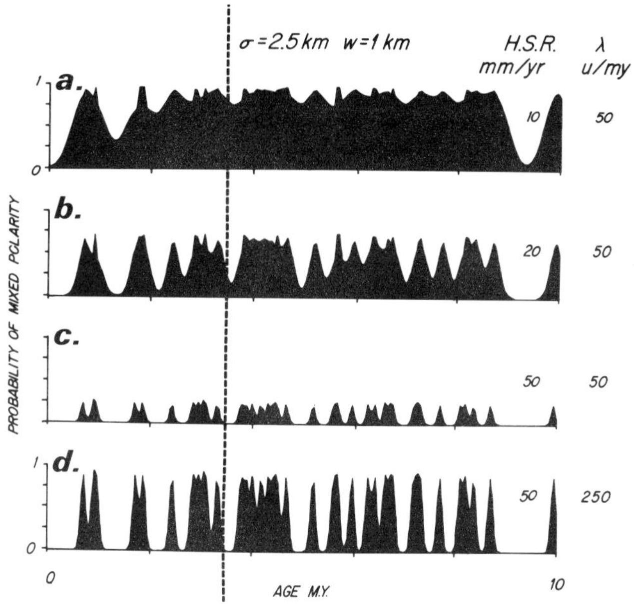

Fig. 2. Probability of striking a polarity reversal in a deep basement drill core, plotted as a function of age. A σ = 2.5 km situation is depicted in this figure for various emplacement parameters. The dashed line marks 3.5 Myr, which was used in the Figure 1 example.

The function $P_o(x)$ is given in Equation 1. The ratio (λ/hsr) serves to convert flows per time (λ) to flows per distance via the half-spreading-rate hsr. The resultant β then is used in Equation 2, which is the conventional Poisson probability for at least one success (i.e. relevant lava).

From knowing the probability P_N of a hit by one or more of the j normal polarity (time) intervals, and also the probability P_R of a hit by one or more of the k reversed intervals, we can merge them to obtain the probability P_{Mixed} for at least one of each:

$$P_N = 1 - \sum_{i=1}^{j} (1 - P_{N_i}) \quad (4)$$

$$P_R = 1 - \sum_{i=1}^{k} (1 - P_{R_i}) \quad (5)$$

$$P_{Mixed} = P_N P_R \quad (6)$$

Probabilities for the Figure 1 example are shown at the lower right. It is clear from Appendix I that no distinction needs to be made between polarity intervals from the old-side and from the young-side of the drillsite. The separation in Figure 1 serves only to show the relative influence of each side for that particular example. The results we will now describe are for full, two-side calculations.

Results

The probabilities of mixed-polarity for various parameter combinations are shown in Figures 2 and 3 for the past 10 million years of the geomagnetic polarity timescale (LaBrecque et al., 1977). We believe that the Atlantic system is most appropriately modeled in curve 2a, where the zone of emplacement and/or tectonic stretching is some 10 kilometers wide (σ = 2.5 km), the half-spreading-rate is 10 mm/yr, and the extrusion half-rate is about 50 flows/my. These values are supported by actual geological observations (Ballard and van Andel, 1977; Macdonald and Luyendyk, 1977). For the Pacific system, curve 3d is the most appropriate, with its narrower accumulation zone, faster spreading, and presumably higher frequency of volcanic activity. The narrower zone of accumulation is deduced from the greater sharpness of magnetic anomaly transitions in the Pacific (Atwater and Mudie, 1973; Blakely, 1974). More frequent volcanism is inferred from submersible observations (R. Ballard, personal communication, 1978) that the Atlantic and Pacific extrusive bodies are similar in spacing, implying more frequent activity in the Pacific to account for the faster spreading.

In the Atlantic model, high probabilities for mixed polarity dominate almost everywhere rather than only close to the inferred crustal polarity boundaries. For the past 10 my, noteable exceptions occur over the long Brunhes interval, the middle Matuyama, and the anomaly 5 region.

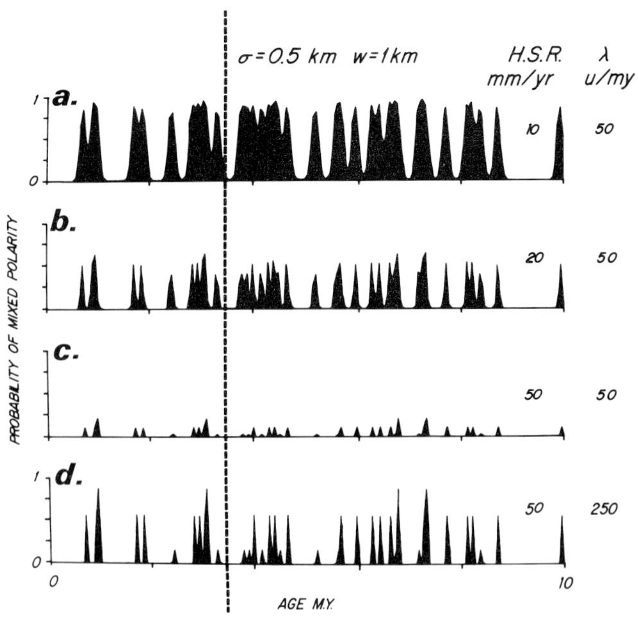

Fig. 3. Same as Figure 2, for a σ = 0.5 km situation.

The Pacific model is very different, with near-zero probabilities everywhere except close to polarity transitions.

The key features which distinguish the Atlantic and Pacific models are the spatial distribution of lava emplacement, and the spreading rate. The temporal rate of extrusive activity has a secondary importance: it strongly influences the local probabilities for mixed polarity, but it does not extend the region of influence laterally very effectively. This finding should be compared with the conclusion of Schouten and Denham (1979) that the rate of extrusive activity is very important to the quality of the lineated seafloor spreading magnetic anomalies.

Acknowledgements. We very much appreciate the contribution which Robert Detrick and Christopher Tapscott made toward this work during early discussions. We thank Kim Klitgord for his critique and suggestions for improvement of the initial manuscript. And we thank Janet Bear for her outstanding typing of the initial and final versions.

The research was fully supported by the Ocean Science and Technology Division of the Office of Naval Research, Contract N00014-74-C0262; NR083-004. Woods Hole Oceanographic Institution Contribution 4279.

Appendix I. The Stretching Function

The apparent instantaneous center of seafloor spreading is the position at which spreading is actually occurring. If we consider the average spreading center to be fixed at the ridge axis, and further if we assume that the spreading is symmetrically away from the instantaneous center at the half-spreading rate hsr, then the average velocity $\overline{V}(x_0)$ of motion of material at distance x_0 away from the ridge axis can be expressed as a function of the spatial probability density $D_s(x)$ of a combination of dike injection, open fissures, and normal faulting.

$$\overline{V}(x_0) = hsr \int_0^{x_0} D_s(x) \cdot dx \qquad (A.1)$$

The average age of material at distance x_0 is then

$$T(x_0) = \frac{1}{2} \int_0^{x_0} \frac{1}{\overline{V}(x)} \cdot dx \qquad (A.2)$$

If an initial spatial probability density of extrusives $D_e(x)$ were allowed to spread according to the above plan, then the final probability density D_f would be given by

$$D_f(x) = \frac{1}{dT(x)/dx} \cdot D_e(x) \qquad (A.3)$$

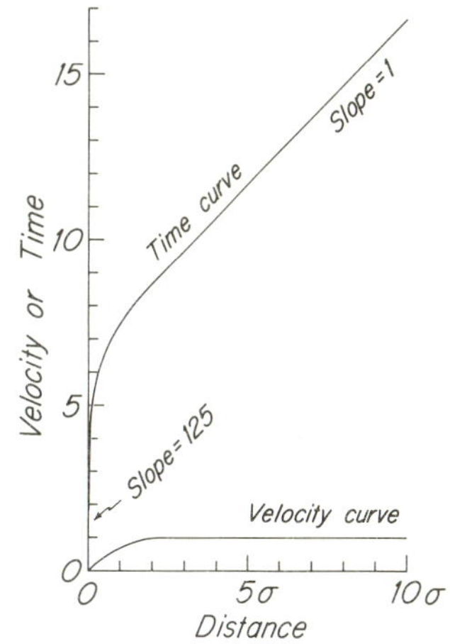

Figure A1. Gaussian stretching curves for $\sigma_s = 1$ in a pure dike emplacement model. The average spreading velocity and the average age of crustal material are shown as functions of distance away from the ridge axis (given in units of σ_s). The straight portion of the time curve is described by $t = x + 6.67$. The curves are calculated by text Equations 1 and 2.

The slope of the age curve (i.e. dT(x)/dx) is the stretching function.

The magnetization curve after an arbitrarily long time T_0 is given by plotting $D_f(x)$ as a function of hsr · $(T(x) + T_0)$, where hsr is the half-spreading-rate, assumed constant at great distance from the ridge axis. Since T(x) is essentially linear beyond a few standard deviations, this curve shows the relation between the distance and the density of lavas of a certain age.

Appendix Figure A1 shows the velocity and time (age) curves for Gaussian stretching in which the extrusive standard deviation σ_e and the stretching standard deviation σ_s are equal (unity in the graph). It is readily apparent that the spreading rate is constant beyond about 2σ, and that the average age of material at any x is slightly greater than would be expected for volcanic emplacement confined to the exact center of the ridge crest only. The reason is that close to the ridge axis, the material undergoes a biased random walk back-and-forth, making slower progress away from the axis than does material already much farther away. For $\sigma_s = \sigma_e = 1$, the relation at large x is

$$T(x) = x + 6.67 \qquad (A.4)$$

Appendix Figure 2A shows the stretching which would be experienced by an initial Gaussian lava pile. The final result is non-Gaussian, with skewness toward the ridge axis for $\sigma_e \lessapprox 2.2\sigma_s$ and skewness away from the ridge axis for $\sigma_e > 2.2\sigma_s$. Integration of those stretched-Gaussian curves yields the magnetization curve for a

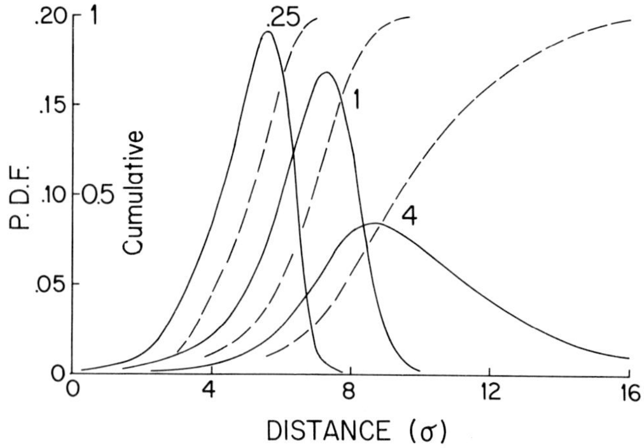

Fig. A2. Stretched Gaussian curves for various values of σ_e/σ_s, calculated by text Equation 3. Their integrals (dashed) show the shapes of the corresponding magnetic polarity transition zone.

polarity transition region. Several examples are shown.

To quantify the probability of striking polarity reversals in deep crustal drill holes, the final probability density curve of the lava pile (i.e. the stretched-Gaussian here) is the one which should be convolved with the magnetic polarity timescale (converted to a spatial scale via the spreading rate) represented as a (+1, -1) telegraph wave. For $\sigma_s = \sigma_e$, Appendix Figure A3 shows the comparison between the final curve and the original Gaussian, with both curves centered at the peak-value and normalized to unit area. The differences between them are not great. Both of them have about 95% of their area within $\pm 2\sigma_e$ of the median, and the peak of the stretched

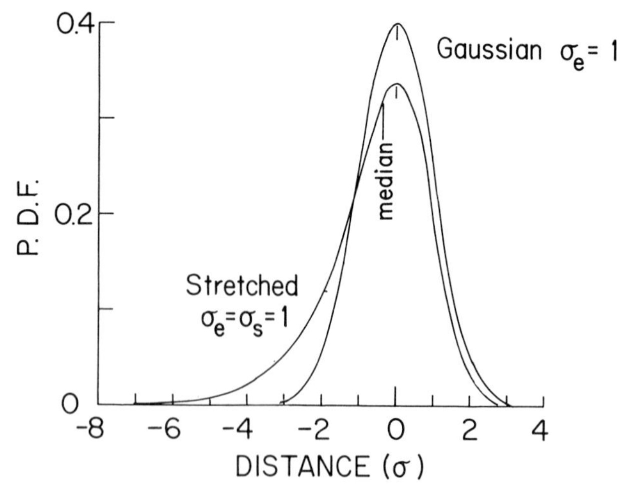

Fig. A3. Comparison of Gaussian and stretched Gaussian curves for $\sigma_e = \sigma_s = 1$. The curves are centered on their peaks. The median of the stretched curve lies 0.3σ units to the left of the peak.

curve is only $0.3\sigma_e$ away from the median position. We took advantage of this similarity by using a Gaussian approximation of the stretched-Gaussian for constructing text Figures 2-3. When $\sigma_e = \sigma_s$, the differences are negligible for our purposes.

Appendix II. Average Probability

Consider a time-interval which has been divided into N partitions, with probability $\frac{a}{N}$ that an attempt will occur within any particular one of them. Furthermore, assign probability P_i for a success within the i-th partition if an attempt were actually to occur at that time. The probability that no success will occur in the i-th partition is then $(1 - \frac{aP_i}{N})$. (A.5)

The probability that no successes will occur in any of the partitions is

$$\prod_{i=1}^{N} \left(1 - \frac{aP_i}{N}\right) \qquad (A.6)$$

In the limit as $N\to\infty$, this expression is equivalent to

$$\lim_{N\to\infty}\left[1 - \frac{1}{N}\sum \frac{aP_i}{N}\right]^N \qquad (A.7)$$

Note how only the average P ($\bar{P}$) is needed here. The probability of at least one success during the whole time interval is the one's complement of the latter expression i.e.:

$$[1 - e^{a\bar{P}}] \qquad (A.8)$$

References

Atwater, T. and J.D. Mudie, Detailed near-bottom geophysical study of the Gorda Rise, J. Geophys. Res., 78, 8665-8686, 1973.

Ballard, R.D. and Tj. H. van Andel, Morphology and tectonics of the inner rift valley at lat 36°50'N on the Mid-Atlantic Ridge, Geol. Soc. Amer. Bull., 88, 507-530, 1977.

Blakely, R.J., Geomagnetic reversals and crustal spreading rates during the Miocene, J. Geophys. Res., 79, 2979-2985, 1974.

Cox, A. and G.B. Dalrymple, Statistical analysis of geomagnetic reversal data and the precision of potassium-argon dating, J. Geophys. Res., 72, 2603-2614, 1967.

Hall, J.M., Major problems regarding the magnetization of oceanic crustal layer 2. J. Geophys. Res., 81, 4223-4230, 1976.

LaBrecque, J.L., D.V. Kent, and S.C. Cande, Revised magnetic polarity timescale for late Cretaceous and Cenozoic time, Geology, 5, 330-335, 1977.

Macdonald, K.C. and B.P. Luyendyk, Deep-tow studies of the structure of the Mid-Atlantic Ridge crest near lat 37°N, Geol. Soc. Amer. Bull., 88, 621-636, 1977.

Opdyke, N.D., D.V. Kent, and W. Lowrie, Details of magnetic polarity transitions recorded in a high deposition rate deep-sea core, Earth Planet. Sci. Lett., 20, 315-324, 1973.

Schouten, H. and C.R. Denham, Modeling the oceanic magnetic source layer, in Implications of Deep Drilling Results in the Atlantic Ocean: Ocean Crust, Proc. Second Ewing Mem. Symp., this volume, 1979.

A MODEL FOR THE STRUCTURAL STATE OF THE UPPER HALF KILOMETER OF NORTH ATLANTIC OCEANIC LAYER 1

J. M. Hall

Department of Geology, Dalhousie University, Halifax, Nova Scotia, Canada B3H 3J5

Abstract. A new model is developed to explain indications of considerable tectonic rotations within the upper half kilometer of oceanic layer 2 in the North Atlantic. The model invokes several processes that are likely to occur in the inner median valley of the Mid Atlantic Ridge. It is suggested that the overall process of structural evolution should be viewed as an integral part, with vulcanicity and geothermal activity, of the process of heat loss from the asthenosphere.

Introduction

This short paper summarizes a discussion of structural state of oceanic layer 2 given in a review of the main results of North Atlantic oceanic basement drilling (Hall and Robinson, to appear in Science). At the Second Ewing Symposium it was suggested several times that oceanic layer 2 consisted largely of little rotated (but perhaps extensively faulted) material. It is considered valuable to present here an alternative view--that large (≥ 40°) tectonic rotations are not uncommon. The evidence for the model is given below.

Evidence

A number of lines of evidence indicate that uppermost ocean crust is likely to be tectonically disrupted:

(i) The location of earthquake epicentres along active spreading centers.

(ii) The folding and faulting of ophiolites.

(iii) Fissuring, faulting and minor rotation of the upper surface of layer 2 as observed from submersibles.

(iv) Deviations of several kinds from dipole paleomagnetic directions in drill cores from layer 2.

(v) Distinct lithological and magnetic differences between drill holes separated by 50 to 450 m.

(vi) The occurrence of tectonized plutonic sequences in contact with pillow sequences in upper layer 2.

No single line of evidence indicates directly the structural state of the upper half kilometer of layer 2. Focal mechanism solutions can only be expected to approximate in a gross way to the details of local strain relief. It is often impossible to distinguish between original-ocean-floor and emplacement structures in ophiolites. Surface observations of layer 2 structures can only provide free surface boundary conditions. Paleomagnetic data can often be interpreted in terms of either geomagnetic or geological history. Lateral changes could be accounted for by the limited lateral extent of submarine volcanic units. Additional difficulties are that core recovery in oceanic basement is usually low, particularly so far as fragile horizons are concerned, and that the common pillow basalt sequences of oceanic basement do not contain conventional tilt indicators such as bedding planes.

Since the argument given here in favour of extensive tectonic disruption depends largely on the paleomagnetic evidence it will be necessary to give this in more detail. Three aspects of the paleomagnetic results are thought to be significant.

1. A number of long vertical intervals of up to 580 m extent are characterized by stable magnetic inclinations that are significantly different from ideal dipole inclinations. Significant here means that divergence between measured and ideal inclinations is greater than the expected range of secular variation. (Holes 332, 333A, 396B, 417D and perhaps in some of the shallow holes in the 407 to 413 sequence).

2. Continuous change of stable magnetic inclination with depth, usually within one lithological-geochemical unit, is observed in Holes 395, 410A and 418A.

3. One or more reversals within a vertical sequence occur in Holes 332A, 334, 395, 396B, 407, 410 and 418A.

---

[1] For young crust (< 20 my) ideal dipole inclinations are calculated from $I = \tan^{-1} (2 \tan \Theta)$ where Θ is the Site latitude. For older crust the latitude is with respect to a paleomagnetic pole as determined by study as a continent coupled by a passive margin to the ocean crust in question.

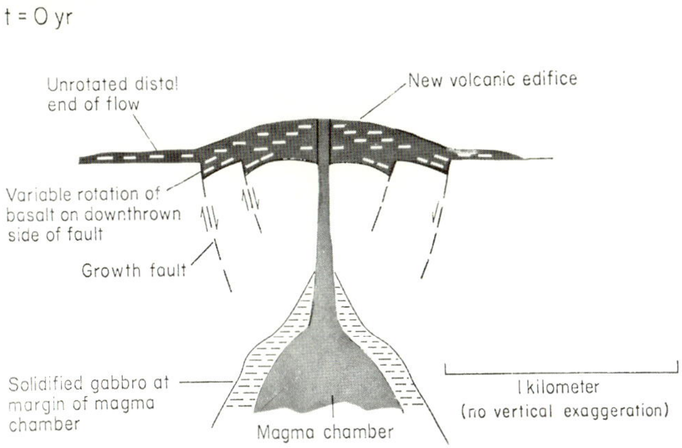

Fig. 1. Structural consequences of the formation of a new volcanic edifice on the floor of the inner median valley of the Mid Atlantic Ridge.

The Structural Model

This structural model is based on the assumption that deviations of stable paleomagnetic inclination from ideal dipole[1] are solely the result of tectonic rotations. If this is true tectonic rotations of at least 40° are quite common. This is an extreme assumption since it does not allow for secular variation. It is suggested, however, that it is a justifiable assumption as the present state of knowledge does not allow any other explanation for the phenomena. Thus, stable magnetization directions are demonstratably rather good indicators of original magnetizing field directions, and, there is no reason to believe that the geomagnetic field is essentially different over the oceans and continents.

The model is best explained by observing the evolution of newly formed crust within the inner median valley of the Mid Atlantic Ridge. Figure 1 illustrates the valley at the time of formation of a new volcanic edifice of the form of Mts. Pluto and Venus of the FAMOUS study area. Note that growth faults are active during the accumulation of the thicker parts of the edifice. This feature is based on the observation that the crust directly in front of thick submarine flows yields by normal faulting (Ballard and Van Andel, 1977). This indicates that the strength of very young oceanic crust is less than the stress exerted by loading it with a single thick flow. If a hinge line occurs between a growth fault and another part of the edifice then continuous rotation of the accumulating flow will occur, with the maximum rotation at the base of the flow. It is suggested that this process can account for the continuous change of stable magnetic inclination within a single unit, as noted above.

The other features of the paleomagnetic results can be explained by reference to Figure 2. Here the edifice of Figure 1 has now been transported to the margin of the inner valley. During transport it has subsided by several hundred meters until it is almost awash in sediment, rubble and the distal ends of younger flows. It is suggested that the subsidence is accompanied by large scale rotation and fracturing into a number of fairly large segments. Rotation could be the result of either or both mechanical instability of the edifice or of lateral change in strength of the underlying valley floor. It is during this slow downlimb convection, consisting of transport and subsidence, which may be viewed as part of the process of heat loss from the asthenosphere, that the pattern of large tectonic tilts and rapid lateral lithological and paleomagnetic variability is acquired. It is suggested that the pattern is essentially frozen-in by this position so that the distal ends of flows from newer edifices, which may now cap the subsided older edifice, will be unrotated. It requires from 1 to 2 x 10^5 years at a spreading rate of $\sim$ 1 cm.yr^{-1} for an edifice to be sufficiently subsided to be covered by these younger flows and for at least the Neogene this interval will often contain a reversal of geomagnetic polarity. In this way a cap of a different polarity, with close to ideal dipole stable paleomagnetic inclination, can be expected to be present in some crustal sections.

Implications and Weaknesses of the Model

If the model is correct a number of implications follow:

1. That slow spreading, Atlantic type crust consists of a large scale tectonic mélange to a depth of at least half a kilometer.
2. That reliable plate motion estimates are unlikely to be obtained from Atlantic type crustal sections. This follows since most of a section is likely to be considerably tectonically rotated, without the prospect of unique tilt correction,

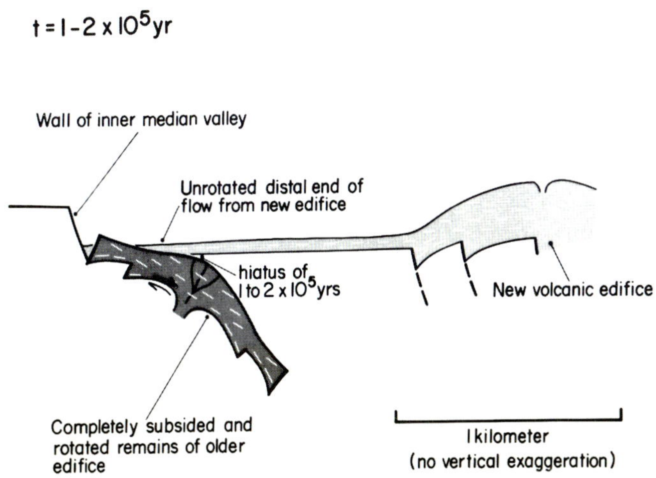

Fig. 2. Structural consequences of the translation of the volcanic edifice of Figure 1 to the margin of the inner median valley.

while unrotated caps where they occur, will contain too little paleomagnetic information to allow secular variation effects to be removed by averaging.

The main weakness of this model is the present difficulty in checking the basic assumption--that deviations from ideal dipole paleomagnetic inclinations are solely the result of tectonic tilting, by an independent means. It will be of great value in furthering our understanding of the structure of oceanic layer 2 if another tilt indicator can be identified. It will also be of value if original (ocean floor) and emplacement structural features of ophiolite sequences can be distinguished.

Acknowledgments. Most of the ideas presented here have developed as the result of discussions since DSDP Leg 37 (1974) with Paul Robinson. Others who have added to and productively criticized the model are Peter Lonsdale, Matthew Salisbury, Ian Gibson and Clinton Milligan.

References

Ballard, R.D., and Tj.H. Van Andel, Morphology and tectonics of the inner rift valley at lat. 36°50'N on the Mid Atlantic Ridge, Geol. Soc. Amer. Bull., 88, 507-530, 1977.

Hall, J.M., and P.T.R. Robinson, Deep crustal drilling in the North Atlantic ocean: A progress report, (to appear in Science).

LOW TEMPERATURE ALTERATION OF THE MAGNETIC MINERALS IN OCEAN FLOOR BASALTS

Nikolai Petersen

Institut fur Allgemeine and Angewandte Geophysik, Ludwig-Maximilians-Universitat Munchen, Munich, FRG

Peter Eisenach

Institut fur Allgemeine and Angewandte Geologie, Ludwig-Maximilians-Universitat Munchen, Munich, FRG

Ulrich Bleil

Institut fur Geophysik, Ruhr-Universitat Bochum, Bochum-Querenburg, FRG

Abstract. The correlation between submarine weathering and magnetic properties of ocean floor basalt has been studied on samples from 12 different oceanic sites: 10 sites in the Atlantic, 1 in the Pacific and 1 in the Mediterranean sea. In agreement with earlier studies it has been found that the natural magnetization of ocean floor basalts is carried by small grains of titanomagnetites with average volume content of 1%.

These titanomagnetites are oxidized grandually under suboceanic conditions and form thereby cation deficient spinels, the so-called titanomaghemites. This oxidation proceeds at low temperatures, most likely at bottom water temperature.

The primary composition of oceanic titanomagnetites varies only within narrow limits. From the samples studied in this paper a primary composition of $Fe^{2+}_{1.50}Fe^{3+}_{0.77}Ti_{0.58}Al_{0.07}Mg_{0.06}Mn_{0.02}O_4$ is deduced. Microprobe analyses of titanomagnetites in combination with Curie temperature measurements indicate that the dominating mechanism of titanomagnetite oxidation under oceanic conditions is iron-migration: Fe migrates out of the spinel lattice and is either incorporated in the surrounding clay minerals, or transported elsewhere by sea water. The process of Fe-migration is limited by the amount of Fe^{2+} left in the spinel lattice; for each Fe-ion leaving the crystal another Fe^{2+} is converted to Fe^{3+}. Fe-migration causes the Fe/Ti ratio of titanomagnetite to change during oxidation.

A correlation between the Curie temperature T_c and the oxidation parameter z has been determined from the analysis of

$$Fe_2O_3/Fe_2O_3 + FeO$$

of separated titanomagnetites. This correlation does not agree with the corresponding one derived from the data of Readman and O'Reilly (1972).

The Curie temperature of ocean floor basalts increases with increasing age of the oceanic crust. Although there is considerable scatter in the data, it seems that T_c increases proportionally with the logarithm of age. The upper 600 m of the predominantly pillow basalts of the Atlantic ocean floor are alterd more or less uniformly with respect to the magnetic minerals. However within single lithological units there is a slight decrease of oxidation rate with depth.

The saturation magnetization of oceanic basalts first decreases with increasing low temperature oxidation until the oxidation parameter reaches z = 0.60. Beyond this z-value, for further increasing oxidation, it increases again. This variation of magnetization can be explained qualitatively by a simple model assuming Fe-migration first from the octahedral spinel lattice sites and then from the tetrahedral sites. The high amplitudes of the mesozoic marine magnetic anomalies could possibly be caused by this process. The jurassic magnetic quiet zone could then be explained by inversion of the completely oxidized titanomagnetites to non-magnetic mineral phases.

The direction of the natural remanent magnetization of ocean floor basalts is not changed by sea floor alteration.

Introduction

Studies of cores from the ocean crust recovered by the Deep Sea Drilling Project's ship Glomar Challenger have made it clear that the ocean floor is permeable to seawater. These studies suggest that all the crust down to at least 600m depth, the deepest crustal penetration sofar, has been alterd to some degree by cold, rather than hot, seawater. Seawater has percolated into the crust over millions of years and slowly reacted with the rocks, producing chemical changes quite different from those ocurring at high temperatures.

This process, generally termed submarine weathering, is characterized by marked increases of the K_2O, H_2O^+ and H_2O^- contents, and the Fe_2O_3/FeO ratio, while the MgO, CaO and, to a lesser extent, the SiO_2 contents decrease (S. Hart 1969, 1971; Miyashiro et al. 1969; R. Hart 1970, 1973 a, b; Hekinian 1971; Mathews; Shido et al. 1974; Honnorez et al. 1978). The changes with respect to Al_2O_3, MnO, total Fe, TiO_2 and Na_2O are generally much smaller and quite variable; the various authors present contradictory data for these oxides.

Also the magnetic minerals, predominantly titanomagnetites, are affected by the submarine weathering which consequently influences the magnetic properties of the ocean floor basalts. Various characteristics of the marine magnetic anomalies are ascribed to the process of submarine weathering of the titanomagnetites, in particular the prominence of the central anomaly and the gradual decrease in amplitude of the anomalies with increasing age on both sides of an active spreading ridge (Irving 1970).

Compared to the studies of the alteration of silicate minerals the effect of submarine weathering of the opaque, magnetic minerals has been somewhat neglected. It is our aim to contribute to the question of how submarine weathering and magnetic properties of ocean floor basalts are correlated.

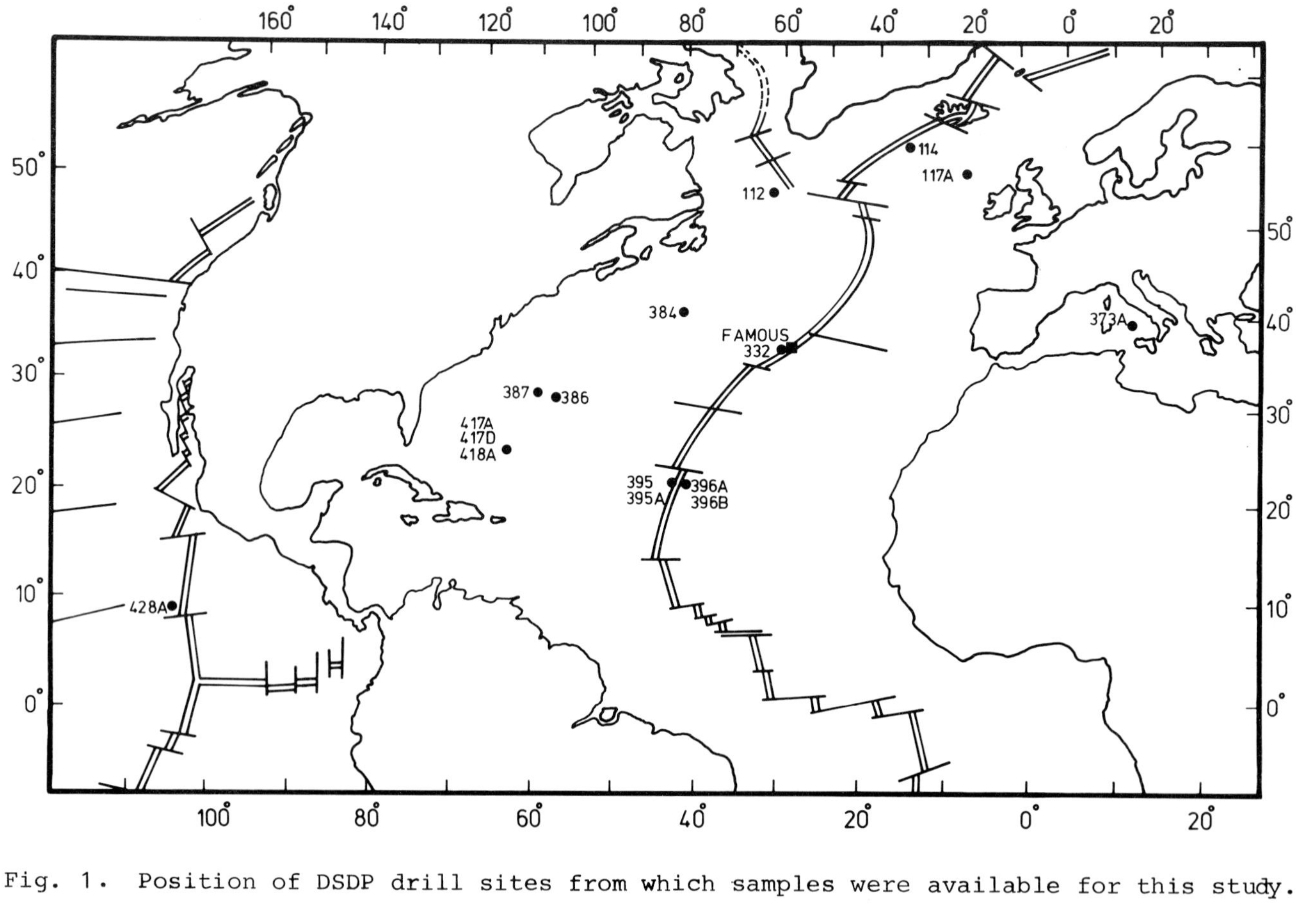

Fig. 1. Position of DSDP drill sites from which samples were available for this study.

Samples

All ocean floor basalt samples studied in this paper can be very broadly termed oceanic tholeiites. They come from 11 different Deep Sea Drilling Project (DSDP) sites and from the FAMOUS area. The locatities are displayed in fig. 1. The majority of the samples comes from the atlantic ocean. The samples from sites 332 A and B, 395 A, 396 B and 417 A are from the deepest DSDP sites drilled so far in the atlantic, with basement penetrations between about 200 and 500 m. The other sites used in this study have much shallower penetrations, apart from the Mediterranean site 373 A (see table 1).

TABLE 1. Location of Samples Used in this Study

DSDP Leg	Site	Position	Water Depth (m)	Basalt Penetration (m)	Age (m.y.)
12	112	54°01.00'N 46°36.24'W	3667	3	66
12	114	59°56.00'N 26°48.00'W	1937	5	12
37	332A	36°52.72'N 33°38.46'W	1806	333	3.5
37	332B	36°52.72'N 33°38.46'W	1806	583	3.5
42	373A	39°43.68'N 12°59.56'E	3517	187	5
43	384	40°21.65'N 51°29.80'W	3920	5	105
43	386	31°11.21'N 64°14.94'W	4783	10	104
43	387	32°19.20'N 67°40.00'W	5118	3	135
45	395	22°45.35'N 46°04.90'W	4484	92	7
45	395A	22°45.35'N 46°04.90'W	4484	571	7
45	396A	22°58.88'N 43°30.95'W	4450	96	9
46	396B	22°59.14'N 43°30.90'W	4459	255	9
51	417A	25°06.69'N 68°02.82'W	5480	219	109
51	417D	25°06.69'N 68°02.82'W	5480	188	109
52	417D	25°06.69'N 68°02.82'W	5480	169	109
54	428A	09°02.77'N 105°26.14'W	3236	53	2
FAMOUS Area		36°49.00'N 33°16.00'W	2600	-	0.12-0.48

Experimental methods

Reflected light microscopy:

Polished sections were studied under reflected light using a Leitz Ortholux Pol microscope. Volume content of the opaque phases and their mean grain diameter were measured on a Quantimet image analyzing device. Conventional point counting was also done on some samples for comparison. The results deviate systematically, point counting giving values higher by a factor of 1.5-2.5, depending on the grain size and the number of grains analyzed per unit area.

To aid the identitication of the magnetic minerals and their variable stages of maghemitization magnetic colloid has also been used.

Some of the samples were treated by ionic etching (Soffel and Petersen 1971) to enhance the optical contrast for the observation of the size and structure of volume change cracks ("shrink cracks") in the titanomagnetite grains. The volume change cracks have also been observed under a scranning electro microscope (Stereoscan, Cambridge Instr.).

X-ray measurements:

Lattice constants of separated Fe-Ti spinels were determined by the conventional Debye-Scherrer technique. The uncertainty in the determinations is of the order of 0.005 Å.

Microprobe analysis:

The instrument used during this study was an ARL - SEMQ electron microprobe equipped with 5 spectrometers, located at the Department of Physics, University of Lausanne-Dorigny, Switzerland, and also a Geoscan, Cambridge Instruments electron microprobe equipped with 2 spectrometers and located at the School of Physics, University of Newcastle, England.

The titanomagnetite grains selected for the microprobe analyses had to be homogeneous and larger than 10 microns in diameter. The latter criterion limited the number of analyzed samples to 23. The 23 samples were prepared as polished sections and coated with a thin carbon layer. Accerleration voltage of the ARL instrument was 15 kV, and of the Geoscan 20 kV. Specimen current was 40nA. These conditions gave an analysis spot area of about 1-2 microns. The counting period was chosen as 20 sec. All titanomagnetite grains were analyzed for Fe, Ti, Al, Mg, Mn, Cr and Si. The last element was measured to estimate the contribution of the surrounding silicate matrix to the analysis, and we rejected analyses giving apparently high Si contents, assuming that titanomagnetite does not contain Si. As reference for the analyses synthetic spinels were used with the following compositions:

$Fe_{2.24}Ti_{0.64}Al_{0.05}Mg_{0.07}Mn_{0.01}V_{0.01}O_4$

$Fe_{2.29}Ti_{0.64}Al_{0.03}Mg_{0.04}O_4$

$Fe_{2.1}Ti_{0.64}Al_{0.03}Mg_{0.04}O_4$

The analyses were corrected for dead time, drift, x-ray absorption and x-ray fluorescence. The results are given as weight percent of the elements, instead of the usually given oxides, as oxygen cannot be measured with the microprobes used in this study.

Chemical analysis:

The FeO/Fe_2O_3 ratio has been determined for separated titanomagnetites and for the corresponding bulk rocks. For the determination of ferrous iron the rock powder was dissolved in a mixture of hydrofluoric and sulfuric acid. The FeO content was measured by titration with potassium bichromate, the total iron content by atomic absorption. The Fe_2O_3 content was derived by subtracting the titrated FeO from total iron. For the separated titanomagnetites the silicon content was also measured by atomic absorption to determine the degree of purity.

Separation of magnetic minerals:

The magnetic minerals(predominantly titanomagnetite) were separated from a number of samples. The separation was carried out in two steps: first by a hand magnet from a acetone slurry of finely crushed rock powder, then by centrifuging the material in Clerici´s solution. Due to the fine grain size and the skeletal shape of the titanomagnetites the purity of the extracted material was on average only about 50%. The degree of purity has been determined by a chemical measurement of the silicon content of the separate.

Thermomagnetic curves:

Thermomagnetic curves (temperature dependence of strong field induced magnetization) were measured with a Forrer-type automatically recording translation balance, the sample environment being argon.
For the Curie temperature measurement a magnetic field of 1800 Oe was chosen to reduce the influence of paramagnetism. From the thermomagnetic curves the Curie temperatures were determined by a graphical method following Grommé et al. (1969). The uncertainty in the absolute value of the magnetization is about 3%.

The Néel curve-type of the magnetic spinel phase has been determined by measuring the magnetization of the rock samples in the temperature range between liquid nitrogen temperature and the Curie temperature in a magnetic field of 10.100 Oe. In this case the paramagnetism of the silicates and the super-paramagnetism of very fine titanomagnetite grains has to be taken into account. Their contribution was determined from the linear field dependence of the magnetization, obtained after saturation of the ferrimagnetic mineral phase, and measured at constant temperatures.

Natural remanent magnetization:

The natural remanent magnetization (NRM) was measured with a Digico Spinner Magnetometer. The "stable" remanence direction was determined after removing the unstable magnetization components by stepwise alternating field demagnetization.

Opaque minerals contained in ocean floor basalts.

The opaque minerals in ocean floor basalt are predominantly titanomagnetites. They are initially stoichiometric members of the solid solution series $Fe_{3-x}Ti_xO_4$ ($0 \leq x \leq 1$) with magnetite (x=0) and ulvöspinel (x=1) endmembers (fig.2).

As we will discuss further below, the range of compositions of the primary titanomagnetites in ocean floor basalts, prior to any alteration, is very limited, with most of the x-values between 0.58 and 0.65.

The titanomagnetite volume content in ocean floor basalts varies between 0.1 and 5%. A volume content of 1% is typical for the pillow basalts from the Atlantic ocean floor.

Titanomagnetite is the main carrier of remanent magnetization, the contribution of the other opaque minerals can be neglected normally. Figs. 3 and 4 show titanomagnetite grains in oceanic tholeiite, of typical skeletal shape. This is a general feature in ocean floor tholeiites. Titanomagnetite skeletons even occur in thick flows with low cooling rates in the center parts,

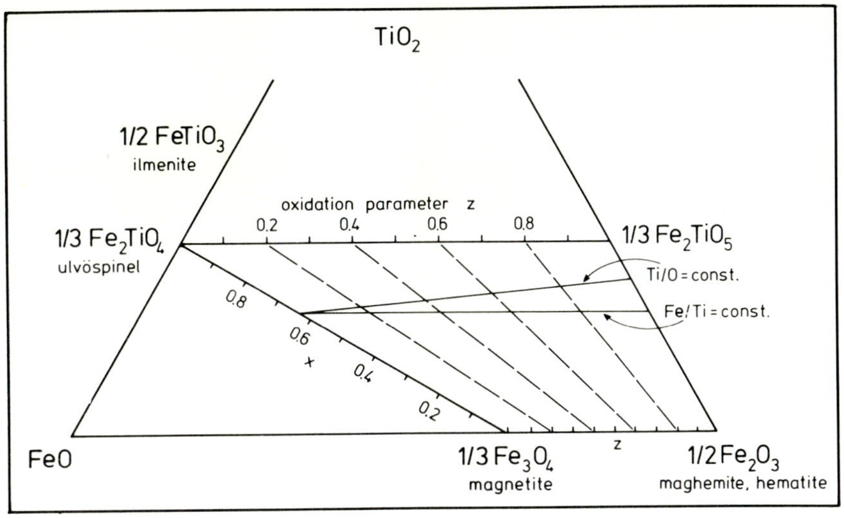

Fig. 2. Ternary diagram FeO-Fe_2O_3-TiO_2. The composition of titanomagnetites can conveniently be represented in this diagram. Low temperature oxidation of titanomagnetite is commonly thought to proceed along the line Fe/Ti = const. The oxidation model proposed in this study proceeds along the line Ti/O = const.which represents a process of Fe-migration.

with grain sizes upto 1 mm in length. However such large grain sizes are the exception in the Atlantic ocean floor, where pillow basalts predominate. Here the titanomagnetite grain size is usually very small, between submicroscopic (i.e. sub-micron) and a few tens of microns (μm). (The most detailed description of skeletal forms of titanomagnetite in basalts is still the early work of Lindley 1926). The skeletal form indicates either rapid cooling of the liquid basalt or very late crystallization of the titanomagnetites due to low oxygen fugacity.

In the coarser grained basalts two size fractions of titanomagnetite are observed normally, a larger generation up to 150 microns grainsize, usually with skeletal to subhedral forms, and a smaller, down to the limit of visibility. Iron sulfides are also commonly present, but their volume content is normally less than 10% of the titanomagnetite phase.

Contrary to extrusive subaerial basalts sulfides can be observed in nearly all ocean floor basalts (grain size between 1 and 100 microns, volume content on average 0.1%). They consist of a primary phase (pyrrhotite) and a secondary phase (pyrite and chalcopyrite). The primary sulfide phase consists of small droplets resulting from the immiscibility of silicate and sulfide melt (fig. 4). The roundish droplets can be considered as indices of restricted outgassing of submarine basalts (Ade-Hall et al 1976). The main portion of the sulfides consists of pyrrhotite and intergrowths of pyrrhotite with pyrite and subordinate chalcopyrite. Fig. 5 shows the oxidation of initial pyrrhotite to pyrite and magnetite, and simultaneous dissolution into chalcopyrite and pentlandite. In many other cases iron hydroxides give evidence for the original existence of sulfides. Other opaque phases which are sometimes also present are chromium spinels, ilmenite, hematite and iron-hydroxide, the later two phases only is strongly altered rocks.

Euhedral cubic chromium spinel (grain size between 5 and few hundred microns) is present in many ocean floor basalts as a rare phenocryst (volume content about 0.01%). It seems to be more common in phyric than in aphyric basalts (Graham et al. 1978) although there is no significant difference in the overall chromium content between these two types of basalts. Chromium spinels totally surrounded by clinopyroxenes are optically homogeneous. On the other hand crystals surrounded by the groundmass show margins of resorption and corrosion and transition to a more iron rich chromium spinel. Fig. 6 shows an example, but here it is difficult to decide whether the graphical texture is actually caused by corrosion or whether it is a primary phenomenon. When ilmenite is

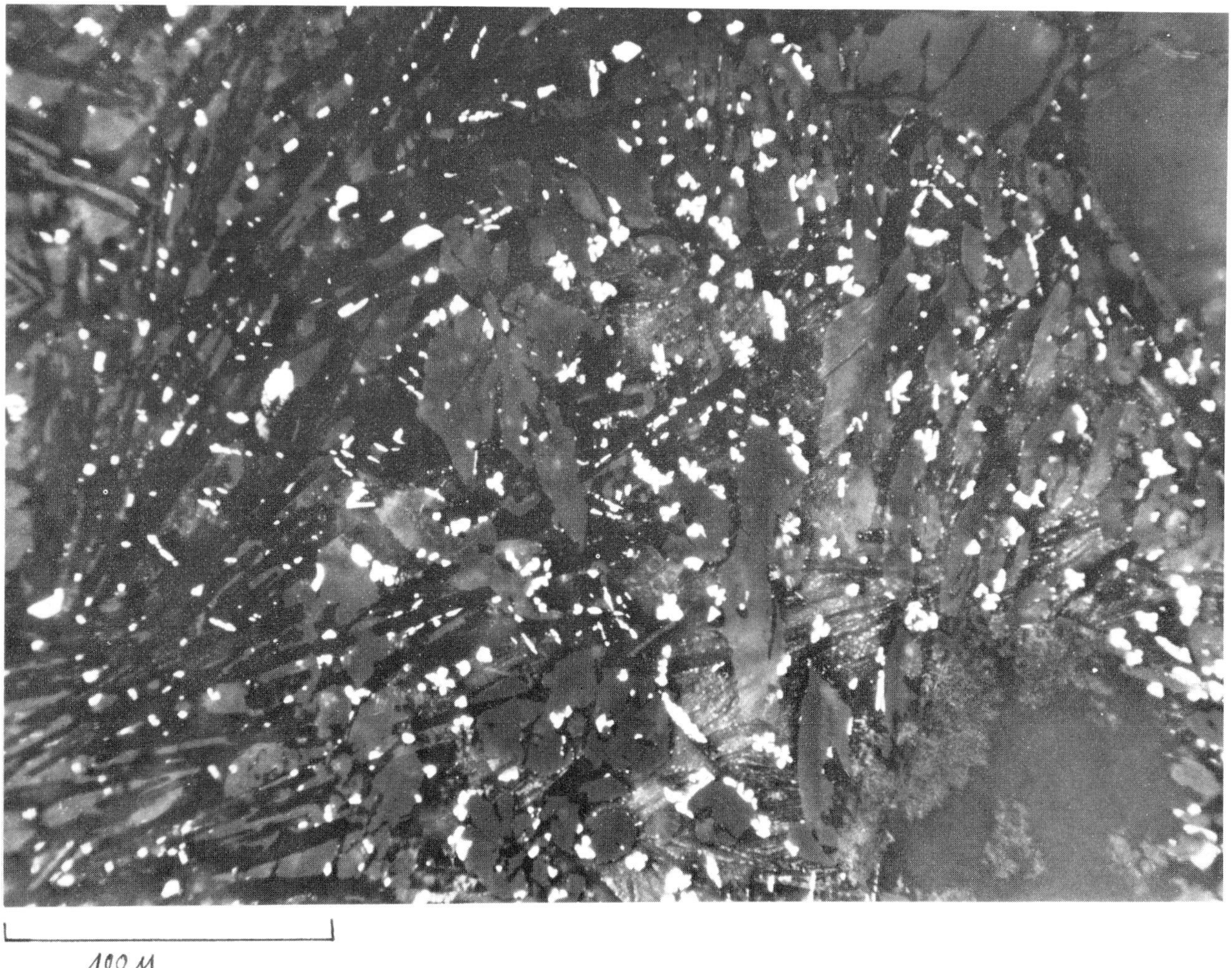

Fig. 3. Skeletal titanomagnetite grains (white) in ocean floor basalt. Sample: 395A-11-1(56-66).

present as primary mineral, it normally forms separate laths (grain size between 5 and 30 microns, volume content between 0 and 0.5%). But also pseudomorphic growth can be observed. An example is shown in fig. 7 where it is easily distinguished from similarly shaped titanomagnetite by the treatment with magnetic colloid, ilmenite being non-magnetic at room temperature. Another widespread phenomenon are ilmenites mantled by titanomagnetite (fig. 8).

Ilmenite also occurs as a secondary mineral produced by high temperature oxidation of titanomagnetite, when it forms the typical ilmenite "exsolution" lamellae along the (111) planes of the host titanomagnetite (fig. 9). However compared to subaerial basalts secondary ilmenite is a rare phenomenon in ocean floor basalts, restricted to the center parts of thick flows.

Oxidation of titanomagnetite

Oxidation is the most important form of alteration of the primary Fe-Ti oxides under natural conditions. According to Verhoogen (1962) the equilibrium oxygen partial pressure of magnetite and ulvöspinel at 298°K is $10^{-68.2}$ atm and $10^{-80.3}$ atm respectively, which means that there exists a strong tendency for oxidation under conditions at the ocean floor. Whether the deep ocean oxidation is caused by dissolved oxygen in seawater or by the dissocia-

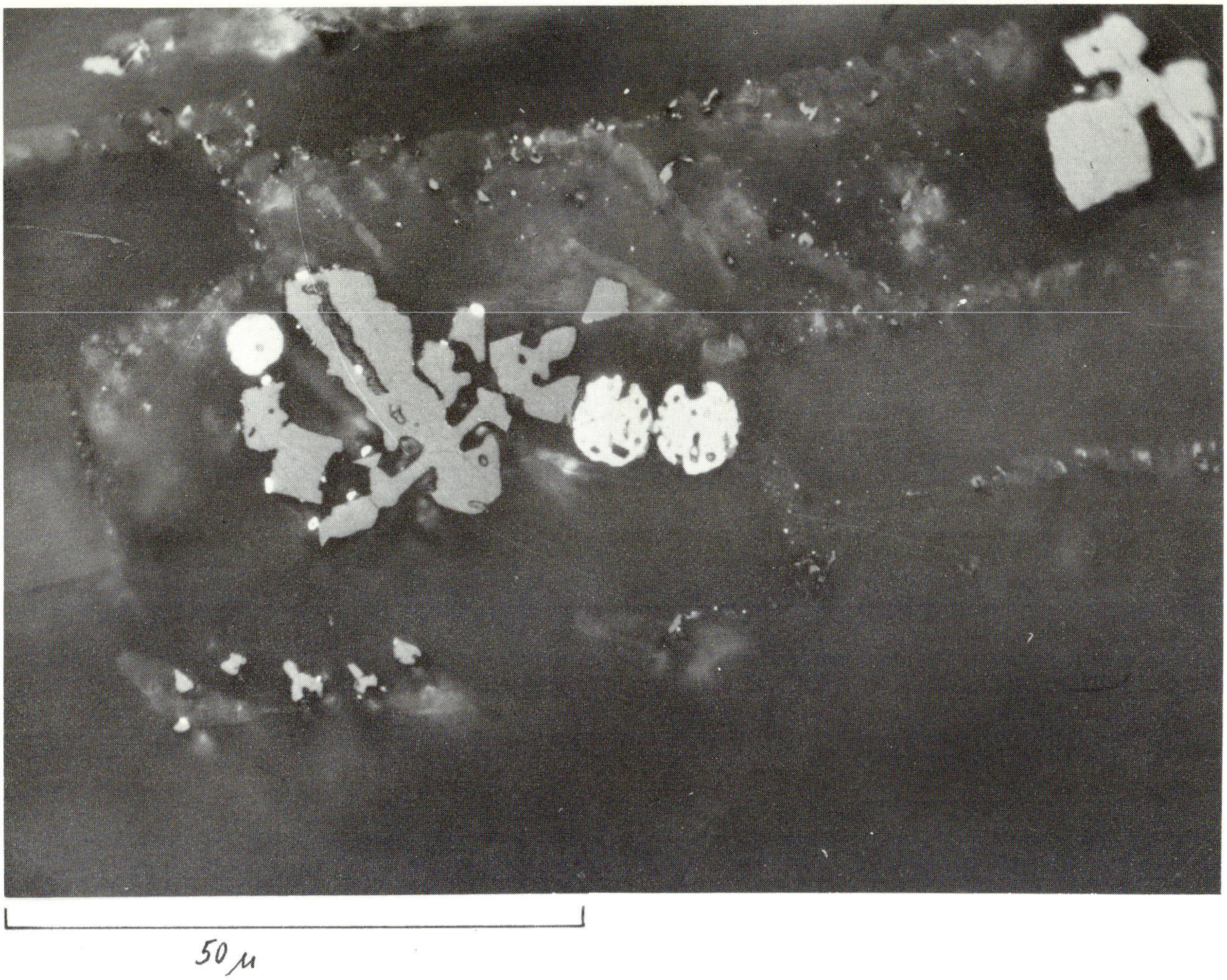

Fig. 4. Skeletal titanomagnetite bordered by small roundish sulfides. The larger sulfide particles, so-called globules, are composed of a mixture of pyrrhotite and pyrite. Sample: 395-11-2(100-106).

tion of water is still a matter of controversy and will not be discussed here.

The way titanomagnetite oxidizes depends critically on the temperature where it takes place. We may distinguish schematically between high temperature and low temperature oxidations. The following gives a simplified picture of these two types of oxidation:

a) high temperature oxidation ($T > 500^{\circ}$ C):

titanomagnetite → ilmenite → rutile (or pseudobrookite if oxidation takes place above 700°C)

ilmenite → hematite

titanomagnetite → magnetite → hematite

This form of oxidation is most common in subaerial basalts. It is rare in ocean floor basalts, where it is only observed in the center parts of thick flow units (Ade-Hall et al. 1976). The well known regular intergrowths of ilmenite lamellae in magnetite is typical for this kind of high temperature deuteric oxidation (see for example fig. 4). The very restricted development of deuteric high temperature oxidation in ocean floor basalts is consistent with a suggestion of Sato and Wright (1966) that hydrogen from dissociated water vapor must first be lost from a flow for deuteric oxidation to be able to proceed. The high confining pressure of the deep ocean environment must considerably reduce gas loss from

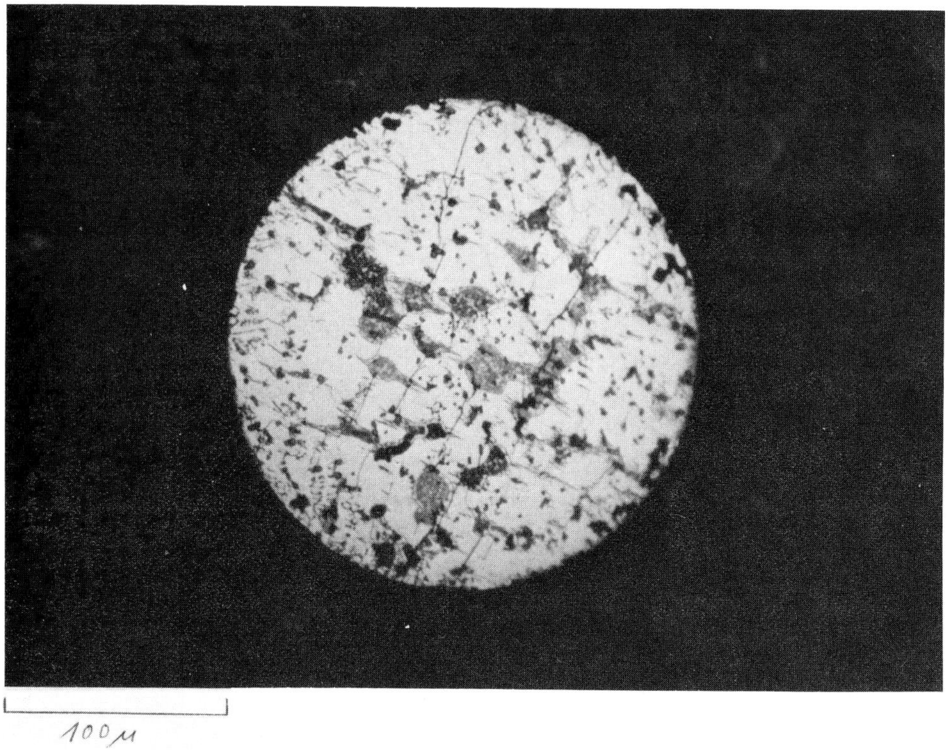

Fig. 5. Large, perfectly spherical sulfide globule consisting of a complex intergrowth. Extremely fine magnetite grains dispersed in the sulfide particle indicates low temperature oxidation of initial pyrrhotite to pyrite and magnetite, with simultaneous dissolution to chalcopyrite and pentlandite. Sample: FAMOUS, Alv 518-3-1.

a cooling flow and hence inhibit this process to occur. Another fact supporting the restricted outgassing of submarine extrusives is the presence of primary sulfides.

The magnetic consequences of high temperature oxidation of titanomagnetite are the increase of Curie temperature (Curie temperatures between 500 and 550°C are typical after deuteric oxidation), of magnetization intensity and of magnetic stability (Chevalier and Pierre 1932, Petersen 1962 and 1966, O'Reilly and Banerjee 1967, Lewis 1968, Ade-Hall et al. 1968).

b) low temperature oxidation ($T < 200°C$):

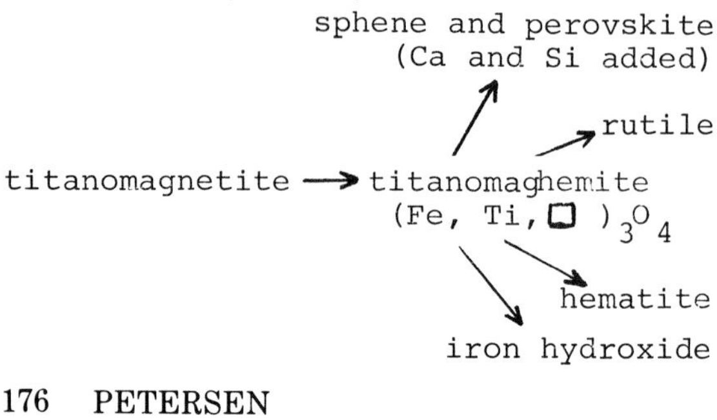

This is the dominant form of titanomagnetite alteration in ocean floor basalts where it most likely proceeds at ambient temperatures. The result is the production of increasing cation deficiency in the titanomagnetite spinel lattice with increasing low temperature oxidation. The product becomes more and more unstable as the cation deficiency increases.

The monophasic non-stoichiometric spinel products of this low temperature oxidation are referred to as titanomaghemites and the oxidation process itself as magnetization.

Low temperature oxidation of titanomagnetites has been thoroughly studied on synthetic samples by Readman and O'Reilly (1972). From their experiments it can be seen that Curie temperature increases with oxidation, as also does the magnetic stability; the magnetization intensity and the lattice constant, on the other hand, decrease. Using the results of Readman and O'Reilly and assuming that the influence of impurities like Mg, Al and other cations that normally occur in natural samples, can be neglected,

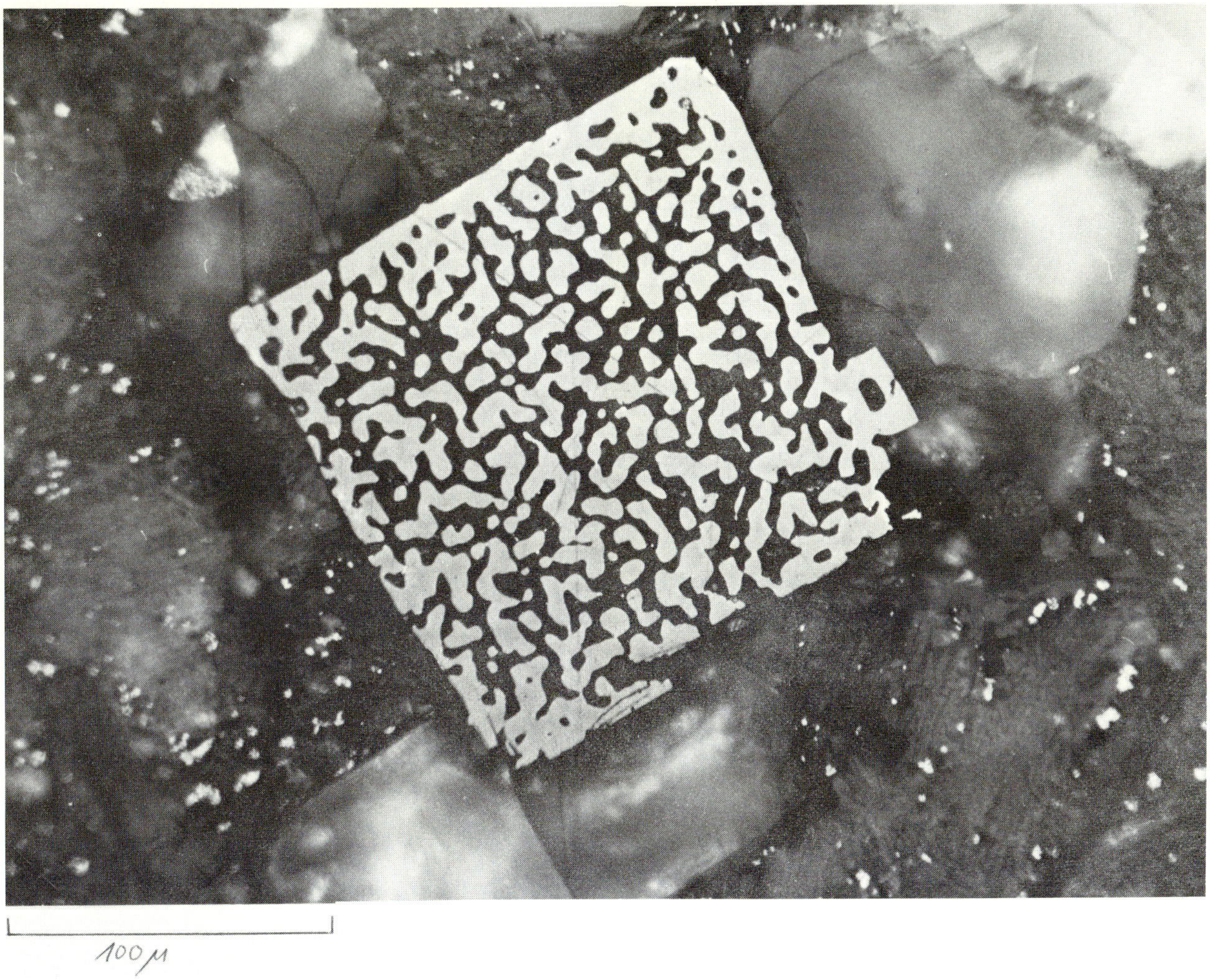

Fig. 6. Euhedral, cubic chromium spinel with graphical textures. It is difficult to decide whether this texture is caused by corrosion or whether it is a primary crystallization phenomenon. Sample: 395A-18-2(35-37).

it is possible to determine the composition and oxidation state of titanomagnetites from the Curie temperature and the lattice constant measurements.

It is commonly assumed that low temperature oxidation of titanomagnetite does not alter the Fe/Ti-ratio, which means that the oxidation product proceeds along a straight line Fe/Ti= const. (see fig. 2), the so called oxidation line (e.g. Akimoto et al. 1957, Readman and O´Reilly 1972, Petersen 1976, Ade-Hall et al. 1976). If this assumption is true, it should be possible to determine the original titanomagnetite composition prior to any alteration by reducing the sample in the appropriate gas atmosphere.

We have carried out such experiments heating a number of ocean floor basalts samples up to 900°C in an oxygen atmosphere of 10^{-10}atm which is close to the equilibrium condition for a titanomagnetite composition of $Fe_{2.4}Ti_{0.6}O_4$ (the corresponding Curie temperature is 160°C).

We will see later that such a composition comes close to the natural original titanomagnetite composition in ocean floor basalts. To our surprise we found that the composition of the thus reduced titanomagnetite samples varied in a systematic way: the higher the original oxidation state of the sample, the higher was the final Ti/Fe ratio. This is reflected in the

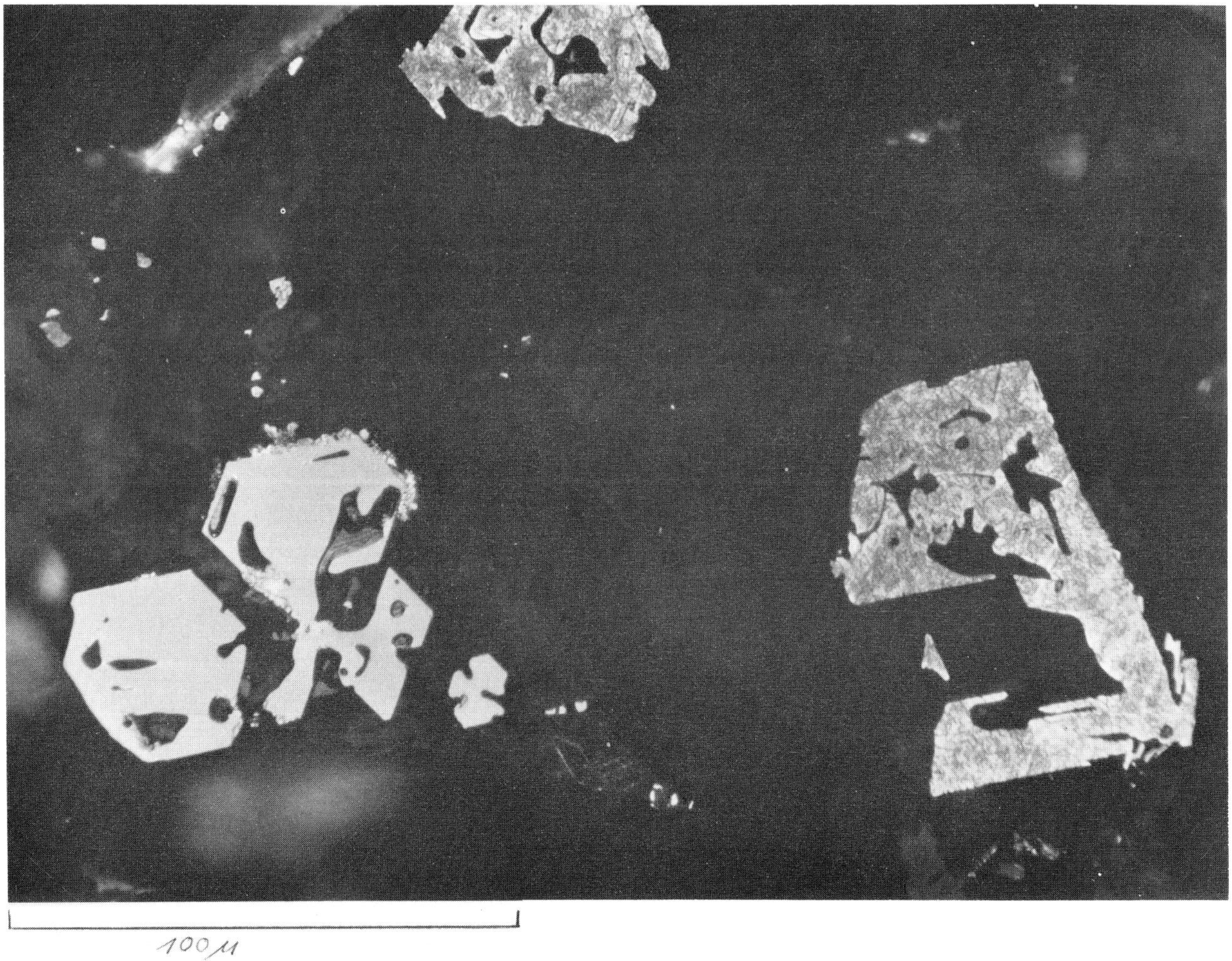

Fig. 7. Pseudomorphic growth of ilmenite (not covered by magnetic colloid) after skeletal titanomagnetite. Also present skeletal titanomagnetite (covered by magnetic colloid). Sample: 395A-14-2(125-134).

Curie temperature of the reduced samples: the samples that were most oxidized yield a lower Curie temperature, far below room temperature. As the Curie temperatures of the titanomagnetite solid solution series $Fe_{3-x}Ti_xO_4$ decrease with increasing x (Pouillard 1950), this made us suspicious that the mechanism of low temperature ocean floor oxidation of titanomagnetite might involve a process of titanium enrichment relativ to iron, or a iron depletion relative to titanium.

Also in the literature there is growing evidence that low temperature oxidation of titanomagnetite involves a diffusion of Fe-cations out of the original spinel lattice into the surrounding silicates, which would result in an increase of the Ti/Fe ratio for the remaining titanomagnetite phase (Prévot et al. 1976, Johnson and Melson 1978, Marshall and Cox 1972). Before we pursue this question, the microscopic characteristics of low temperature oxidation are treated in the following section.

Ore microscopy of altered titanomagnetites in ocean floor basalts

The first indication of low temperature oxidation of titanomagnetites under the microscope is a slight change in colour, a transition from the original redish brown to a more greyish colour, eventually becoming bright grey-

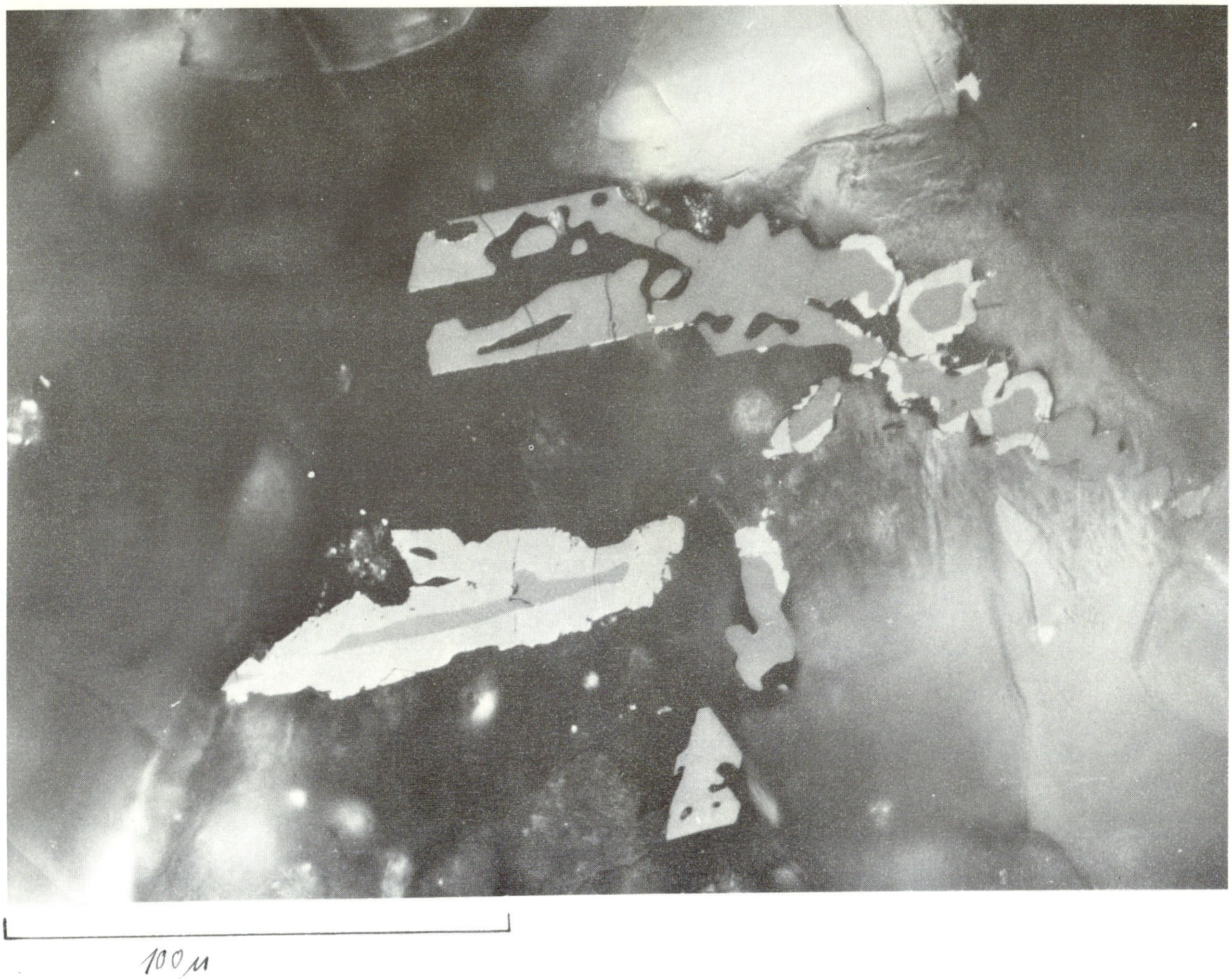

Fig. 8. Ilmenite grains (darker grey) mantled by titanomagnetite (light grey). The grain in the lower left corner may possibly represent a sandwich texture and not be primary ilmenite. Sample: 395A-14-2(125-134).

ish. The grain shown in fig. 10 a and b gives an example of this transition. When covered with magnetic colloid it becomes evident that the oxidized zones have a lower magnetization intensity than the unoxidized zones and therefore attract less of the colloid. However, the main characteristic of low temperature oxidized titanomagnetites in ocean floor basalts is the formation of irregular, curved cracks, also visible in the grain of fig. 10. These cracks have already been observed in mildly metamorphosed continental basalts (Katsura and Kushiro 1961).

Larson and Strangway (1969) were the first to interpret them as due to the volume change associated with the process of magnetization. In ocean floor basalts they were first described by Cockerham and Ade-Hall (1976).

The basic process underlying this cracking is the reduction of lattice constant with increasing oxidation (see fig. 26). If we assume that the kinetics of low temperature oxidation of titanomagnetites is the same as the oxidation of magnetite to maghemite it can be explained in the following way. Feitknecht and Gallagher (1970) investigating the mechanism of the oxidation of magnetite assume a diffusion of Fe^{2+} out of the crystal to the surface of the magnetite grain where it combines with oxygen to be oxidized to Fe^{3+}, thus forming lattice vacancies, but leaving the internal oxygen structure unchanged.

Fig. 9. High temperature oxidation of euhedral titanomagnetite with ilmenite "exsolution" lamellae also along the margins of the grain. Sample: 395A-63-1(0-10).

For each Fe^{2+} ion diffusing out the magnetite lattice, two more are converted to Fe^{3+} to maintain constant total cationic charge within the structure. Feitknecht and Gallagher further maintain that during oxidation a high concentration gradient of cation holes would develop in the monocrystalline particles, leading to a gradient in the lattice constant, thus developing strain, whereby the effect on large particles must be greater than on small particles.

We observe the cracks mainly in larger titanomagnetite grains which have already undergone some degree of low temperature oxidation and interpret them consequently as due to shrinkage of the lattice constant. They are called "shrink cracks" in the following parts of this paper.

There is a correlation between the amount of shrink cracks and Curie temperature as can be seen clearly in fig. 11 a, b, c. The expanding cracks of the grain of fig. 11 c produce a sponge-like structure. Grains of this advanced low temperature oxidation state are often surrounded by a halo of red stained silicates which is most likely caused by Fe^{3+} that migrated out of the titanomagnetite crystal. This question will be discussed below in detail.

Most interesting is the titanomagnetite grain shown in fig. 12 where the shrink cracks are filled with secondary pyrite. This provides evidence that the cracks are open to solution and very likely also to sea water.

Fig. 13 a and b show volume change cracks of submicroscopic size. The picture was taken by a scanning electron microscope.

Primary composition of titanomagnetites

As outlined above, the major magnetic mineral in almost all submarine basalts recovered to date is a member of the titanomagnetite solid solution series that has commonly undergone various stages of posteruptive low temperature alteration. Only relatively seldom has it retained its original unoxidized composition.

Two variables, x and z, are needed in order to define the composition of this type of magnetic mineral (fig. 2). Upon initial eruption, the mineral that is originally formed is a stoichiometric member of the solid solution series between magnetite and ulvöspinel, which is defined by "x" in the formula $Fe_{3-x}Ti_xO_4$. Normally the natural titanomagnetites contain a minor amount of metal ions other than Fe and Ti, mainly Mg and Al. The second variable "z" must be determined in order to characterize the degree of oxidation of titanomagnetite to titanomaghemite. The degree of oxidation, the so-called oxidation parameter "z" ($0 \leq z \leq 1$) is defined as the fraction of the initial Fe^{2+} ions converted to Fe^{3+} ions (Readman and O'Reilly 1971).

For all considerations of alteration of titanomagnetites it is essential to know the original composition prior to any alteration. The first step in the present consideration shall be therefore the evaluation of the "x"-value. To date there does not exist a large number of analyses of titanomagnetites contained in ocean floor basalts. Johnson and Hall (1978) give an average value of $x = 0.65 \pm 0.2$. This value agrees very closely with the value of $x = 0.64$ obtained from continental tholeiite basalts by Petersen (1976).

Table 2 gives a compilation of the primary composition of titanomagnetites in tholeiite basalts, both continental and oceanic, as found in the literature. In this table the x-values have been obtained assuming that low temperature oxidation of titanomagnetites does not change the cation-ratio. The only exception are the data of Johnson and Melson (1978) who argued that progressive low temperature oxidation causes a reduction in the iron content and consequently a decrease in the Fe/Ti-ratio. In this case the simple averaging technique of both unoxidized and oxidized samples would lead to x-values that are too high.

Prevot (1968), analyzing a partially maghemized titanomagnetite grain in basalt, was the first to suggest that the process of low temperature oxidation involved a change in the Fe/Ti-ratio.

For our consideration of the chemical composition of titanomagnetite we use the respective Curie temperature as reference value, as it is a convenient indicator of the degree of progressive low temperature oxidation, increasing Curie temperature indicating increasing oxidation. The results of our microprobe analyses are shown in table 3 and figs. 14, 15, 16 and 17.

Apart from Fe and Ti the analyzed titanomagnetites contain Al, Mg and Mn. The amount of other cations is much lower compared to these and hence they are not represented in this compilation. We have taken care not to include samples with signs of deuteric high temperature oxidation (visible under the microscope as ilmenite exsolution).

We observe a distinct decrease of the Fe-content and also of the Fe/Ti-ratio with increasing low temperature oxidation. Obsiously the spinel lattice becomes iron depleted with respect to the other cations. We interpret this phenomenon as Fe-migration out of the original titanomagnetite lattice, probably into the surrounding silicates. This agrees with the frequent microscopic observation of red staining of the silicates in the vicinity of oxidized titanomagnetite grains (Ade-Hall et al. 1976).

The process of Fe-migration is also well documented in the single titanomagnetite grain of fig. 18 and table 4 that is only partially maghemized. We observe the distinct iron depletion in the oxidized part of the grain (also characterized by lower magnetization compared to the unoxidized part as documented by the attraction of magnetic colloid) whereas titanium remains more or less constant. There may even be some slight increase in titanium, but this may be only an apparent increase due to the decrease of the lattice parameter associated with maghemitization (see also fig. 26). Although titanium is rather mobile at

Fig. 10 a. Titanomagnetite grain with different stages of maghemitization. The central parts of unoxidized titanomagnetite (grey) grade into titanomaghemite (light grey) that forms along the grain margins and also along irregularly formed cracks, so-called "volume change cracks" or "shrink cracks".

higher temperatures (Petersen 1970), we did not observe any indication of Ti-migration at low temperature oxidation in ocean floor basalts.

The same trend as for iron is visible for Mg, although not so clear (fig. 16). There is further indication that Ti and Al are replacing each other. We also observe a slight increase of the cation sum Ti+Al+Mg+Mn with oxidation. Very conspicuous is the increase of Mn in the strongly oxidized samples (fig. 17), a result that has also been observed by Prévot (1968).

From these results it is evident that the primary titanomagnetite composition cannot be deduced by simply averaging the Fe/Ti-ratios of the different samples.

The primary composition has been determined by taking into account the observed trends and making a best fit of all the data to produce a stoichiometric spinel. The composition thus derived is indicated in figs. 14 and 15 and also given in table 2 and 5. Taking into account the oxidation trend it is remarkable how small the scatter of data is although the samples come from very different localities.

For comparison the synthetic equivalent of the above inferred primary titanomagnetite was then produced. The appropriate amounts of Fe_2O_3, TiO_2, MgO, Al_2O_3 and MnO have been sintered for 10 h at 1300°C in an appropriate gas mixture of H_2 and CO_2 corresponding to an oxygen fugacity of $fo_2=$

Fig. 10 b. The same grain as in fig. 10 a covered by magnetic colloid. The maghemized parts have lower magnetization and attract less magnetic colloid therefore. Sample: 395A-32-1(140-150).

10^{-8}atm. After the first sintering process, the material was reground, pressed and then reheated under the same conditions. In this way two samples were prepared independently from each other which had in the end identical properties. Curie temperature, saturation magnetization at room temperature and lattice parameter were then determined. These values are listed in table 5. The actually measured Curie temperature of 125°C falls well into the range of Curie temperatures which we would expect (Richards et al. 1973) from the chemical analyses (see figs. 14 and 15).

Mechanism of low temperature oxidation

From the chemical trends described in the foregoing chapter we deduce the following model of low temperature oxidation of titanomagnetites in ocean floor basalts: The process of oxidation starts from an original titanomagnetite composition close to

$$Fe_{2.27}Ti_{0.58}Al_{0.07}Mg_{0.06}Mn_{0.02}O_4$$

The basic process underlying the oxidation is the migration of Fe-ions, out of the spinel lattice of the titanomagnetite, thus producing cation-vacancies. In this model the oxygen lattice of titanomagnetite remains untouched by the process of oxidation. It is interesting to note that Hoernes

Fig. 11a

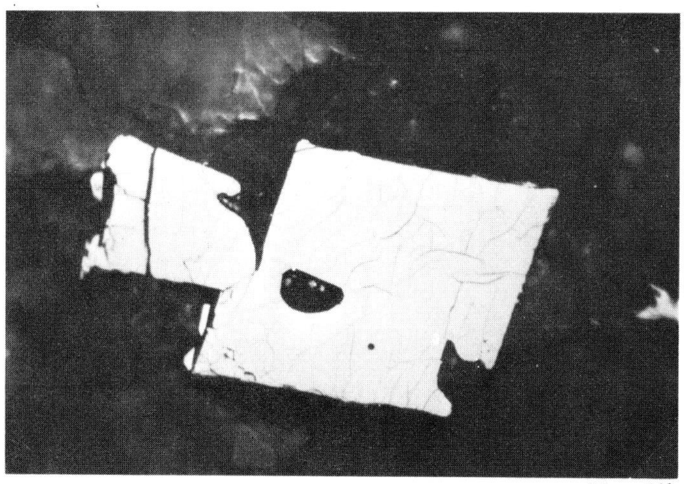

Fig. 11b

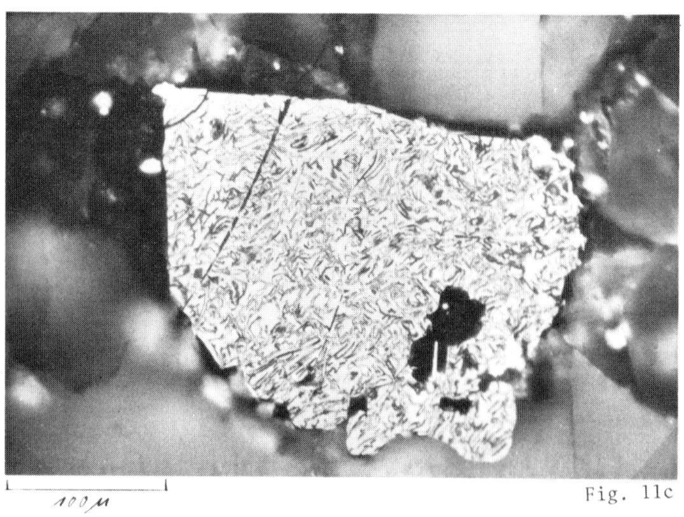

Fig. 11c

and Friedrichsen (1977) from measurements of the oxygen isotope ratios of altered titanomagnetites from DSDP-Leg 37 come independently to the same conclusion.

To maintain the charge balance in the spinel crystal the process requires the transformation of the appropriate number of Fe^{2+} into Fe^{3+}. The process of Fe-migration comes to an end when all the available Fe^{2+} has been exhausted by this transformation.

This model of iron-emigration can be characterized by

$$Fe^{2+} \rightarrow (1-z)Fe^{2+} + \frac{2z}{3}Fe^{3+} + \frac{z}{3}\square + \boxed{\frac{2z}{3}e^{-} + \frac{z}{3}Fe^{2+}}\uparrow$$

or

$$Fe^{2+} \rightarrow (1-z)Fe^{2+} + \frac{2z}{3}Fe^{3+} + \frac{z}{3}\square + \boxed{ze^{-} + \frac{z}{3}Fe^{3+}}\uparrow$$

where z is the oxidation parameter defined by O'Reilly and Banerjee (1966) as the fraction of the initial Fe^{2+} content converted to Fe^{3+}. To make this definition of z applicable also to the first of the above two equations we treat the migrating Fe^{2+} formally like a converted Fe^{2+} ion.

The final composition of the cation deficient spinel is independent of whether iron migrates as Fe^{2+} or Fe^{3+}. On the other hand we have no means of deciding between these two types of Fe-migration. But judging from the ionic radius (Fe^{2+} = 0.81 Å, Fe^{3+} = 0.67 Å, Ti^{4+} = 0.69 Å) a higher mobility of Fe^{3+} and consequently Fe^{3+} rather than Fe^{2+}-migration seems to us the more likely mechanism. This model of titanomagnetite low temperature oxidation due to iron migration is different from the commonly accepted concept of constant Fe/Ti-ratio during the process of oxidation (e.g. Akimoto et al. 1957, Sakamoto et al. 1968, Ozima and Larson 1970, Ozima and Sakamoto 1971, Readman and O'Reilly 1971 and 1972, Bleil and Petersen 1976, Ade-Hall et al. 1976, O'Donovan and O'Reilly 1977).

In the following discussion we will call the two models as "Fe-migration model" and "Readman - O'Reilly model" respectively.

The Readman-O'Reilly model, which pertains to experiments with artificial materials, assumes that Fe^{2+}-ions diffuse out of the original oxide lattice and are combined with external oxygen at the crystal surface forming the same spinel lattice as the inside material. The other cations are thereby rearranged resulting in no net change in the Fe/Ti ratio.

The Readman-O'Reilly model can be characterized by

$$Fe^{2+} + \frac{z}{2}O \longrightarrow (1-z)Fe^{2+} + zFe^{3+} + \frac{z}{2}O^{2-} + \frac{3z}{8}\square$$

where z is the oxidation parameter again.

In the ternary system FeO-Fe_2O_3-TiO_2 (fig.2) the Readman-O'Reilly model can be represented by lines of Fe/Ti = const, the Fe-migration model by lines of Ti/O = const.

Correlation between Curie temperature and oxidation parameter z

The oxidation parameter z of titanomagnetites in ocean floor basalts is usually determined by Curie temperature and/or lattice parameter measurements using the Readman and O'Reilly (1972) contour lines in the ternary system FeO-Fe_2O_3-TiO_2, obtained from synthetic cation deficient titanomagnetites (e.g. Ade-Hall et al. 1976, Bleil and Petersen 1976, Grommé and Mankinen 1976). To avoid this indirect method of z-determination we extracted the titanomagnetites from powdered rock samples and measured directly the oxidation ratio $Fe_2O_3/(Fe_2O_3+FeO)$ of the extracted material by chemical analysis.

As the grain size of titanomagnetites in most of the samples is very fine (between 1 and 20 microns) an

Fig. 11. Different titanomagnetite grains with increasing amount of "shrink cracks". Such shrink cracks are characteristic for low temperature oxidized titanomagnetites. A correlation exists between the amount of shrink cracks and Curie temperature T_C.

a. Large titanomagnetite grain with curvelined and ramifying shrink cracks. Also visible in this grain are two single ilmenite lamellae, product of a high temperature oxidation. T_C = 240°C. Sample: 417D-69-2(35-37).
b. Euhedral titanomagnetite with typical shrink cracks, T_C = 250°C. Sample: 417D-68-5(89-92).
c. The expanding cracks produce a sponge-like structure in the highly oxidized titanomagnetite grain. T_C = 320°C. Sample: 417D-68-3(64-66).

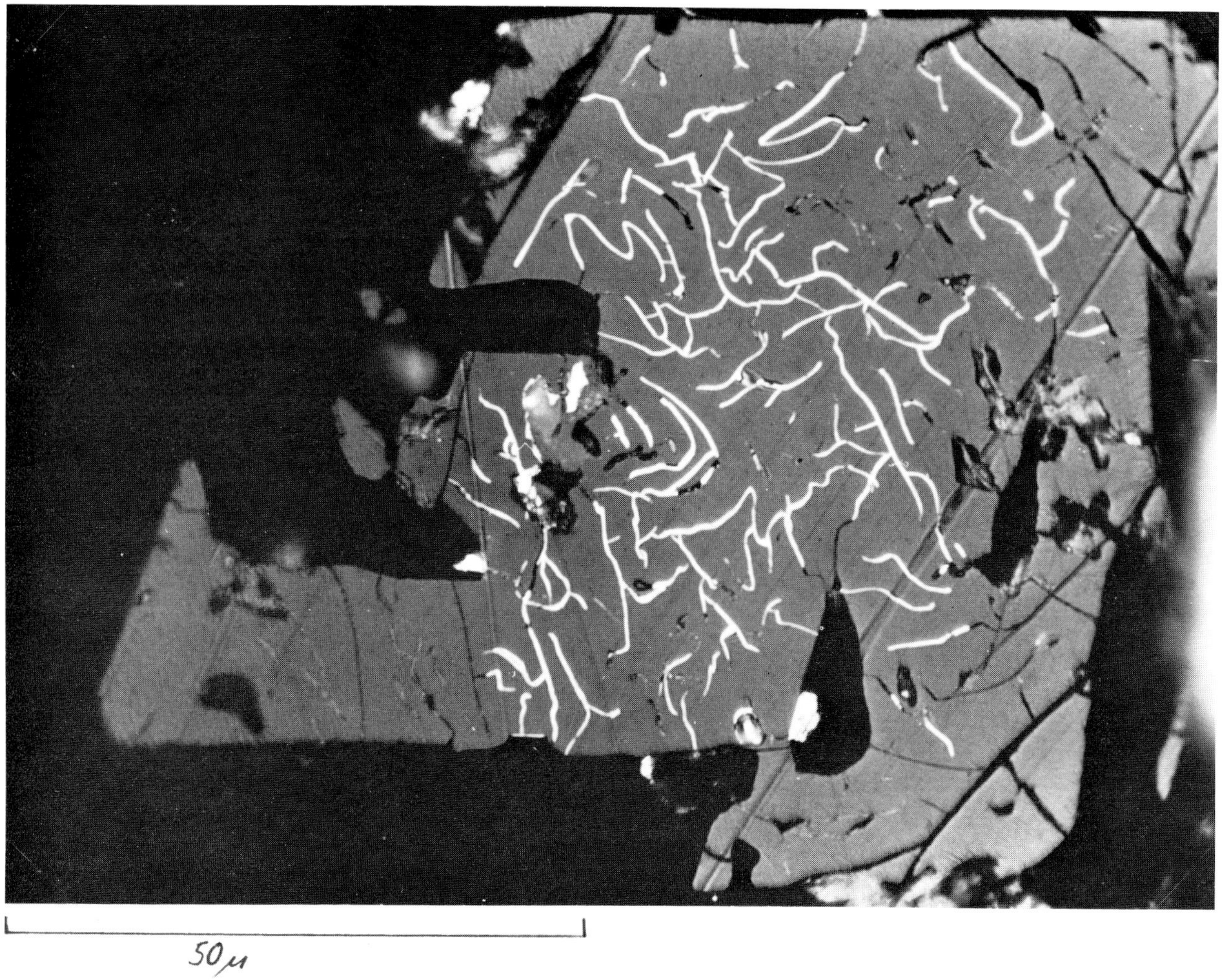

Fig. 12. Titanomagnetite grain with shrink cracks partly filled by sulfide (pyrite?). Sample: 373A-OB 12.

appreciable amount of silicate impurity was associated with the extracted material, even after centrifuging in Clerici´s solution. The Fe^{2+}/Fe^{3+} ratio of the silicate impurities is normally higher than that of the pure titanomagnetite grains. As the degree of impurity of most of our extracted material is as high 50% we had to correct the measured oxidation ratio of the extracted material for the silicate impurities. For such a correction it is necessary to know the total iron content of the whole rock and of the extracted material, the microscopically determined titanomagnetite content in the extracted titanomagnetites (determined from a chemical Si analysis). The thus corrected $Fe_2O_3/(Fe_2O_3+FeO)$ values together with the uncorrected values are given in table 6 and fig. 19, where they are plotted against the Curie temperature. For comparison we have also included in fig. 19 the data of Ozima et al (1974) which show considerable scatter, but fall into the same range. Our corrected $Fe_2O_3/(Fe_2O_3+FeO)$ values fit well a smooth curve.

Knowing the initial $Fe_2O_3/(Fe_2O_3+FeO)$ ratio of 0.36 (corresponding to z=0) from the above determined primary titanomagnetite composition, and taking into account the process of Fe-migration we can then convert the $Fe_2O_3/(Fe_2O_3+FeO)$ ratios into z-values.

We thus obtain the Curie temperature versus z curve shown in fig. 20. It can be seen from this figure that none of the investigated samples has reached complete oxidation.

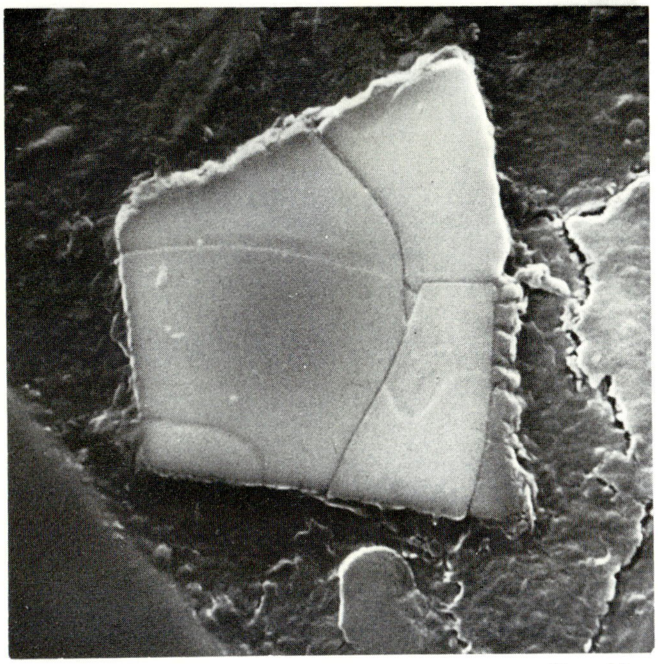

Fig. 13a

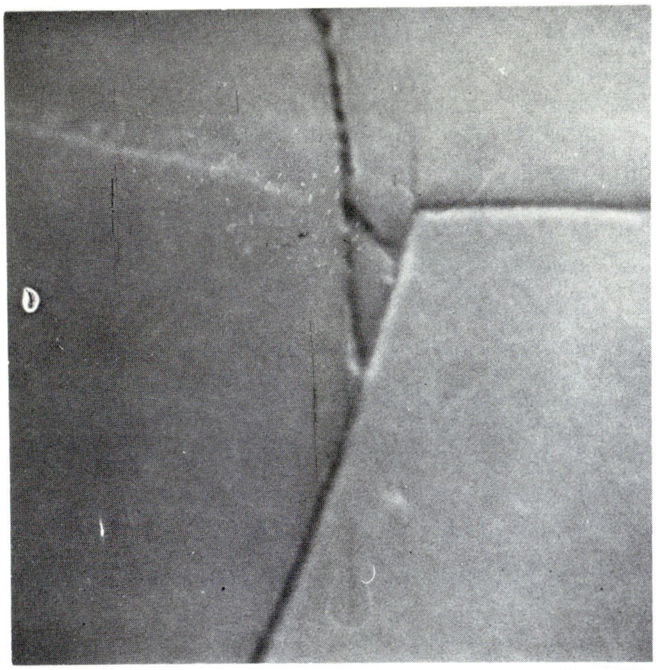

Fig. 13b

Fig. 13 a and b. Small euhedral titanomagnetite grain with few shrink cracks. Picture taken by a scanning electron microscope. Sample: 387-50-1(22-25).

TABLE 2. Compilation of Literature Data of the Primary Composition of Titanomagnetites in Basalts

Rock Type and Locality	Reference	$Fe_{3-x-y-z-a}Ti_xAl_yMg_zMn_aO_4$ x	y	z	a	Curie Temperature °C†	Technique Used	Number of Analyses
Continental Tholeiites	Petersen (1976)	0.64	0.03	0.04	-	103	Microprobe	49
Ocean Floor Tholeiites Atlantic Ocean DSDP Leg 45	Johnson and Melson (1978)	0.63±0.05	0.03±0.01	0.05±0.02	0.02±0.002	110	Microprobe	6
Ocean Floor Tholeiites Atlantic Ocean DSDP Leg 37	Bleil and Petersen (1977)	0.62±0.02	0.09±0.01	0.03±0.01	0.01	89	Microprobe	4
Ocean Floor Tholeiites Atlantic Ocean FAMOUS Area	Prêvot and Lecaille (1976)	0.61±0.02	0.11±0.01	0.03±0.01	0.02±0.01	89	Microprobe	10
Ocean Floor Tholeiites Atlantic and Pacific Ocean, DSDP Legs 16, 34, 38 and MOHOLE EM-7	Johnson and Hall (1978)	0.65±0.02	-	-	-	117	Microprobe	60
Ocean Floor Basalts Atlantic and Pacific Ocean, Dredged and MOHOLE EM-7	Ozima et al. (1974)	0.73±0.11	-	-	-	55	Wet Chemical	5
Ocean Floor Tholeiites Pacific Ocean DSDP Leg 34	Mazullo and Bence (1976)	0.68±0.05	0.07±0.02	0.03±0.01	0.02±0.002	55	Microprobe	14
Ocean Floor Tholeiites Pacific Ocean DSDP Leg 34	Ade-Hall et al. (1976)	0.62±0.02	-	-	-	140 138⁺	Curie Temperature	8
Ocean Floor Tholeiites Atlantic, Pacific, Mediterranean Sea	This Study	0.58±0.02	0.07±0.02	0.06±0.02	0.02±0.005	131 125ˣ	Microprobe	24

†) Curie Temperature of Synthetic Equivalent Calculated According to Richards et al. 1973
+) Curie Temperature Measured on Natural Samples
×) Curie Temperature Measured on Synthetic Equivalent

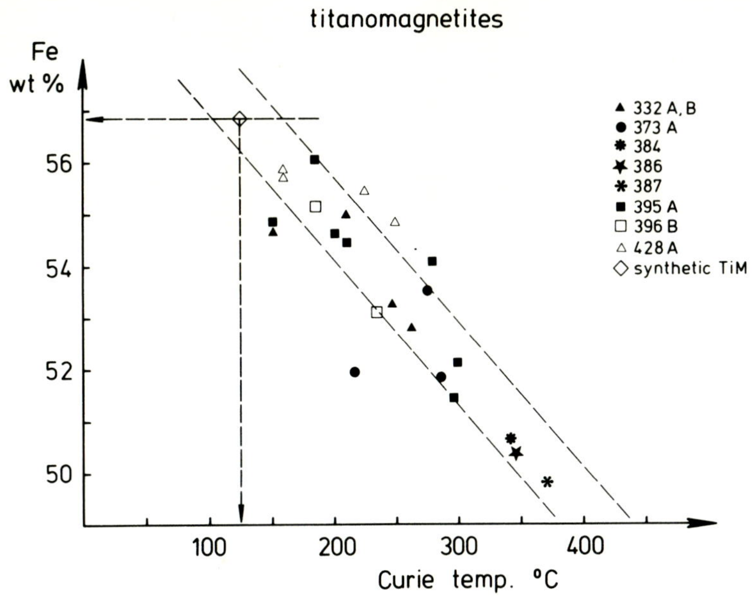

Fig. 14 . Microprobe analyses of titanomagnetites contained in ocean floor tholeiites plotted against Curie temperature T_C. The Curie temperature indicates the degree of low temperature oxidation of the titanomagnetites: As higher T_C, as higher the oxidation state. Samples with indication of high temperature titanomagnetite oxidation are excluded. A distinct iron depletion with increasing oxidation is observed. Also indicated in this diagram is the composition and Curie temperature of a synthetic spinel with a composition equivalent to the primary titanomagnetite composition as deduced from the microprobe analyses.

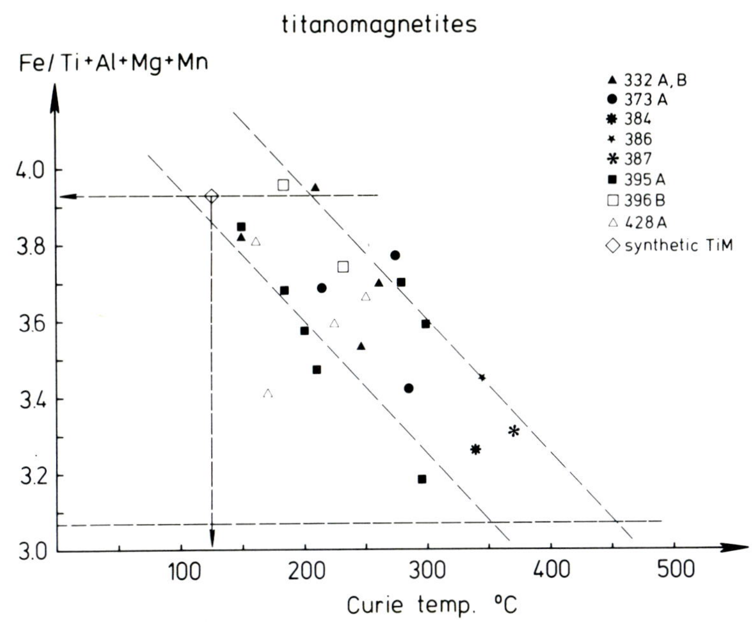

Fig. 15. Microprobe analyses of titanomagnetites contained in ocean floor tholeiites. A distinct variation of the cation ratio with low temperature oxidation is observed. As in fig. 14 also the synthetic equivalent of the primary titanomagnetite composition is indicated. The lower dashed line represents the minimum value of the Fe/(Ti + Al + Mg + Mn) ratio to be expected according to the model of Fe-migration (see text) when complete oxidation is reached.

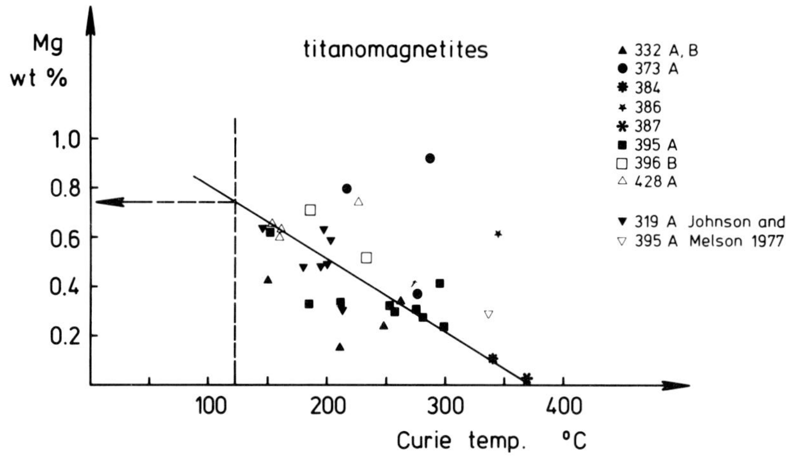

Fig. 16. Microprobe analyses of titanomagnetites contained in ocean floor tholeiites. Magnesium shows a similar trend of depletion as iron.

We will use this curve in the following section for other samples to determine the oxidation parameter z from Curie temperature measurements.

Finally, a comparison is attempted between the thus determined Curie temperature versus z curve of fig. 20 and the corresponding data of Readman and O'Reilly (1972) obtained from synthetic samples:

First we take account of the impurities in our natural samples by selecting the equivalent pure titanomagnetite composition having the same Curie temperature of 125°C as our natural primary spinel composition. The equivalent pure titanomagnetite composition is x=0.63.

We have seen from the microprobe analyses that our investigated samples when transferred into the ternary system FeO-Fe_2O_3-TiO_2 follow a line Ti/O = const (Fe-emigration). We therefore compare our T_c versus z curve with the Curie temperature and z-values taken from Readman and O'Reilly's (1972)

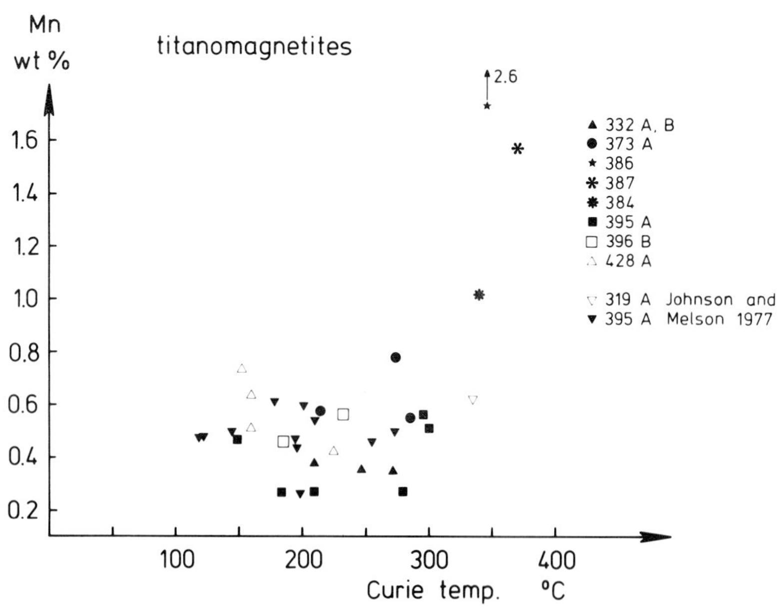

Fig. 17. Microprobe analyses of titanomagnetites contained in ocean floor tholeiites. There exists a conspicuous increase of manganese content in the strongly oxidized samples.

TABLE 3. Microprobe Analyses of Titanomagnetites

Leg	Site	Sample	Age m.y.	Fe	Ti	Al	Mg	Mn	Cr	Total	Rel.Total[x] Atomic %	$\frac{Fe}{Ti+Al+Mg+Mn}$	T_c °C
				wt.%									
42	373A	7-2,143-146	5	53.51	12.41	0.63	0.37	0.78	0.02	67.27	94.2	3.77	275
42	373A	7-3,105-110	5	51.85	12.27	1.41	0.92	0.55	0.04	67.04	95.4	3.31	285
42	373A	12-OB	5	51.93	12.97	0.71	0.80	0.58	0.03	67.02	94.1	3.45	215
43	384	22-1,118-120	105	50.65	13.14	1.25	0.11	1.02	0.05	66.22	93.1	3.26	340
43	384	22-2,56-58	105	51.56	12.74	0.35	0.21	0.49	0.02	65.37	90.8	3.74	575
43	384	22-CC,146-148	105	49.77	14.89	0.90	0.35	0.38	0.05	66.34	94.3	3.01	550
43	386	66-CC,3	104	50.34	12.00	1.35	0.62	2.62	0.02	66.95	94.7	3.03	345
43	387	50-2,134-137	135	49.81	13.73	0.75	0.00	1.58	0.04	65.91	91.9	3.10	370
45	395A	14-1,87-99	7	54.48	14.33	0.78	0.34	0.26	0.01	70.20	98.3	3.47	210
45	395A	14-2,125-134	7	56.08	14.03	0.63	0.33	0.25	0.01	71.33	99.5	3.68	185
45	395A	32-1,140-150	7	54.67	13.84	0.77	0.49	0.23	0.02	70.02	98.2	3.57	200
45	395A	61-1,140-150	7	54.13	13.17	0.90	0.28	0.26	0.00	68.74	96.2	3.71	280
45	395A	63-1,0-10	7	54.44	12.79	0.75	0.31	0.22	0.02	68.53	95.6	3.87	255
45	395A	63-4,115-122	7	51.46	14.13	1.05	0.42	0.56	0.02	67.64	95.2	3.18	295
45	395A	64-1,137-142	7	55.56	13.38	0.70	0.71	0.49	0.03	70.97	99.6	3.64	270
45	395A	64-2,109-112	7	52.13	12.81	0.97	0.24	0.52	0.01	66.68	93.5	3.59	300
45	395	11-2,100-106	7	54.86	12.09	1.10	0.61	0.46	0.04	69.16	97.2	3.85	150
46	396B	15-2,120-122	9	55.15	11.38	1.39	0.71	0.46	0.04	69.13	97.8	3.96	185
46	396B	15-4,59-61	9	53.07	12.09	1.01	0.52	0.56	0.02	67.27	94.2	3.74	233
54	428A	5-2,5D	2	55.85	12.52	0.91	0.63	0.51	0.03	70.45	98.8	3.83	160
54	428A	5-2,6A	2	54.87	12.81	0.96	0.60	0.64	0.03	69.91	98.3	3.66	160
54	428A	5-4,6	2	55.73	13.92	1.02	0.65	0.74	0.04	72.10	101.6	3.41	153
54	428A	7-2,6	2	55.49	12.98	1.30	0.74	0.42	0.03	70.96	100.4	3.59	225
54	429A	3-1,6	5	54.76	12.85	1.07	0.74	0.46	0.02	69.90	98.6	3.62	86

[x]) Atomic % with Respect to the Total Cations of the Stoichiometric Titanomagnetite $Fe_{2.27}Ti_{0.58}Al_{0.07}Mg_{0.06}Mn_{0.02}O_4$ It Gives a Measure of the Degree of Cation Deficiency in the Respective Spinel. This Representation Has First Been Described by Prévot et al. (1968).

TABLE 4. Microprobe Analyses of an Oxidized Titanomagnetite

Sample	395A,32-1,140-150										
Analyses Points	1	2	3	4	5	6	7	8	9	10	11
Elements wt.%											
Fe	52.52	54.39	55.79	56.45	56.25	56.46	55.91	55.48	54.42	53.75	53.52
Ti	13.58	13.82	13.53	13.44	13.44	13.41	13.64	13.42	13.72	13.75	13.79
Al	0.76	0.83	0.78	0.87	0.83	0.87	0.88	0.92	0.90	0.91	0.92
Mg	0.55	0.48	0.52	0.45	0.48	0.49	0.53	0.51	0.47	0.49	0.46
Mn	0.23	0.20	0.21	0.23	0.22	0.20	0.24	0.25	0.22	0.22	0.24
Cr	0.00	0.05	0.00	0.00	0.00	0.00	0.05	0.04	0.00	0.00	0.00
Total	67.64	69.77	70.83	71.44	71.22	71.43	71.25	70.62	69.73	69.12	70.93
+Atomic % with Respect to the Total Cations of the Stoichiometric Titanomagnetite $Fe_{2.27}Ti_{0.58}Al_{0.07}Mg_{0.06}Mn_{0.02}O_4$											
Fe	69.8	72.4	74.2	75.1	74.8	75.1	74.4	73.8	72.4	71.5	71.2
Ti	21.0	21.4	21.0	20.9	20.9	20.8	21.2	20.8	21.2	21.3	21.4
Al	1.0	1.1	1.0	1.2	1.1	1.2	1.2	1.3	1.3	1.3	1.3
Mg	1.7	1.5	1.6	1.4	1.5	1.5	1.6	1.6	1.4	1.5	1.4
Mn	0.3	0.3	0.3	0.3	0.3	0.3	0.3	0.3	0.3	0.3	0.3
Cr	0.0	0.0	0.0	0.0	0.0	0.0	0.0	0.0	0.0	0.0	0.0
Rel.Total	93.8	96.8	98.1	98.9	98.6	98.9	98.7	97.8	96.6	95.9	95.6

+ A Representation First Described by Prévot et al. (1968)

fig. 7 b correspondingly along the line Ti/O=const, starting from an original composition x=0.63. This comparison makes visible a distinct discrepancy between the two curves, which is much greater than the uncertainty of our measurements and calculations. On the other hand we find remarkable agreement when we compare Readman and O´Reilly´s Curie temperature and z-values taken along the line Fe/Ti=const. instead. This result, also shown in fig. 20, is contrary to what we would expect.

From this comparison we can draw the conclusion that either we have misinterpreted our microprobe data, the Fe/Ti ratios actually being constant, or that Readman and O´Reilly have misinterpreted their data, or that natural and synthetic samples cannot be compared in this way.

We think that the microprobe data give overwhelming evidence for the iron-migration model and against a process that proceeds along Fe/Ti=const, at least so far as the present samples are concerned. It seems to us to be more likely that in Readman and O´Reilly´s synthetic samples Fe-migration had actually occured as oxidation mechanism. If this is the case, their samples cannot have been single phase oxidation products, the migrated iron must have formed a second mineral, most likely thin films of Fe_2O_3. As the amount of emigrating Fe is relatively small (a maximum of about 10 wt% for x-values around 0.60) it can easily be overlooked in x-ray diffraction data. However, at this time this is merely a suggestion and has not been substantiated by any direct measurements.

Variation of Curie temperature with age

Given a certain grain size of the titanomagnetites in submarine basalt, the process of maghemitization increases with increasing age of the rock. Ozima et al. (1974) assumed a first order rate process of the following form

$$z = 1 - e^{-t/\tau}$$

where z denotes the oxidation parameter, t the age of the rock and π the time constant of the process. From their data they derive a time constant of $\tau = 5 \times 10^7$ y.

The actual maghemitization process in ocean floor basalts must be very complicated depending on various factors such as grain size, chemical composition, depth below basement, temperature etc. Hence the actual maghemitization is probably much more complicated than that described in the above expression of Ozima et al. (1974).

If we plot the Curie temperature (T_C) as a measure of maghemitization against age for a number of Atlantic sites, we observe actually an increase with age for the mean values of T_C

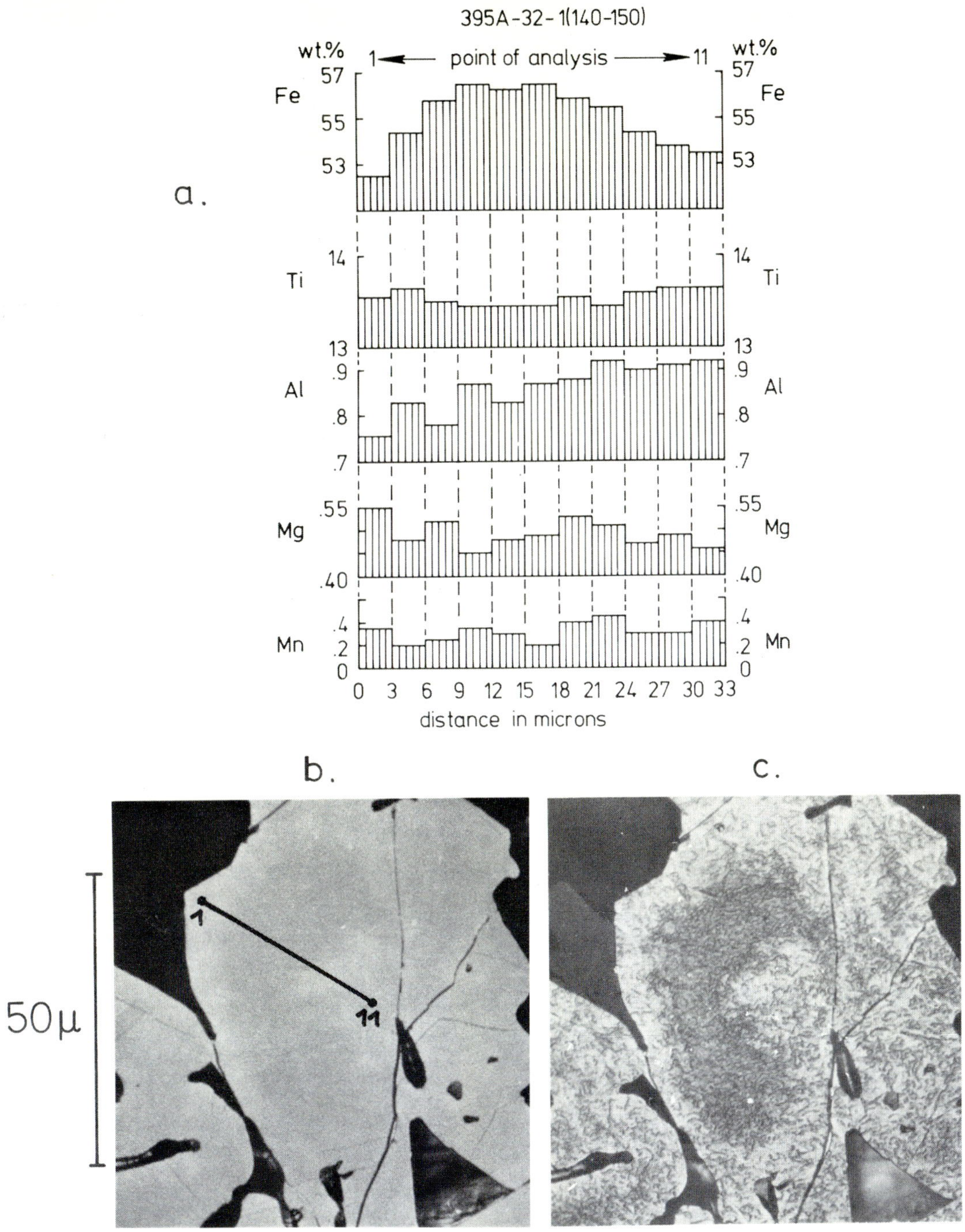

Fig. 18. Section of the titanomagnetite grain fully shown in fig. 10. The grain displays different stages of maghemitization.

a. Microprobe analyses along a profile shown in b). There is iron depletion in the oxidized outer parts of the grain with respect to the unoxidized central part. Titanium remains more or less constant over the whole profile.

b. Reflected light microphotograph of the analyzed titanomagnetite grain, showing the profile of microprobe analysis.

c. Titanomagnetite grain covered by magnetic colloid. The oxidized parts (with Fe-deficiency) are characterized by lower magnetization compared to the unoxidized central part as documented by the attraction of colloid.

TABLE 5. Synthetic Spinel

$Fe_{2.27}Ti_{0.58}Al_{0.07}Mg_{0.06}Mn_{0.02}O_4$

Equivalent to the Average Titanomagnetite Composition in Oceanic Basalts as Derived in this Study

Curie Temperature °C	Saturation Magnetization at Room Temperature Gauss·cm³/g	Lattice Constant Å
125±2	26±2	8.475±0.003

(fig. 21). However, the scatter of individual Curie temperatures within a site is considerable. The most important factor causing this scatter seems to be the grain size. Inspite of a large scatter in the data fig. 22 shows clearly that there is an inverse relationship between grain size and Curie temperature. Nevertheless it seems that such factors average out if the number of samples for a certain site is large enough. However our data cannot be fitted to the above relationship of Ozima et al. (1974).

In fig. 23 are plotted the mean values of Curie temperature against the logarithm of age. The maximum Curie temperature that can be obtained by maghemitization prior to any inversion of the cation deficient spinel phase is probably close to 400°C. This can be taken from fig. 15, where the theoretically possible minimum value of Fe/Ti is reached at $T_c = 400°C$. Therefore the age - T_c relationship of fig. 23 leads to the suggestion that ocean floor basalt of jurassic age is liable to the inversion of the magnetic spinel phase into non magnetic or only weakly magnetic phases like iron hydroxide or hematite. The so-called jurassic quiet zone could therefore be the result of this mineral instability.

Variation of maghemitization with depth below basement

Direct evidence about the depth variation of maghemitization exists only from the few holes with deeper penetrations of the oceanic crust. The depth limit to date is about 600 m. Judging from the few data we have available, it seems that the whole portion of the upper 600 m basalt has been maghemitized more or less uniformly, at least so far as the pillow basalts are concerned which obiously predominate in the upper 600 m of the atlantic ocean crust. Fig. 24 shows the downhole Curie temperature variation in the Atlantic DSDP hole 396B.

No overall decrease of the Curie temperature is observed here. However, within individual lithological units there seems to be a tendency of slight decrease of T_c. For lithologic unit 1 of site 396B we determine a Curie temperature decrease of

$$0.57\ °/m$$

Similar Curie temperature gradients have been determined by Bleil (1976) from the sites 332 A and B of DSDP Leg 37.

Variation of magnetization intensity

Irving (1970) and Irving et al. (1970) were the first to study different magnetic and mineralogical properties of ocean floor basalt samples taken along a profile across the Midatlantic Ridge at 45°N, extending about 150 km away from the ridge axis. They found a drastic decrease of the NRM intensity with increasing distance from the ridge axis and correlated this with the similarly increasing whole rock oxidation ratio. The explanation for this correlation is given by experimental results obtained from synthetic titanomagnetites (Sakamoto et al. 1968, Ozima and Larson 1970, Readman and O'Reilly 1972) that clearly demonstrate a decrease of magnetization intensity with increasing low temperature oxidation. The decrease of magnetization intensity, both NRM and saturation magnetization, with increasing low temperature oxidation, has consequently been used for an explanation for the general decrease of the amplitudes of marine magnetic anomalies away from a ridge axis.

We too have tested for this relationship in our samples, excluding those samples which indicate that high temperature deuteric oxidation took place.

For this purpose the titanomagnetite content was determined microscopically using a Quantimet image analyzing device. The magnetization intensity was then referred to the magnetic mineral. In fig. 25 the specific saturation magnetization of the magnetic mineral at room temperature is plotted against Curie temperature. From this diagram we find actually a decrease of magnetization with increasing oxidation, but only up to a z-value of about z = 0.6 (corresponding to a Curie temperature

TABLE 6. $Fe_2O_3/FeO+Fe_2O_3$, z-Parameter, Lattice Constant a_o and Curie Temperature T_c of Separated Titanomagnetites

Leg	Site	Sample	Age m.y.	$\frac{Fe_2O_3}{FeO+Fe_2O_3}$	k	z	a_o Å	T_c + °C	T_c × °C
FAMOUS		Alv 521-4-2	0.1-0.5	0.425	0.51	0.11	8.483±0.002	165	167
42	373A	7-3,105-110	5	0.761	0.77				285
43	384	22-1,118-120	105	0.779	0.83	0.69	8.416	323	340
43	386	66-CC,3	104	0.763	0.78	0.67	8.437	313	345
43	387	50-2,134-137	135	0.822	0.88	0.76	8.411		370
46	396B	15-4,59-61	9	0.628	0.66	0.46	8.444	233	233
54	428A	1-3,3C	1.2-2.5	0.498	0.53	0.24	8.477	150	160
54	428A	5-4,6	1.2-2.5	0.430	0.44	0.12	8.476	147	153

k - $Fe_2O_3/FeO+Fe_2O_3$ Corrected for the Impurity of the Separated Titanomagnetites
z - Oxidation Parameter Derived from the Corrected $Fe_2O_3/FeO+Fe_2O_3$
+) Curie Temperature of Separated Titanomagnetites
×) Curie Temperature of Whole Rock

of 250°C). For higher z-values the magnetization increases again. Part of the scatter in fig. 25 is probably due to errors in the optical determination of the magnetic mineral content. It is interesting to note that the lattice parameters a_o of the titanomagnetites when plotted against the Curie temperature T_c (see fig. 26 and table 7) also display an anomaly at T_c = 250°C.

If the observed trend of magnetization increase for z-values higher than 0.6 is a general feature of the oceanic crust, we may speculate if this would cause the high amplitudes of the mesozoic M-anomalies.

Two explanations seem possible

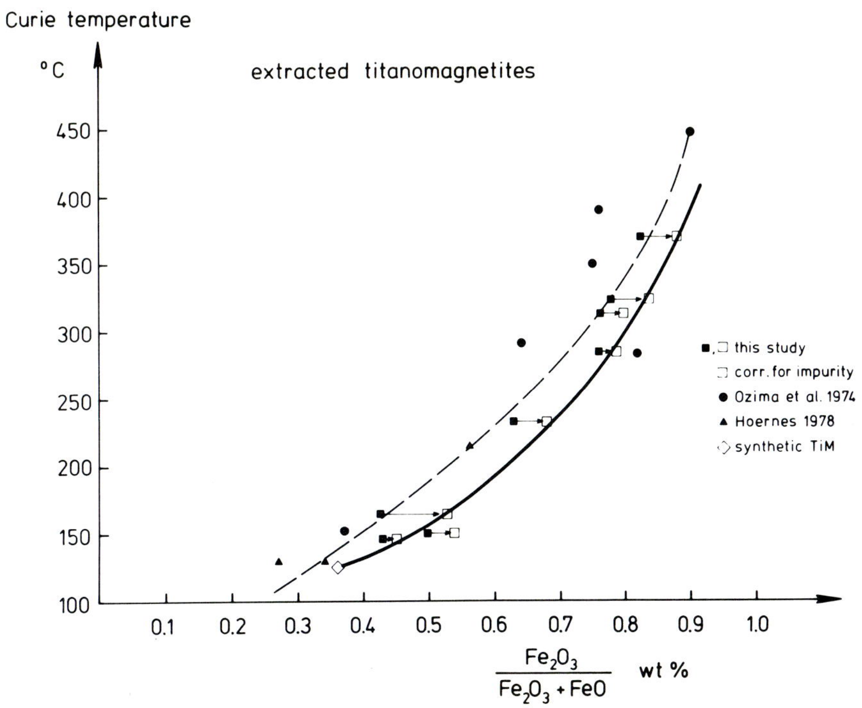

Fig. 19. Curie temperature T_c versus oxidation ratio Fe_2O_3/Fe_2O_3 + FeO of extracted titanomagnetites. The dashed curve gives the uncorrected oxidation ratios where the influence of silicate impurities adhering to the titanomagnetite grains is not eliminated. The solid line is the curve for the corrected Fe_2O_3/Fe_2O_3 + FeO values. Also included are the data of Ozima et al. (1974) which show considerable scatter, but fall into the same range. They are not corrected for silicate impurities.

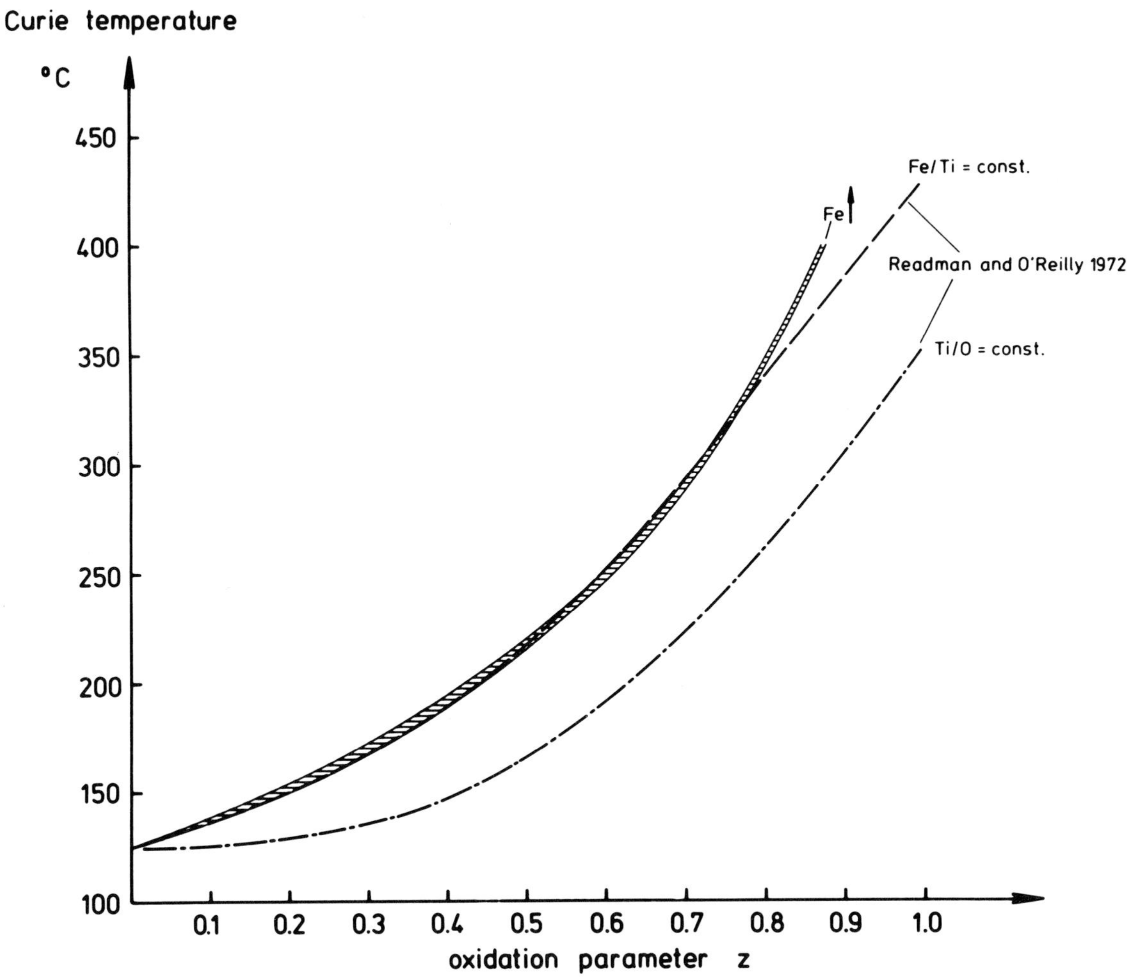

Fig. 20. Curie temperature T_c versus oxidation parameter z. The shaded curve has been derived from the solid curve of fig. 19 by conversion of the Fe_2O_3/Fe_2O_3 + FeO values of the extracted titanomagnetites into z-values (see text). This curve represents an oxidation mechanism due to Fe-migration which is equivalent to a process whereby Ti/O is kept constant. Also shown in this diagram are two curves with T_c and z-values taken from Readman and O'Reilly's (1972) fig. 7b along the line Ti/O = const (dash-dotted) and along the line Fe/Ti = const (dashed). Assuming Fe-migration for the samples studied in this paper the shaded curve would consequently be expected to follow the dash-dotted curve Ti/O = const instead of the curve Fe/Ti = const. This obvious discrepancy is discussed in the text.

for the magnetization increase at high z-values:

1. Inversion of the metastable cation deficient titanomagnetite to an intergrowth of magnetite and other stable Fe-Ti oxides (Readman and O'Reilly 1970). This process is realized probably only in a few of the investigated samples, for example in 112-17-1(99-100) (see fig. 27). However the majority of the samples plotted in fig. 25 and 26 with Curie temperatures higher than 250°C seem to be cation deficient single phase titanomaghemites as for example 332A-12-1(81-83) and 114-9-1(8-9) in fig. 27. This conclusion can be drawn from the shape of their thermomagnetic curves and their x-ray diffraction pattern. Also their lattice parameters are much too small for being magnetite (see table 7 and fig. 26).

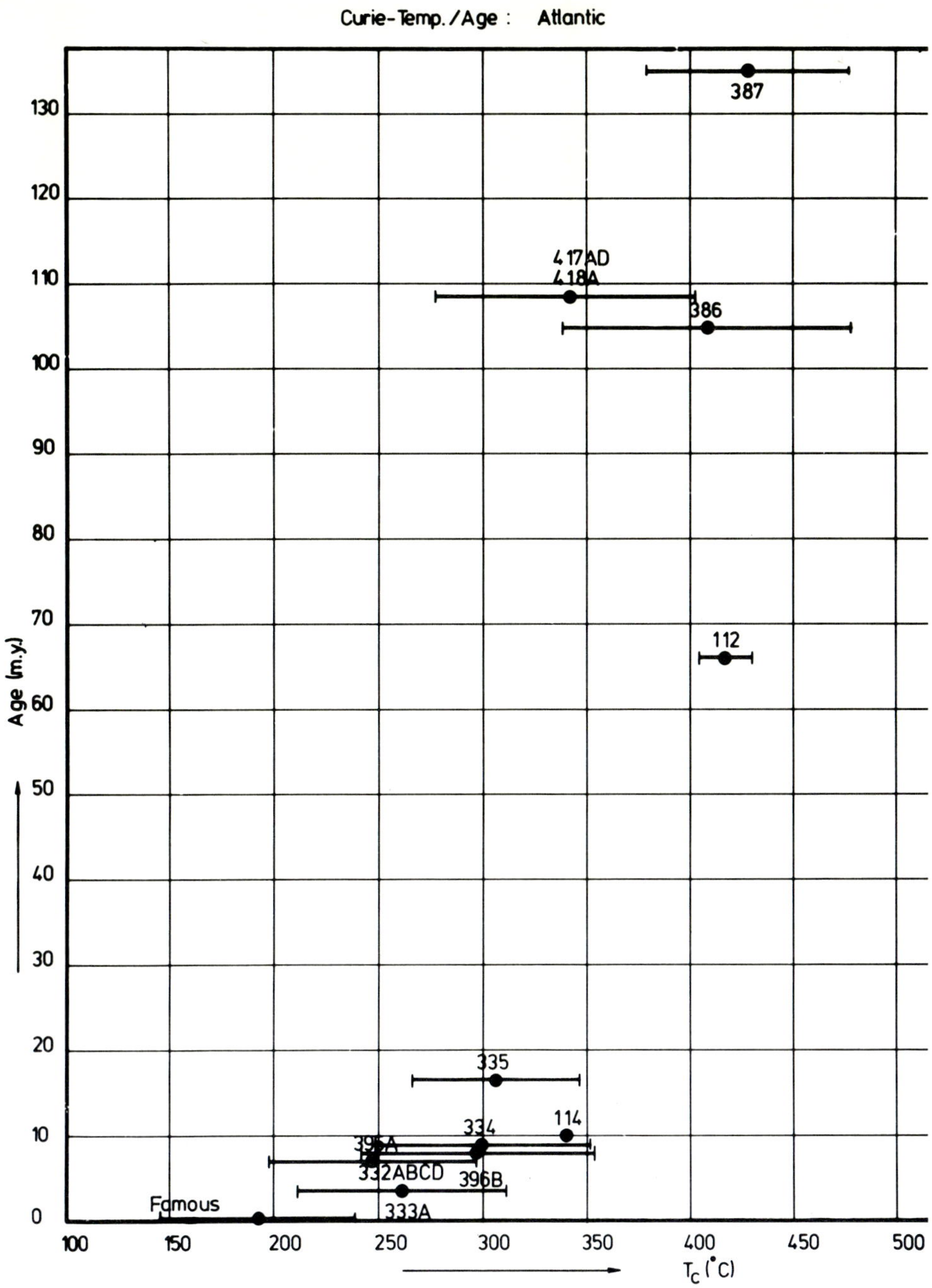

Fig. 21. Curie temperature T_C versus age of ocean floor basalts from different Atlantic sites.

2. Formation of cation-vacancies in the spinel lattice due to Fe-migration.

   a) We first consider the case where vacancies are formed only on the octahedral sites of the spinel lattice. This results in a decrease of the net magnetic moment first, but eventually the magnetic moment reverses sign and then grows in the opposite direction. (This is correct only at $0^{O}K$, at room temperature the situation may be different, de-

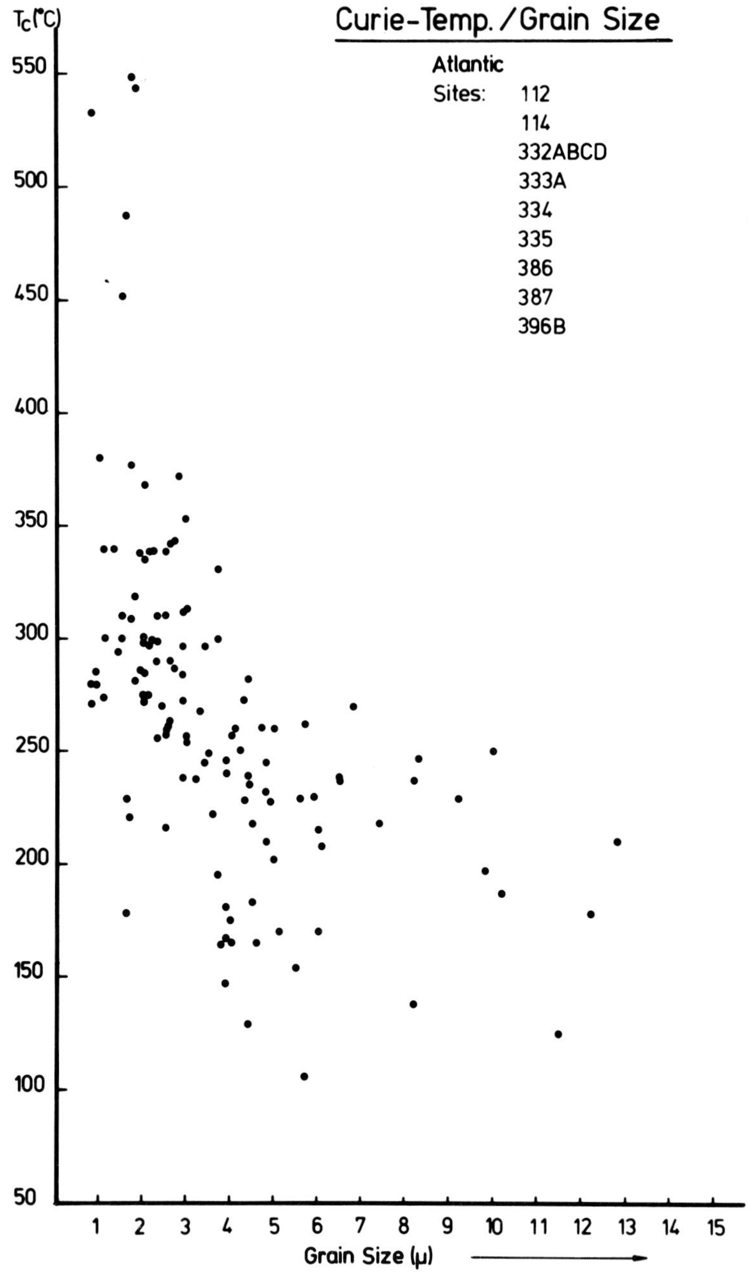

Fig. 22. Curie temperature T_C versus mean grain size of titanomagnetites contained in ocean floor basalts. An inverse relationship between the two variables seems to exist.

pending on the Néel type curve. See for example fig. 27).

This case is demonstrated in table 8 for our primary titanomagnetite composition assuming a cation distribution according to the O'Reilly and Banerjee model (1966). We also make the simplifying assumption that Ti, Al, Mg, Mn and cation vacancies are located on octahedral sites only.

The reversal occurs at z=0.60 which corresponds to a Curie temperature of 250°C. This actually coincides with the minimum of saturation magnetization of the natural titanomaghemites as shown in fig. 25 For such a model we would theo-

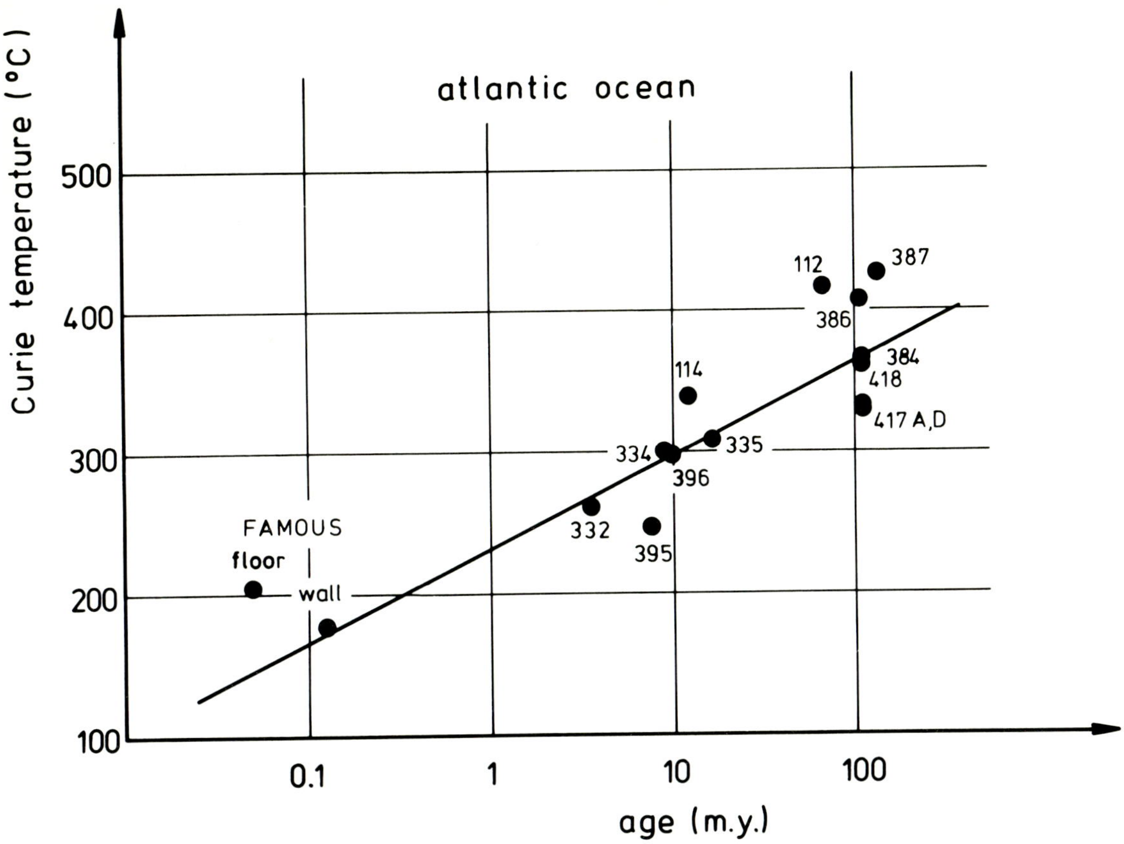

Fig. 23. Mean values of Curie temperatures taken from fig. 21 plotted against the logarithm of age of the rocks. As deduced from fig. 15 extreme degree of maghemitization is reached at a Curie temperature around 400°C. Ocean floor basalt of jurassic age is therefore liable to an inversion of the cation deficient magnetic titanomagnetites to non-magnetic phases.

retically expect a sequence of Q-P-L-N-Q Neel curves with increasing oxidation parameter z (Ozima and Sakamoto 1971, Readman and O´Reilly 1972, Schult 1978). This prediction is tested for a number of samples in fig. 27. There the saturation magnetization has been measured in a magnetic field of 10.100 Oe between liquid nitrogen temperature and the Curie temperature of the sample. From fig. 27 it can be taken that we cannot observe the expected sequence of Neel curves; instead the measured curves let us suggest the sequence

P - L - P

b) In table 9 we consider the case that vacancies are formed on octahedral sites first, until z = 0.60 is reached, and then, beyond this z-value, vacancies shall be formed on tetrahedral sites only.

This model also gives a decrease of magnetization until z = 0.60 is reached, and then for higher z-values, an increase of magnetization again. For such a model we expect theoretically a sequence of

P - L - P Néel-curves

what we actually observe (fig. 27). O´Donovan and O´Reilly (1977) observed a similar sequence of Néel curves measuring the oxidation sequence of a synthetic titanomagnetite of the composition

$Fe_{2.35}Mg_{0.05}Ti_{0.60}O_4$

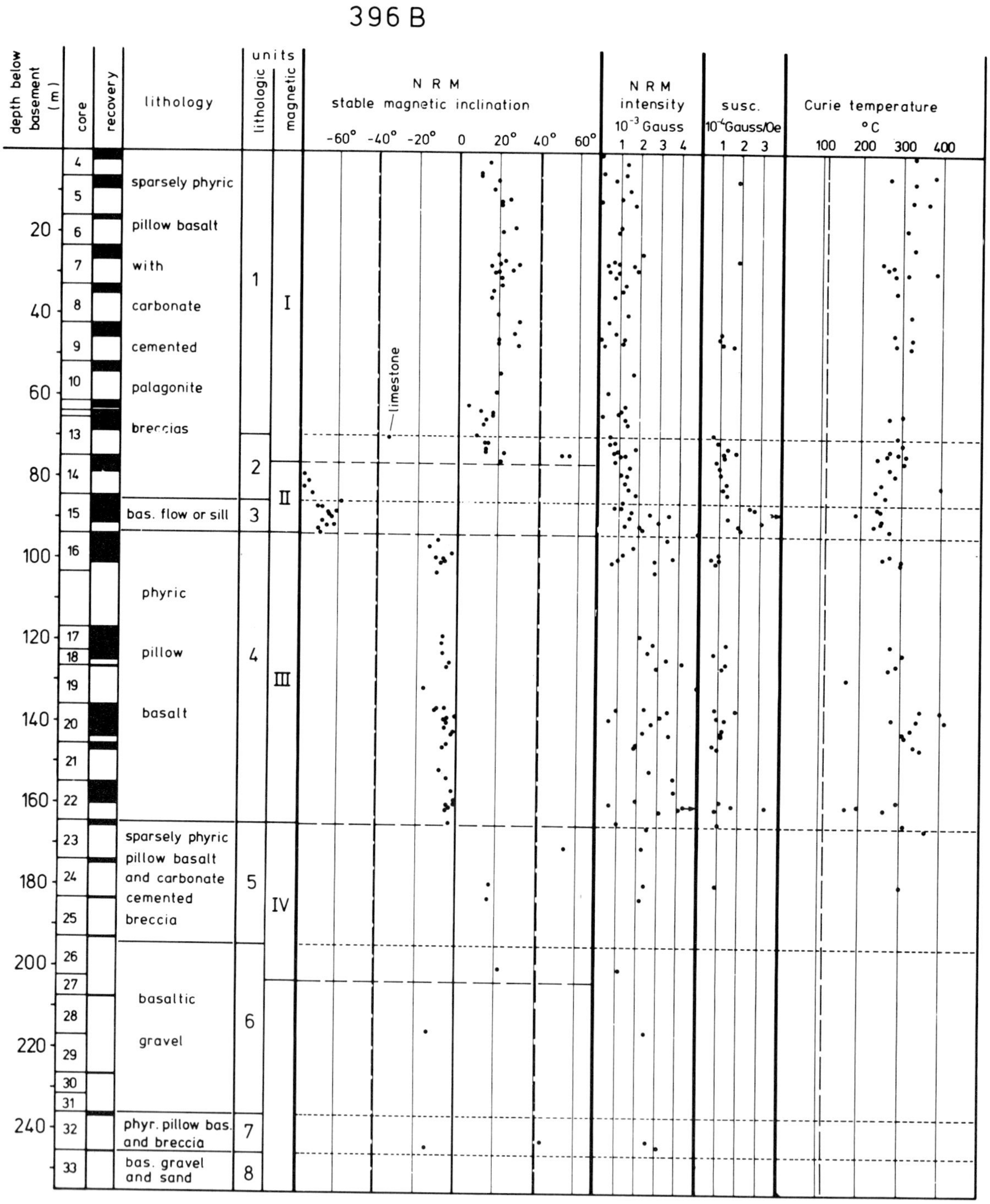

Fig. 24. Downhole variation of different magnetic parameters in DSDP hole 396B. The dashed line in the Curie temperature downhole plot indicates the Curie temperature of the unalterd, primary magnetic mineral component. There is a slight decrease of T_C within individual lithologic units.

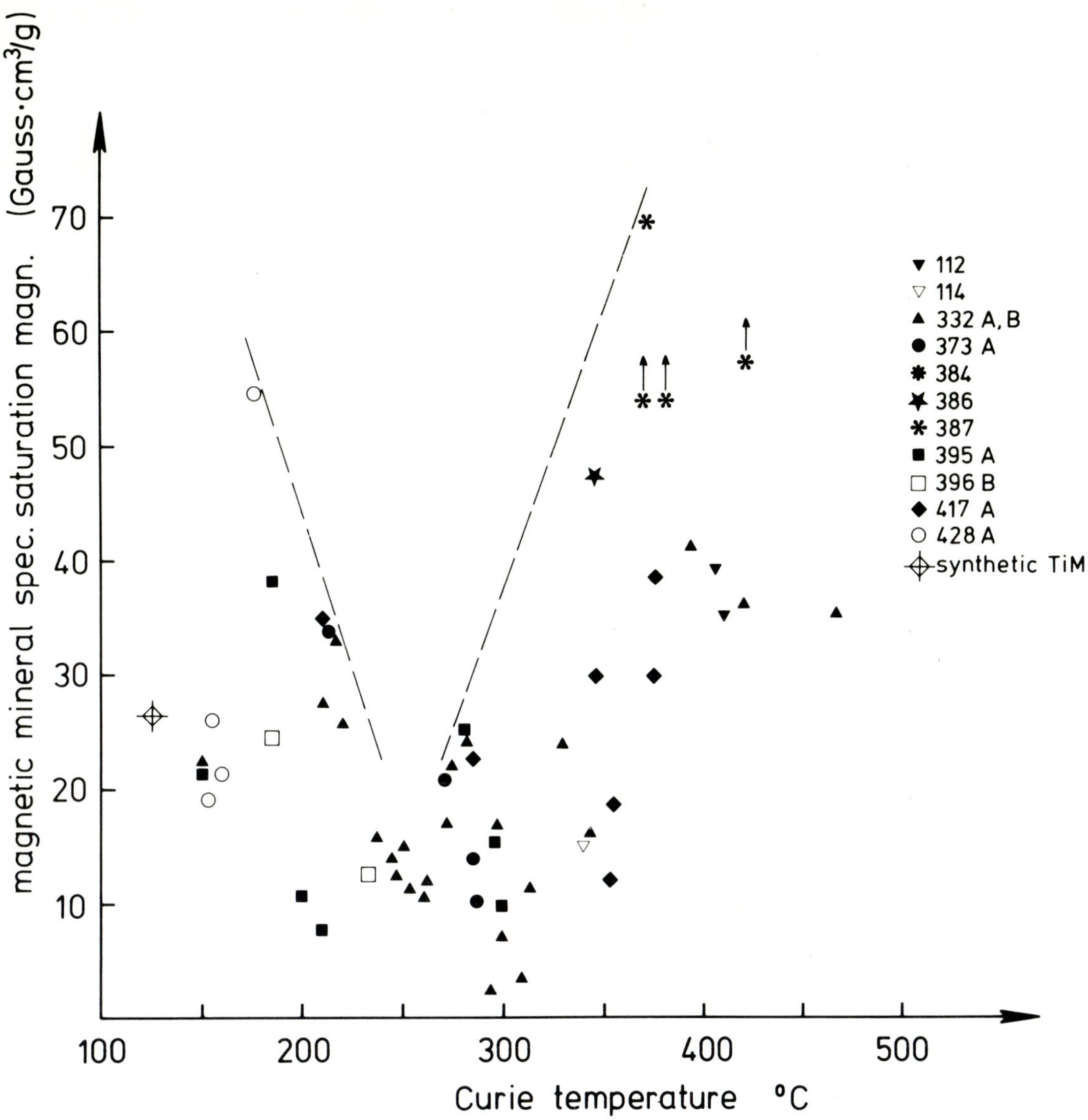

Fig. 25. Specific saturation magnetization of the magnetic minerals plotted against Curie temperature. The specific mineral magnetization has been determined from the specific rock magnetization and the optically measured titanomagnetite content. For the conversion a rock density of 3 and titanomagnetite density of 5 has been assumed: spec. mineral magnetization = spec. rock magn.•(3/5)•(100/timag. vol%).

They report P-type behaviour for this composition that becomes more pronounced with increasing z until it is almost L-type at z = 0.59 and then, beyond this z-value, becomes less pronounced P-type again. This agrees well with our observations.

However, we are well aware that our simple "model b" is by far not the only one that can suffice the experimental data. The actual mechanism of Fe-migration may be much more complicated in nature. But we wanted to demonstrate that it is easily possible to develop a physically feasible model that yields the main characteristics observed in our samples.

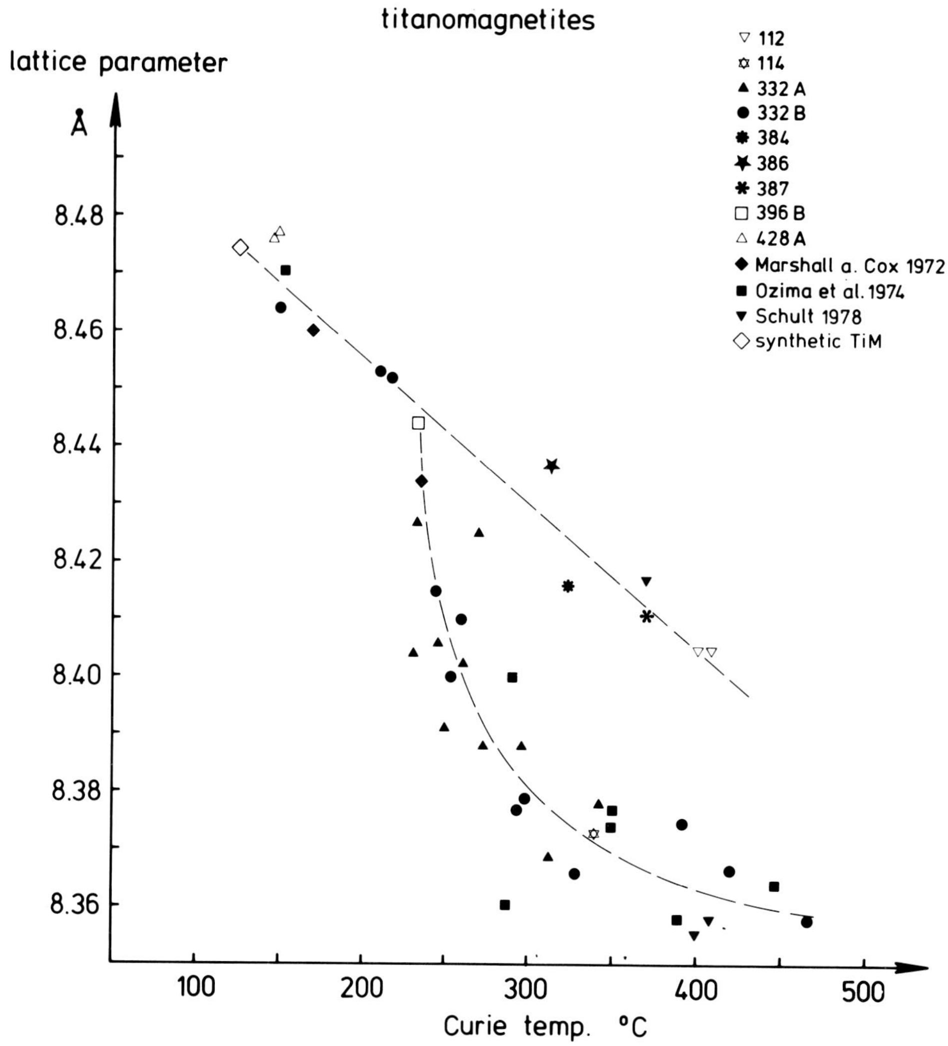

Fig. 26. Lattice parameter a_o of separated titanomagnetites versus Curie temperature T_c. There is a general decrease of a_o with increasing Curie temperature. The anomaly at T_c = 250°C corresponds to a z-value of 0.6. It is not clear if we observe two different trends for T_c-values higher than 250°C or if the scatter becomes so large although the accuracy of the individual measurements is by far better.

Variation of magnetization direction with low temperature oxidation

Most of the magnetic parameters measured on ocean floor basalts change with increasing degree of maghemitization of the contained titanomagnetites. In the context of interpretation of the marine magnetic anomalies it is important to know if also the direction of natural remanent magnetization is affected by low temperature oxidation.

Following Vine and Matthews (1963) it seems reasonable to assume that the unaltered oceanic crust close to the axial rift, containing primary titanomagnetite, carries a thermoremanent magnetization acquired soon after deposition of the rock. However the primary titanomagnetite is substantially changed chemically by the process of maghemitization that constantly proceeds in the ocean floor environment at temperatures far below the Curie temperature. One may argue therefore that this oxidation gives rise to a secondary chemical remanent magnetization (CRM) acquired in the ambient earth magnetic field long after the deposition of the rock, with a magnetization direction in general different from the original thermoremanence.

CRM forms in small grains of magnetic

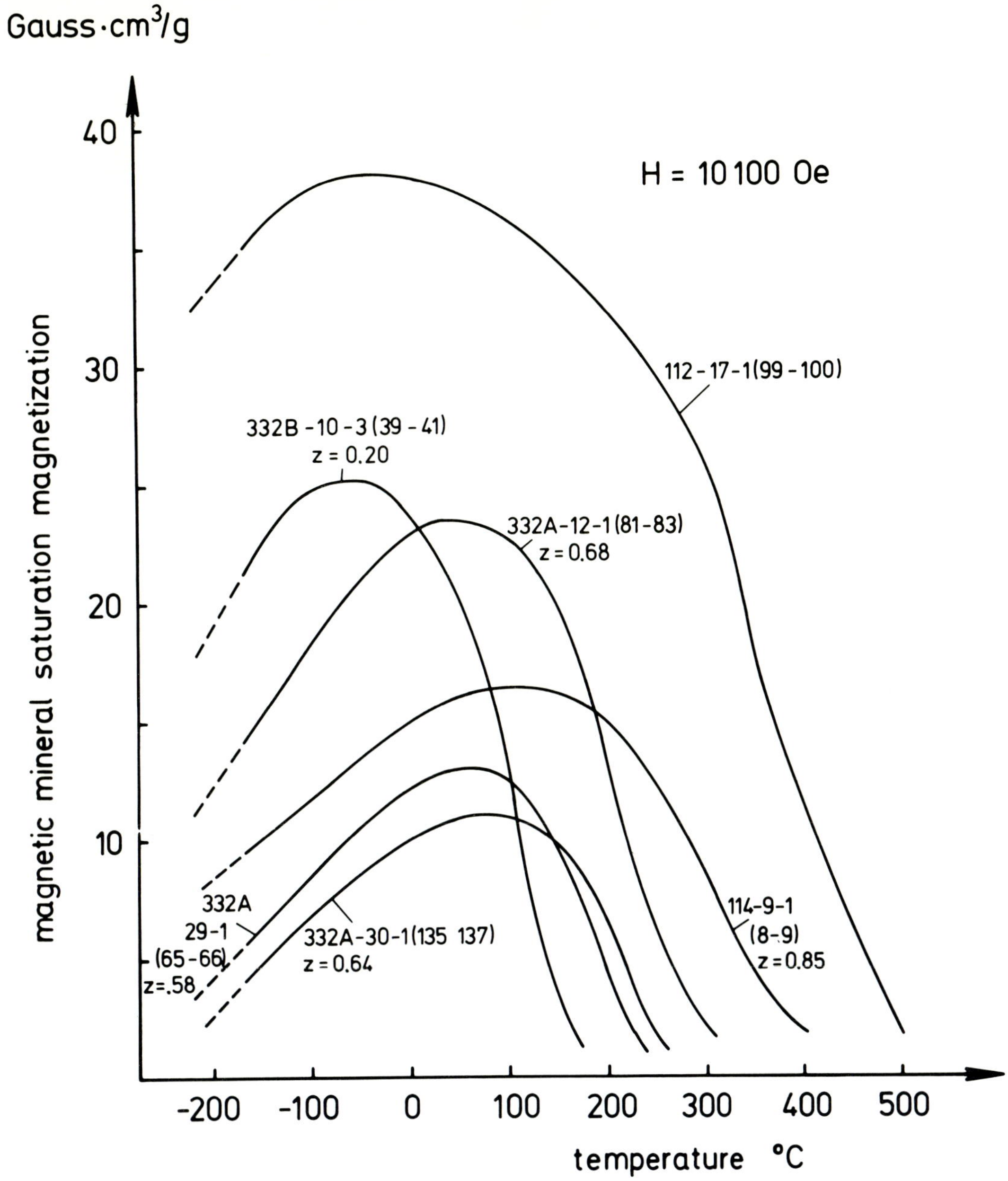

Fig. 27. Thermomagnetic curves of natural titanomagnetites measured in a magnetic field of 10.100 Oe.

material that grow from small nuclei through the blocking volume in the presence of the geomagnetic field (Kobayashi 1959).

When we consider under this point of view the Fe-migration model as the basic process of ocean floor titanomagnetite alteration, it seems unlikely that a chemical remanence should develop, because no new magnetic material is growing by this process, the magnetic grains are only losing iron that either is incorporated in non-magnetic clay minerals (Honnorez et al. 1978) or transported away by sea water. Laboratory experiments by Marshall and Cox (1971) also come to the conclusion that the original remanence direction of ocean floor basalts is not destroyed by oxidation if the oxidation occurs below the original Curie temperature of the titanomagnetite. On the other hand, Butler (1973) has argued that low temperature oxidation of oceanic basalts will cause a significant portion of the titanomagnetite grains, originally stable

TABLE 7. Curie Temperature T_c, Oxide-Content, Saturation Magnetization J_s, NRM, and Lattice Parameter a_o of Titanomagnetites

Leg	Site	Sample	T_c °C	Vol.% Oxide	J_s(Rock) x10^{-2} emu/g	J_s(Min.) emu/g	NRM x10^{-3} Gauss	a_o Å
12	112	17-1,99-100	405	1.9	124		9.17	8.405
12	112	17-1,118-119	410	2.1	123		12.21	8.405
12	114	9-1,8-9	340	1.2	30		4.20	8.373
37	332A	6-2,107-109	343	0.6	16	16	2.56	8.378
37	332A	8-1,50-52	313	1.3	24.5	11	7.89	8.369
37	332A	12-1,81-83	273	1.0	37	22	3.69	8.388
37	332A	17-1,50-52	250	1.1	27.4	15	0.69	8.391
37	332A	29-1,65-66	247	1.3	27	12	1.76	8.406
37	332A	30-1,135-137	262	1.4	27.6	12	1.89	8.403
37	332A	32-2,28-30	271	0.6	17	17	3.77	8.425
37	332A	36-1,21-23	297	1.0	28.5	17	1.20	8.338
37	332A	37-1,101-103	237	1.5	39.5	16	5.37	8.427
37	332A	40-3,35-37	282	0.4	16	24	0.60	8.404
37	332B	1-5,57-60	393	0.2	13.8	41	1.62	8.375
37	332B	2-4,122-125	467	0.2	11.8	35	1.52	8.358
37	332B	3-1,27-30	420	0.2	12	36	1.95	8.367
37	332B	6-1,111-114	253	1.2	22.5	11	7.98	8.400
37	332B	9-1,88-99	244	1.2	28	14	1.18	8.415
37	332B	10-3,39-41	150	1.6	58.5	22	2.68	8.464
37	332B	15-1,103-105	260	1.4	24.8	11	2.65	8.410
37	332B	16-2,63-65	328	0.3	12	21	10.29	8.366
37	332B	25-1,106-113	298	0.8	9.5	7	4.92	8.379
37	332B	31-1,116-119	293	1.3	4.5	2	1.70	8.377
37	332B	44-5,124-126	217	1.3	71.5	33	0.81	8.452
37	332B	46-2,58-61	210	1.7	79	28	2.54	8.453
42	373A	3-3,40-45	287	0.9	15.5	10	2.44	
42	373A	6-CC	270	0.8	18	21	1.68	
42	373A	7-2,51-54	295		34		3.28	
42	373A	7-2,143-146	275		31		3.54	
42	373A	7-3,105-110	285	1.2	28	14	1.61	
42	373A	12-OB	215	1.3	73	34	3.53	
43	384	22-1,118-120	340	1.2	67.5	34	2.93	8.416

Leg	Site	Sample	T_c °C	Vol.% Oxide	J_s(Rock) x10^{-2} emu/g	J_s(Min.) emu/g	NRM x10^{-3} Gauss	a_o Å
43	386	66-1,1	360		60		3.94	
43	386	66-2,14	410		110		0.64	
43	386	66-CC	345	1.1	87.2	48	3.59	8.437
43	387	50-1,35-38	370	0.9	105	70	0.95	
43	387	50-1,98-101	420	0.2?	95.5	29?	3.24	
43	387	50-2,30-33	380	0.1?	90	540?	4.15	
43	387	50-2,48-51	390		77		4.19	
43	387	50-2,134-137	370	0.3?	90	180?	2.62	8.411
45	395A	11-2,100-106	150	1.6	58	22		
45	395A	14-1,87-99	210	2.5	32	8	9.19	
45	395A	14-2,125-134	185	1.3	83	38	3.33	
45	395A	32-1,140-150	200	2.6	46	11		
45	395A	61-1,140-150	280	0.6	25.5	26		
45	395A	63-4,115-122	295	1.2	31	15		
45	395A	64-2,109-112	300	1.2	20	10	7.44	
46	396B	15-2,120-122	185	1.6	65	24	1.58	
46	396B	15-4,59-61	233	1.6	33.5	13	1.28	
51	417A	37-3,23-25	355	0.7	22	31	6.46	
51	417A	38-2,28-31	375	0.3	15	50	5.32	
51	417A	39-1,37-40	345	2.5	13	50	5.32	
51	417A	41-1,103-105	285	2.0	76	38	12.96	
51	417A	41-2,68-70	355	1.0	20	20	8.20	
51	417A	43-4,27-30	210	1.5	88	59	21.94	
51	417A	44-1,52-55	375	1.8	115	64	23.07	
52	417D	63-2,12-14	230	1.6	104		40.52	
52	417D	68-3,64-66	320	5.4	110		17.72	
52	417D	68-5,89-92	250	3.7	127		31.93	
52	417D	69-2,35-37	240	4.1	118		38.52	
54	428A	1-3,3C	155	2.4	104	26	6.08	8.483
54	428A	2-1,68-71	177	0.8	73	55	8.88	
54	428A	5-2,85-88	160	3.1	110	21	3.70	
54	428A	5-4,89-91	153	3.1	102	20	5.55	8.486

J_s(Rock) - Whole Rock Saturation Magnetization Measured at Room Temperature in 1800 oe
J_s(Min.) - Magnetic Mineral Saturation Magnetization

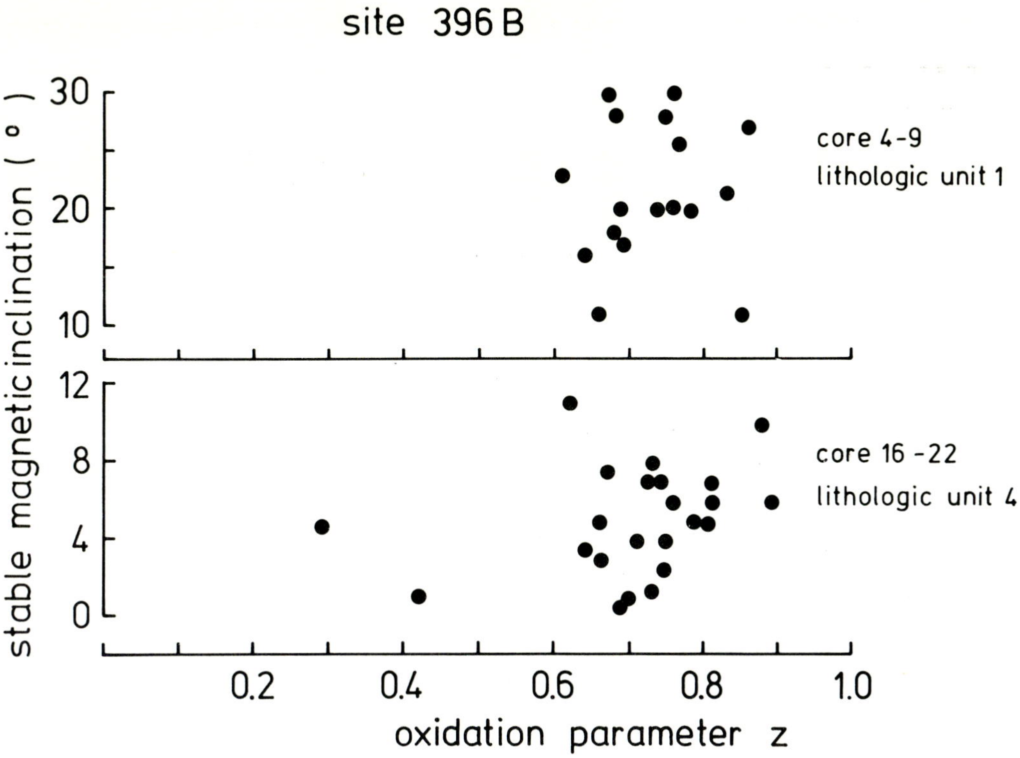

Fig. 28. Stable magnetic inclination values of two lithological units from atlantic DSDP site 396B, plotted against the titanomagnetite oxidation parameter z. The inclination data are the same as in the downhole plot of fig. 24.

TABLE 8. Formation of Cation-Vacancies on Octahedral Sites of the Spinel Lattice

z = 0: net magnetic moment: +0.61 $_B$

$$Fe^{3+}_{0.62}\ Fe^{2+}_{0.38}\ (Fe^{2+}_{1.12}\ Fe^{3+}_{0.15}\ Al^{3+}_{0.07}\ Mg^{2+}_{0.06}\ Mn^{2+}_{0.02}\ Ti^{4+}_{0.58})\ O^{2-}_{4}$$

the brackets denote the cations on octahedral sites

$Fe^{3+}_{0.50}$ maximum amount of Fe-migration

z = 1: net magnetic moment: -1.15 $_B$

$$Fe^{3+}_{1.00}\ (Fe^{3+}_{0.77}\quad 0.50\ Al^{3+}_{0.07}\ Mg^{2+}_{0.06}\ Mn^{2+}_{0.02}\ Ti^{4+}_{0.58})\ O^{2-}_{4}$$

carriers of the remanence, to become superparamagnetic and thus lose their initial magnetization. Increase of the spontaneous magnetization at high z-values would then shift the superparamagnetic particles back into the field of stable single domain particles which thereby would actually acquire a new chemical remanence.

We have tested this question by studying the variation of NRM directions within single lithological units of the oceanic crust, where it can safely be assumed that all the samples acquired

TABLE 9. Formation of Cation-Vacancies on Octahedral and Tetrahedral Sites of the Spinel Lattice

z = 0: net magnetic moment: +0.61 $_B$

$$Fe^{3+}_{0.62}\ Fe^{2+}_{0.38}\ (Fe^{2+}_{1.12}\ Fe^{3+}_{0.15}\ Al^{3+}_{0.07}\ Mg^{2+}_{0.06}\ Mn^{2+}_{0.02}\ Ti^{4+}_{0.58})\ O^{2-}_{4}$$

the brackets denote the cations on octahedral sites

$Fe^{3+}_{0.30}$ amount of Fe-migration

z = 0.60: net magnetic moment: +0.01 $_B$

$$Fe^{3+}_{0.62}\ Fe^{2+}_{0.38}\ (Fe^{2+}_{0.22}\ Fe^{3+}_{0.75}\ \ 0.30\ Al^{3+}_{0.07}\ Mg^{2+}_{0.06}\ Mn^{2+}_{0.02}\ Ti^{4+}_{0.58})\ O^{2-}_{4}$$

$Fe^{3+}_{0.20}$ amount of Fe-migration

z = 1: net magnetic moment: +0.86 $_B$

$$Fe^{3+}_{0.80}\ \ 0.20\ (Fe^{3+}_{0.97}\ \ 0.30\ Al^{3+}_{0.07}\ Mg^{2+}_{0.06}\ Mn^{2+}_{0.02}\ Ti^{4+}_{0.58})\ O^{2-}_{4}$$

their original thermoremanent magnetization within a short time interval. The samples should correspondingly posses the same original magnetization direction.

Fig. 28 shows the stable inclination values for two lithological units from the atlantic DSDP site 396B, plotted against the oxidation parameter z. Although there is a wide spectrum of z-values, particularly within unit 1, we cannot observe any corresponding variation of the magnetization direction. In particular no self-reversal is observed when passing through the critical z-value of 0.6. Hall (1977) describes similar studies and thereby comes to the same conclusion: The initial direction of thermoremanent magnetization is not significantly altered by sea floor oxidation.

Conclusions

1. Carrier of the remanent magnetization of ocean floor basalts are small grains of titanomagnetites which have varying degrees of cation deficiency.

2. The primary titanomagnetite composition in oceanic tholeiites varies only within narrow limits and is close to

   $$Fe^{2+}_{1.50}Fe^{3+}_{0.77}Ti_{0.58}Al_{0.07}Mg_{0.06}Mn_{0.02}O_4$$

   The corresponding primary Curie temperature T_c = 125 ± 3°C, the lattice parameter a_o = 8.475 ± 0.003 A

3. The dominating mechanism of titanomagnetite sea floor oxidation is Fe-migration. Contrary to the oxidation model of Readman and O'Reilly (1971) this assumes a constantly changing Fe/Ti ratio in the spinel lattice with oxidation. The migrated iron may either be incorporated in surrounding clay minerals or be transported elsewhere by sea water.

4. The correlation between Curie temperature T_c and oxidation parameter z of natural oceanic titanomagnetites is different from the corresponding correlation derived from the data of Readman and O'Reilly (1972).

5. The Curie temperature of ocean floor basalts increases with increasing age of the oceanic crust. Although there is considerable scatter, it does seem that T_c increases proportionally with the logarithm of age.

6. The upper 600 m of the predominantly pillow basalts of the Atlantic ocean floor are more or less uniformly altered so far as the magnetic minerals are concerned. However within single lithological units there is a slight decrease of the oxidation state with depth.

7. The magnetization intensity of oceanic basalts first decreases with low temperature oxidation until the oxidation

parameter reaches z = 0.6. For z-values higher than 0.6 it increases again. The variation of saturation magnetization can be explained qualitatively by a simple model assuming Fe-migration first form octahedral sites and then from tetrahedral sites.

The high amplitude of the mesozoic marine magnetic M-anomalies could possibly be caused by this process. The jurassic magnetic quiet zone could then be explained by inversion of the completely oxidized titanomagnetites to non-magnetic mineral phases.

8. The direction of the natural remanent magnetization of ocean floor basalts is not changed by sea floor alteration.

Acknowledgments. This study is a contribution of the Institute of Geophysics and the Institute of Geology, both University of Munich, and the Institute of Geophysics, Ruhr-University, Bochum. We thank Prof. G. Angenheister, Prof. D. Klemm and Prof. H. Baule for generously making available the facilities of the above named institutions and for their support throughout the study.

We would like to thank Prof. L. Rinderer (Lausanne, Switzerland) and Dr. W. O'Reilly (Newcastle, England) for making possible the use of the electron microprobes of their institutions. We also thank Dr. A. Stephenson, W. Davidson and J. Burri for help during the microprobe measurements. Dr. G. Schönharting made part of the thermomagnetic measurements. His assistance is gratefully acknowledged. We also want to express our thanks to Dr. J. Heirtzler, Dr. W. Bryon, Dr. F. Fabrizius and Dr. P. Vogt who made it possible for us to obtain samples from the Famous area, and from DSDP Legs 42 and 43. We also acknowledge discussions with other members of our research group, in particular Dr. J. Pohl, Dr. E. Schmidbauer, Dr. A. Schult and Prof. H. Soffel. The financial support for this project was supplied by the Deutsche Forschungsgemeinschaft.

References

Ade-Hall, J.M., Khan, M.A., and R.L. Wilson, A detailed opaque petrological and magnetic investigation of a single tertiary lava from Skye, Scotland: Part 1. irontitanium oxide petrology, Geophys. J., 16, 374, 1968.

Ade-Hall, J.M., Johnson, H.P., and J.L.Rhall, Rockmagnetism of basalts, Leg 34, Initial Reports of the Deep Sea Drilling Project, Volume 34, Washington, 459, 1976.

Akimoto, S., Katsura, T., and M. Yoshida, Magnetic properties of the Fe_3O_4-Fe_2TiO_4 System and their change with oxidation, J.Geomagn.Geoelectr., 9, 165, 1957.

Bleil, U., Notes on the magnetic structure of the oceanic crust, Bull, Soc. géol. France, 4, 913,1976

Bleil, U., and N. Petersen, Magnetic properties of basement rocks, Leg 37, Site 332, Initial Reports of the Deep Sea Drilling Project, Volume 37, Washington, 449, 1977.

Butler, R.F., Stable single domain to superparamagnetic transition during low-temperature oxidation of oceanic basalts, J. Geophys. Res., 78, 6868, 1973.

Chevallier, R., and J. Pierre, Propriétés thermomagnétiques des roches volcaniques, Ann. de Phys., série 10, 17, 390, 1932.

Cockerham, R.S., and J.M. Hall, Magnetic properties and paleomagnetism of some DSDP, leg 33, basalts and sediments and their tectonic implications, J. Geophys. Res., 81, 4207, 1976

Feitknecht, W., and K.J. Gallagher, Mechanisms for the oxidation of Fe_3O_4, Nature, 228, 548, 1970.

Graham, A.,m Symes, R.F., Bewan, I.C., and V.K.Din, Chromium bearing spinels in some rocks of Leg 45: pase chemistry, zoning and relation to hostbasalt chemistry, Initial Reports of the Deep Sea Drilling Project, Volume 45, Washington, 1979 (in press).

Grommé, C.S., Wright, T.L., and D.L. Peck, Magnetic properties and oxidation of iron-titanium oxides in Alae and Makaopuhi lava lakes, Hawaii, J. Geophys. Res., 74, 5277, 1969.

Grommé, C.S., and E. Mankinen, Natural remanent magnetization, magnetic properties and oxidation of titanomagnetite in basaltic rocks from DSDP, Leg 34, Initial Reports of Deep Sea Drilling Project, Volume 34, Washington, 485, 1976.

Hall, J.M., Does TRM occur in oceanic Layer 2 basalts?, J. Geomagn. Geoelectr., 29, 411, 1977.

Hart, R., Chemical exchange between seawater and deep ocean basalts, Earth Planet, Sci. Lett., 9, 269, 1970.

Hart, R., Geochemical and geophysical implications of the reaction between seawater and the oceanic crust, Nature, 243, 76, 1973.

Hart, R., A model for chemical exchange in the basalt-seawater system of oceanic - Layer II, Canadian J. Earth Sci., 10, 799, 1973b.

Hart, S.R., K/Rb, Cs contents and K/Rb K/Cs ratios of fresh and altered submarine basalts, Earth Planet, Sci. Lett., 6, 295,1969.

Hart, S.R., K, Rb, Cs, Sr and Ba contents and Sr isotope ratios of ocean floor basalts, Phil. Trans. Roy. Soc. London, A268, 473, 1971.
Hekinian, R., Chemical and mineralogical differences between abyssal hill basalts and ridge tholeiites in the Eastern Pacific Ocean, Marine Geol., 11, 77, 1971.
Hoernes ,S., and H. Friedrichsen, Oxygen isotope investigation of rocks of Leg 37, Initial Reports of the Deep Sea Drilling Project, Volume 37, Washington, 603, 1977.
Honnorez, J., Böhlke, J.K., and B.M. Honnorez-Guerstein, Petrographical and geochemical study of the low-temperature submarine alteration of basalt from hole 396B. Leg 46, Initial Reports of the Deep Sea Drilling Project, Volume 46, Washington, 1979 (in press).
Irving, E., The mid-atlantic ridge at 45°N XIV, oxidation and magnetic properties; review and discussion, Can. J. Earth Sci., 7, 1528, 1970.
Irving, E., Park, J.K., Haggerty, S.E., Aumento, F., and B. Loncarevic, Magnetism and opaque Mineralogy of basalts from the midatlantic ridge at 45°N, Nature, 228, 974, 1970.
Johnson, H.P., and J.M. Hall, A detailed rock magnetic and opaque mineralogy study of the basalts from the Nazca plate, Geophys. J. Roy, astr. Soc., 52, 45, 1978.
Johnson, H.P., and W.G. Melson, Electron microprobe analyses of some titanomagnetite grains from hole 395A, Initial Reports of the Deep Sea Drilling Project, Volume 45, Washington, 1979 (in press).
Katsura, T., and Kushiro, Titanomaghemite in igneous rocks, American Mineralogist, 46, 134, 1961.
Kobayashi, K., Chemical remanent magnetization of ferromagnetic minerals and its application to rock magnetism, J. Geomag. Geoelectr., 10, 99, 1959.
Larson, E. E., and D.W. Strangway, Magnetization of the Spanish Peak dike swarm, Colorado and Shipwreck Dike, New Mexico, J. Geophys. Res., 74, 1505, 1969.
Lewis,M., Some experiments on synthetic titanomagnetites, Geophys. J.R. astr. Soc., 16, 295, 1968.
Lindley, H.W., Mikrographie der Eisenerzmineralien oberhessischer Basalte, N.Jb. Min. Geol. Paläont., 53 Beilagen Bd., 324, 1926.
Marshall, M., and A. Cox, Effect of oxidation on the natural remanent magnetization of titanomagnetite in suboceanic basalt, Nature, 230, 28, 1971.
Marshall, M., and A. Cox, Magnetic changes in pillow besalts due to sea floor weathering, J. Geophys. Res.,77, 6459, 1972.
Mathews, D.C., Altered basalts from Swallow Bank, an abyssal hill in the NE atlantic, and from nearby seamount, Phil. Trans. Roy. Soc. London, A268, 551, 1971.
Mazullo, L.J., and A.E. Bence, Abyssal tholeiites from DSDP, Leg 34: The Nazca Plate, J. Geophys. Res., 81, 4327, 1976
Miyashiro, A., Shido, F., and M. Ewing, Diversity and origin of abyssal tholeiite from the Mid-Atlantic Ridge near 24° and 30° North, Latitude, Contrib. Mineral. Petrol., 23, 38, 1969.
O'Donovan, J.B., and W.O'Reilly, The preparation, characterization and magnetic porperties of synthetic analoques of some carriers of the paläomagnetic record, J.Geomagn. Geoelectr., 29, 331, 1977.
O'Reilly, W., and S.K. Banerjee, Oxidation of titanomagnetites and self-reversal, Nature, 221, 26, 1966.
O'Reilly, W., and S.K. Bannerjee, The mechanism of oxidation in titanomágnetites: a magnetic study, Min. Mag., 36, 29, 1967.
O'Reilly, W., and P.W. Readman, The preparation and unmixing of cation deficient titanomagnetites, Z. Geophys., 37, 321, 1971.
Ozima, M., and E.E. Larson, Low- and high-temperature oxidation of titanomagnetite in relation to irreversible changes in the magnetic properties of submarine basalts, J. Geophys. Res., 75, 1003, 1970.
Ozima, M., and N. Sakamoto, Magnetic properties of synthesized titanomaghemite, J. Geophys. Res., 76, 7035, 1971.
Ozima, M., Joshima, M., and H. Kinoshita, Magnetic Properties of submarine basalts and the implications on the structure of the oceanic crust, J. Geomagn. Geoelectr., 26, 335, 1974.
Petersen, N., Untersuchungen magnetischer Eigenschaften von Titanomagnetiten in Basalten des Rauhen Kulm (Operpfalz) in Verbidnung mit elektronenmikrokopischer Beobachtung, Z. Geophys., 28, 79, 1962.
Petersen, N., Beobachtungen einiger mineralogischer und magnetischer Eigenschaften dreier Basaltproben nach unterschiedlicher thermischer Behandlung, J. Geomagn. Geoelectr., 18, 463,1966.
Petersen, N., Calculation of diffusion coefficient and activation energy of of titanium in titanomagnetite, Phys. Earth Planet. Interiors, 2, 175, 1970.
Petersen, N., Notes on the variation of magnetization within basalt lava flows and dikes, Pageoph, 114, 177, 1976.
Pouillard, E., Sur le comportement de

l'alumine et de l'oxyde de titane vis-a-vis des oxydes de fer, Ann. de Chim., série 12, 5, 164, 1950.

Prévot, M., Remond, G., and R. Caye, Etude de la transformation d'une titanomagnétite en titanomaghémite dans une roche volcanique, Bull, Soc. fr. Minéral. Cristallogr., 91, 65, 1968.

Prévot, M., and A. Lecaille, Sur le caractère épisodique du functionnement des zones d'accréation: Critique des arguments géomagnetiques, Bull. Soc. Géol. France, 18, 903, 1976.

Readman, P.W., and W.O'Reilly, The synthesis and inversion of non-stoichiometric titanomagnetites, Phys. Earth Planet. Interiors, 4, 1970.

Readman, P.W., and W.O'Reilly, Magnetic properties of oxidized (cation deficient) titanomagnetites (Fe, Ti, $)_3O_4$, J. Geomagn. Geoelectr., 24, 69, 1972.

Richards, J.C.W., O'Donovan, J.B., Hauptman, Z., O'Reilly, W., and K.M. Creer, A magnetic study of titanomagnetite substituted by magnesium and aluminium, Phys. Earth Planet. Inst., 7, 121, 1973.

Sakamoto, N., Ince, P.I., and W.O'Reilly, The effect of wet-grinding in the oxidation of titanomagnetites, Geophys. J.R. astr. Soc., 15, 509, 1968.

Sato, M., and T.L. Wright, Oxygen fugacities directly measured in magnetic gases, Science, 153, 1103, 1966.

Schult, A., Die Selbstumkehr der remanenten Magnetisierung beobachtet in einigen tertiären Alkali-Basalten, Habilitationsschrift, Ludwig-Maximilians-Universität München, 1978.

Schult, A., Self-reversal above room temperature due to N-type magnetization in basalt, J. Geophys., 42, 81, 1976.

Shido, F., Miyashiro, A., and M. Ewing, Compositional variation in pillow lavas from the Mid-Atlantic Ridge, Marine Geol., 16, 177, 1974.

Soffel, H., and N. Petersen, Ionic etching of titanomagnetite grains in basalts, Earth Planet. Sci. Lett., 11, 312, 1971.

Verhoogen, J., Oxidation of iron-titanium oxides in igneous rocks, J. Geol., 70, 168, 1962.

Vine, F.J., and D.H. Matthews, Magnetic anomalies over oceanic ridges, Nature, 199, 947, 1963.

MAGNETISM OF THE MID-ATLANTIC RIDGE CREST NEAR 37°N FROM FAMOUS AND DSDP RESULTS: A REVIEW

Michel Prévot[1], Alaîn Lecaille

Laboratoire de Géomagnétisme, Université Paris VI, 4 Avenue de Neptune, 94 100 Saint-Maur, France

Roger Hekinian[2]

Centre Océanologique de Bretagne BP 337, 29273 Brest Cedex, France

Abstract. Comparison of magnetic properties of pillow basalts from the Rift Valley (FAMOUS area) and from the Crest Mountains (site 332 of DSDP Leg 37) of the Mid-Atlantic Ridge near 37°N shows that major differences in magnetic structure and properties of the oceanic crust occur between these two sites.

Mean effective intensity of magnetization of the pillow basalt layer drops from 1.4 x 10^{-2} emu cm^{-3} (14 A m^{-1}) at the axis to about 0.4 x 10^{-2} emu cm^{-3} (4 A m^{-1} at site 332. While the magnetic anomaly over the spreading center may be accounted for by a magnetized layer less than 500 m thick, deeper seated material contributes to the magnetic anomalies over the Crest Mountains.

The large reduction of the intensity of magnetization of the magnetized layer appears to be a general feature resulting from low-temperature oxidation (maghemitization) of titanomagnetite. The effect of maghemitization is a reduction by 2/3 to 3/4 of the remanence intensity, accompanied by a comparable decrease of the Q ratio. Correlative increase of Curie temperature and decrease of saturation magnetization agree with experimental data for synthetic titanomagnetites. However, contrary to previous conclusions deduced from magnetic studies on submarine basalts, almost no changes in susceptibility and coercivity parameters result from maghemitization. It is also pointed out that maghemitization of the top of layer 2A runs over too long a time constant to account for the occurrence of the "magnetization high" found in the FAMOUS Rift Valley. This feature corresponds to eruptions of fine-grained olivine-rich basalts, which are on the average twice as magnetic as the surrounding plagioclase or pyroxene-rich basalts. The high remanence of olivine-rich basalts results from their smaller magnetic oxide grain size. A similar relationship between petrology and magnetism is also observed for Leg 37 basalts at site 332.

The thickness of the highly magnetized layer found at the ridge axis is less than the approximately 1 km thick pillow basalt layer (2A). This implies that some process strongly reduces the intensity of magnetization of the pillows at depth. A moderate elevation in temperature resulting in a rapid and intense maghemitization seems to be the best explanation. At Leg 37 sites the isotherm corresponding to the mean Curie temperature (300°C) is at least at 2 km depth, which indicates that the whole layer 2 can contribute to the magnetic anomalies.

Within the FAMOUS Rift Valley the mean inclination of magnetization fits the dipole inclination, while anomalously shallow inclinations are commonly observed at DSDP and IPOD sites from the Mid-Atlantic Ridge Crest. This is in agreement with tectonic models of rifted mid-oceanic rises, which assume block tilting within or at the edges of the rift valley as a result of normal faulting. However, tectonic rotations required for explaining paleomagnetic result for Cretaceous North Atlantic IPOD sites may indicate a more complex structural evolution of the upper part of the oceanic crust.

Introduction

FAMOUS (French American Mid-Ocean Undersea Study) and DSDP (Deep Sea Drilling Program) Leg

---

[1]Now at U.S. Geological Survey, Menlo Park, California, U.S.A.

[2]Contribution number 592 of the Départment Scientifique, Centre Océanologique de Bretagne B.P. 337, 29273 Brest France.

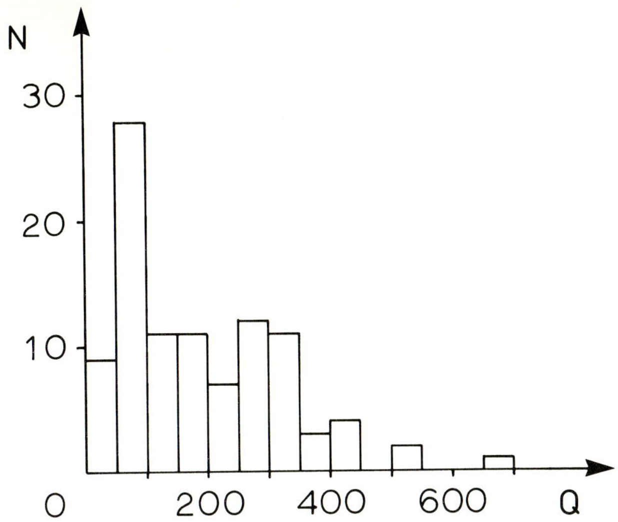

Fig. 2. Histogram of the Q ratio of 103 pillow basalts of the FAMOUS rift valley.

37 yielded large collections of pillow basalts that have been studied in great detail. The extensive magnetic results obtained provide a unique opportunity to compare the magnetism of the upper part of the oceanic crust at and near a ridge axis, in the present-day accretion zone and at 30 km from the spreading center.

The FAMOUS program was mainly devoted to the study of the Rift Valley near 37°N. This area was extensively sampled either by dredging (Needham and Francheteau, 1974) or by using manned submersible (Bellaiche et al., 1974; Arcyana, 1975; Ballard et al., 1975). Figure 1 shows the location of the dredge stations where most of the rocks studied magnetically by the French team (Lecaille et al., 1974; Prévot et al., 1976; Prévot and Lecaille, 1976) were recovered. A total number of about 160 samples was studied, including 120 large samples about 300 g in weight. Experiments include weak-field susceptibility and natural remanent magnetization (NRM) measurements, alternating field treatment, magnetic hysteresis studies, thermomagnetic analysis and electron microprobe analysis (carried out on only 10 samples with titanomagnetite crystals large enough for such studies). Detailed experimental procedures and results will be given elsewhere (Lecaille and Prévot, 1979). Fission-track ages obtained by Storzer and Selo (1974, 1976) indicate these rocks are less than 0.1 m.y. old.

Another magnetic study has been carried out by Johnson and Atwater (1977) on rock samples from the Rift Valley, most of them recovered by submersibles. Their results are restricted to small specimens with mean magnetic grain size larger than 1 μm and include NRM and susceptibility measurements, alternating field treatment and thermomagnetic analysis. This sampling bias results in a systematic rejection of the specimens with mean grain size below the single-domain to pseudosingle domain threshold (about 1 μm for titanium-rich titanomagnetites, e.g., Day, 1977). There is therefore some doubt that the mean values they obtain for grain size-related magnetic parameters (intensity of remanence, susceptibility, Koenigsberger ratio, median destructive field) are really representative of the overall FAMOUS pillow basalts.

Regarding the Leg 37 cores magnetically studied by several authors (in Aumento, Melson et al., 1977) and in greatest detail by Hall and Ryall (1977), we will consider the data from holes 332A and 332B, the nearest site to the ridge axis, for comparison with FAMOUS results. Site 322 is located at 37°N, only 30 km west of the present-day plate boundary. Estimated age of the crust at this site is about 3.5 m.y. (Aumento, Melson et al., 1977).

The comparison of these two magnetically well documented sets of submarine basalts will be made in regard to three problems: (1) the contribution of pillow basalts to the observed magnetic anomalies; (2) the magnetic effects of the low-temperature oxidation (maghemitization) of titanomagnetites and (3) the relationship between petrology and magnetism of the basaltic layer.

Pillow basalts as a source of magnetic anomalies

The magnetic anomalies produced by pillow basalts depend on several magnetic parameters: Firstly, the Koenigsberger (or Q) ratio, the ratio of the intensity of remanence to the intensity of magnetization induced by the geomagnetic field at the site. As a first approximation, it may be considered that the Q ratio must be larger than one for the Vine and Matthews (1963) interpretation of magnetic anomalies to be valid. Secondly, we have to consider the characteristics of the remanent magnetization itself: intensity, polarity and direction.

1. FAMOUS pillow basalts

The Q ratio is quite high (figure 2) with a geometric mean value of 151 (table 2). Individual Q ratios range from 20 to 680.

The mean Q value found by Johnson and Atwater (1977) for their specimens from the Rift Valley is about one order of magnitude smaller, due to their higher susceptibility (3.5 x 10^{-3} emu cm^{-3}), about 10 times that listed in table 2. Such a high susceptibility is difficult to accept for unoxidized pillow basalts. It is several times larger than the susceptibility of unoxidized massive flows (Johnson and Hall, 1978). However, the magnetic susceptibility of the latter should be higher, due to both a larger grain size and a higher content of magnetic minerals. We note also that the susceptibilities given in the same paper by Johnson and Atwater (1977) for samples dredged near DSDP site 332 are similarly one order of magnitude higher than the mean value for the pillow basalts drilled at this site (table 2).

There is therefore no doubt that the Q value is quite high and that the magnetic anomaly over this spreading center is not due to lateral changes of the induced magnetization carried by the pillow lava sequence.

It has been pointed out by Harrison (1976) that the arithmetic mean (and not the geometric mean) has to be used to estimate the intensity of remanence as a source of magnetic anomalies. The arithmetic mean value of the intensity of remanence of the pillow basalts is $1.9 \pm 0.3 \times 10^{-2}$ emu cm^{-3} (19 A m^{-1}). The histogram (figure 3) shows only one maximum, indicating that probably this mean value is meaningful. Johnson and Atwater (1977) found approximately the same value for a less extensive collection of rocks from the median valley.

Polarity and inclination of remanence have been determined for 18 partially oriented pillow basalts from the Rift Valley (Prévot et al., 1976; Johnson and Atwater, 1977). For most of the samples, the criterion used for orientation was the occurrence of frozen-in lava levels resulting in the formation of ledges assumed to be horizontal. During the diving operation in the FAMOUS area of the Mid-Atlantic Ridge near 36°50', field observations have shown (Ballard et al., 1975; Heirtzler, 1975) the existence of a series of ledges or shelves on the walls of collapsed feeder tubes. The outer surfaces of these tubes or pillows are glassy and the interior ledges or shelves are cryptocrystalline and not as glassy as the exteriors. Hence it is believed that cold sea water did not enter into the cavities defined by the ledges. These shelves and cavities are formed when the feeding lava is cut off intermittently at its source, permitting partial drainage of the remaining flow and quenching of lower lava levels within the cooling pillow (Ballard and Moore, 1977). The upper surface of the ledges is smooth while the bottom surface has a rugose texture due to the presence of septa and lava stalactites (Prévot et al., 1976). The cavities between shelves vary in width from a few centimeters to about twenty centimeters and are often elongated in a preferential direction (Fig. 4). However, the cavities are not always continuous but may be interrupted by internal walls (Bideau et al., 1977). Usually several levels of cavities may be seen in a single pillowed flow. In the interior of the pillows, a series of elongated cavities with a smoother floor and a more rugose roof gives an indication of the vertical at the time of cooling. Hence in order to orient a fragment of such a pillow basalt showing cavities, the vertical at the time of cooling is taken to be normal to the flattish ledge or normal to a line of small elongated cavities. In both cases the polarity of the vertical is deduced from the relative position of the smooth and rugose surfaces of the cavities (Prévot et al., 1976).

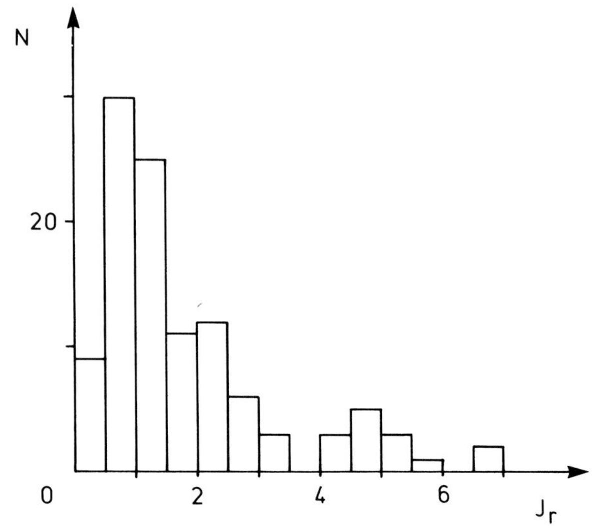

Fig. 3. Histogram of the intensity of magnetization J_r (natural values) of 111 pillow basalts of the FAMOUS rift valley (10^{-2} emu cm^{-3}).

The difficulties with this method of rock reorientation are related to the facts that the cavities are often irregular in shape, that the ledges may be missing, and/or the cavities may be scarce. Thin sections made across the pillow fragments from the surface to the interior show only textural variations accompanied by an increasing degree of crystallinity towards the inner part of the flow. However, the most significant variations are found within the first 3 centimeters from the surface. Examples of reorientation using the method of the cavities were tested on various types of basalts recovered by dredging and by submersibles from the ocean floor. Plagioclase-pyroxene-rich basalts (ARP 73-10-02, ARP 74-17-40), olivine basalt (ARP 74-11-17, CH 31-DRI-111), and plagioclase-rich basalts were used (Fig. 4).

This criterion has been used for 13 samples. For the remaining samples, two were collected by Alvin divers from outcrop, then reoriented according to diver description after the return to the surface. The other three samples were drilled and reoriented assuming the drilling barrel was vertical and the pillows did not move after cooling. Figure 5 shows the location and polarity of oriented samples.

The nine samples studied by Prévot et al. (1976), broadly distributed within the median valley, are normally magnetized. The mean inclination is equal to 56° ± 10° (s.d.), which corresponds to the expected inclination for an axial centered dipole at this latitude. Similarly, Johnson and Atwater (1977) found a mean inclination equal to 53° ± 8°. But a major difficulty arises from the fact that two of their nine samples, all collected within the median valley, exhibit a negative inclination.

One of the two reversely magnetized samples was oriented from cooling levels as described above. This criterion provides an unequivocal determin-

Fig. 4. Selected pillow flows showing oriented cavities with ledges. Samples ARP 74-11-17, and CH 31-DR 1-111 are olivine basalts. Samples ARP 73-10-02 and ARP 74-17-40 are plagioclase-pyroxene-rich basalts. The vertical is taken to be normal to a line defined by the elongation of the cavities. All the samples are from the inner floor of the Rift Valley near 36°50'N except for sample ARP 74-17-40 which was collected in transform fault "A".

ation of the horizontal plane but the determination of up and down is less sure, being based only upon differences in rugosity between supposed downward and upward-facing walls of the cavities. The second reversely magnetized sample was drilled. In the absence of direct visual inspection of the site, the possibility that this sample was recovered from a talus pile cannot be ruled out. Other possible interpretations are self-reversal or recording of some geomagnetic excursion.

In view of the indirect methods which have to be used for reorienting submarine pillow basalts, the occurrence of 16 normally magnetized samples from a total of 18, strongly suggests that in this part of the FAMOUS Rift Valley (Rift Valley 2, according to Macdonald 1977) the pillow basalts are normally magnetized, in agreement

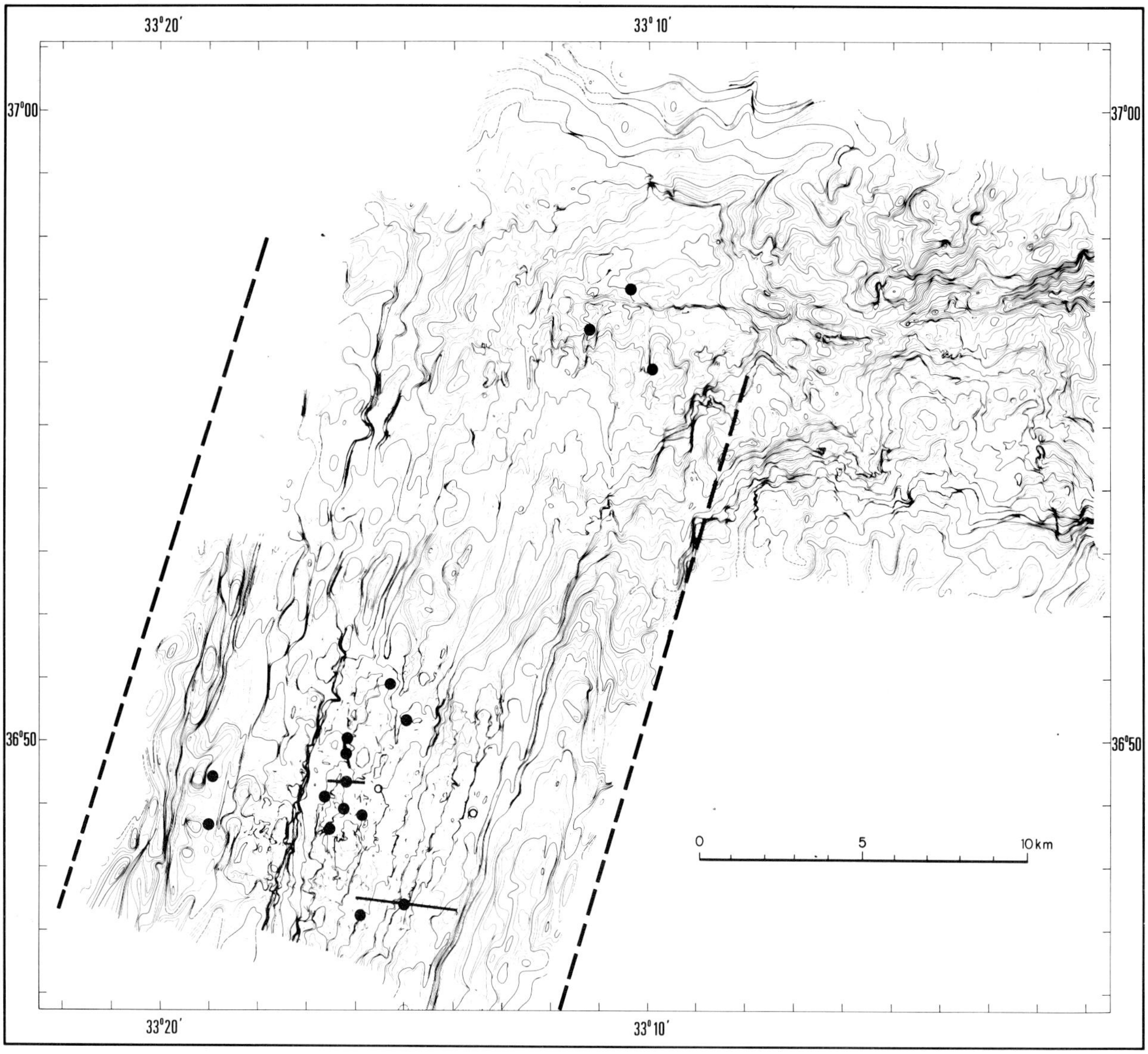

Fig. 5. Location and polarity of oriented samples from the median valley (from Prévot et al., 1976, and Johnson and Atwater, 1977). Black dots: positive inclination; circles: negative inclination. Heavy lines with black dots correspond to dredge hauls samples. Boundaries of the crust formed during the Brunhes epoch are indicated by heavy dashed line.

with their Brunhes age. Similarly, paleomagnetic studies of the Afar rift valley--which is interpreted as an accreting plate boundary--showed that all samples are normally magnetized (Harrison et al., 1977; Smith and Sichler, personal communication).

Assuming that the fraction of non-magnetic material is 0.2 at the ridge axis (Salisbury, personal communication) and that the standard deviation of the directions of NRM is 20° (Marshall and Cox, 1971), the overall mean intensity of magnetization of the pillow-lava layer is 1.4×10^{-2} emu cm^{-3}. A magnetized layer less than 500 m thick is enough (Lecaille et al., 1974) to account for the amplitude of the axial anomaly at 37°N (Needham and Francheteau, 1974). Hence the pillow basalt layer is responsible for the magnetic anomaly observed at the axis of the Mid-Atlantic Ridge near 37°N.

2. Leg 37 pillow basalts

The numerous magnetic studies carried out on Leg 37 basalts (Hall and Ryall, 1977a and b; Bleil and Petersen, 1977; Dunlop and Hale, 1977; Plessard and Prévot, 1977; Kent and Lowrie, 1977) showed large differences in magnetic properties with respect to the FAMOUS rocks.

The relatively high Q ratio of Leg 37 basalts (approximately equal to 35 according to Ryall et al., 1977) seems to indicate that--as found for FAMOUS rocks--the remanence dominates over induced magnetization. However, it must be noted that Leg 37 basalts are often magnetically viscous (Dunlop and Hale, 1977; Evans and Wayman, 1977; Kent and Lowrie, 1977; Plessard and Prévot, 1977) in contrast to FAMOUS basalts (Lecaille et al., 1974). Magnetic viscosity results in the fact that not only the remanence (viscous remanent magnetization or VRM) but also the induced magnetization (viscous induced magnetization or VIM) increases significantly with t, the time of application of the field. If we neglect short-lived geomagnetic excursions, the acquisition time to consider in nature is 0.7 m.y., the beginning of the Brunhes polarity epoch, except for younger rocks. The measurements carried out by Hall and Ryall (1977) using a high frequency susceptibility bridge, yield an induced magnetization definitely smaller than the induced magnetization carried in situ by the specimen. The actual mean Q value is therefore smaller than 35. Plessard and Prévot (1977) showed that the VIM of Leg 37 basalts increased almost linearly with log t. According to a tentative estimation of the intensity of induced magnetization for t = 0.7 m.y., they concluded nevertheless that for a large majority of their samples, the induced magnetization carried in situ is noticeably smaller than the NRM.

The properties of NRM are also quite different from those observed in the FAMOUS rift valley. The magnetic polarity is mostly reversed (Ryall et al., 1977) in accordance with interpretation of sea-surface magnetic profiles suggesting that site 332 is within the 3.32-3.78 m.y. old negative polarity block (Aumento, Melson et al., 1977). The approximately 100 m thick normally magnetized layer occurring at the top of hole 332A may correspond to a normal polarity event preceding the Gilbert-Gauss boundary (Aumento, Melson et al., 1977). The mean inclination is shallower by a much as 25°-30° from the expected dipole value suggesting considerable tectonic tilting (Ryall et al., 1977; Harrison and Watkins, 1977). The effective intensity of the drilled basaltic layer is only 0.37×10^{-2} emu cm^{-3} (Ryall et al., 1977), about four times weaker than that calculated for the rift valley. Hence, contrary to our conclusion for the FAMOUS Rift Valley, the layer responsible for magnetic anomaly at site 332 probably exceeds the accepted thickness of 2 km for layer 2 (Ryall et al., 1977).

The low-temperature oxidation of titanomagnetites

1. Magnetic effects of maghemitization: comparison of FAMOUS and Leg 37 pillow basalts.

Low-temperature oxidation of titanomagnetites (maghemitization) is probably the main process explaining the decrease of the contribution of the pillow lava sequence to magnetic anomalies outwards from the ridge axis (Irving, 1970).

All evaluations of the magnetic effects of naturally occurring maghemitization are based on the assumption that the two specimens or the two rock units under comparison were identical before alteration. This can never be demonstrated. It may be hoped that a favorable situation for comparing two collections of pillow basalts would occur when the rocks were erupted at the same latitude from the same accreting plate boundary and close to each other in time. The FAMOUS and Leg 37 collections meet these conditions. Moreover, many petrographical similarities were noted between Leg 37 basalts and those in the FAMOUS area (Bryan and Thompson, 1977). We shall restrict the comparison between the FAMOUS and Leg 37 collections to pillow basalts. Hence, for Leg 37, we shall consider only the magnetic data of the petrological units identified as pillow lava sequences by Aumento, Melson et al., (1977). Because NRM intensity and Q ratio have a log normal distribution, we shall compare their geometric mean values.

Comparison of electron-microprobe and thermomagnetic data indicates that the FAMOUS pillow basalts contain only slightly oxidized titanomagnetites (table 1). This is demonstrated by the absence of significant difference between measured and calculated Curie points for most of the samples. The latter Curie point is deduced from chemical analysis, assuming that no oxidation occurred and using Curie point data for synthetic titanomagnetites (see review by Smith and Prévot, 1977, for pure titanomagnetite, and experimental data by Richards et al., 1973 for substituted titanomagnetites). Note that estimated uncertainties about calculated values are ± 30°C (Smith and Prévot, 1977).

Sample DR06-104A is certainly oxidized, and perhaps also samples DR02-102A_2, DR03-103 and DR03-105 B_3. Moreover, the mean calculated Curie point of the samples is about 40° lower than the mean measured Curie point for the entire collection (table 2). This suggests a slight maghemitization of the FAMOUS pillow basalts, the mean oxidation index z being about 0.2 according to Readman and O'Reilly's (1972) data for synthetic titanomaghemite.

The Leg 37 pillow basalts are largely maghemitized as demonstrated by the irreversibility of the thermomagnetic curves of most of the samples (Bleil and Petersen, 1977; Dunlop and Hale, 1977; Hall and Ryall, 1977b; Schwarz and Fujiwara, 1977; Murthy et al., 1977; Kent and Lowrie, 1977) and the high mean Curie point (table 2).

For most of the magnetic parameters (except Curie temperature and saturation magnetization), the estimation of the magnetic effects of maghemitization requires the two suites of rocks under consideration to contain titanomagnetites with the same grain size. The condition seems fulfilled here, as shown by the absence of significant difference in the H_{rc}/H_c ratio (remanent coercive force to coercive force) for FAMOUS and Leg 37 pillow basalts (table 2). The J_{rs}/J_s and H_{rc}/H_c ratios found may

Table 1 Average chemical composition and Curie point of titanomagnetite from FAMOUS pillow basalts (rift valley)

Sample number	n	Fe	Ti	Al	Mg	Mn	Cr	Si	x	Θ	Θ_c
DR 02-102 A_2	23	53.0 ± 0.1	13.4 ± 0.1	1.5 ± 0.2	0.4 ± 0.2	0.4 ± 0.1	0.1 ± 0.1	0.2 ± 0.1	0.64	105 ± 15	80 ± 30
DR 03-103	5	54.6 ± 0.7	11.5 ± 0.1	1.2 ± 0.2	0.3 ± 0.2	0.1 ± 0.01	0.01± 0.01	0.6 ± 0.1	0.56	180 ± 20	160 ± 30
DR 03-105 B_3	5	54.0 ± 0.4	12.7 ± 0.5	1.2 ± 0.1	0.5 ± 0.2	0.4 ± 0.1	0.1 ± 0.01	0.5 ± 0.1	0.61	172 ± 22	130 ± 30
DR 03-112 B	16	52.8 ± 1.1	13.0 ± 0.6	1.4 ± 0.2	0.4 ± 0.1	0.4 ± 0.1	0.1 ± 0.1	0.5 ± 0.2	0.63	100 ± 15	95 ± 30
DR 06-100 G_4	20	53.3 ± 0.9	12.5 ± 0.4	1.5 ± 0.1	0.3 ± 0.2	0.4 ± 0.1	0.1 ± 0.1	0.7 ± 0.3	0.61	113 ± 13	105 ± 30
DR 06-102 A	20	52.1 ± 0.8	12.8 ± 0.7	1.1 ± 0.1	0.3 ± 0.1	0.5 ± 0.4	0.03± 0.01	0.3 ± 0.1	0.64	117 ± 13	105 ± 30
DR 06-104 A	23	54.2 ± 0.3	13.1 ± 0.3	1.1 ± 0.2	0.2 ± 0.1	0.4 ± 0.2	0.03± 0.02	0.6 ± 0.1	0.63	212 ± 12	115 ± 30
DR 06-106	16	55.4 ± 0.8	12.4 ± 0.9	1.1 ± 0.2	0.2 ± 0.1	0.4 ± 0.1	0.3 ± 0.2	0.5 ± 0.2	0.59	144 ± 18	145 ± 30
DR 09-111 A	20	55.5 ± 0.5	12.7 ± 0.5	1.2 ± 0.1	0.2 ± 0.1	0.3 ± 0.2	0.01± 0.01	0.4 ± 0.1	0.60	122 ± 14	130 ± 30
DR 10-100 B_2	22	53.9 ± 0.3	13.2 ± 0.5	1.2 ± 0.2	0.4 ± 0.1	0.5 ± 0.1	0.01± 0.01	0.5 ± 0.2	0.63	99 ± 15	110 ± 30

n, Number of analyzed grains (three spots from each grain); mean metal content is given in weight percent with standard deviation;

x, molecular percentage of ulvöspinel of the pure and non-oxidized equivalent titanomagnetite (Prévot and Mergoil, 1973); Θ and Θ_c are respectively the measured and calculated Curie points in °C (calculation method is given in the text).

Table 2 Comparison of average magnetic properties of FAMOUS (rift valley) and Leg 37 (holes 332A and 332B) pillow basalts

	FAMOUS	Leg 37
Θ (°C)	159 ± 9 (N=86)	294 ± 14 (N=69)
z	≃ 0.2	0.7 ± 0.04 (N=69)
J_s (emu cm^{-3})	0.87 ± 0.11 (N=53)	0.40± 0.04 (N=85)
J_r (10^{-2} emu cm^{-3})	1.44 {1.23, 1.68} (N=103)	0.366 {0.312, 0.428} (N=85)
k (10^{-4} emu cm^{-3})	2.99 ± 0.34 (N=103)	2.41 ± 0.22 (N=84)
Q	1.51 {129, 178} (N=103	39.5 {32.9, 47.6} (N=84)
J_{rs}/J_s	0.41 ± 0.03 (N=53)	0.54 ± 0.12 (N= 8)
H_{rc}/H_c	1.43 ± 0.07 (N=51)	1.41 ± 0.14 (N=20)
H_c (Oe)	297 ± 42 (N=53)	253 ± 31 (N=20)
H_{rc} (Oe)	407 ± 74 (N=51)	353 ± 50 (N=20)
MDF (Oe)	359 ± 51 (N=53)	303 ± 25 (N=85)

Θ, Curie point; z, degree of oxidation; J_s, saturation magnetization; J_r, natural remanent magnetization; k, initial susceptibility; Q, Koenigsberger ratio ($\frac{J_r}{kH}$ with H=0.42 Oe); J_{rs}, saturation remanent magnetization; H_{rc}, remanent coercive force; MDF, Median destructive field of NRM. Error bars indicate the 95% confidence interval for the average (brackets for log normally distributed parameters). Leg 37 data from Bleil and Petersen (1977), Dunlop and Hale (1977), Hall and Ryall (1977a and b) and Murthy et al. (1977). Note that the MDF of the FAMOUS pillow basalts is about 80% higher than that found by Johnson and Atwater (1977) for a collection restricted to samples with a mean titanomagnetite grain size larger than 1 μm (see text).

be interpreted as corresponding to small pseudo-single domain grains (Lecaille and Prévot, 1979).

Magnetic effects of maghemitization are as follows:

i) Large decreases of saturation magnetization, remanence and Q ratio occur (table 2).

The decrease in spontaneous magnetization is compatible with the trend observed at 0°C for synthetic titanomaghemites (Prévot, 1973).

It must be pointed out that J_r and Q changes are overestimated because the shallow-reversed primary remanent magnetization of Leg 37 basalts is often reduced in intensity by overprinted normally direc-

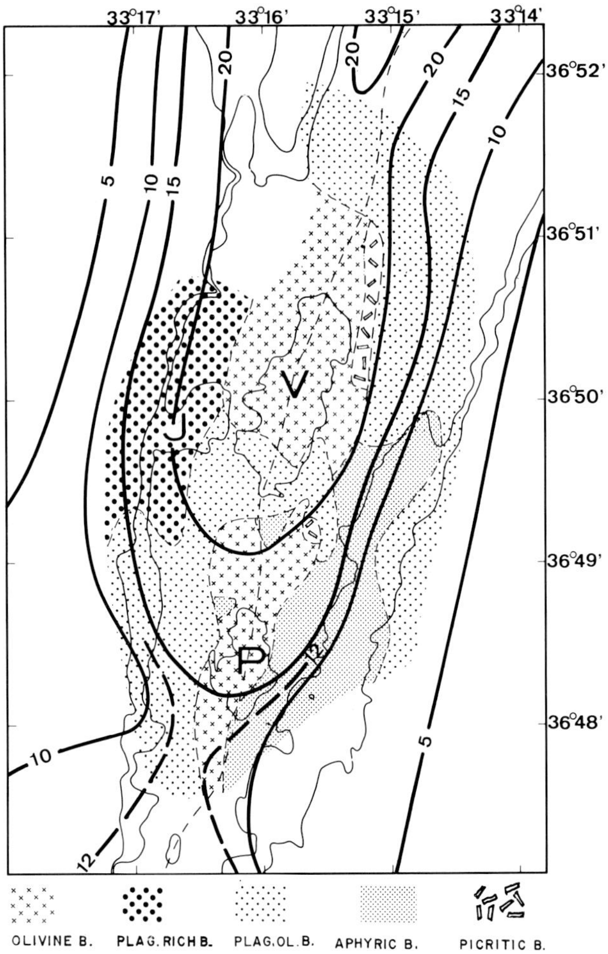

Fig. 6. Generalized distribution of five basalt types in the FAMOUS rift valley (from Hekinian, Moore and Bryan, 1976) with superimposed magnetization calculated from the inversion solution of near-bottom magnetic profiles (5 A/m contour interval; from Macdonald, 1977). Letters indicate topographic highs: V, Mount Venus; P, Mount Pluto; J, Mount Jupiter. The fields of plag. ol. basalts (plagioclase-olivine-pyroxene basalts) and of the aphyric basalts refers here to the pyroxene-rich basalts discussed in the text. The picritic and the olivine basalts correspond to the olivine-rich basalts.

ted VRM (Bleil and Petersen, 1977; Ryall et al., 1977). Prévot and Grommé (1975) showed that for subaerial basalts with a viscosity index (Thellier and Thellier, 1944) comparable to that of Leg 37 basalts (Plessard and Prévot, 1977) the intensity of the VRM acquired since the beginning of the Brunhes epoch is, on average, equal to 1/5 of the intensity of the primary remanence.

ii) Quite small changes in susceptibility and coercivity parameters (H_c, H_{rc} and MDF) are observed (table 2).

It can be demonstrated from theory (Prévot et al., 1979) that maghemitization of single-domain grains results in a large decrease in coercivity and a large increase in susceptibility whereas the opposite trends are expected for true multidomain particles. The later trends have been observed by Ade-Hall et al., (1976) for coarse-grained basalts from Leg 34.

The small changes observed here in coercivity-related parameters are probably a characteristic of small pseudo-single domain particles.

2. Development with time of maghemitization

At some spreading centers, and in particular within the FAMOUS rift valley 2, deep tow magnetic profiles indicate the occurrence of a "central anomaly magnetization high" (Klitgord, 1976; Macdonald, 1977). In the rift valley 2 it is described as a sharp, narrow maximum in crustal magnetization occurring near the axis of the inner floor and coinciding with large volcanic features such as Mount Pluto and Mount Venus (Fig. 6). According to Klitgord and Macdonald the magnetization high indicates the youngest eruptions on the sea-floor. The less magnetized surrounding lava flows are supposed to be older, their weaker magnetization being interpreted as a result of weathering. In other words, maghemitization is supposed to occur intensively within the rift valley. Only direct magnetic measurements can verify this hypothesis. From a comparison of preliminary magnetic data for "axial basalts" (dredged within 1 km from the axis: dredge hauls 1, 4, 5, 8, 9, 11 and 12) and for "more distant basalts" (dredge hauls 2, 3, 6 and 10), Prévot and Lecaille (1976) concluded that the magnetization high cannot result from an alteration process. The more complete data now available (Table 3) confirms their conclusions:

1) The axial basalts are about twice as magnetic as the more distant basalts. In other words, direct magnetic measurements are compatible with intensity variations deduced from deep-tow profiles.

2) This change in intensity of remanence cannot be explained by any maghemitization process because of the absence of change in Curie temperature and the larger saturation magnetization (which decreases through maghemitization) of the more distant basalts.

It is obvious therefore that such a magnetization high is not necessarily a marker of the most recent extrusions on the sea floor and may not be useful in detecting present plate boundaries.

Relationship between petrology and magnetism of the basaltic layer

1. FAMOUS pillow basalts

The idea that the magnetization high could be due to petrological variations of the magma at the time of eruption came from the observation that highly

Table 3 Comparison of average magnetic properties of the axial basalts and the more distant basalts from the median valley of the Mid-Atlantic ridge (FAMOUS area)

Location	Θ	J_s	k	J_r	Q
Axial basalts	160 ± 10°C	0.755 ± 0.145	2.50 ± 0.40	1.98 { 2.39 / 1.64	241 { 277 / 210
	(N=57)	(N=30)	(N=57)	(N=57)	(N=57)
More distant basalts	155 ± 14°C	1.10 ± 0.15	3.63 ± 0.53	0.978 { 1.18 / 0.813	78.2 { 94.7 / 64.6
	(N=46)	(N=23)	(N=46)	(N=46)	(N=46)

Legend: see table 2

magnetic hills found in the inner floor (Fig. 6) of the FAMOUS Rift Valley correspond mainly to olivine-rich basalt while plagioclase and pyroxene-rich basalt occurs on the eastern and western walls (Bougault and Hekinian, 1974; Hekinian and Hoffert, 1975; Hekinian, Moore and Bryan, 1976; Arcyana, 1977; Bryan and Moore, 1977).

To test this hypothesis, thin sections from each of the 120 large samples studied magnetically were examined and half of the samples were chemically analyzed. We compare the magnetic properties of two groups of basalts:

- olivine-rich basalts (Bougault and Hekinian, 1974) which include both olivine and plagioclase as early formed mineral phases. These are aphanitic, aphyric or slightly phyric basalts. The picritic basalts, which are less common, will be considered here as olivine-rich basalts.
- pyroxene-rich basalts (plagioclase-pyroxene-olivine basalts) and plagioclase-rich basalts (moderately phyric and highly phyric plagioclase basalts). The pyroxene-rich basalt may contain small amount of olivine.

Results listed in table 4 show that the intensity of remanence of olivine-rich basalts is statistically twice as large as other basalt types. This does not result from any difference in the chemical composition of the titanomagnetite, as attested by the absence of any difference in the mean Curie temperature. Neither can the difference in remanence intensity be due to changes in the magnetic oxide content. On the contrary, titanomagnetites are more abundant within the group with the smallest intensity of remanence as shown by their higher saturation magnetization (which is proportional to the oxide content). The difference in intensity of remanence may result either from changes in the intensity of the paleofield or from a variation in the grain size of the magnetic minerals. For titanomagnetites, the intensity of the thermoremanent magnetization decreases within the pseudosingle domain range when particle size increases, the total diminution in intensity corresponding to a factor of ten (Day, 1977). We think that the difference in intensity of remanence between olivine-rich basalts and pyroxene-rich or plagioclase-rich basalts is mainly due to a statistical grain size difference, the smallest magnetic oxide grains being found in the olivine-rich basalts. This view is substantiated by the trends of the grain size-dependent magnetic parameters (H_c, H_{rc}, MDF and Q) which all indicate that, statistically, the finest titanomagnetites occur within the olivine-rich basalts (Table 4). Microscopic observations lead to the same conclusion.

It must be pointed out that this correlation between magnetism and petrology is only valid at a statistical level. Moreover, it is not an unequivocal relationship: within each of the two magnetically different groups, basalts which are petrologically quite distinct coexist. The differences in magnetic properties observed are in fact related to the textural features of the rocks rather than to their petrology as a whole.

Figure 7 shows a ternary diagram giving the relative amount (vol %) of phenocrysts, matrix minerals and mesostasis in the rock. The term mesostasis as used here refers to the glassy portion and the cryptocrystalline (< 0.02 mm in length) material not identified under a high power microscope. The matrix minerals range between 0.02 and 0.9 mm in length and include plagioclase laths, pyroxene, and olivine granules. The phenocrysts consist mainly of cumulate plagioclases having a grain size larger than 1 mm in length. On the same diagram the value of the coercive force H_c is superimposed for each sample. It is apparent that the largest coercive forces (corresponding to the finest grains) are found for the least crystalline rocks, which are the olivine-rich basalts.

Table 4 Average magnetic properties of olivine-rich pillow basalts and plagioclase or pyroxene-rich pillow basalts from the FAMOUS area (median valley)

	olivine-rich basalts	plagioclase or pyroxene-rich basalts
Θ	162 ± 12 (N=35)	159 ± 13 (N=45)
J_s	0.690 ± 0.150 (N=28)	1.05 ± 0.13 (N=23)
Q	234 { 193 / 284 (N=52)	98.5 { 88.3 / 121 (N=45)
H_c	368 ± 76 (N=28)	250 ± 55 (N=23)
H_{rc}	502 ± 113 (N=28)	314 ± 80 (N=22)
MDF	393 ± 69 (N=28)	323 ± 80 (N=23)
J_r	2.01 { 1.64 / 2.48 (N=52)	0.995 { 0.818 / 1.21 (N=45)

Legend: see table 2

2. Leg 37 pillow basalts

Petrographic studies of Leg 37 basalts led to the recognition of petrological types similar to those found in the FAMOUS area (Bryan and Thompson 1977; Dimitriev and Aumento, 1977). Using these studies and geochemical data from Aumento, Melson et al. (1977) we were able to classify Leg 37 pillow basalts from hole 332B into the two groups previously defined: olivine-rich basalts and plagioclase or pyroxene-rich basalts.

Table 5 shows that the relative variation of the magnetic parameters between these two groups are the same as those found for the FAMOUS pillow basalts; olivine-rich basalts are about twice as magnetic as plagioclase and pyroxene-rich basalts. Again, this cannot result from variations of oxide content (oxide content is lower in the olivine-rich basalts) and has to be attributed to the smaller magnetic oxide grain size (attested by a larger median destructive field) in the olivine-rich basalt.

The fact that low-temperature oxidation affects these two groups of basalts to the same degree is demonstrated by the absence of any significant difference between the two mean Curie points. Hence maghemitization does not modify the relative contribution of olivine-rich basalts and of plagioclase or pyroxene-rich basalts to magnetic anomalies.

IV Conclusions

Extensive magnetic studies carried out at the Mid-Atlantic Ridge Crest near 37°N show that the magnetic properties of the pillow basalt layer are

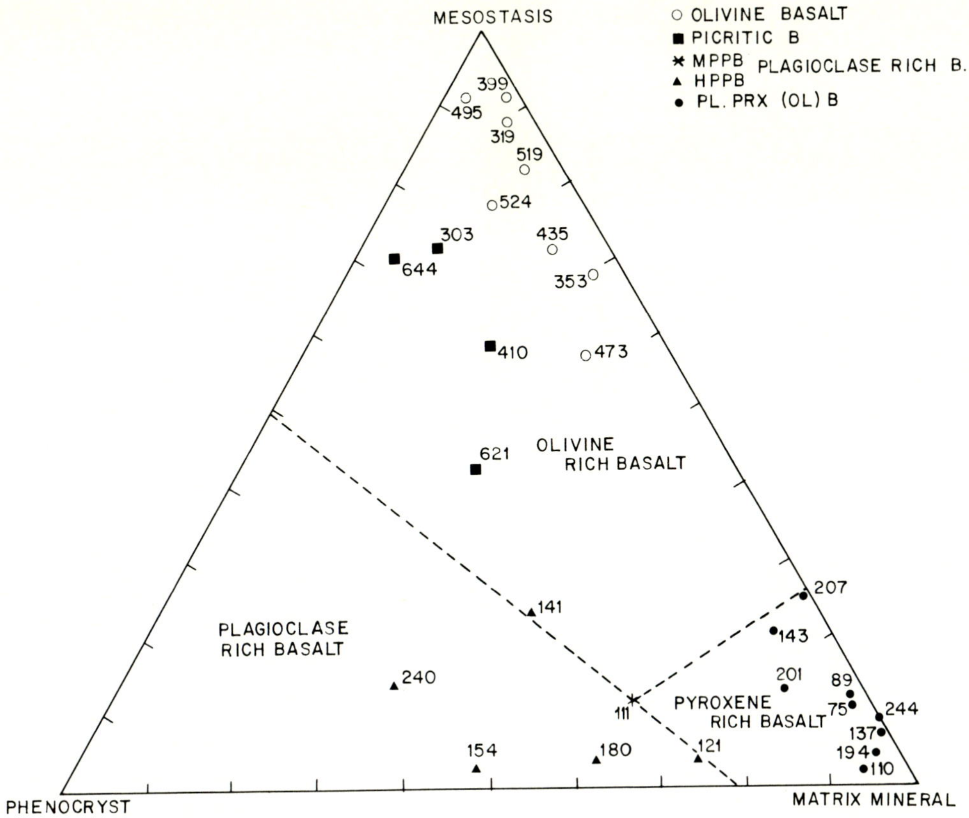

Fig. 7. Ternary diagram of the relative amount (vol %) of phenocrysts, matrix mineral and mesostasis (see text for precise definitions). Number indicates coercive force of the sample (Oe). Olivine basalts and picritic basalts are olivine-rich basalts. Plagioclase-rich basalts are divided into moderately phyric plagioclase basalts (MPPB) and highly phyric plagioclase basalts (HPPB). PL. PRX (OL)B are called pyroxene-rich basalts in the text. See text for references about these different types of basaltic rocks found in the rift valley.

petrology dependent and change notably with time as a result from sea-floor weathering and tectonic tilting. In addition the thickness of the magnetized layer increases rapidly outwards from the ridge axis. All these features seem to be typical of the behavior of the basaltic layer of the oceanic crust.

1. A relationship between petrology and magnetism was first suggested by Vogt and Johnson (1973) in order to explain high-amplitude anomalies zones located close to some postulated "hot spots". Chemical and magnetic studies of rocks from the Reykjanes Ridge (de Boer, 1973; Day et al., 1979), the Galapagos spreading center (Vogt and de Boer, 1976) and high-amplitude zones in the Juan de Fuca area (Vogt and Byerly, 1976) seem to support this conjecture. The highest magnetizations of these Fe-Ti enriched basalts are supposed to result from a larger titanomagnetite concentration, contrary to our observations in the FAMOUS area. However, the rock magnetic study carried out on seven pillow fragments from the Galapagos spreading center showed a poor correspondence between the titanomagnetite abundance and the remanence intensity (Anderson et al., 1975). These workers suggest that in addition to having twice as much titanomagnetite, the samples from the Galapagos rift zone have five times more intense remanence than the samples from the Costa Rica ridge because the former have a greater proportion of smaller grains. As it is true for the magnetization high in the FAMOUS rift valley, variations in the grain size of the magnetic minerals seem essential to interpret the high-amplitude anomalies zones found in other parts of the oceans.

2. Maghemitization of titanomagnetite is probably the major cause for the decrease in the amplitude of magnetic anomalies outwards from the ridge axis. However, according to the Curie temperatures of the rocks collected within the FAMOUS Rift Valley, the maghemitization of the top of the basement is a progressive process. Samples collected as far as 4 km west of the axis yield Curie temperatures near 150°C (Johnson and Atwater, 1977), which is not significantly different from the mean value at the axis (table 4). This confirms our

Table 5 Average magnetic properties of olivine-rich pillow basalts and plagioclase or pyroxene-rich pillow basalts from Leg 37 (hole 332B)

	olivine-rich basalts	plagioclase or pyroxene-rich basalts
Θ	294 ± 9 (N=65)	305 ± 14 (N=78)
J_s	0.305 ± 0.023 (N=65)	0.429 ± 0.066 (N=78)
Q	$74.5 \left\{ \begin{matrix} 63.6 \\ 87.2 \end{matrix} \right.$ (N=50)	$34.7 \left\{ \begin{matrix} 28.6 \\ 42.2 \end{matrix} \right.$ (N=49)
MDF	370 ± 31 (N=49)	254 ± 30 (N=49)
J_r	$0.488 \left\{ \begin{matrix} 0.423 \\ 0.561 \end{matrix} \right.$ (N=50)	$0.270 \left\{ \begin{matrix} 0.222 \\ 0.327 \end{matrix} \right.$ (N=49)

Magnetic data from Hall and Ryall (1977a and b), petrological classification from Dmitriev and Aumento (1977). Legend: see table 2.

conclusion that the magnetization high found in the FAMOUS Rift Valley 2 does not result from an alteration process and cannot be used *a priori* as a marker of the present plate boundary.

From the FAMOUS area to DSDP 332 site, the mean intensity of NRM decreases by a factor of 4. Taking into account the reducing of NRM of reversely magnetized Leg 37 basalts due to VRM overprinting, the reduction of the intensity of the primary remanent magnetization must be slightly less, probably between 2/3 and 3/4. There is an excellent agreement between intensities of remanence measured and intensities calculated from direct modeling of near-bottom magnetic anomalies at 37°N (Macdonald, 1977). In both cases the intensity decreases from 2×10^{-2} emu cm^{-3} right at the axis (effective average intensity of remanence of axial basalts) to about 0.4×10^{-2} emu cm^{-3} for 3.5 m.y. old crust (holes 332).

The 2/3 to 3/4 reduction due to maghemitization is the same as that found by Prévot and Grommé (1975) for dredged basalts from the whole North Atlantic basement. The larger decrease observed at 45°N (Irving, 1970) probably results from some combination of maghemitization and local variations in magnetic oxide content and/or size, between basalts from the median valley and from the flanks (Prévot, 1973). It is worth noting that moderately magnetized pillow basalts have also been recovered within the median valley at 45°N (Ade-Hall et al., 1973).

3. Tectonic rotations due mainly to normal faulting at the edges of the median valley have been suggested by Van Andel (1968) and Van Andel and Bowin (1968). Another geologic model by Moore et al. (1974) assumes that the subsidence of the ridge center induces a substantial tilting (of 20° to 40°) towards the rift-valley axis. The paleomagnetic data so far obtained in the North Atlantic strongly support such models. It has been frequently observed at DSDP and IPOD sites that, even after cleaning, the mean magnetic inclination of the magnetic units differs from the value expected from paleolatitude: sites 332A, 332B, and 333A (Ryall and Hall, 1977); sites 395 and 396 (Johnson, 1979), site 396B (Petersen, 1979); sites 407, 408, and 410A (Luyendyk, Cann et al., 1977), site 317D (Donnelly, Francheteau et al., 1977) and site 418A (Flower, Salisbury

et al., 1977). Expected and anomalous inclinations, which generally differ by 20 to 30°, have been observed several times in a single hole.

It is difficult to explain these anomalous inclinations by any non-dipole behavior of the geomagnetic field (secular variation or geomagnetic excursions or transitions) at the time of cooling because the probability of recording such directions is very low (Harrison and Watkins, 1977). Decisive geological and paleomagnetic observations in favor of tectonic rotations of crustal blocks have been reported recently (Donnelly, Francheteau et al., 1977; Flower, Salisbury et al., 1977).

The axis of spreading of the North Atlantic Ridge, trending approximately N-S, is more or less parallel to the horizontal component of the Upper Tertiary and Quaternay geomagnetic field. Rotations around the horizontal axis parallel to the ridge axis postulated by the two geological models reported above would reduce the magnetic inclination. The unexpectedly low inclinations frequently observed along the North Atlantic Ridge Crest (Ryall and Hall, 1977, Johnson, 1979; Petersen, 1979; Luyendyk, Cann et al., 1977) can therefore be readily explained by tectonic rotations.

Qualitatively different inclination discrepancies have been obtained for older oceanic crust. At the Cretaceous North Atlantic sites (IPOD megaleg 51-52-53) magnetic units with a mean inclination larger than expected dipole inclination were found, the difference being about 25°-35° (Donnelly, Francheteau et al., 1977; Flower, Salisbury et al., 1977) In such cases, horizontal rotation axes have to be more or less oblique with respect to the horizontal component of the paleofield. Taking into account the Cretaceous pole position for North America (Mankinen, 1978) and the 10-20° East strike of the ridge crest, rotations about axes parallel to the ridge crest can, at the most, increase the expected dipole 32° inclination to 43°-50°. This seems insufficient to account for the observed discrepancy. It may be suggested that large faulting occurred along directions oblique with respect to the spreading axis strike or that some rotations are due to large scale slumps rather than to normal faults (Francheteau, personal communication).

Another unsolved question is whether the postulated rotations parallel to the spreading center axis take place within (Moore et al., 1974) or at the edges of the median valley (van Andel and Bowin, 1968). The paleomagnetic data so far obtained for rocks collected within the median valley cannot settle this point because the magnetic inclinations calculated are in most cases relative to the horizontal plane at the time of cooling rather than to the present horizontal plane. Drilling within the rift valley is probably the only way to solve this problem.

An argument against the interpretation of anomalous inclinations in terms of tectonic rotation is that it would require extreme tilting of the blocks (Lowrie, 1977). Assuming that the rotation axis was in the plane of the paleofield vector, Harrison and Watkins (1977) found that the amount of rotation needed to account for the excessively shallow inclinations at the Leg 37 sites is 55°. However, the tectonic studies of the Afar accretion zone (Morton and Black, 1975) show that in several large areas extension tectonics resulted in normal faulting with blocks strongly tilted. Dip angles up to 60° and locally up to 90° have been observed. Probably as a result of successive stages of normal faulting, the faults were sometimes observed dipping at only shallow angles. In view of these results, there is no doubt that the tectonic complexity of the oceanic crust has been so far underestimated.

4. Thickening of the magnetized layer outwards from the ridge axis. It is now well established (Harrison, 1976) that the mean magnetic intensity of DSDP cores is several times less than the values calculated from magnetic anomaly profiles assuming the magnetized layer is only 500 m thick. DSDP sites recently studied along the MAR (Johnson, 1979, Petersen, 1979) confirms Leg 37 results (Ryall et al., 1977) that the magnetized layer is several kilometers thick away from the ridge axis. On the other hand, the high average magnetic intensity of the FAMOUS pillow basalts seems to indicate that the magnetized layer is less than 500 m within the median valley (Lecaille et al., 1974).

Three processes can result in a decrease of the intensity of remanence with depth at the MAR axis:

1. transition from pillow basalts to possibly less magnetic intrusive units.
2. geothermal gradient.
3. intense maghemitization at depth.

Regarding the first interpretation, the main transition between intrusive and intrusive rocks is supposed to correspond to the 2A/2B interface within layer 2 (Talwani et al., 1971). Fowler (1976) showed that beneath the median valley at 37°N layer 2A is about 1 km thick. Similarly, in a detailed geophysical study of the Reykjanes Ridge Crest Talwani et al. (1971) found that the thickness of the pillow basalts varies from 0.6 to 1.1 km in the crestal zone. This figure is to compare with the 400 m thickness they assign to the magnetized layer.

The second interpretation is that the bottom of the magnetized layer corresponds to the Curie point isotherm (160°C at the axis). This is supported by some thermal models of the oceanic crust. Sleep's (1975) study of the thermal structure of mid-ocean ridge axes suggests that the 300°C isotherm would be as shallow as 300 m at the axis of the MAR near 37°N. According to Sclater and Francheteau's (1970) cooling model of lithospheric plate, the variation of heat flow q with age t of the site is given approximately by (Francheteau, personal communication):

$$q = \frac{12}{\sqrt{t}}$$

where q is in Heat Flow Unit (1 HFU=10^{-6} cal sec^{-1} cm^{-2}) and t in m.y. The thermal conductivity of pillow basalts being 4.0 mcal °C^{-1} cm^{-1} sec^{-1} (Hyndman and Jessop, 1971; Hyndman et al., 1977) the depth of the 160°C isotherm can be calculated. This depth is approximately 400 m for t=0.5 m.y. and 500 m for t=1 m.y. At site 332, the calculated depth of the relevant 300°C Curie point isotherm is approximately 2 km, in agreement with Ryall et al.'s (1977) suggestion that the entire layer 2 contributes to the magnetic anomaly. As the plate cools the lower part of layer 2 becomes progressively magnetized. Such a cooling by conduction results in the formation of blocks of alternating polarity with broad sloping transition zones as described in Cande and Kent's (1976) magnetic model of layer 3. Such thermal models provide therefore a straightforward explanation of the thickening of the magnetized layer near the ridge axis.

There are however two major objections regarding this interpretation. The first one deals with the importance of nonconductive processes, in particular hydrothermal circulation, in the oceanic crust. Percolation by sea water extends down to several kilometers in depth (Ribando et al., 1976) and may cause significant loss of heat near the ridge axis (e.g. Talwani et al., 1971), which is not considered in Sclater and Francheteau's and Sleep's models. This idea led Cande and Kent (1976) and Kusznir and Bott (1976) to suggest that layer 2 as a whole is at 0°C at ridge axes. However, according to a numerical study of hydrothermal circulation in the oceanic crust (Ribando et al., 1976), percolation alters the spatial distribution of surface heat flow but not its mean value, unless circulating water leaves the porous crust as extensive hot springs with a temperature in excess of 50°C.

The second objection is that any rise in temperature will increase the reaction rate of maghemitization. It has been shown by Readman and O'Reilly (1970) and by Ozima (1971) that the reaction rate of maghemitization observed for 0.1 μm size titanomagnetites heated in air yields a time constant of the order of 10^6 years at 0°C, in agreement with the time dependence of maghemitization near the MAR axis (Irving, 1970; Johnson and Atwater, 1977). Pillow basalts at ridge axes are highly vesicular and are buried with trapped water, as shown by their very low seismic velocity (Talwani et al., 1971). As is true for the top of the oceanic basement, maghemitization at depth can proceed in the presence of sea water. Calculations following Ozima's (1971) procedure yield time constants of about 10^3 years at 50°C and a few years at 100°C for completion of maghemitization. Hence the lower boundary of the magnetized layer corresponds probably to a decrease in magnetization due to an intense maghemitization at depth at ridge axes. A further contribution comes from the decrease of the spontaneous magnetization as temperature increases. The geothermal gradient within the rift valley is therefore less important than the gradient needed to interpret the lower boundary of the magnetized layer in terms of the Curie point isotherm. However, depending upon the hydrothermal circulation pattern, and the depth of the eventually transient magma chamber (Nisbet and Fowler, 1978) at the ridge axis, the possibility remains that the lower part of layer 2A is heated at some places above its Curie point and becomes remagnetized as the plate moves away from the ridge axis.

Acknowledgments. Discussion with R. S. Coe, J. Francheteau, S. Grommé, E. A. Mankinen and J. G. Moore were very helpful as were the reviews of C. G. A. Harrison and H. P. Johnson. We thank also B. Bregman for typing the manuscript. This work was supported by the CNRS, the Université de Paris 6, and the CNEXO.

References

Ade-Hall, J. M., H. P. Johnson, and P. J. C. Ryall, Rock magnetism of basalts Leg 34, Initial Rep. DSDP, 34, 459-468, 1977.

Ade-Hall, J. M., F. Aumento, P. J. C. Ryall, and R. E. Gerstein, The Mid-Atlantic Ridge near 45°N. XX1. Magnetic results from basalt drill cores from the median valley, Can. J. Earth Sci., 10, 679-696, 1973.

Anderson, R. N., D. A. Clague, K. D. Klitgord, M. Marshall, and R. K. Nishimori, Magnetic and petrologic variations along the Galapagos spreading center and their relation to the Galapagos melting anomaly, Geol. Soc. Amer. Bull., 86, 683-696, 1975.

Arcyana, Transform fault and rift valley from bathyscaph and diving saucer, Science, 190, 108-116, 1975.

Arcyana, Rocks collected by bathyscaph and diving saucer in the FAMOUS area of the Mid-Atlantic rift, valley: petrological diversity and structural setting, Deep-Sea Res., 24, 565-589, 1977.

Aumento, F., W. E. Melson et al., Initial Rep. DSDP, 37, Washington, 1977.

Ballard, R. D., W. B. Bryan, J. R. Heirtzler, G. Keller, J. G. Moore and Tj Van Andel, Manned submersible observations in the FAMOUS area, Mid-Atlantic Ridge, Science, 190, 103-108, 1975.

Ballard, R. D., and J. G. Moore, Photographic Atlas of the Mid-Atlantic Ridge Rift Valley, 114 p., 1977.

Bellaiche, G., J. L. Cheminée, J. Francheteau, R. Hekinian, X. Le Pichon, H. Needham, and R. Ballard, Inner floor of the rift valley:

first submersible study, Nature, 250, 558-560, 1976.

Bideau, D., R. Hekinian, and J. Francheteau, Orientation of ocean floor basaltic rocks at time of cooling: a general method, Contrib. Mineral Petrol., 65, 19-28, 1977.

Bleil, U., and N. Petersen, Magnetic properties of basement rocks, Leg 37, site 332, Initial Rep. DSDP, 37, 449-456, 1977.

Bougault, H. and Hekinian, R., Rift valley in the Atlantic Ridge near 36°50'N. petrology and geochemistry of basaltic rocks, Earth Planet. Sci. Lett., 24, 249-261, 1974.

Bryan, W. B. and J. C. Moore, Compositional variations of young basalts in the Mid-Atlantic Ridge rift valley near 36°49'N., Geol. Soc. Am. Bull., 88, 556-570, 1977.

Bryan, W. B. and G. Thompson, Basalts from DSDP Leg 37 and the FAMOUS area: compositional and petrogenetic comparisons, Can. J. Earth Sci., 14, 875-885, 1977.

Cande, S. C. and D. V. Kent, Constraints imposed by the shape of marine magnetic anomalies on the magnetic source, J. Geophys. Res., 81, 4157-4162, 1976.

Day, R., TRM and its variation with grain size, a review, J. Geomagn. Geoelect., 29, 233-266, 1977.

Day, R., S. Halgedahl, M. Steiner, K. Kobayashi, T. Furuta, T. Ishii, and A. Faller, Magnetic properties of basalts from DSDP Leg 49, Initial Rep. DSDP, 49, (in press), 1979.

de Boer, J., Variations in the magnetic properties of post-glacial pillow lavas along the Reykjanes ridge, J. Geophys. Res., 80, 3769-3776, 1975.

Dimitriev, L. and F. Aumento, Geochemical characteristics and evolution of basalts from DSDP site 332, Can. J. Earth Sci., 14, 899-902, 1977.

Donnelly, T. W., J. Francheteau et al., Mid-ocean ridge in the Cretaceous, Geotimes, 22, 6, 21-23, 1977.

Dunlop, D. J. and C. J. Hale, Simulation of long-term changes in the magnetic signal of oceanic crust, Can. J. Earth Sci., 14, 716-744, 1977.

Evans, M. E. and M. L. Wayman, Magnetic properties of igneous samples, Leg 37, Initial Rep. DSDP, 37, 471-473, 1977.

Flower, M. F. J., M. H. Salisbury et al., Cretaceous crust sought, Geotimes, 22, 9, 20-22, 1977.

Fowler, C. M. R., Crustal structure of the Mid-Atlantic ridge crest at 37°N, Geophys. J. R. astr. Soc., 47, 459-491, 1976.

Hall, J. M. and P. J. C. Ryall, Paleomagnetism of basement rocks, Leg 37, Initial Rep. DSDP, 37, 425-448, 1977a.

Hall, J. M. and P. J. C. Ryall, Rock magnetism of basement rocks, Leg 37, Initial Rep. DSDP, 37, 489-501, 1977b.

Harrison, C. G. A., Magnetization of the oceanic crust, Geophys. J. R. astr. Soc., 47, 257-284, 1976.

Harrison, C. G. A., L. Stieltjes and E. Tarasiewicz, Paleomagnetism of samples from the axial zone of the Afar depression, Earth Planet. Sci. Lett., 34, 273-283, 1977.

Harrison, C. G. A. and N. D. Watkins, Shallow inclinations of remanent magnetism in Deep-Sea Drilling Project igneous core, geomagnetic field behavior or postemplacement effects? J. Geophys. Res., 82, 4869-4877, 1977.

Heirtzler, J. R., Project FAMOUS, first voyages down to the Mid-Atlantic Ridge, where the Earth turns inside out, Nat. Geogr. Mag., 147, 586-603, 1975.

Hekinian, R. and M. Hoffert, Rate of palagonitization and manganese coating on basaltic rocks from the rift valley in the Atlantic ocean near 36°50'N, Marine Geol., 19, 91-109, 1975.

Hekinian, R., J. G. Moore and W. B. Bryan, Volcanic rocks and processes of the Mid-Atlantic Ridge rift valley near 36°40'N, Contrib. Mineral. Petrol., 58, 83-110, 1976.

Hyndman, R. D., and A. M. Jessop, The Mid-Atlantic Ridge near 45°N. XV. Thermal conductivity of dredge and drill core samples, Can. J. Earth Sci., 8, 391-393, 1971.

Hyndman, R. D., R. P. Von Herzen, A. J. Erickson and J. Jolivet, Heat-flow measurements DSDP Leg 37, Initial Rep. DSDP, 37, 344-362, 1977.

Irving, E., The Mid-Atlantic Ridge at 45°N. XIV. Oxidation and magnetic properties of basalt, review and discussion, Can. J. Earth Sci., 7, 1528-1538, 1970.

Johnson, H. P. and T. Atwater, Magnetic study of basalts from the Mid-Atlantic Ridge, lat 37°N, Geol. Soc. Amer. Bull., 88, 637-647, 1977.

Johnson, H. P., and J. M. Hall, A detailed rock magnetic and opaque mineralogy study of the basalts from the Nazca plate, Geophys. J. R. astr. Soc., 52, 45-64, 1978.

Johnson, H. P., Paleomagnetism of igneous rock samples, DSDP Leg 45, Initial Rep. DSDP, 45, (in press), 1979.

Kent, D. V. and W. Lowrie, VRM studies in Leg 37 igneous rocks, Initial Rep. DSDP, 37, 525-529, 1977.

Klitgord, K. D., Sea-floor spreading: the central anomaly magnetization high, Earth Planet. Sci. Lett., 29, 201-209, 1976.

Kusznir, N. J., and M. H. P. Bott, A thermal study of the formation of oceanic crust, Geophys. J. R. astr. Soc., 47, 83-95, 1976.

Lecaille, A. and M. Prévot, Rock-magnetism of basalts from the FAMOUS rift valley, in preparation, 1979.

Lecaille, A., M. Prévot, J. C. Tanguy and J. Francheteau, Intensité d'aimantation de basaltes dragués dans le rift médio-atlangique vers 36°50'N, C. R. Acad. Sci. Paris, 279, sér B., 617-620, 1976.

Lowrie, W., Intensity and direction of magnetization in oceanic basalts, J. Geol. Soc., 133, 61-82, 1977.

Luyendyk, B., J. R. Cann et al., "Young and hot" drilling, Geotimes, 22, 3, 25-28, 1977.

Macdonald, K. C., Near-bottom magnetic anomalies, asymmetric spreading, oblique spreading, and tectonics of the Mid-Atlantic Ridge near lat. 37°N, Geol. Soc. Amer. Bull., 88, 541-555, 1977.

Mankinen, E. A., Paleomagnetic evidence for a late Cretaceous deformation of the great valley sequence, Sacramento Valley, California, J. Res. U.S. Geol. Survey, 6, 383-390, 1978.

Marshall, M. and A. Cox, Magnetism of pillow basalts and their petrology, Geol. Soc. Amer. Bull., 82, 537-552, 1971.

Moore, J. G., H. S. Fleming and J. D. Phillips, Preliminary model for extrusion and rifting at the axis of the Mid-Atlantic Ridge, 36°48'N, Geology, 8, 437-440, 1974.

Morton, W. H. and R. Black, Crustal attenuation in Afar, in The Afar region of Ethiopia and related rift problems, ed. A. Pilger and A. Rosler, Stuttgart, 55-65, 1975.

Murthy, G. S., E. R. Deutsch, R. R. Pätzold, Inferences on the magnetic domain state of Leg 37 basalt, Initial Rep. DSDP, 37, 515-523, 1977.

Needham, H. D. and J. Francheteau, Some characteristics of the rift valley in the Atlantic ocean near 36°48'N, Earth Planet. Sci. Lett., 22, 29-43, 1976.

Nisbet, E. G. and C. M. R. Fowler, The Mid-Atlantic Ridge at 37 and 45°N: some geophysical and petrological constraints, Geophys. J. R. astr. Soc., 54, 631-660, 1978.

Ozima, M., Magnetic processes in oceanic ridge, Earth Planet. Sci. Lett., 13, 1-5, 1971.

Petersen, N., Rock and paleomagnetism of basalts from site 396B, Leg 46, Initial Report DSDP, 46, (in press), 1979.

Plessard, C. and M. Prévot, Magnetic viscosity of submarine basalts, Deep Sea Drilling Project, Leg 37, Initial Rep. DSDP, 37, 503-506, 1977.

Prévot, M., Oxydation basse température des titanomagnétites et variation de l'intensité d'aimantation des basaltes océaniques au voisinage de l'axe de la dorsale médio-atlantique, C. R. Ac. Sci., Paris, sér. D, 1101-1104, 1973.

Prévot, M. and Grommé, S., Intensity of magnetization of subaerial and submarine basalts and its possible change with time, Geophys. J. R. astr. Soc., 40, 207-224, 1975.

Prévot, M. and A. Lecaille, Comments on paper by K. D. Klitgord, "Sea-floor spreading: the central anomaly magnetization high", Earth Planet. Sci. Lett., 33, 166-168, 1976.

Prévot, M., A. Lecaille, J. Francheteau and R. Hekinian, Magnetic inclination of basaltic lavas from the Mid-Atlantic Ridge near 37°N, Nature, 259, 649-653, 1976.

Prévot, M. and J. Mergoil, Crystallization trend of titanomagnetites in an alkali basalt from Saint Clément (Massif Central, France), Miner. Mag., 39, 474-481, 1973.

Prévot, M., A. Lecaille and E. A. Mankinen, Magnetic effects of maghemitization of the oceanic crust, in preparation.

Readman, P. W. and W. O'Reilly, The synthesis and inversion of non-stoichiometric titanomagnetites, Phys. Earth Planet. Interiors, 4, 121-128, 1970.

Readman, P. W. and W. O'Reilly, Magnetic properties of oxidized (cation-deficient) titanomagnetites, J. Geomagn. Geoelect., 24, 69-90, 1972.

Renard, V., B. Schrumpf and J. C. Sibuet, Bathymétrie détaillée d'une partie de vallée du rift et de faille transformante près de 36°50'N dans l'Océan Atlantique, CNEXO, Paris, 1975.

Ribando, R. J., K. E. Torrance and D. L. Turcotte, Numerical models for hydrothermal circulation in the oceanic crust, J. Geophys. Res., 81, 3007-3012, 1976.

Richards, J. C. W., J. B. O'Donovan, Z. Hauptan, W. O'Reilly and K. M. Creer, A magnetic study of titanomagnetite substituted by magnesium and aluminium, Phys. Earth Planet. Inter., 7, 437-444, 1973.

Ryall, P. J. C., J. M. Hall, J. Clark and T. Milligan, Magnetization of oceanic crustal layer 2: results and thoughts after DSDP Leg 37, Can. J. Earth Sci., 14, 684-706, 1977.

Schwarz, E. J. and Y. Fujiwara, Remanent magnetization and magnetic mineralogy in DSDP Leg 37 oceanic basalts, Initial Rep. DSDP, 37, 507-509, 1977.

Sclater, J. G. and J. Francheteau, The implications of terrestrial heat flow observations on current tectonic and geochemical models of the crust and upper mantle of the Earth, Geophys. J. R. astr. Soc., 20, 509-542, 1970.

Sleep, N. H., Formation of oceanic crust: some thermal constraints, J. Geophys. Res., 80, 4037-4042, 1975.

Smith, B. M. and M. Prévot, Variation of the magnetic properties in a basaltic dyke with concentric cooling zones, Phys. Earth Planet. Inter., 14, 120-136, 1977.

Storzer, D. and M. Selo, Ages par la méthode des traces de fission de basaltes prélevés dans la vallée axiale de la dorsale médio-atlantique aux environs de 37°Nord, C. R. Acad. Sci., Paris, 279, sér. D., 1649-1651, 1974.

Storzer, D. and M. Selo, Uranium content and fission track ages of some basalts from the FAMOUS area, Bull. Soc. Géol. France, 18, 807-810, 1976.

Talwani, M., C. C. Windisch and M. G. Langseth, Jr., Reykjanes ridge crest: a detailed geophysical study, J. Geophys. Res., 76, 473-517, 1971.

van Andel, T. H., The structure and development of rifted mid-oceanic rises, J. Marine Res., 26, 144-161, 1968.

van Andel, T. H. and C. O. Bowin, Mid-Atlantic Ridge between 22° and 23° North latitude and the tectonics of mid-oceanic rises, J. Geophys. Res., 73, 1279-1298, 1968.

Vine, F. J. and D. H. Matthews, Magnetic anomalies over ocean ridges, Nature, 196, 947-950, 1963.

Vogt, P. R. and G. R. Byerly, Magnetic anomalies

and basalt composition in the Juan de Fuca-Gorda ridge area, Earth Planet. Sci. Lett., 33, 185-207, 1976.

Vogt, P. R. and J. de Boer, Morphology, magnetic anomalies and basalt magnetization at the ends of the Galapagos high-amplitude zone, Earth Planet. Sci. Lett., 33, 145-163, 1976.

Vogt, P. R. and G. L. Johnson, Magnetic telechemistry of oceanic crust? Nature, 245, 373-375, 1973.

METAMORPHISM IN THE OCEAN CRUST

Johnson R. Cann

Department of Geology, The University,
Newcastle upon Tyne, NE1 7RU, England

Abstract. Ocean floor rocks can be assigned to five mineral assemblage facies, brownstone, zeolite, greenschist, amphibolite and gabbro. The rocks characteristically preserve igneous textures and are little deformed. Changes in bulk chemistry are controlled by congruent mineral reactions, in which one phase is transformed into another. Overall bulk crustal changes cannot easily be calculated. Metamorphism to high grades follows on very shortly after crustal formation, during an intense period of hydrothermal activity as the ridge-crest-parallel faults form. Brownstone alteration continues as the crust ages during less intense, lower temperature hydrothermal episodes.

Introduction

Metamorphosed igneous rocks are recovered relatively frequently from the ocean floor by dredging, and are to be expected as the solid components of the rock-water hydrothermal systems that have been predicted on geophysical grounds (Lister, 1972), observed indirectly (Williams et al., 1974), and recently observed directly as emerging jets of hot water (Weiss et al., 1977). However, metamorphic rocks have very rarely been recovered by drilling, and none of the deep holes in the ocean crust, such as those on legs 37 and 45, have shown any clear-cut increase in the degree of alteration with depth. This paper sets out to examine this paradox, by first giving an outline account of the nature of the metamorphic rocks from the ocean crust, and then examining the processes by which they form.

Characteristics of Ocean-floor Metamorphics

All of the metamorphic rocks so far described from the ocean floor are metamorphosed igneous rocks, either basalts, gabbros and related basic rocks, or peridotites and related ultramafic rocks. Metamorphic deformational fabrics are only poorly developed. Though these rocks are described as belonging to, for example, the greenschist or amphibolite facies, this refers to mineral assemblage only and does not imply that the greenschists are schistose, or that the amphibolite facies rocks are made up of lineated lustrous black prisms of hornblende. Almost all of the metamorphics preserve a well-developed igneous texture, with a characteristic lath shape to the plagioclase demonstrating an origin by crystallisation from a liquid. Chilled margins preserve variolitic and skeletal crystal forms, now pseudomorphed by other minerals, clearly indicating very rapid cooling and rapid crystal growth. Zones of shearing may cut across areas where primary textures are well-developed, but these make up only a small proportion of the bulk of most rock samples.

Table 1 gives a summary of the mineral assemblages observed in ocean floor metamorphics for both basic and ultramafic compositions, over the range of metamorphic conditions encountered. Specific features are discussed in the following paragraphs.

The brownstone facies, newly defined here, is the facies of low temperature ocean floor weathering or cool hydrothermal alteration. It is the facies of most older oceanic crust, and persists through most of the drilled sections of ocean crust. Its products have characteristically a yellowish brown tint (from which the name is derived) when conditions are, as they usually are, oxidising, but are a dark bluish grey under reduction. Secondary minerals are closely related to particular primary phases, and cannot be demonstrated to form equilibrium assemblages. The most widespread mineral in oxidising conditions is a K-rich dioctahedral iron illite, resembling celadonite in composition, but not always showing a true illite X-ray diffraction pattern (Kempe, 1974; Bass, 1976; Robinson et al., 1977; Scarfe and Smith, 1977; Andrews, 1977). This mineral replaces olivine, lines and fills vesicles, replaces interstitial glass, and eventually, under extreme alteration, replaces augite. Under reducing conditions its place is taken by a magnesium-rich trioctahedral smectite, a true saponite, which gives a good X-ray diffraction pattern characteristic of this mineral. This phase fulfils the same mineralogical functions as the illite, but with a very different chemical character. Where both phases occur in the same rock, the saponite appears usually to be earlier than the illite. The plagioclase of the basalt remains little affected during the early stages of this alteration, except for some replacement by the appropriate clay, but at more advanced stages of alteration becomes replaced by potassium feldspar, probably best described as adularia (Mellis, 1952; Matthews, 1962, 1971). Secondary pyrite is very characteristic of the reduced zones in this facies, developing as irregular cross cutting stringers.

Where basalt glass is abundant, as in the glassy rinds of lava pillows, a rather different style of alteration is found. Orange palagonite, a mineral resembling in composition the illitic mineral described above, but very poorly crystalline, forms from the glass in association with phillipsite or other low temperature zeolites and calcium carbonate. At the upper limit of the brownstone facies, the celadonitic phase becomes very rare, so that it is seldom seen in the zeolite facies rocks, and phillipsite disappears. New minerals that appear at or near the bottom of the zeolite facies are zeolites characteristic of rather higher temperatures, such as analcite and natrolite (Miyashiro, Shido and Ewing, 1971). To judge from similar, but not identical, mineral transitions in the Icelandic geothermal fields (Krismannsdottir, 1975), the temperature of the brownstone-zeolite facies boundary would be in the region of 50-100°C.

The zeolite facies has been well described by Miyashiro, Shido and Ewing (1971). As is the case with the brownstone facies, it is characterised by the presence of relict igneous phases which have only rarely been completely replaced by alteration phases. The most characteristic mineral of this facies is a magnesium-rich trioctahedral saponite or saponite-chlorite mixed layer mineral. This is in many ways similar to the mineral of the reduced assemblage of the brownstone facies, but is usually more coarse-grained, and may contain chlorite interlayers (Miyashiro, Shido and Ewing, 1971). It occurs replacing olivine and glass, and filling vesicles and cracks. It is commonly found partly replacing plagioclase, which is also often replaced by zeolites. Augite is usually unreplaced in this facies, retaining its igneous appearance clear and unclouded even when secondary minerals replace nearly all of the rest of the rock.

The upper limit of the zeolite facies is marked by the disappearance of zeolites and saponite and the appearance of albite and chlorite. These changes happen nearly synchronously, though in some rocks near the border the assemblages albite + saponite or zeolites + chlorite may be found (R.A.Heath, personal communication). Again the Icelandic analogy suggests temperatures of about 250-300°C for this transition (Sigvaldason, 1962; Tomasson and Kristmannsdottir, 1972). This is supported by preliminary results from Hawaii (Stone and Fan, 1978).

Nothing is yet known of the mineralogical characteristics of ultramafic rocks such as peridotites and serpentinites under brownstone or zeolite facies conditions. It is possible that particular forms of serpentine may be recognisable, or that Mg-rich clays such as saponite may be formed as in rocks of basaltic composition. Preliminary investigations (Prichard, personal communication) suggest that both types of mineral may be present, though their relative conditions of formation are not yet clear.

With the incoming of greenschist facies mineralogy characterised by the assemblage albite + chlorite ± actinolite ± epidote + sphene, the extent of alteration of the primary assemblage increases. Usually less than half of the primary phases remain, and complete replacement by the secondary phases is not uncommon. In contrast with the situation in the brownstone and zeolite facies, here it is possible to conclude from the close spatial juxtaposition and wide development of secondary phases that they represent some sort of equilibrium assemblage. Augite is the phase most commonly appearing as a relict, and sometimes it occurs as water-clear crystals which have apparently been completely unaffected while the rest of the rock is extensively replaced. It is often, however, closely associated with bluish green needles of actinolite which may replace it completely. Numerous veins are characteristic of greenschist facies material, containing usually some of the secondary minerals present in the rock, often in association with quartz. Within the rock there is a close association of individual secondary phases with particular primary phases, so that albite replaces calcic plagioclase, which may also be replaced by chlorite. Chlorite may also replace olivine and glass, and fill vesicles. This relationship is described in more detail by Cann (1969).

The upper limit of the greenschist facies is marked by the disappearance of albite, chlorite and actinolite, and the appearance of green aluminous hornblende associated with calcic plagioclase. Miyashiro, Shido and Ewing (1971) suggested that the assemblage actinolite + calcic plagioclase might form an intermediate assemblage in this complex change, on the basis of some of their observations from the Atlantis collection, and this has been supported by observations of Heath (personal communication) on Palmer Ridge metamorphics (Cann, 1971).

The amphibolite facies, which succeeds the greenschist facies, also shows well developed igneous textures, but instead of those of the basalts and dolerites of the preceding three facies, these are of dolerites and gabbros, demonstrating its development deeper in the ocean crust, where lavas no longer occur. Even dolerites are rare in the amphibolite facies, suggesting that the full development of the facies begins near the level at which the doleritic part of the ocean crust gives way to gabbros. The mineralogy of this facies is very simple, consisting of hornblende + calcic plagioclase + titanomagnetite ± epidote. The plagioclase is frequently zoned in a style exactly analogous with the plagioclase in the corresponding unmetamorphosed rocks, suggesting that its composition has been little affected by the metamorphism in general, though the plagioclase must have contributed some Al to the hornblende. Ocean floor gabbros sometimes show the assemblage hornblende + calcic plagioclase as a clearly igneous product (Prinz et al., 1976), though when the assemblage is clearly igneous, with euhedral hornblende associated with euhedral plagioclase, the hornblende is chestnut brown and titaniferous. However, amphibolite facies assemblages clearly persist in this environment into the igneous regime, where they are perhaps best called diorites. Every gradation is seen within such rocks from hornblendes that are clearly primary, through varieties that are pale brown or brownish green that are probably late magmatic, to green hornblendes with textural relationships that show them clearly to be secondary. Such a gradational relationship demonstrates clearly that metamorphism has followed very closely upon the crystallisation of the gabbros from the melt, and hence that it takes place very early in the history of the ocean crust.

Many metamorphosed gabbros appear to have been through no such period of primary crystallisation of hornblende. In these, clearly secondary green hornblende develops at the expense of primary anhydrous phases, producing a uralitic product, where primary plagioclase survives relatively unaffected and ferromagnesian minerals are replaced entirely by green hornblende (Cann, 1971).

The most important metamorphic event to affect the ultramafic rocks is serpentinisation. Most serpentinisation probably occurs under greenschist facies conditions, since the upper temperature limit for the formation of serpentine is about 500°C (Kitahara et al., 1966; Scarfe and Wyllie, 1967). In ocean floor serpentinites, lizardite is the first polymorph to form, and is followed by chrysotile at a later stage, after serpentinisation is essentially complete (Prichard, personal communication). Serpentinisation seems to take place with a considerable increase of volume so that low density serpentinite can intrude upwards through the ocean crust as diapirs, often bringing with them fragments of the lower oceanic crust (Cann and Funnell, 1967; Barrett and Aumento, 1970). Other mineral phases are known from altered peridotites, including carbonates, talc, tremolite and brucite (Cann, 1971; Aumento and Loubat, 1971), but the controls on their appearance have yet to be precisely established.

Facies nomenclature

Objections have been raised to the continued use of the terms "greenschist" and "amphibolite" to refer to the products of ocean floor metamorphism, since the greenschists are not schistose, nor the amphibolites the typical amphibolites of orogenic belts. In the case of the greenschists, the mineral assemblage is exactly the same, as far as can be determined, as that of the schistose greenschists, and it would certainly be difficult to find a means of distinguishing the two on mineralogical grounds. Such a distinction would seem to be essential if a new facies name is to be used. It could be argued that the distinction could be made on textural grounds, reserving the term "greenschist" for schistose materials. This would, however, lead to the objectionable result that, in orogenic environments, the sheared margins of massive metabasalt units would be assigned to the greenschist facies, while adjacent unsheared material, metamorphosed under identical conditions, would receive a different facies name. I would argue that the problem here is a psychological one, not a scientific one, and that petrologists should reconcile themselves to the proper application of facies terms on the basis of mineral assemblages as indicating conditions of metamorphism (Miyashiro, 1973, p. 67), and refrain from worrying about implications of names which are merely the result of historical accident.

The amphibolites are a slightly different case. There is a definite, though not very useful, mineralogical difference involved, and there is one alternative name of respectable pedigree available (with another possibility, too). The mineralogical distinction from conventional amphibolites is that almandine garnet is not seen in ocean floor amphibolite facies rocks. The difficulty of applying this distinction is of knowing how hard you have to look before saying that almandine garnet is absent from a particular suite of rocks, especially as garnet is not present in all normal amphibolite facies metabasites. This distinction is one of those (the others involve pelitic compositions) by which a hornblende hornfels facies has been distinguished from the amphibolite facies (Turner and Verhoogen, 1960, p. 511). However, the use of this facies term has never been very widespread, possibly because rocks belonging to it are not common, but most probably because of the difficulty of making the distinction clearly and reliably. In addition, the use of a facies name which includes the term "hornfels" raises just the problem that we are trying to avoid, since if the ocean floor rocks are not amphibolites, neither are they hornfelses.

The other possibility is to use some term such as the adjective "uralitic" (or the Scottish equivalent noun "epidiorite") for metamorphic rocks with hornblende-plagioclase mineralogy and igneous texture. But this is not a facies term, and runs into the same problems as were considered in connexion with the greenschist facies, above. I conclude, therefore, that there are no good grounds for changing the application of facies names to ocean floor rocks, and that "greenschist" and "amphibolite" should continue to be used. However, petrologists should be careful, if there is any chance of confusion, to qualify their descriptions with rock name terms such as spilite (for the greenschist facies), or epidiorite (or the use of "uralite" for the hornblende) for the amphibolite facies, so as to leave no room for confusion.

Congruence and element fluxes

As has already been indicated in the descriptions of individual facies, the primary mineralogy exerts a very strong, perhaps dominant, effect on the occurrence and distribution of the secondary minerals, and hence on the bulk chemical composition of the rock. Such an effect has been appropriately called "congruence" by Amstutz (1968). The way in which congruence affects mineral relationships in the greenschist facies has been described by Cann (1969). It leads to the replacement of plagioclase by albite, olivine by chlorite, basalt glass by chlorite, titanomagnetite by iron ore and augite by actinolite. The result is a clear change in composition from basaltic to spilitic, with a decrease in Ca, Sr and K, and an increase in Na and perhaps Si. The mobility of elements implied by the effects of congruence has been used by several authors to calculate elemental fluxes into and out of the ocean crust, and may, as in the case of magnesium, make up a source or sink that is significant on a global scale (Hart, 1970; Humphris and Thompson, 1978; Wolery and Sleep, 1976). From this point of view congruence has an important role to play, but interpretation is not always as simple as some authors would wish.

Fig. 1 gives a summary of the different congruent relationships for primary basaltic minerals in the different metamorphic facies. This is rather simplified by comparison with the real world, but it conveys its essential features. Note that there is not a one to one correspondence between primary and secondary minerals. Some primary phases may be replaced under different circumstances by more than one secondary phase, and more than one primary phase in a particular rock may alter to a single secondary phase, so that, for example, a basalt may become largely chloritised in the greenschist facies. The composition of the resulting material may depend critically on the relative proportions of the different primary phases present, and particularly on the proportion of basaltic glass present (Cann, 1969). Glassy pillow rinds and interstitial hyaloclastite may end up with a very different composition from holocrystalline pillow cores, and the overall composition of a pile of pillows may depend critically on the relative proportion of glass in the pile as a whole. It is not, therefore, easy to make deductions about the magnitude of elemental fluxes in particular cases from the analysis of a few samples unless these samples are truly representative, and unless the results are weighted by the relative proportions of the different materials in the original rock body.

When elemental fluxes are sought for a typical piece of oceanic crust, the problem becomes more complex. Now, in addition to the requirements for a particular body of rock, some estimate is required of the relative proportions of the different facies present in a typical piece of ocean crust. That this is crucial can be seen from Fig. 1. Brownstone facies alteration leads typically to a bulk increase in K in the rock, while greenschist facies metamorphism releases K. Is the net flux of K into or out of the crust? Similarly, zeolite facies alteration usually involves uptake of Mg by the rock, but extreme brownstone facies alteration releases Mg into solution. The net flux can be changed by altering the proportion of the different facies in the crust.

A few years ago, it seemed that this question was not too difficult to answer. The relative proportions of rocks of different facies in dredge hauls could be estimated, a correction applied for the chances of recovering material from progressively deeper in the oceanic crust, and the two combined to give an answer that was not very precise, but probably good enough for its purpose. Such a calculation was not made, probably because a very similar answer was derived at the same time from ophiolite complexes, in particular from the Troodos complex (Gass and Smewing, 1973, 1978; Stern et al., 1976). Here, detailed field work led to the view of metamorphic facies boundaries lying more or less parallel to the original ocean floor, with the brownstone facies (not specifically recognised as separate) forming a layer a few hundred metres thick,

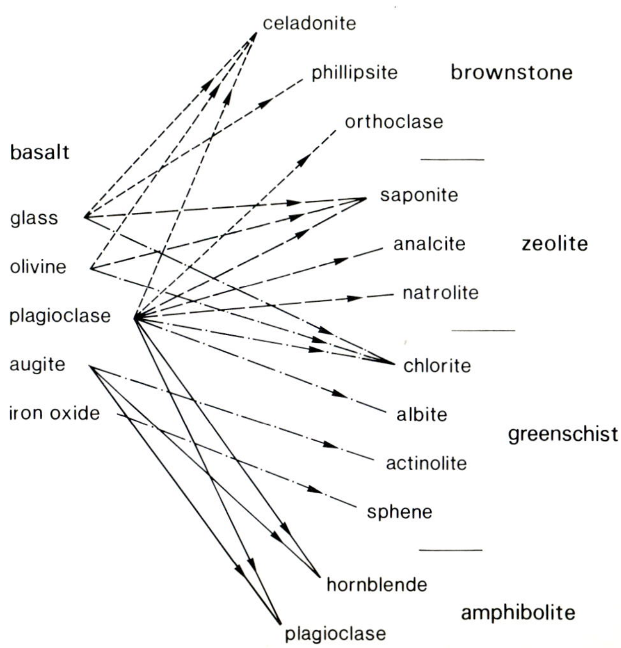

Fig. 1. Congruent mineral reactions in different facies of alteration in the ocean crust, shown schematically. The different reactions in different facies lead to considerable variation in the elemental flux from facies to facies.

the zeolite facies extending nearly to a kilometre, and the greenschist facies forming a layer about one kilometre thick beneath. The degree of alteration is more or less uniform, and extensive. Such a pattern seems characteristic of ophiolites, where they are unaffected by later metamorphism and undeformed, so that it can be recognised today.

However, drilling in the ocean crust has given a very different picture. Here, some holes have now reached depths of several hundred metres into the crust, such as at site 332 on leg 37 (Aumento and Melson, 1977), site 395 on leg 45 (Rabinowitz et al., 1976) and sites 417 and 418 on legs 51-53 (Donnelly et al., 1977; Flower et al., 1977). In all of these holes, the extent of alteration has been very small. Fresh basaltic glass persists, even though not in great amounts, to the bottom of the holes, and the proportion of primary phases is rarely less than 50%, except where brownstone facies alteration is highly developed in parts of the core from legs 51-53. No sign is found of transition to a uniform, highly altered, greenschist facies layer which, if the ophiolite model is correct, must lie only a short distance below the bottoms of the holes. This evidence gives a very different picture of the proportion of the different metamorphic facies that make up the ocean crust, and, if it proved characteristic of the crust as a whole, would have a drastic effect on the calculation of elemental fluxes. A useful understanding of the problem would require some reconciliation of the evidence from dredging and from drilling, and an integration of this with that from ophiolites. The last section of this paper attempts the first of these, but not the second.

One final point must be made in relation to the calculation of elemental fluxes. The congruent replacement reactions summarised in Fig. 1 are classical metasomatic reactions, and it has long been commonplace in metasomatic calculations that knowledge of the composition of the initial and final states is not enough in itself to calculate the elemental flux into or out of the system. Some third piece of knowledge is necessary, usually expressed as the volume of the initial system relative to the final system (Turner and Verhoogen, 1960, p. 563) or that one or more components has remained inert during the transformation(Loughnan, 1969, p. 51). Thus, if quartz (SiO_2) is transformed to wollastonite ($CaSiO_3$), Ca has certainly entered the system, but whether Si has come in, remained constant, or left the system is not contained in the bare statement that one phase has transformed to the other.

One common way of making the necessary bridge is to assume that volume remains constant during the transformation. Certainly when primary textures are well preserved, as with ocean floor metamorphics, then the assumption seems well justified. A classic application of this approach is seen in the serpentinisation of peridotite (Turner and Verhoogen, 1960, p. 318), where the constant volume assumption is justified by arguments from the preservation of primary textures. However, constant volume serpentinisation of peridotite requires the removal of large quantities of relatively insoluble Mg and Si from the system, and thus quantities of water orders of magnitude greater than that necessary for the formation of the serpentine minerals themselves. If about 40% expansion in volume is allowed, then the amount of element transport is much reduced, and the result seems altogether more reasonable in terms of water-rock geochemistry. Preservation of texture alone, then, is not enough to justify a constant volume assumption without further evidence.

A more useful approach, especially for the low temperature reactions of the brownstone and zeolite facies, is to assume that some elements have been virtually immobile during the transformation. Cann (1970), Pearce and Cann (1973), Wood et al. (1976) and Humphris and Thompson (1978) have shown, by looking at altered and unaltered rocks, that elements with a high formal ionic charge, such as Ti, Zr, P, Nb, Ta and Hf, are least mobile in such an alteration situation. This is confirmed by analyses of thermal waters, which are characteristically very low in such elements, and also in Al (Rittenhouse et al., 1969). Such waters must have formed in association with hot, altering rock, and their poverty in these elements suggests a low solubility in hydrothermal fluids at low temperatures.

Adopting this approach does change radically ones view of elemental fluxes, as is shown by comparing the results of Andrews (1977), who adopts a constant volume approach to the formation of palagonite, with those of Pritchard et al. (1978), using very similar data, by approaching the problem using immobile elements.

Modelling metamorphic processes

In this last section I want to construct a model of metamorphic processes within the oceanic crust at mid-ocean ridge crests. I shall first discuss the constraints on any such model, and then put them together to form a single pattern.

The fundamental constraint for such a model must be the relationship between the occurrence of metamorphism and the cooling of the upper oceanic lithosphere by penetrative hydrothermal convection (Elder, 1965; Lister, 1972). Though some of the water for the metamorphic hydration reactions may come from magmatic water, the amount available from that source is strictly limited (c. 0.35%; Moore and Schilling, 1973; Bryan and Moore, 1977), and cannot supply the several percent needed by metamorphic rocks where the degree of alteration is high. Most of this must come from the ocean, some, perhaps from water trapped between basaltic pillows during the initial eruption, but more,especially that reaching the lower crust and upper mantle, from hydrothermal circulation.

The pattern of such circulation will be controlled by the thermal gradient before it

Table 1. Mineral Assemblage Facies In Outline

Facies	Basalt	Peridotite
Brownstone	celadonite + phillipsite + palagonite + saponite	?
Zeolite	saponite + mixed layers + analcite + natrolite	?
Greenschist	chlorite + albite + actinolite + epidote + sphene	lizardite + chrysotile + magnetite
Amphibolite	hornblende + plagioclase + iron oxide	tremolite + olivine + enstatite
Gabbro	augite + plagioclase + olivine + hypersthene + iron oxide	olivine + enstatite + diopside + chromite

starts, the structure of permeability within the oceanic crust, and the effective thermal diffusivity of the less permeable parts of the crust. The convection will be most vigorous along zones of brecciation, such as faults, where the permeability will be orders of magnitude greater than through the regions between faults. Heat will be transferred to the fault planes by subsidiary more sluggish convection cells or by conduction. Not only does the ocean crust have a very inhomogeneous structure of permeability, but the structure is also very anisotropic. The most abundant ridge-crest faults are parallel to the ridge crusts, with a spacing of about 1 km (Searle and Laughton, 1977; Klitgord and Mudie, 1974; Macdonald and Luyendyk, 1977), so that the main convection should be in thin, vertical zones paralleling ridge crests. Transform faults are less abundant, occurring every few tens of kilometres, and should contribute relatively little to the total pattern of hydrothermal circulation.

Lister (1972) has pointed out that hydrothermal convection of this kind is likely to be so efficient at removing heat that it will probably be episodic, with short episodes of active convection alternating with longer episodes of repose while the thermal gradient necessary to restart convection is restored. Another factor affecting the occurrence of convection in time is that permeability changes with time. Thus, in very young crust, faults have yet to develop, and, when faults have been inactive for some time, continued convection will result in the reduction of permeability by the precipitation of materials from solution in the voids of the fault planes. Convection will thus be most active when active faulting is taking place, renewing the highly permeable zones as fast as they are clogged. Such an effect is clearly seen in the hydrothermal veins of the continents (see for example Park and McDiarmid, 1975, p. 122).

Further constraints on metamorphic models come from estimates of water-rock ratios based on isotopic measurements (Heaton and Sheppard, 1977; Spooner et al., 1974, 1977; Muehlenbachs and Clayton, 1972) and hydrothermal experiments (Bischoff and Dickson, 1975; Hajash, 1975). There has been some problem in understanding the meaning of the idea of water-rock ratio in this context, and the confusion seems an essential part of fitting a simple concept to the complexity of a real world. The only application that seems to be reproducible and meaningful is where the ratio refers to the bulk water-rock ratio in a whole geothermal system, averaged over all of it. The changes in formulation suggested by Spooner and Fyfe (1973), while giving a better feel for the quantity of water seen by parts of the system, has the defect of being dependent on the unit of length used for measurement, since as formulated it is not a dimensionless ratio. The results of both isotopic and experimental approaches suggest bulk water-rock ratios of about 1-10, but with considerable variation, and with much larger effective ratios in the more permeable parts of the system.

Elder (1965) introduced another distinction, between open hydrothermal systems (single pass systems), which draw in cold sea water, and vent hot hydrothermal fluids into the ocean, and sealed systems (multi-pass systems) where circulation is sealed beneath a sedimentary cover, and the same water circulates repeatedly. Sealed hydrothermal fluids are likely to be better agents of metamorphism than open system fluids, primarily because the mean temperature of the circulating system is higher, but also because of the lower water-rock ratio and the more reducing conditions.

Observations on ophiolites provide further constraints, though this evidence is not easy to interpret, because of the uncertainty of how far ophiolites represent oceanic crust. However, on Troodos, hydrothermal circulation produced at times massive sulphide deposits above the outwelling limbs of the systems. These deposits were formed close enough to the axis of spreading to be covered by later lavas flows (Gass and Smewing, 1973; Spooner, 1977; Kidd, 1977), probably within 1-2 km, and thus very early in the life of the crust. It is not clear, as I said earlier, how far the metamorphic structure of ophiolites can be extrapolated to the ocean crust, especially in view of the drilling results, and this information will not be used as a constraint.

Some of the evidence from the ocean crust itself is direct, other indirect. Dredging has recovered metamorphic rock from the walls of the median valley in the Mid-Atlantic Ridge (Aumento et al., 1971; Melson et al., 1968; Bonatti et al., 1975), showing that metamorphism occurs very soon after the formation of the crust, as suggested by the evidence from the ophiolitic sulphide deposits. Confirmation of this comes from the very high geothermal gradient at the time of metamorphism shown by the telescoping of metamorphic facies covering several hundred degrees within the upper half of the oceanic crust (Cann, 1968, 1974), the gradational transition from magmatic to metamorphic mineralogy in the ocean floor gabbros (see above), and the observation of hot water emerging very close to the ridge crest at the Galapagos spreading centre (Weiss et al., 1977).

The connexion of the ridge-crest-parallel faults with metamorphism suggested circumstantially in the discussion of permeability above, receives strong support from observations in the mounds area just south of the Galapagos Spreading Centre. The mounds are composed of hydrothermally deposited manganese oxides and clay minerals (Corliss et al., 1978) and clearly represent the vents of hydrothermal systems, perhaps extinct, perhaps intermittently active. They form clusters, aligned along faults of this kind, identified by deep-tow survey (Klitgord and Mudie, 1974), and the hydrothermal circulation that gives rise to them must be associated with the faults.

Very similar faults have been extensively studied by Whitmarsh and Laughton (1976) and Searle and Laughton (1977) in the Atlantic at several places using GLORIA, a long-range side-scan sonar. Their data suggest that the faults initiate within the median valley, about 1 km from the median line, are active during the rift valley faulting phase, but become inactive after the crust passes the upper rim of the median valley. The mean spacing between these faults is about 1.5 km and individual faults may stretch for as far as 20 km parallel to the ridge crest. A very similar spacing and continuity can be seen in maps of the Galapagos Spreading Centre (Klitgord and Mudie, 1974). How such faults form is controversial, but this is not important here: they appear to be a constant feature of the spreading process, whether the crust becomes rifted eventually or not.

With this basis, it is possible to discuss the metamorphic process relatively exactly. As the ocean crust forms, by some such injection -effusion mechanism as proposed by Kidd (1977), it is at first unmetamorphosed, and hot. Beneath the crust as it forms is a source of magma - perhaps a small isolated magma chamber at slow spreading rates, perhaps a continuous one at higher rates. After about 1 km of spreading the crust fractures with the first ridge-crest-parallel fault. This fracturing terminates downwards in the plastic, nearly molten cap to the magma chamber. Within the fault plane, highly permeable and very hot, intense hydrothermal circulation is initiated, cooling the crust and producing the main metamorphic effect. This is most complete near to the fault, where the availability of water is greatest, and is much less well-developed in the centres of the blocks, which are relatively impermeable and from which the heat is most easily extracted by conduction. Local water-rock ratios vary from very low at the centres of the blocks to very high near the faults in which the convection is most active.

After this initial very rapid convective overturn, there is a pause while conductive equilibrium is re-established, and during this time the brittle zone is extended into the mantle. Then, renewed convection is possible, leading to little change in the initial mineral assemblage, but to serpentinisation of the upper part of the mantle followed, very often, by diapiric intrusion of serpentinite to high levels in the crust, especially along faults (Barrett and Aumento, 1970; Cann and Funnell, 1967). Further episodes of convection are milder. They are active in transporting heat out of the crust, and extend the low temperature metamorphic effects into the centres of the less permeable blocks, producing brownstone or zeolite facies alteration there. This low temperature alteration is a slow business, however, and does not become complete for tens of millions of years.

On such a model, the high temperature metamorphic imprint is imposed on the crust very rapidly early in its history, but is largely limited to the edges of crustal blocks, especially along the ridge-crest-parallel faults. Such a picture can help to reconcile the apparently contradictory results of dredging and drilling, since dredging by its nature selectively samples the edges of blocks and the more highly fractured, more permeable rocks, while drilling selectively recovers the massive flow units and less permeable horizons, losing brecciated rock as cuttings that do not enter the core barrel. This picture receives some support from the observations of the unusually complete recovery at sites 417 and 418 in 105 m.y. old Atlantic crust. Here brecciated zones had become cemented and softened, and could be recovered, and from these zones Pritchard (in press) has reported the presence of chlorite, indicating a higher temperature alteration than any seen previously on the Deep Sea Drilling Project.

References

Amstutz, G.C., Spilites and spilitic rocks, in Hess, H.H. and Poldervaart, A., Basalts, Interscience, New York, 737-754, 1968.

Andrews, A.J., Low temperature fluid alteration of oceanic layer 2 basalts, DSDP leg 37, Can. J. Earth Sci., 14, 911-926, 1977.

Aumento, F., Loncarevic, B.D.,and Ross, D.I., Hudson Geotraverse: geology of the Mid-Atlantic Ridge at 45°N, Phil. Trans R. Soc. Lond., A268, 623-650, 1971.

Aumento, F.,and Loubat, H., The Mid-Atlantic Ridge near 45°N, XVI Serpentinised ultramafic intrusions, Can. J. Earth Sci., 8, 631-663, 1971.

Aumento, F., and Melson, W.G., Initial Reports of the Deep Sea Drilling Project, 37, Washington (US Government Printing Office), 1977.

Barrett, B.D., and Aumento, F., The Mid-Atlantic Ridge near 45°N, XI, Seismic velocity, density and layering of the crust, Can. J. Earth Sci., 7, 1117-1124, 1970.

Bass, M.N., Secondary minerals in oceanic basalt, with reference to leg 34, Deep Sea Drilling Project, in Yeats, R.S., Hart, S.R. et al., Initial Reports of the Deep Sea Drilling Project, 34, Washington (US Government Printing Office), 393-432, 1976.

Bischoff, J.L., and Dickson, F.W., Sea water-basalt interaction at 200°C and 500 bars: implications for origin of sea-floor heavy-metal deposits and regulations of sea-water chemistry, Earth Planet. Sci. Lett., 25, 387-397, 1975.

Bonatti, E., Honnorez, J., Kirst, P., and Radicati, F., Metagabbros from the Mid-Atlantic Ridge at 6°N: contact-hydrothermal-dynamic metamorphism beneath the axial valley, J. Geol., 83, 61-78, 1975.

Bryan, W.B., and Moore, J.G., Compositional variations of young basalts in the Mid-Atlantic Ridge rift valley near lat. 36°49'N, Geol. Soc. Amer. Bull., 88, 556-570, 1977.

Cann, J.R., and Funnell, B.M., Palmer Ridge: a section through the upper part of the ocean crust?, Nature, 213, 661-664, 1967.

Cann, J.R., Geological processes at mid-ocean ridge crests, Geophys. J. R. astr. Soc., 15, 331-341, 1968.

Cann, J.R., Spilites from the Carlsberg Ridge, Indian Ocean, J. Petrol., 10, 1-19, 1969.

Cann, J.R., Rb, Sr, Y, Zr and Nb in some ocean floor basaltic rocks, Earth Planet. Sci. Lett., 10, 7-11, 1970.

Cann, J.R., Petrology of basement rocks from Palmer Ridge, NE Atlantic, Phil. Trans R. Soc. Lond., A 268, 605-618, 1971.

Cann, J.R., A model for oceanic crustal structure developed, Geophys. J. R. astr. Soc., 39, 169-187, 1974.

Corliss, J.B., Lyle, M., Dymond, J., and Crane, K., The chemistry of hydrothermal mounds near the Galapagos Rift, Earth Planet. Sci. Lett., 40, 12-24, 1978.

Donnelly, T.W. et al., Mid-ocean ridge in the Cretaceous, Geotimes, 22, no. 6, 21-23, 1977.

Elder, J.W., Physical processes in geothermal areas, Monogr. Ser. Amer. geophys. Union, 8, 211-239, 1965.

Flower, M.F.J. et al., Cretaceous crust sought, Geotimes, 22, no. 9, 20-22, 1977.

Gass, I.G., and Smewing, J.D., Intrusion, extrusion and metamorphism at constructive margins: evidence from the Troodos massif, Cyprus, Nature, 242, 26-29,1973.

Gass, I.G., and Smewing, J.D., Ophiolites: obducted oceanic lithosphere, in Emiliani, C., The Sea, vol. 7, 1978.

Hajash, A., Hydrothermal processes along mid-ocean ridges: an experimental investigation, Contrib. Miner. Petrol., 54, 208-226, 1975.

Hart, R., Chemical exchange between sea water and deep ocean basalts, Earth Planet. Sci. Lett., 9, 269-279, 1970.

Heaton, T.H.E., and Sheppard, S.M.F., Hydrogen and oxygen isotope evidence for sea-water-hydrothermal alteration and ore deposition, Troodos complex, Cyprus, Sp. Publ. geol. Soc. Lond., 7, 42-57, 1977.

Humphris, S.E., and Thompson, G., Hydrothermal alteration of oceanic basalts by sea water, Geochim. Cosmoch. Acta, 42, 107-125, 1978.

Humphris, S.E. and Thompson, G., Trace element mobility during hydrothermal alteration of oceanic basalts, Geochim. Cosmoch. Acta, 42, 127-136, 1978.

Kempe, D.R.C., The petrology of the basalts, leg 26, in Davies, T.A., Luyendyk, B.P. et al., Initial Reports of the Deep Sea Drilling Project, 26, US Government Printing Office,Washington, 465-504, 1974.

Kidd, R.G.W., A model for the process of formation of the upper oceanic crust, Geophys. J. R. astr. Soc., 50, 149-183, 1977.

Kitahara, S., Takenouchi, S., and Kennedy, G.C., Phase relations in the system $MgO-SiO_2-H_2O$ at high temperatures and pressures, Amer. J. Sci., 264, 223-233, 1966.

Klitgord, K., and Mudie, J., The Galapagos spreading center: a near-bottom geophysical survey, Geophys.J. R. astr. Soc., 38, 563-586, 1974.

Kristmannsdottir, H., Hydrothermal alteration of basaltic rocks in Icelandic geothermal areas, Proc. Second United Nations Symp. on the Development and Use of Geothermal Resources, 1, 441-445, 1975.

Lister, C.R.B., On the thermal balance of a mid-ocean ridge, Geophys. J. R. astr. Soc., 39, 575-535,1972.

Loughnan, F.C., Chemical weathering of the silicate minerals, American Elsevier, New York, 1969.

Macdonald, K.C.,and Luyendyk, B.P., Deep-tow studies of the structure of the Mid-Atlantic Ridge crest near lat. 37°N, Geol. Soc. Amer. Bull., 88, 621-636, 1977.

Matthews, D.H., Altered lavas from the floor of the eastern North Atlantic, Nature, 194, 368-369, 1962.

Matthews, D.H., Altered basalts from Swallow Bank, an abyssal hill in the NE Atlantic and from a nearby seamount. Phil. Trans R. Soc. Lond., A 268, 551-572, 1971.

Mellis, O., Replacement of plagioclase by orthoclase in deep-sea deposits, Nature, 169, 624, 1952.

Melson, W.G., Thompson, G., and van Andel, Tj.H., Volcanism and metamorphism in the Mid-Atlantic Ridge, 22°N latitude, J. geophys. Res., 73, 5925-5941, 1968.

Miyashiro, A., Shido, F., and Ewing, M., Metamorphism in the Mid-Atlantic Ridge near 24° and 30°N, Phil. Trans R. Soc. Lond., A 268, 589-604, 1971.

Miyashiro, A., Metamorphism and metamorphic belts, London, Allen and Unwin, 1973.

Moore, J.G., and Schilling, J.-G., Vesicles, water and sulfur in Reykjanes Ridge basalts, Contr. Miner. Petrol., 41, 105-118, 1973.

Muehlenbachs, K., and Clayton, R.N., Oxygen isotope composition of the oceanic crust and its

bearing on seawater, J. geophys. Res., 81, 4365-4369, 1972.
Park, C.F., and MacDiarmid, R.A., Ore deposits, Freeman, San Francisco, 3rd edition, 1975.
Pearce, J.A., and Cann, J.R., Tectonic setting of basic volcanic rocks determined using trace element analysis, Earth Planet. Sci. Lett., 19, 290-300, 1973.
Prinz, M., Keil, K., Green, J.A., Reid, A.M., Bonatti, E., and Honnorez, J., Ultramafic and mafic dredge samples from the equatorial Mid-Atlantic Ridge and fracture zones, J. geophys. Res., 81, 4087-4103, 1976.
Rabinowitz, P. et al., Challenger Drills on leg 45, Geotimes, 21, no. 4, 20-23, 1976.
Rittenhouse, G., Fulton, R.B., Grabowski, R.J., and Bernard, J.L., Minor elements in oil-field waters, Chem. Geol., 4, 189-209, 1969.
Robinson, P.T., Flower, M.F.J., Schmincke, H.-U., and Ohnmacht, W., Low temperature alteration of oceanic basalts, leg 37, in Aumento, F., Melson, W.G. et al., Initial Reports of the Deep Sea Drilling Project, 37, Washington(US Government Printing Office), 775-794, 1977.
Scarfe, C.M., and Wyllie, P.J., Serpentine dehydration curves and their bearing on serpentinite deformation in orogenesis, Nature, 215, 945-946, 1967.
Scarfe, C.M., and Smith, D.G.W., Secondary minerals in some basaltic rocks from DSDP leg 37, Can. J. Earth Sci., 14, 903-910,1977.
Searle, R.C., and Laughton, A.S., Sonar studies of the Mid-Atlantic Ridge and Kurchatov Fracture Zone, J. geophys. Res., 82, 5313-5328, 1977.
Sigvaldason, G.E., Epidote and related minerals in two deep geothermal drill holes, Reykjavik and Hveragerdi, Iceland, U.S. Geol. Surv. Prof. Paper, 450-E, 77-84, 1962.
Spooner, E.T.C., and Fyfe, W.S., Sub-sea-floor metamorphism: heat and mass transfer, Contr. Miner. Petrol., 42, 287-304, 1973.
Spooner, E.T.C., Beckinsale, R.D., Fyfe, W.S., and Smewing, J.D., O^{18} enriched ophiolitic metabasic rocks from E. Ligurin (Italy), Pinlos (Greece), and Troodos (Cyprus), Contr. Mineral. Petrol., 47, 41-62, 1974.
Spooner, E.T.C., Chapman, H.J., and Smewing, J.D., Strontium isotope contamination and oxidation during sea-floor hydrothermal metamorphism of the ophiolitic rocks of Troodos, Cyprus, Geochim. Cosmochim. Acta, 41, 873-890, 1977.
Spooner, E.T.C., Hydrothermal model for the origin of the ophiolitic cupriferous pyrite ores of Cyprus, Sp. Publ. geol. Soc. Lond., 7, 58-71, 1977.
Stern, C., de Wit, M.J., and Lawrence, V.R., Igneous and metamorphic processes associated with the formation of Chilean ophiolites and their implication for ocean-floor metamorphism, seismic layering and magnetism, J. geophys. Res., 81, 4370-4380, 1976.
Stone, C., and Fan, P.-F., Hydrothermal alteration of basalts from Hawaii Geothermal Project Well-A, Kilanea, Hawaii, Geology, 6, 401-404, 1978.
Tomasson, J., and Kristmansdottir, H., High temperature alteration minerals and thermal brines, Reykjanes, Iceland, Contrib. Miner. Petrol., 36, 123-127, 1972.
Turner, F.J., and Verhoogen, J., Igneous and metamorphic petrology, McGraw Hill, New York, 1960.
Weiss, R.F., Lonsdale, P.F., Lupton, J.E., Bainbridge, A.E., and Craig, H., Hydrothermal plumes in the Galapagos Rift, Nature, 267, 600-603, 1977.
Whitmarsh, R.B., and Laughton, A.S., A long-range sonar study of the Mid-Atlantic Ridge crest near 37°N (FAMOUS area) and its tectonic implications, Deep Sea Res., 23, 1005-1023, 1976.
Williams, D.L., Von Herzen, R.P., Sclater, J.G., and Anderson, R.N., The Galapagos spreading center: lithospheric cooling and hydrothermal circulation, Geophys. J. R. astr. Soc., 38, 587-608, 1974.
Wolery, T.J., and Sleep, N.H., Hydrothermal circulation and geochemical flux at mid-ocean ridges, J. Geol., 84, 249-275, 1976.
Wood, D.A., Gibson, I.L., and Thompson, R.N., Elemental mobility during zeolite facies metamorphism of the Tertiary basalts of eastern Iceland, Contrib. Miner. Petrol., 55, 241, 1976.

TECTONIC AND IGNEOUS EMPLACEMENT OF CRUST IN OCEANIC TRANSFORM ZONES

Enrico Bonatti

Lamont-Doherty Geological Observatory of Columbia University, Palisades, New York 10964

Andy Chermak and Jose Honnorez

Rosenstiel School of Marine and Atmospheric Science, University of Miami, Miami, Florida 33149

Abstract. The following questions are briefly reviewed: (1) Does tectonic emplacement of crust occur in the large oceanic fracture zones? (2) Does igneous emplacement of crust occur in the same? (3) Can fracture zones expose thick sections of oceanic crust?

On the basis of topographic, geophysical and lithological data, a "secondary tectonized domain" is defined for the large fracture zones. This domain, located on one or both sides of the "main transform domain" (deeper part of the main transform valley), consists of a zone of faulted blocks with intense vertical tectonism which might be related to convergent wrenching in the transform zone. Tectonic emplacement of crust, incorporating also slivers of mantle-derived material, occurs in the secondary tectonized domain. Basaltic rocks with alkaline affinities, found at the St. Paul, Romanche and Owen Fracture Zones may represent examples of igneous emplacement of crust in fracture zones, related mainly to divergent wrenching. The possibility that relatively undisturbed sections of oceanic crust are exposed in some fracture zones is discussed. Possible examples of such sections are given from the Romanche and the Owen fracture zones.

Introduction

Among the transform faults which intersect the ocean ridge system a distinction has been made between (a) "boundary" or "initial" transforms, i.e., transforms with long (>100 kms) offset and which constitute an important, permanent or semi-permanent boundary between lithosphere of different age, and (b) small, second generation transforms constituting transient features affecting young, hot lithosphere and probably related only to the geometry of crustal spreading (Gilliland and Meyer, 1976). Small (<50 km) type (b) offsets have been estimated to occur on the average every 50 kms or less along the Mid-Atlantic Ridge (Fox *et al.*, 1976) while the large "boundary" fracture zones are spaced at irregular intervals generally of several hundred kms or more.

Some of the sharpest topographic relief of the ocean floor is associated with the large "boundary" fracture zones. For instance, at the Romanche Fracture Zone (F.Z.) between 18° and 16°W the depth changes within a few tens of km from greater than 7 km below sea level (bottom of the main fracture valley) to less than 1 km below sea level (summit of the transverse ridge on the north side of the fracture valley), with a vertical relief of over 6 km. Thus, thick crustal sections are exposed along the large oceanic fracture zones. The following two questions pertaining to the geology of fracture zones are reviewed briefly in this paper:

1. Do the large oceanic fracture zones expose sections of "normal" oceanic crust [where by "normal" crust we mean crust generated at an accretionary plate boundary (mid ocean ridge)], thereby exposing lower crust? Our direct knowledge of the composition of the lower crust is presently very scant, since drilling has been achieved only through the top few hundred meters of the igneous crust beneath the ocean.

2. Is new crust emplaced along the large oceanic fracture zones by tectonic or igneous processes? This question is related to the first because, if processes of crustal generation are ubiquitous along fracture zones, the probability of their exposing "normal" oceanic crust is low. Moreover, if, in addition to accretionary plate boundaries, crust is generated also along fracture zones, it is of interest to understand the mechanisms and the products of such crustal generation.

Tectonic Emplacement of Crust in Oceanic Fracture Zones

The ridges commonly found parallel and adjacent to one or both sides of fracture zone valleys (transverse ridges) are clearly caused

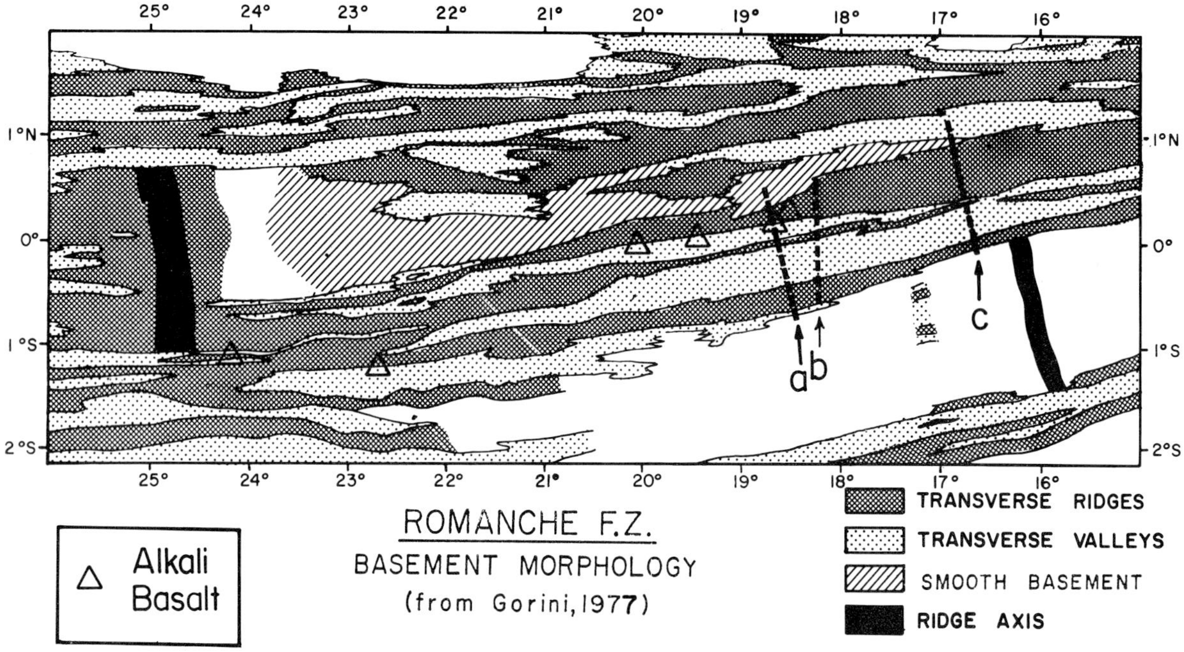

Fig. 1. Basement morphology in the Romanche F.Z. region. Location of profiles discussed in this paper is shown: a = profile of figure 4; b = profile of figure 7; c = profile of figure 2. Triangles indicate locations where basaltic rocks with alkaline affinity were recovered.

by fracture zone-related processes, i.e., either tectonic activity or igneous activity or both.

Geophysical data and detailed sampling of sections exposed across some of the large Atlantic and Indian Ocean fracture zones indicate that the fracture zone transverse ridges are generally uplifted blocks of oceanic crust and/or uplifted blocks of oceanic upper mantle.

(a) Uplifted Oceanic Crust. Two examples are shown in Figures 1, 2 and 3, one at the Romanche F.Z. (equatorial Atlantic) and one at the Owen F.Z. (western Indian Ocean).

The north-south section of Figure 2 shows locations where samples were obtained on the transverse ridge on the north side of the Romanche fracture valley at about 17°W. The summit of the transverse ridge is over 2 km above the level of "normal" crust, presumably of the same age, immediately to the north. Rocks obtained on the lower level of the ridge were described by Melson and Thompson (1970) and interpreted as part of a gabbro-ultramafic basic layered complex. These rocks are identical to what generally accepted models predict for constituents of the lower oceanic crust. Basalt, of a type normally encountered in the upper part of the oceanic crust, is the prevalent material recovered at about half-slope on the transverse ridge. Limestones were recovered on, and close to, the summit of the ridge. These limestones, described in Bonatti et al. (1977), were deposited at sea level about 5 million years before the present. It appears, therefore, that the Romanche F.Z. transverse ridge is in this area a block of crust which has been subjected to intense vertical motion, having reached sea level 5 million years ago and having subsided since at an average rate about one order of magnitude faster than that of crust of equivalent age (Bonatti et al., 1977).

A subbottom reflection is visible about 100 meters below the summit of the transverse ridge in a seismic reflection profile taken where the limestones were dredged. This reflection has been interpreted as marking the base of the limestone cap (Chermak, 1978). Similar reflections appear in profiles taken across the summit of the transverse ridge about 60 km to the east and 40 km to the west of the profile taken where the limestones were dredged. It is possible, therefore, that the vertical motions documented at the section of figure 2 have affected the transverse ridge for considerable distance.

Figure 3 shows a section across the Owen F.Z., where the transverse ridge on the western side of the sediment-filled fracture valley exposes prevalently lower crustal metagabbroic rocks overlain by upper crustal basalts. This lithology is consistent with the hypothesis that the Owen transverse ridge along this profile constitutes an uplifted block of oceanic crust.

(b) Uplifted Oceanic Mantle. In some sections across the large oceanic fracture zones the transverse ridges have been shown to expose exclusively ultramafic rocks. A spectacular example is at the Romanche F.Z. on the southern wall of the fracture valley between 18°W and

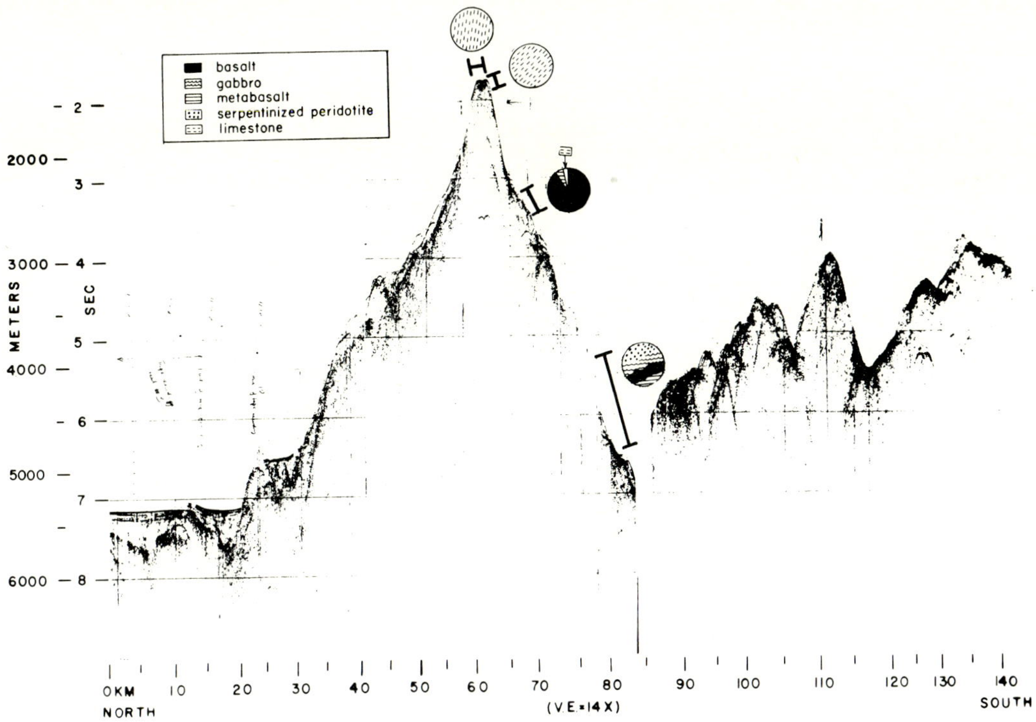

Fig. 2. Seismic reflection profile across the Romanche F.Z. at about 16°45'W (profile C in figure 1). Rock types recovered along this section are indicated.

19°W (figure 4). Here ultramafic rocks were recovered almost exclusively, suggesting they outcrop for a vertical thickness of over 4 km (Bonatti and Honnorez, 1976).

Another example is provided by the St. Peter Paul Rocks at the St. Paul F.Z., also in the equatorial Atlantic, which constitutes the subaerial summit of a prevalently ultramafic block (Melson et al., 1972).

An additional example is at the Owen F.Z., close to its intersection with the Mid-Indian (Sheba) axis of spreading (figure 5), where ultramafic rocks, mainly spinel lherzolites, have been shown to be the only material outcropping from the base to the summit of the transverse ridge for a thickness of over 2 km (Bonatti and Hamlyn, 1978).

In the three cases cited above the ultramafic massifs can be interpreted as blocks of uplifted upper mantle. The large size of these bodies makes a lower crustal origin for them unlikely because ultramafic rocks are considered to comprise only a minor constituent of the lower crust. In addition, the texture of the rocks and pressure/temperature estimates based on pyroxenes compositions indicate a mantle derivation in all three cases (Printz et al., 1976; Melson et al., 1971; Bonatti and Hamlyn, 1978). The Owen F.Z. ultramafic protrusion occurs in very young crust (only about 40 km from the axis of accretion). Its rate of ascent in the fracture zone has been estimated at being of the order of 1 mm/year (Bonatti and Hamlyn, 1978).

Igneous Emplacement of Crust in Oceanic Fracture Zones

Basalts and gabbros have been recovered abundantly from the scarps of fracture zone valleys. These rocks generally have a composition similar to that of basalts from accretionary plate boundaries (ocean ridges). They can be interpreted as having originated at the ocean ridge axis, close to its intersection with the transform zone, and having subsequently become exposed on the scarps of the fracture valley during the process of crustal spreading at the edge of the transform zone. The northern side of the Vema fracture valley (Atlantic at 11°N) provides an example of this situation: basalts sampled along the fracture valley scarp at various distances from the axis of spreading show some indication of increase in age with distance from the axis of spreading, and are chemically and mineralogically very similar to samples collected from the axis of spreading, except that they tend to be more altered (Bonatti and Honnorez, 1976; Bonatti et al., 1977).

However, at several sites along some of the

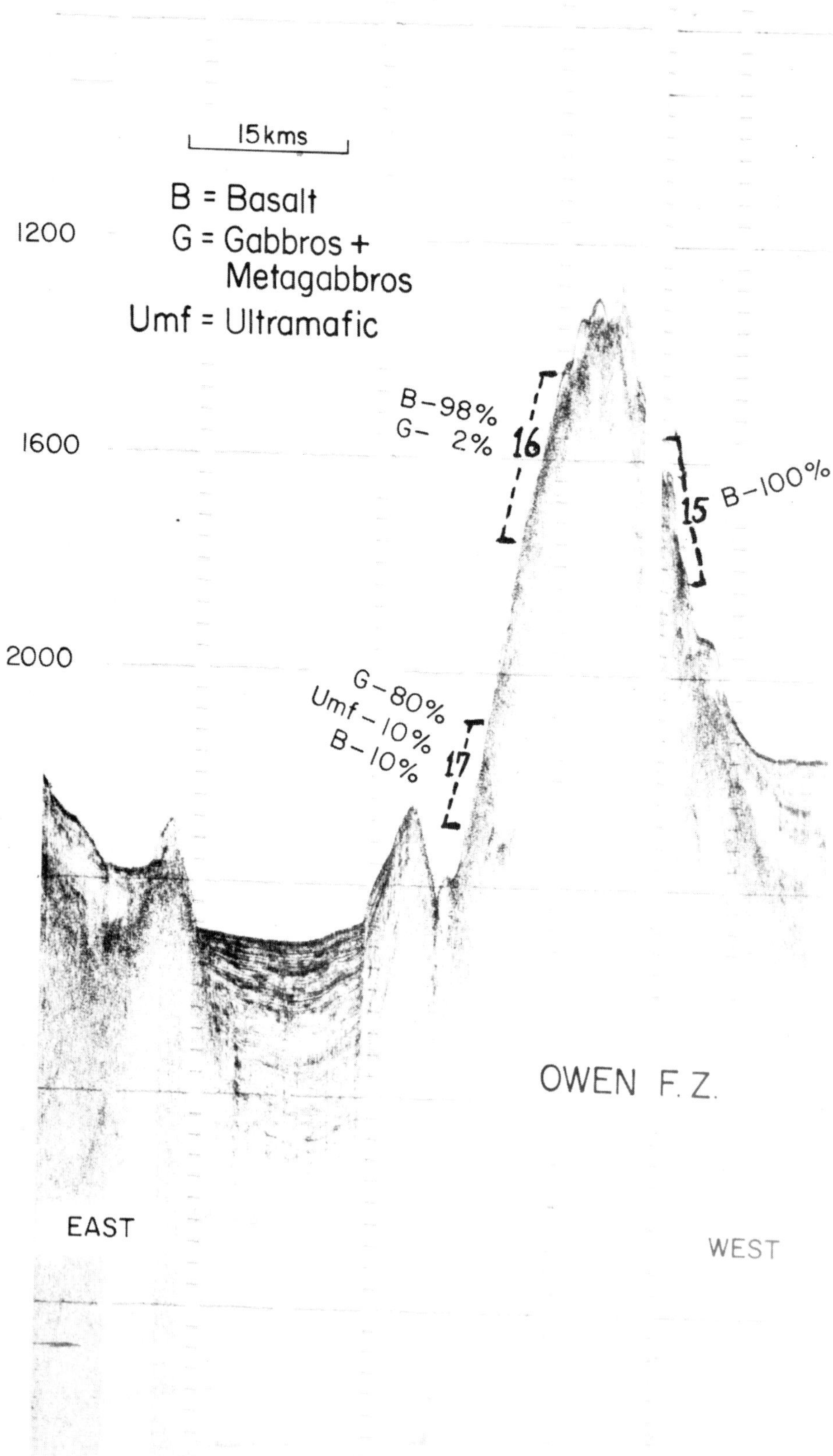
15 kms
B = Basalt
G = Gabbros + Metagabbros
Umf = Ultramafic
1200
1600
2000
B-98%
G- 2%
16
15
B-100%
G-80%
Umf-10%
B-10%
17
OWEN F. Z.
EAST
WEST

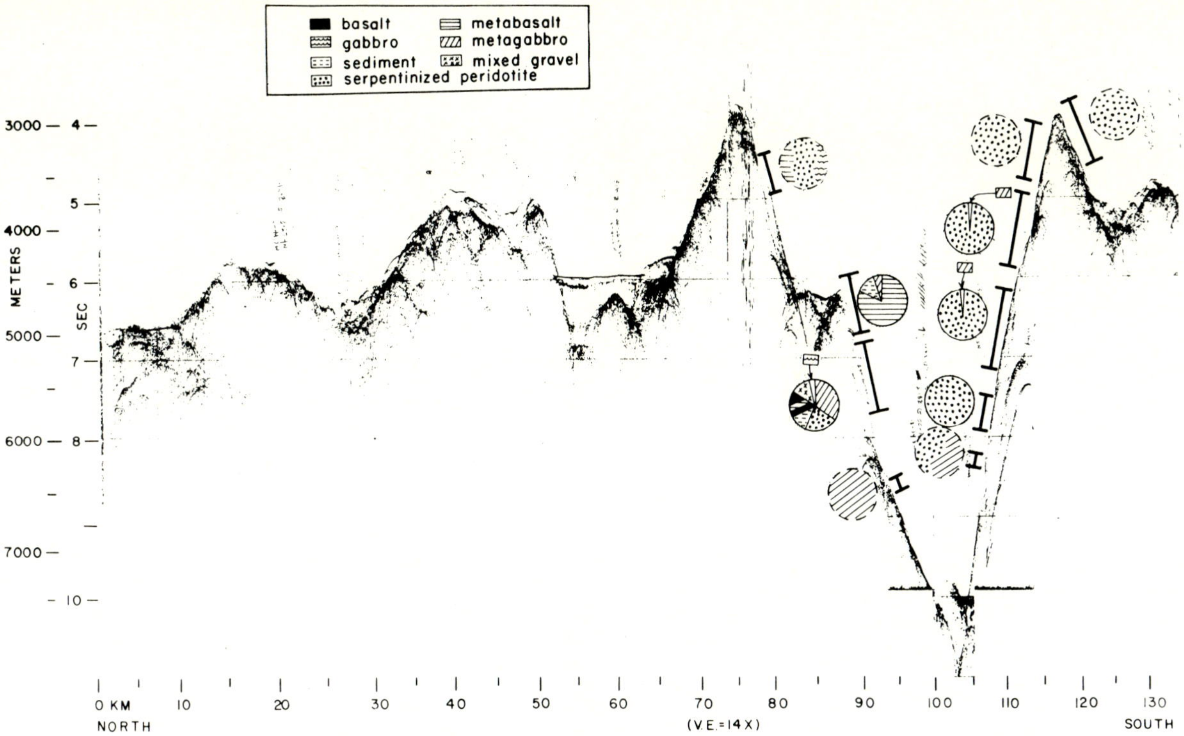

Fig. 4. Seismic reflection profile across the Romanche F.Z. at about 18°45'W (profile a in figure 1). Rock types recovered along this section are indicated.

large oceanic fracture zones ignoeus rocks have been recovered with a different petrochemistry from that of basalts from oceanic accretionary plate boundaries. Basalt with alkali affinities and with a light rare earth elements (REE) enriched pattern was collected at the St. Paul F.Z., close to the St. Peter-Paul ultramafic massif (Melson et al., 1967; Frey, 1970). An alkali gabbro with modal nepheline was described from the Romanche F.Z. (Honnorez and Bonatti, 1970).

Similar basalts with alkaline affinities have been recovered at several sites in the Romanche F.Z. (figure 1) and the Owen F.Z. They are characterized by: (a) a normative composition which is displaced into or close to the nepheline-normative field relative to the population of ridge-axis basalts which are olivine tholeiites; (b) a clinopyroxene composition different from that of the population of ridge-axis basalts; for instance, a higher Ti/Al ratio (figure 6); (c) a concentration of several lithophile elements (such as Ba, Sr and Zr) substantially higher than in ridge-axis basalts; (d) a flat or light REE enriched pattern (chondrite-normalized) in contrast to ridge axis basalts which generally display a light REE depleted pattern.

These data, as yet unpublished, suggest that a magma source, different from that which produces oceanic crust at accretionary plate boundaries, is active in the large oceanic transform zones. Thus, igneous emplacement of crust probably does occur in the large fracture zones, even though the volume of crust produced by this process is minor relative to that produced along accretionary boundaries.

Structure of Oceanic Fracture Zones

Several authors have pointed out that the large oceanic transforms are more complex than just two crustal blocks sliding passively by each other. Dewey (1975) stressed the fact that the large transform faults constitute major plate boundaries, where compressional and tensional stresses of the upper part of the lithosphere may be relieved. Bonatti (1978)

Fig. 3. Seismic reflection profile across the Owen F.Z. (profile b). Location of profile is shown in figure 5. Rock types recovered along this section are indicated.

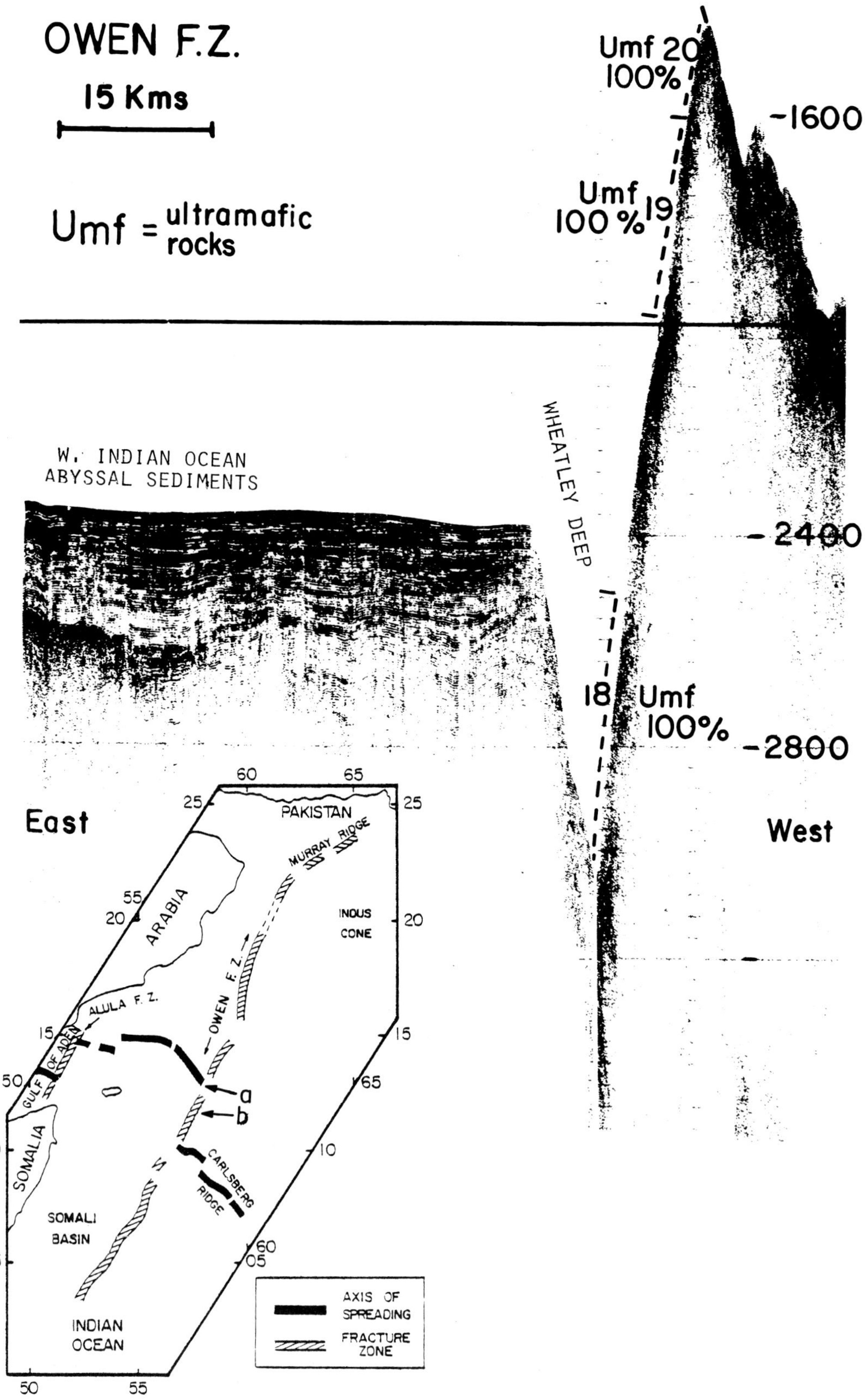

OWEN F.Z.
15 Kms
Umf = ultramafic rocks
Umf 20 100%
Umf 19 100%
Umf 18 100%
-1600
-2400
-2800
W. INDIAN OCEAN ABYSSAL SEDIMENTS
WHEATLEY DEEP
East
West
PAKISTAN
MURRAY RIDGE
ARABIA
INDUS CONE
OWEN F.Z.
ALULA F.Z.
GULF OF ADEN
SOMALIA
SOMALI BASIN
CARLSBERG RIDGE
INDIAN OCEAN
a
b
AXIS OF SPREADING
FRACTURE ZONE

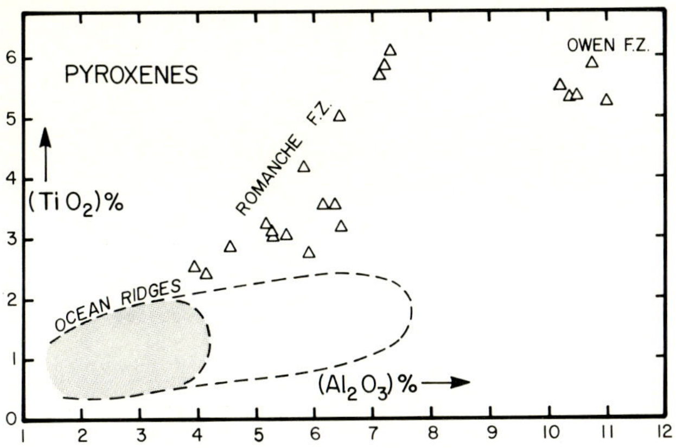

Fig. 6. Al_2O_3/TiO_2 ratio in clinopyroxene crystals from basaltic rocks of alkaline affinity from the Romanche F.Z. (see figure 1) and from the Owen F.Z. Shown for comparison is the field for pyroxenes from basalts of oceanic axial zones of spreading and DSDP sites, the bulk of which falls in the shaded area.

discussed the occurrence of anomalous topographic highs (transverse ridges) along many of the large oceanic fracture zones, and the factors determining intense vertical tectonism on these structures. Recent detailed topographic and seismic reflection surveys of large fracture zones such as the Charlie-Gibbs F.Z. (Searle, 1978) and Romanche F.Z. (Chermak, 1978) have shown that these features have a complicated structure with a highly tectonized zone extending for considerable distance on one or both sides of the main transform boundary.

The detailed topographic coverage made possible by the "Gloria" side-scan sonar system allowed Searle (1978) to detect, among other features, sets of oblique faults, striking about 45° from the transform direction, on the margins of the Charlie-Gibbs main transform fault boundary. Searle (1978) pointed out the similarity of these oblique structures to normal faults and tension fractures commonly associated with the early stages of crustal shearing (Wilcox et al., 1973). In fact, Searle (1978) stressed that the topography and structural framework of a long-offset oceanic fracture zone such as the Charlie-Gibbs F.Z. are similar to those of large continental strike-slip fault zones.

Chermak (1978), on the basis of topographic data and of conventional seismic reflection data, also stressed the similarity of the eastern portion of the Romanche F.Z. to large continental strike-slip shear zones such as the San Andreas fault system. Chermak has shown that the transverse ridge on the north side of the Romanche F.Z. main transform fault boundary (corresponding to the deepest part of the Romanche F.Z. valley) has a complicated structure, with a complex system of faults, some subparallel to the main transform fault direction, some trending obliquely to that direction. This system of faults includes high-angle normal faulting on the summit and southern, internal (i.e., facing toward the deep valley) flank of the transverse ridge, reverse and thrust faulting on its northern, external flank. Both Searle (1978) and Chermak (1978) interpret the complicated structures associated with the large oceanic fracture zones in terms of the concepts of "wrench tectonics" as derived from laboratory experimental models and from observation of large continental shear zones (Moody and Hill, 1956; Wilcox et al., 1963). In their interpretation, the fracture zone transverse ridges consist of a mosaic of faulted blocks created in response to compressional and tensional stresses resulting from the shear motion at the transform boundary. A model on the evolution of the Romanche F.Z. based on the concepts of wrench tectonics has been discussed by Chermak (1978).

The strong vertical tectonic motions documented for fracture zones (see Bonatti et al., 1977) were ascribed by Bonatti (1978) to several factors, the principal being compressional and tensional stresses related mainly to small changes in the spreading direction of the crust adjacent to the transform boundary. This interpretation can be combined with the "wrench tectonics" interpretation of Chermak (1978). In fact, Wilcox et al. (1973) and Emmons (1969) discuss cases of convergent and divergent wrenching on the basis both of clay model experiments and field occurrences. Convergence or divergence can develop locally within a wrench fault where the direction of motion of adjacent blocks is parallel on a regional scale; or they can occur due to non-parallel displacements of adjacent crustal blocks on a regional scale. Both cases are probably common in large oceanic transforms, the latter case resulting from small changes in spreading direction. Convergent wrenching on whatever scale enhances compressive structures, including folding, reverse faulting and thrusting, while divergent wrenching enhances tensional structures, mainly normal faults (Wilcox et al., 1973). Experiments on clay models suggest that even with a small, 2° convergence, important secondary structures are created (Wilcox et al., 1973). Thus, convergent wrenching is probably most effective in causing vertical motions related to the formation of fracture zone transverse ridges.

These considerations lead to a model of the

Fig. 5. Seismic reflection profile across the Owen F.Z. (profile a). Rock types recovered along this section are indicated. Profile b is shown in figure 3.

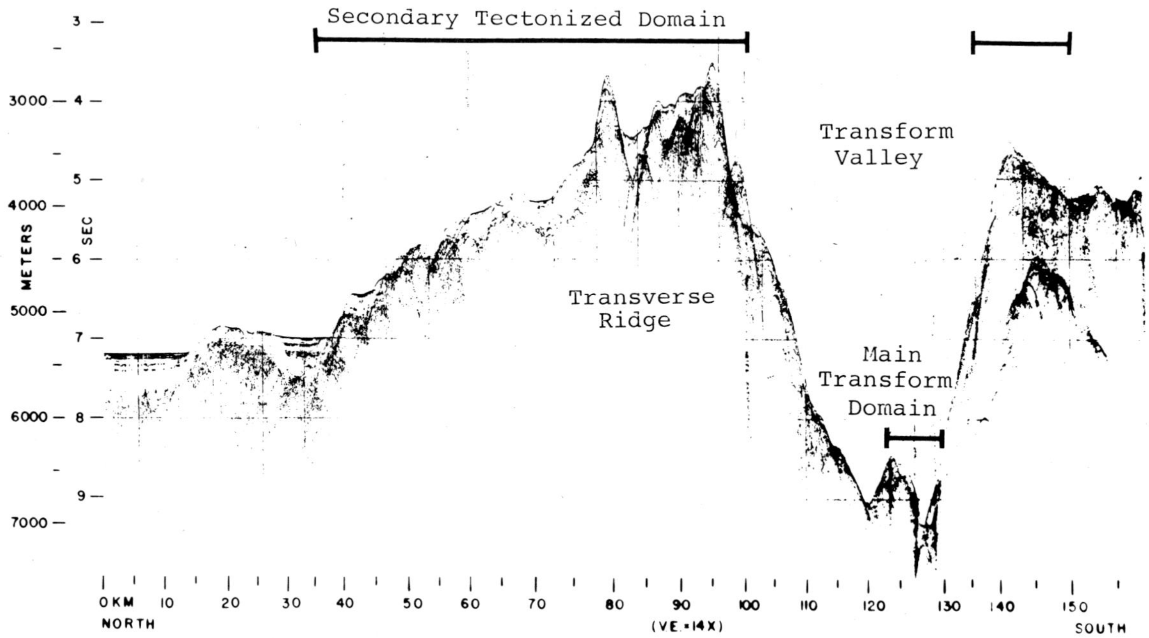

Fig. 7. Structural scheme of large oceanic fracture zones, based on a section across the Romanche F.Z. at about 18°10'W (section b in figure 1). For explanation see text.

structure of the large oceanic fracture zones, as shown in figure 7. The main strike-slip fault ("main transform domain") corresponds to the axial, deeper zone of the main fracture valley. On one or both sides of the main fracture valley a "secondary tectonized domain" exists, corresponding generally to the transverse ridges, and consisting of a mosaic of faulted blocks subjected not only to strike-slip motions but to considerable vertical motions. The width of the secondary tectonized domain appears to be, very roughly, related to the age of the plate, i.e., its maximum width in a given transform zone is roughly proportional to the length of the transform offset. However, in some cases the secondary tectonized domain appears to be well developed in very young crust (i.e., see section of figure 4 at the Owen F.Z.). In other cases the secondary tectonized domain is well developed on one side of the main transform valley (i.e., south side of the Vema F.Z.) but is absent on the other side (i.e., north side of the Vema F.Z.).

Conclusions

On the basis of the considerations of the structure of oceanic fracture zones outlined in the previous section, and of the lithology of samples recovered from them, some partial answers to the questions posed in the introduction of this paper can be attempted.

Tectonic Emplacement of Crust in Fracture Zones

The "secondary tectonized domain" is interpreted by us essentially as produced by tectonic emplacement of crust (incorporating also upper mantle blocks), involving considerable vertical motions. Compression related to convergent wrenching in the transform zone is probably a major cause of this vertical tectonism. The reasons for mantle-derived ultramafics to frequently reach high levels in this domain relative to other rock types have been discussed by Bonatti and Honnorez (1976), Bonatti (1976), Francheteau et al. (1976) and Bonatti and Hamlyn (1978).

Igneous Emplacement of Crust in Fracture Zones

Divergent wrenching on a regional scale along a transform zone may lead to extension in a direction perpendicular to the strike of the transform zone (as in the concept of "leaky" transform faults of Van Andel et al. (1971) and Thompson and Melson (1972)), as well as to enhanced tensional structures in the secondary tectonized domain. Igneous activity may occur in zones of extension, preferentially along the main transform fault which probably extends deeper in the lithosphere than the tensional features in the secondary tectonized domain. The alkaline affinity of the basaltic magmas emplaced along the fracture zones may reflect a small degree of partial melting in the source due to limited upwelling, and differentiation during the relatively slow ascent of the liquid.

Both convergent and divergent wrenching can occur commonly along any one long-offset transform, one or the other being prevalent at different times or in different areas. Thus, both tectonic emplacement of crust (favored by convergent wrenching) and igneous emplacement of crust (favored by divergent wrenching) can occur in the same fracture zone at different times and/or locations.

Sections of Oceanic Crust in Fracture Zones

Whether or not relatively thick sections of "normal" oceanic crust can become exposed in the large fracture zones has been discussed by Bonatti and Honnorez (1976) and Francheteau et al. (1976). Two situations can be envisaged where relatively thick sections of oceanic crust might be exposed in fracture zones.

(1) If the secondary tectonized domain is absent on one side of a transform valley due, for instance, to reduced shear related to slight divergence of the plates on opposite sides of the transform boundary, windows exposing relatively untectonized crust might exist. The north side of the Vema F.Z. might provide a case of this situation, as discussed by Bonatti and Honnorez (1976). Francheteau et al. (1976) argue that the walls of transform valleys consist of many faults each with small vertical offsets, which would allow exposure only of thin sections of the upper crust. This is probably true in transforms with small offset, such as fracture zone A in the FAMOUS area. However, single scarps over several hundred meters high have been observed at the Vema F.Z. by side-sonar methods in a recent cruise of the French vessel Charcot, and by submersible observation in the Cayman Trough area (De Long et al., 1978). In addition, it has been suggested that layer 2 may be thinner than normal in the vicinity of fracture zones because a smaller quantity of basaltic lava is injected at an axis of spreading close to a cold transform boundary. If so, the chemical composition of these basalts could be expected to differ from that of basalts from the axis of spreading away from the fracture zone. However, the composition of basalts from the north side of the Vema F.Z. is identical to that of basalts from the axis of spreading at various distances from the fracture zone (Bonatti et al., 1977), suggesting a normal thickness of layer 2 on the north side of this fracture zone.

(2) If the secondary tectonized domain is well developed in a fracture zone, it is possible that slivers of crust might be uplifted which preserve, more or less, their original stratigraphy. Possible examples of such uplifted crustal blocks are the section of figure 2 from the northern side of the Romanche main transform valley, and the section of figure 3 from the Owen F.Z. Detailed field work in these promising areas is required to test this interpretation.

Acknowledgments. Research supported by ONR contract N001475-C0210 and NSF grant OCE 76-01755.

Contribution from the Rosenstiel School of Marine and Atmospheric Sciences, and No. 2785 from Lamont-Doherty Geological Observatory.

References

Bonatti, E., Serpentinite protrusions in the oceanic crust, Earth Planet. Sci. Lett., 32, 107-113, 1976.

Bonatti, E., Vertical tectonism in oceanic fracture zones, Earth Planet. Sci. Lett., 37, 369-379, 1978.

Bonatti, E., and P.R. Hamlyn, Mantle uplifted block in the western Indian Ocean, Science, 201, 249-251, 1978.

Bonatti, E., K. Hartman, F. Innocenti, and R. Kay, Basalt drilled at the Vema fracture zone, DSDP Leg 39, DSDP Initial Reports, 39, 507-511, 1977.

Bonatti, E., and J. Honnorez, Sections of the Earth's crust in the equatorial Atlantic, Jour. Geophys. Res., 81, 4104-4116, 1976.

Bonatti, E., M. Sarnthein, A. Boersma, M. Gorini and J. Honnorez, Neogene crustal emersion and subsidence at the Romanche fracture zone, equatorial Atlantic, Earth Planet. Sci. Lett., (1977).

Chermak, A., A structural study of the Romanche Fracture Zone based on geophysical data, MS Thesis, University of Miami, Miami, Florida, 1978.

DeLong, S., P.J. Fox, F. Malcolm, E. Schreiber, C. Sondergeld, R. Spydell and J. Stroup, Plutonic rocks from the rift valley wells of the Cayman spreading center: Geologic implications for the plutonic foundation of the oceanic crust, Abstracts, Ewing Symposium, 16-17, 1978.

Dewey, F.J., Finite plate motions: some implications for the evolution of rock masses at plate margins, Amer. Jour. Sci., 275-A, 260-284, 1975.

Emmons, R.C., Strike-slip rupture patterns in sand models, Tectonophysics, 7, 71-87, 1969.

Fox, P.J., E. Schreiber, H. Rowlett and K. McCamy, The geology of the Oceanographer fracture zone: a model for fracture zones, Jour. Geoph. Res., 81, 4117-4128, 1976.

Francheteau, J., P. Choukroune, R. Hekinian, X. Le Pichon and H.D. Needham, Oceanic fracture zones do not provide deep sections in the crust, Can. Jour. Earth Sci., 13, 1223-1235, 1976.

Frey, F.A., Rare earth and potassium abundances in St. Paul's Rocks, Earth Planet. Sci. Lett., 7, 351-360, 1970.

Gilliland, W.N., and G.P. Meyer, Two classes of transform faults, Geol. Soc. Amer. Bull., 87, 1127-1130, 1976.

Gorini, M., The tectonic fabric of the equatorial Atlantic and adjoining continental margins. Ph.D. Thesis, Columbia University, New York, 1977.

Honnorez, J., and E. Bonatti, Nepheline gabbro from the Mid-Atlantic Ridge, Nature, 228, 850-852, 1970.

Melson, W.G., E. Jarosewich, R. Cifelli, and G. Thompson, Alkali olivine basalt dredged near St. Paul's Rocks, Mid-Atlantic Ridge, Nature, 215, 381-382, 1967.

Melson, W.G., S.R. Hart, and G. Thompson, St. Paul's Rocks, equatorial Atlantic: Petrogenesis, radiometric ages and implications on sea floor spreading. Geol. Soc. Amer. Memoir, 132, 241-272, 1972.

Moody, J.D., and M.J. Hill, Wrench-fault tectonics, Geol. Soc. Amer. Bull., 67, 1207-1246, 1956.

Prinz, M., K. Keil, J.A. Green, A.M. Reid, E. Bonatti, and J. Honnorez, Ultramafic and mafic dredge samples from the equatorial Mid-Atlantic Ridge and fracture zones, Jour. Geophys. Res., 81, 4087-4103, 1976.

Searle, R.C., Side-scan sonar studies of North Atlantic fracture zones, Quat. Jour. Geol. Soc. London, in press, 1978.

Thompson, G., and W.G. Melson, The petrology of oceanic crust across fracture zones in the Atlantic ocean: evidence of a new kind of sea floor spreading, Jour. Geol., 80, 526-538, 1972.

Van Andel, T.H., R.P. von Herzen, and J. Phillips, The Vema fracture zone and tectonics of transverse shear zones in oceanic crustal plates, Mar. Geophys. Res., 1, 261-283, 1971.

Wilcox, R.E., T.P. Harding, and D.R. Seely, Basic wrench tectonics, Am. Assoc. Petrol. Geol., 57, 74-96, 1973.

BASALTIC GLASS ERUPTED ALONG THE MID-ATLANTIC RIDGE BETWEEN 0-37° N: RELATIONSHIPS BETWEEN COMPOSITION AND LATITUDE

William G. Melson and Timothy O'Hearn

Department of Mineral Sciences, Smithsonian Institution, Washington, D. C. 20560

Abstract. The basaltic rocks from the upper part of the basement of the Atlantic between the Equator and 37°N have major and minor element signatures which correlate with large latitude intervals. 204 compositional basalt types based on electron microprobe analyses of the glassy rinds of pillow lavas fall mainly into two groups of Recent to Eocene age: a 0-29°N group characterized by higher TiO_2, Na_2O, P_2O_5/K_2O, and lower CaO than a 36-37°N group Basalt cores from Cretaceous and Jurassic DSDP drill sites suggest migration of the 0-29°N group to the north through time. The higher K_2O contents of the 36-37°N group may reflect the influence of a southward longitudinal migration of the postulated Azore's 'thermal plume' magmas. If so, the migration is dammed by the Oceanographer Fracture Zone, which marks the southern limit of this group. High K_2O lavas reappear in the vicinity of the Equator, perhaps reflecting on-going activity of the proposed Sierra Leone or St. Paul's Rocks 'thermal plume'.

Introduction

Studies of the Mid-Atlantic Ridge are in part aimed at a clear understanding of the process of lithospheric formation at a slowly accreting plate margin. This paper focuses on submarine volcanism, and on but one facet of it; the variation in space and time of the major and minor element composition of the glassy rinds of pillow lavas. It is restricted to the publication and preliminary interpretation of results on samples dredged, drilled, or collected by submersibles between the Equator and 37°N. Samples range in age from Recent to about 150 m.y. (late Jurassic). The data, in particular, are examined within the context of the evolution of the proposed Azores and St. Paul's Rocks 'thermal plumes' during the on-going opening of the Atlantic Ocean Basin. Sample ages and regional distribution present a long sought opportunity to examine these and the many other problems addressed in this volume and so appropriately dedicated to Maurice Ewing.

Analyses of glassy pillow rinds are rapidly and accurately done using the electron microprobe. Numerous investigators are now routinely using such analyses in studies of volcanic glasses from land and beneath the sea. The glass analyses reported here are an extension of work reported previously (Melson, 1973; Melson et al.,1976; Byerly et al.,1977; Melson et al.,1977). The analyses are for Si, Al, Fe, Mg, Ca, Na, K, Ti and P. Table 1 gives the results for Atlantic abyssal basalts between 37°N latitude and the Equator.

Comparison of Field Sampling Methods

The analyzed samples come from dredges, drill cores and submersibles. Each method involves a different sampling capability and are complementary. Dredging involves least control on where samples are precisely collected and least knowledge of the geometric relationships of rock units. Nonetheless, dredging is the least expensive way of obtaining exposed deep sea basement rocks and remains useful in reconnaissance sampling. Also, as shown here, basalt from extensive regions of the Mid-Atlantic Ridge have 'chemical signatures' which were revealed by dredge samples and later substantiated by drill cores and samples obtained by submarines. In and very near the median valley a single dredge has yielded up to 5 compositional basalt types (locality A25.40N, table 1) which are probably derived from 5 separate flows. Normally, though, 1 to 2 types of basalt are recovered in a single dredge haul.

Submarines give samples carefully selected in regard to field relationships. Flows of different ages are recognizable and there is excellent control in regard to bottom morphologic features (Heirtzler and Van Andel, 1977). Except in fissures and along fault scarps, the sampling is essentially two-dimensional.

Core drilling of the Deep Sea Drilling Project gives vertical sequences and can provide basement sequences beneath thick sedimentary sections, a capability essential in verifying current plate tectonic theory in regard to sea-floor spreading. Ar present, drilling requires about 40 meters or more of sediment to spud in, a severe limitation

Table 1. Chemical types of deep sea volcanic glasses from the oceanic basement of the Atlantic between the Equator and 37°N. latitude based on electron microprobe analyses. Key is at end of table.

GROUP	LONG.	N	R	SIO2	AL203	FEO*	MGO	CAO	NA20	K20	TIO2	P205	SUM	S	T	A	E	SO
A 0.33N1	16.95W	1	0	50.88	14.85	11.27	6.77	10.70	2.92	0.11	2.06	0.18	99.74	FZ	D	Q	E	W
A 0.91N1	30.28W	7	0	50.96	15.96	9.54	7.82	11.83	2.43	0.23	1.28	0.12	100.17	FZ	D	Q	E	W
A 0.93N1	29.37W	8	0	50.76	15.18	9.69	6.30	11.14	2.61	0.61	1.85	0.21	98.35	FZ	D	Q	E	W
A 0.93N2	29.37W	1	1	50.61	15.13	10.39	5.38	10.44	2.92	0.85	2.55	0.27	98.54	FZ	D	Q	E	W
A 5.92N1	33.25W	1	0	51.45	15.60	9.71	6.96	10.67	3.05	0.15	1.58	0.17	99.34	R	D	<10	E	M
A 6.01N1	33.28W	3	0	51.69	15.14	10.91	7.40	10.29	2.99	0.14	1.73	0.16	100.45	R	D	<1	E	M
A 7.15N1	33.97W	1	0	52.19	15.19	10.50	6.75	10.68	3.12	0.16	1.89	0.21	100.69	R	D	<1	E	M
A 7.82N1	38.04W	1	0	51.22	15.68	10.56	6.86	11.12	3.09	0.17	1.73	0.20	100.63	R	D	<1	E	M
A 9.60N1	40.65W	27	0	51.16	15.28	10.36	7.02	10.80	2.95	0.12	1.84	0.15	99.68	FZ	D	<10	E	W
A11.02N1	43.67W	1	0	51.39	15.56	10.40	7.79	11.14	2.72	0.12	1.64	0.08	100.84	FZ	D	<10	E	W
A11.02N2	43.67W	7	0	50.69	14.82	11.21	6.63	10.81	2.95	0.12	2.07	0.16	99.46	FZ	D	<10	E	W
A11.07N1	44.22W	2	0	51.28	14.72	10.88	6.97	11.49	3.02	0.10	1.80	0.16	100.42	FZ	D	<10	E	W
A11.22N1	43.60W	7	0	51.27	15.25	10.43	7.29	11.32	2.85	0.08	1.68	0.13	100.30	FZ	D	<1	E	M
A11.40N1	43.50W	1	0	50.50	15.18	10.16	8.02	10.83	2.81	0.10	1.74	0.17	99.51	R	D	<10	E	W
A11.42N1	43.58W	1	0	51.36	14.96	9.74	7.49	11.73	2.75	0.07	1.57	0.14	99.81	R	D	<1	E	M
A11.48N1	43.63W	5	0	50.47	15.65	10.29	7.62	10.98	2.89	0.09	1.58	0.14	99.71	R	D	<1	E	M
A11.52N1	43.60W	3	0	50.80	15.53	9.65	7.67	11.50	2.78	0.08	1.56	0.13	99.70	R	D	<1	E	M
A11.97N1	43.70W	5	0	51.19	15.94	9.41	8.03	11.52	2.71	0.08	1.50	0.13	100.51	R	D	<1	E	M
A22.17N1	45.25W	10	1	50.55	15.69	10.12	7.30	11.12	2.94	0.15	1.76	0.15	99.78	R	D	<1	E	M
A22.17N2	45.25W	2	0	51.27	14.95	10.03	7.16	12.32	2.56	0.12	1.36	0.10	99.87	R	D	<10	E	V
A22.17N3	45.25W	2	0	50.78	15.08	10.89	6.93	10.80	3.04	0.17	1.91	0.14	99.74	R	D	<10	E	V
A22.24N1	45.02W	18	3	50.53	15.49	10.20	7.54	10.96	2.90	0.13	1.71	0.14	99.60	R	D	<10	E	V
A22.36N1	45.20W	4	0	50.76	15.23	11.06	6.79	10.96	3.11	0.14	1.97	0.16	100.18	R	D	<10	E	V
A22.52N1	45.05W	24	0	50.82	15.51	10.04	7.36	11.46	2.86	0.13	1.73	0.14	100.05	R	D	<10	E	V
A22.52N2	45.05W	4	0	50.49	15.13	10.72	7.26	10.87	2.97	0.14	1.95	0.15	99.68	R	D	<10	E	V
A22.52N3	45.05W	1	0	50.61	16.56	9.69	7.97	11.52	2.76	0.09	1.44	0.12	100.76	R	D	<10	E	V
A22.76N1	46.08W	6	0	50.49	15.57	10.94	7.21	10.72	3.00	0.11	1.66	0.16	99.86	R	395	7	E	1
A22.76N2	46.08W	1	0	51.48	14.33	9.72	6.90	12.15	2.67	0.11	1.49	0.17	99.02	R	395	7	E	1
A22.76N1	46.08W	1	0	50.24	15.10	9.87	8.06	11.02	2.72	0.11	1.65	0.14	98.91	R	395A	7	E	1
A22.76N2	46.08W	1	0	49.64	15.33	10.98	8.01	10.53	2.95	0.11	1.52	0.13	99.20	R	395A	7	E	1
A22.76N3	46.08W	1	0	50.58	15.85	11.07	6.97	10.89	2.96	0.10	1.68	0.16	100.26	R	395A	7	E	1
A22.76N4	46.08W	3	0	51.62	14.95	9.78	7.50	12.30	2.46	0.10	1.40	0.15	100.26	R	395A	7	E	1
A22.76N5	46.08W	5	0	51.49	14.77	9.61	7.09	11.88	2.80	0.11	1.52	0.14	99.41	R	395A	7	E	1
A22.76N6	46.08W	1	0	50.63	15.22	9.63	6.91	11.60	3.08	0.12	1.62	0.14	98.95	R	395A	7	E	1
A22.76N7	46.08W	8	0	50.92	15.49	9.86	7.25	11.20	2.90	0.13	1.73	0.17	99.65	R	395A	7	E	1
A22.76N8	46.08W	9	0	50.05	15.52	10.22	8.05	11.01	2.84	0.11	1.61	0.14	99.55	R	395A	7	E	1
A22.76N9	46.08W	1	0	50.39	15.82	10.22	7.38	11.24	2.92	0.12	1.68	0.14	99.91	R	395A	7	E	1
A22.76N10	46.08W	1	0	50.32	15.54	10.42	7.96	10.81	2.83	0.10	1.62	0.19	99.79	R	395A	7	E	1
A22.80N1	45.20W	2	0	50.00	16.37	9.81	7.56	11.01	2.91	0.12	1.71	0.11	99.60	R	D	<10	E	V
A22.80N2	45.20W	2	0	50.38	15.53	10.60	6.87	11.02	2.93	0.15	1.89	0.15	99.52	R	D	<10	E	V
A22.98N1	43.52W	5	0	50.89	15.57	9.53	7.22	11.31	2.93	0.13	1.65	0.16	99.39	R	396	13*	E	1
A22.98N2	43.52W	11	0	50.55	16.26	8.73	7.94	11.62	2.81	0.11	1.36	0.14	99.52	R	396	13*	E	1
A22.98N3	43.52W	6	0	51.15	15.16	9.90	7.44	11.65	2.70	0.11	1.53	0.16	99.80	R	396	13*	E	1
A23.02N1	44.96W	4	0	50.59	15.86	9.88	7.53	11.22	2.86	0.11	1.54	0.16	99.75	R	D	<10	E	W
A23.02N2	44.96W	19	1	50.77	15.45	10.41	7.08	10.96	3.01	0.16	1.72	0.17	99.73	R	D	<10	E	W
A23.03N1	44.91W	22	1	50.92	15.21	10.21	7.15	11.05	3.10	0.13	1.73	0.17	99.67	R	D	<10	E	W
A23.03N1	45.03W	13	3	50.48	15.39	10.25	7.24	10.88	3.02	0.12	1.79	0.20	99.37	R	D	<10	E	W
A23.07N1	45.17W	2	0	51.74	15.08	10.57	7.28	10.80	3.00	0.16	1.81	0.16	100.60	R	D	<10	E	W
A23.22N1	44.92W	3	0	48.53	17.82	9.87	8.41	10.82	3.09	0.11	1.11	0.09	99.85	R	D	<10	E	W
A23.22N2	44.92W	2	0	50.25	15.29	10.52	7.23	11.09	3.27	0.16	1.90	0.17	99.88	R	D	<10	E	W
A23.42N1	52.03W	1	0	49.23	17.73	8.62	8.86	11.58	2.84	0.22	1.11	0.13	100.32	R	D	50	E	W
A23.42N2	52.03W	1	0	51.75	15.63	9.75	7.26	11.40	2.84	0.09	1.56	0.16	100.44	R	D	50	E	W
A23.52N1	44.82W	2	0	50.18	15.86	9.56	7.92	11.12	3.04	0.10	1.48	0.14	99.40	FZ	D	<1	E	W
A23.60N1	45.03W	3	0	50.73	15.60	9.49	7.66	11.54	2.77	0.11	1.52	0.14	99.56	FZ	D	<1	E	W
A23.88N1	46.26W	5	0	50.90	14.78	10.65	6.66	11.49	3.03	0.13	1.78	0.16	99.58	FZ	D	<1	E	W

Table 1. cont.

GROUP	LONG.	N	R	SIO2	AL203	FEO*	MGO	CAO	NA20	K20	TIO2	P205	SUM	S	T	A	E	SC
A23.92N1	46.24W	1	0	51.04	15.18	10.04	6.83	11.43	3.06	0.16	1.78	0.16	99.68	FZ	D	<1	E	W
A23.92N2	46.24W	1	0	50.89	14.65	11.01	6.28	11.10	3.19	0.13	1.96	0.16	99.37	FZ	D	<1	E	W
A24.01N1	46.19W	2	0	50.86	15.60	9.38	7.66	11.28	2.90	0.12	1.54	0.14	99.48	FZ	D	<1	E	W
A25.04N1	68.06W	12	0	51.15	14.36	10.31	7.67	12.39	2.21	0.10	1.26	0.11	99.56	R	418A	109	E	1
A25.04N2	68.06W	7	0	51.00	14.72	10.33	7.60	12.21	2.26	0.08	1.33	0.11	99.64	R	418A	109	E	1
A25.04N3	68.06W	11	0	50.27	15.50	9.71	8.28	12.60	2.21	0.06	1.13	0.10	99.86	R	418A	109	E	1
A25.04N4	68.06W	20	0	50.13	15.28	10.17	8.00	12.40	2.24	0.06	1.24	0.11	99.63	R	418A	109	E	1
A25.04N5	68.06W	10	0	50.45	15.25	10.37	7.97	12.34	2.26	0.08	1.23	0.11	100.06	R	418A	109	E	1
A25.04N6	68.06W	1	0	50.58	14.21	11.46	7.21	11.70	2.31	0.11	1.57	0.12	99.27	R	418A	.09	E	1
A25.04N7	68.06W	3	0	50.83	14.55	10.59	7.49	12.22	2.32	0.11	1.38	0.13	99.62	R	418A	109	E	1
A25.04N8	68.06W	19	0	50.60	14.32	11.65	6 98	11.62	2.42	0.12	1.67	0.14	99.52	R	418A	109	E	1
A25.04N9	68.06W	6	0	50.84	14.73	10.76	7.49	12.20	2.36	0.09	1.37	0.13	99.97	R	418A	109	E	1
A25.04N10	68.06W	18	0	51.03	14.19	11.31	7.13	11.86	2.34	0.11	1.54	0.14	99.65	R	418A	109	E	1
A25.04N11	68.06W	1	0	50.73	14.34	10.59	7.37	11.72	2.39	0.11	1.44	0.12	98.81	R	418A	109	E	1
A25.04N12	68.06W	1	0	52.14	13.92	11.69	7.21	11.84	2.42	0.08	1.55	0.15	101.00	R	418A	109	E	1
A25.04N13	68.06W	1	0	49.79	15.18	9.98	7.92	12.30	2.25	0.08	1.21	0.12	98.83	R	418A	109	E	1
A25.04N14	68.06W	1	0	50.76	14.31	10.16	7.41	12.11	2.23	0.12	1.32	0.16	98.58	R	418A	109	E	1
A25.04N1	68.06W	1	0	50.93	14.73	9.99	7.50	12.36	2.29	0.10	1.26	0.11	99.27	R	418B	109	E	1
A25.11N1	68.04W	10	0	50.46	14.46	11.27	7.20	11.73	2.33	0.10	1.59	0.14	99.28	R	417D	109	E	1
A25.11N2	68.04W	19	0	50.70	14.59	11.28	7.46	11.86	2.28	0.09	1.53	0.13	99.92	R	417D	109	E	1
A25.11N3	68.04W	15	0	50.71	14.32	11.61	7.14	11.81	2.31	0.10	1.64	0.14	99.78	R	417D	109	E	1
A25.11N4	68.04W	2	0	50.60	14.28	11.10	7.30	12.08	2.27	0.09	1.50	0.14	99.36	R	417D	109	E	1
A25.11N5	68.04W	7	0	50.70	14.04	11.74	7.04	11.81	2.34	0.10	1.69	0.14	99.60	R	417D	109	E	1
A25.11N6	68.04W	10	0	50.64	13.95	11.97	6.84	11.61	2.36	0.08	1.80	0.15	99.40	R	417D	109	E	1
A25.11N7	68.04W	1	0	51.08	14.65	11.26	7.34	11.72	2.27	0.12	1.45	0.13	100.02	R	417D	109	E	1
A25.11N8	68.04W	15	0	50.72	14.32	11.58	7.13	11.64	2.34	0.09	1.62	0.14	99.58	R	417D	109	E	1
A25.40N1	45.30W	8	0	51.05	15.85	9.45	7.62	11.22	2.93	0.10	1.60	0.13	99.95	R	D	<10	E	L
A25.40N2	45.30W	9	0	50.55	16.30	9.44	7.67	11.22	2.88	0.10	1.61	0.14	99.91	R	D	<10	E	L
A25.40N3	45.30W	6	1	51.32	15.36	9.69	7.34	11.27	2.89	0.08	1.54	0.13	99.62	R	D	<10	E	L
A25.40N4	45.30W	3	0	50.36	16.20	8.63	8.85	11.53	2.66	0.07	1.29	0.11	99.70	R	D	<10	E	L
A25.40N5	45.30W	2	0	49.75	15.33	10.34	7.33	10.81	2.88	0.13	1.87	0.17	98.61	R	D	<10	E	L
A25.77N1	45.16W	1	0	51.32	15.82	9.25	7.83	11.08	2.95	0.18	1.57	0.14	100.14	R	D	<10	E	T
A27.20N1	43.55W	1	0	50.94	15.46	9.38	8.60	11.39	2.48	0.10	1.47	0.10	99.92	R	D	<10	E	W
A28.56N1	44.83W	2	0	49.75	15.33	10.35	7.33	10.81	2.89	0.14	1.88	0.18	98.66	R	D	<10	E	L
A28.90N1	43.32W	22	0	48.80	17.75	9.30	8.63	11.62	2.49	0.05	0.91	0.08	99.63	R	D	<10	E	W
A28.90N2	43.32W	10	0	51.08	15.85	9.91	7.45	11.44	2.69	0.10	1.50	0.14	100.16	R	D	<10	E	W
A29.71N1	42.65W	1	0	50.21	16.41	10.39	7.46	11.27	2.94	0.07	1.46	0.15	100.36	R	D	<10	E	S
A29.81N1	42.73W	1	0	51.04	16.40	9.69	7.29	10.82	2.65	0.13	1.35	0.11	99.48	R	D	<10	E	S
A29.86N1	42.77W	1	0	50.08	15.60	11.62	7.66	10.69	2.92	0.34	1.93	0.23	101.07	R	D	<10	E	S
A32.87N1	52.22W	2	0	50.30	16.38	9.22	7.51	12.05	2.42	0.45	1.52	0.18	100.03	Q	10	–	I	1
A34.90N1	61.17W	7	0	50.82	15.24	9.67	8.47	12.26	1.88	0.05	1.00	0.07	99.46	Q	105	150*	IQ	1
A36.40N1	33.77W	1	0	51.51	14.68	10.59	7.26	12.00	2.36	0.23	1.30	0.16	100.09	R	D	<1	E	W
A36.42N1	33.65W	4	0	51.39	15.06	9.94	7.47	12.01	2.25	0.15	1.31	0.15	99.73	R	D	<1	E	W
A36.43N1	33.66W	1	0	51.31	14.95	9.51	7.84	12.44	2.21	0.12	1.12	0.12	99.62	R	D	<1	E	W
A36.43N1	33.68W	1	0	51.66	14.54	9.76	7.10	12.14	1.97	0.18	1.13	0.14	98.62	R	D	<1	E	W
A36.44N1	33.64W	6	0	52.19	14.12	10.83	6.48	11.56	2.19	0.15	1.32	0.15	98.99	R	D	<1	E	W
A36.44N2	33.64W	7	0	51.31	14.35	12.17	6.23	10.62	2.57	0.20	1.84	0.20	99.49	R	D	<1	E	W
A36.44N1	33.68W	2	0	51.44	14.88	9.12	7.56	12.55	1.90	0.13	0.96	0.10	98.64	R	D	<1	E	W
A36.46N1	33.58W	4	0	50.82	15.41	8.81	8.02	12.92	2.00	0.14	1.01	0.12	99.25	R	D	<1	E	W
A36.49N1	33.65W	1	0	51.17	15.24	9.22	8.40	12.98	2.20	0.08	0.96	0.10	100.35	R	D	<1	E	W
A36.49N2	33.65W	3	0	51.64	14.99	9.92	7.69	12.18	2.22	0.20	1.12	0.12	100.08	R	D	<1	E	W
A36.54N1	33.52W	1	0	51.62	14.77	9.68	7.84	12.59	2.01	0.14	1.11	0.12	99.88	R	D	<1	E	W
A36.58N1	33.60W	1	0	51.99	13.96	12.11	6.34	11.06	2.50	0.18	1.74	0.22	100.10	R	D	<1	E	W
A36.58N1	33.52W	1	0	50.02	15.93	9.55	8.92	12.42	2.00	0.13	0.85	0.11	99.93	R	D	<1	E	W
A36.60N1	33.46W	5	0	51.58	15.21	8.71	8.31	12.52	1.95	0.25	1.10	0.13	99.76	FZ	D	<1	E	W

Table 1. cont.

GROUP	LONG.	N	R	SIO2	AL203	FEO*	MGO	CAO	NA20	K20	TIO2	P205	SUM	S	T	A	E	SO
A36.64N1	33.47W	1	0	51.14	14.64	10.44	7.45	11.26	2.30	0.33	1.66	0.19	99.41	R	D	<1	E	W
A36.64N1	33.46W	5	0	51.19	15.29	8.84	8.06	12.46	2.04	0.21	1.07	0.13	99.29	R	D	<1	E	W
A36.66N1	33.34W	5	1	51.32	14.95	9.46	7.87	12.79	2.05	0.17	1.16	0.12	99.89	R	D	<1	E	W
A36.66N2	33.34W	8	0	51.36	14.63	10.18	7.31	12.05	2.14	0.22	1.46	0.16	99.51	R	D	<1	E	W
A36.70N1	33.29W	1	0	51.01	14.87	9.32	8.02	12.80	1.97	0.13	1.06	0.13	99.31	R	D	<1	E	W
A36.70N2	33.29W	6	0	50.75	14.86	10.27	7.72	11.92	2.14	0.17	1.28	0.14	99.25	R	D	<1	E	W
A36.70N3	33.29W	1	0	51.01	14.66	10.92	7.17	11.63	2.21	0.20	1.46	0.19	99.45	R	D	<1	E	W
A36.72N1	33.33W	4	0	51.01	15.07	9.24	8.31	13.03	2.04	0.11	1.02	0.11	99.94	R	D	<1	E	W
A36.72N2	33.33W	1	0	52.17	13.84	11.91	6.52	11.21	2.45	0.20	1.70	0.18	100.18	R	D	<1	E	W
A36.72N1	33.34W	5	0	50.61	14.99	9.58	7.71	12.81	2.11	0.19	1.27	0.17	99.44	R	D	<1	E	W
A36.72N2	33.34W	2	0	51.09	14.96	9.33	7.96	13.30	2.03	0.12	1.07	0.12	99.98	R	D	<1	E	W
A36.73N1	33.31W	13	0	50.94	15.02	9.56	7.83	12.56	2.09	0.14	1.18	0.13	99.45	R	D	<1	E	W
A36.73N2	33.31W	3	0	50.57	14.89	10.16	7.92	12.14	2.29	0.16	1.24	0.14	99.51	R	D	<1	E	W
A36.75N1	33.29W	1	0	51.16	15.10	9.26	7.82	12.90	2.00	0.14	1.14	0.11	99.63	R	D	<1	E	W
A36.76N1	33.31W	5	0	50.82	14.90	10.07	8.00	12.34	2.22	0.16	1.18	0.14	99.83	R	D	<1	E	W
A36.76N1	33.34W	10	0	51.46	15.02	9.42	7.77	12.65	2.04	0.15	1.09	0.13	99.73	R	D	<1	E	W
A36.79N1	33.27W	2	0	50.95	14.78	9.52	7.50	12.57	2.05	0.18	1.16	0.12	98.83	R	S	<1	E	W
A36.80N1	33.27W	2	0	49.24	16.21	9.05	9.32	12.39	1.98	0.12	0.95	0.12	99.38	R	S	<1	E	W
A36.80N1	33.27W	1	0	50.04	15.29	10.39	7.88	11.34	2.33	0.18	1.30	0.14	98.89	R	S	<1	E	W
A36.80N1	33.26W	1	0	49.54	16.62	8.74	9.43	12.64	2.08	0.09	0.74	0.08	100.16	R	S	<1	E	W
A36.81N1	33.27W	1	0	49.00	16.37	8.88	9.55	12.38	1.96	0.10	0.86	0.09	99.19	R	S	<1	E	W
A36.81N1	33.27W	2	0	51.32	14.93	9.89	7.63	12.21	2.17	0.18	1.30	0.14	99.77	R	S	<1	E	W
A36.81N1	33.26W	1	0	48.68	15.89	9.16	9.46	12.38	2.13	0.11	0.88	0.09	98.78	R	S	<1	E	W
A36.81N1	33.26W	2	0	49.45	16.30	9.00	9.28	12.49	2.11	0.12	0.85	0.10	99.70	R	S	<1	E	W
A36.81N1	33.25W	1	0	50.17	15.32	10.56	7.91	11.60	2.38	0.24	1.40	0.16	99.74	R	S	<1	E	W
A36.81N1	33.25W	2	0	50.84	15.06	10.21	7.80	11.60	2.28	0.25	1.42	0.16	99.62	R	S	<1	E	W
A36.81N1	33.25W	1	0	50.11	15.16	10.37	8.02	11.54	2.40	0.23	1.45	0.18	99.46	R	S	<1	E	W
A36.81N1	33.27W	1	0	49.77	15.95	9.08	8.82	12.32	1.98	0.12	0.90	0.08	99.02	R	S	<1	E	W
A36.81N1	33.26W	4	0	48.79	16.84	8.99	9.41	12.70	2.05	0.07	0.66	0.07	99.58	R	S	<1	E	W
A36.81N1	33.26W	1	0	49.52	16.04	8.96	9.86	12.39	2.07	0.17	0.85	0.09	99.95	R	S	<1	E	W
A36.81N1	33.28W	8	0	51.17	14.83	9.95	7.80	12.53	2.05	0.18	1.07	0.12	99.70	FZ	D	<1	E	W
A36.81N2	33.28W	1	0	51.49	14.57	10.51	6.95	11.52	2.29	0.23	1.48	0.15	99.19	FZ	D	<1	E	W
A36.81N1	33.27W	3	0	48.36	16.07	9.33	9.61	12.50	2.05	0.11	0.77	0.08	98.88	R	S	<1	E	W
A36.81N1	33.27W	1	0	49.71	16.20	9.10	8.94	12.53	1.92	0.11	0.81	0.11	99.43	R	S	<1	E	W
A36.81N1	33.27W	1	0	50.74	14.68	10.29	7.40	12.27	2.21	0.23	1.20	0.16	99.18	R	S	<1	E	W
A36.81N1	33.26W	1	0	49.08	16.48	9.08	9.20	12.61	2.01	0.09	0.77	0.09	99.41	R	S	<1	E	W
A36.81N1	33.26W	2	0	48.81	16.44	9.13	9.52	12.51	2.07	0.11	0.78	0.09	99.46	R	S	<1	E	W
A36.82N1	33.26W	3	0	49.45	16.52	9.17	9.15	12.53	2.00	0.09	0.77	0.10	99.78	R	S	<1	E	W
A36.82N1	33.26W	2	0	49.05	16.20	9.25	9.63	12.53	2.03	0.11	0.78	0.08	99.66	R	S	<1	E	W
A36.82N1	33.27W	3	0	49.14	16.37	9.16	9.37	12.56	2.01	0.11	0.76	0.08	99.56	R	S	<1	E	W
A36.82N1	33.26W	3	0	49.90	16.69	9.13	9.10	12.71	1.92	0.08	0.76	0.08	100.37	R	S	<1	E	W
A36.82N1	33.24W	1	0	51.56	14.64	9.17	7.93	13.01	1.96	0.14	1.02	0.11	99.54	R	S	<1	E	W
A36.82N1	33.25W	1	0	50.72	14.72	10.37	7.76	11.36	2.33	0.24	1.52	0.16	99.18	R	S	<1	E	W
A36.82N1	33.25W	1	0	50.16	15.55	9.88	8.45	11.92	2.20	0.16	1.12	0.11	99.55	R	S	<1	E	W
A36.82N1	33.25W	1	0	50.40	14.79	10.36	8.06	11.61	2.30	0.22	1.45	0.16	99.35	R	S	<1	E	W
A36.82N1	33.25W	3	0	49.82	16.11	8.48	9.11	13.00	1.85	0.13	0.85	0.10	99.45	R	S	<1	E	W
A36.82N1	33.25W	1	0	49.97	15.52	9.38	8.81	11.77	2.19	0.14	1.13	0.13	99.04	R	S	<1	E	W
A36.82N1	33.26W	1	0	50.67	16.16	9.37	8.77	11.99	2.23	0.10	1.04	0.10	100.43	R	S	<1	E	W
A36.82N1	33.26W	1	0	50.17	15.27	9.80	8.88	11.59	2.39	0.16	1.11	0.14	99.51	R	S	<1	E	W
A36.82N1	33.27W	2	0	51.14	16.28	8.22	8.88	12.37	2.25	0.20	0.95	0.11	100.40	R	S	<1	E	W
A36.82N1	33.26W	3	0	50.34	16.61	9.05	8.98	11.99	2.19	0.08	0.85	0.10	100.19	R	S	<1	E	W
A36.82N1	33.27W	1	0	51.91	14.58	10.07	7.26	12.34	2.08	0.22	1.14	0.13	99.73	R	S	<1	E	W
A36.82N1	33.28W	1	0	51.06	14.59	10.81	6.87	11.40	2.52	0.27	1.47	0.16	99.15	R	S	<1	E	W
A36.82N1	33.26W	4	0	51.75	15.15	9.96	7.35	11.82	2.28	0.15	1.28	0.13	99.87	R	S	<1	E	W
A36.82N1	33.24W	1	0	49.67	14.80	10.40	7.94	11.64	2.36	0.24	1.43	0.15	98.63	R	S	<1	E	W

Table 1. cont.

GROUP	LONG.	N	R	SIO2	AL203	FEO*	MGO	CAO	NA20	K20	TIO2	P205	SUM	S	T	A	E	SO
A36.82N1	33.24W	1	0	49.44	14.99	10.44	7.74	11.73	2.42	0.24	1.45	0.15	98.60	R	S	<1	E	W
A36.83N1	33.25W	1	0	49.72	14.47	10.37	7.71	11.45	2.37	0.36	1.48	0.18	98.11	R	S	<1	E	W
A36.83N1	33.25W	1	0	50.24	15.57	9.30	8.45	12.24	2.29	0.24	1.16	0.14	99.63	R	S	<1	E	W
A36.83N1	33.27W	1	0	50.89	16.21	8.29	9.40	12.26	2.06	0.16	0.89	0.12	100.28	R	D	<1	E	W
A36.84N1	33.67W	4	0	50.88	14.86	8.83	8.95	12.59	1.85	0.18	0.93	0.11	99.18	R	333A	3.5	E	1
A36.84N2	33.67W	1	0	50.62	14.41	9.43	8.15	12.34	2.01	0.20	1.19	0.12	98.47	R	333A	3.5	E	1
A36.84N3	33.67W	1	0	50.82	15.46	10.77	6.71	12.72	2.48	0.13	1.28	0.11	100.48	R	333A	3.5	E	1
A36.84N4	33.67W	8	0	51.33	14.91	10.72	7.05	12.15	2.12	0.21	1.26	0.14	99.89	R	333A	3.5	E	1
A36.84N5	33.67W	1	0	51.67	14.92	9.41	7.88	13.15	1.86	0.13	0.95	0.12	100.09	R	333A	3.5	E	1
A36.84N6	33.67W	2	0	51.29	15.56	10.64	7.11	12.08	2.08	0.20	1.25	0.14	100.35	R	333A	3.5	E	1
A36.85N1	33.32W	1	0	51.11	14.70	9.99	7.83	12.60	2.20	0.17	1.18	0.12	99.90	R	D	<1	E	W
A36.85N1	32.97W	1	0	51.05	15.36	8.71	8.08	13.22	1.87	0.14	0.95	0.11	99.49	R	D	<1	E	O
A36.88N1	33.64W	1	0	52.31	14.68	10.86	6.85	11.61	2.29	0.19	1.31	0.13	100.23	R	332A	3.5	E	1
A36.88N2	33.64W	3	0	51.51	14.97	10.59	6.91	11.82	2.24	0.20	1.13	0.10	99.47	R	332A	3.5	E	1
A36.88N3	33.64W	6	0	51.30	14.91	10.00	7.46	12.38	2.11	0.16	1.04	0.09	99.45	R	332A	3.5	E	1
A36.88N4	33.64W	3	0	51.84	14.69	10.93	6.66	11.45	2.31	0.18	1.18	0.11	99.35	R	332A	3.5	E	1
A36.88N5	33.64W	5	0	51.11	14.80	10.58	7.47	11.90	2.11	0.22	1.25	0.13	99.57	R	332A	3.5	E	1
A36.88N6	33.64W	1	0	51.28	14.69	9.82	7.45	12.33	2.11	0.15	1.10	0.09	99.02	R	332A	3.5	E	1
A36.88N7	33.64W	1	0	51.89	14.68	9.81	8.05	12.25	2.04	0.16	1.18	0.12	100.18	R	332A	3.5	E	1
A36.88N8	33.64W	3	0	51.43	14.75	10.44	7.35	11.97	2.17	0.16	1.06	0.10	99.43	R	332A	3.5	E	1
A36.88N9	33.64W	2	0	51.22	14.60	11.22	6.95	11.45	2.20	0.19	1.21	0.10	99.14	R	332A	3.5	E	1
A36.88N1	33.64W	4	0	51.58	14.93	9.93	7.71	12.39	2.03	0.18	1.16	0.11	100.02	R	332B	3.5	E	1
A36.88N2	33.64W	8	0	50.19	16.43	9.30	8.90	12.35	2.06	0.09	0.77	0.06	100.15	R	332B	3.5	E	1
A36.88N3	33.64W	1	0	50.15	16.52	9.25	9.44	11.84	2.24	0.09	0.73	0.06	100.32	R	332B	3.5	E	1
A36.88N4	33.64W	1	0	50.74	16.66	9.54	8.95	12.84	1.93	0.10	0.80	0.07	101.63	R	332B	3.5	E	1
A36.88N5	33.64W	13	0	49.83	16.76	9.14	9.03	12.61	2.03	0.07	0.66	0.05	100.18	R	332B	3.5	E	1
A36.88N6	33.64W	1	0	50.56	16.09	10.03	7.87	12.02	2.41	0.08	0.90	0.07	100.03	R	332B	3.5	E	1
A36.88N7	33.64W	11	0	50.60	16.07	10.01	7.98	11.78	2.33	0.08	0.89	0.06	99.80	R	332B	3.5	E	1
A36.88N8	33.64W	1	0	50.95	15.04	10.13	7.39	12.41	2.08	0.17	1.13	0.10	99.40	R	332B	3.5	E	1
A36.88N9	33.64W	1	0	50.41	16.20	9.14	8.84	12.59	2.07	0.08	0.71	0.08	100.12	R	332B	3.5	E	1
A36.88N10	33.64W	5	0	49.78	15.79	8.75	8.69	13.83	1.90	0.07	0.67	0.05	99.53	R	332B	3.5	E	1
A36.88N11	33.64W	1	0	51.84	14.23	10.57	7.55	12.37	2.10	0.18	1.13	0.10	100.07	R	332B	3.5	E	1
A36.88N12	33.64W	1	0	49.69	16.68	9.16	8.80	12.65	2.03	0.05	0.66	0.06	99.78	R	332B	3.5	E	1
A36.88N13	33.64W	2	0	50.83	14.75	10.56	7.20	12.23	2.17	0.16	1.05	0.10	99.05	R	332B	3.5	E	1
A36.88N1	33.64W	2	0	51.66	14.71	10.71	7.19	11.75	2.18	0.20	1.18	0.12	99.70	R	332D	3.5	E	1
A36.92N1	33.13W	3	0	51.39	14.39	10.23	6.89	12.06	2.11	0.22	1.30	0.16	98.75	R	D	<1	E	W
A37.04N1	34.41W	2	0	51.98	14.76	9.94	7.41	12.10	2.05	0.21	0.95	0.13	99.53	R	334	10	E	1
A37.04N2	34.41W	4	0	52.03	14.53	9.96	7.91	12.72	1.81	0.09	0.83	0.07	99.95	R	334	10	E	1
A37.30N1	35.20W	40	0	50.59	15.68	9.43	8.01	11.94	2.39	0.17	1.14	0.10	99.45	R	335	13	E	1

Explanation of column headings: Group consists of latitude collected (A=Atlantic) and chemical type designation, e.g. A22.76N3 is sampled at 22.76°N and is the third chemical type from this locality. Long.=longitude of the type; N=number of samples which were averaged for the chemical type; R=number of analyses very close to but outside the chemical discriminant limits used in defining the chemical type. These are rejected in defining the type. S=probable geologic setting: R=present or ancient Mid-Atlantic Ridge Rift Valley, FZ=fracture zone. T=sampling method: D=dredge, S=submersible, and for DSDP cores, DSDP hole number. A=probable age of extrusive(E) from magnetic anomaly age or from age of oldest overlying sediment, the latter indicated by an asterisk. E=mode of emplacement as eruptive(E) or inrusive(I). SO=Institution or individual who contributed samples- see acknowledgements for key to source codes. For drill cores, sequence of listing is sequence of coring, i.e., from top to bottom.

Table 2. Average of analyses for various basalts groups (table 1). N = number of chemical types

Group	N	SiO_2	Al_2O_3	FeO*	MgO	CaO	Na_2O	K_2O	TiO_2	P_2O_5
0-29°N	69	50.73	15.56	10.08	7.45	11.19	2.88	0.12	1.63	0.15
36-37°N	108	50.73	15.29	9.79	8.04	12.24	2.14	0.16	1.10	0.12
K-25°N	23	50.73	14.53	10.91	7.42	12.01	2.30	0.09	1.45	0.13
SITE 105	1	50.82	15.24	9.67	8.47	12.26	1.88	0.05	1.00	0.07
0-93°N	2	50.69	15.16	10.04	5.84	10.79	2.77	0.73	2.20	0.24
SITE 10	1	50.30	16.38	9.22	7.51	12.05	2.42	0.45	1.52	0.18
TOTAL[1]	204	50.72	15.29	10.01	7.75	11.84	2.42	0.15	1.33	0.13
TOTAL[2]	--	50.73	15.51	10.02	7.56	11.39	2.75	0.13	1.53	0.14

*ALL Fe as FeO

1 Average, all analyses

2 Average of 0-29°N and 37-38°N weighted by inferred along-axis distribution.

to combining the complementary advantages of drilling, diving, and dredging in the same area. The 37°N region is the only one sampled by all methods and has yielded the largest data set.

Regional Setting and Sampling Density

With the exception of drill cores, most samples were collected from the median rift between

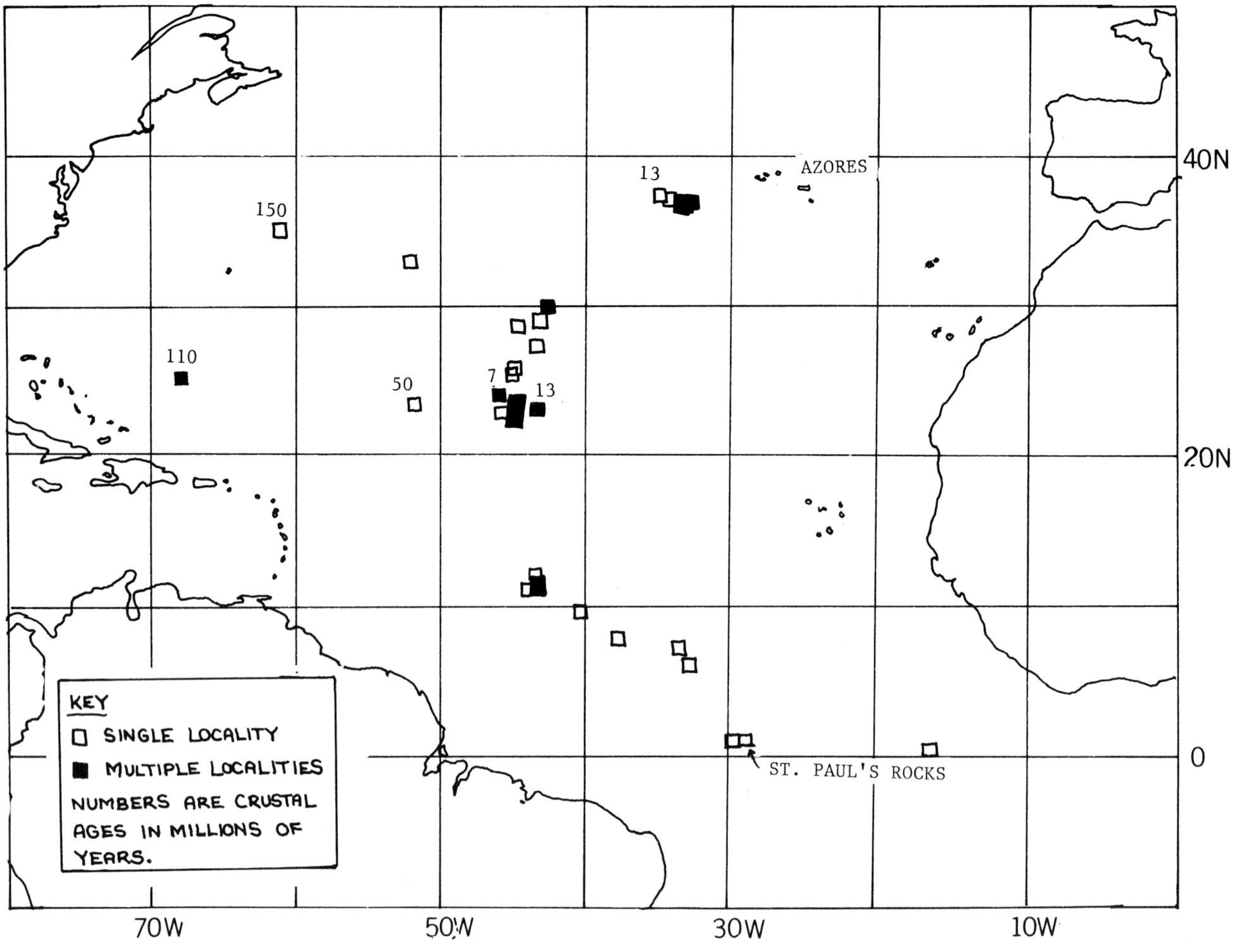

Fig. 1. Sample locations

the equator and 30°N latitude, a distance of about 4100 km (figure 1). Sample density between the equator and 15°N is poor - 17 stations over a distance of nearly 2000 km. 9 of these 17 are within fracture zones, and the most fractionated samples (A0.93N1 and N2, table 1) in the entire collection are from one of these near St. Paul's Rocks. The largest gap in sampling, some 1150 km., occurs between 15°N and 22°N. Sample density is high between 22°N and the Kane fracture zone around 24°N: 17 stations over an along-the-axis distance of about 230 km. Between the Kane and Atlantis fracture zone, 25-29°N, sample density is again low: 8 stations over about 550 kilometers of ridge axis. Moving northward, a gap of about 600 km. exists between the Atlantis and Oceanographer fracture zones. North of the Oceanographer fracture zone, centered around 37°N and including the FAMOUS area, station density reaches a maximum: 67 stations over about 230 km. of ridge axis.

Relations To Previous Work on Basaltic Glasses from 0-37°N

The bottom samples obtained by ALVIN dives have been previously described in regard to their major element compositions by Bryan and Moore (1977). These samples are reanalyzed here to eliminate between-lab differences. The drill core glass analyses from Leg 37 sites near 37°N. are extensive and permit a look at thick sections of basalt at a single site. Individual analyses of these have been previously reported (Aumento et al., 1977) but here are compiled into averages for individual cooling units. Byerly and Wright (1978) have integrated Leg 37 glass analyses with whole rock analyses and recognized 'chemical types', individual units with compositions different from overlying and underlying units, and 'magma batches', inferred magma compositions serving as a parental liquid for a number of 'chemical types'. Partly using glass analyses as guides to liquid compositions, they correct analyses of slightly altered basalt back to unaltered compositions. Of special significance is their model which sorts out crustal fractionation from high pressure processes and points to an important role for clinopyroxene as an intermediate level fractionating phase or as a residual phase during mantle partial melting.

Bryan and Moore (1977) using analyses of basaltic glass as well as other data put forth a model involving a zoned magma chamber at 37°N to account for the abundance of "primitive" liquids near the center of the median valley and of more fractionated lavas on the margins. The Leg 37 drill cores show an alternation of "primitive" and "fractionated" lavas (table 1) a view arguing against this model (Byerly and Wright, 1978) for crust formed in Pliocene time.

Basalt from the Leg 37 and FAMOUS sites have been compared (Bryan and Thompson, 1977). The wider range in composition of the dredged and dive samples compared to Leg 37 basalt is pointed out, a conclusion also evident in the analyses of table 1. Variation trends are shown to be similar for both groups, and a parental liquid for the site 332 basalt is suggested to have been like a young basalt from the median rift. Variations within all the 37°N group are modeled in a preliminary way using plagioclase, olivine, and clinopyroxene phenocryst compositions and crystal-liquid fractionation. Plagioclase is shown to be a dominant fractionating phase in most calculations, a view in accord with the abundance of plagioclase phenocrysts in many samples. On the other hand, olivine is normally a more common and more abundant phenocryst phase than clinopyroxene in hand specimens, and yet fractionation models relating one liquid to another require that the amount of separated clinopyroxene phenocrysts be greater than that of separated olivine. Bryan and Thompson (1977) suggest a stagnated magma chamber that was isolated from a primitive magma as a source for the site 332 basalt in contrast to the steady state magma chamber postulated to account for wider diversity and more highly fractionated marginal basalt of the present rift valley. Their different models are controversial. For example, 'primitive' and 'fractionated' basalt alternate in the deep site (332B), with fractionated basalt occuring both shallow and deep. This can be seen by inspection of the 332B glass analyses (table 1), which are arranged in "stratigraphic" sequence, top to bottom (A36.88N1, Site 332B=top; A36.88N13=bottom).

Phenocrysts whose compositions indicate that they crystallized in "primitive" liquids have been found in evolved, fractionated liquids in basalt cores from the Miocene 22°N drill cores (Dungan et al., 1978). This observation, combined with inclusions of "primitive" liquid compositions in phenocrysts (also Byerly and Sinton, 1979), suggest magma mixing beneath the mid-Atlantic Ridge, or crystal sorting in a zoned magma chamber. This phenomena is also observed at 37°N and appears to be widespread beneath the mid-Atlantic rift valley.

The deep holes drilled in Cretaceous Crust during Legs 51, 52 and 53 provided a surprisingly large amount of fresh basaltic glasses. Their analyses and interpretations have been summarized by Byerly and Sinton (1979). They conclude that these baslats are more similar to the FAMOUS basalts than to the recent Miocene basalts from 22°N, the approximate region from which they spread in Cretaceous time. They conclude that intermediate to high pressure processes involving clinopyroxene as either a fractionating phase or mantle residual phase accounts for differences in composition observed between their Cretaceous basalt and the recent to Miocene basalt from 22°N and 37°N

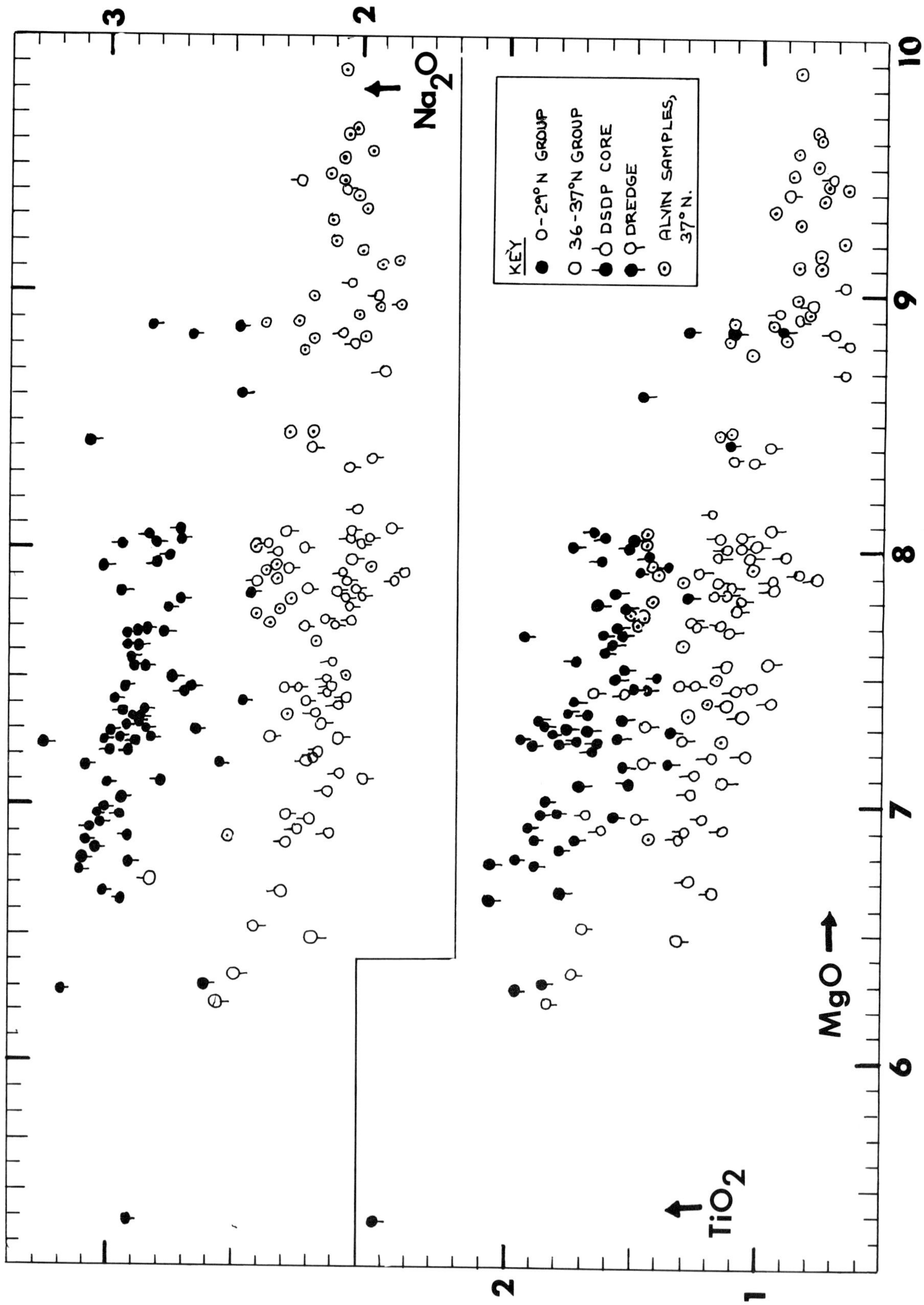

Fig. 2. Differences between the 0-29°N and 36-37°N groups in terms of MgO-TiO_2-Na_2O.

Characteristics of the Data Base

The analyses are limited to major and a few minor elements: Si, Al, Fe, Mg, Ca, Na, K, Ti and P, all reported as oxides, with all Fe reported as FeO.

Analyses are of glass chips from pillow lava rinds performed by an ARL 9-channel electron microprobe. A defocused beam of about 50-microns diameter at 15 kilovolts and 30-microamp beam current was used for all analyses. The large beam size practically eliminates volatilization effects. Correction and normalization procedures and an examination of homogeneity and carbon coating effects are in Melson el al. (1976) and Byerly et al. (1977).

Analyses from a given dredge station, drill site, or submersible station normally fall into one or more groups. Analyses within each group do not differ by more than the analytical precision or by pre-determined tolerances. These group recognition criteria are described in Melson et al. (1975) and are here in modified only to better handle the analyses of glasses in cored pillow lava sequences. For these, each defined chemical type differs from its overlying type in consistently different non-overlapping values for one or more of the oxides.

The data base has a distinct advantage over most in that all analyses were done in the same laboratory using identical analytical procedures. Thus, biases which typically result from different laboratories are eliminated. The number of analysed samples has increased greatly since the inception of IPOD drilling and Project FAMOUS. From 33 chemical groups in 1975, representing 176 analyses, the base has grown to 204 groups, representing 848 analyses (table 1). New relationships have emerged and old ones have been reaccessed. Phenocryst abundances and compositions are not described here. This important area of study is at present incomplete.

The specimen "library" on which the data of table one are based consist of 2.5 cm. diameter lexan discs containing ten peripheral 4 mm diameter holes, each containing polished glass chips of one unknown mounted in epoxy. A central 4 mm hole contains similarly mounted chips of one or more standards run every eleventh analysis. One of these (VG-2, USNM 111240/52) serves as an internal standard used to "normalize" each of the 10 preceding unknowns in a given run. The discs containing the samples analysed in table 1 and in the collections of the U. S. National Museum are available on loan for ion probe, additional electron microprobe and other minimally destructive surface analyses. Researches interested in obtaining chips of the analyzed samples should write those investigators who contributed the samples (see source, table 1 and acknowledgements).

The Two Major Cenozoic Groups

Studies of basalt between 0-37°N have so far dealt mainly with intra-region variations. The recent publication of such studies and the data of table 1 permit a look at inter-region variations. For table 1 data, the analyses fall into two major Cenozoic groups which correlate with location. Of particular importance in discrimination of these groups are Mg, Ca, Na and Ti. Relationships between Mg, Ti and Na are in figure 2. Starting at the Equator and extending to around 29°N Recent to Eocene basalt is characterized by low Mg and Ca and high Na and Ti. This group, termed here the 0-29°N group, includes dredge samples and drill cores to at least Miocene age on both sides of the median valley at 22°N (DSDP sites 395 and 396) and to at least Eocene age on the west side of the median valley (A23.42N1 and N2, table 1).

The second major group, termed the 36-37°N group, is characterized by higher Mg, Ca, K, and lower Ti and Na than the 0-29°N group. This group carries this signature to at least Miocene time in this area (DSDP site 335), and includes all dredge, dive and drill samples. Table 2 includes the average compositions of these two groups.

We have no samples of basaltic glass from the MAR between 29°N (Atlantis Fracture Zone) and 35°N (Oceanographer Fracture Zone). This region is the contact zone between the two above groups and thus the nature of this contact zone is uncertain based on the data presented here.

Previous work suggested that Ti and Na increased gradually between 37°N and the Equator (Shido and Miyashiro, 1973; Melson et al., 1976). Looking closely at the 0-29°N group, we find that samples from 0-15°N have subtle but significant differences compared to samples from 22-29°N. These include the highest Na (greater than 3% in types A5.92N1, A7.15N1 and 7.82N1) and highest Ti (greater than 2% in types A0.33N1, A0.93N2 and A11.02N2) contents in the entire data set. Thus, there is a suggestion of a gradient-like increase in these two elements between the Equator and 37°N.

The analyses of Shibata and Fox (1975) show that basalt from the Oceanographer fracture zone is mainly high in Na and Ti and low in Ca, that is, belongs in the 0-29°N group. However, two samples of the 54 analyzed have low Ti and Na and high Ca and are close to some members of the 36-37°N group. From this work, it looks like the change between the 36-37°N and 0-29°N groups is abrupt and occurs at the Oceanographer Fracture Zone.

At about 0.93°N on the north flank of St. Paul's Rocks basalt low in MgO and high in K_2O, P_2O_5 and TiO_2 occurs (A0.93N1 and A0.93N2, table 1). These highly fractionated basalt samples are unique among the 0-37°N samples and will be discussed more fully in relationship to the Sierra Leone 'hot spot'.

Petrogenesis

The focus of this paper is mainly on the definition in space and time of basalt erupted along the Mid-Atlantic Ridge. Nonetheless, a few suggestions are in order concerning genetic re-

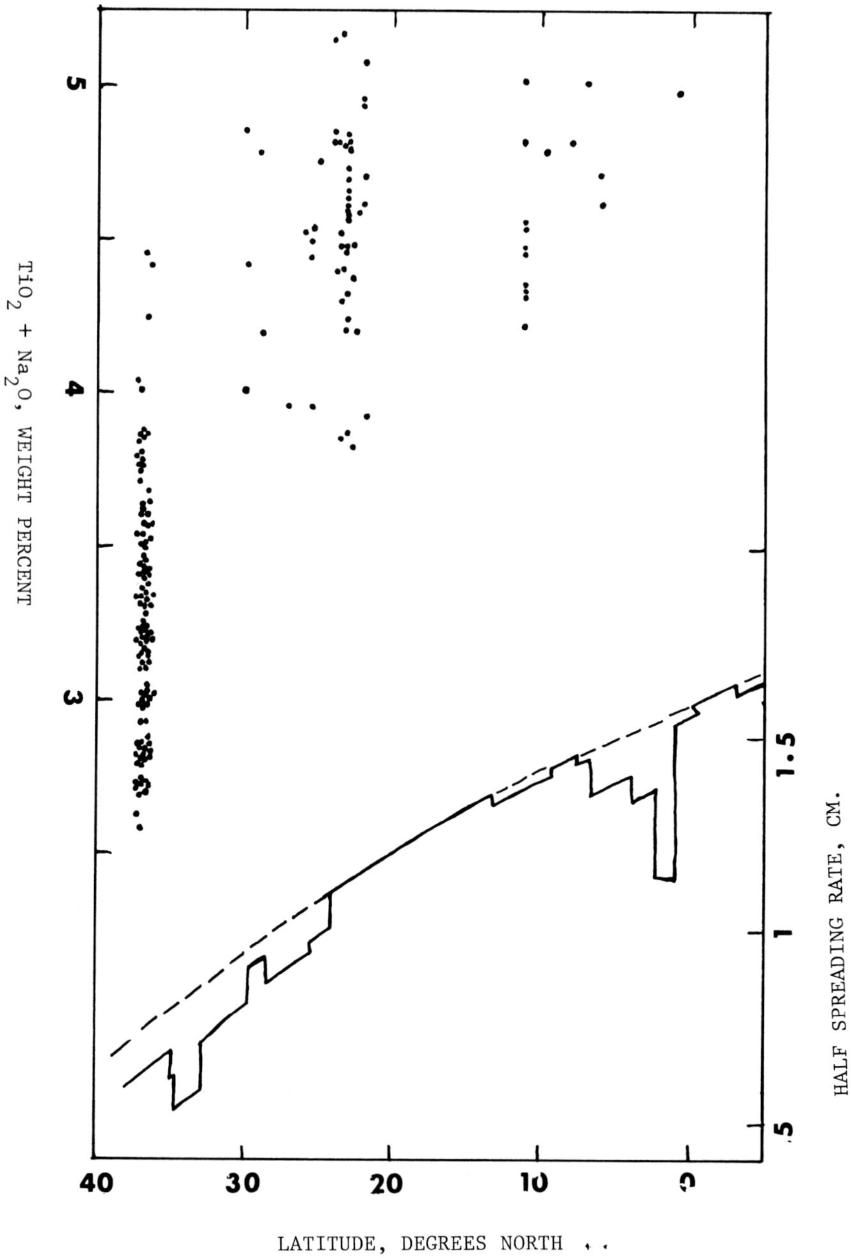

Fig. 3. Relationship between half-spreading rate (Morgan, 1967) and $Na_2O + TiO_2$.

lationships between the various basalt groups just described and listed in tables 1 and 2.

A scan of MgO variation diagrams two of them are reproduced in figure 2- show the effects of plagioclase and olivine fractionation in creating much of the compositional diversity. However, even allowing for clinopyroxene fractionation, we find that more than one parent liquid composition must be invoked and that the mantle source compositions for the two major Cenozoic groups must differ. In regard to this latter point, we conclude this from the average lower K_2O and higher P_2O_5 contents of the 0-29°N group compared to the 36-37°N group. This feature rules out fractionation of plagioclase, olivine or clinopyroxene as accounting for the differences between these two groups because such fractionation cannot account for the different K_2O/P_2O_5 ratios: neither K nor P occur in the fractionating phases. It is noteorthy, too, that the higher K_2O content of the 36-37°N basalt is not coherent with the major element differences between the groups: the 36-

37°N group is mainly higher in Ca and Mg and has lower Fe/Fe+Mg than than the 0-29°N group; i.e., is more primitive.

Some of the differences in major element composition between the 0-29°N and 36-37°N groups can be accounted for in terms of a higher diopside component in the later. Presnal (1978) studied the simplified basalt system diopside-anorthite-forsterite from 1 atm. to 7 kilobars. In this interval the first liquids in equilibrium with the 3 phases become less diopsidic as pressure increases: from 30% to 50% diopside. From this line of reasoning, it looks like the higher diopsidic component in the 36-37°N group reflects lower pressure melting of a mantle source similar in major element composition to that of the 0-29°N group. Or alternatively, more effective re-equilibration of a rising liquid in the 36-37°N region.

The 0-29°N group contains a distinctive high alumina(Al_2O_3 greater than 17%), high Mg, and low Na, K, Ti and Si subgroup of 'primitive' aspect (A23.22N1, A24.42N1, and A28.90N1, table 1). These differences lead to a very high normative anorthite+forsterite/diopside ratio compared to all other chemical types in table 1. The high alumina type A28.90N1 was previously reported by G. D. Nicholls (1965) and attributed to partial melting of a hornblende peridotite, a view unlikely in the now documented very low H_2O content of abyssal basalt from the Mid-Atlantic Ridge. We are left with explanations involving different source mantle composition (higher Al), different partial melting pressures and temperatures and/or different crystal fractionation during ascent- e.g. rapid ascent leading to retention of much of the plagioclase and olivine which are normally lost by near surface crystal settling and/or flotation.

The high alumina glasses from 23°N (A23.22N1, A23.42N1) are paradoxical in that although'primitive' appearing in their high MgO and low TiO_2 (1.11%), suggesting a high degree of partial melting, they are abnormally high in K_2O, one containing 0.22%, the highest value among the 22--24°N samples. This seeming paradox also applies to the higher content of K_2O in the 37°N group compared to the 0-29°N group.

Relationships to 'hot spots'

'Hot spots' as used here are synonymous with major, centralized or elongate active volcanic regions containing unusually thick piles of lavas compared to adjacent regions. Of special importance in the context of plate tectonics are those 'hot spots' which appear to have remained relatively'anchored' or fixed compared to lithospheric plates, leaving traces of abnormal accumulations of lavas outward from the active volcanic zone. In regard to the data at hand, the Azores, Colorado Seamounts, and Sierra Leone Rise - moving successively south along the Mid-Atlantic Ridge-are of interest. These three 'hot spots' may be reflected, for example, in the higher K_2O contents in nearby abyssal basalt described here.

Basalt from 'hot spots' have a number of compositional characteristics which distinguish them from abyssal basalt erupted at considerable distances from 'hot spots'(e.g. White et al.,1978). Of interest here is K_2O, which, along with other large ion lithophile elements, is enriched in most basalt from 'hot spots'. The high K_2O content of the 36-37°N group compared to the 0-29°N group may be attributed to a southward flow of Azores basaltic magma, mixing with 'normal','depleted' abyssal basaltic liquids. Such a mixing process has been speculated to be be important in this region by a number of authors, and to be reflected in light rare-earth element enrichment and in higher Sr^{87}/Sr^{86} ratios. For this case, it looks like a 'damming' of the Azore's liquids occurs at the Oceanographer fracture zone.

The proposed Colorado Seamount 'hot spot' is extrapolated to 34°N along the current ridge axis(Burke et al.,1973), the region for which we have no samples. The data of White et al.,(1978) do indicate, though, that some of the lavas from the 34--35°N region do have'hot spot' chemical characterisitics. It may be significant too that our most northerly sample in the 0-29°N group (A29.86N1, table 1) has the highest K_2O content (0.34%) of this group.

The high K_2O content of basalt dredged near St. Paul's Rocks (A0.91N1,0.93N1 and N2, table 1), and the light rare earth element enrichment (Frey, 1970) and mainly high $Sr^{87/86}$ (Melson et al.,1971) of St. Paul's Rocks itself may be in some way connected with the Sierra Leone Rise. Previously, the alkaline nature of the igneous rocks associat-with St. Paul's Rocks were attributed to small degrees of melting beneath a 'leaky' transform fault zone (Thompson and Melson, 1972). Such an interpretation cannot explain the high Sr^{87}/Sr^{86} values. The 'thermal plume' models, which in some versions involves tapping of largely undifferentiated deep mantle with high Rb/Sr, can account for this feature. St. Paul's Rocks is on the same roughly east-west trending fracture zone high which, to the east of St. Paul's Rocks,intersects the Sierra Leone Rise. St. Paul's Rocks may well include among its light-rare earth enriched, high Sr^{87}/Sr^{86} peridotites candidates for 'hot spot' magmas.

Secular Variations

Two groups were erupted in Jurassic(DSDP site 105, table 2) and Cretaceous time (K-25°N, table 2) from Mid-Atlantic Ridge segments currently between the Equator and 37°N and erupting the just described two major compositional groups. The Cretaceous basalt cored on Legs 51, 52 and 53 is distinct from both the 36-37°N group and the 0-29°N groups but has affinities with the former. This Cretaceous basalt is readily distinguishable from the 0-29°N group by lower Ti and Na, which fall in the range of the 36-37°N group. This basalt was cored on sea-floor spreading flow lines now extending to about 19°N and thus show that a distinct secular change occurred in basalt composition between Eocene and Cretaceous time. There are no drill cores from basement between these ages and thus the character of this change - whether abrupt, transitional, alternating, or some other form - remains unknown. It is now clear that a great deal can be learned about the chemical signature of basalt from a given region by even a single bit or shallow hole into basement. Thus, any future DSDP Atlantic drilling could readily define the nature of this transition.

Basalt cored in Jurassic crust at DSDP site 105 may be either intrusive or extrusive. Its composition falls in the 36-37°N group. If it is an extrusive, it was erupted from the MAR segment currently erupting the 0-29°N group. This observation and the similarity of the 25°N Cretaceous group (K-25°N, table 2) to the 36-37°N group indicate, albeit weakly at present, that the high Na, Ti, low Ca, Mg 0-29°N group has successively migrated northward along the ridge axis since Mesozoic time. Additional basalt drill cores are needed to verify or disprove this suggestion.

Relationships with Spreading Rate

The current range in sea-floor spreading half rate is about 0.6 cm/yr at 37°N to 1.5 cm/yr at the Equator around a seemingly coherent pole of opening around 62°N, 36°W (Morgan, 1972). Correlation coefficients were calculated between individual oxide values and latitude, which correlates well with spreading rate. This exercise reveals that there is no correlation. A more interesting result is obtained by plotting the sum of TiO_2 and Na_2O, two major discriminants for the two major Cenozoic groups, against latitude and spreading rate (figure 3). If we delete the high alumina basalt of the 0-29°N group and the high K_2O basalt dredged near St. Paul's Rocks, we find a suggestion of a correlation between the minimum TiO_2+Na_2O and spreading rate. Finally, it is true that the region of the 0-29°N group, with the various already described chemical discriminants, is from the region with an average higher spreading rate than for the 36-37°N group.

Summary

(1) Basalt from 0-37°N from Recent to Miocene age show a diversity of compositions reflecting well-documented effects of different parental mantle source compositions, high to low pressure fractional crystallization of plagioclase, clinopyroxene and olivine, and magma mixing. Nonetheless, the basalt major and minor element compositions fall into two distinctive groups which correlate with latitude intervals. Chemicals discriminants among the nine elements reported here are:

0-29°N: High Na_2O, TiO_2, low CaO, high P_2O_5, low K_2O

36-37°N: Low Na_2O, TiO_2, P_2O_5, high CaO, K_2O, MgO

(2) The contact zone between these two groups appears to be at the Oceanographer fracture zone (35°N) and may be abrupt. This contact may have migrated northward from at least 22°N since the Jurassic based on basalt core compositions from the western Atlantic.

(3) Basalt in the deep drill cores in Cretaceous crust on flow lines from 0-29°N form yet another distinctive group. This group is closer to the 36-37°N group but has lower MgO. Compared to the 0-29°N, it has higher CaO and lower TiO_2 and Na_2O.

(4) Higher K_2O (and possibly coherent U and Th values) may reflect the effect of the Azores 'plume' in the 37°N group. The otherwise more primitive aspect of this group compared to the 0-29°N may reflect more extensive melting due to a higher abundance of radiogenic heat sources in the mantle beneath the Azores platform.

(5) Maximum K_2O values in abyssal basalt occur near St. Paul's Rocks and may be a reflection of ongoing activity of the postulated Sierre Leone, on St. Paul's Rocks 'thermal plume'.

(6) Each magma type in table 1 is the end result of a long chain of processes including: (1) parent mantle composition and pressure and temperature intervals and times of partial fusion, (2) subsequent rise of partial melt and consequent "wall rock" interactions and high to intermediate pressure crystal-liquid fractionation, and (3) low pressure near surface (magma chamber) crystal-liquid fractionation, and (4) mixing with magmas with different histories and compositions at any stage in the processes. Thus it is expected and observed that most individual glasses from single erupted cooling units are commonly unique in regard to one or more elements, to combinations of elements, or to radiogenic isotope ratios. Thus, it not surprising that the search for simple explanations for the diversity of compositions is a frustrating one. It is nonetheless important that the search continue if we are to understand the processes of magma origins and their relationships to the mechanics of spreading centers.

Acknowledgements

Volcanic glass samples were contributed by numerous institutions and individuals as follows and indicated in table 1: W = Woods Hole Oceanographic Institution through Geoffrey Thompson, Wilfred Bryan and Vaughan Bowen; O = Office of U. S. Naval Research through Peter Vogt and Leonard Johnson, V = from Tjeerd Van Andel now at Stanford University, I = Deep Sea Drilling Project; L = Lamont-Doherty Geological Observatory; S = Anatol Sharaskin; M = University of Miami through Elba and Jose Honnorez, Enrico Bonatti, and Martin Perlmutter; T = University of Texas through Robert Scott. At the U. S. National Museum, Eugene Jarosewich, Charles Obermeyer and Joe Nelen assisted with the electron microprobe analyses and wet chemical analyses of standards and Richard Johnson prepared the polished grain mounts. We gratefully acknowledge all this assistance.

References

Aumento, F., Melson, W.G., et al., Initial Reports of the Deep Sea Drilling Project, Vol. 37, Washington (U.S. Government Printing Office), 1008p., 1977.

Burke, K., Kidd, W.S.F., and Wilson, J.T., Relative and latitudinal motion of Atlantic hot spots, Nature 245, 133-137, 1973.

Bryan, W.B., and Thompson, G., Compositional variations of young basalts in the Mid-Atlantic Rift Valley near 36°49'N, Geol. Soc. Am. 88, 556-570, 1977.

Bryan, W.B., and Thompson, G., Basalts from Leg 37 and the FAMOUS area: compositional and petrogenetic comparisons, Can. Jour. Earth Sci 19, 875-885, 1977.

Byerly, G.R., Melson, W.G., Nelen, J.A., and Jarosewich, e., Abyssal basaltic glasses as indiators of magma compositions, Smithsonian Contr. Earth Sci.,19, 22-29, 1977.

Byerly, G.R., and Sinton, J.M., Compositional trends in natural basaltic glasses from DSDP holes 417D and 418A, Initial Reports of the Deep Sea Drilling Project 51, 52, 53, in press, 1979.

Byerly, G.R., and Wright, T.L., Origin of major element trends in DSDP Leg 37 basalts, Mid-Atlantic Ridge, Jour. Vol. Geotherm. Res. 3, 229-279, 1978.

Frey, F.A., Rare earth and potassium abundances in St. Paul's Rocks, E.P.S.L., 7, 351- 360, 1970.

Heirtzler, J.R., and Van Andel, Project FAMOUS: Its origin, programs, and setting, Geol. Soc. Am., 88, 481-487, 1977.

Melson, W.G., Basaltic glasses from the Deep Sea Drilling Project: chemical characteristics, compositions of alteration products, and fission track 'ages', EOS,54, 1011, 1973.

Melson, W.G., Byerly, G.R., Nelen, J.A., O'Hearn, T., Wright, T.L., and Vallier, T., A catalog of the major element chemistry of abyssal volcanic glasses, Smithsonian Contr. Earth Sci., 19, 31-60, 1977

Melson, W.G., Vallier, T., Wright, T.L., Byerly, G., and Nelen, J., Chemical diversity of abyssal volcanic glass erupted along Pacific, Atlantic, and Indian Ocean sea-floor spreading centers, AGU Geophys. Mon. 19, 351-368, 1976.

Melson, W.G., Hart, S.R., and Thompson, G., St. Paul's Rocks, Equatorial Atlantic: petrogenesis, radiometric ages, and implications on sea-floor spreading, Geol. Soc. Am. Mem. 132, 241-272, 1972.

Morgan, W.J., Rises, trenches, great faults, and crustal blocks, Jour. Geophys. Res. 73. 6,1967.

Nichols, G.D., Basalts from the deep ocean floor, Min. Mag. 34, 373-388, 1965.

Presnal, D.C., Dixon, S.A., Dixon, J.R., O'Donnell, T.H., Brenner, N.L., Shrock, R.L., and Dycus, D.W., Liquidus phase relations on the join diopside-forsterite-anorthite from 1 atm to 20 kilobars: their bearing on the generation and crystallization of basaltic magma: Contr. Min. Pet. 66, 203-220, 1978.

Shibata, T., and Fox, P.J., Fractionation of abyssal tholeiites: samples from the Oceanographer fracture zone, EPSL 27, 62-67.

Shido, F., and Miyashiro, A., Compositional differences between abyssal tholeiites from north and south of the Azores, Nature Phys. Sci. 245, 59-60, 1973.

Thompson, G., and Melson, W.G., The petrology of oceanic crust across fracture zones in the Atlantic Ocean: evidence of a new kind of seafloor spreading, Jour. Geol. 80, 535, 1972.

White, W.M., and Schilling, J., The nature and origin of geochemical variation in Mid-Atlantic Ridge basalts from the central north Atlantic, Geochim. Cosmochim. Acta 42, 1501-1516, 1978

THE EVOLUTION OF OCEAN-FLOOR BASALTIC MAGMAS

J. M. Rhodes

Department of Geology, University of Massachusetts, Amherst, Massachusetts 01003

M. A. Dungan

Department of Geological Sciences, Southern Methodist University, Dallas, Texas 75275

Abstract. The spatial and temporal control provided by recent drilling into the volcanic basement of the Atlantic Ocean, in combination with integrated petrologic and geochemical studies, has afforded new insights into the volcanic processes occurring along mid-ocean ridges. Several lines of evidence indicate that a specific, rather well-defined, "primitive" magma type, characterized by high atomic Mg/(Mg+Fe) values (0.70-0.72), high CaO (12.0-13.5%) and MgO (10%) contents, and low TiO_2 (0.6-0.9%) Na_2O (< 2.0%) and magmaphile trace element (Sm 6-8xch.) abundances, may well be a primary, mantle-derived melt, parental to other, more abundant, evolved ocean-floor basalts. Basalts such as these have been sampled directly at only a few localities, but there is abundant indirect evidence, such as phenocryst compositions, melt inclusion data, and glass and whole-rock compositional trends, attesting to their importance in ocean-floor basalt petrogenesis. We believe that more evolved basalts are derived from these "primitive" melt compositions by a combination of crystal fractionation and magma mixing. "Primitive" magma is repeatedly injected into a magma chamber and mixed with a more evolved, consanguineous magma that has fractionated from previous episodes of injection and mixing. This "steady-state" model results in the generation of voluminous moderately evolved tholeiites, without the necessity of large magma chambers, and accounts for several enigmatic features of ocean-floor basalts not readily explained by crystal fractionation processes alone. These include: difficulties frequently encountered in attempts to relate spatially associated basalts to a common parental magma; the necessity to invoke pyroxene-dominated fractionation, when pyroxene is a minor component or is absent from the phenocryst assemblage; magmaphile element abundances in the evolved basalts that are higher than those predicted solely from crystal fractionation processes; and nickel values that are too high for the amount of fractionation deduced from magmaphile element abundances. This model successfully predicts that both "primitive" and highly differentiated basalts and silicic derivatives will be rare and that moderately evolved basalts should dominate layer 2 of the oceanic crust. Gabbroic derivatives of these evolved magmas should be volumetrically important in layer 3, whereas olivine-rich cumulates, troctolites and highly iron-enriched gabbros would be less significant.

Introduction

Petrological and geochemical studies of ocean-floor basalts have benefited substantially from an energetic sampling program within the last decade, and in particular from the spatial and temporal control provided by a combination of drilling into the oceanic basement and detailed sampling by dredge and submersibles along segments of mid-ocean ridges. It is now widely recognized, following O'Hara (1968), that the vast majority of ocean-floor basalts sampled to date are of an evolved nature. That is, their compositions have been modified from those of primary magmas generated by partial melting within the mantle, both en-route to, and during emplacement on the sea floor. Evidence for this is provided by: (1) petrographic observations, supported by experimental studies, showing that most ocean-floor basaltic liquids are multiply saturated, and that their compositions are controlled by the low-pressure olivine-plagioclase cotectic, or olivine-plagioclase-clinopyroxene peritectic (e.g., O'Hara, 1968; Kay et al., 1970; Shido et al., 1971; Frey et al., 1974; Bryan et al., 1976; Dungan et al., 1978a; Bender et al., 1978). Mg'-values* for basaltic liquids clustering about 0.60 (Figs. 1,2) that are too low for inferred primary mantle derived melts (Green, 1970; O'Hara et al., 1975) and imply substantial olivine fractionation; (3) Ni abundances commonly less than 200 ppm, again indicating olivine fractionation (Sato,

*Mg'-value = mol. prop. Mg/(Mg+Fe) after adjusting Fe^{+3}/total Fe as FeO = 0.1.

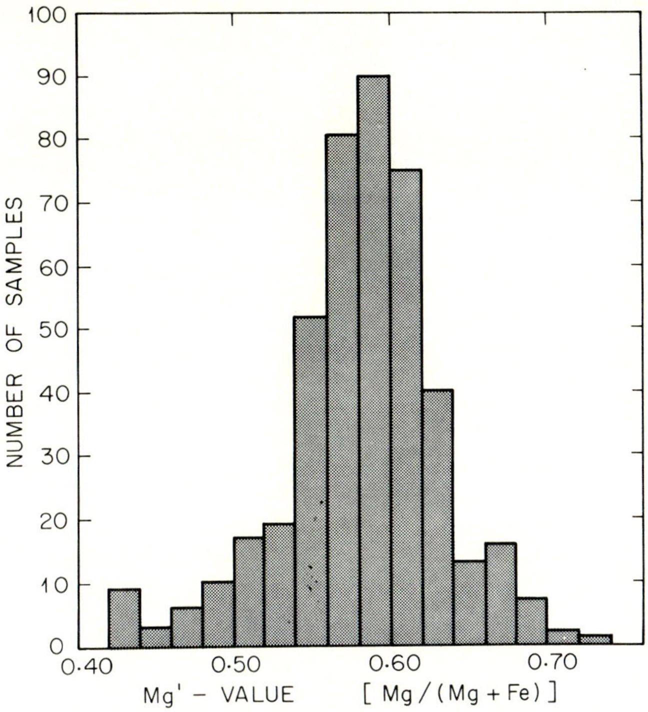

Figure 1. Histogram of Mg'-values for ocean-floor basaltic glasses. Data from Frey et al., 1974; Melson et al., 1976; Langmuir et al., 1977.

1977); (4) the presence of small negative Eu anomalies in some basalts, especially those with high magmaphile element abundances (Kay and Hubbard, 1978), suggesting extensive (>25%) plagioclase fractionation.

The nature of the evolutionary process is not yet clearly understood, and is a fundamental problem in ocean-floor basalt petrogenesis, with important implications concerning the nature and composition of the oceanic crust and volcanic processes occurring along mid-ocean ridges. For example, consider the following potential but not mutually exclusive possibilities: (1) the evolved basalts that dominate the samples obtained from the oceanic basement may be derived from more "primitive," possibly primary basalts, by crystal fractionation at shallow levels in the crust. If this is the case, then as much as 40 to 50 percent of the magma must solidify to produce the commonly observed varieties of ocean-floor basalts (Bender et al., 1978), with the consequence that huge volumes of "primitive" magma are required beneath spreading ridges, either in large magma chambers or within vast interconnecting conduits of dykes, sills and laccoliths, etc. A further requirement of this model is that the cumulates so formed will be dominated by early crystallizing olivine and plagioclase. If layer 3 of the oceanic crust is composed in part of the coarsly crystalline derivatives of basaltic magmas, and their metaphorphosed equivalents (e.g., Dewey and Kidd, 1977), then dunites and troctolites, with mineral chemistries reflecting the "primitive" nature of the parental magma, should form a substantial component of this layer. (2) If, however, fractionation occurs within the mantle close to the source, possibly at depths of around 25-30 km, there is no compelling reason for the presence of large magma chambers beneath the ridges. The evolved basalts could be derived directly from mantle depths, or could have accumulated in small magma chambers and conduits prior to eruption. If this is the case, then layer 3 should differ substantially from that proposed for the previous model, being dominated by gabbros containing plagioclase, olivine and pyroxene with compositions reflecting the evolved nature of the parental magma. (3) A third possibility involves magma mixing, where "primitive" mantle-derived magma is repeatedly injected into a shallow magma chamber and mixed with an evolved, consanguineous magma that has fractionated from previous episodes of injection and mixing (O'Hara, 1977; Rhodes et al., 1978). Again, there is no necessity for large magma chambers beneath the ridges, and gabbroic derivatives are expected to dominate in layer 3 rather than the dunitic and troctolitic derivatives of the first model.

It is the purpose of this paper to explore these evolutionary models and their implications, emphasizing in particular the role of magma mixing. Although magma mixing has been largely ignored by petrologists in recent years (c.f. Carmichael et al., 1974) it is becoming increasingly evident, following the work of Wright and colleagues (Wright and Fiske, 1971; Wright, 1973), Anderson (1976) and O'Hara (1977) that it is an intrinsically plausible phenomenon in volcanic processes. In view of the tectonic regime of ocean-ridge volcanism, where magma is being constantly generated and replenished along a narrow zone of activity, it appears likely that new batches of magma will be intruded prior to the solidification of previous magma batches, and therefore that magma mixing is to be expected. Furthermore, we believe that magma mixing may explain several enigmatic features of ocean-floor basalt petrogenesis not readily understandable in terms of crystal fractionation processes alone.

The Nature of Primary Magmas

Before evaluating the processes of basalt evolution it is desirable to discuss the nature of the primary, mantle-derived melts from which the evolved ocean-floor basalts might be derived. Inspection of Figures 1 and 2 shows that, although the majority of ocean floor basaltic glasses have Mg'-values of about 0.60, a few have higher values, between 0.70 and 0.73. These are the most "primitive" basaltic <u>liquid</u> compositions identified to date. There are no basaltic glasses with Mg'-values higher than 0.73, and rocks with higher values invariably show signs of olivine accumulation. Mg'-values between 0.70 and 0.73 are compatible with the most forsteritic olivine

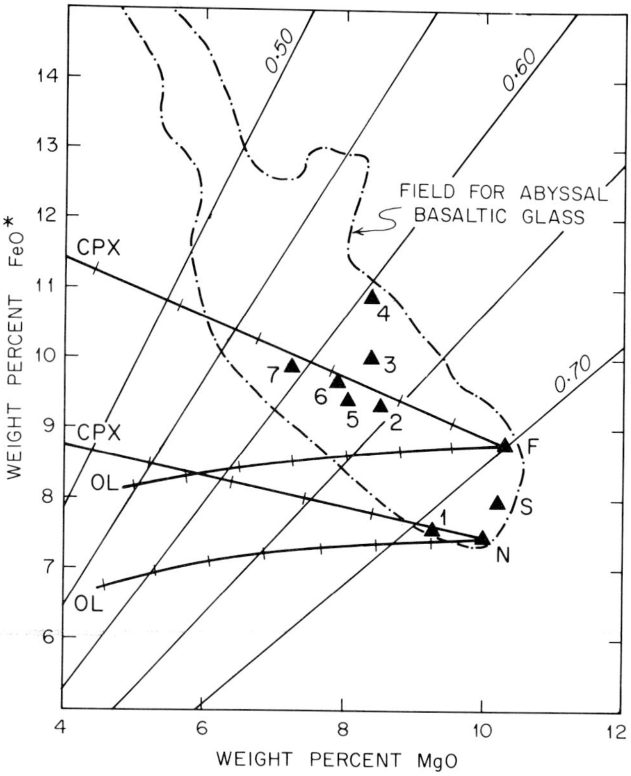

Figure 2. MgO-FeO* relationships showing the field for ocean-floor basaltic glasses and selected basalts discussed in the text. Samples F, S, N are primitive basalts respectively from the FAMOUS area (Langmuir et al., 1977), South Atlantic (Frey et al., 1974), North Atlantic (Rhodes and Rodgers, 1972). Samples 1, 2, 3 and 4 are aphyric basalts with olivine on the liquidus, and samples 5, 6 and 7 are also aphyric but with both olivine and plagioclase on the liquidus. Samples 1 and 2 are from the North Atlantic, 45°N (Rhodes and Rodgers, 1972), 3 and 4 from DSDP Site 395 (Rhodes et al., 1978), and 5, 6 and 7 from DSDP Site 396 (Dungan et al., 1978b). The dashed lines indicate Mg'-values from 0.5 to 0.7. Olivine and clinopyroxene fractionation trends from two "primitive" basalts (F,N) are shown in 2 and 10 percent intervals respectively. Plagioclase fractionation will follow the constant Mg'-value lines. Data for the basaltic glass field are taken from Frey et al., 1974; Melson et al., 1976; Bryan and Moore, 1977; Langmuir et al., 1977.

phenocryst cores (Fo_{90}) found in ocean-floor basalts, assuming K^D between 0.27 and 0.33 for olivine-melt partitioning (Roeder and Emslie, 1970), and correspond closely with values predicted for primary magmas derived by partial melting of model mantle compositions (Green, 1970; O'Hara et al., 1975; Bender et al., 1978).

Sato (1977) and Hanson and Langmuir (1978) predict that primary ocean-floor basaltic magmas in equilibrium with olivine-bearing model mantle compositions should contain about 240 to 300 ppm Ni. This is precisely the range found for ocean-floor basalts with Mg'-values between 0.70 and 0.73 (Bougault and Hekinian, 1974; Frey et al., 1974; Langmuir et al., 1977; Rhodes, unpubl. data). On the other hand, the majority of evolved basalts contain less than 200 ppm Ni (Kay et al., 1970), and those with over 300 ppm show signs of olivine accumulation (Blanchard et al., 1976; Rhodes unpubl. data) and do not reflect liquid compositions.

An important aspect of these "primitive" basaltic glass compositions is that, in addition to the high Mg'-values, between 0.70 and 0.73, they have many compositional characteristics in common, and it appears possible to place relatively tight constraints on such "primitive" ocean-floor basalt magma compositions (Rhodes and Dungan, 1977). All are characterized by low TiO_2 (0.6-0.9%), Na_2O (1.7-2.1%) and magmaphile trace element abundances (e.g., Sm = 6-8xch.), and by high MgO (≃10%) and CaO (12.0-13.5%) contents and CaO/Al_2O_3 (0.75-0.85) ratios. Figure 3 illustrates the way in which the TiO_2 contents of basaltic glasses converge at these high Mg'-values towards a well-defined compositional apex. This is an important observation, bearing on subsequent discussion of basalt evolution, and it must be emphasized that there are no recorded examples of ocean-floor basaltic liquid compositions with Mg'-values greater than 0.70 and TiO_2 contents above 1.0 percent.

Basalts with the above compositions are rare, but have been identified at several dispersed localities in the Atlantic Ocean (Rhodes and Rodgers, 1972; Frey et al., 1974; Bougault and Hekinian, 1974; Bryan and Moore, 1977), and typical analyses are given in Table 1. Although

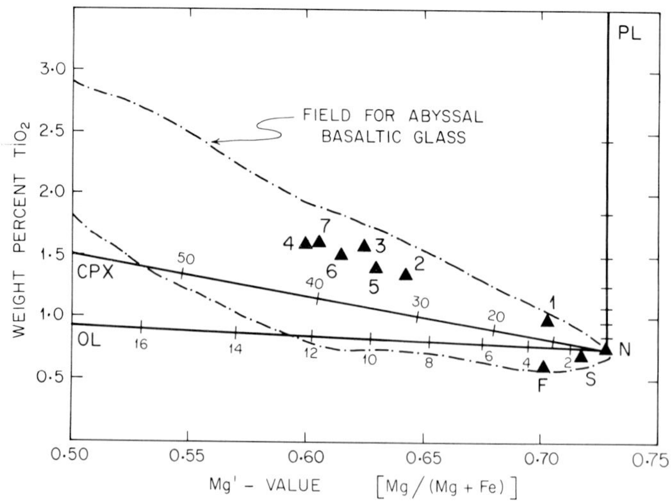

Figure 3. The relationship between Mg'-value and TiO_2 content for ocean-floor basalts. Symbols and sources of data are the same as in Figure 2.

TABLE 1. Chemical Characteristics of Some Potential Primary Ocean-Floor Basaltic Magmas (recalculated to 100% on an anhydrous basis)

	1	2	3	4	5	6
SiO_2	49.7	49.5	49.1	50.8	50.0	50.5
TiO_2	0.72	0.81	0.62	0.62	0.65	0.76
Al_2O_3	16.4	15.7	16.5	14.4	14.5	15.6
FeO*	7.89	7.45	8.78	6.89	8.26	7.3
MnO	0.12	0.15	0.15	0.10	n.d.	0.2
MgO	10.1	10.0	10.3	12.0	11.7	9.4
CaO	13.2	13.0	12.4	13.6	13.0	13.7
Na_2O	2.00	1.95	1.92	1.43	1.71	2.1
K_2O	0.01	0.17	0.07	0.07	0.03	0.4
P_2O_5	--	0.08	0.06	--	--	--
Cr_2O_3	0.07	0.14	0.06	0.10	--	--
Mg'-value	0.72	0.73	0.70	0.77	(0.72)	0.72
CaO/Al_2O_3	0.81	0.82	0.75	0.94	0.90	0.88
Ni ppm	320	249	232	--	--	--
Sm ppm	1.60	1.95	1.37	--	--	--
La/Sm	0.63	1.81	1.57	--	--	--

1. Frey et al. (1974), 3-14.
2. Rhodes (unpubl. data) Chain 43 #23, 45°N.
3. Langmuir et al. (1977), FAMOUS, 527-1-1.
4. Donaldson and Brown (1977), Avg. Melt inclusion in spinel.
5. Dungan and Rhodes (1978) interpolated from melt inclusion data.
6. Watson (1976) interpolated from melt inclusion data.

uncommon, there is indirect corroborative evidence in phenocryst and melt inclusion data attesting to their importance in ocean-floor basalt petrogenesis. As noted earlier, the maximum observed forsterite content of Fo_{90} for olivine phenocrysts in ocean-floor basalts is appropriate for crystallization from a primary magma having Mg'-values between 0.70 and 0.75. Similarly, Sato (1977) showed that the maximum Ni content of 0.41 percent observed in olivine phenocrysts from ocean-floor basalts is compatible with crystallization from a primary melt immediately following segregation from mantle peridotite. Olivines with such Ni abundances should crystallize from basalt magmas containing about 270 ppm Ni (Hart and Davis, 1978; Irving, 1978), providing, as suggested above, the MgO content is about 10 percent. Similar relationships also hold for plagioclase phenocrysts. Microphenocryst data from basaltic glasses, together with experimental melting studies of appropriate compositions, and theoretical calculations based on plagioclase-melt equilibria, all indicate that plagioclase in equilibrium with common ocean-floor basaltic melts rarely exceeds An_{80}. Yet many phyric ocean-floor basalts contain plagioclase phenocrysta with complex zoning patterns that are frequently accompanied by compositional breaks and cores that are too anorthitic (An_{84-92}) to be in equilibrium with melts appropriate to the whole-rock composition (Donaldson and Brown, 1978; Rhodes et al., 1978). We infer that these refractory cores crystallized from more "primitive" melts, such as the primary magma compositions under discussion, prior to episodes of magma mixing. Evidence supporting this argument comes from experimental melting studies on a single sample of this type (Bender et al., 1978), and from calculations of plagioclase compositions in equilibrium with the proposed primary magma compositions (Table 1), using the expression developed by Drake (1976) for plagioclase-melt equilibria and crystallization temperatures extrapolated from melting studies. The range varies from An_{82} to An_{87} and is in reasonable agreement with the observed phenocryst compositions.

Several authors (Watson, 1976; Donaldson and Brown, 1978; Dungan and Rhodes, 1978) have attempted to deduce the composition of primary ocean-floor basaltic magmas from compositions of melt inclusions in olivine, plagioclase, pyroxene and spinel phenocrysts. A variety of approaches, ranging from direct measurement to extrapolations attempting to compensate for entrapment at various stages of magmatic evolution, and for reaction with the host mineral, reveal certain well-defined characteristics. All the proposed primary magma compositions have high Mg'-values, about 0.7 or greater. All have low TiO_2 (<1.0%) and Na_2O (<2.0%) contents and are high in CaO (>12%), with high CaO/Al_2O_3 ratios. In other words, they have strong compositional similarities to the primary magma compositions discussed earlier in this section. Several examples of these melt inclusion compositions are also included in Table 1.

Thus, on the basis of a variety of information, including phenocryst compositions, melt inclusion data, and glass and whole-rock chemistry, it appears that the most "primitive" basaltic compositions identifiable to date from the ocean floor have many striking characteristics in common (Table 1). These include high Mg'-values (0.70-0.73), Ni concentrations between 250 and 300 ppm, high CaO contents (12.0-13.5%), and low TiO_2 (<0.9%), and Na_2O (<2.18%) and magmaphile element abundances. There is compelling evidence to believe that such compositions, perhaps modified by small amounts of olivine fractionation, are close to those of primary ocean-floor basaltic magmas. This is not to imply that there is only one such composition. Others have been proposed (Bryan, 1978), and it is abundantly clear that, although there is close similarity in major element chemistry, there is no such coherence in trace element and isotopic abundances (Table 1). Such differences are presumed to reflect heterogeneity in the mantle source, or in the magma generating processes (Langmuir et al., 1977). Despite these caveats, we believe that the wide geographic distribution of these "primitive" magma compositions, coupled with the diverse lines of evidence for their existence, attests to their importance in ocean-floor basalt petrogenesis, and makes them a suitable starting point for the discussion of the evolution of ocean-floor basalts.

Evolution of Ocean-Floor Basalts

Once a basaltic melt has segregated from its mantle source, the composition may be modified by a variety of processes, including crystal fractionation within the mantle at high to moderate pressure, and crystal fractionation at shallow depths in magma chambers, conduits, or during extrusion on the sea floor. In addition to these, there is the added complication that successive magma batches may be mixed, probably in shallow magma chambers beneath mid-ocean ridges.

Petrographic and compositional data provide strong evidence for low-pressure crystal fractionation, dominated by olivine and plagioclase, in ocean-floor basalts. These minerals constitute the prevailing phenocryst phases (Shido et al., 1971), and basaltic glass compositions follow a narrow trend, apparently controlled by the olivine-plagioclase cotectic (Bryan et al., 1976). Experimental melting studies at 1 atm. confirm these observations, demonstrating the importance of olivine and plagioclase as the early crystallizing phases, and the significance of the olivine-plagioclase cotectic to ocean-floor basalt petrogenesis (Kushiro, 1973; Fujii and Kushiro, 1977; Dungan et al., 1978a; Bender et al., 1978). Again supporting petrographic evidence, these studies show that pyroxene crystallizes late, anywhere from 40-130°C below the liquidus, depending on the bulk composition. Trace element data are broadly commensurate with the petrographic and experimental conclusions, in particular the depletion in Ni with increasing differentiation and the development of negative Eu-anomalies in otherwise unfractionated rare-earth patterns (Kay et al., 1970; Schilling, 1971; Frey et al., 1974). Figures 2 and 3 illustrate the necessity for coprecipitation of plagioclase and olivine to generate the trends found for ocean-floor basaltic glasses. However, it is important to note that extremely large amounts of crystallization are required for a primary magma of the type discussed earlier to achieve typical ocean-floor basalt compositions solely by crystal fractionation. This varies from about 40-60 percent for the majority of basalts, to 60-80 percent for the high FeO-TiO_2 basalts (FeTi basalts of Melson et al., 1976) common in the Pacific.

In early studies of ocean-floor basalts there was no *a priori* reason why samples from dredge hauls or shallow drill holes in the volcanic basement, separated by several or by hundreds of km, should be related to common parental magmas or even similar magma-types. Consequently, it was difficult to evaluate the evolution of these basaltic magmas, except in a general sense.

The advent of crustal drilling during the IPOD program, and commensurate detailed studies of a segment of the mid-Atlantic ridge by the FAMOUS project, enabled more in-depth studies to be undertaken. From these studies it is clear that magma is erupted onto the sea-floor in discrete, chemically distinct units. They may be of wide areal extent (Bryan and Moore, 1977), and can attain thicknesses of up to 200 m (Rhodes et al., 1978). These magmatic units may contain many separate individual cooling units, yet they are generally chemically homogeneous, with little evidence of internal differentiation.

Since, at a specific sampling locality, these magmatic units are closely associated in space and time, have probably derived from the same mantle source, and presumably shared the same plumbing system, it might reasonably be expected that they are comagmatic, having episodically erupted onto the sea-floor during continuing differentiation of a common parental magma. Detailed studies show that this is not often obviously the case (e.g., Blanchard et al., 1976; Hekinian et al., 1976; Rhodes et al., 1976; Bryan and Moore, 1977; Langmuir et al., 1977; Rhodes et al., 1978). The dilemma is that it is rarely possible to relate associated magmatic units to one another, to a common parental magma, or to hypothetical primary magmas of the type discussed earlier solely by crystal fractionation processes employing the observed phenocryst phases. Frequently, the magmaphile element content of the more evolved unit is higher than that predicted by crystal fractionation models, or, contrary to petrographic evidence, extensive clinopyroxene fractionation appears to be necessary.

The Clinopyroxene Problem

We have noted earlier that petrographic evidence, supported by 1 atm. melting studies, indicates that clinopyroxene is subordinate to plagioclase and olivine during low-pressure fractionation of ocean-floor basalts. However, clinopyroxene figures prominently in crystal fractionation models for ocean-floor basalts (Clague and Bunch, 1976), and in order to relate associated magmatic units by crystal fractionation, it is frequently necessary to include clinopyroxene as a major fractionating component, along with plagioclase and/or olivine, even though pyroxene is not present, or is clearly subordinate in the phenocryst assemblage (Thompson et al., 1976; Bryan et al., 1976; Bryan, 1978). This is because there is a tendency for the CaO/Al_2O_3 ratio to decrease with increasing differentiation (Figure 4). Two specific examples are shown in Figure 4, one from DSDP Site 396 (Dungan et al., 1978b) and the other from a single dredge haul at 45°N on the mid-Atlantic ridge (Rhodes and Rodgers, 1972). The 396 samples are successive aphyric basalt units with olivine and plagioclase microphenocrysts, whereas the 45°N

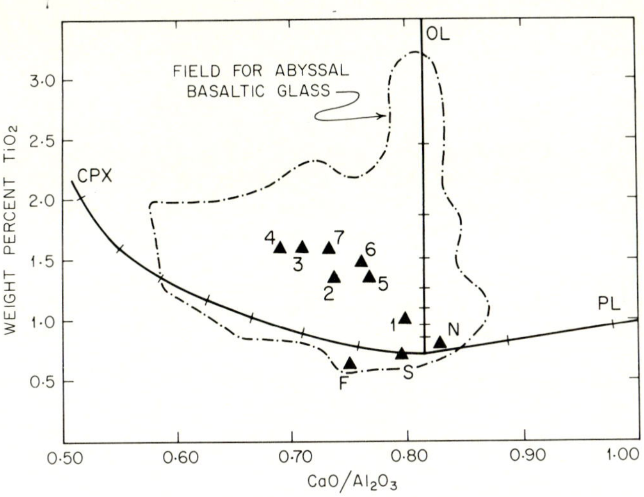

Figure 4. The relationship of CaO/Al_2O_3 and TiO_2 content of ocean-floor basalts to crystal fractionation. Symbols and source of data are the same as in Figure 2. Olivine, plagioclase and clinopyroxene fractionation trends from a hypothetical primary magma composition (see text) are shown in 10 percent intervals.

samples are aphyric basalts with only olivine on the liquidus. Since only clinopyroxene can effect a reduction in the CaO/Al_2O_3 ratio, it follows that both examples require substantial clinopyroxene fractionation in order to derive the evolved units from the more primitive ones. Thus, it appears that, despite petrographic evidence to the contrary, chemical arguments demand that clinopyroxene has participated in the fractionation process, if the associated basalts are indeed comagmatic, and not derived from separate primary magmas that have retained their integrity through subsequent fractionation, emplacement and eruption.

How can these seemingly conflicting arguments be resolved? One possibility is that clinopyroxene fractionation may occur within the mantle at moderate to high pressure where it will be the liquidus phase (Kushiro, 1973; Fujii and Kushiro, 1977; Bender et al., 1978), followed by olivine and plagioclase fractionation at or near the surface. Donaldson and Brown (1978) describe pyroxene megacrysts which may be unstable relicts of former clinopyroxene crystallization, and Bender et al. (1978) suggest that clinopyroxene microphenocrysts or xenocrysts that are minor (<1%) but widespread constituents of ocean-floor basalts (Bryan, 1978), and are characteristically lower in the ferrosilite component than groundmass pyroxene, may be evidence of a prior history of high pressure pyroxene fractionation. Consequently, the ocean floor basalt trends in Figures 2-4 could reflect an initial stage of clinopyroxene fractionation in the mantle shortly after segregation, followed by low-pressure cotectic fractionation of olivine and plagioclase in the crust, which is followed in turn by peritectic fractionation of olivine, plagioclase and clinopyroxene in the more evolved magmas. Although this process is appealing as a mechanism for producing an imprint of clinopyroxene fractionation in magmas that have only olivine or olivine + plagioclase as liquidus phases, Sc abundances in ocean-floor basalts (Frey et al., 1974), commonly in the range of 35-45 ppm, seem to preclude significant amounts of high-pressure clinopyroxene fractionation. The specific examples plotted in Figures 2-4 have Sc concentrations of 38 ± 2 ppm, thus placing even tighter constraints on clinopyroxene fractionation. The effective distribution coefficient for Sc in ocean-floor basalts appears therefore to be close to unity, suggesting that the clinopyroxene fractionation necessary to account for the decrease in CaO/Al_2O_3 with differentiation has occurred at low-pressures, accompanied by co-precipitation of plagioclase and olivine.

Rhodes et al. (1978) have suggested that the compositional imprint of clinopyroxene fractionation on basaltic magmas crystallizing only olivine and plagioclase can be produced by magma mixing. This process is illustrated schematically in Figure 5 by means of the ternary system forsterite-anorthite-diopside. Consider a primary magma (P) fractionating within a shallow magma chamber. Olivine will crystallize first, followed by olivine and plagioclase along the cotectic, until plagioclase, olivine and clinopyroxene crystallize together from an evolved magma at the peritectic (E). If a new batch of primary magma is injected into the magma chamber, it will mix with the evolved magma to produce a magma such as M, somewhere on the line P-E. The mixed magma (M) will crystallize olivine until it reaches the plagioclase-olivine cotectic, at which point it will follow a crystallization path similar to that above, and the whole cycle may be repeated many times. The important point is that although the mixed magma (M) has only olivine on the liquidus, its bulk chemistry will bear the imprint of the preceding episode, or episodes of clinopyroxene, plagioclase and olivine fractionation. According to this model, the examples shown in Figures 2-4 are mixed magmas resulting from the injection of "primitive" melts, close to the primary magma compositions, into magma chambers containing highly fractionated magmas similar to high FeTi basalts.

The Magmaphile Element Problem

A common problem in modeling the crystal fractionation of an assumed parental basaltic magma to produce its evolved derivatives is that the magmaphile element content (TiO_2, P_2O_5, K_2O and trace elements) of the more evolved basalt is frequently higher than the values predicted by the fractionation model using linear least squares mixing calculations (e.g., Bryan et al., 1976; Bryan and Moore, 1977). In other words, there is a distinct discrepancy between the extent of fractionation calculated from major element data

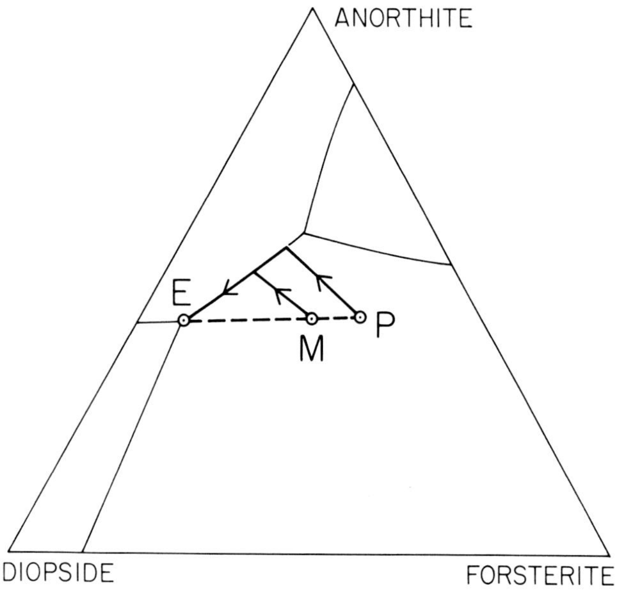

Figure 5. Schematic representation in the system forsterite-diopside-anorthite of fractionation of a primary magma (P) to an evolved magma (E), followed by magma mixing, and the subsequent fractionation path of the mixed magma (M).

and that deduced from the increase in magmaphile elements in the evolved basalt relative to their abundance in the parental basalt (Anderson and Greenland, 1969). Where discrepancies occur, the magmaphile element abundances nearly always provide the higher estimate.

In part, this is due to the calculation used; the magmaphile element abundances are logarithmically related to fractionation and will tend to be higher than the estimates based on linear mixing calculations. However, this is not the entire explanation, and we believe the excess in magmaphile elements in the more evolved basalt results from magma mixing processes (Rhodes et al., 1978). This is shown in Figure 6, where TiO_2 content is plotted against the amount of crystallization necessary for a hypothetical primary magma, similar to those proposed earlier (SiO_2 = 49.9, TiO_2 = 0.7, Al_2O_3 = 16.1, FeO* = 7.7, MgO = 10.1, CaO = 13.1, Na_2O = 1.9, K_2O = 0.1), to produce a derivative magma composition. Crystal fractionation curves for olivine, plagioclase and clinopyroxene have been calculated using the TiO_2 distribution coefficients of Bougault and Hekinian (1974). It is readily apparent from this diagram that a mixed magma involving, for example, the primary melt and an evolved basalt resulting from 70 percent fractionation, will have a much higher TiO_2 content than an equivalent basalt derived solely by crystal fractionation processes. This diagram also shows the average Mid-Ocean Ridge Basalt composition given by Melson et al. (1976), and the field for ocean-floor basaltic glass compositions taken from the same source (see Rhodes et al., 1978 for computational details).

The mid-ocean ridge basalt average plots well above the crystal fractionation curves, as do many of the abyssal basaltic glasses, suggesting either that many ocean-floor basaltic compositions result from a combination of crystal fractionation and magma mixing, or that the primary magma composition used in the model is inappropriate for the majority of ocean floor basalts. We do not believe the latter to be the case, for reasons discussed earlier, and also because: (a) the lower boundary of the basaltic glass field closely mimics the fractionation curves, a necessary condition if this model is realistic; (b) the basaltic glass field converges towards the hypothetical primary magma composition, tending to confirm the composition used; and (c) "primitive" basaltic glass compositions with TiO_2 contents greater than 1.0 percent are not known.

An analogous but more acute example of the discrepancy in estimates of crystal fractionation is found in attempting to compare estimates of crystal fractionation based on magmaphile elements with those based on a highly compatible element such as Ni. The most striking discrepancies are found in those basalts that appear to have undergone only olivine fractionation, with Ni abundances much higher than the magmaphile element-based estimates of olivine fractionation would suggest. Consider three basalt types, samples from a single dredge haul on the mid-Atlantic ridge at 45°N, with the following chemical characteristics:

	Mg'-value	TiO_2 (%)	Ni (ppm)
Type III	0.73	0.81	249
Type II	0.71	1.00	240
Type I	0.64	1.34	197

The Type III basalts have the chemical characteristics of primary magma, and since all three basalt types have only olivine as the liquidus phase, it seems reasonable to suggest that they may be comagmatic, related only by olivine fractionation. Inspection of the data shows that this cannot be, the increase in TiO_2 ($D_{TiO_2}^{oliv/liq}$ = 0.04) requires about 40 percent crystallization, whereas the decrease in Ni is commensurate with only 2 percent olivine fractionation ($D_{Ni}^{oliv/liq}$ = 12). This problem is illustrated in a general sense in Figure 7, where the 45°N data are plotted with respect to four fractionation curves calculated from a hypothetical primary magma. Also plotted are average analyses of aphyric basalts containing only olivine microphenocrysts from DSDP Site 395 (Rhodes et al., 1978) and examples of potential primary magmas. The position of the sample points relative to the lowermost olivine fractionation curve illustrates the problem, indicating that the 45°N samples cannot be interrelated nor can any of the basalt types be derived from the hypothetical primary magma, simply by olivine fractionation. Thus, either each basalt type is derived solely by olivine fractionation

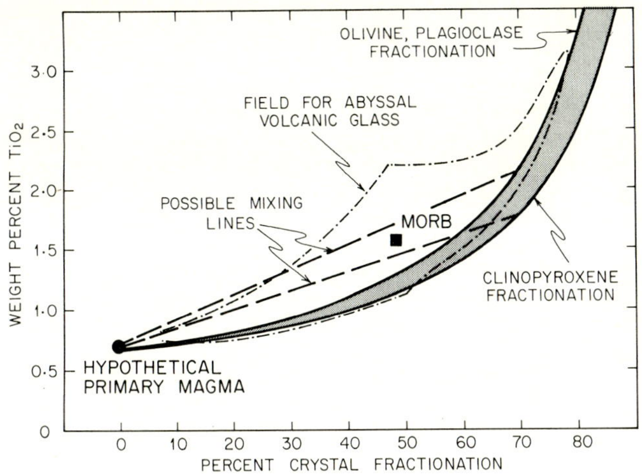

Figure 6. Effects of crystal fractionation and magma mixing on the TiO_2 content of ocean-floor basalts. Olivine, plagioclase and clinopyroxene fractionation curves for a hypothetical primary magma are shown, together with possible mixing trends. The average mid-ocean ridge basalt (MORB) is taken from Melson et al. (1976) and the field for abyssal volcanic glasses from the same sources as Figure 2.

from its own distinct primary magma, or these basalts result from mixing of "primitive" or primary magmas with more evolved magmas in the manner illustrated in Figure 5.

If these basalts are derived from distinct primary magmas directly by olivine fractionation, inspection of Figures 2, 3 and 7 shows that they must possess unique compositions, characterized by high TiO_2 (1.0-1.5%) and FeO (9.5-11.2%) contents. We emphasize that "primitive" basalts (i.e., Mg'-values $\geq$0.70) with these characteristics have yet to be identified. It is unlikely that such compositions will exist because, as was shown by Bender et al. (1978), varying amounts of partial melting of a suitable mantle source will produce magmas with relatively constant Mg'-values (0.74-0.76) and with FeO and TiO_2 rarely above 9.0 and 1.0 percent, respectively. Furthermore, if the site 395 and 45°N Type I basalts are derived directly by olivine fractionation from high FeO-TiO_2 primary melts with Mg'-values close to 0.70 and Ni contents of about 300 ppm, the amount of olivine crystallization required (10% and 7%, respectively) is too much for the observed Ni content (Figures 2, 7).

Consider, then, the mixing of a hypothetical primary magma (Figure 7) with basalts derived from a similar magma by a previous episode of crystal fractionation. In addition to the olivine trend, three other fractionation trends are shown in Figure 7. The uppermost one reflects plagioclase and olivine coprecipitation in a ratio of 4:1, with an effective Ni distribution coefficient of 2.4. Below this are two, more realistic models in which olivine crystallizes first, and is joined by plagioclase, again in the ratio 4:1, after 4 and 8 percent of olivine crystallization. The effects of clinopyroxene fractionation once the peritectic is reached have not been considered, in part because of uncertainty concerning the onset of pyroxene crystallization, and in part because of discrepancies in appropriate values for $D_{Ni}^{cpx/liq}$ (Irving, 1978). Since $D_{Ni}^{cpx/liq}$ is likely to be low, between 1-3, it is unlikely that coprecipitation of clinopyroxene with plagioclase and olivine will result in drastically different interpretations of Figure 7. The point of these fractionation trends is that for reasonably realistic fractionation models Ni should be rapidly depleted, with the consequence that basalts with both high Ni and TiO_2 abundances are unlikely to have been produced solely by crystal fractionation, and can be regarded as mixed magmas. This is shown schematically in Figure 7, where all the basalts under discussion could have been produced by mixing the hypothetical primary magma with evolved basaltic magmas resulting from about 60-70 percent crystallization of a similar primary magma. Although the evolved end-member will most likely be multiply saturated with olivine, plagioclase and pyroxene (E, Figure 5), and be compositionally similar to FeTi basalts, the mixed magma may only have olivine on the liquidus (M, Figure 5). In reality, a single step process such as this seems unlikely. Repeated mixing of small volumes of magma, followed by moderate fractionation prior to the next mixing episode (O'Hara, 1977) will ultimately produce compositions similar to a single-step process involving a highly-evolved end member. Also, repeated episodes of mixing and fractionation will tend to buffer the Ni content at a high level, while continually increasing the magmaphile element content.

Discussion

In the preceding section we have attempted to review some of the processes that may modify the composition of an ocean-floor basaltic magma following segregation from its mantle source. The arguments given are to some extent model-dependent, relying heavily on our inferences concerning the compositional characteristics of a primary magma. Several lines of evidence, including theoretical constraints, phenocryst data, the composition of melt inclusions, and bulk compositions of glasses and rocks, suggest that a specific, rather well-defined primary magma-type, although relatively rare, may be of widespread importance in ocean-floor basalt petrogenesis. Consequently, it is upon this compositional type, examples of which are given in Table 1, that we base our evaluation of the processes that modify ocean-floor basaltic magma compositions.

Although the importance of crystal fractionation in ocean-floor basalt petrogenesis has long been recognized, it has become increasingly apparent, through detailed studies made possible by improved sampling, that magmatic processes at mid-

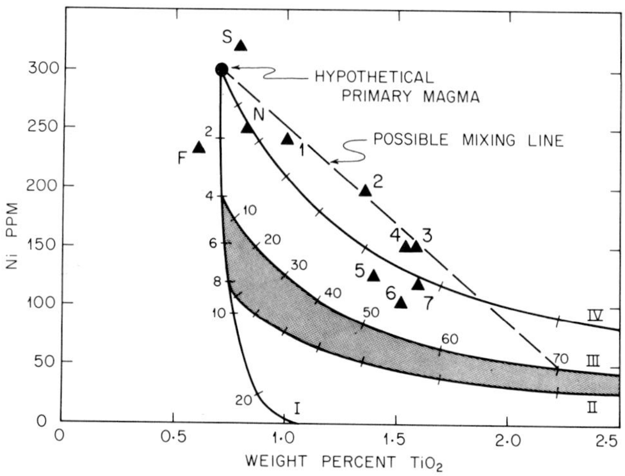

Figure 7. Effects of crystal fractionation and magma mixing on the TiO_2 and Ni content of ocean-floor basalts. Source of data are the same as in Figure 2. Curve I--olivine fractionation only; Curves II, III--initial olivine fractionation followed by olivine-plagioclase fraction in the ratio 1:4; Curve IV--olivine-plagioclase fractionation in the ratio 1:4 ($D_{Ni}^{Pl/liq} = 0.01$; $D_{TiO_2}^{Pl/liq} = 0.04$; $D_{Ni}^{ol/liq} = 12$; $D_{TiO_2}^{ol/liq} = 0.04$).

ocean ridges are more complex than earlier studies suggested. Several commonly occurring enigmatic features cannot be explained solely in terms of crystal fractionation. These include: difficulties in relating spatially associated basalts to a common parental or primary magma; magmaphile element abundances in evolved basalts that are higher than those predicted solely from fractionation processes; Ni concentrations that appear to be too high for the amount of fractionation deduced from magmaphile element abundances; and the necessity to invoke clinopyroxene dominated fractionation, when pyroxene is a minor component or is absent from the phenocryst assemblages. The clinopyroxene anomaly may be produced by clinopyroxene fractionation in the mantle at moderate to high pressures, but the evidence is not compelling, and high pressure fractionation may only provide part of the answer. Most of these features are, however, readily understandable in terms of a combination of crystal fractionation and magma mixing, where "primitive," possibly primary magmas are repeatedly injected into magma chambers containing evolved magmas that have fractionated from previous episodes of injection and mixing (O'Hara, 1977; Rhodes et al., 1978). Such a process explains why it is commonly difficult to relate spatially associated, and apparently comagmatic, basalt-types by crystal fractionation processes. First, differences in residence time and thermal history of the fractionating magma in the magma chamber prior to the next mixing episode will give rise to evolved magmas with differing compositions. Mixing of these magmas with the incoming "primitive" magma, which also may be more or less fractionated than the previous batch, will produce mixed magmas that cannot be related by simple fractionation models. Second, as is evident from Figure 6, the mixed magmas will always be higher in magmaphile element abundance than magmas of similar major element chemistry produced from the same primary magma strictly by crystal fractionation. Third, the Ni contents will also tend to be higher in mixed magmas, and the Ni and TiO_2 abundances will be "decoupled" (Figure 7). Fourth, the necessity to invoke fractionation models dominated by clinopyroxene when pyroxene is rarely a major phenocryst phase is readily explained. Clinopyroxene fractionation has taken place, but in the evolved magma prior to mixing with the "primitive" magma, and the resulting mixed magma bears the chemical, but not the mineralogical, imprint of this earlier process.

If magma mixing, accompanying crystal fractionation, is, as we believe, a commonly occurring process of fundamental importance to the evolution of ocean-floor basaltic magmas, what are the implications concerning volcanic processes occurring along mid-ocean ridges, and the nature and composition of the oceanic crust?

A major problem of ocean-floor basalt petrogenesis is to reconcile the apparent conflict between the vast volumetric significance of these basalts and their evolved characteristics. Traditional models involving partial melting followed by crystal fractionation fail to resolve this discrepancy since one would expect relatively "primitive" basalts to be volumetrically more important. On the other hand, a "steady state" combination of magma mixing, of the type discussed previously, together with crystal fractionation provides an effective mechanism for buffering the magma composition (O'Hara, 1977; Rhodes et al., 1978), the dominant product of which will be moderately evolved basalts. Both "primitive" basalts and highly differentiated basalts and their silicic derivatives should be rare, and observations of ocean-floor basalts (Figure 1) confirm this prediction. Because of this buffering process, there is no necessity for the existence of large magma chambers, or large magma volumes, beneath mid-ocean ridge crests: a conclusion that is apparently in accord with recent interpretations of geophysical data (Rosendahl et al., 1976), and with models proposed for the development of the oceanic crust (Sleep, 1975; Dewey and Kidd, 1977). A further consequence of this model is that the crystalline products of this "steady state" mixing and fractionating process will reflect the evolved nature of the parental magma. For most of its history the magma will be multiply saturated, crystallizing olivine, plagioclase and clinopyroxene. It is only immediately following injection of "primitive" magma, and then only for a short part of the fractionating history, that olivine alone will crystallize (Figure 5). If layer 3 of the oceanic crust is produced in part by the crystallization of magmas beneath mid-ocean ridges, it follows that this component of layer 3 will be dominated by gabbros, rather than the troctolites and dunites that would result from substantial crystallization of more "primitive" basalts. This suggestion is compatible with the ophiolite-based models of the ocean crust (e.g., Dewey and Kidd, 1977), and reconciles field studies with laboratory based observations and experiments.

References

Anderson, A.T. Magma mixing--petrological processes and volcanologic tool. J. Volc. Geotherm. Res., 1, 3-33, 1976.

Anderson, A.T. and Greenland, L.P. Phosphorus fractionation diagram as a quantitative indicator of crystallization differentiation of basaltic liquids. Geochim. Cosmochim. Acta, 33, 493-505, 1969

Bender, J.F., Hodges, F.N. and Bence, A.E. Petrogenesis of basalts from the FAMOUS area: Experimental studies from 0 to 15 kb. Earth Planet. Sci. Lett., (in press).

Blanchard, D.P., Rhodes, J.M., Dungan, M.A., Rodgers, K.V., Donaldson, C.H., Brannon, J.C., Jacobs, J.W. and Gibson, E.K. The chemistry and petrology of basalts from Leg 37 of the Deep Sea Drilling Project. J. Geophys. Res., 81, 4231-4346, 1976.

Bougault, H. and Hekinian, R. Rift valley in the Atlantic ocean near 36°50' N: petrology and geochemistry of basaltic rocks. Earth Planet. Sci. Lett., 24, 249-261, 1974.

Bryan, W.B. Regional variation and petrogenesis of basalt glasses from the FAMOUS area, Mid-Atlantic Ridge. J. Petrol., (in press).

Bryan, W.B., Thompson, G., Frey, F.A., and Dickey, J.J. Inferred settings and differentiation in basalts from the Deep Sea Drilling Project. J. Geophys. Res., 81, 4285-4304, 1976.

Bryan, W.B. and Moore, J.G. Compositional variations of young basalts in the Mid-Atlantic Ridge rift valley near lat 36°49' N. Bull. Geol. Soc. Amer., 88, 556-570, 1977.

Carmichael, I.S.E., Turner, F.J. and Verhoogen, J. Igneous Petrology, p. 67, McGraw-Hill, New York, 1974.

Clague, D.A. and Bunch, T.E. Formation of ferrobasalt at East Pacific midocean spreading centers. Jour. Geophys. Res., 81, 4247-4256, 1976.

Dewey, J.F. and Kidd, W.J.F. Geometry of plate accretion. Bull. Geol. Soc. Amer., 88, 960-968, 1977.

Donaldson, C.H. and Brown, R.W. Refractory megacrysts and magnesium-rich melt inclusions within spinel in oceanic tholeiites: indicators of magma mixing and parental magma composition. Earth Planet. Sci. Lett., 37, 81-89, 1977.

Drake, M.J. Plagioclase-melt equilibria. Geochim. Cosmochim. Acta, 40, 457-466, 1976.

Dungan, M.A. and Rhodes, J.M. Magma mixing in basalts from the mid-Atlantic ridge near 22°N: The combined use of residual glass and glass melt inclusion compositions to determine differentiation process in MORB. Contr. Mineral. and Petrol., (in press).

Dungan, M.A., Long, P.E., and Rhodes, J.M. The petrography, mineral chemistry and one-atmosphere phase relations of basalts from site 395-Leg 45 D.S.D.P., in Initial Reports of the Deep Sea Drilling Project, V. 45, Washington (U.S. Government Printing Office), (in press).

Dungan, M.A., Rhodes, J.M., Long, P.E., Blanchard, D.P., Brannon, J.C. and Rodgers K.V. The petrology and geochemistry of basalts from Site 396-Legs 45 and 46 of the Deep Sea Drilling Project, in Initial Reports of the Deep Sea Drilling Project, V. 46, Washington (U.S. Government Printing Office), (in press).

Frey, F.A., Bryan, W.B. and Thompson, G. Atlantic Ocean floor: Geochemistry and petrology of basants from legs 2 and 3 of the Deep Sea Drilling Project. J. Geophys. Res., 79, 5507-5527, 1974.

Fujii, T. and Kushiro, I. Melting relations and viscosity of an abyssal tholeiite. Carnegie Inst. Washington Yr. Bk., 76, 461-465, 1977.

Green, D.H. The origin of basaltic and nephelenitic magmas. Trans. Leicester Lit. Philos. Soc., 64, 28-54, 1970.

Hanson, G.N. and Langmuir, C.H. Modelling of major elements in mantle-melt systems using trace element approaches. Geochim. Cosmochim. Acta, 42, 725-742, 1978.

Hart, S.R. and Davis, K.E. Nickel partitioning between olivine and silicate melt. Earth Planet. Sci. Lett., 40, 203-219, 1978.

Hekinian, R., Moore, J.G. and Bryan, W.B. Volcanic rocks and processes of the Mid-Atlantic Ridge rift valley near 36°49' N. Cont. Mineral. and Petrol., 58, 83-110, 1976.

Irving, A.J. A review of experimental studies of crystal/liquid trace element partitioning. Geochim. Cosmochim. Acta, 42, 743-770, 1978.

Kay, R., Hubbard, N.J. and Gast, P.W. Chemical characteristics and origin of oceanic ridge volcanic rocks. J. Geophys. Res., 75, 1585-1613, 1970.

Kay, R. and Hubbard, N.J. Trace elements in ocean ridge basalts. Earth Planet. Sci. Lett., 38, 95-116, 1978.

Kushiro, I. Origin of some magmas in oceanic and circum-oceanic regions. Tectonophysics, 17, 211-222, 1973.

Langmuir, C.H., Bender, J.F., Bence, A.E. and Hanson, G.N. Petrogenesis of basalts from the FAMOUS area: Mid-Atlantic Ridge. Earth Planet. Sci. Lett., 36, 133-156, 1977.

Melson, W.G., Vallier, T.L., Wright, T.L., Byerly, G. and Nelen, J. Chemical diversity of abyssal volcanic glass erupted along Pacific, Atlantic, and Indian ocean sea-floor spreading centers, in The Geophysics of the Pacific Ocean basin and its margins, pp. 351-367. American Geophysical Union, Washington, D.C., 1976.

O'Hara, M.J. Are ocean floor basalts primary magmas? Nature, 220, 683-686, 1968.

O'Hara, M.J. Geochemical evolution during fractional crystallization of a periodically refilled magma chamber. Nature, 286, 503-507, 1977.

O'Hara, M.J., Saunders, M.J. and Mercy, E.L.P. Garnet-peridotite, primary ultrabasic magma and eclogite: interpretation of upper mantle processes in kimberlite. Phys. Chem. Earth, 9, 571-604, 1975.

Rhodes, J.M. and Rodgers, K.V. Chemical diversity of mid-ocean ridge basalts from a single dredge haul (abstract). Eos Trans. AGU, 53, 1133, 1972.

Rhodes, J.M., Blanchard, D.P., Rodgers, K.V., Jacobs, J.W. and Brannon, J.C. Petrology and chemistry of basalts from the Nazca plate: Part 2--major and trace element chemistry, in Initial Reports of the Deep Sea Drilling Project, V. 34, pp. 239-244 (U.S. Government Printing Office), Washington, D.C., 1976.

Rhodes, J.M. and Dungan, M.A. The nature of primary ocean-floor basalts, in Papers presented to the second inner-team meeting, Basaltic Volcanism Study Project, 50-52, Lunar Science Institute, Houston, 1977.

Rhodes, J.M., Blanchard, D.P., Dungan, M.A., Rodgers, K.V. and Brannon, J.C. Chemistry of basalts from Leg 45 of the Deep Sea Drilling Project, in Initial Reports of the Deep Sea Drilling Project, V. 45, Washington (U.S. Government Printing Office) (in press).

Rhodes, J.M. Dungan, M.A., Blanchard, D.P. and Long, P.E. Magma mixing at mid-ocean ridges: Evidence from basalts drilled near 22°N on the mid-Atlantic ridge. Tectonophysics (in press).

Roeder, P.L. and Emslie, P.F. Olivine-liquid equilibrium. Contr. Mineral. and Petrol., 29, 275-289, 1970.

Rosendahl, B.R. Evolution of oceanic crust, 2. constraints, implications and inferences. J. Geophys. Res., 81, 5305-5314, 1976.

Sato, H. Nickel content of basaltic magmas: identification of primary magmas and a measure of the degree of olivine fractionation. Lithos, 10, 113-120, 1977.

Schilling, J.G. Sea floor evolution: Rare-earth evidence. Phil. Trans. Roy. Soc. London, Ser. A, 268, 663-706, 1971.

Shido, F.A., Miyashiro, A. and Ewing, M. Crystallization of abyssal tholeiites. Contrib. Mineral. Petrol., 31, 251-266, 1971.

Sleep, N.H. Formation of oceanic crust: Some thermal constraints. J. Geophys. Res., 80, 4037-4042, 1975.

Thompson, G., Bryan, W.B., Frey, F.A., Dickey, J.S. and Sven, C.J. Petrology and geochemistry of basalts from DSDP Leg 34, Nazca plate, in Initial Reports of the Deep Sea Drilling Project, V. 34, Washington (U.S. Government Printing Office), 215-226, 1976.

Watson, E.B. Glass inclusions as samples of early magmatic liquid: determinative method and application to a South Atlantic basalt. J. Volcanol. Geotherm. Res., 1, 73-84, 1976.

Wright, T.L. Magma mixing as illustrated by the 1959 eruption, Kilauea Volcano, Hawaii. Bull. Geol. Soc. Amer., 84, 849-858, 1973.

Wright, T.L. and Fisk, R.S. Origin of the differentiated and hybrid lavas of Kilauea Volcano, Hawaii. J. Petrol., 12, 1-65, 1971.

PETROLOGIC CHARACTER OF THE ATLANTIC CRUST FROM DSDP AND IPOD DRILL SITES

Wilfred Bryan

Geoffrey Thompson

Woods Hole Oceanographic Institution Woods Hole, Massachusetts 02543

Frederick Frey

Department of Earth and Planetary Science Massachusetts Institute of Technology Cambridge, Massachusetts 02139

Abstract. Basement rocks recovered by DSDP and IPOD drilling in the Atlantic are predominantly pillowed or massive basalt lava flows which resemble modern Mid-Atlantic Ridge (MAR) basalts in their range of chemical composition and petrographic characteristics. Basalts from the oldest sites drilled (70-150 m.y.) generally resemble modern basalts presently being erupted at the locations on the MAR where these older basalts should have originated; all of these represent "normal" ridge segments unaffected by "mantle plume" activity. The Leg 37 transect drilled opposite the Azores "mantle plume" or "blob" shows evidence of some fluctuation in geochemical parameters, possibly indicating short-term fluctuations in plume activity.

Only two sites (334 and 395) have penetrated layer 3 plutonic rocks, but it is likely that these are tectonically emplaced, and the true thickness of layer 2 has not been defined by drilling. In general, the transition from sediment to basalt basement is sharp. Excellent core recovery at Sites 417 and 418 shows that pillowed and massive flows are interbedded; dikes and sills are rare at these and other sites, but probably are hard to recognize unless core recovery is very good. Drilling at Site 417 suggests that intense low temperature alteration may be limited to topographic highs which are not immediately buried by sediment. Hydrothermal alteration and metamorphism are not encountered in the sites so far drilled.

Most DSDP and IPOD sites have not been well-placed to indicate the persistence of "mantle plumes" with time, although north-south geochemical variations in continental triassic basalts resemble those documented for modern basalts associated with the Azores plume and suggest that such plumes may have initiated spreading in the Atlantic. Future drilling should be designed to define the persistence of such plumes in time and space, and also should attempt to penetrate the lower part of layer 2 and the upper part of layer 3 in a crustal section of normal thickness.

Introduction

In an earlier paper (Bryan et al., 1976), we summarized results of the Deep Sea Drilling Project (DSDP) drilling in the three major ocean basins, with special emphasis on geochemical variation in basalt glasses. At that time, surprisingly little drilling had been completed in the Atlantic, and with the exception of the Leg 37 holes, the sites had been selected with predominantly non-petrologic objectives. Recently, deep crustal penetrations have been attempted on Leg 45, Leg 46 and the Leg 51-52-53 "megaleg". While none of the recent drilling succeeded in penetrating the layer 2 -layer 3 boundary, some excellent penetration into layer 2 has been accomplished with above-average core recovery. These new sites have added considerably to our knowledge of the stratigraphy of layer 2 and have produced examples of both the freshest and also the most altered deep ocean crust so far known among the sites drilled in crust over 100 m.y. old.

In this paper we attempt to present a summary picture of the layer 2 basement rocks in the Atlantic in terms of their variation in composition and petrographic character as a function of space and time. We have deliberately avoided extended speculation on magmatic processes within the crust and upper mantle which may contribute to this variation. Overall, the situation does not seem to warrant serious departure from the position we took in our first paper on Atlantic DSDP basalts (Frey et al., 1974) in which arguments were presented both for significant high-level fractional crystallation and also for melting of contrasted mantle sources in order to explain all of the geochemical variability observed in a rather limited selection of samples. The greater variety of compositions now known from the Atlantic drill holes makes it even more advisable to retain multiple working hypotheses, and enhances the probability that several processes operate at any given site. Indeed, the relative importance of each process is likely to vary from case to case, and probably can be quantitatively evaluated only through detailed study of specific sample sets, a task (or series of tasks) beyond the scope of this paper. However, we will allude to these processes as appropriate within the rather broad and qualitative perspective we prefer to maintain.

Modern dredged basalts show a considerable degree of geochemical variability which is most directly expressed in variations in some of the large-ion-lithophile (LIL) elements in basalts which otherwise are very similar in their major element composition. High concentrations of LIL-elements are centered on certain geographic positions along the Mid-Atlantic Ridge, and may be marked by islands (Schilling, 1973; Hart et al., 1973; Schilling et al., 1976). Of these, the best known centers are Iceland, the Azores and Bouvet; other islands in the South Atlantic such as Ascension and Tristan da Cunha are suspected of marking similar centers. At least one other center (at 45°N latitude) is known which has no special physiographic expression. Geochemical parameters vary gradationally away from these centers, and the degree of light rare earth enrichment and strontium isotope ratios are especially characteristic (White et al., 1976; Schilling, 1975). Although the physical significance of these centers remains obscure, they are widely believed to mark the location of upwelling deep mantle which is then partially melted to yield the basalts forming the elevated ridges and islands. Because this mantle probably lies below the level from which magmas are usually extracted, it is believed to contain a much higher proportion of LIL-elements and other components which are preferentially partitioned into the magma during mantle melting, while the uppermost mantle is presumed to have been depleted in these components because of its continued involvement in magma production during earlier spreading events.

Because of these demonstrated along-ridge compositional variations, compositional variation as a function of time can only be evaluated properly if basalts at the various drill sites can be compared directly to basalts presently being erupted at the same point on the Mid-Atlantic Ridge from which these older basalts were derived. Alternatively, basalts from drill holes lying along a common spreading "flow line" may be compared. In such comparisons, allowance must be made for the effects of varying degrees of high-level crystallization or random variations in degree of mantle melting; such comparisons are most easily made in terms of parameters not seriously changed by moderate differences in degree of melting or crystallization, such as isotopic ratios, ratio of light to heavy rare earths, or ratios between certain LIL elements. At present, only Leg 3 and Leg 37 have sampled systematically along spreading flow lines, and only Leg 37 data can be related directly back to a thoroughly sampled modern ridge segment. Although Legs 45, 46, 51, 52 and 53 were supposed to drill on a common flow line related to a well-sampled modern ridge segment at 22°-23°N, operational considerations forced a shift of the Leg 51, 52 and 53 sites to the south, so that they correspond more closely to a poorly known ridge segment at 18°-20°N. Although we believe variations along the modern ridge are sufficiently well known to permit valid interpolations of modern compositions through this and other poorly sampled ridge segments, the limitations of such extrapolations must be recognized.

Chemical data reported here are necessarily drawn from a variety of sources, and for the most recently drilled sites (395, 396, 417, 418) it has been necessary to rely heavily on shipboard x-ray fluorescence analyses and unpublished data. We have deliberately excluded altered material and, where possible, have used aphyric samples or glasses in order that the data may represent magmatic liquids as closely as possible. To simplify graphical presentations, we have commonly plotted only site averages; but in a few cases, where within-site variation has seemed especially significant, compositions of specific magmatic types within sites have been plotted. Most of the trace element data also represent aphyric rock or glass and obviously altered material has been excluded. Wherever possible, these data have been taken from our own papers already published or in press, and these references should be consulted for details of analytical procedures and precision and accuracy.

Compositional Variation in Layer 2

The locations of Atlantic DSDP sites for which significant basalt penetration has been reported are indicated in Figure 1. However, for many of these sites definitive compositional data are lacking, generally because the samples are intensely weathered or because of very limited core recovery. Also, data for sites drilled on the Reykjanes Ridge and in the FAMOUS area by Leg 49 were not available to us at the time this paper was prepared. Although the distribution of these sites implies a rather comprehensive sampling of the Atlantic sea floor, the data are, in fact, much more limited. In the South Atlantic, only Sites 14, 15, 18, 19 and 20 have basalt of sufficient freshness for geochemical and mineralogical characterization. In the western Atlantic, data for older sea floor is similarly limited to Sites 10, 100, 105, 417 and 418. None of the sites in an old crust in the eastern Atlantic have provided basalt samples of adequate freshness. The only definitive near-ridge sites

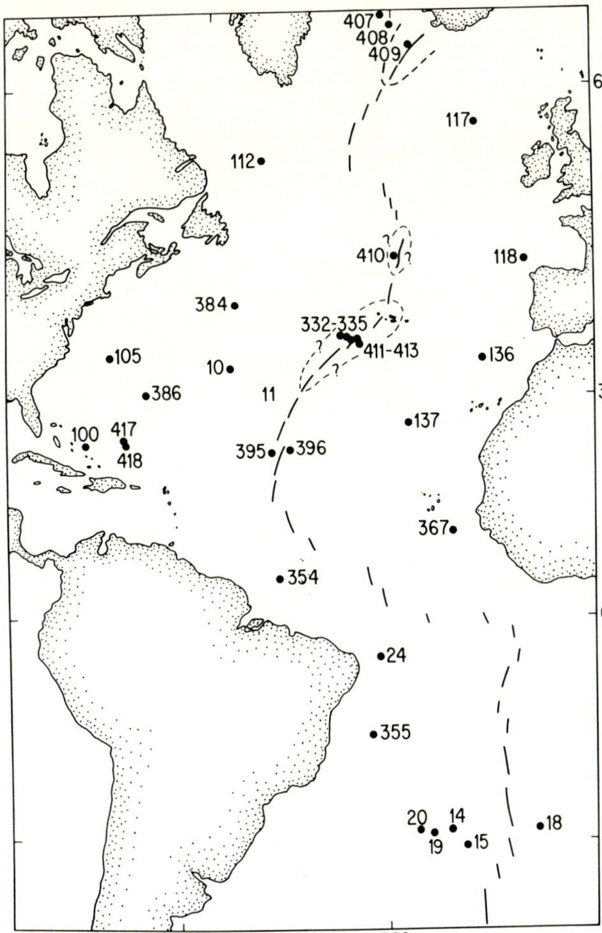

Figure 1. Location of DSDP and IPOD drill sites in the Atlantic Ocean which penetrated significant amounts of basement rock. Many of these samples are badly altered or poorly recovered and provide little reliable compositional data; see text for discussion.

lie near 22°N, 31°N, and 37°N. Sampling of the deep crust is equally limited; only Sites 332, 333, 395, 396, 417, and 418 penetrated more than 200 m into layer 2. The limitations posed by this restricted sampling must be kep in mind in the discussion which follows.

Major element data can be summarized concisely in a plot of FeO^*/MgO vs. TiO_2, where FeO^* represents total Fe as FeO (Figure 2). We prefer FeO^*/MgO to the popular molecular $Mg/(Mg + Fe^{2+})$ ratio for graphical portrayals because the XRF and microprobe data do not discriminate Fe^{3+} and Fe^{2+}, and because of the wider numerical range of the FeO^*/MgO ratio. TiO_2 and FeO^*/MgO also are the most variable of the major element parameters, and generally correlate well with other important major elements such as CaO, Al_2O_3, K_2O, and Na_2O. In this plot the data tend to lie within a rough parallelogram-shaped area, with the longest diagonal representing an approximately one to one variation in FeO^*/MgO with TiO_2. The large dashed field encloses basalt glass data from one small area (the FAMOUS area) on the Mid-Atlantic Ridge, while the smaller dotted field and stars represent basalt glasses from the Mid-Atlantic Ridge at 23° and 22°N, respectively. Several points are made clear by this graphical plot:

1) The limited samples recovered by Legs 2 and 3 (Sites 10, 14, 15, 18 19, 20) span most of the range in TiO_2 and FeO^*/MgO represented by all the rest of the Atlantic data.

2) The low-TiO_2, low FeO^*/MgO compositions from Sites 14 and 18 are relatively rare among DSDP basalts, but correspond to the more mafic compositions in the FAMOUS area and at Site 332B.

3) Sites 395 and 396 differ somewhat in their average compositions even though they were drilled in approximately symmetrical positions either side of the Mid-Atlantic Ridge; Site 295 seems to correspond more closely to the modern ridge compositions than does 396.

4) The Leg 37 sites tend toward higher FeO^*/MgO than do FAMOUS area basalts of comparable TiO_2 content.

5) The 150 m.y. old basalts from Sites 100 and 105 generally correspond to the more mafic compositions from Leg 37 and the FAMOUS area.

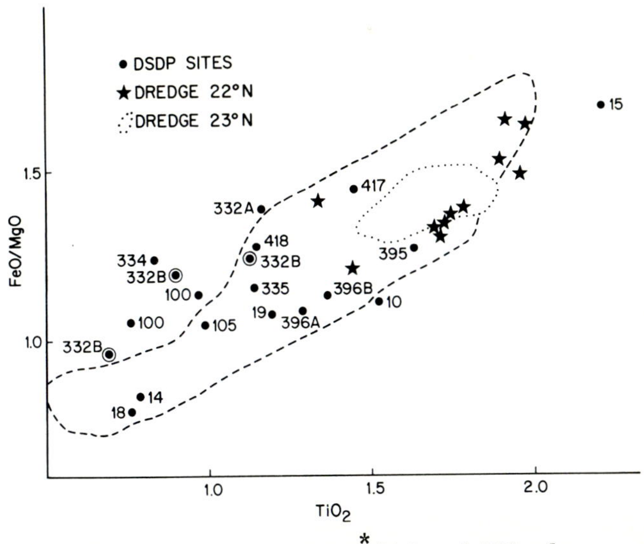

Figure 2. Variation in FeO^*/MgO and TiO_2 for fresh basalt and glass from Atlantic DSDP and IPOD drill sites. Data sources for Sites 10-335 as in Bryan et al., 1976; data for Sites 395, 396, 417 and 418 are provisional, based on shipboard XRF analyses. Dredge data from Melson et al., 1968, and Bryan, 1969, outlined by short dashes; dredge data is from Bryan and Sargent, 1978.

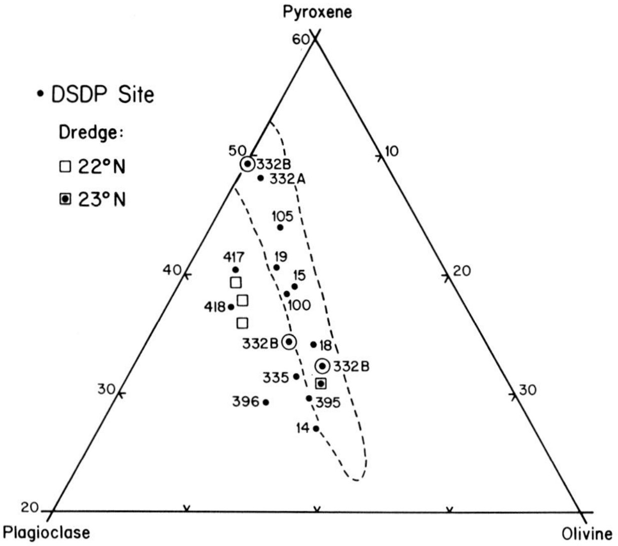

Figure 3. Variation of total normative pyroxene, olivine, and plagioclase for Atlantic DSDP and IPOD sites. Data sources as in Figure 2.

6) Site 417 and 418 basalts tend to lie at the high end of the FeO^*/MgO range compared to other basalts of comparable TiO_2 content.

7) Overall, the compositions of basalts recovered by drilling tend to lie within or close to the range of compositions represented by submersible and dredge samples from two very limited portions of the Mid-Atlantic Ridge at 36-37°N and 22-23°N.

Remembering that these sites represent a very limited sampling of the Atlantic layer 2, the data nevertheless do suggest that these basalts vary in composition within fairly well-defined limits. The restricted range of compositions recovered by drilling at any one site are apparently subsets of a more diverse assemblage which can be generated within a short period of time and limited space at an active spreading center.

This relationship can be portrayed in an even more striking way in the normative pyroxene-plagioclase-olivine ternary (Figure 3) in which, again, most of the data lie within the field of the FAMOUS glasses. The narrow elongated composition field is suggestive of the low-pressure binary cotectic in the system diopside-anorthite-forsterite (Osborn and Tait, 1952) and implies composition control by low-pressure crystal-melt equilibria. It is interesting that very few of these basalts contain either normative nepheline or normative quartz; again, the range of compositions seems to be constrained to those which lie along this cotectic. In an earlier summary (Bryan et al., 1976) we showed that only a few glass compositions lie appreciably off this cotectic. Those few exceptions represented seamounts or other off-ridge extrusions, and were offset into the plagioclase field. This same effect is shown here by ridge-derived basalts from 22°N and from Sites 417, 418, and 396. Such an offset might result from a shift in the cotectic due to higher volatile contents or to higher confining pressure (Dick and Bryan, 1978; Thompson, Bryan and Melson, in preparation).

We have previously shown (Bryan et al., 1976) that basalts of otherwise similar major element composition may differ considerably in concentration of large-ion-lithophile elements. This contrast can be demonstrated for the Atlantic data, in plots of Ba vs. FeO^*/MgO or in relative enrichments of light vs. heavy rare earth elements (Figure 4a, b, c).

Such large differences in LIL elements such as Ba, REE, or in Sr isotopes (Figure 4d) are generally attributed to differences in the composition of the mantle source rocks. However, as noted at individual sites (Bryan and Moore, 1977; Bryan et al., 1978; Langmuir et al., 1977) whilst incompatible and major elements vary in a systematic manner LIL elements often show a wider range in content than can be attributed to high level fractional crystallization processes. This apparent decoupling of major elements from LIL trace elements is generally attributed to variations in degree of partial melting of the mantle. Such a relation can be expected to arise because LIL elements are very strongly partitioned into liquid relative to mantle mineral phases, and the first few percent of liquid formed will contain most of these components. Further melting serves mainly to increase the proportion of major elements and heavy transition metal trace elements and effectively dilutes the concentration of LIL elements.

Petrography and Mineralogy

Water-quenched basalts generally are too fine-grained for definitive study of mineral phases other than those occurring as phenocrysts. However, on several occasions drilling has penetrated relatively thick cooling units which display a holocrystalline, doleritic or gabbroic texture. Good coarsely-crystalline plutonic gabbro and peridotite were recovered at Site 334. These are the best examples of "layer 3"-type material so far documented by the drilling project (Hodges and Papike, 1976). These rocks also are unusual in containing amphibole, which was considered to be a secondary phase. Relatively fresh peridotite and gabbro also have been recovered beneath basalt at Site 395.

The majority of the basement rocks recovered consist of pillowed or massive basalt with varying proportions of plagioclase and olivine phenocrysts. Clinopyroxene is a significant phenocryst phase in basalts of relatively high TiO_2 and FeO^*/MgO content. It is relatively conspicuous, for

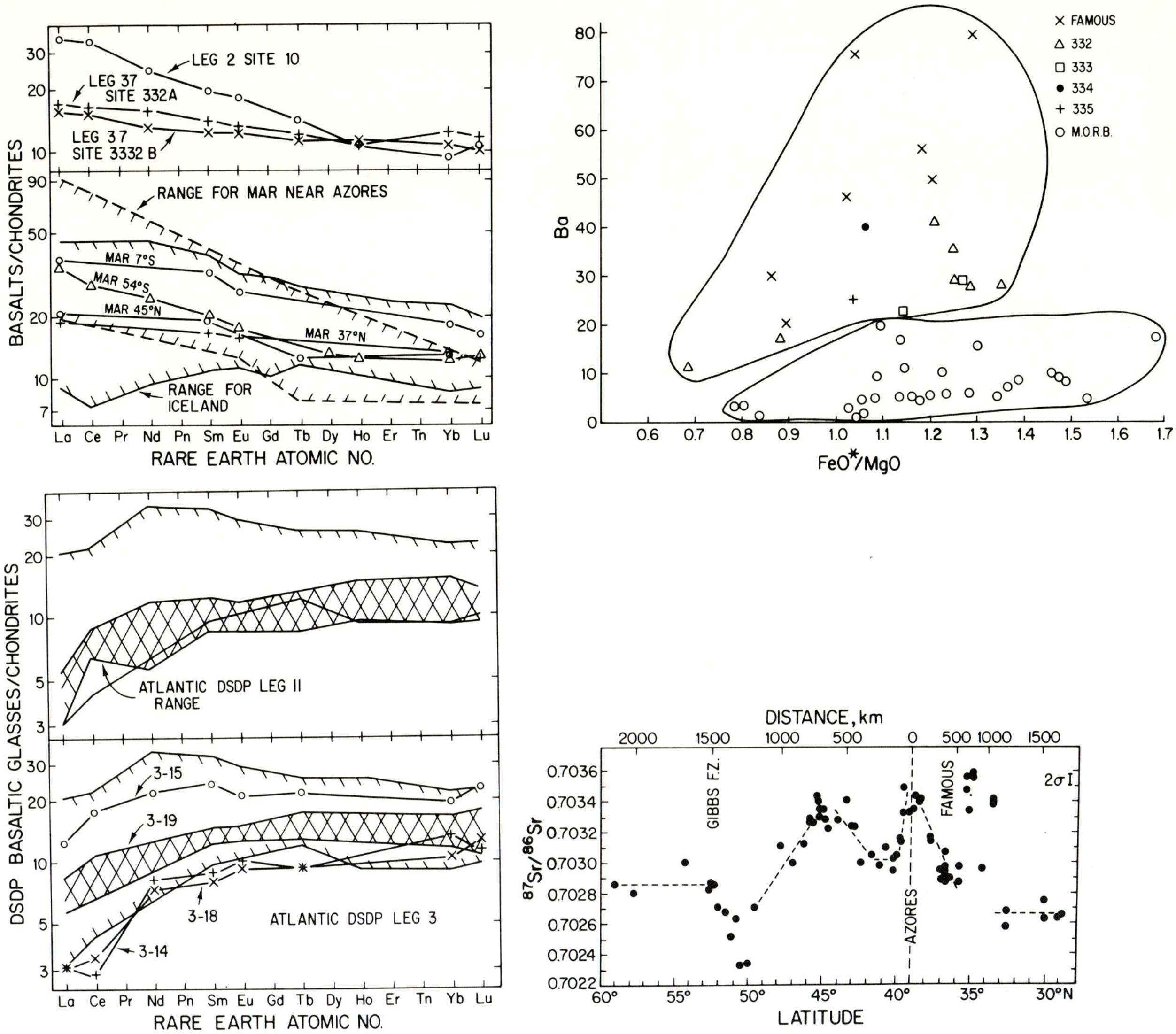

Figure 4. A. Variation in chondrite-normalized rare earth data for DSDP and dredged basalts related to mantle plumes; B. Variation in chondrite-normalized rare earth elements and dredged basalts (open hatched field) from normal spreading ridges; C. Variation of Ba vs. FeO*/MgO for Leg 37 and FAMOUS plume-related basalts (upper field) and normal ocean ridge basalts (lower field); D. Variation in $^{87}Sr/^{86}Sr$ as a function of latitude in the North Atlantic. Data for A and B after Frey et al., 1974; Bryan et al., 1976, 1977a, b; for C, Bryan and Thompson, 1977; Bryan and White, 1977; Frey et al., 1974; Bryan et al., 1977a, b; and for D, White et al., 1976.

example, at Sites 11, 15, 332A, 417 and 418, and is sparingly present in some basalts at Site 395. By contrast, the low TiO_2, low FeO*/MgO basalts lack pyroxene, but commonly contain small phenocrysts of spinel.

Several studies have shown a tendency for the relative proportions of phenocrysts to fall into fairly distinct groupings, which also can be related to overall phenocryst abundance. Dick and Bryan (1978) showed that the nearly aphyric basalts from Site 396 with less than 5% phenocrysts tended to contian olivine as the dominant phenocryst, while the phyric basalts with 10% or more phenocrysts have plagioclase as the dominant phase. This relationship is very similar to that reported by Bryan and Moore (1977) for the FAMOUS area. Staudigel et al. (1979a) showed that phenocryst assemblages at Sites 417 and 418 tend to fall into two distinct groups, one in which plagioclase is almost the only phenocryst, and a second in which significant amounts of olivine accompany the plagioclase. The presence of plagioclase- or olivine-enriched phenocryst assemblages tends to cause whole rock compositions to fall to one side or the other of the inferred cotectic line in the normative ternary (Figure 3). Hence, the cotectic may be computed as a discriminant function between the fields of plagioclase or olivine basalt whole rock data, or may be computed as a regression line from basalt glass data. These relations seem to hold equally well for dredged basalts, as originally noted by Miyashiro et al., (1969).

Plagioclase, pyroxene, and olivine appear as euhedral microphenocrysts in basalt glasses at Site 332 (Bryan et al., 1977a). The three minerals may be closely associated in glomeroporphyritic clusters and show no evidence of a reaction relation either with the liquid or with each other. Bryan (1978) has shown that most olivine and pyroxene compositions from Site 332 and the FAMOUS area correlate with the FeO^*/MgO ratio of their enclosing liquid as predicted from experimental data, which also indicates that both minerals are in equilibrium with the liquid. This suggests that the three-phase eutectic in the experimental system diopside-anorthite-forsterite is actually the piercing point of a three-phase cotectic in the real basalt system. Orthopyroxene and pigeonite are rare or absent in most DSDP basalts, but hypersthene and pigeonite are present as lamellae in augite in the plutonic rocks at Site 334. Hypersthene also occurs as individual crystals. Detailed description and analyses of these hypersthenes are given by Hodges and Papike (1976). Most of the clinopyroxenes in both the basalts and plutonic rocks are diopsidic augites containing moderate amounts of TiO_2 and Al_2O_3. Aluminous chrome diopside is occasionally observed; an analysis of one of these from a basalt at Site 334 has been given by Bryan et al. (1977a). Analyses of chromian diopside at Site 11 are given by Frey et al. (1974).

Olivines typically show a limited range of composition, between Fo_{90} and Fo_{80}. In the more mafic basalts, olivine usually contains inclusions of aluminous chrome spinel, and may be intergrown with plagioclase, forming the "knot" in a "bow-tie" intergrowth. Analyses of olivines and spinels from various Atlantic sites have been given by Frey et al. (1974), Hodges and Papike (1976), and Sinton and Byerly (1979).

Plagioclase phenocrysts generally fall in the range An_{80}-An_{60} and usually contain significant amounts of iron and magnesium. Bryan (1974) showed that Mg and Fe^{2+} must be accommodated in tetrahedral sites. Many of the plagioclases are sector-zoned, suggesting rapid, possibly disequilibrium growth. Good examples have been described at Sites 11, 15 and 105, as well as in modern dredged basalts. Hodges and Papike (1976) found that in the plutonic plagioclase at Site 334 there is almost no magnesium and the iron appears to substitute as Fe^{3+} for Al. This again suggests that the Fe^{2+} and Mg substitution may be metastable and typical of quickly cooled pillow lavas. Thus the presence of sector-zoning and Mg substitution in plagioclase could be important criteria for distinguishing extrusive basalt from intrusive sills or dikes in drill cores.

Although basaltic compositions predominate in all recovered Atlantic drill cores, Sinton and Byerly (1979) have reported trondjhemitic segregations within massive basalt at Site 418. These consist of sodic plagioclase, quartz, and apatite, and may have associated pigeonite. Their mode of occurrence implies that these patches are extreme differentiates of normal ocean ridge basalt.

Stratigraphy of Layer 2

Two Atlantic sites (334 and 395) have penetrated gabbro and peridotite, thought to be typical of layer 3, beneath a relatively thin basalt capping. At Site 334, gabbro and peridotite were encountered beneath 50 meters of basalt, and at Site 395, similar lithologies were encountered beneath 57 meters of basalt. These very shallow occurrences of layer 3-type material probably are atypical, and may reflect localized diapiric intrusion of layer 3 material into layer 2 along fault planes. Such an interpretation is supported by the relatively large amount of breccia associated with these plutonic rocks (Hall and Robinson, in preparation).

Excluding these apparently anomalous cases, surprisingly little plutonic rock or intrusive dikes and sills have been encountered at other sites. The three deepest holes at Sites 332, 395 and 418 all penetrated greater than 500 m without encountering layer 3 material, and only a small amount of the basalt recovered is likely to represent intrusive sills or dikes. In general, there is a rather abrupt transition from overlying sediment to basalt; a few layers of volcanic ash or basalt breccia may appear in the sediment above the basalt contact, and small amounts of indurated sediment may be intercalated between basalt pillows. Interlayered sediment was encountered at Site 332 up to 200 m below the top of the basalt sequence. Possibly such sediment is more common than the recovery indicates, due to a tendency for it to be washed out of the core. However, distinct sediment intervals interbedded with basalt were very rare at Sites 417 and 418, where core recovery averaged about 70% in basalt basement.

Recognition of intrusive basalt units is generally based on evaluation of textural criteria, physical relations (contacts) with adjacent rock units, and comparison of chemical composition and magnetic polarity with adjacent rock units. A massive basalt unit, lacking internal quenched glassy intervals and showing an ophitic or gabbroic texture as its center is approached, may be suspected of being a shallow intrusive unit. This interpretation is enforced if the unit is chemically distinct from basalt units above and below, and if it also differs from them in magnetic polarity or inclination. Very few of the massive basalts encountered in the drilling meet all of these criteria. Probably most represent thick

TABLE 1. Comparison of Geochemical Variation in Dredged Basalts From the Mid-Atlantic Ridge, 29-40°N, With Average "Normal" Mid-Atlantic Ridge Basalt (After White and Bryan, 1977 - Table 2). Maximum Values are Typical Near 40°N, Minimum Values are Typical Near 29°N.

	29°-40°N Max.	29°-40°N Min.	Normal M.A.R. Average
K (ppm)	9050	430	855
Rb (ppm)	22.5	0.22	0.87
Cs (ppm)	0.283	0.0014	0.012
Sr (ppm)	298	55	103
Ba (ppm)	268	1.08	8.3
La (ppm)	30.7	0.53	2.7
(La/Sm) E.F.	3.6	0.24	0.62
Rb/K	0.0031	0.0005	0.0009
Ba/K	0.036	0.003	0.007
Rb/Sr	0.080	0.002	0.008
$^{87}Sr/^{86}Sr$	0.70349	0.70258	0.70266

flows, rapidly extruded and ponded on the sea floor. Dikes are very difficult to identify unless core recovery is very high and the drill fortuitously follows the dike margin for some distance. A few dikes have been recognized on this basis in Holes 417D and 418A. Possibly, massive basalts recovered in these and other holes represent the central portions of relatively thick dikes which have been followed for some distance by the drill. At Sites 417 and 418, where core recovery is sufficiently good to evaluate these possibilities, most contacts between massive basalt and pillow basalt were at a low angle and other features such as crystal settling are more compatible with a tabular, near-horizontal cooling unit. Most of the massive basalts at Sites 417 and 418 have been interpreted as thick flows because of their generally fine-grain size and chemical and magnetic compatibility with adjacent pillow lavas (Bryan, Robinson et al., 1977).

The available evidence then suggests that layer 2 consists mainly of interbedded pillowed and massive basalt flows at least down to a depth of 500 m. At this depth some dikes begin to appear, based on very limited evidence at Sites 417 and 418. The upper several hundred meters of layer 2 (layer 2A?) may include interbedded sediment and may be more fractured and brecciated than the deeper units (layer 2B?). At the sites drilled in older crust (417, 418) induration of sediment and cementation of fractures by calcite effectively eliminates the physical distinction between 2A and 2B (Robinson and Hall, in preparation). Until deeper penetrations are achieved, we can only speculate on the nature of the crust beyond 600 m depth; but because of the persistence of extrusive basalt to the greatest depths drilled (583 m at Site 332) it is not unreasonable to suppose basalt persists to depths of 1.5 km or more, with an increasing proportion of intrusive dikes and sills below 500 m.

Alteration

Previous studies of dredged pillow basalts (Hart, 1969; Hart, 1970; Thompson, 1973) indicate that the reaction of basalts with seawater at ambient bottom water temperatures results in uptake by the basalt from seawater of H_2O, K, P, Rb, Cs, U, B, and Li, and occasionally Ba and Sr. Elements leached from the rock are Si, Ca, and Mg. Elements such as Ti, Al, Zr, Y, REE are not affected unless extreme alteration occurs.

Studies of the DSDP basalts (Thompson, 1973; Frey et al., 1974) indicate that similar effects occur at depth in the basaltic layer 2. However, as indicated by such workers as Muehlenbachs (1976), Robinson et al. (1977), Aumento et al. (1977), and Bryan et al. (1977b), basalts recovered by the DSDP are not always completely altered, and fresh glass is commonly found. Recently, studies of the alteration products (Hart et al., 1979; Staudigel et al., 1979b; Muehlenbachs, 1970) indicate that alteration into the upper few hundreds of meters of layer 2 takes place predominantly in the first 10-15 million years by penetration of a limited amount of cold sea water. After that period, the system is relatively closed and only minor amounts of seawater penetrate along restricted fractures. The result is limited alteration of the upper 0.5 km of basaltic crust at low temperature. No large scale hydrothermal alteration at elevated temperatures (>100°C) of the basaltic crust has been seen in DSDP cores (although such hydrothermally altered materials are commonly dredged).

Although low temperature alteration of the oceanic crust apparently is not a major process in providing chemical fluxes between seawater and igneous basement, at least based on the observations of the majority of DSDP recovered cores, Site 417A is an exception. Donnelly et al. (1979) have shown

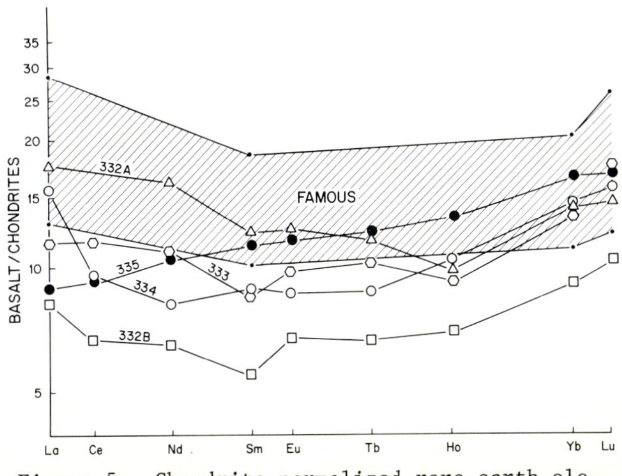

Figure 5. Chondrite-normalized rare earth elements for FAMOUS area (diagonal ruled field) and Leg 37 DSDP sites (Bryan et al., 1977a). Note anomalous LREE depletion at Site 335.

that alteration here proceeded in the same direction as low temperature alteration of dredged basalts and other DSDP basalts, but at slightly elevated temperatures - 30-40°C. The net effect is to increase the rate of alteration with extreme exchanges, involving large uptake of K, Rb, B and Li and concomitant loss of Mg and Ca. If Site 417A is representative of only a small percentage of the oceanic crust, then such reactions would have a major effect in chemical budgets and control of seawater composition. More drilling of oceanic crust up to 15 million years old that exists as distinct topographic highs is required to test this hypothesis.

Chemical Variations in Space and Time

Some of the trace element characteristics and ranges of the Mid-Atlantic Ridge and the segment 20-40°N are shown in Table 1. Typically the LIL-enriched basalts are not greatly different in major element composition or transition metal trace element contents than normal mid-ocean ridge basalts (MORB) (Bryan et al., 1976). Some of the critical diagnostic characteristics of MORB (K, Rb, Cs, Ba, U, and Sr isotopes) can be drastically affected by low-temperature alteration by seawater. The resultant uptake of these ions from seawater gives the basalt the characteristics of a LIL element-enriched basalt. The rare earth elements (REE) are not greatly affected by alteration and can be used as a discriminant, particularly the chondrite-normalized La/Sm ratio (Schilling, 1975).

Geochemical and petrographic data are now available for four sites in crust over 100 m.y. old. Site 417 and 418 are in crust about 107 m.y. old, and Sites 100 and 105 are in crust about 150 m.y. old. These latter sites approach the age of the initial opening of the North Atlantic (about 180 m.y.). In addition, basalts from Leg 3 lie along a spreading "flow line" ranging in age up to about 70 m.y., and basalts from Leg 37 also are arranged along a "flow line" originating in the FAMOUS area and range in age up to about 15 m.y. These data allow some conclusions to be drawn about both short-term and long-term variations in composition with time. Also, the Leg 37 transect is related to a ridge segment influenced by the Azores "mantle plume", while the Leg 3 transect appears to originate at a ridge segment free of such "plume" influence. The other DSDP sites in the Atlantic typically show basalts that are characteristic of MORB except those associated with volcanic chains alkalic in character (i.e. Site 385, Leg 43) or those that occur as sills or mid-plate eruptives, such as Sites 10 and 138 which are LIL element-enriched in character (Frey et al., 1974; Wright et al., 1972). These basalts are apparently not related to spreading center activity.

Frey et al. (1974) compared the chemical and petrographic characteristics of basalts from the Leg 3 Sites 14, 15, 18, 19 and 20 and showed that, if reasonable allowance is made for high-level fractionation and alteration effects, all of the basalts are LIL element-depleted tholeiites which are typical of the modern spreading center at that latitude (30°S). Our own data for Sites 417 and 418 (Staudigel et al., 1979a) show that these sites also are relatively fractionated, LIL element-depleted basalts, characteristics which they share with basalts from Sites 395 and 396 (Rhodes et al., 1978). These characteristics are also shared with basalts at 22-23°N (Melson et al., 1968; Ludden and Thompson, 1978; Thompson et al., in preparation) and suggest that no major changes in geochemical characteristics have occurred over the past 100 m.y. at this latitude.

The geochemistry and mineralogy of basalts from Sites 100 and 105 have been discussed in considerable detail by Ayuso et al. (1976) and Bryan et al. (1977b). These basalts are mineralogically and geochemically comparable to modern ocean ridge basalts erupted beyond the influence of "mantle plumes", a relation that is compatible with a projection of these sites along spreading flow lines to their points of origin on the Mid-Atlantic Ridge. Bryan et al. (1977b) showed that these oldest Atlantic basalts are contrasted geochemically with the continental basalts of nearly the same age on the adjacent eastern margin of North America. These latter basalts appear to be systematically enriched in LIL-elements relative to modern sea floor basalts, probably due to to interaction with continental crust or sub-continental mantle. Thus, the geochemical characteristics of layer 2 basalt appear to have been established very early in the history of the opening of the Atlantic, and the most important variations appear to be related not to age, but to geographic proximity to "mantle plumes". Unfortunately, no drill sites have been placed in critical locations to test the persistence of these plumes as a function of time. Bryan et al. (1977b) noted, however, that the overall range and geographic distribution of geochemical variation in the continental basalts in eastern North America resembles that which has been documented along the Mid-Atlantic Ridge (Table 1) south of the Azores (Schilling, 1975; White and Bryan, 1977; White and Schilling, 1978) and suggests that "plume" activity may have initiated opening of the Atlantic.

Data from Leg 37 drill sites suggests that there may be at least short-term fluctuations in LIL-element enrichments on a scale of 10 m.y. or less. This is especially evident in variations in the ratio of light to heavy rare earths and in overall enrichment in LILE (Schilling et al., 1977; Bryan and Thompson, 1977). Basalts at Site 335 are light rare earth depleted, while those at Site 332 have relatively flat patterns, and basalts in the FAMOUS area show systematically greater rare earth enrichments than Site 332 (Figure 5). The interpretation of these differences in rare earth enrichments is further complicated by additional data from the FAMOUS area, where basalts closely associated in

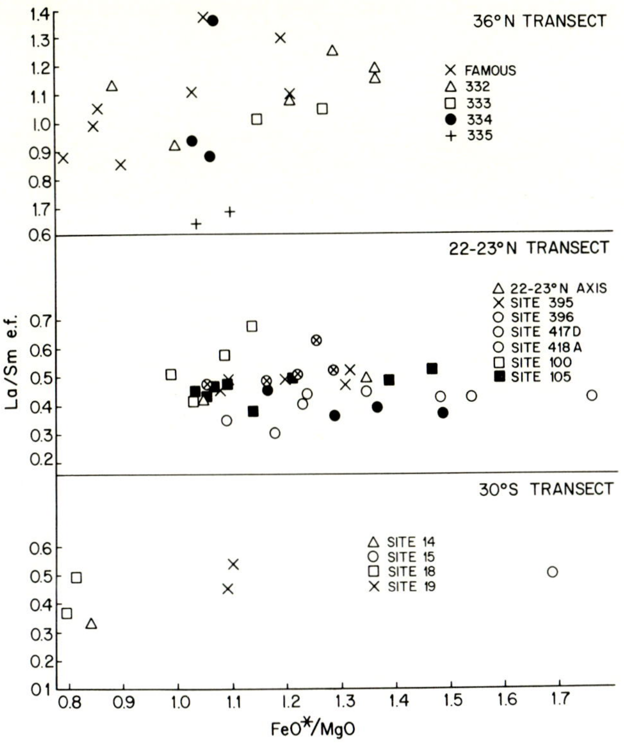

Figure 6. Chondrite-normalized La/Sm ratios vs. FeO*/MgO for three DSDP-IPOD transects. The 36°N transect is related to the Azores plume; the 22-23°N and 30°S transects are related to normal ridge segments. Note anomalously low values for Site 335, and general lack of correlation of La/Sm with FeO*/MgO. Data from Bryan et al., 1977a, b; Frey et al., 1974; Rhodes et al., 1978; Bryan and White, 1977; Langmuir et al., 1977; Schilling et al., 1977.

space and time show distinctly different rare earth patterns. These differences have been attributed to variations in degree of melting of a homogeneous mantle source rather than to fluctuations in "plume" activity (Langmuir et al., 1977), a relation supported by the consistently high Ba and Sr^{87}/Sr^{86} ratio in the LREE-depleted FAMOUS basalts (White and Bryan, 1977).

In Figure 6 we show the La/Sm enrichment factor as a function of FeO*/MgO for the three transects. Although representing a wide range of differentiation, the MORB of 30°S and 23°N show chondrite-normalized La/Sm ratios of 0.5 or less. However, the 37°N basalts of 0, 3, and 9 million years age show high (>0.8) La/Sm. Site 335 at 16 million years is apparently not affected by the Azores 'hot spot' activity suggesting that such events are not continuous in time. Figure 7 shows the La/Sm values plotted as a function of time for all the DSDP sites. At 23°N the DSDP basalts from this transect suggest the Atlantic Ocean has been erupting basalts of very similar composition since its opening in Jurassic times. These three transects also show differences in their REE relative abundances (Figure 4). The typical MORB is depleted in the light rare earth elements, while basalts from 37°N, at least for the last 9 million years, are LREE enriched. Available data for fresh basalts from the 37°N transect suggests that they are also higher in Sr^{87}/Sr^{86} (White and Bryan, 1977) and in Ba. They are not greatly different in major element composition, transition metal trace element contents, or in LILE such as Zr, and Y.

These compositional differences with respect to REE, Sr isotopes, Ba, etc. may be related to fundamental differences in mantle sources for the two kinds of basalt. There are also small but significant compositional differences between basalts at any one site. These variations are commonly explained as due to different relative melting effects, mixing of slightly different magmas, and magmatic differentiation (Bryan et al., 1978; Frey et al., 1974; Langmuir et al., 1977; Rhodes, et al., 1978). The nature and relative importance of these latter processes still require some detailed investigation.

Conclusions and Suggestions for Future Drilling

1) Basalts recovered by drilling show the same range of composition, mineralogy, and textures as modern basalts from the Mid-Atlantic Ridge. The youngest sites (<20 m.y.) include examples of both "normal" and "plume-related" basalts; the oldest sites (>50 m.y.) include only examples of "normal" basalts.

2) Layer 2 is composed predominantly of extrusive basalt with minor intercalated sediment to a depth of at least 600 m. Massive and pillowed lava flows are interbedded, and dikes and sills are rare. The few examples of layer 3 material recovered are at anomalously shallow depths and probably were tectonically emplaced. These plutonic rocks consist of gabbro and peridotite as commonly inferred for layer 3. The "normal" thickness of layer 2 is not constrained by existing drilling data, but is inferred to be at least 1.5 km.

3) The observed chemical and mineralogical characteristics of layer 2 were established very early in the history of opening of the Atlantic and show no major variations as a function of time. Short-term variations at the scale of 10 m.y. or less are, however, implied by variations in LIL-element enrichments in Leg 37 sites compared with one another to the FAMOUS area.

4) Systematic compositional variation with position along the Mid-Atlantic Ridge has been well documented among modern spreading center basalts, but drill holes in old crust have not been well placed to document the persistence of this modern pattern in crust older than 15 m.y.

5) Low-temperature weathering appears to have

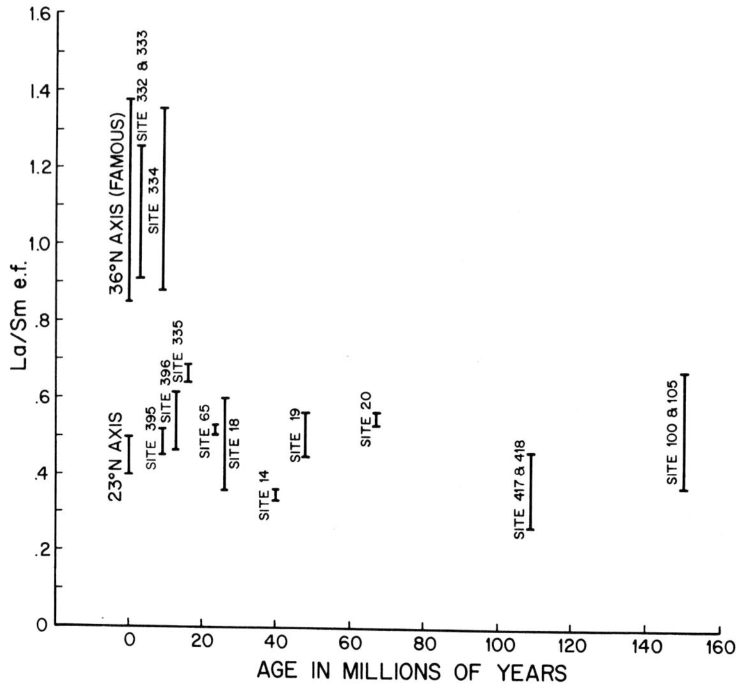

Figure 7. Chondrite-normalized La/Sm ratios as a function of age for DSDP-IPOD sites; range of values is indicated by length of bars at each site. FAMOUS area and Sites 332, 333, 334 and 335 are related to Azores mantle plume; Site 335 appears anomalous. Other sites are related to normal ridge segments and show no significant variation with age. Data sources as in Figure 6.

been concentrated in basalts exposed on local highs, while it shows very limited effects in basalts covered by sediment almost immediately after extrusion. There is little evidence of metamorphism or of high-temperature hydrothermal alteration even in the deepest levels penetrated. In particular, absence of metamorphic effects at Sites 395 and 396 is surprising in view of the abundant greenstones recovered in the median valley at 22°N (Melson et al., 1968). These observations suggest that both intense low temperature weathering and metamorphism may be more localized than has often been assumed, with obvious additional implications for chemical budgets arising from rock-water interactions.

Because of its relatively simple spreading history, and because of the presence of several successful deep holes in both young and old crust, we consider the North Atlantic to be an ideal location for additional deep drilling to define the stratigraphy of the deeper part of layer 2 and upper part of layer 3, and to further define both short-term and long-term geochemical variations with time to achieve a better mapping of along-ridge variations in old crust. The simplest way to achieve these objectives would appear to be to continue the Leg 37 transect out to crust of about 110 m.y. age and to place several deep drill holes in 110 m.y. old crust between the end of the northern transect and the existing holes at Sites 417 and 418. Several additional holes should be drilled along a spreading flow line based either on Sites 395 or 417-418 in order to complete a second transect in "normal" crust unrelated to a mantle plume.

Acknowledgements

We are especially grateful to W. G. Melson, P. T. Robinson, and J. M. Rhodes for discussion of recent IPOD sites and for providing advance copies of data and manuscripts for Sites 395 and 396. Work summarized in this paper has been supported by the National Science Foundation under Grant DES 76-15858. This is WHOI Contribution No. 4189.

References

Aumento, F. W. S. Mitchell and M. Fratta (1976). Interaction between seawater and oceanic layer 2 as a function of time and depth. 1. Field evidence. Canad. Mineralogist, 14: 269-290.

Ayuso, R. A., A. E. Bence and S. R. Taylor (1976). Upper Jurassic tholeiitic basalts from DSDP Leg 11. J. Geophys. Res., 81: 4305-4325.

Bryan, W. B. (1974). Fe-Mg relationships in sector-zoned submarine basalt plagioclase. Earth Planet. Sci. Letts., 24: 157-165.

Bryan, W. B. (1978). Regional variations and petrogenesis of basalt glasses from the FAMOUS area, Mid-Atlantic Ridge. J. Petrology, in press.

Bryan, W. B., G. Thompson, F. A. Frey and J. S. Dickey (1976). Inferred geologic settings and differentiation in basalts from the Deep Sea Drilling Project. J. Geophys. Res., 81: 4285-4304.

Bryan, W. B., G. Thompson, F. A. Frey, J. S. Dickey and S. Roy (1977a). Petrology and geochemistry of basement rocks recovered on Leg 37, DSDP. In: Initial Reports of the Deep Sea Drilling Project, 37: 695-703.

Bryan, W. B., F. A. Frey and G. Thompson (1977b). Oldest Atlantic sea floor: Mesozoic basalts from western North Atlantic margin and eastern North America. Contrib. Mineral. Petrol., 64: 223-242.

Bryan, W. B. and J. G. Moore (1977). Compositional variations of young basalts in the Mid-Atlantic Ridge rift valley near 36°49'N. Geol. Soc. Amer. Bull., 88: 556-570.

Bryan, W. B., P. T. Robinson et al. (1977). Studying oceanic layer 2. Geotimes, 22: 22-26.

Bryan, W. B. and D. Sargent (1978). Basalts from 22-23°N, Mid-Atlantic Ridge median valley. In: Initial Reports of the Deep Sea Drilling Project, 45, in press.

Bryan, W. B. and G. Thompson (1977). Basalts from DSDP Leg 37 and the FAMOUS area: compositional and petrogenetic comparisons. Can. J. Earth Sci., 14: 875-885.

Bryan, W. B., G. Thompson and P. M. Michael (1978). Compositional variation in a steady state zoned magma chamber: Mid-Atlantic Ridge at 36°50'N. Tectonophysics, in press.

Dick, H.J.B. and W. B. Bryan (1978). The variation of basalt phenocryst mineralogy and rock compositions in DSDP Hole 396B. In: Initial Reports of the Deep Sea Drilling Project, 46, in press.

Donnelly, T. W., G. Thompson and M. Salisbury (1979). The chemistry of altered basalts at Site 417A, DSDP Leg 51. In: Initial Reports of the Deep Sea Drilling Project, 51, in press.

Frey, F. A., W. B. Bryan and G. Thompson (1974). Atlantic Ocean floor: geochemistry and petrology of basalts from Legs 2 and 3 of the Deep Sea Drilling Project. J. Geophys. Res., 79: 5507-5527.

Hart, S. R. (1969). K, Rb, Cs contents and K/Rb and K/Cs ratios of fresh and altered submarine basalts. Earth Planet. Sci. Letts., 6: 295-303.

Hart, S. R. (1970). Chemical exchange between seawater and deep ocean basalts. Earth Planet. Sci. Letts., 9: 269-279.

Hart, S. R., J. G. Schilling and J. L. Powell (1973). Basalts from Iceland and along the Reykjanes Ridge: Sr isotope geochemistry. Nature, 246: 104-107.

Hodges, F. N. and J. J. Papike (1976). DSDP Site 334: magmatic cumulates from oceanic layer 3. J. Geophys. Res., 81: 4135-4151.

Langmuir, C. H., J. F. Bender, A. E. Bence, G. N. Hanson and S. R. Taylor (1977). Petrogenesis of basalt from the FAMOUS area, Mid-Atlantic Ridge. Earth Planet. Sci. Letts., 36: 133-156.

Ludden, J. N. and G. Thompson (1978). The behavior of the rare earth elements during submarine weathering of tholeiitic basalts. Nature, in press.

Melson, W. G., G. Thompson and Tj. van Andel (1968). Volcanism and metamorphism in the Mid-Atlantic Ridge, 22°N latitude. J. Geophys. Res., 73: 5925-5941.

Miyashiro, A., F. Shido and M. Ewing (1969). Diversity and origin of abyssal tholeiite from the Mid-Atlantic Ridge near 24° and 30°N. Contrib. Mineral. and Petrol., 23: 38-52.

Muehlenbachs, K. (1976). Oxygen isotope geochemistry of Leg 34 basalts. In: Initial Reports of the Deep Sea Drilling Project, 34: 337-339.

Muehlenbachs, K. (1979). The alteration and aging of the basaltic layer of the sea floor: O_2 evidence from DSDP Legs 51, 52 and 53. In: Initial Reports of the Deep Sea Drilling Project, 52, in press.

Osborn, E. J. and D. B. Tait (1952). The system diopside-forsterite-anorthite. Amer. J. Sci., Bowen Volume, 413-433.

Rhodes, J. M., M. A. Dungan, D. P. Blanchard and P. E. Long (1978). Magma mixing at mid-ocean ridges: evidence from basalts drilled near 22°N on the Mid-Atlantic Ridge. Tectonophysics, in press.

Robinson, P. T., M.F.J. Flower, H. V. Schmincke and W. Ohnmacht (1977). Low temperature alteration of oceanic basalt. In: Initial Reports of the Deep Sea Drilling Project, 37: 775-793.

Schilling, J. G. (1973). Iceland mantle plume: geochemical evidence along the Reykjanes Ridge. Nature, 242: 565-571.

Schilling, J. G., (1975). Azores mantle blob: rare earth evidence. Earth Planet. Sci. Letts., 25: 103-115.

Schilling, J. G., R. N. Anderson and P. R. Vogt (1976). Rare earth, Fe, and Ti variations along the Galapagos spreading center, and the relationships to the Galapagos mantle plume. Nature, 261: 108-113.

Schilling, J. G., R. Kingsley and M. Bergeron (1977). Rare earth abundances in DSDP Sites 332, 334 and 335, and inferences on the Azores mantle blob activity with time. In: Initial Reports of the Deep Sea Drilling Project, 37: 591-597.

Sinton, J. M. and G. R. Byerly (1979). Mineral compositions and crystallization trends in DSDP Holes 417D and 418A. In: Initial Reports of the Deep Sea Drilling Project, 52, in press.

Staudigel, H., W. B. Bryan and G. Thompson (1979a). Chemical variation in glass-whole rock pairs from individual cooling units in Holes 417D and 418A. In: Initial Reports of the Deep Sea Drilling Project, 52, in press.

Staudigel, H., F. A. Frey and S. R. Hart (1979b). Rare earth elements and Sr isotopes in basalts and glass from Sites 417 and 418: fresh rock composition and alteration effects. In: Initial Reports of the Deep Sea Drilling Project, 52, in press.

Thompson, G. (1973). A geochemical study of the low temperature interaction of seawater and oceanic igneous rocks. EOS, Trans. Amer. Geophys. Union, 54: 1015-1019.

Thompson, G., M. Rivers, S. Henrichs and D. C. Bankston. The low temperature weathering of oceanic basalts by seawater. In preparation.

Thompson, G., W. B. Bryan and W. G. Melson. Geochemical variation and petrogenesis of basalt glasses from the Cayman Trench spreading center. In preparation.

White, W. M., J. G. Schilling and S. R. Hart (1976). Evidence for the Azores mantle plume from strontium isotope geochemistry of the central North Atlantic. Nature, 263: 659-663.

White, W. M. and W. B. Bryan (1977). Strontium isotope, K, Rb, Cs, Sr, Ba, and rare earth geochemistry of basalts from the FAMOUS area. Geol. Soc. Amer. Bull., 88: 571-579.

White, W. M. and J. G. Schilling (1978). The nature and origin of geochemical variation in Mid-Atlantic Ridge basalts from the central North Atlantic. Geochim. Cosmochim. Acta, in press.

Wright, T. L., W. E. Benson, W. G. Melson and S. R. Hart (1972). Petrology of basaltic rocks collected from Leg 14. In: Initial Reports of the Deep Sea Drilling Project, 14: 767-772.

NATURE OF MANTLE HETEROGENEITY IN THE NORTH ATLANTIC :

EVIDENCE FROM LEG 49 BASALTS

J.Tarney[1], D.A.Wood[1], J. Varet[2], A.D.Saunders[1], and J.R.Cann[3]

Abstract. Previous deductions concerning the nature of mantle heterogeneity in the North Atlantic have been based largely on samples dredged from near the ridge axis. Leg 49 added two further dimensions to this array of data - variations with depth and with time - in three critical areas where information was already available from dredging. Three holes at 63°N provided information on the compositions of basalts erupted at the Reykjanes segment of the ridge for almost 36 m.y., two holes at 45°N up to 10 m.y. and four holes at 36°N (combined with data from Leg 37) up to 16 m.y. Major and trace element data on more than 150 basalt samples have been used to ascertain compositional variations resulting from fractional crystallization, batch melting and dynamic melting and to define more closely geochemical differences in the mantle source regions in space and time. The results demonstrate the following features. (1) An unusually wide range of basalt compositions may be erupted at a single site, although in most cases it is possible to relate these basalts to a single source through partial melting and fractional crystallization. (2) Basalts from different sites along the same mantle flow line have remarkably consistent ratios of the more hygromagmatophile trace elements, indicating that the mantle source feeding a particular ridge segment was relatively uniform with respect to these elements for long time periods. (3) There are inter-site differences along a particular mantle flow-line with respect to Rb/Sr, Ce/Yb and $^{87}Sr/^{86}Sr$ which may reflect smaller scale heterogeneities in the source. (4) There are large differences in hygromagmatophile trace element ratios between different segments of the ridge, but in each case these regional differences are maintained for long time periods. Basalts from 45°N and 36°N have similar hygromagmatophile element ratios (but different abundances) and these ratios are different from those at 63°N and on Iceland, while both are different from the ratios in 'depleted' MORB from 22°N.

The data are difficult to reconcile with mantle plume models invoking mixing of two contrasted 'enriched' and 'depleted' sources or magmas. Rather there are several sources each with a characteristic composition, or range of compositions, feeding the Mid Atlantic Ridge. While compositional and isotopic differences between North Atlantic basalts may partly reflect earlier mantle depletion processes, such as crust extraction, later enrichment events may also have contributed to the observed chemical diversity, particularly in the 45°N region.

[1]Department of Geological Sciences, University of Birmingham, England

[2]Bureau de Recherches Geologiques et Minieres, Orleans, France

[3]Department of Geology, University of Newcastle-upon-Tyne, England

Introduction

That geochemical differences exist in basalts sampled from different sections of the Mid-Atlantic Ridge is now well established (e.g. Schilling, 1975b). Because these geochemical differences correlate to some extent with variations in Sr isotopes (Hart et al, 1973; O'Nions and Pankhurst, 1974; O'Nions et al, 1976; White et al, 1975, 1976), Pb isotopes (Sun and Jahn, 1975; Sun et al, 1975) and Nd isotopes (O'Nions et al, 1977, 1978; Carter et al, 1978a) there is a concensus of opinion that these differences reflect some sort of geochemical heterogeneity in the mantle source.

The generalised picture emerging from previous studies suggests that the Mid-Atlantic Ridge consists of 'normal segments' together with 'anomalous hot spot areas' (such as Iceland and the Azores) and with transitional zones between (e.g. the Reykjanes Ridge and in the 39°-35°N region south of the Azores). To explain these features, Schilling and co-workers, (Schilling, 1973, 1975a, b, 1976; Schilling and Noe-Nygaard, 1974; Hart et al, 1973; Hermes and Schilling, 1976; White et al, 1975) have proposed models involving mixing of an incompatible element 'depleted' mantle source, feeding the normal ridge segments, with that of an incompatible element 'enriched' mantle source feeding oceanic

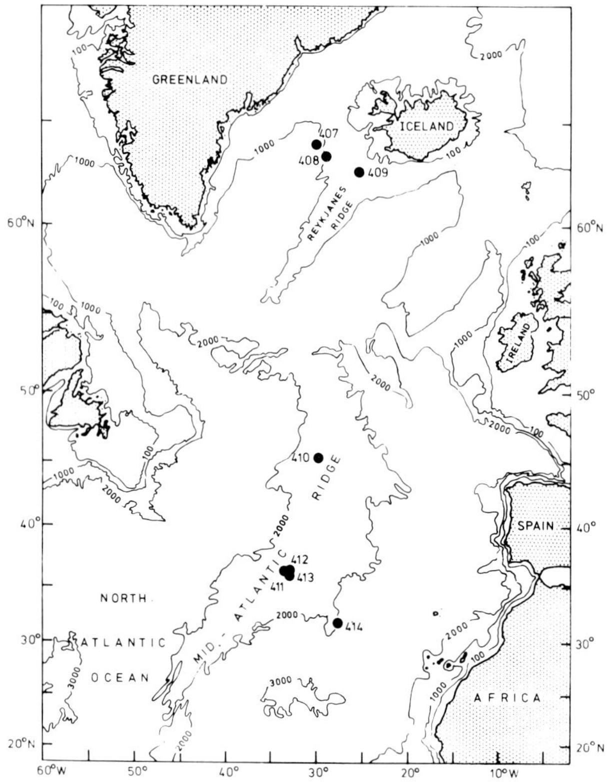

Figure 1. Location of Sites 407 to 413 drilled by Leg 49.

islands and rising up in the form of plumes (Morgan, 1971) or blobs in hot spot regions.

Inevitably there have been arguments concerning the extent to which the observed compositional differences can be accounted for by the processes associated with basalt generation (O'Hara, 1977). At one extreme it would be possible to regard each basalt composition as representing the partial melting product of a unique mantle source; while at the other extreme various partial melting and dynamic melting mechanisms (e.g. Langmuir et al, 1977) combined with open and closed fractional crystallization (e.g. O'Hara, 1977) and perhaps contamination, are capable in theory of producing a wide variety of magma compositions from a uniform source. The true situation inevitably lies somewhere between these two extremes, and it is important to try to establish which element or element ratio variations can be accounted for by observed fractionation processes within the normal range of basalt generation, and then to consider residual differences in terms of inhomogeneity of the source.

Isotopic studies of course are invaluable in establishing time-integrated differences in parent/daughter ratios in the mantle source regions, and while correlations between $^{143}Nd/^{144}Nd$ and $^{87}Sr/^{86}Sr$ (O'Nions et al, 1977) suggest a degree of covariance between Sm/Nd and Rb/Sr and therefore perhaps other element pairs too, it is nevertheless vital to establish whether in fact there are significant regional differences in these other elements. This may have a bearing on the processes which have given rise to heterogeneity.

There is however still considerable uncertainty as to the nature, degree and extent of mantle heterogeneity. With the simplistic mantle plume mixing model for example, the spectrum of basalt compositions along the Mid-Atlantic Ridge should reflect the mixing of the two mantle sources once the effects due to partial melting and fractional crystallization have been accounted for. Not all geochemists believe that the data can be explained so simply (O'Nions et al, 1976; Tatsumoto, 1978). Langmuir et al. (1978) have applied mixing equations to the Iceland - Reykjanes Ridge data and have argued that these data are not consistent with either mixing of magmas or of an enriched ocean island source with a depleted ocean ridge source. They concluded that there are either two separate trends in the Iceland - Reykjanes Ridge region, or that the basalts have been derived from a multiplicity of sources which have been isolated for hundreds of millions of years.

These concepts have generally been discussed using analytical data for a few elements and/or isotopes on a limited number of samples — and, moreover, samples located or dredged near the ridge axis. Leg 49 added two further dimensions to this array of data — (a) variations with time and (b) variations with depth — in critical areas of the North Atlantic where a considerable body of data was already available. Nine holes were drilled at seven sites (Fig. 1) in three different transects across the Mid-Atlantic Ridge at 63°N (Reykjanes Ridge), 45°N (Hudson Geotraverse Area) and 36°N (FAMOUS area). Three sites (407, 408 and 409) were located along the same mantle flow line transverse to the Reykjanes Ridge at 63°N, at approximately the mid-point of the supposed geochemical gradient south of Iceland, and sampled basalts erupted over a 34 m.y. interval between 36 m.y. and 2.3 m.y. ago. Two holes at Site 410 penetrated 10 m.y. old crust west of the ridge axis at 45°N and permitted comparisons with young basalts dredged from the Median Valley at 45°N. Three sites (411, 412, 413) in the FAMOUS area near 36°N were drilled in young (1 m.y. to 3.5 m.y.) basaltic crust and complement earlier drilling on Leg 37 at Sites 332 to 335 (3.5 to 16.5 m.y.).

The recovery at most of these sites was good, and produced a surprisingly wide variety of basalt types. In order to provide a comprehensive body of data as a framework for petrogenetic discussion, over 150 samples have been analysed for up to 40 major and trace elements by X-ray fluorescence and instrumental neutron activation techniques in four different laboratories. Several elements have been analysed on the same samples in different laboratories, and inter-laboratory agreement has been found to be good. Nevertheless, to avoid possible bias in comparing basalts from different sites, all the diagrams presented make use of data from one laboratory for each element. The complete petrological, mineralogical and geochemical data is listed in the D.S.D.P. Initial Reports (Tarney et al., in press; Wood et al., in press) and only 17 analyses representing the main lithological units are presented here (Table 1).

Much of the discussion will center around 'incompatible' trace elements (i.e. those with a solid/liquid distribution coefficient, D, less than 1). Nevertheless this term covers elements with a wide range of D values (Fig. 7), and for greater clarity we will use the term 'hygromagmatophile' (HYG) for elements with D significantly less than 1, qualified by 'less-HYG' for elements with D near 0.1 and 'more-HYG' for those with $D \leqslant 0.01$.

Mid-Atlantic Ridge 63°N: Sites 407, 408 and 409

These three sites were chosen to lie along a mantle flow line from a single segment of the Reykjanes Ridge, and provide a record of the chemistry of basalts produced during almost 40 m.y. of spreading. Moreover, this transect corresponds with the mid-point of the geochemical gradient south of Iceland suggested by dredged basalts from the Reykjanes Ridge (Schilling, 1973; Hart et al., 1973). Recent work on Iceland has shown that basalts, even in quite restricted areas, may show a wide range of trace element and isotopic compositions (O'Nions et al., 1976; Sigvaldason et al., 1976; Wood, 1978), thus it was important to establish whether the geochemical gradient, based on relatively few samples from each latitude, would remain apparent after more detailed sampling in depth and with time.

Hole 407 (63° 56.52'N; 30° 34.56'W), on magnetic anomaly 13 (35-40 m.y.), penetrated almost 160m. of aphyric or slightly phyric (plagioclase, olivine, minor clinopyroxene) basalts. Geochemical studies have revealed four distinct units, each made up of several flow units. Units 2 and 3 are separated by a major change in magnetic polarity and Units 3 and 4 by a sedimentary horizon. The degree of alteration increases downhole.

Hole 408 (63° 22.66'N; 28° 54.71'W), sited on a topographic high within magnetic anomaly 6 (20 m.y.), penetrated 33m. of vesicular aphyric (or sparsely phyric) basalts. Although there is some geochemical variation within the hole, the basalts belong essentially to one geochemical unit. Alteration is widespread.

Hole 409 (62° 36.98'N; 25° 57.17'W), sited close to the crest of the Reykjanes Ridge, drilled young 2.3 m.y. old crust on anomaly 2' and penetrated 240m. of vesicular basalt, mostly aphyric or sparsely phyric, and relatively fresh. There are four geochemical units, but the major break occurs between Units 2 and 3. A thin interval within Unit 2 consisted of much more evolved basalt.

All the fresh basalts from this transect are hypersthene normative, and follow a distinct trend of iron-enrichment on an AFM diagram. Some of the units, particularly at Site 407, are also Ti-rich (Table 1). A plot of Ce against Y (Fig. 2a) shows that samples from each of the geochemical units fall in discrete and separate fields, each with a constant Ce_N/Y_N ratio (within the limits of Ce precision by XRF). Chondrite-normalised rare-earth patterns of representative samples from each unit are shown in Figure 3 and confirm that there is a wide variation in Ce_N/Yb_N ratio in basalts from this transect, from light-RE depleted to light-RE enriched. Even within one hole (407) there is an appreciable variation in Ce_N/Yb_N.

A plot of Y versus Zr (Fig. 2b) serves to distinguish between fractional crystallization and partial melting processes in the petrogenesis of these basalts. The Y/Zr ratio changes only very slightly during fractional crystallization, but the differences in partition coefficient between the two become significant during moderate to low degrees of partial melting, with Y being progressively retained in the source at lower degrees of partial melting (Saunders and Tarney, in press) It can be seen that samples from each unit have constant Y/Zr ratios and plot along lines with zero intercept suggesting that within-unit variation may be attributed to limited closed system fractional crystallization. Units 407-1 and 407-3 have similar Y/Zr and Ce/Y ratios and (although separated by a magnetic polarity reversal) 407-1 can be derived from a liquid similar to 407-3 by fractional crystallization. A least squares solution of the major element oxides for the fractionation of the observed phenocryst mineral compositions using the method of Wright and Doherty (1970) suggests some 40% crystal fractionation of plagioclase + clinopyroxene with minor olivine and titanomagnetite (Wood et al, in press). This is supported by the lower Sr, Cr, and Ni in 407-1, whereas the other incompatible elements increase in proportion to Zr. The more evolved flow unit (Sample 10-2, 22-24cm) at Site 409 can similarly be related by fractional crystallization to the main group of basalts (Unit 409-2).

Attempts to relate the other geochemical units at Site 407 or the two main units at Site 409 by crystal fractionation processes, open or closed, were unsuccessful. In each case the amount of fractional crystallization of olivine, clinopyroxene or plagioclase demanded by the increase in incompatible elements is not matched by the predicted reductions in Cr, Ni or Sr expected by applying reasonable partition coefficient data. Nor can differences in Ce/Y, Ce/Yb, Zr/Y ratios between these units be accounted for. In Hole 409 for instance, Unit 409-3 has higher

Table 1. Major element oxide (wt %) and trace element (p.p.m) analyses of the 17 major lithostratigraphic units recovered by IPOD Leg 49

Latitude	63°N								45°N				36°N				
Unit	407-1	407-2	407-3	407-4	408	409-1	409-2	409-3	410-1	410-2	410A-1	410A-2	411-1	411-2	412A	413-1	413-2
SiO_2	49.79	48.98	48.32	48.30	49.00	50.40	49.91	49.30	51.46	50.16	50.46	49.24	48.39	50.6	50.38	49.86	47.48
TiO_2	3.91	1.59	2.87	1.96	1.53	1.30	1.17	1.74	1.41	1.51	1.40	1.46	0.71	0.96	1.00	1.58	0.70
Al_2O_3	12.97	15.03	12.89	13.96	14.75	13.97	14.09	14.01	15.84	17.71	16.16	16.44	15.31	16.93	15.00	14.11	16.26
Fe_2O_3	6.58	3.41	5.18	4.59	5.67	3.13	2.46	2.94	2.15	3.59	3.84	4.50	1.29	1.22	2.42	3.45	3.85
FeO	8.17	7.93	9.11	7.93	4.42	7.66	8.14	8.28	6.18	4.36	4.86	3.83	7.73	7.03	6.96	5.54	6.47
MnO	0.19	0.15	0.22	0.19	0.13	0.18	0.17	0.19	0.14	0.17	0.12	0.13	0.17	0.16	0.18	0.15	0.16
MgO	5.02	7.45	6.32	6.60	8.50	8.34	8.15	8.09	6.95	3.77	8.38	8.45	12.20	7.28	7.48	10.64	10.51
CaO	8.96	11.71	10.74	11.27	11.60	11.85	11.89	11.65	11.33	10.07	11.36	11.26	11.97	12.96	12.97	11.22	10.88
Na_2O	2.90	2.47	2.50	2.16	2.38	2.20	2.21	2.30	3.00	3.98	2.70	2.50	1.88	2.50	2.18	2.30	2.44
K_2O	0.54	0.17	0.35	0.38	0.21	0.10	0.07	0.18	0.64	1.76	0.42	0.48	0.05	0.20	0.24	0.49	0.15
P_2O_5	0.38	0.14	0.33	0.32	0.18	0.13	0.14	0.22	0.23	0.35	0.22	0.27	0.11	0.17	0.12	0.28	0.08
H_2O+	0.49	0.84	0.91	1.70	1.44	0.32	0.69	0.83	1.06	1.26	0.66	1.21	0.46	0.24	0.18	1.48	1.36
CO_2	0.01	0.07	0.04	0.08	0.11	0.03	0.15	0.01	-	0.66	0.11	0.41	0.10	0.13	0.15	0.20	0.04
Total	99.91	99.94	99.78	99.44	99.92	99.61	99.24	99.74	100.39	99.35	100.69	100.18	100.37	100.41	99.26	101.30	100.38
Mg/Mg+Fe	0.389	0.547	0.450	0.494	0.614	0.587	0.584	0.569	0.604	0.470	0.643	0.657	0.710	0.615	0.594	0.687	0.654
FeO	14.09	11.00	13.77	12.06	9.52	10.48	10.35	10.93	8.12	7.59	8.32	7.88	8.89	8.13	9.14	8.64	9.93
Sc	43.7	-	39.6	40.0	42.4	44.6	44.3	45.3	-	26.7	31.2	29.1	39.6	37.5	39.8	33.2	34.9
V	530	-	406	357	319	340	328	377	265	260	295	261	202	235	262	265	-
Cr	24	233	100	274	316	160	161	330	189	206	201	258	532	162	89	708	580
Co	54	-	53	51	49	51	51	49	35	32	39	38	54	47	47	50	135
Ni	19	84	57	97	98	76	75	98	86	108	107	141	227	55	50	277	354
Zn	130	81	109	85	77	78	72	87	69	72	70	68	57	63	71	67	64
Ga	23	19	22	18	19	17	18	18	17	23	18	18	14	17	17	14	16
Rb	10	1	4	8	2	1	1	3	7	33	5	5	1	3	4	6	1
Sr	201	160	222	168	185	83	83	101	226	470	217	300	80	121	120	234	75
Y	53	24	36	29	27	26	23	35	24	23	22	21	18	19	21	22	19
Zr	263	83	185	119	105	65	57	110	115	206	113	142	39	55	57	102	33
Nb	26	6	19	12	11	5	3	10	25	55	25	32	5	8	10	25	5
Cs	-	-	0.06	0.17	0.02	-	-	-	-	0.81	0.26	0.25	-	0.04	0.12	0.15	-
Ba	172	49	123	75	70	34	38	70	172	447	175	191	37	55	56	124	12
La	17.7	-	12.9	9.3	8.8	3.1	2.4	6.1	-	42.2	14.1	19.9	2.4	6.3	5.7	16.0	1.9
Ce	47.2	14.9	35.6	23.1	26.2	6.9	6.0	18.6	32.6	68.2	31.8	39.3	6.3	13.7	12.5	27.8	4.5
Nd	31.8	10.0	24.7	16.1	18.5	6.7	5.9	13.9	16.0	26.9	15.4	19.5	4.9	8.6	8.9	14.8	4.5
Sm	8.89	3.45	6.86	5.02	5.88	2.11	2.09	4.71	3.99	5.40	3.67	3.96	1.79	2.43	2.34	3.41	1.69
Eu	2.89	1.32	2.32	1.71	1.79	0.89	0.78	1.59	1.35	1.73	1.32	1.40	0.72	0.91	0.92	1.24	0.67
Gd	10.0	-	-	5.1	4.9	3.0	2.8	5.3	-	-	4.6	4.4	2.2	2.5	3.2	4.0	2.6
Tb	1.62	0.73	1.18	0.93	0.79	0.57	0.50	0.97	0.70	0.76	0.64	0.66	0.47	0.54	0.49	0.62	0.48
Tm	0.8	0.3	0.5	0.5	0.4	0.4	0.3	0.6	0.4	0.4	0.4	0.4	0.3	0.4	0.4	0.3	0.3
Yb	4.76	2.65	3.49	3.21	2.93	2.54	2.14	3.94	2.30	2.10	2.36	2.11	2.17	2.18	2.21	2.03	2.32
Lu	0.77	0.43	0.57	0.55	0.50	0.41	0.33	0.67	0.39	0.38	0.37	0.30	0.39	0.37	0.40	0.33	0.35
Hf	6.6	2.2	4.8	3.2	3.4	1.6	1.4	3.1	2.5	4.0	2.3	2.8	1.0	1.5	1.5	2.4	1.0
Ta	1.88	-	1.40	0.87	0.86	0.27	0.24	0.68	-	4.61	1.59	2.18	0.23	0.62	0.57	1.80	-
Th	1.85	0.53	1.27	0.67	0.96	0.34	0.29	0.59	2.12	5.82	2.07	2.92	0.35	0.73	0.62	1.52	0.22
U	0.5	-	0.44	0.16	0.11	-	-	0.14	-	1.84	0.28	0.58	-	0.20	0.21	0.24	-

HYG element abundances than Units 409-1 and 409-2, but also has higher Cr and Ni abundances and more normative olivine, although lower Mg/Fe ratio. A similar relationship obtains between Units 407-2 and 407-4 at Site 407.

In Figure 2b possible partial melting relationships between these units are indicated. For each hole individual units lie close to a single partial melting curve, those with higher Zr also having slightly higher Fe/Mg ratios, but also lower Y/Zr ratios, suggesting retention of Y in the source by a phase such as clinopyroxene. There are however three separate curves, indicating that the source mineralogy or composition at the three sites is different. A small amount of garnet in the source at Site 408 for example would account for the lower Y/Zr ratio (and higher Ce/Y and Ce/Yb ratios) of Site 408 basalts. Unit 407-3 basalts are too low in Cr and Ni and their Mg/Fe ratios are also too low for primary mantle melts. Allowing 35% pre-eruptive crystallization (ol, plag, cpx) would bring the compositions of these basalts closer to those of Hole 408 and to the more Mg-rich basalts from eastern Iceland (Wood, 1978), all of which have similar Ce/Y ratios. That all the units cannot be related by equilibrium partial melting of a similar mantle source, irrespective of its mineralogy is demonstrated by the strontium isotope data for these sites

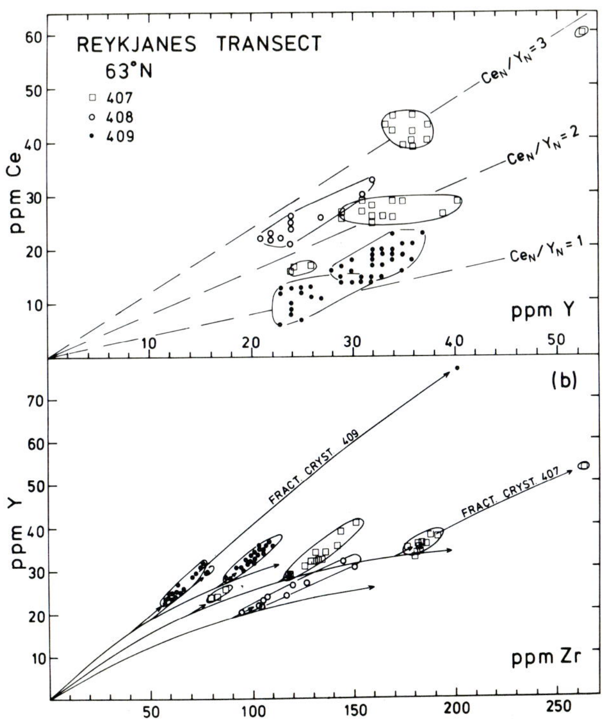

Figure 2.(a) Ce versus Y and (b) Y versus Zr for 63°N basalts. In (b) fractional crystallization and possible partial melting relationships between the units are indicated (see text).

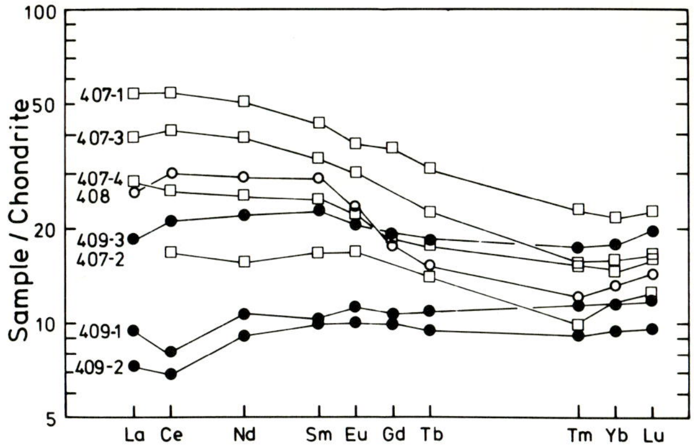

Figure 3. Chondrite-normalized rare earth patterns for representative 63°N basalts.

(Wood et al., in press). There is a significant positive correlation between $^{87}Sr/^{86}Sr$ ratio and enrichment in Sr and with Rb/Sr and Ce_N/Yb_N ratios. There is however a much greater difference in $^{87}Sr/^{86}Sr$ ratio between Hole 407 and Hole 408 basalts (0.7034) and Hole 409 basalts (0.7029) than there is between the individual units of a single hole, which generally have the same $^{87}Sr/^{86}Sr$ ratio ($\pm$ 0.0001). The Sr isotope data do therefore suggest that there are inhomogeneities, at least on an inter-site scale, in the mantle feeding a single segment of the Mid-Atlantic Ridge. Such inhomogeneities are on a similar scale to those previously described in Iceland (O'Nions et al, 1976).

We conclude that, although much of the compositional variability in basalts from the 63°N Reykjanes transect results from fractional crystallization and especially partial melting processes, there are at least two different mantle source compositions which have had a separate time-integrated Rb-Sr history. Note however that none of the basalt types related to these sources has the geochemical characteristics of 'MORB'. All are in fact more radiogenic and are enriched in hygromagmatophile elements relative to MORB. This is readily demonstrated by plotting any pair of incompatible elements such as Zr-Nb, Zr-Ce, Zr-Ba, Tb-Ta, Hf-Th etc (see Figs. 8 to 10, and Tarney et al., in press; Wood et al., in press), where basalts from all 63°N sites form a linear array either with zero intercept or with a small intercept on the axis of the less hygromagmatophile element; basalts of 'MORB' type have distinctly different ratios of these elements. In fact it is the consistency of these element ratios in basalts produced over almost 40 m.y. of spreading which is the more remarkable, and implies that the ratios of the more-hygromagmatophile elements in the mantle source are relatively constant. Differences have however developed in the ratios of more-HYG to less-HYG elements and are reflected in Sr-isotope variations.

The data are difficult to reconcile with the mantle plume mixing model for the Iceland-Reykjanes Ridge area (Vogt, 1972; Schilling, 1973; Schilling and Noe-Nygaard, 1974). Evidently there is not a relative enrichment in HYG elements or $^{87}Sr/^{86}Sr$ ratios towards the ridge crest, as predicted by the plume model, but rather a decrease. Moreover the most 'enriched' basalts were erupted (at Site 407) some 35 to 40 million years ago when the postulated plume activity was supposed to be at a minimum on the Iceland segment of the ridge. Our data thus corroborate the conclusions of O'Nions et al. (1976) and of Langmuir et al. (1978) who showed that binary mixing of depleted 'MORB' magmas or sources and enriched ocean island material did not satisfy the data.

Mid-Atlantic Ridge 45°N: Site 410

The area around the Mid-Atlantic Ridge at 45°N has been the subject of detailed investigation by the Canadian Hudson Geotraverse project (Aumento et al., 1971) in which samples have been dredged from 40 sites. Although this is a bathymetrically normal segment of the ridge, a variety of tholeiitic, transitional and alkalic basalts with some high-alumina basalts and ferro-basalts have been recovered (Muir and Tilley, 1964; Aumento, 1968) The basalts from this area are well known for their high levels of hygromagmatophile elements, high $^{87}Sr/^{86}Sr$ ratios (~0.7033) and light RE enriched REE patterns (Frey et al., 1968; Hart, 1971a, b; Aumento, 1971; Erlank and Kable, 1976; White et al., 1975). Leg 49 drilled west of the median valley on magnetic anomaly 5 (ca 10 m.y.).

Hole 410 (45° 30.51'N, 29° 28.56'W) penetrated 47m of basement. The upper 26m was a breccia of sparsely phyric (ol, plag, cpx) basalt while the lower unit consisted of a pillow breccia of mugearitic composition (MgO~4% ne-normative) with up to 10% phenocrysts of titaniferous clinopyroxene, olivine and plagioclase (An_{45}).

Hole 410A(120m south of Hole 410) penetrated 49m of sparsely phyric pillow basalts similar to those in the upper unit in Hole 410.

Site 410 samples range from transitional to mildly alkaline mugearites, and clearly form part of a geochemically coherent group, but trend towards alkali enrichment rather than Fe-enrichment on an AFM diagram. Dredged rocks from 45°N do however include some ferrobasalt differentiates (Aumento et al., 1971) and even plagiogranites (Aumento, 1969). 49°N drilled basalts have very different compositions from those at 63°N (Table 1). They have lower FeO and TiO_2 at equivalent MgO values, but much higher levels of K_2O and other hygromagmatophile elements. Ratios of more-HYG to less-HYG elements are also very different from those in 63°N basalts and in 'MORB', but there are many similarities in trace element relationships between 45°N and 36°N (FAMOUS area) basalts and to emphasise this, samples from the two areas are included on the same diagrams.

A Ce versus Y plot (Fig. 4a) demonstrates that Ce_N/Y_N ratios of Unit 410-2 basalts (>6) are different from those of Unit 410-1 and 410A basalts (~4). This is confirmed by the chondrite-normalised REE patterns of representative samples of the three units (Fig. 5), which all show strongly light-RE enriched concave-upwards patterns, very different from those of Reykjanes basalts. The samples analysed by Frey et al. (1968) from 45°N have similar REE patterns and lie just below those for Unit 410-1. An interesting feature of the REE patterns is that there is little variation in HREE levels. A plot of Y versus Zr (Fig. 4b) also demonstrates that while the Zr concentration almost doubles (from 115 ppm to 206 ppm) Y remains constant at between 24 and 21 ppm. A Ti versus Zr plot (Fig. 6) exhibits the same feature. This relationship contrasts with that in Reykjanes basalts, and in fact in most ocean basalts (Pearce and Cann, 1973), where there is normally a good linear relationship between Y - Zr and between Ti - Zr.

To account for the compositional variation at Site 410 by fractional crystallization presents many difficulties. Separation of a Ti-rich clinopyroxene could probably explain the lack of Ti-enrichment, but although published partition coefficient data for LREE/HREE between clinopyroxene and liquid (Schnetzler and Philpotts, 1970; Grutzeck et al., 1974) are less than unity, appreciable separation of clinopyroxene would be required to increase Ce/Yb ratios to the required values. Moreover to prevent Y and HREE levels from rising would require that $K_{DY+HREE}^{cpx/L}$ be greater than unity, which is three or more times the published values. The Cr, Ni and Sr contents of the mugearites (ca. 200, 100 and 470 ppm respectively) relative to the values in the basalts (190, 90 and 226 ppm respectively) severely restricts the amount of clinopyroxene, olivine and plagioclase fractionation which could have occurred, yet the increase in Zr levels indicates that at least 50% fractional crystallization would be required (and the increase in some of the more HYG elements even more). These conflicting factors argue against fractional crystallization being the main mechanism relating Site 410 lavas. A partial melting relationship is much more likely. The compatible behaviour of Ti, Y and the heavy REE (their concentrations remaining constant in liquids produced by differing degrees of partial melting) indicates that the bulk distribution coefficients for these elements must be close to unity. Moreover since the Ti/Y ratio in Site 410 basalts is invariant and is the same as in other Leg 49 basalts it implies that any refractory phase which is holding Y and the heavy REE is retaining Ti in the same proportion. The efficiency of garnet in retaining Y and the

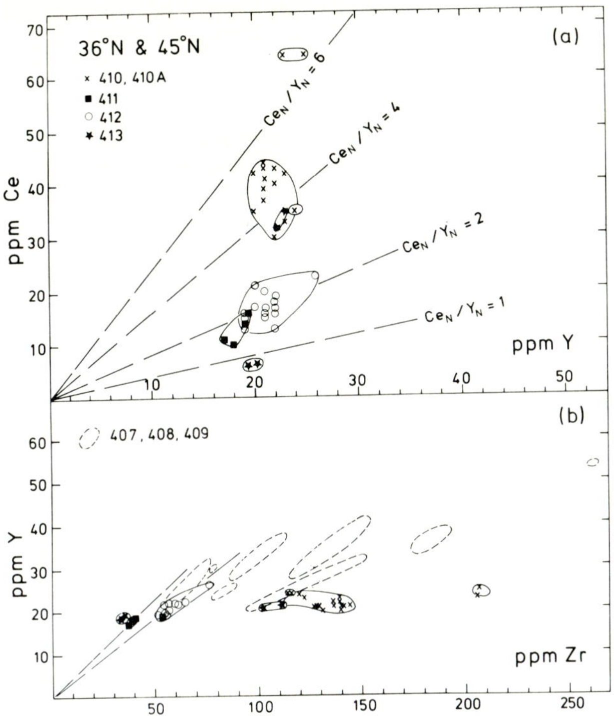

Figure 4.(a) Ce versus Y and (b) Y versus Zr for basalts from 36°N (Sites 411, 412 and 413) and 45°N (Site 410). In (b) fields for 63°N basalt units are shown for comparison.

heavy REE is well known, but there is little data available on Ti partitioning into garnet under mantle conditions. Both titanian and zirconian garnets are known (Dowty, 1971; Huggins et al., 1977) and garnets in ultramafic nodules in kimberlite contain as much or more Ti than coexisting clinopyroxenes (Bishop et al., 1978). That garnet may be a residual phase is also suggested by the lower P/Zr ratios of the more fractionated lavas (Table 1): phosphorus can substitute in mantle garnet (Thompson, 1975; Bishop et al., 1978). Note also that Sc and V are low in 45°N lavas. Other phases which might retain Ti are hornblende (see Basu and Murthy, 1977) and titan-phlogopite (Forbes and Flower, 1974). Note however that the latter would need to melt in significant quantities to account for the high K, Rb and Ba contents and low K/Rb ratios at 45°N.

Trace element relationships at 45°N are very different from those at 63°N. Plots of a more-HYG element against a less-HYG element form a linear array and generally have a small intercept on the axis of the less-HYG element (e.g. Figs. 8 to 10). This produces a small range of ratios, and probably reflects fractionation of HYG element ratios at small degrees of partial melting. However these differences are small compared with the changes in ratios such as Ce/Y, Ti/Zr, Y/Zr which are more strongly dependent on mineral fractionation during partial melting.

Thus ratios such as Zr/Nb, Ce/Zr, Ba/Zr, K/Zr, Th/Hf at 45°N are very different from those at 63°N and in MORB. Trace element ratios determined by Erlank and Kable (1976) on dredged basalts from near the median valley at 45°N are very similar to those at Site 410 and confirm that 'enriched' basalts have been erupted at this segment of the ridge for at least 10 m.y.

Mid-Atlantic Ridge 36°N: Sites 411, 412 and 413 (FAMOUS area)

The rift valley at 36°N has been extensively studied by dredging and by submersibles during the FAMOUS project (Ballard et al., 1975). In addition, DSDP Leg 37 drilled basement ranging in age from 3.5 to 16.5 m.y. at four sites west of the ridge. The results have been presented in a series of publications (Hekinian et al., 1976; Aumento, Melson, et al., 1977; Bryan and Moore, 1977; White and Bryan, 1977; Lambert and Holland, 1977; Flower et al., 1977) and have demonstrated that basalts with a similar range of composition have been erupted at this segment of the ridge for up to 16 m.y. Leg 49 drilling at this latitude not only extended the sampling to very young crust, but also to basalts erupted adjacent to nearby fracture zones.

Schilling (1975a) and White et al. (1975) have suggested, on the basis of dredged samples, that there may be a geochemical gradient south of the Azores, similar to that along the Reykjanes Ridge but extending over a much larger distance (ca. 1000 km). They have interpreted the gradient in terms of overspill from the supposed Azores mantle plume and mixing with a depleted MORB source. The FAMOUS area at 36°N lies at an intermediate position on this geochemical gradient. Erupted basalts include many near-primitive high-Mg compositions with low incompatible element levels, but with moderately high $^{87}Sr/^{86}Sr$ ratios

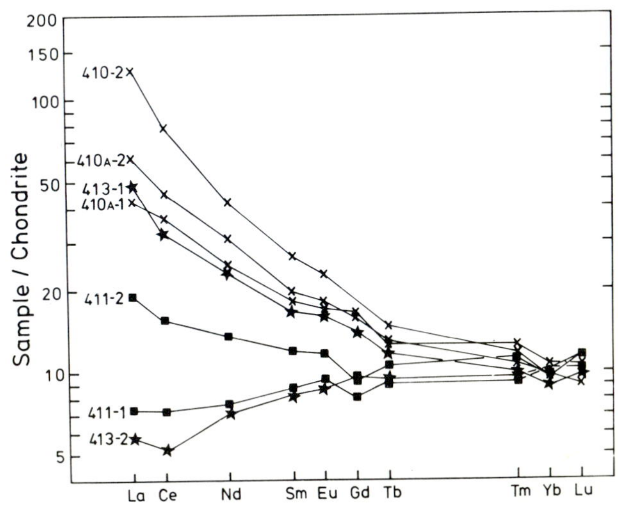

Figure 5. Chondrite-normalized rare-earth patterns for representative 36°N and 45°N basalts.

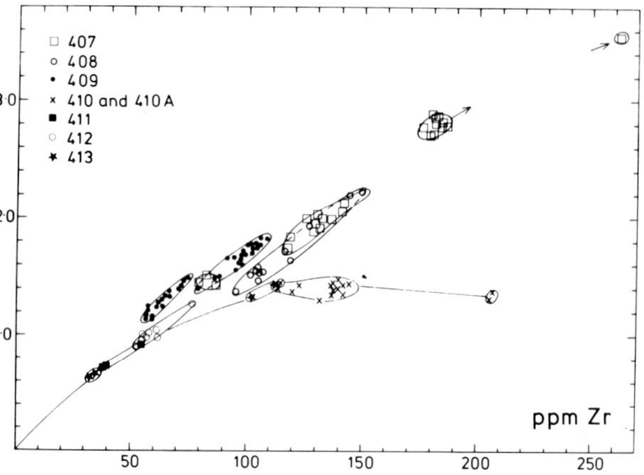

Figure 6. TiO_2 versus Zr for all Leg 49 basalts, illustrating suppression of Ti-enrichment in the more fractionated basalts from 36^oN and 45^oN.

(~0.7030; O'Nions et al., 1976) and REE patterns which range from slightly light-RE enriched to light-RE depleted (see Langmuir et al., 1977).

Hole 411 (36° 45.97'N; 33° 23.30'W), sited only 10 km west of the inner rift valley, penetrated 45 m of 1 m.y. old basaltic crust. There are two geochemical units, the upper unit comprising primitive (high-Mg) aphyric or sparsely phyric olivine basalts, and the lower one containing basalts with 10-20% plagioclase megacrysts and associated olivine and clinopyroxene phenocrysts.

Hole 412 (36° 33.74'N; 33° 09.96'W) penetrated 18 m of 1.6 m.y. old crust east of the rift valley and just south of the northern valley wall of Fracture Zone B. The basalts belong to one geochemical unit but vary from aphyric to coarsely phyric with plagioclase megacrysts and clinopyroxene and olivine phenocrysts.

Hole 412A (same locality as Hole 412) penetrated 50 m of basalt similar to Hole 412 and a further 80 m of sparsely phyric to aphyric basalts. They all belong to one geochemical unit but there are 3 cycles of Fe-enrichment represented by the flow units within the core. A least squares solution for the major elements (Wright and Doherty, 1970) suggests that the compositional variation could be accommodated by 25% fractional crystallization of plagioclase, clinopyroxene and minor olivine.

Hole 413 (36° 32.6'N; 33° 10.5'W) was drilled only 2 km south of Site 412, but into 3.5 m.y. old crust close to and on the south side of Fracture Zone B. The 39 m of basalt penetrated formed two very distinct geochemical units, the upper being a glassy to microcrystalline basalt with ~10% olivine phenocrysts and the lower unit a sparsely olivine phyric basalt.

Basalts recovered from the FAMOUS area by Leg 49 have many of the chemical characteristics (high MgO, Cr, Ni, Mg/Fe; low TiO_2 and HYG elements) of other dredged and drilled basalts previously recovered from this area. Many are close to primary magma compositions. Glasses from Units 411-1 and 413-2 (Wood et al., 1978) are in fact among the most primitive liquid compositions sampled on the ocean floor (Frey et al., 1974) with SiO_2 = 49%, Mg/Mg + Fe ≐ 0.65 high Al_2O_3 (> 16%), and low TiO_2 (0.8%) and P_2O_5 (0.05%). Langmuir et al. (1977) have described similar primitive samples from the FAMOUS area. The sparsely phyric basalts from Unit 413-1 are also very primitive, with Mg/Mg + Fe = 0.70 but have lower Al_2O_3 (~14%) and significantly higher TiO_2 (1.5%), P_2O_5 (0.25%) and HYG element contents than basalts from Units 411-1 and 413-2. Aphyric basalts from Units 411-2, 412 and 412A are intermediate between these two extremes in terms of HYG element contents and are rather less primitive in having Mg/Mg + Fe = 0.6 and lower Cr and Ni. They also have appreciably less normative olivine and plot close to the Di-Hy join on the Ol-Di-Hy normative diagram, some samples in fact being quartz-normative. Two of these groups of basalts (411-1, 413-2 and 411-2, 412, 412A respectively) correspond fairly closely with two magma types previously recognised from study of Leg 37 samples (Flower et al., 1977); magma types A and B respectively. A magma corresponding to Unit 413-1 was not represented by any of the samples recovered by Leg 37. Basalts from this unit have however many compositional similarities with the more primitive basalts from 45^oN (e.g. Unit 410-1).

A plot of Ce versus Y (Fig. 4a) shows that basalts from the three groups fall into three distinct fields with widely differing Ce_N/Y_N ratios. Representative chondrite-normalised REE patterns for the different geochemical units (Fig. 5) confirm the large variation in Ce_N/Yb_N ratios, from light-RE depleted to light-RE enriched. The variation in REE distributions within the short section cored at Site 413 in fact encompass almost the total Ce_N/Yb_N range in samples previously recovered from the FAMOUS area. The REE patterns, like those of 45^oN basalts, are concave upwards, show little variation in heavy REE abundance and have unusually high La/Ce ratios (see also Langmuir et al., 1977). A plot of Y versus Zr (Fig. 4b) demonstrates that Y remains almost invariant at about 20 ppm in the more primitive lavas despite a three fold increase in Zr, a feature also seen at 45^oN. Only in fact in Site 412 basalts is there a colinear relationship between Y and Zr, indicative of approximately 25% fractional crystallization, as noted above. Considering 36^oN and 45^oN basalts together, the lack of variation in Y and the HREE compared with the six-fold increase in Zr abundance is unusual. Titanium shows a similar, if not identical, relationship with Zr as does yttrium (Fig. 6). In the more primitive basalts with low Zr, TiO_2 varies in sympathy with Zr and appears to be acting incompatibly. However Unit 412-1 basalts, although having higher TiO_2, in fact have a much lower TiO_2/Zr ratio which is identical to that of the more primitive basalts from 45°N. Thus the more primitive basalts from the FAMOUS area cluster tightly about a curve (Fig. 6) which apparently continues through the 45°N samples. The more evolved basalts in Unit 412 rise off this curve with a constant TiO_2/Zr ratio.

Consideration of major element compositions and trace element abundances indicates that it is impossible to derive these more primitive basalt magma types (with different Ce/Yb ratios) from a single parent magma by open or closed system fractionation. For instance the difference in zirconium abundance between Units 411-1 and 411-2 would demand ca. 35% removal of crystals. The marked reduction in Ni and Cr could be accomplished by extensive separation of olivine and Cr-spinel, which might account for the observed increase in SiO_2 and the trend towards quartz-normative compositions in Unit 411-2. However, in spite of the fact that Unit 411-2 basalts are strongly plagioclase-phyric, little plagioclase must have been removed because Sr is 50% higher than in Unit 411-1. Moreover removal of olivine and spinel should produce a comparable enrichment in all incompatible elements: but in fact they are enriched to widely varying extents. The difference in Ce/Y and Ce/Yb ratios between the two units would necessitate removal of substantial amounts of clinopyroxene, yet this could not account for the major element variation. Even larger problems are apparent in trying to relate Units 413-1 and 413-2 by fractional crystallization. Both are magnesian, high-Cr and high-Ni basalts, but the differences between the levels of many incompatible elements in these two units could only be reconciled with up to 90% fractional crystallisation, which is clearly inconceivable without changing transition metal levels drastically. Thus either they are related by some process involving partial melting or they are derived from two compositionally different sources.

Arguments favouring heterogeneity in the source regions of FAMOUS area basalts are severely constrained by the fact that Sr-isotope ratios are relatively uniform (O'Nions and Pankhurst, 1976; White and Bryan, 1977) and also that the ratios of the more-HYG elements are very similar (Bougault et al., 1978). Indeed, the two extreme basalt types from Hole 413 have similar ratios of La/Th and Ta/Th. Such considerations indicate that the fundamental differences between the main geochemical units are determined during the processes of basalt generation. Simple batch partial melting of a normal peridotite mineral assemblage cannot however explain the REE data, particularly the lack of variation of the heavy REE. Faced with similar problems in explaining the crossing REE patterns of primitive basaltic glasses from the FAMOUS area, Langmuir et al., (1977) have developed a more complex dynamic melting model. This integrates varying degrees of batch melting with incremental melting, and makes use of an observation by Dick (1977) on the Josephine peridotite indicating that melt is not always completely extracted. They propose that melting at ocean ridges may take place over a considerable depth range, but with an increasing degree of melting as the mantle diapir rises. Melt can be extracted at any point but it is suggested that a small proportion of melt always remains in equilibrium with the mantle residue (sufficient at least to remain fertile enough in HYG elements to yield further basalts). More extensive melting of mantle which has already yielded up an earlier liquid fraction at greater depth may thus produce magmas with much lower concentrations of incompatible elements and in particular lower more-HYG/less-HYG element ratios. Thus a liquid with ratios of more-HYG/less-HYG elements lower than the initial source can be produced, and it is not necessary to invoke an initial source composition which is LREE depleted to produce basalts such as 413-2. Such basalts should normally be accompanied by basalts with more LREE-enriched characteristics. Note that ratios of the more-HYG elements (e.g. Th/Ta) are not significantly changed by this process, although their abundances may vary widely.

The dynamic melting model is inherently very flexible and capable of reconciling quite diverse basalt compositions to a uniform source. Of the extreme basalt compositions at Site 413, Unit 413-1 would represent approximately a 3% equilibrium batch melt of the initial source, whereas Unit 413-2 would be an incremental remelt of such a source after some basalt had been extracted. The fact that basalt 413-1 occurs near Fracture Zone B in the FAMOUS area, may be a consequence of more efficient melt segregation in this tectonic environment. Basalts with similar geochemical characteristics are common at 45°N, but it seems unlikely that the whole range of basalt compositions at 36°N and 45°N could be accommodated by dynamic melting processes (as opposed to invoking source compositional differences, perhaps involving some more subtle mineralogical control, differences in P_{H_2O} etc) because present evidence indicates[2] that 45°N basalts are significantly more radiogenic with respect to Sr and Pb isotopes than those in the FAMOUS area (White et al., 1975; Mattinson, in press) and indeed have isotopic compositions similar to ocean island basalts. Nevertheless it is interesting that on almost all inter-element plots there is a continuum between 36°N and 45°N basalts and that their HYG-element ratios are almost identical and very different from those at 63°N and in 'MORB'.

The two extreme primitive magma types, 411-1 and 413-2, from the FAMOUS area are close to primary magmas and, assuming both have been derived

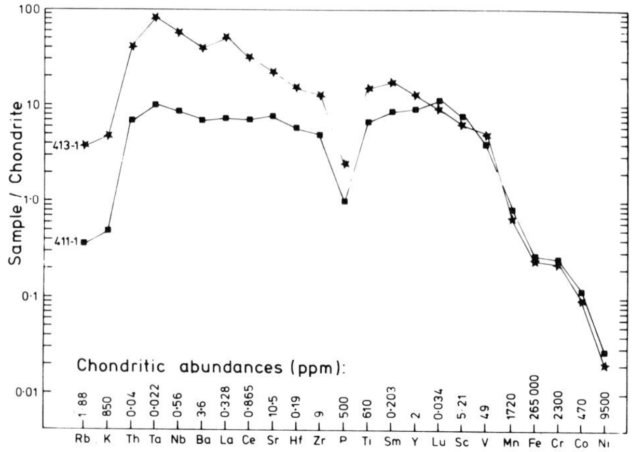

Figure 7. Chondrite-normalized abundances of 23 elements in Units 411-1 and 413-1 (FAMOUS area), arranged approximately in order of increasing bulk mantle/basalt liquid partition coefficient.

by differing degrees of equilibrium partial melting of a uniform source, can be used to compare trace element behaviour during partial melting. Abundances of 23 elements in these two basalts, normalised to chondrites, are shown in Fig. 7 and have been arranged approximately in order of increasing D by comparing enrichment factors. Note that the respective enrichments are specific to the mantle mineralogy in the FAMOUS area and may be complicated by a mineral phase (such as that holding Ti and Y) being consumed at higher degrees of melting. The diagram shows that ratios such as Zr/Y, Ce/Y and La/Sm are greatly modified by the widely differing degrees of partial melting and that ratios such as Zr/Nb, Th/Hf and Ba/Sr are affected to a lesser extent. However ratios of the more HYG elements, K/Rb, Th/Ta, Th/La etc., are essentially unmodified. Ratios between the elements Th, Nb, Ba and La are also chondritic in both basalts and indicate that the mantle has chondritic ratios for these elements, although Ta/Th and Ta/Nb ratios are higher than chondritic. Anomalies are apparent in the abundances of K and Rb (the Rb/Sr ratio of the bulk earth is much lower than chondritic: O'Nions et al., 1977) and in phosphorous (possibly due to entry into the earth's core: Sun and Nesbitt, 1977). Transition metals Cr, Co and Ni are systematically fractionated with respect to chondrites to a greater degree than would be expected from differences in their distribution coefficients (see Langmuir et al., 1977).

Discussion

Variables controlling basalt compositions

It is clear that the processes of partial melting and fractional crystallization can give rise to wide compositional differences in erupted basalts, and many of these processes we have recognised in basalts recovered by Leg 49. Fractional crystallization is certainly responsible for much of the within-unit variation, and some between-unit relationships, but seems incapable of accounting for the wider trace element differences. O'Hara (1977) has suggested that open system fractionation in a continuous but periodically refilled magma chamber can produce a wider range of trace element levels, but as Pankhurst (1977) has pointed out, this is still incapable of accounting for the wide variation in trace element ratios such as we have observed in Leg 49 samples. Fractional crystallization thus seems subordinate to partial melting, and may reflect the fact that continuous sub-axis magma chambers are not an important feature of the slow-spreading Mid-Atlantic ridge; the data are more easily reconciled to a series of short-lived magma chambers. Not all the products of sub-axis crystal differentiation may be erupted into the lava sequence - indeed in ophiolite complexes, plagiogranite differentiates are mostly confined below the sheeted dyke unit (Saunders et al., 1978).

Polybaric partial melting processes are capable of producing much more chemical diversity (Hanson, 1977), owing to variations in degree of partial melting, nature of residual phases and whether or not they are

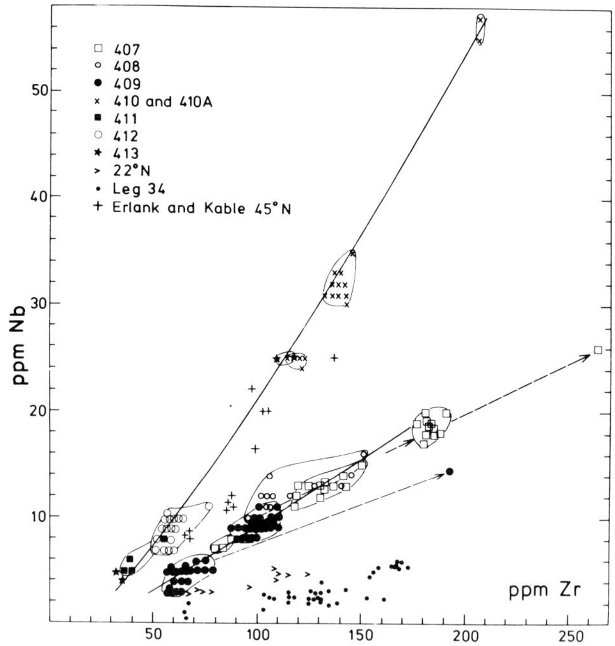

Figure 8. Nb versus Zr for all Leg 49 basalts, illustrating regional differences in Zr/Nb ratios. Dashed lines indicate fractional crystallization vectors and solid lines partial melting (see text). 45°N data from Erlank and Kable (1976) and data on 'MORB' type basalts from 22°N and Nazca Plate (Leg 34) shown for comparison.

consumed, variation in P_{H_2O}, fO_2, etc. controlling the stability of mineral phases. Dynamic melting (Langmuir et al., 1977) provides an added variable in that magma compositions need not directly reflect the composition of the primary mantle source. Modelling back to the initial source is always easier when high-magnesian primary liquids are available (Sun and Nesbitt, 1977), but alas this is the exception rather than the norm. Primary liquids with a greater chance of being erupted are those generated at high levels in a rising diapir, or perhaps those adjacent to fracture zones. Magmas generated deep enough to be in equilibrium with garnet may suffer much more crystal fractionation, zone refining, even mixing before eruption (O'Hara, 1965, 1968, 1973). The larger number of samples from a given locality recovered by drilling, as opposed to dredging, provides more information necessary to assess these processes, but there are still many uncertainties. As we have seen however, hygromagmatophile element ratios are much less affected by the range of partial melting processes associated with basalt generation, and can be used, like radiogenic isotope ratios, to decipher differences in the source.

Regional Comparisons

The three different regions studied are characterised by basalts of distinct compositions. Reykjanes basalts show a trend toward Fe- and Ti-enrichment similar to that on Iceland, even at Site 407 (36 m.y. ago). Basalts from 45°N instead trend more towards alkali enrichment, whereas FAMOUS basalts include many high-MgO compositions and the trend towards Fe-enrichment is less marked than in the Reykjanes samples. Both at 45°N and 36°N the trend toward Ti-enrichment in the more fractionated basalts is strongly inhibited, and basalts from these two areas include more high-alumina types than at 63°N. Some of these differences could reflect differences in the mantle mineralogy and/or differences in P_{H_2O} during basalt generation. This is suggested also by the differences in shape of the REE patterns (Figs. 3 & 5), all those from 45°N and 36°N having distinctive concave-upwards patterns with high La/Ce ratios.

It is with the more hygromagmatophile elements that regional differences are accentuated. Erlank and Kable (1976) noted significant differences in Zr/Nb ratio between Atlantic basalts. Leg 49 data are plotted on a Zr versus Nb plot in Fig. 8 together with Erlank and Kable's data from 45°N, with data on 'MORB' type basalts from 22°N drilled by Legs 45 and 46, and with data on 'MORB' basalts from Leg 34, Nazca Plate (Rhodes et al., 1976; Cann and Heath, 1976). Basalts from different regions group into distinct linear arrays with different abundances and ratios of these elements. The data of Erlank and Kable (1976) from 45°N plot close to Leg 49 data from this area. Note that Zr/Nb ratios are not constant because the correlation lines intercept the Zr axis. In fact there is a linear correlation between Zr/Nb and Ce/Y at 45°N and 36°N (and a similar but different one at 63°N too) indicating that, at least in part, this results from partial melting. Basalts from 45°N and 36°N have different abundances of Zr and Nb, but lie on the same line with Zr/Nb~5. Reykjanes basalts have distinctly higher ratios (Zr/Nb~10) but it is clear that none of the North Atlantic basalts sampled by Leg 49 has the geochemical characteristics of 'MORB'. Nor, since these ratios are relatively constant in basalts erupted along mantle flow lines for periods in excess of 36 m.y., is it likely that a basalt with the major and trace element characteristics of 'MORB' or MAR 63°N basalts could ever be erupted at 36°N or 45°N segments of the ridge, or vice versa. Preliminary data from 110 m.y. old crust drilled by Legs 51 and 53, which lies approximately along the same mantle flow line from 22°N suggests that basalts with 'MORB' geochemistry have been erupted here since the opening of the Atlantic. Considering that crust along the Mid-Atlantic Ridge is being generated at much the same rate (1-2 cm yr^{-1}), and that basalts from each region studied represent the products of a wide range of partial melting and fractional crystallization processes, it would appear that the observed Zr/Nb ratios (blurred to some extent by these fractionation processes) are characteristics of the mantle source regions. Moreover the volumes of mantle with these Zr/Nb ratio characteristics must be large. It is also apparent that mixing of a high Zr/Nb ratio 'MORB'-type source with a low Zr/Nb ratio 'ocean island'-type source, in the manner envisaged by the plume hypothesis, should produce a scatter of intermediate Zr/Nb ratios rather than the consistent trends observed.

Similar regional differences are observed in equivalent plots of other HYG elements against Zr, such as Ce-Zr, Ba-Zr (Fig. 8) and Sr-Zr — differences in the latter two remaining apparent despite the known mobility of Ba and Sr under conditions of mild alteration. Large regional differences are also found in the ratios of the most-HYG elements — yet these ratios, as we have seen, are unaffected by the range of partial melting associated with basalt generation and can only be changed at very small degrees of partial melting. A plot of Th versus La (Fig. 10a) shows that each region, 63°N, 45°N, 36°N and 22°N, has a distinct Th/La ratio (or small range of ratios). MORB basalts from 22°N have Th/La ratios significantly lower than chondrites, 63°N basalts rather less so, whereas 45°N basalts have Th/La ratios slightly higher than chondritic. Plots of Nb and Ta against Th show similar features. There is less variation in inter-element plots of La, Nb and Ta of course (e.g. Fig. 10b) owing to the similarity of their D

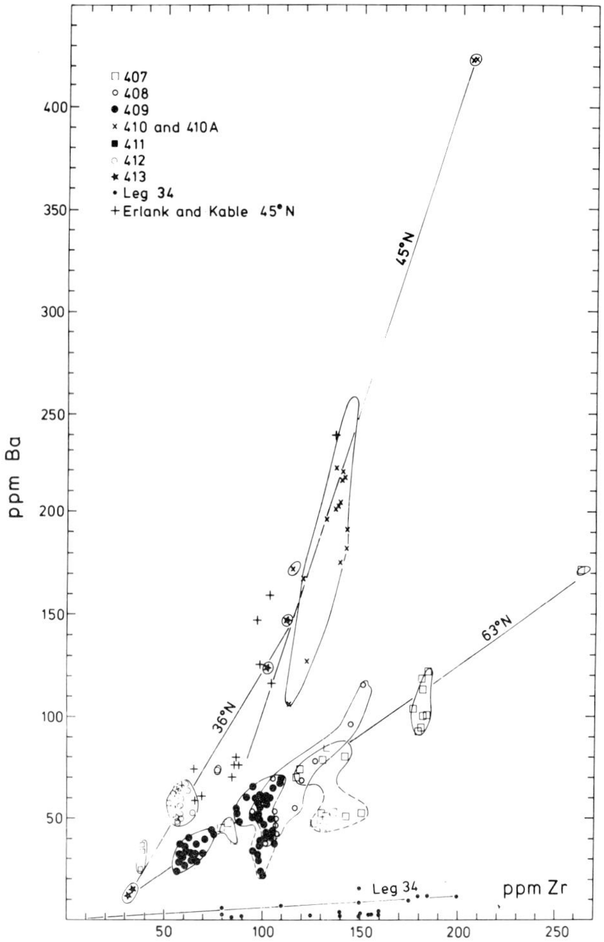

Figure 9. Ba versus Zr for Leg 49 basalts. The greater scatter compared with Fig. 8 is due to the mobility of Ba during alteration, but large regional differences in Ba/Zr ratio are apparent. 45°N data from Erlank and Kable (1976) and data on 'MORB' type basalts from Leg 34 (Nazca Plate) shown for comparison.

values, but significant regional differences do exist. To emphasise the fine detail in the differences between the elements Th, Ta, Nb and La, the abundances in representative basalts from the different regions have been normalised in Fig. 11 to those in a basalt from Unit 409-2, which has the lowest abundances of the samples studied. Two 'MORB'-type basalts from Hole 348 (north east of Iceland) and Hole 395 (at 22°N) are also included. It is clear that 45°N and 36°N basalts have a very similar ratios of these elements irrespective of their abundance, and their sources must in some way be similar. The two 'MORB'-type basalts also have very similar ratios of these elements, but in going from 45°N, 36°N and 63°N to MORB there is a progressive depletion of Th relative to La.

The fact that basalts from 45°N and 36°N have similar element ratios on these and almost all other inter-element plots is particularly interesting in view of the fact that the abundances of incompatible elements at 45°N are all higher than at 36°N. Thus when two different regions with different abundances yield the same ratios this gives added confidence that element ratios differences in other regions provide a valid indication of differing source compositions.

Nature and Causes of Regional Heterogeneities

We may ask what the processes were that changed the element ratios in the source regions of Atlantic basalts? A solution to this problem would in itself provide some constraints on the nature of heterogeneity. It is evident that the partial melting processes associated with basalt generation do not significantly alter ratios of the more-HYG elements, although where smaller degrees of partial melting are involved, as in the production of alkalic basalts at 45°N, some changes in these ratios are seen. Elements with similar D values have ratios which vary only slightly and may be close to chondritic. Because the volumes of mantle with differing HYG element ratio characteristics appear to be large, it is suspected that a major process (or processes), probably involving crystal-liquid equilibria, has produced the observed chemical heterogeneties.

To change the more-HYG element ratios significantly by partial melting requires very small degrees of melting (because of the very low D's involved). This could be accomplished by a single process, or by a series of processes culminating in the extraction or addition of a small amount of melt. Fig. 11 reveals that the more-HYG elements have behaved independently during such processes to generate the observed variations: i.e. there is not a systematic variation in Th/La or La/Ta ratios between the regions and each region is characterized by REE patterns with distinctive shapes. It would be difficult to account for these features by a single process. Moreover the fact that there seems to be significant inter-site $^{87}Sr/^{86}Sr$ variation in basalts along a particular mantle flow line while the ratios of the more-HYG elements remain relatively constant also suggests that more than one process may be involved, perhaps operating on different scales, and at different times.

Evidence from Sr and Nd isotopes (DePaulo and Wasserburg, 1976; O'Nions et al, 1977, 1978) indicates that the source regions of oceanic basalts have been variably depleted in Rb/Sr and Nd/Sm relative to the primordial bulk earth on a time-integrated basis, but this does not necessarily imply only a single depletion event. Multi-stage depletion-enrichment events are possible and, according to O'Nions et al. (1978) do not necessarily conflict with the Sr and Pb isotopic data on basalts from oceanic regions which yield pseudo-isochrons with ages of 1.5-2.0 b.y. for the major fractionation event (Sun and Hanson, 1975; Brooks et al., 1976).

The growth of continental crust may be regarded as a major mantle depletion event, with the effects being integrated over 3.8 b.y. depending on the extent of mantle convection. Processes in subduction zones are capable of significant chemical fractionation, in that material already fractionated as a result of ridge magmatism is subject to further melting with the extraction of lithophile element enriched silicic liquids. In effect the continental material extracted represents a

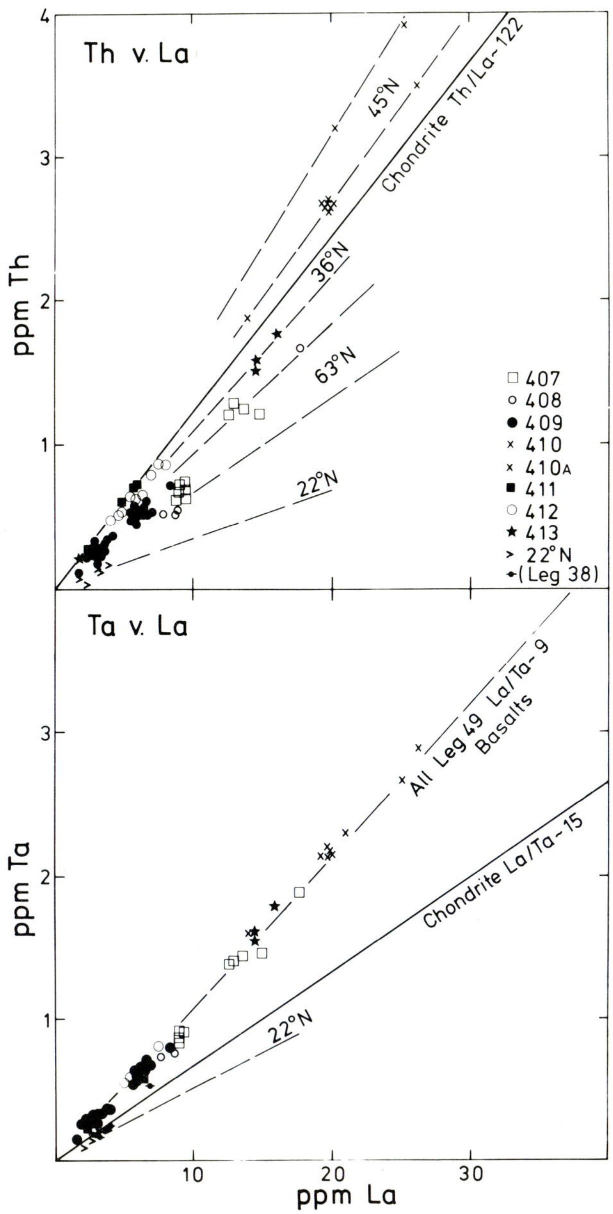

Figure 10. (a) Th versus La and (b) Ta versus La for Leg 49 basalts, with additional data from 22°N and from Leg 38.

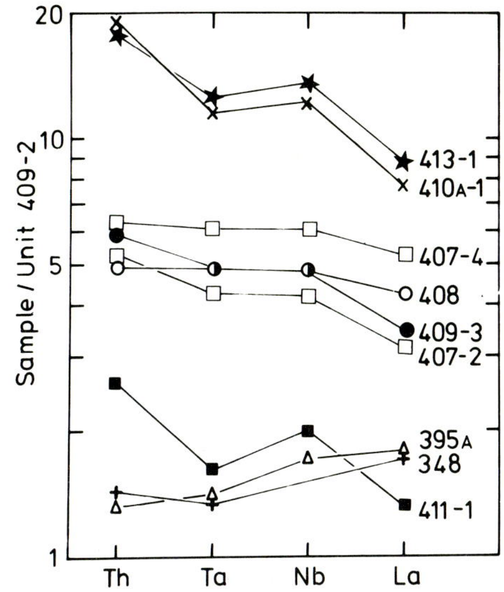

Figure 11. Abundances of Th, Ta, Nb and La in selected Leg 49 basalts normalized to the most primitive basalt in Unit 409-2 (Table 1). Two 'MORB' type basalts from Site 348 and Site 395A are included.

very small degree of partial melting of the initial mantle source. Trace element data on Andean calc-alkaline and island arc igneous rocks indicate a marked decoupling in the behaviour of large lithophile ions (K, Rb, Cs, Th, U) and smaller highly charged ions (Ta, Nb, Ti, Zr, P, etc.), with the former group being enriched in intermediate to silicic compositions whereas the latter group is impoverished. This probably arises because minor phases such as ilmenite, rutile, zircon and apatite, in addition to hornblende, pyroxene and garnet, are residual phases during magma genesisin subduction zones and retain the highly charged ions (which are normally incompatible during basalt genesis). Element ratios in continental crust are thus very different from those in 'MORB'. Archaean crust in particular has high Ce/Y and Ce/Yb ratios (Tarney and Windley, 1977) and hence a low Sm/Nd ratio rendering it unradiogenic with respect to Nd isotopes (Carter et al., 1978b). Likewise continental material has high Rb/Sr ratios and high present day $^{87}Sr/^{86}Sr$ ratios. In both respects these features complement the high $^{143}Nd/^{144}Nd$ and low $^{87}Sr/^{86}Sr$ ratios of oceanic basalts (O'Nions et al., 1977).

Depleted material returned to the mantle in subduction zones would mix to varying extents with undepleted pristine mantle, producing both small scale and large scale compositional heterogeneties which may, by the time this material is recycled to younger oceanic spreading centers, give rise to isotopic and trace element differences in erupted basalts. Moreover since the volcanic and plutonic products above subduction zones are themselves very variable (e.g. andesitic island arc or granitic continental margin) the material returned to the mantle may itself be variably depleted.

Such an explanation may account in part for the depleted nature of the mantle source for MORB, in which case the source regions at 63°N, 36°N and 45°N could be regarded as progressively less fractionated relative to the primordial mantle. However it is apparent that basalts from these regions have much higher concentrations of elements such as Ta and Nb relative to the values in MORB, in which case it is equally feasible that their source regions have been enriched in these and other HYG elements, irrespective of whether they had been involved in earlier depletion events. Incipient melting processes, producing very small amounts of liquid rich in HYG elements (and perhaps H_2O and CO_2) may have been responsible. These liquids need not have been removed from the mantle, but may just have migrated vertically from one region to another. At very small degrees of melting their compositions may have been controlled more by minor mineral phases than by major phases and would thus produce the necessary fractionation of HYG elements. Lloyd and Bailey (1975) suggested that mantle metasomatism may occur in the sub-continental mantle, and similar processes have been invoked to explain the trace element geochemistry of suites of ultramafic nodules in continental basalts (Frey and Green, 1974; Frey and Prinz, 1978). Although the mantle regions underlying the ridge were indeed sub-continental before the opening of the Atlantic ocean, there is no reason why such processes should be restricted to the sub-continental mantle. Hanson (1977) favours a veined mantle in view of the common occurrence of pyroxenite veins in mantle nodules and in ultramafic components of ophiolites.

A mantle region variably veined by HYG-element enriched material does offer some advantages in interpreting the geochemical variations in N. Atlantic basalts. The HYG-element enriched material in the veins would, at mantle temperatures, be in partial isotopic equilibrium with the surrounding depleted mantle to an extent depending on the number and spacing of the veins and the effectiveness of diffusion. At small degrees of melting the veins would contribute significantly to the HYG element content of the melt and HYG element ratios would be governed by those in the veins. However at higher degrees of partial melting the more-HYG element depleted host mantle would be involved and HYG element abundances in the melt would be lower and more-HYG / less-HYG element ratios transitional toward those of the host mantle. This could explain why there is an intercept on the axis of the less-HYG element on most regional inter-element plots. However, ratios of the more-HYG elements would be determined by those in the veins except where the proportion of veins was very small.

Such a model may account for the fact that basalts at 45°N and 36°N have similar ratios of the more-HYG elements, but display systematically varying ratios of more-HYG to less-HYG elements. Bearing in mind that the latent heat of fusion exerts a strong buffering effect on the amount of liquid produced during mantle melting, it would be unrealistic to interpret the wide range of HYG element abundances observed in the two areas as simply being due to differing degrees of partial melting. Instead, these features could be reconciled with a higher proportion of veins in the source regions at 45°N. This would allow isotopic differences to develop in proportion to the amount of HYG-element enriched vein material while still maintaining consistent ratios of the more HYG elements.

Conclusions

Previous deductions concerning mantle heterogeneity in the North Atlantic have been based very largely on samples dredged from the ridge axis. Leg 49 drilling has added two further dimensions (variations with depth and with time) to this array of data and has demonstrated several important features.

1. Basalts recovered from a single drill hole may show a surprisingly wide range of composition, particularly with respect to REE, Ti/Zr ratios, etc., thus making deductions concerning heterogeneity based on these parameters alone of doubtful value. Nevertheless it is possible in most cases to derive all basalt compositions at a single site from the same mantle source through partial melting processes (including dynamic melting) and fractional crystallization. While fractional crystallization is an important cause of compositional variation within a particular geochemical unit, several such units may occur at any one site that can only be related by partial melting processes. This suggests that long-lived sub-axial magma chambers are not an important feature of the slow-spreading Mid-Atlantic Ridge.

2. Basalts from different sites along the same mantle flow line (e.g. 63°N) do show some compositional and isotopic differences which cannot readily be accommodated by fractional crystallization and partial melting. These probably reflect local heterogeneities in the mantle source feeding the ridge. However, in terms of the more-HYG element ratios of basalts at these same sites, it is the constancy (or consistency) of these ratios which is the more striking, implying that the mantle feeding particular ridge segments was uniform with respect to these HYG elements for at least 35 m.y. (at 36°N) and probably since the opening of the Atlantic.

3. By contrast, ratios of the more-HYG elements vary markedly along the Mid-Atlantic Ridge, but in each case these regional differences are maintained at a particular spreading axis for long time periods. HYG-element ratios in the Iceland-Reykjanes Ridge area are different from those at 45°N and at 36°N, and both are different from the ratios at 22°N and in typical MORB. It is apparent that the volumes of mantle with these distinctive element ratios are large, and moreover that there

are not just two contrasted ('depleted' and 'enriched') sources mixing and feeding the Mid-Atlantic Ridge, but rather several sources with their own HYG-element character.

The results suggest that the mantle may be heterogeneous on different scales for different elements, in which case the plume-mixing model would seem to be insufficiently flexible to account for the observed variations. A viable alternative model, bearing in mind isotopic constraints, is to consider the data in terms of depletion/enrichment processes operating throughout earth history. The processes of crustal extraction are capable of causing both depletion and significant fractionation of HYG elements in the residual mantle source. This may not however be the only process. Mantle variably veined with HYG element enriched material offers distinct advantages in interpreting the observed patterns of geochemical and isotopic variation. Further drilling in critical areas of the N. Atlantic would greatly assist in understanding the nature and causes of mantle heterogeneity.

Acknowledgements: We are grateful to the Natural Environment Research Council and the Royal Society for financial support. The helpful comments of W. B. Bryan, C. Langmuir and S-S. Sun are greatly appreciated.

References

Aumento, F. The Mid-Atlantic Ridge near 45°N. II. Basalts from the area of Confederation Peak, Can. J. Earth Sci., 5, 1-21, 1968

Aumento, F. Diorites from the Mid-Atlantic Ridge at 45°N, Science, 165, 1112-1113, 1969.

Aumento, F. Uranium content of mid-oceanic basalts, Earth Planet. Sci. Lett., 11, 90-94, 1971.

Aumento, F., B.D. Loncarevic and D.I. Ross. Hudson Geotraverse: geology of the Mid-Atlantic Ridge at 45°N, Phil. Trans. R. Soc. Lond., A268, 623-650, 1971

Aumento, F., W.G. Melson, et al. Initial Reports of the Deep Sea Drilling Project (Washington, U.S. Government Printing Office), 37, 1977.

Ballard, R.D., W.B. Bryan, J.R. Heirtzler, G. Keller, J.G. Moore and Tj.H.van Andel, Manned submersible observations in the FAMOUS area, Mid-Atlantic Ridge, Science, 190, 103-108, 1975

Basu, A.R. and V.R. Murthy. Kaersutites, suboceanic low-velocity zone and the origin of mid-oceanic ridge basalts, Geology, 5, 365-368, 1977.

Bishop, F.C., J.V. Smith and J.B. Dawson. Na, K, P and Ti in garnet, pyroxene and olivine from peridotite and eclogite xenoliths from African kimberlites, Lithos, 11, 15-173, 1978.

Bougault, H., M. Treuil and J.L. Joron, Trace elements: fractional crystallization and partial melting process, heterogeneity of the upper mantle material, Tectonophysics (in press), 1978.

Brooks, C., S.R. Hart, A. Hofmann and D.E. James, Rb-Sr mantle isochrons from oceanic regions, Earth Planet. Sci. Lett., 32, 51-61, 1976.

Bryan, W.B., and J.G. Moore, Compositional variations of young basalts in the Mid-Atlantic Ridge rift valley near lat 36° 49'N, Geol. Soc. Amer. Bull., 88, 556-570, 1977.

Cann, J.R. and R. Heath, Some trace elements in basalts from Leg 34, in Initial Reports of the Deep Sea Drilling Project, (Washington, U.S. Government Printing Office), 34, 289-292, 1976.

Carter, S.R., N.M. Evenson, P.J. Hamilton and R.K. O'Nions. Continental volcanics derived from enriched and depleted source regions, Earth Planet Sci. Lett., 37, 401-408, 1978a.

Carter, S.R., N.M. Evenson and R.K. O'Nions, Contamination of basalts with continental crust: Nd- and Sr- isotope evidence, EOS, Trans. Am. Geophys. Union, 59, 393, 1978b.

DePaulo, D.J. and G.J. Wasserburg, Nd isotopic variations and petrogenetic models, Geophys. Res. Lett., 3, 249-252, 1976.

Dick, H.D., Partial melting in the Josephine peridotite, I, Amer. J. Sci., 277, 801-832, 1977

Dowty, E., Crystal chemistry of titanian and zirconian garnet: I. Review and spectral studies, Amer. Mineral., 56, 1983-2009, 1971.

Erlank, A.H. and E.J.D. Kable. The significance of incompatible elements in Mid-Atlantic Ridge basalts from 45°N with particular reference to Zr/Nb, Contr. Mineral. Petrol., 54, 281-291, 1976

Flower, M.J.V., P.T. Robinson, H.U. Schmincke and W. Ohnmacht, Magma fractionation systems beneath the Mid-Atlantic Ridge at 36°-37°N, Contrib. Mineral. Petrol., 64, 167-195, 1977.

Forbes, W.C. and M.F.J. Flower, Phase relations of titan-phlogopite, $K_2Mg_4TiAl_2Si_6O_{20}(OH)_4$: a refractory phase in the upper mantle, Earth Planet. Sci. Lett., 22, 60-66, 1974.

Frey, F.A. and D.H. Green, The mineralogy, geochemistry and origin of lherzolite inclusions in Victorian basanites, Geochim. Cosmochim. Acta, 38, 1023-1059, 1974.

Frey, F.A. and M. Prinz, Ultramafic inclusions from San Carlos, Arizona: Petrologic and geochemical data bearing on their petrogenesis, Earth Planet. Sci. Lett., 38, 129-176, 1978.

Frey, F.A., W.B. Bryan and G. Thompson, Atlantic ocean floor: geochemistry and petrology of basalts from Legs 2 and 3 of the Deep Sea Drilling Project, J. Geophys. Res., 79, 5507-5527, 1974.

Frey, F.A., M.A. Haskin, J.A. Poetz and L.A. Haskin, Rare-earth abundances in some basic rocks, J. Geophys. Res., 73, 6085-6097, 1968.

Grutzeck, M.W., S.J. Kridelbaugh and D.F. Weill, The distribution of Sr and REE between diopside and silicate liquid, Geophys. Res. Lett., 1, 273-275, 1974.

Hanson, G.N., Geochemical evolution of the sub-oceanic mantle, J. Geol. Soc. Lond., 134, 235-253, 1977.

Hart, S.R., The geochemistry of basaltic rocks, Carnegie Inst. Washington Yearbook, 70, 353-355, 1971a.

Hart, S.R., K, Rb, Cs, Sr and Ba contents and Sr isotope ratios of ocean floor basalts, Phil. Trans. Roy. Soc. Lond., A268, 573-588, 1971b.

Hart, S.R., J.G. Schilling and J.L. Powell, Basalts from Iceland and along the Reykjanes Ridge: Sr isotope geochemistry, Nature, Lond. 246, 104-107, 1973.

Hekinian, R., J.G. Moore and W.B. Bryan, Volcanic rocks and processes of the Mid-Atlantic Ridge rift valley near 36^{c} 49'N, Contrib. Mineral. Petrol., 58, 83-110, 1976.

Hermes, O.D. and J.G. Schilling, Olivine from Reykjanes Ridge and Iceland tholeiites, and its significance to the two-mantle source model, Earth Planet. Sci. Lett., 28, 345-355, 1976.

Huggins, F.E., D. Virgo and H.G. Huckenholz, Titanium-containing silicate garnets, I. The distribution of Al, Fe^{3+} and Ti^{4+} between octahedral and tetrahedral sites, Amer. Mineral., 62, 475-490, 1977.

Lambert, R.St.J. and J.G. Holland, Trace elements and petrogenesis of DSDP 37 basalts, Can. J. Earth Sci., 14, 809-836, 1977.

Langmuir, C.H., J.F. Bender, A.E. Bence, G.N. Hanson and S.R. Taylor, Petrogenesis of basalts from the FAMOUS area: Mid-Atlantic Ridge, Earth Planet. Sci. Lett., 36, 133-156, 1977.

Langmuir, C.H., R.D. Vocke and G.N. Hanson, A general mixing equation with applications to Icelandic basalts, Earth Planet.Sci. Lett. 37, 380-392, 1978.

Lloyd, F.E. and P.K. Bailey, Light element metasomatism of the continental mantle: The evidence and consequences, in Phys. Chem. Earth, 9, 389-416, 1975.

Mattinson, J.M., Lead isotope studies of basalts from IPOD Leg 49, in Initial Reports of the Deep Sea Drilling Project. (Washington, U.S. Government Printing Office) 49, (in press), 1978.

Morgan, W.J., Convection plumes in the lower mantle, Nature, Lond., 230, 42, 1971.

Muir, I.D. and C.E. Tilley, Basalts from the northern part of the rift zone of the Mid-Atlantic ridge, J. Petrol., 5, 409-434, 1964.

O'Hara, M.J., Primary magmas and the origin of basalts, Scott. J. Geol., 1, 19-40, 1965.

O'Hara, M.J., The bearing of phase equilibria studies in synthetic and natural systems on the origin and evolution of basic and ultrabasic rocks, Earth-sci. Rev., 4, 69-133, 1968.

O'Hara, M.J., Non-primary magmas and the dubious mantle plume beneath Iceland, Nature, Lond., 243, 507-508, 1973.

O'Hara, M.J., Geochemical evolution during fractional crystallization of a periodically refilled magma chamber, Nature, Lond., 266, 503-507, 1977.

O'Nions, R.K. and R.J. Pankhurst, Petrogenetic significance of isotope and trace element variations in volcanic rocks from the mid-Atlantic, J. Petrol., 15, 603-634, 1974.

O'Nions, R.K. and R.J. Pankhurst, Sr-isotope and rare-earth geochemistry of DSDP Leg 37 basalts, Earth Planet. Sci. Lett., 31, 255-261, 1976.

O'Nions, R.K., R.J. Pankhurst and K. Grönvold, Nature and development of basalt magma sources beneath Iceland and the Reykjanes Ridge, J. Petrol., 17, 315-338, 1976.

O'Nions, R.K., P.J. Hamilton and N.M. Evenson, Variations in $^{143}Nd/^{144}Nd$ and $^{87}Sr/^{86}Sr$ ratios in oceanic basalts, Earth Planet. Sci. Lett., 34, 13-22, 1977.

O'Nions, R.K., N.M. Evenson, P.J. Hamilton and S.R. Carter, Melting of the mantle past and present: isotope and trace element evidence, Phil. Trans. Roy. Soc. Lond., A258, 547-559, 1978.

Pankhurst, R.J., Open system crystal fractionation and incompatible element variation in basalts, Nature, Lond., 268, 36-38, 1977.

Pearce, J.A. and J.R. Cann, Tectonic setting of basic volcanic rocks determined using trace element analyses, Earth Planet. Sci. Lett., 19, 290-300, 1973.

Rhodes, J.M., D.P. Blanchard, K.V. Rodger, J.W. Jacobs and J.C. Brannon, Petrology and chemistry of basalts from the Nazca Plate: Part 2 - Major and trace element chemistry, in Initial Reports of the Deep Sea Drilling Project. (Washington, U.S. Government Printing Office), 34, 239-244, 1976.

Saunders, A.D. and J. Tarney, Geochemistry of basalts from the back-arc spreading centre in the Scotia Sea, Geochim. Cosmochim. Acta (in press).

Saunders, A.D., J. Tarney, C.R. Stern and I.W.D. Dalziel, Geochemistry of Mesozoic marginal basin floor igneous rocks from Southern Chile. Geol. Soc. Amer. Bull. (in press), 1978.

Schilling, J.-G., Iceland mantle plume: geochemical study of Reyjkanes Ridge. Nature, Lond., 242, 565-575, 1973.

Schilling, J.-G., Azores mantle blob: rare-earth evidence, Earth Planet. Sci. Lett., 25, 103-115, 1975a.

Schilling, J.-G., Rare-earth variations across 'normal segments' of the Reykjanes Ridge, 60°N-53°N, Mid-Atlantic Ridge, 29°S, and East Pacific Rise, 2°-19°S, and evidence on the composition of the low-velocity layer, J. Geophys. Res. 80, 1459-1473, 1975b.

Schilling, J.-G., Rare earth, Sc, Cr, Fe, Co and Na abundances in DSDP Leg 38 basement basalts: some additional evidence on the evolution of the Thulean Volcanic Province, in Initial Reports on the Deep Sea Drilling Project (Washington, U.S. Government Printing Office), 38, 741-750, 1976.

Schilling, J.-G. and A. Noe-Nygaard, Faroe-Iceland Plume: rare-earth evidence, Earth Planet Sci. Lett., 24, 1-14, 1974.

Schnetzler, C.C. and J.A. Philpotts, Partition coefficients of rare earth elements between igneous matrix material and rock-forming mineral phenocrysts, II, Geochim. Cosmochim. Acta, 34, 331-340, 1970.

Sigvaldason, G.E., S.Steinthorsson, P. Imsland and N. Oskarsson. The simultaneous production of basalts enriched and depleted in large lithophile trace ions (LIL) within the same fissure swarms in Iceland, Bull. Soc. Geol. France, 18, 863-867, 1976.

Sun, S.-S. and G.N. Hanson, Evolution of the mantle: geochemical evidence from alkali basalts, Geology, 3, 297-302, 1975.

Sun, S.-S. and Jahn, B.-M., Lead and strontium isotopes in post-glacial basalts from Iceland, Nature, Lond., 255, 527-530, 1975.

Sun, S.-S. and R.W. Nesbitt, Chemical heterogeneity of the Archaean mantle, composition of the earth and mantle evolution, Earth Planet. Sci. Lett., 35, 429-448, 1977.

Sun, S.-S., M.Tatsumoto and J.-G. Schilling, Mantle plume mixing along the Reykjanes Ridge axis: lead isotopic evidence, Science, 190, 143-147, 1975.

Tarney, J. and B.F. Windley, Chemistry, thermal gradients and evolution of the lower continental crust, J. Geol. Soc. Lond., 134, 153-172, 1977.

Tarney, J., A.D. Saunders, S.D. Weaver, N.C.B. Donellan and G.L. Hendry, Minor element geochemistry of basalts from Leg 49, North Atlantic Ocean, in Initial Reports of the Deep Sea Drilling Project (Washington, U.S. Government Printing Office), 49 (in press) 1978.

Tatsumoto, M., Isotopic compositions of lead in oceanic basalt and its implication to mantle evolution, Earth Planet. Sci. Lett., 38, 63-87, 1978.

Thompson, G., W.B. Bryan, F.A. Frey, J.S. Dickey and C.J. Suen, Petrology and geochemistry of basalts from DSDP Leg 34, Nazca Plate, in Initial Reports of the Deep Sea Drilling Project. (Washington, U.S. Government Printing Office), 34, 215-226, 1976.

Thompson, R.N., Is upper-mantle phosphorus contained in sodic garnet? Earth Planet. Sci. Lett., 26, 417-424, 1975.

Vogt, P.R., Evidence for global synchronism in mantle plume convection, and possible significance for geology, Nature, Lond., 240, 338-342, 1972.

White, W.R., W.B. Bryan, Sr isotope, K, Rb, Cs, Sr, Ba and rare earth geochemistry of basalts from the FAMOUS area, Geol. Soc. Amer. Bull., 88, 571-576, 1977.

White, W.R., S.R. Hart and J.-G. Schilling, Geochemistry of the Azores and the Mid-Atlantic Ridge: 29°N to 60°N, Carnegie Inst. Wasington Yearbook, 74, 224-234, 1975.

White, W.R., J.-G. Schilling and S.R. Hart, Evidence for the Azores mantle plume from strontium isotope geochemistry of the central North Atlantic, Nature, Lond., 263, 659-663, 1976.

Wood, D.A., Major and trace element variations in the Tertiary lavas of eastern Iceland with respect to the Iceland geochemical anomaly. J. Petrol. (in press), 1978.

Wood, D.A., J. Varet, H. Bougault, O. Corre, J.L. Joron, M. Treuil, H. Bizouard, M.J. Norry, C.J. Hawkesworth and J.G. Roddick, The petrology, geochemistry and mineralogy of North Atlantic basalts: a discussion based on IPOD Leg 49, in Initial Reports of the Deep Sea Drilling Project (Washington U.S. Government Printing Office), 49 (in press), 1978.

Wright, T.L. and P.C. Doherty, A linear programming and least squares computer method for solving petrologic mixing problems, Geol. Soc. Amer. Bull., 81, 1995-2008, 1970.

THE PRIMARY MELT OF THE OCEANIC THOLEIITE AND THE UPPER MANTLE COMPOSITION

L. V. Dmitriev, A. V. Sobolev, and N. M. Suschevskaja

Vernadsky Institution of Geochemistry of the USSR Academy of Sciences

Introduction

The chemical variations of pillow lavas from seismic layer two of the world oceans are relatively limited and markedly distinguished from their continental volcanic counterparts. This has led to the following assumptions: stable and specific conditions of partial melting accompanies the formation of typical oceanic crust, and a specific and uniform upper mantle composition exists at the depth where oceanic tholeiite magma appears. These conditions have characterized the ocean crust formation during the last 150 m.y. All of these considerations support the concept of the oceanic tholeiites from seismic layer two occupying a single worldwide basaltic province (Dmitriev, Sharaskin, and Garanin, 1976).

Nevertheless alternative viewpoints based on the diversity of pillow lava compositons and also on experimental petrology have been discussed in the literature. These postulate the existence of two or more parental tholeiitic magmas derived from a heterogeneous upper mantle.

For instance, experimental data by Kushiro (1972) show that the minimal changes in the melt composition lead to the appearance of different phases along the liquidus (Figure 1). Composition (T-87) has 9.2 wt % MgO and the stable olivine along the liquidus from 1 atm. to 7.5 kilobars pressure. The last corresponds to four-phase equilibrium, whilst the lower (T-89) has 7.4 wt. % MgO yet stable plagioclase along the liquidus within the same interval in pressure. In the last case olivine appeared on the liquidus only with about 2 wt. % H_2O. In this regard recent data on the homogenization of melt-microinclusions in olivine, plagioclase, and spinel occurring in oceanic tholeiites indicate temperatures of formation in the range 1240°-1280°C (Clocchiatti, 1977; Shugurova and Dolgov, 1976; Karpuchin, Sobolev, Vernadsky Institute of Geochemistry, U.S.S.R. Academy of Science). In light of these data it seems likely that oceanic tholeiites originated under the dry condition. Therefore, the Kushiro and Thompson experiment requires the simultaneous existence of lherzolitic and pyroxenitic mantle as the source of magma types with compositions (T-87) and (T-89) and are actually primary tholeiitic melts. This leads to serious consequences relating to global geology, tectonics, petrology, etc.

However, a more or less exact estimate of the real primary (or parental) oceanic tholeiite melt composition is required, and this should be obtained from chemical variations of the quenched basaltic glasses formed on the surface of submarine undersea lava flows. The compositions of these glasses are expected to vary in composition from the primary melt (in cases where phenocrysts did not have time to appear) to the residual melt (when porphyric rocks formed). In both cases the glass composition does not depend upon the mechanical enrichment or depletion in phenocrysts.

The significant deviation of the glass composition from the primary or residual liquid should only be due to the dissolution (or remelting) of phenocrysts. However, this needs a remarkable and gradual change in conditions of the thermodynamic equilibrium during the eruptive process.

More exact information about the primary melt can be obtained through the determination of the composition of microinclusions in phenocrysts homogenized at their trapping temperatures. There are few available published data pertaining to this topic.

Thus the present paper is based largely on the numerous glass composition data available with new information about microinclusions in phenocrysts.

The chemical variation amongst 1000 analyses of volcanic glasses from the world's oceans have been statistically assessed (the original data from W.G. Melson, et al., Catalog, Smithsonian Institution, 1977).

Figure 2 shows the average glass composition for each of 68 Sites from the Atlantic, 47 Sites from the Pacific, and 17 Sites from the Indian Ocean plotted in the olivine-diopside-plagioclase ternary system. Some characteristic compositions and the field of pillow lava compositions are also plotted here.

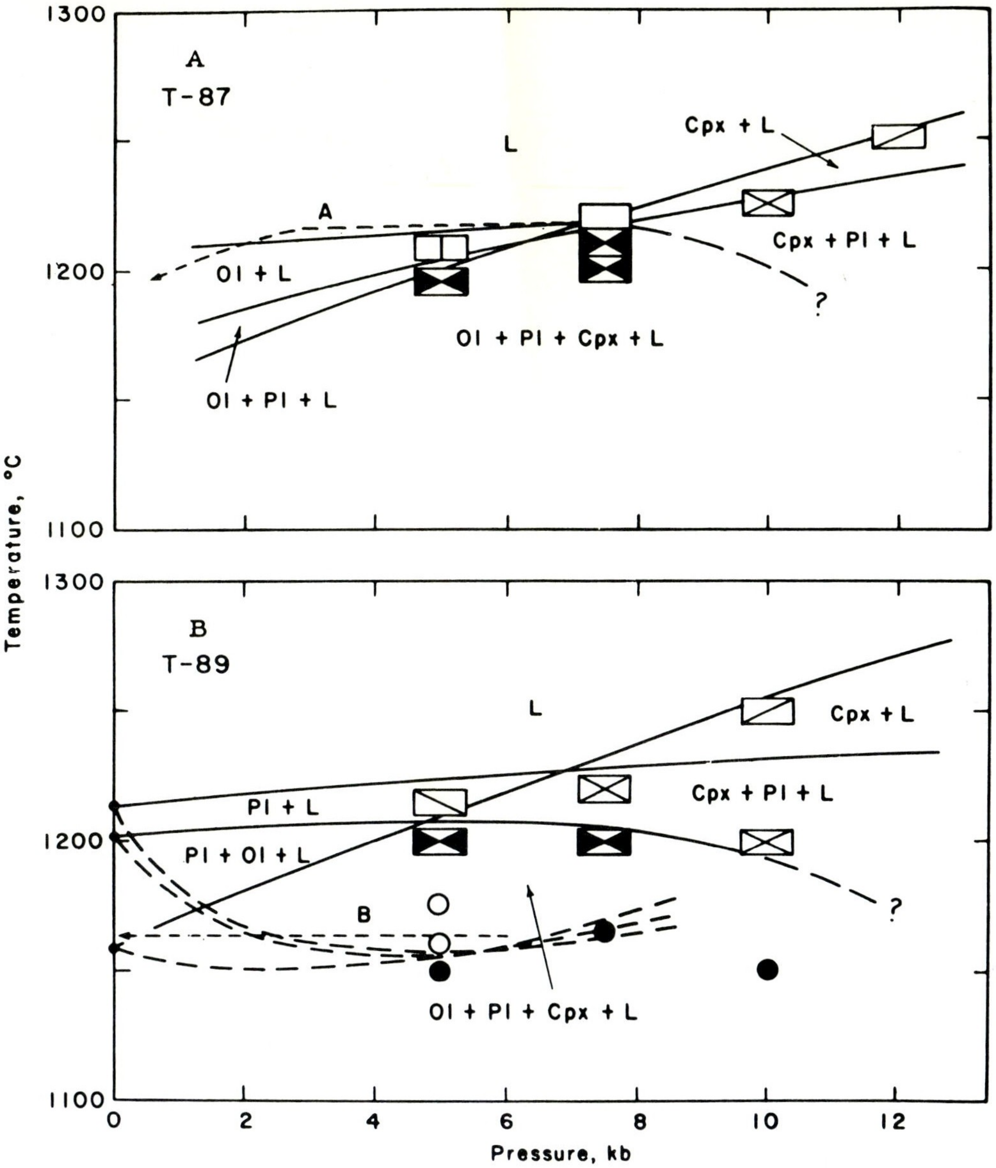

Fig. 1. Melting relations for two abyssal tholeiites (T-87) and (T-89) after I. Kushiro (1972). For compositions, see Table 2.

Figure 2 clearly illustrates that (1) chemical variations of tholeiitic glasses are limited in comparison with one for pillow lavas, and (2) all the glasses compositions occupy a compact area bordered by the five-phase olivine-tholeiite point. The point representing alumina-rich tholeiite with the plagioclase as the liquidus phase (T-89) lies outside of the area of the glass compositions.

These observations lead to the following main conclusions.

1. Only one kind of parental (or primary) oceanic tholeiite melt exists. It corresponds to the magnesium-rich tholeiite coexisting with lherzolitic mantle at 5-11 kilobars of pressure. There are no primary melts of alumina-rich tholeiite.

2. The chemical diversity of pillow lavas are explained by various degrees of their mechanical saturation in phenocrysts: olivine, plagioclase, and pyroxene or combinations of all.

3. The restricted range of the natural glass compositions observed does not permit significant remelting or dissolution of the phenocrysts.

Data based on Site 332 B (Dmitriev, 1977; Aumento and Kempe, 1977) confirm these con-

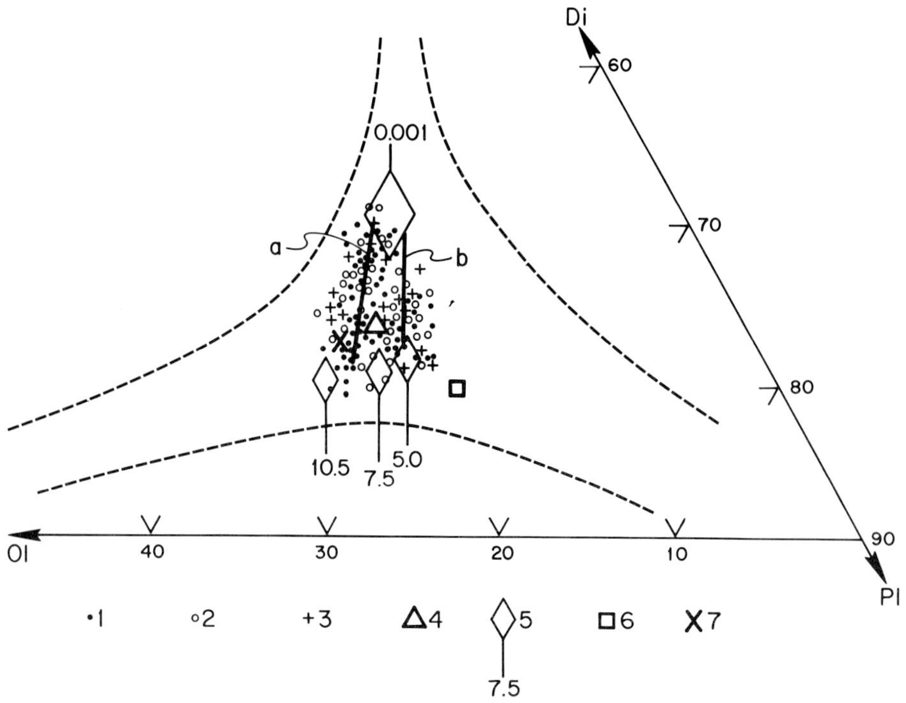

Fig. 2. Average glass compositions for Sites from world oceans plotted in ternary diagram Olivine (Ol)-Diopside (Di)-Plagioclase (Pl).
1. Atlantic Ocean (68 Sites). 2. Pacific Ocean (47 Sites). 3. Indian Ocean (17 Sites). 4. Average M.O.R.B. (Dmitriev et al., 1976). 5. Points of four-phase equilibrium for 10.5 kilobars (Bence, et al., 1978), 7.5 kilobars (Kushiro, 1972), 5.0 kilobars (extrapolated from Kushiro, 1972), and 1 atmosphere (Tilley, et al., 1965; Tilley, et al., 1966; Brown and Shairer, 1966; Emslie and Lindsley, 1967). 6. Alumina-rich tholeiite (T-89) (Kushiro, 1972). 7. Composition of melt microinclusion in olivine, Site 332 B (see text) "a" and "b" lines estimated for olivine-plagioclase cotectic boundaries for real compositions with normative plagioclase An_{70} and An_{50} respectively (wt. %).
The compositions of pillow lavas are inside the dashed area.

clusions totally (see Figure 3). The glass compositions are very restricted occupying a field distinct from the point of alumina-rich basalt.

The individual glass compositions for the Atlantic are plotted on the same ternary system (293 single points for all sites, Figure 4). It is seen here that they form two discrete groups. The first group (crosses) occupies the area between the 7.5-10.5 kilobars five-phase points and the 1 atmosphere point. The second group stretches from the 5 kilobars point to the 1 atmosphere point also. Group N 1 is relatively depleted in TiO_2, while group N 2 is markedly enriched in TiO_2.

The existence of these two single groups was confirmed by a statistical assessment study of the correlation between major elements. Some examples of this statistical operation are demonstrated in Figures 5 and 6. It is seen here that both groups are characterized by similar covariation but differ one from another in the level of element concentrations. It is important to point out that these are significant coefficients or correlations for single groups while there are no correlations between the same elements for joined groups (see Table 1).

The more detailed statistical treatment of glass compositions for the Atlantic permits a recalculation of the variation lines for both groups and to determine the compositions of the terminal points on the ends of these lines. These compositions are placed into Table 2 together with primary melt composition estimated by A. Bence, et al. (1978), F. Frey, et al. (1974), and M. Flower, et al. (1977). Composition of the melt inclusion in olivine from Site 332 B (our data, see below), olivine tholeiite (T-82) (Kushiro, 1972), plagioclase tholeiite (T-89) (Kushiro, 1972), and average M.O.R.B. (Dmitriev, Sharaskin, and Garanin, 1976) are also placed in this table. All these compositions and glass variation lines are plotted on Figure 7 as a function of MgO content.

Figure 7 shows the following. All primary melts considered lie inside the oval area.

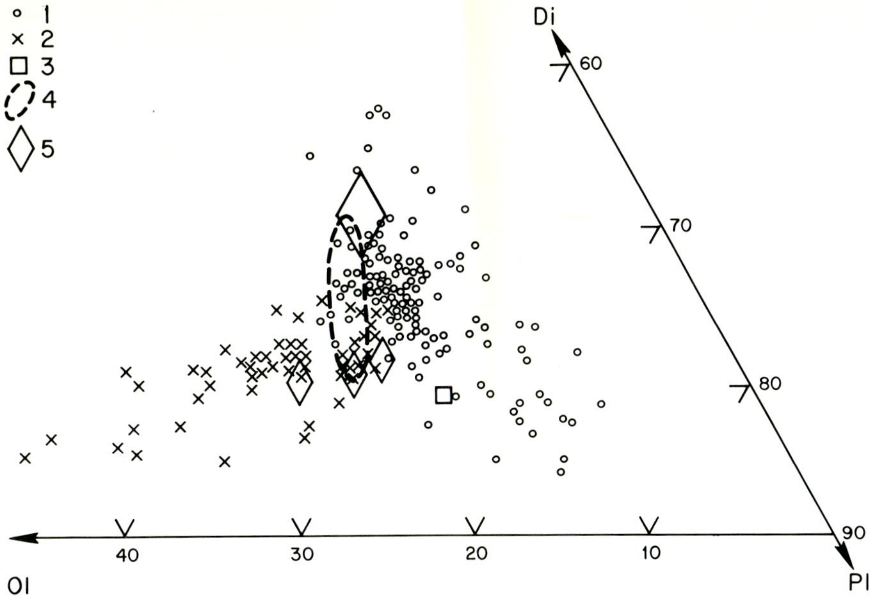

Fig. 3. The compositions of pillow lavas and glasses for Site 332 B (Dmitriev, 1977; Aumento and Kampe, 1977) in ternary diagram Ol-Di-Pl.

1. Plagioclase-phyric basalts
2. Olivine-phyric basalts
3. Alumina-rich basalt (T-89) (Kushiro, 1972)
4. Glass compositions for all rocks from Site 332 B
5. Four-phase points for various pressure (see Fig. 2)

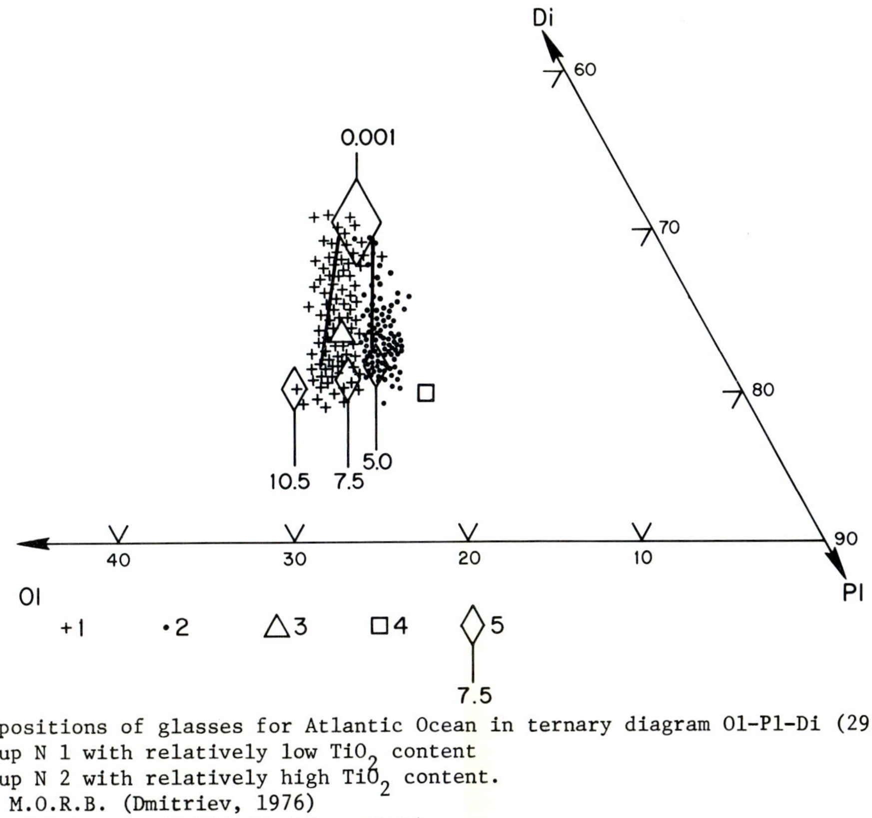

Fig. 4. The compositions of glasses for Atlantic Ocean in ternary diagram Ol-Pl-Di (293 single analyses)

1. The group N 1 with relatively low TiO_2 content
2. The group N 2 with relatively high TiO_2 content.
3. Average M.O.R.B. (Dmitriev, 1976)
4. Alumina-rich basalt (T-89) (Kushiro, 1972)
5. Four-phase points for various pressures (see Fig. 2). Solid lines see also Fig. 2.

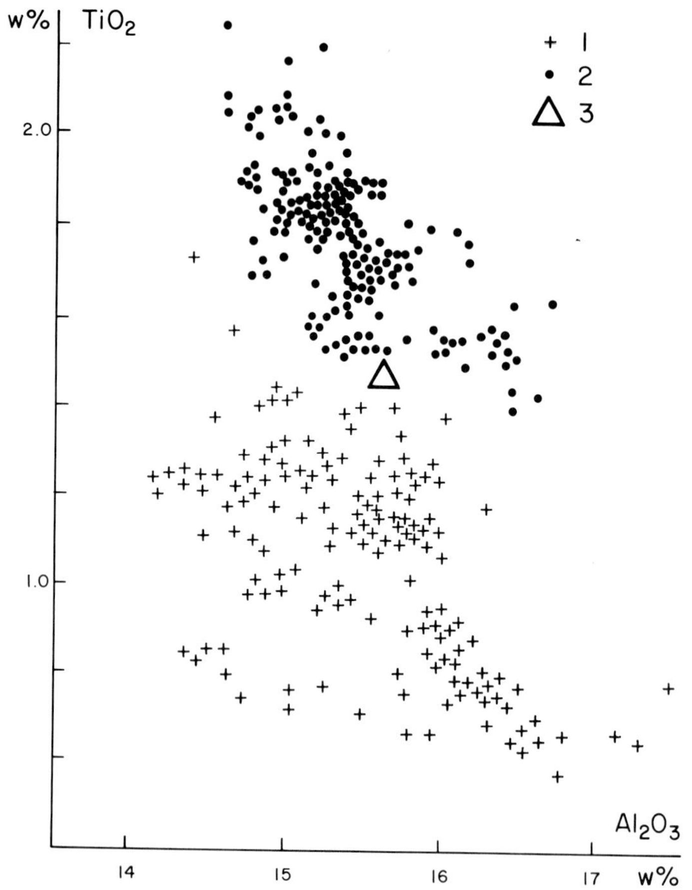

Fig. 5. Correlation between Al_2O_3 and TiO_2 contents in glasses from Atlantic (wt. %).
1. The group N 1 (125 analyses)
2. The group N 2 (168 analyses)
3. Average M.O.R.B. (Dmitriev, et al., 1976)

Their compositions change depending on pressure as indicated by the arrow. The dashed (1-2) and dotted (3-4) lines most probably reflect the change of residual melt composition due to crystallization of the primary melts. The latter point is confirmed by simple calculations (Figure 8). The same variation lines are plotted in the Ol-Pl-Di ternary system: the phenocrysts assemblages which must be subtracted from the primary melts (1) and (3) to get the residual compositions (2) and (4), respectively, are indicated. In both cases the extracted phenocrysts make up about 40% wt. of the original primary melts. The orientation of variation lines (1-2) and (3-4) coincide with the directions of the cotectic lines shown on Figure 2. Olivine to plagioclase ratio of the phenocrysts for the first group is markedly higher than for the second group. This could be due to differences in Ca/Ca + Na which can be responsible for the shift of the Ol-Pl cotectic boundary.

The observations concerning Figure 7 and 8 above allow us to conclude that the primary melt formed under decreasing pressure and crystallizatin of this melt led to the similar trends in variation of the glass composition. SiO_2, TiO_2 (FeO), and Na_2O contents increase while CaO and MgO decrease during this process. Change of pressure does not significantly influence Al_2O_3 and K_2O contents in the primary melt whilst the residual melt becomes depleted in Al_2O_3 and enriched in K_2O due to crystallization differentiation.

Average MORB, olivine (T-87) and plagioclase (T-89) tholeiites are located mostly outside the field of primary melts and glass variation. It is that all of these compositions are saturated in phenocrysts to various degrees.

Figure 9 shows a combination of the diagrams used previously for various petrochemical constructions (Dmitriev, 1969; Dmitriev, 1974). Two histograms of "A" and "S" components for oceanic ultramafic rocks are seen in the lower left part. About 150 analyses of these rocks were used to construct these histograms

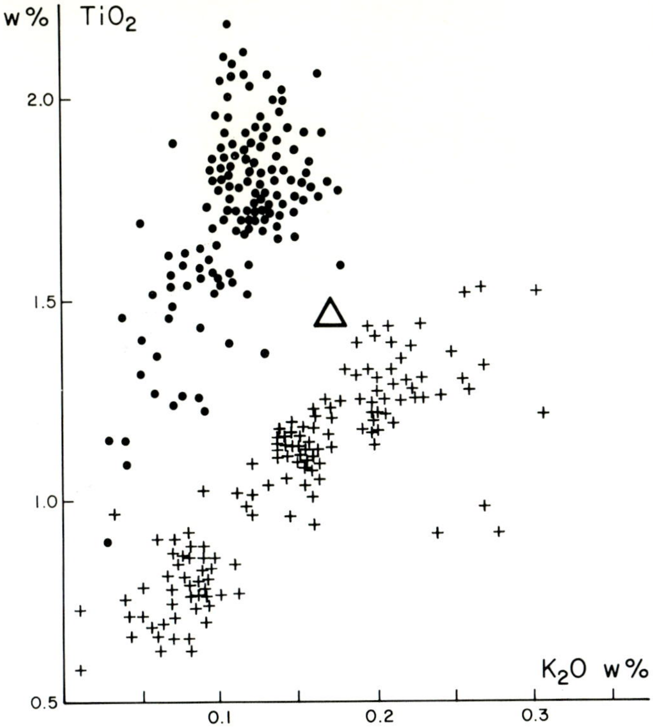

Fig. 6. Correlation between K_2O and TiO_2 contents in glasses from Atlantic (wt. %). For symbols, see Fig. 5.

(Dmitriev, 1969). The bimodal character corresponds to the existence of two main groups of oceanic ultramafic rocks: harzburgite (H) and lherzolite (L). The calculated primary melts of tholeiites for a pressure of 8-10 k bars (point 1) and for 3-5 k bars (point 3) are located in the upper left part of the diagram. Points (1) and (3) corresponding to average lherzolite and harzburgite lie practically along the one straight line. This allows a test of the petrological model of partial melting of the tholeiites from lherzolite to leave harzburgite as residual material (Dmitriev, 1969). The result of these evaluations is shown in Table 3. It is seen that the real average oceanic lherzolite and lherzolites calculated from real harzburgite and primary melts (1) and (3) have practically the same compositions. We consider this result as confirmation of our estimates for the primary melt composition and also the petrological model for the origin of tholeiite melt from oceanic mantle by 13% of partial melting.

It has been indicated above that very important information about the primary melt can be obtained through the melt microinclusion studies. The technique for the homogenization and measurement of melt microinclusion compositions in olivine and spinels from oceanic tholeiites has been updated. Figure 10 shows the microphotograph of a polyphase inclusion in olivine (Site 332 B) before and after homogenization. The inclusions were heated at 1270 ± 10°C within a CO_2 - CO atmosphere for one hour and then quenched. The compositions

TABLE 1. Correlation Between Al_2O_3-TiO_2 and TiO_2-K_2O for Groups 1, 2 and for Both Groups

	Group N 1 (125 analyses)			Group N 2 (168 analyses)			Both Groups (293 analyses)		
	m	s	r	m	s	r	m	s	r
Al_2O_3	15.65	0.691		15.619	0.474		15.633	0.578	
			−0.726			−0.546			−0.450
TiO_2	1.046	0.241		1.716	0.219		1.433	0.400	
			+0.838			+0.727			+0.290
K_2O	0.139	0.070		0.121	0.039		0.129	0.054	

m - mathematical expectation
s - root mean square deviation
r - correlation coefficient
(Standard statistical coefficients)

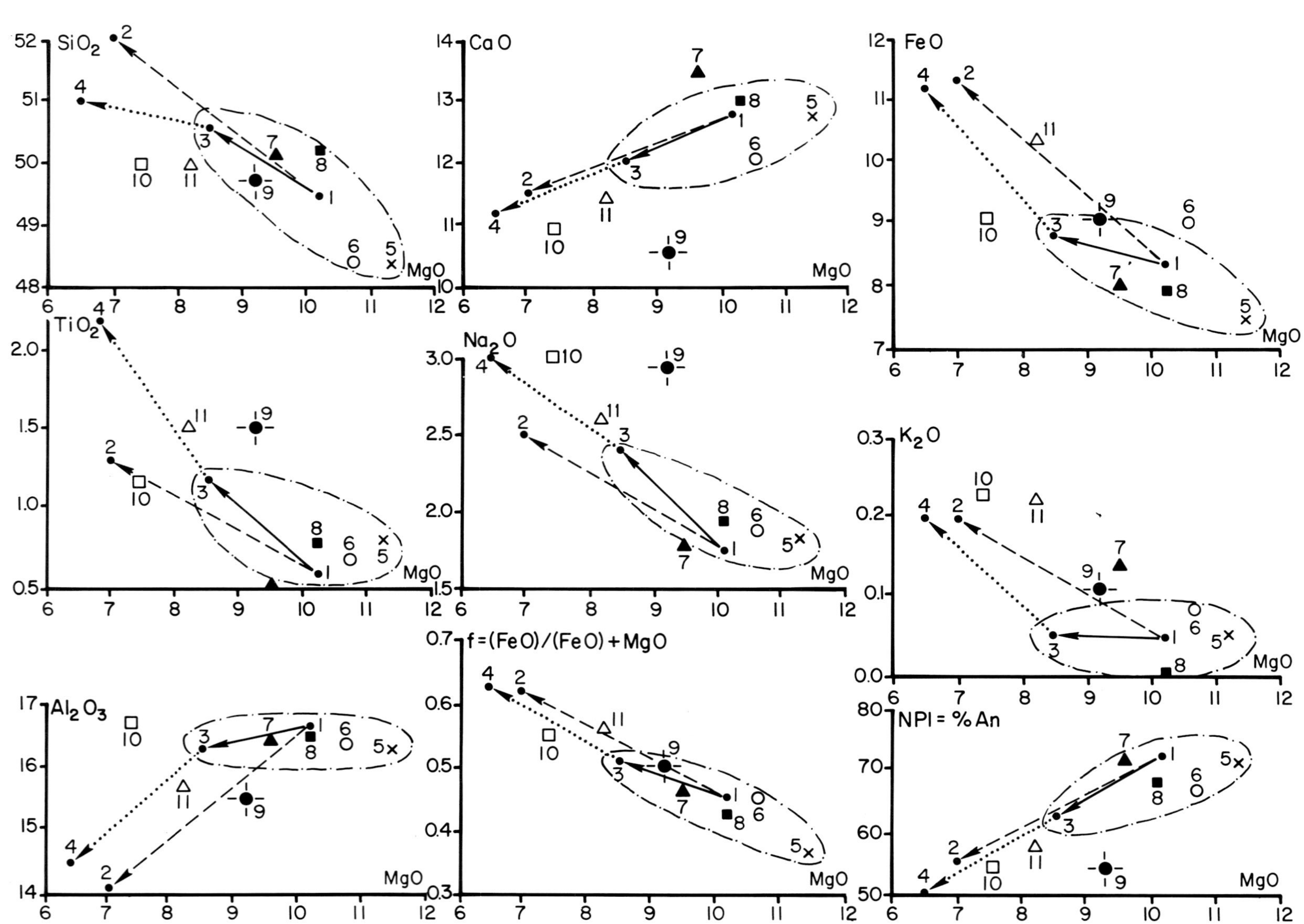

Fig. 7. The calculated variation lines for glass composition from Atlantic as function MgO content. For explanations see text and Table 2. Numbers of points on this diagram correspond to numbers of columns in Table 2.

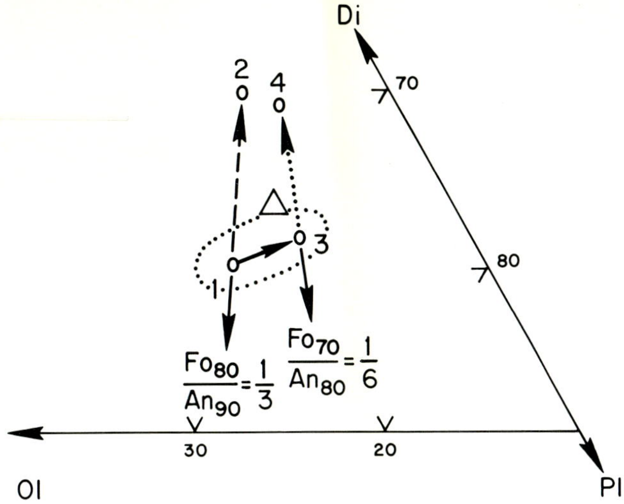

Fig. 8. The calculated variation lines for glass composition from Atlantic in ternary diagram Ol-Di-Pl. For explanation, see text.

of homogenized and quenched microinclusions were determined by JXA-5A microprobe. The average of determinations is seen in Table 2 (Column 5). The partition coefficient for Fe and Mg between olivine and the inclusion K_d = 0.31. Following Roeder and Emsley (1970) this quantity indicates that the equilibrium between olivine and melt occurred at a temperature of T=1280°C, practically the same as the temperature of homogenization. It is important

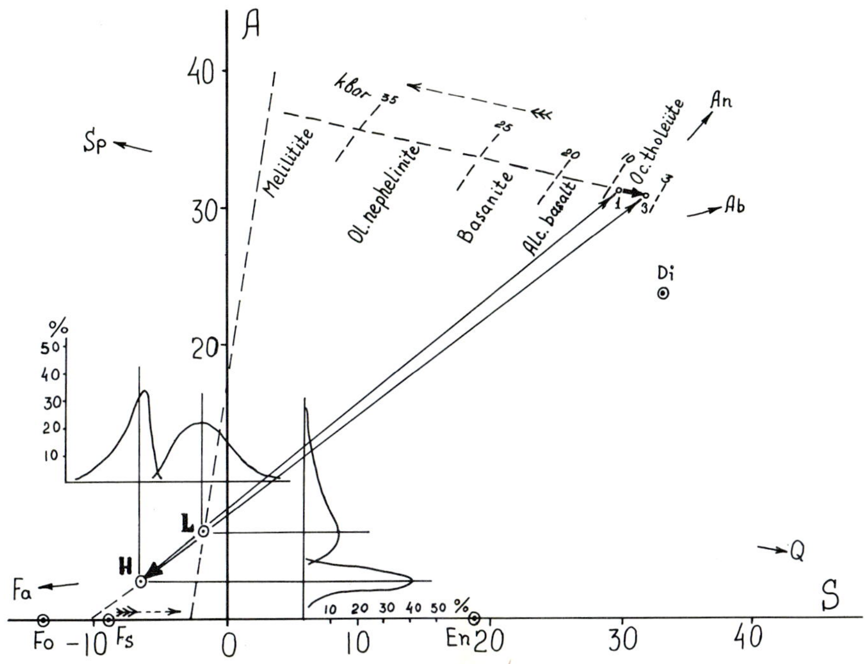

Fig. 9. Primary tholeiitic melts and oceanic ultramafic rocks compositions on the A-S diagram
$A = Al_2O_3 + CaO + Na_2O + K_2O$; $S = SiO_2 - ((FeO) + MgO + MnO + TiO_2 + Cr_2O_3 + NiO)$
Solid arrows show direction to some minerals compositions.
H - average oceanic harzburgite
L - average oceanic lherzolite
1 - primary tholeiitic melt under 8-10 k bar
3 - primary tholeiitic melt under 3-5 k bar
For additional explanation see text.

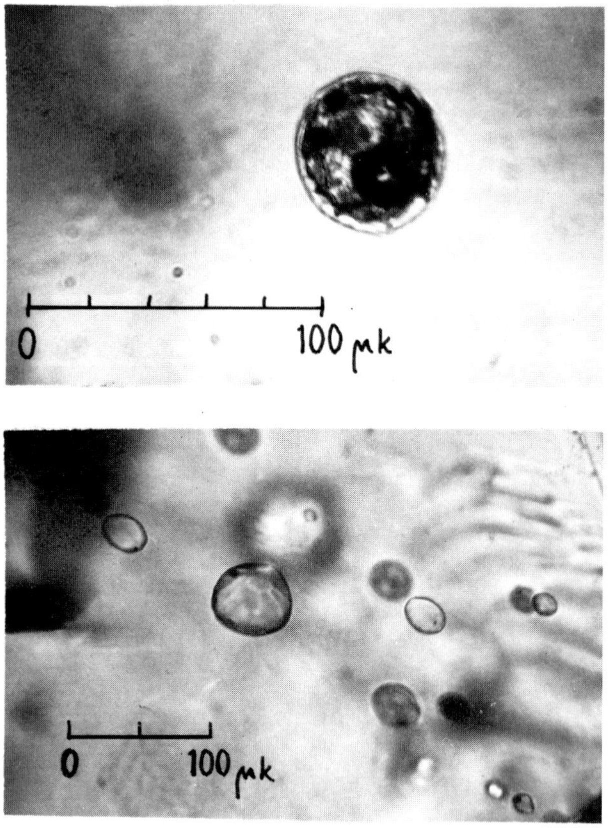

Fig. 10. Polyphase melt microinclusion in olivine (Site 332 B).

a. Before homogenization

b. After homogenization and quenching

to note the composition of melt microinclusions determined also corresponded to the equilibrium with olivine from oceanic harzburgite (Fa=9.5) which is considered to be the residual material after extraction of tholeiitic melt (Dmitriev, 1969). The temperature calculated for this equilibrium under dry conditions and 10 k bar pressure is 1320°C (Vaganov and Kusnetsov, 1976). All of these conditions (temperature, pressure, and composition) are very similar to ones determined recently by A.E. Bence, et al. (1978) (see Figure 11 and Table 2). The maximal temperatures of olivine and plagioclase phenocryst crystallization determined by the homogenization of melt microinclusions in these phenocrysts from basalts of Site 332 B have been plotted on the A. Bence diagram. The results are well in accordance with experimental data and also allow us to assume a

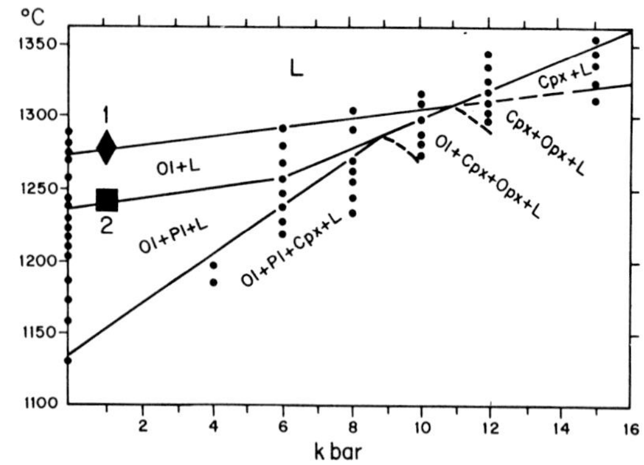

Fig. 11. Melting relations for abyssal tholeiitic glass from FAMOUS, 527-1-1, after Bence, et al. (1978). For composition, see Table 2.
1. Temperature of olivine crystallization (Site 332 B)
2. Temperature of plagioclase crystallization (Site 332 B)

shallow depth of phenocryst appearance in the melt (about 3-4 km) under dry conditions. The existence of two groups of primary tholeiitic melt for the Atlantic is illustrated on the schematical map of the Atlantic Ocean, (Figure 12) from which it is seen that the first group (crosses) occurs north from 36°N south of 25°S along the Mid-Atlantic Ridge. The second group (solid points) is located in the equatorial part of the Ridge.

The geochemical difference between basalts located N and S of 36°N was also noted by W.G. Melson (1978), H. Bougault (1978), and W.B. Bryan (1978) in papers at the Second M. Ewing Symposium and occurring elsewhere in this volume.

If our assumptions about the dependence between primary melt composition and pressure is correct, then the depth of tholeiitic magma generation in the mantle of the Atlantic Ocean decreases from North and South toward the equatorial area, and probably from the western margin to the Mid-Atlantic Ridge (see two points at the western Atlantic on Figure 12). From a petrological point of view this phenomenon is explained as the result of the partial melting of spinel lherzolite to form the first group of primary melt and plagioclase lherzolite to form the second group of melt.

Conclusions

1. Most probably the oceanic upper mantle is quite uniform in composition along the depth interval where the oceanic tholeiites originate. This mantle consists of spinel and plagioclase lherzolite and these are the source of tholeiitic melts and also involves the production of harzburgite as the residual matter after tholeiitic melt extraction. There are no indications of the existence of pyroxenitic mantle.

2. The primary tholeiitic melt is enriched in MgO, CaO, and Al_2O_3 and depleted in (FeO),

TABLE 2. Compositions of the Oceanic Tholeiites (wt. %)

	1	2	3	4	5	6	7	8	9	10	11
SiO_2	49.5	52.0	50.6	51.0	49.3	48.2	50.1	50.3	49.68	49.89	49.96
TiO_2	0.6	1.3	1.2	2.2	0.8	0.73	0.5	0.73	1.55	1.35	1.48
Al_2O_3	16.7	14.0	16.3	14.5	16.1	16.3	16.5	16.6	15.49	16.70	15.67
(FeO)	8.3	11.3	8.8	11.2	6.8	8.92	8.0	7.99	9.08	9.15	10.29
MnO	0.2	0.2	0.2	0.2	0.1	0.25	0.13	0.12	0.18	0.19	0.18
MgO	10.2	7.0	8.5	6.5	11.4	10.70	9.5	10.2	9.17	7.44	8.20
CaO	12.8	11.5	12.0	11.2	12.6	12.0	13.3	13.2	10.61	10.90	11.36
Na_2O	1.7	2.5	2.4	3.0	1.8	1.95	1.76	2.0	2.88	3.01	2.63
K_2O	0.05	0.2	0.05	0.2	0.06	0.09	0.14	0.01	0.11	0.23	0.23
$\frac{(FeO)}{(FeO) + MgO}$	0.45	0.62	0.51	0.63	0.37	0.45	0.46	0.44	0.50	0.55	0.56
Ol	22.9	19.5	19.8	16.7	22.7	25.0	20.2	20.7	21.9	19.2	21.0
Pl	55.2	53.0	57.7	56.2	54.4	54.7	54.9	54.9	57.7	61.4	56.7
Px	21.9	27.5	22.5	27.1	22.9	20.3	24.8	24.4	20.4	19.3	22.3

Explanation for Table 2

1. The glass composition near 8-10 kilobars four-phase point for group N 1.
2. The glass composition near 1 atm. point for group N 1.
3. The glass composition near the 3-5 kilobars four-phase points for group N 2.
4. The glass composition near the 1 atm. point for group N 2.
5. The melt microinclusions in olivine, Site 332 B (average from six determinations).
6. Four-phase equilibrium under P=10.5 kilobars and T=1320°C (Bence, et al., 1978).
7. Primary tholeiitic melt after Flower, et al., 1970.
8. Primary tholeiitic melt after Frey, et al., 1974.
9. Olivine tholeiite (T-87) (Kushiro, 1972).
10. Plagioclase tholeiite (T-89) (Kushiro, 1972).
11. Average M.O.R.B. (Dmitriev, et al., 1976).

(FeO) = 0.9 Fe_2O_3 + FeO, wt. %

TABLE 3. Oceanic ultramafic rocks (wt. %)

	1	2	3	4
SiO_2	45.4	45.9	46.07	45.95
TiO_2	0.1	0.2	0.24	0.17
Al_2O_3	1.7	3.7	3.60	3.69
(FeO)	8.3	8.3	8.35	8.29
MnO	0.1	0.1	0.11	0.11
MgO	42.9	38.6	38.42	38.53
CaO	0.7	2.3	2.17	2.34
Na_2O	0.2	0.3	0.49	0.37
K_2O	0.05	0.05	0.05	0.05
Cr_2O_3	0.3	0.3	0.3	0.3
NiO	0.2	0.2	0.2	0.2

1. Average oceanic harzburgite (75 analyses) (Dmitriev, et al., 1976).
2. Average oceanic lherzolite (64 analyses) (Dmitriev, et al., 1976).
3. Calculated lherzolite as the sum of 87% average harzburgite (Dmitriev, et al., 1976) + 13% of primary tholeiitic melt under 8-10 kilobars (Table 2, column 1).
4. Calculated lherzolite as the sum of 86.6% average harzburgite (Dmitriev, et al., 1976) + 13.4% of primary tholeiitic melt under 3-5 kilobars (Table 2, column 3).

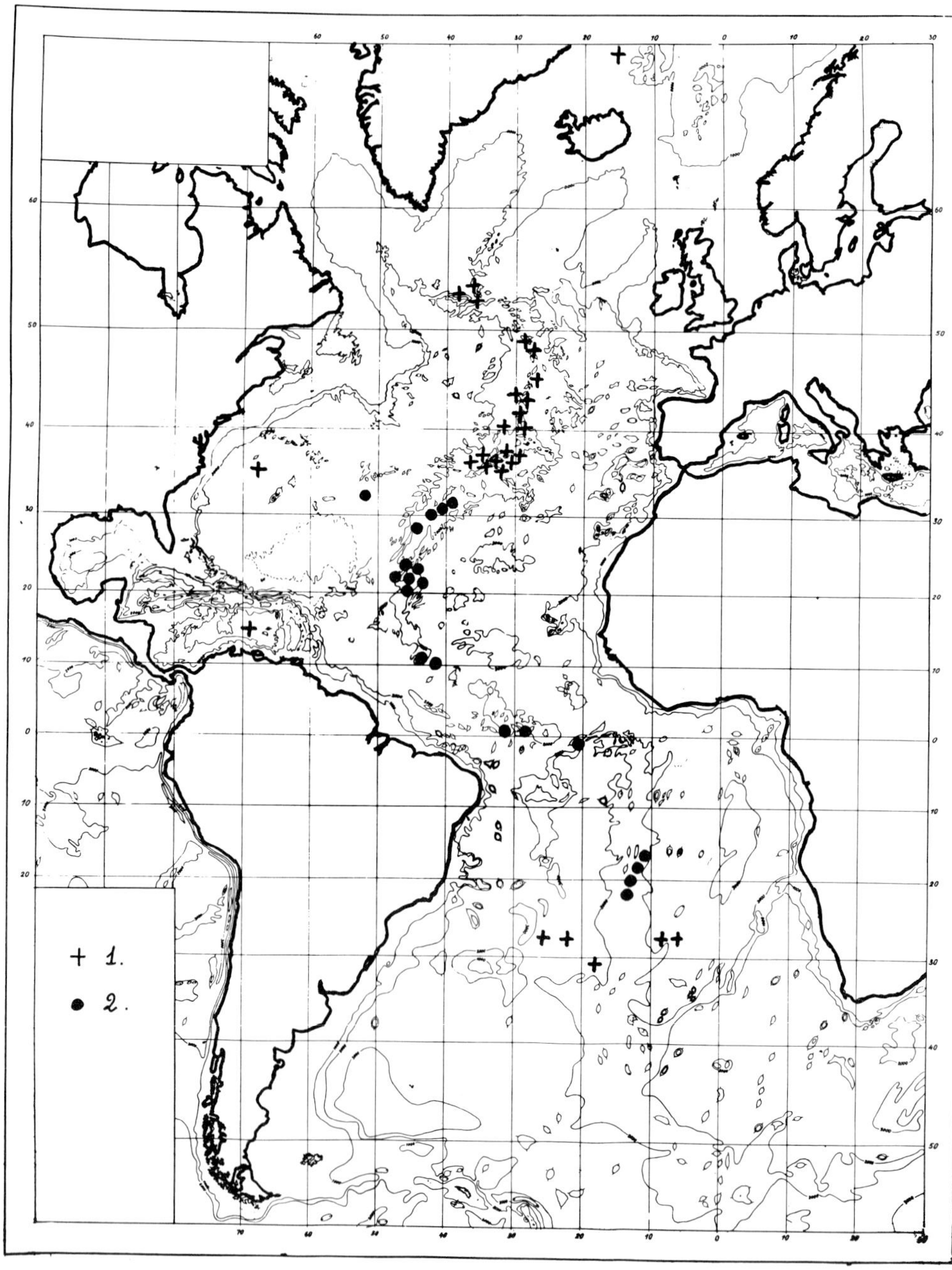

Fig. 12. Two groups of glass composition on the schematical map of Atlantic Ocean.
1. Group N 1 with relatively low TiO_2 content
2. Group N 2 with relatively high TiO_2 content

TiO_2 and alkalies in comparison with tholeiitic pillow lavas. This is due to the significant role of crystallization differentiation during the process of tholeiitic lava formation. It is possible that intermediate magma chambers exist at shallow depth where phenocrysts are formed. The dynamic regime of the eruptive process makes it impossible for significant amounts of the phenocrysts to dissolve in the residual liquid. However, it should be emphasized that the primary melt lost some products by differentiation at the depth in mantle.

3. Two main groups of the primary melts exist in the Atlantic. The first was produced by melting of the spinel lherzolitic mantle at a depth of 35-45 km at 1280-1320°C and dry conditions. The second group was formed by

the partial melting of plagioclase lherzolitic mantle at a depth of 25-30 km at 1250-1300°C and dry conditions. This leads to the geochemical differences between pillow lavas formed by evolution of one or another of these primary melt groups.

The initial crystallization of primary melts probably developed at a shallow depth of about 3-4 km. This is compatible with some geophysical data concerning the existence of the magma chambers at about this same depth.

4. The now established existence of two groups of primary melts in the Atlantic needs to be discussed from the standpoint of global geology and tectonics. Now it is clear, however, that we need additional data about the distribution of different basalts in seismic layer II in various tectonic environments of the ocean floor.

References

Aumento, F., and D.R.C. Kempe, Geochemistry, normative mineralogy, and differentiation trends of basalt glasses from DSDP leg 32, in Initial Reports of Deep Sea Drilling Project, 37, 729-735, 1977.

Bence, A.E., J.F. Bender, and D. Baylis, Major and minor element variations in mid-ocean ridge basalts: The role of partial melting, presented at the Second M. Ewing Symposium, Arden House, Harriman, New York, March, 1978.

Bougault, H., Geochemistry of ocean crust: Trace elements--fractional crystallization, partial melting and heterogeneity of upper mantle from FAMOUS and IPOD results, presented at the Second M. Ewing Symposium, Arden House, Harriman, New York, March, 1978.

Brown, G.M., and J.F. Shairer, Melting relations of some calalcaline volcanic rocks, in Carnegie Institute Washington Yearbook, 66, 460-467, 1966.

Bryan, W.B., Compositional variations in space and time of basalts from Atlantic DSDP Sites, presented at the Second M. Ewing Symposium, Arden House, Harriman, New York, 1978.

Clocchiatti, R., Les liquides silicates pieges dans les cristaux d'olivine, de plagioclase et de pyroxene: Prises d'essai du magma, C. R. Acad. Sc. Paris, 284, 1977.

Dmitriev, L.V., To the question about origination of the ultramaphic rocks from Mid-Indian Ocean Ridge, Geochimia, 10, 1179-1187, 1969.

Dmitriev, L.V., Petrochemical study of the basaltic basement of the Mid-Indian Ridge: Leg 24, Djibouti to Mauritius, in Initial Reports of Deep Sea Drilling Project, 24, 767-780, 1974.

Dmitriev, L.V., Petrochemistry of basalts and plutonic rocks, Leg 37, DSDP, in Initial Report of Deep Sea Drilling Project, 37, 681-693, 1977.

Dmitriev, L.V., A.J. Sharaskin, and A.V. Garanin, The main features of the ocean floor magmatism, in Problems of Petrology, NAUKA, Moscow, 173-189, 1976.

Emslie, R.F., and D.H. Lindsley, Experiments bearing on the origin of anorthosite intrusions, in Carnegie Institute Washington Yearbook, 67, 108-112, 1967.

Flower, M.F.J., P.T. Robinson, H.U. Schmincke, and W. Ohnmacht, Magma fractionation systems beneath the Mid-Atlantic Ridge at 36-37°N, Contrib. Mineral. Petrol., 64, 167-195, 1977.

Frey, F.A., W.B. Bryan, and G. Thompson, Atlantic Ocean floor: Geochemistry and petrology of basalts from legs 2 and 3 of DSDP, Jour. Geophys. Res., 79, 5507-5527, 1974.

Green, D.H., A review of experimental evidence on the origin of basaltic and nephelinitic magmas, Phys. Earth Planet. Interiors, 3, 1970.

Kushiro, I., Origin of some magmas in oceanic and circum-oceanic regions, Tectonophysics, 17, 211-222, 1972.

Melson, W.G., Some highlights of crustal drilling in the North Atlantic, presented at the Second M. Ewing Symposium, Arden House, Harriman, New York, March, 1978.

Melson, W.G., G.R. Byerly, et al., A catalog of the major element chemistry of abyssal volcanic glasses, Smithsonian Insti. Press, Washington, D.C., 31-61, 1977.

Roeder, P.L., and R.F. Emslie, Olivine-liquid equilibrium, Contr. Miner. and Petrol., 29, 275-289, 1970.

Shugurova, N.A., And J.A. Dolgov, The inclusions in minerals from oceanic crust rocks, in Genetical investigation in mineralogy, Novosibirsk, 29-31, 1976.

Tilley, C.E., et al., Melting relations of volcanic rock series, in Carnegie Institute Washington Yearbook, 65, 260-269, 1965.

Tilley, C.E., et al., Melting relations of igneous rocks series, in Carnegie Institute Washington Yearbook, 66, 450-457, 1966.

Vaganov, V.I., and I.E. Kusnetsov, The determination of the temperature of olivine from maphic and ultramaphic rocks, in The Modern Methods of Petrological Investigation, NAUKA, Moscow, 142-156, 1976.

EVOLUTION OF THE 'FAMOUS' OCEAN RIDGE SEGMENT: EVIDENCE FROM SUBMARINE AND DEEP SEA DRILLING INVESTIGATIONS

M. F. J. Flower

Department of Mineral Sciences, Museum of Natural History, Smithsonian Institution, Washington, D. C. 20560

Paul T. Robinson

Department of Geological Sciences, University of California, Riverside, California 92521

Abstract. Data from Leg 37 basement drill sites are compared with the FAMOUS surveys of the Mid-Atlantic Ridge at 36 - 37°N in relation to possible models of crust construction. At the youngest (ca. 3.5 my BP) drill site and in the modern median valley magma compositions range from high CaO to low CaO types of ocean ridge tholeiite. Compositional variation defines several systems of shallow-level fractionation, each characterized by major and lithophile element parameters and each related to separate parent magma batches. Spatial variation of different systems in the median valley corresponds to the volcanic topography and is reconciled to magnetic and chemical stratigraphy at the drill sites. A hypothesis is advanced to explain the chemical zonation and exclusive exposure of low dips in the median valley.

Introduction

Until about 4 years ago little was known regarding ocean floor structure, magnetization of ocean crust, and processes of crustal construction along spreading ridges. In order to investigate these problems basement drilling by the Glomar Challenger was initiated in the Atlantic Ocean and the FAMOUS (French-American Mid-Ocean Undersea Study) project was undertaken, involving a detailed study of the median valley of the Mid-Atlantic Ridge at 36 - 37°N by submersible surveys and other means. Since 1974 seven basement drilling legs have been completed in the Atlantic Ocean, largely as part of the International Phase of Ocean Drilling (IPOD). The first of these, Leg 37, (Aumento, Melson et al., 1977) directly complemented the FAMOUS project by drilling 5 deep holes in oceanic basement along a sea floor spreading flow line extending WNW from the present median valley (Figure 1). Sampling at depths up to 583 m within basement led to several new concepts regarding the ocean crust which have been confirmed and refined by later legs.

Our main objective in this paper is to integrate the chemical, lithologic, and magnetic stratigraphy of crustal rocks drilled during Leg 37 with interpretations of volcanic features and processes in the FAMOUS ridge segment in order to understand the dynamic processes of crustal construction at the spreading axis. Although deep crustal drilling and submersible surveying provide distinctly different types of data, each complements the other to a high degree. Submersible surveying allows study of spatial and morphologic relationships of crustal features but provides little data on eruptive chronology and stratigraphy. In contrast, crustal drilling, although areally restricted, provides a detailed record of crustal stratigraphy and permits a chronologic reconstruction of geologic events. Equivalent investigations have been possible previously only at subaerially exposed spreading centers (e. g. Afar and Iceland) where crustal stratigraphic relationships can be viewed in the context of axial eruptive processes (Walker, 1975a; 1975b; Gibson and Piper, 1972).

One of the basic goals of crustal studies in the Atlantic Ocean is to understand the processes by which the active zone of the Mid-Atlantic Ridge is translated laterally into a rigid section of oceanic basement. Two extreme, though not mutually exclusive, models for this process are gaining increasing attention. The first, influenced by studies of ophiolite complexes and Icelandic crustal sections, suggests that systematic deformation and burial occur as successive eruptive centers migrate away from the spreading axis to form an imbricate crustal section. An underlying sheeted dike complex is believed to result from accretion of buried axial feeder conduits (Palmason and Saemundsson, 1974; Kidd, 1977a; 1977b). The other model, based on morphologic analysis of the median valley in the

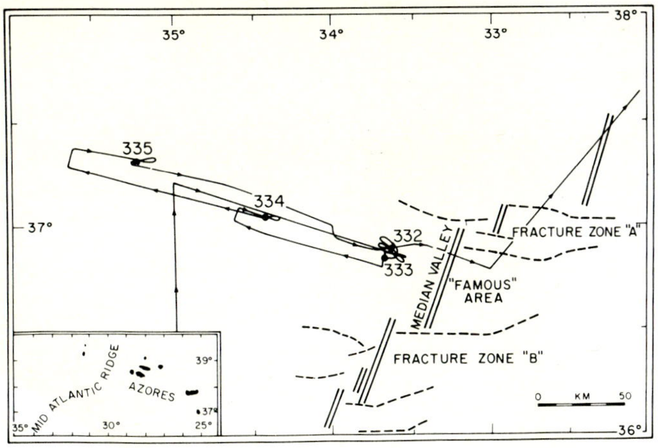

Figure 1. The 'FAMOUS' region of the Mid-Atlantic Ridge showing principal offsets and Leg 37 sites; the 'AMAR' or south rift lies to the south of Fracture Zone B.

FAMOUS area, argues that volcanic edifices once formed in the valley are not buried by later eruptions but are carried laterally to the flanks of the valley to form a 'welded' crustal layer (Atwater, 1978; Macdonald and Atwater, 1978). Each of the models predicts a specific stratigraphic profile for vertical sections of oceanic crust.

In this paper we present interpretations of the Leg 37 drill core and FAMOUS surveys from which we attempt to define the nature of the subrift magma supply and conduit systems, the chronology of crustal accretion, and the deformation of eruptive units. These interpretations are then used to critically evaluate the hypothetical models outlined above.

The 'FAMOUS' Project

Morphology of the Median Valley

The FAMOUS region is a 20-km-long segment of the median valley bounded on the north and south by transform faults, Fracture Zone A and Fracture Zone B (Figure 1). This region has been surveyed in detail by submersibles and deep tow package techniques, and extensively sampled by dredging. The results of these investigations have been published in numerous papers, most of which appeared in a special volume of the Geological Society of America Bulletin (Volume 88, Numbers 4 and 5). Further work is in progress in the AMAR (American Mid-Atlantic Ridge) rift, south of Fracture Zone B (Figure 1). Between the two fracture zones the median valley averages 4 km in width and is about 500 m deep. Within an inner rift valley, 1-2 km wide, linear volcanoes form a set of central topographic highs. Recent eruptive activity is concentrated along this central axis but is also evident at the foot of the eastern fault scarp where young flows overlie older sediment-covered units. Between the central topographic highs and the rift walls the flows are everywhere fractured and deformed, but do not appear to be rotated.

Geomorphic analysis of the valley surface (Ballard and Van Andel, 1977) suggests that rifting occurs along the flanks of the central volcanoes so that as crustal dilation proceeds the axis of spreading migrates back and forth across the rift floor instead of following a static central suture. Lava flows often spread up to 500 meters or more from their vents, and surface measurements suggest that spreading is accompanied by subsidence of up to 100 meters, possibly due to collapse of shallow depleted magma chambers (cf. Wright and Fiske, 1971). Troughs flanking the inner rift and uplifted highs at the outer rift margins are associated with dip-slip faulting. Inward-facing scarps with slopes ranging up to 30° are interpreted by Ballard and Van Andel (1977) as slope 'envelopes' or groups of closely spaced faults with individual dips of 60-90°.

Local spreading rates of 13.4 mm/yr (east) and 7.0 mm/yr (west) result in asymmetric profiles for both inner and outer rifted walls with uplift comparatively compressed to the west (Macdonald, 1977). Macdonald (1977) and Macdonald and Luyendyk (1977) conclude from deep tow magnetic surveys that spreading has been oblique relative to the axial trend for the last 8 m.y. B. P. Ramberg et al. (1977) observe that re-orientation of the oblique spreading pattern has involved complex migration of ridge segments and jumping of shorter ridge segments, resulting in local parallel spreading axes south of 36° 30'N.

The whole FAMOUS-AMAR region is one of intense crustal fracturing. Both Fracture Zones A and B have extensive westward offsets to the south and the median valley merges into these complex zones with increasing width between the confining walls.

Lithology and Eruptive Chronology

As summarized by Hekinian et al. (1976) the major basalt types in the FAMOUS region are: 1) Olivine phyric basalt; 2) picritic basalt; 3) plagioclase-olivine-clinopyroxene-phyric basalt; 4) highly plagioclase-phyric basalt; and 5) aphyric basalt. Types 1 and 2 are believed to represent high temperature, primitive lavas whereas types 3, 4, and 5 are interpreted as lower temperature derivatives formed by fractional crystallization.

Despite the short time span of crustal formation (probably less than 0.5 m.y.) age estimates for the inner rift floor and adjacent walls from manganese and palagonite coatings and fission track measurements suggest three chronologically distinct groups of flows (Bryan and Moore, 1977; Hekinian and Hoffert, 1975; Storzer and Selo, 1974). With decreasing age these are: a) Olivine-rich basalts (lithologic types 1 and 2) forming central hills and more differentiated plagioclase-olivine-clinopyroxene-phyric basalts (lithologic type 3) associated with flank eruptions; b) phyric and aphyric basalts (lithologic types 3 and 5) concentrated on the eastern part of the inner rift floor; and c) highly plagioclase-phyric basalts (type 4) occurring mostly on the western rift wall.

Chemical Variation of Basalts

Several hundred chemical analyses have been made of whole rock and glass samples from the FAMOUS region in order to establish basic compositional trends and to compare these rocks with those from other parts of the Mid-Atlantic Ridge (Bougault and Hekinian, 1974; Bryan et al., 1976; Bryan, in press). Because the composition of whole rock specimens is affected by phenocryst accumulation and by alteration, fresh glass compositions have been used most successfully to establish chemical variations within the FAMOUS region. For any given MgO value, TiO_2 is more variable in north rift (FAMOUS) glass samples and CaO and Al_2O_3 are higher in south rift (AMAR) samples (Figure 2). AMAR and fracture zone basalts are more fractionated than FAMOUS samples but there is no evidence to indicate derivation from chemically different parent magmas. The small segment of the north rift including Mount Pluto and Mont de Venus, studied by the submersible ALVIN, shows a regular spatial pattern of phenocryst assemblages and glass compositions (Bryan and Moore, 1977); Hekinian et al., 1976). These authors found a zonation of several oxides, particularly TiO_2, and parameters such as FeO/MgO between axial and valley flank locations. From this zonation they infer a lateral progression from primitive 'axial' to fractionated 'flank' magmas which they believe to be the surface expression of an extensive sub-rift magma chamber spanning the width of the median valley. This model postulates a continuous process of eruption from, replenishment to, and fractionation within a single large magma chamber. The lateral portions of the chamber are postulated to cool more rapidly and hence undergo more fractional crystallization than the axial portions, thus by virtue of, the spreading of eruptive loci leading to the observed zonation. Bryan and Moore (1977) concluded that fractional crystallization of olivine, plagioclase and clinopyroxene can account for most variations in glass compositions despite some difficulties in reconciling lithophile element contents between proposed parent and daughter magmas. To overcome this problem White and Bryan (1977) suggest that successive magma batches of variable trace element composition may be injected into the chamber during cooling. Incomplete mixing of the batches coupled with fractional crystallization thus gives rise to the observed surface zonation.

We have re-interpreted the FAMOUS glass data keeping in mind the following factors: a)

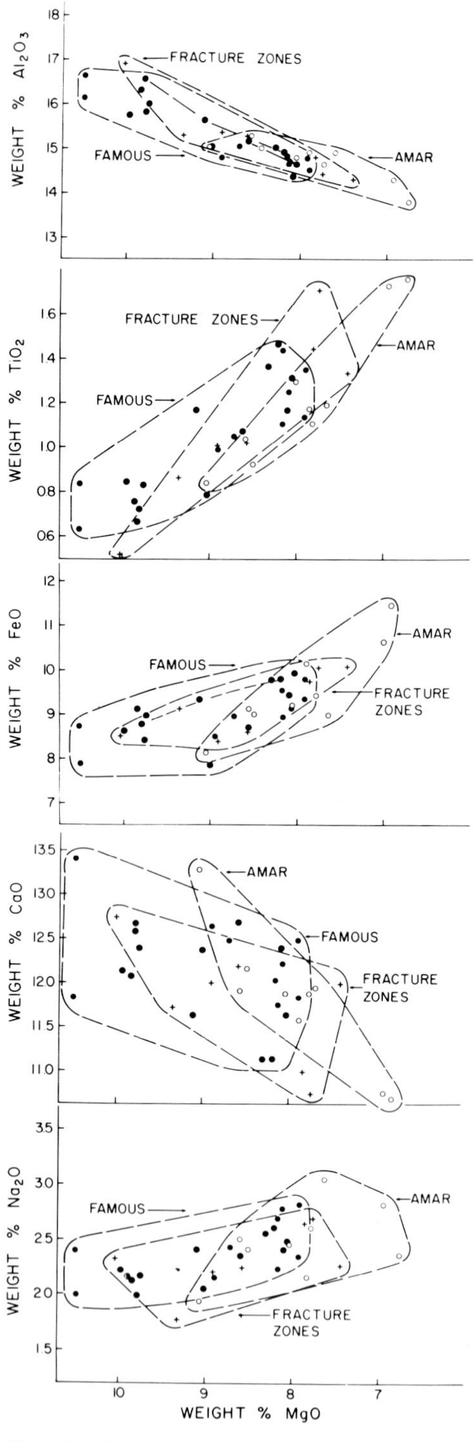

Figure 2. Oxide versus MgO plots for FAMOUS - AMAR glass chemical data;(◉)symbols denote glass group averages for the Mount Pluto - Mont de Venus region of FAMOUS sampled by ALVIN (Bryan and Moore, 1977) (see text); other symbols represent chemical type(●)averages for dredge stations in FAMOUS and AMAR(○) rift segments and Fracture Zone(+)locations, (Bryan, 1978).

Chronologic relations of shallow-level co-genetic magmas; b) spatial relations of co-genetic magmas to topographic features, and thus by implication to eruptive conduit systems; c) the role of median valley tectonics in determining the distribution and extent of burial of eruptive sequences; and d) the constraints placed on the stability of sub-rift magma chambers by such parameters as heat flow and spreading rates.

If eruptions occur along a more or less constant axial location in the median rift the observed chemical zonation should reflect a simple secular trend. If, on the other hand, the zonation reflects coeval eruption of flank and axis magmas the completed crust should have a distinctly different chemical imprint either vertically or horizontally depending on the mode of crustal emplacement. One approach to resolving this question is to model the major element compositions in terms of shallow level fractionation processes. For this purpose we have sorted 50 glass analyses (Bryan and Moore, 1977) into chemical groups by using a set of discriminant oxide tolerances (Table 1) and have tested inter- and intra-group relationships by least squares analysis (Wright and Doherty, 1970). Twelve compositional groups were established by this procedure, the average compositions of which are given in Table 2. Each group is characterized by distinct element parameters such as CaO/Al_2O_3, K_2O/TiO_2 and TiO_2/Al_2O_3. In addition, trace element data from the same glass samples (Langmuir et al., 1977) show that the major element groups have characteristic and distinct ratios of Zr/Y, Ce/Sm and Sm/Yb.

Major element variations indicate that most magma groups cannot be related by simple fractionation models, hence, we infer that they represent discrete magma batches representing independent shallow-level fractionation systems. Exceptions are groups 1, 8 and 12 and groups 7 and 10, each set of which may define a line of liquid descent. Fractionation models relating other groups cannot be accomodated by least squares fits for parent and daughter lavas and fractionating phase compositions because residuals for TiO_2, Na_2O, and K_2O are excessive.

The spatial distribution of these glass groups (Figure 3) shows that most groups have a sub-parallel linear arrangement, generally coincident with the topographic grain, even though sample localities may be up to 500 m from source vents. Some magma groups (e.g. 10 and 11) are split into eastern and western segments separated by sequences believed to be younger because of thinner manganese and palagonite coatings (Hekinian et al., 1976). Based on the same kind of age data, group 11 west appears to be an old inlier completely surrounded by younger basalts. Viewed in this context the observed chemical zonation is the result of a complex combination of fractionated basalts with $TiO_2 > 1$ wt % and $Mg/Mg + Fe^{+2}$ ratios between .61 and .66 (e.g. 7, 8, 9, 10 east, 10 west, 11 east, 11 west, and 12) and primitive basalts with $TiO_2 < 1$ wt % and $Mg/Mg + Fe^{+2}$ ratios > .66 (e.g. groups 1-6) occurring in linear strips. The probable co-genetic groups (1, 8, 12 and 7, 10) are also co-linear with one another and with the topographic grain.

In summary, we believe that the zonation in glass compositions reported by Bryan and Moore (1977) and Hekinian et al. (1976) cannot be due to transverse zonation of a single sub-rift magma chamber. Such a model is precluded by the quantitative incompatibilities of fractionation schemes linking supposedly cogenetic magma groups. This conclusion, coupled with the linear distribution of rocks representing individual magma groups, is compatible with the suggestion of Flower et al. (1977b) that multiple shallow-level fractionation systems co-exist beneath the evolving spreading axis.

DSDP Leg 37

Information derived from basement drilling on Leg 37 (Figure 4) provides a radically different perspective to mid-ocean ridge processes. Most shore-based work on Leg 37 material is now finished (Aumento, Melson et al., 1977) and numerous interpretative papers have been published.

Lithology and Magnetic Stratigraphy

Recovered basement material is mostly pillow basalt and breccia with minor interlayered sediment. Basalts drilled at the youngest sites (332 and 333) are similar to those in the FAMOUS rift valley, ranging from coarsely-phyric to aphyric (Aumento, Melson et al., 1977). An

TABLE 1.

Oxide Tolerances Used in 'Chemsort' grouping of FAMOUS Glass Compositions, (ALVIN sample set).

SiO_2	Al_2O_3	FeO*	MgO	CaO	Na_2O	K_2O	TiO_2
1.10	.60	.40	.50	.50	.20	.04	.10

In any group each analysis differs from every other by less than the stated tolerance.

* total iron oxide

Table 2

	1.	2.	3.	4.	5.	6.
SiO_2	49.16	49.09	49.93	49.34	51.36	49.80
Al_2O_3	16.53	16.30	15.82	16.00	15.03	15.88
FeO^t	8.76	9.05	8.37	8.95	7.86	8.58
MgO	9.80	9.85	9.78	9.74	9.01	9.94
CaO	12.57	12.08	12.67	12.39	12.38	12.12
Na_2O	2.16	2.13	1.97	2.17	2.04	2.22
K_2O	0.05	0.08	0.05	0.09	0.10	0.10
MnO	0.18	0.15	0.15	0.15	0.10	0.14
TiO_2	0.66	0.75	0.72	0.83	0.78	0.84
Cr_2O_3	0.12	0.03	0.08	0.06	0.11	0.07
	99.99	99.51	99.54	99.72	98.77	99.69
n	2	9	1	3	1	9
CaO/Al_2O_3	.760	.741	.801	.774	.824	.763
TiO_2/Al_2O_3	.040	.046	.046	.052	.052	.053
Mg-No.	.689	.682	.698	.682	.693	.696
FeO/MgO	.89	.92	.86	.92	.87	.86
Ti	3955	4494	4314	4937	4674	5033

Table 2 (continued)

	7.	8.	9.	10.	11.	12.
SiO_2	51.14	50.35	51.82	51.36	51.05	50.52
Al_2O_3	15.17	15.63	14.37	14.62	15.01	14.93
FeO^t	8.73	9.20	9.36	9.82	9.74	9.80
MgO	8.61	9.12	8.09	8.04	8.30	8.20
CaO	12.69	11.62	12.41	11.65	11.14	11.14
Na_2O	2.35	2.40	2.39	2.48	2.53	2.60
K_2O	0.12	0.14	0.16	0.20	0.15	0.33
MnO	0.12	0.14	0.17	0.16	0.15	0.15
TiO_2	1.00	1.09	1.16	1.31	1.36	1.47
Cr_2O_3	0.06	0.07	0.03	0.03	0.05	0.05
	99.99	99.76	99.96	99.67	99.48	99.19
n	1	6	2	8	5	3
CaO/Al_2O_3	.837	.743	.864	.792	.742	.746
TiO_2/Al_2O_3	.065	.070	.081	.090	.091	.098
Mg-No.	.661	.662	.631	.617	.627	.624
FeO/MgO	1.01	1.01	1.16	1.22	1.17	1.20
Ti	5992	6531	6951	7850	8149	8808

Glass group averages for ALVIN sample analyses; data from Bryan and Moore (1977) for Mount Pluto - Mont de Venus region of FAMOUS median valley.

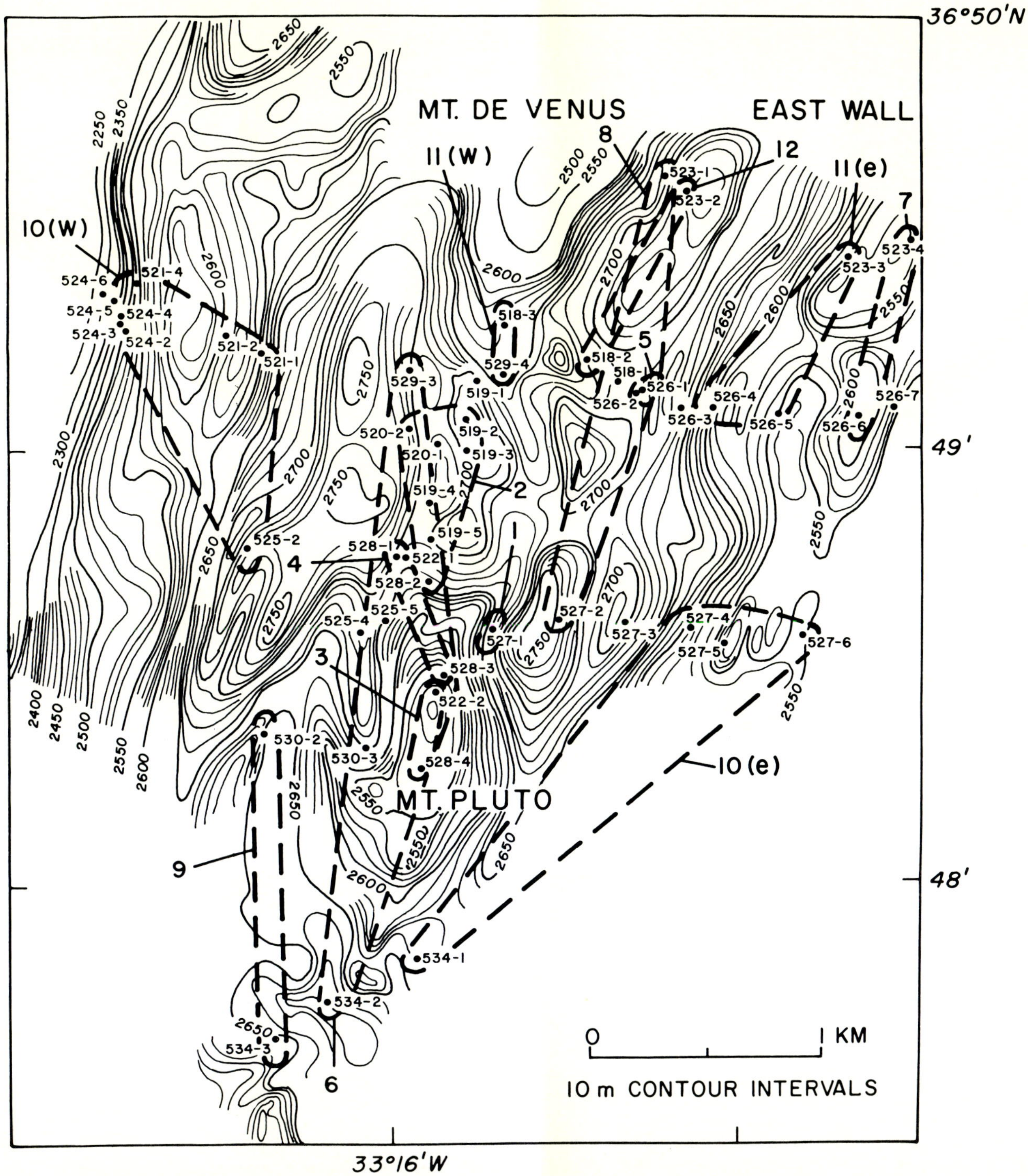

Figure 3. Bathymetric map (contours in meters) showing locations of analyzed samples collected by ALVIN (after Bryan and Moore, 1977). Localities for constituent samples of glass groups defined in this work (see text) are outlined and numbered.

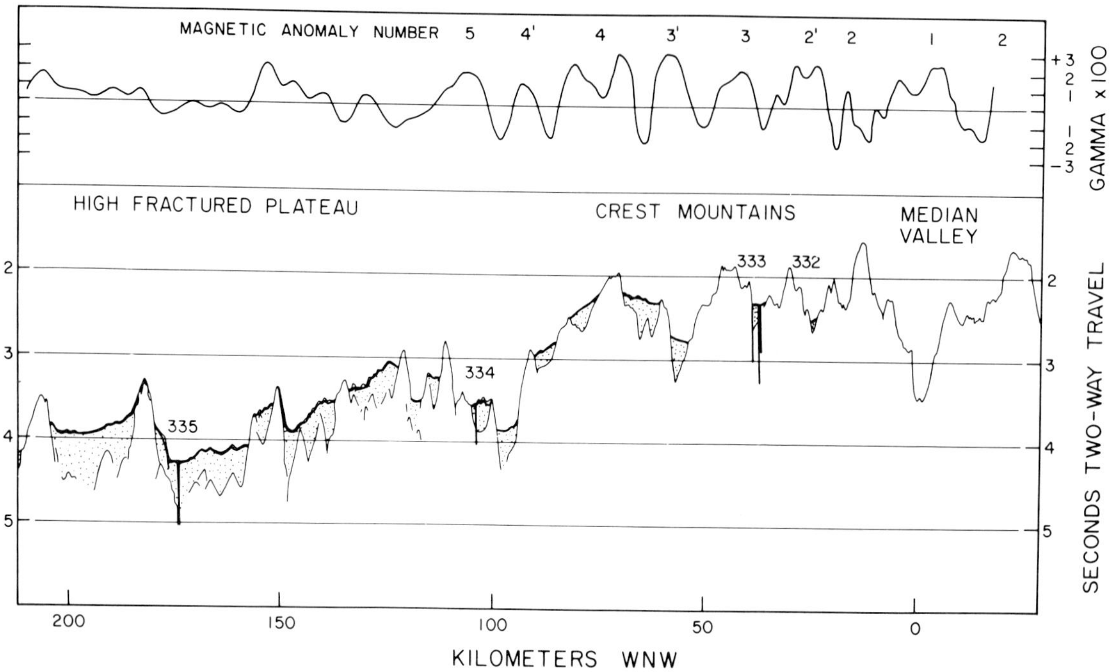

Figure 4. Location of Leg 37 sites in relation to topography and magnetic anomaly profile along the flowline from the median valley.

unusual coarsely-phyric three-phase basalt with phenocrysts of olivine, plagioclase, and clinopyroxene was encountered at Site 332. Basalts at 334 and 335 are sparsely phyric to aphyric, and relatively uniform. A plutonic complex composed of rhythmically interlayered two-pyroxene gabbro, lherzolite, harzburgite, and dunite was encountered at Site 334 beneath about 50 meters of basalt. This complex was interpreted as a cumulate body extruded onto the median valley floor after solidification (Aumento, Melson et al., 1977).

Downhole magnetic polarity reversals and stable inclination changes at Site 332 correspond closely to major lithologic divisions and appear to define single eruptions or sequences of eruptions. Hall (1976; 1978) has interpreted this to indicate a pattern of tectonic rotation as well as secular magnetic field changes. Eruptive sequences at Sites 334 and 335 are less variable magnetically, with expected polarities and with inclinations close to the paleolatitude dipole values.

Chemical Variation of Igneous Rocks

Chemical studies of basement samples have been summarized by Flower et al., (1977a; 1977b), Byerly and Wright (1978) and Bryan and Thompson (1977). Here we shall review only the main chemical aspects of Leg 37 basalts and discuss the relationship of chemical groups to the known lithologic and magnetic stratigraphy.

Flower et al. (1977b) and Byerly and Wright (1978) identified chemical groupings of glass selvedge and whole-rock analyses and used the concept of 'magma systems' to explain compositional lineages resulting from ascent and fractional crystallization of single parent magma batches. From the stratigraphic succession of magma types and the correspondence of such types with eruptive chronologic divisions, we can gain insights into crustal construction mechanisms and the complex nature of sub-rift magma transport.

Figure 5 shows oxide versus MgO variation for Leg 37 glass groups (from Byerly and Wright, 1978) compared to group averages for FAMOUS samples analyzed by Bryan and Moore (1977). Shallow level magma systems inferred by least squares analysis are outlined. At Holes 332A and 332B magmas appear to derive from at least four separate systems (1-4) none of which correspond to a common line of liquid descent. Magmas from a single system are chemically compatible with simple phenocryst removal and accumulation at shallow levels, but those from different systems exhibit 'decoupling' of lithophile elements (e.g. Zr, Y, Ti and REE) and major oxides, such that each system follows a unique compositional trend (cf. Langmuir et al., 1977). The chief example of inter-system element decoupling are:

a) High-TiO_2 systems at Site 332 (1 and 2) are light rare earth element (LREE)-enriched while a similar system at Site 335 is strongly LREE-depleted (Schilling et al., 1977);

b) low TiO_2 systems (3 and 4) at Site 332 are

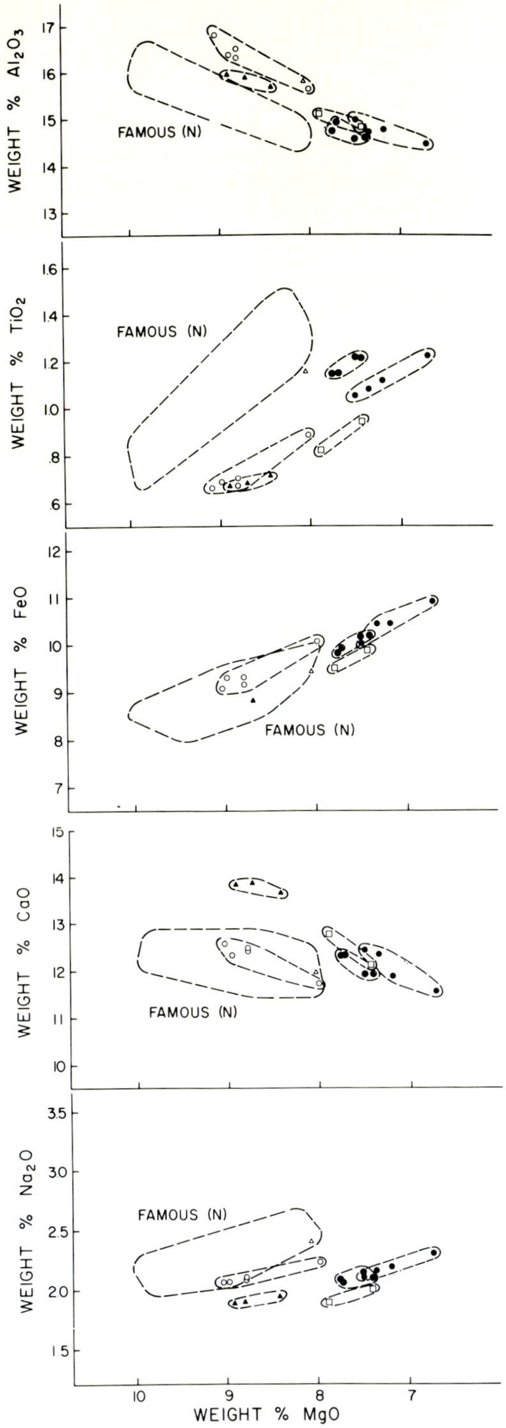

Figure 5. Oxide versus MgO plots for Leg 37 stratigraphic glass group averages (taken from Byerly and Wright, 1978); postulated low pressure fractionation systems (from Flower et al., 1976) are outlined for Site 332 compositions; FAMOUS (north rift) glass compositions (Bryan and Moore, 1977) are also outlined for comparison; ● Site 332-1, ◉ Site 332-2, ○ Site 332-3, ▲ Site 332-4, □ Site 334, △ Site 335.

LREE-depleted while a similar system at Site 334 is LREE-enriched; Na_2O correlates with TiO_2 except for Site 332, group 3 which is Na_2O-rich and TiO_2-poor; and
c) CaO-poor magmas at Site 332 are exclusively LREE enriched, but CaO-rich magmas are both enriched and depleted in LREE and TiO_2.

Generally, Leg 37 glass compositions do not correspond to FAMOUS glass groups which are typically more enriched in lithophile elements and lower in Al_2O_3 and CaO for equivalent MgO content (Bryan and Thompson, 1977). Site 335 glass compositions compare most closely with FAMOUS data but are higher in Al_2O_3 and very different in lithophile element contents. The ranges of CaO, Al_2O_3, Na_2O in Site 332 magmas are interpreted by Flower et al. (1977b) and Byerly and Wright (1978) to reflect buffering of liquid compositions by plagioclase and clinopyroxene in the mantle residue under variable conditions of partial melting. We thus consider low pressure cotectic trends to be superimposed on a range of differing parent magma batches.

Distinct parental magma batches are further defined by lithophile element variation exemplified by REE, Zr, and Y. Zr and Y show partial decoupling from major elements in magma groups at Site 332 and total decoupling between magmas at different sites (Figure 6). Zr/Y ratios are 'polarized', with low Zr/Y ratios characterizing low CaO magmas (Site 332, groups 3 and 4) and high Zr/Y ratios characterizing

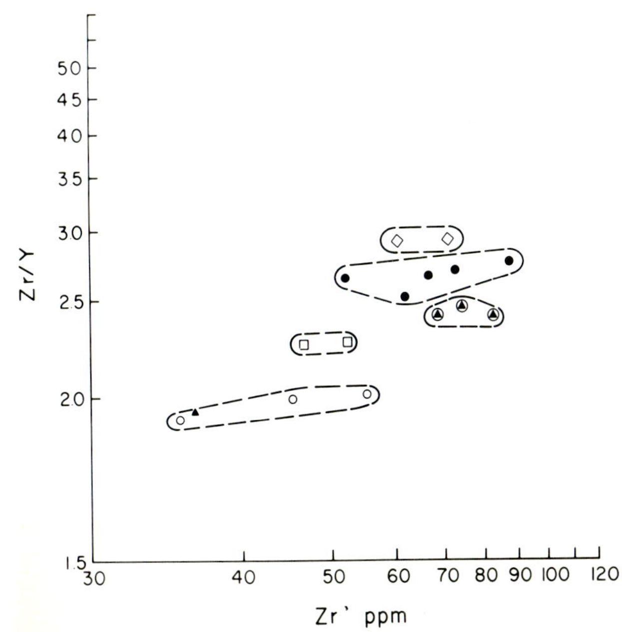

Figure 6. Zr/Y versus Zr (ppm) for sparsely-phyric and aphyric Leg 37 stratigraphic rock group averages defined by Byerly and Wright, (1978); data from Aumento, Melson et al. (1977); ● Site 332-1, 2; ○ Site 332-3; ▲ Site 332-4; ◊ Site 333, □ Site 334; ◬ Site 335.

high CaO magmas (Site 332, groups 1 and 2). Zr/Y values at older sites (334 and 335) show no inter-site correlations with any major element parameter. Polarization and partial decoupling of Ce/Sm ratios from major elements and Zr/Y ratios also occurs at Site 332 (Figure 7) and between sites. Data for FAMOUS glass samples (Langmuir et al., 1977) show a range from chondritic to greater than chondritic Ce/Sm ratios for both low-and high- CaO magmas.

Despite the observed diversity of basalt compositions, Sr and Nd isotope measurements of FAMOUS and Leg 37 samples indicate an isotopically homogeneous mantle between 15 m.y. B.P. and present (O'Nions and Pankhurst, 1977; White et al., 1976). We see from Table 3 that there are no linear changes in Ce/Sm or Zr/Y ratios, or in other parameters insensitive to shallow-level fractionation processes, between FAMOUS and Site 335. Likewise, no linear trends are apparent for major oxides, adjusted to equivalent $Mg/Mg + Fe^{+2}$ ratios. Because the erupted lavas along the FAMOUS - Leg 37 flow line apparently were derived from homogeneous mantle material we attribute the random variations of parent magma compositions to variable degrees of partial melting. The FAMOUS - Leg 37 data do not support the concept of progressive change in mantle source material due to mantle plume activity (Schilling, 1975; Schilling, et al., 1977).

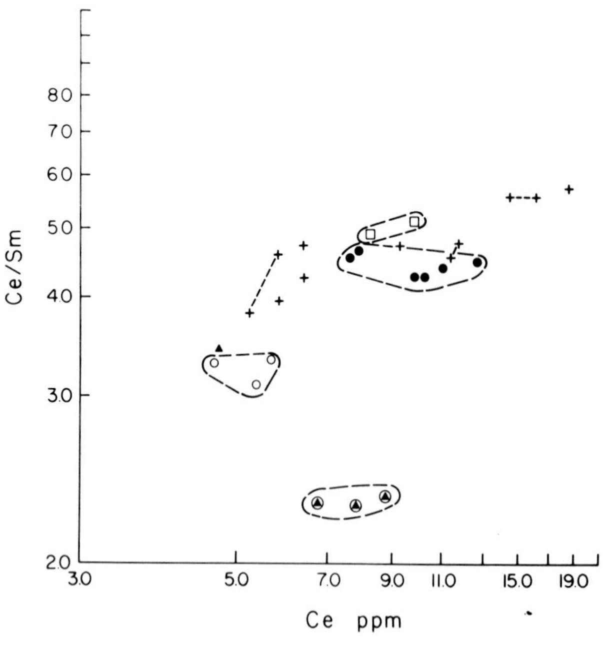

Figure 7. Ce/Sm versus Ce (ppm) for sparsely-phyric or aphyric Leg 37 stratigraphic rock groups (data grouped as for Fig. 6) and FAMOUS glass compositions from Langmuir et al. (1977); symbols same as those in Figure 6; + , FAMOUS glass data.

Interpretation of Leg 37 Sequences

In order to relate chemical and morphologic features of the median valley to the stratigraphy observed in Leg 37 drill holes we now examine Leg 37 sequences in light of possible dynamic models of crust formation.

Integrated models of eruption, burial, and dike intrusion have been postulated from observations on Icelandic crust (Palmason and Saemundsson, 1974) and on ophiolites (Kidd, 1977a; 1977b). Although differing in detail both models postulate: a) A narrow intrusive zone, 1-2 km wide; b) a wider zone of eruption outflow, perhaps 2-10 km wide; c) subsidence of the erupted rocks along curved trajectories, the configuration of which depends on spreading rate and axis width; d) formation of a sheeted dike complex; and e) development of a basal layer of plutonic rocks formed from an accreting magma storage zone.

A schematic projection of these features is shown in Figure 8. The shape of the magnetic polarity changes is marked by lava/dike isochrons which determine the shape of the anomaly blocks (Kidd, 1977a). Kidd extended this approach by modelling anomalies determined by dike spreading trajectories and the Curie point isotherm of trapped magma. According to these models, a drill hole should first penetrate isochron-bounded eruptive sequences, or chemically-defined eruptive phases of 'anomalous' inclination, probably showing one or more polarity reversals. At greater depth the hole should encounter a zone of abundant dikes, finally passing into a plutonic layer.

Alternative 'nonburial' models based on observations in the FAMOUS median rift (Macdonald and Atwater, 1978) predict that a given section of crust is composed of several large volcanic edifices which are transported laterally to the rift wall where they form an accreting eruptive layer. This model postulates that despite at least two episodes of vertical displacement by faulting, tilting of eruptive sequences is minimal, probably not exceeding 4-5° (Atwater, 1978). As a consequence, few if any, down-hole magnetic polarity or inclination changes would be expected and downward chemical variations would be comparatively limited, although compositions might vary laterally from site to site. This model is very difficult to reconcile with data from the crustal sequences drilled by Leg 37 where downhole magnetic, chemical, and lithologic changes are the rule, not the exception. We believe that if downhole changes in magnetic inclination or polarity coincide with compositional and lithologic boundaries of eruptive sequences the magnetic patterns most probably reflect ridge axis tectonic activity rather than faulting of the ocean crust or secular variation of the magnetic field (Hall, 1978; Watkins and Walker, 1977). Such a relationship is strongly supported by Leg 37 observations (Ade-Hall et

Table 3

Magma system(s)	Mt. Pluto - Mt. de Venus	DSDP site 332-1	-2	-3	-4	334	335
age m.y.	present	3.5	----------	----------	----------	9.5	13
TiO_2	0.95 - 1.3	0.75	0.90	0.66	0.65	0.58	0.93
Na_2O	2.0 - 2.5	1.8	1.7	2.05	1.9	1.7	2.25
CaO	11.5 - 12.9	14.2	13.5	12.5	14.2	14.0	13.3
Al_2O_3	15.0 - 16.0	16.0	15.8	16.8	16.2	15.9	16.8
Ce	5.2 - 16.2	6.8	7.5	5.0	5.0	7.0	6.0
Sm	1.37 - 2.60	1.62	1.70	1.56	1.47	1.40	2.61
Yb	1.90 - 3.01	2.0	-	2.2	1.9	2.0	3.3
Zr	31 - 67	50	55	37	37	43	62
Y	24 - 28	18.9	20.4	19.5	19.0	19.1	25.8
Ce/Sm	3.8 - 5.64	4.2	4.4	3.2	3.4	5.0	2.3
Sm/Yb	1.13 - 0.60	0.81	-	0.71	0.77	0.70	0.79
Zr/Y	1.3 - 2.8	2.65	2.70	1.90	1.95	2.25	2.40
TiO_2/Al_2O_3	0.063 - 0.081	0.047	0.057	0.039	0.040	0.036	0.055
CaO/Al_2O_3	0.741 - 0.80	0.888	0.854	0.744	0.877	0.881	0.792
TiO_2/CaO	0.063 - 0.11	0.053	0.067	0.053	0.046	0.041	0.070
$^{87}Sr/^{86}Sr$	.70288 - .70307	.70299	.70298	.70291	.70308	-	.70316

Secular chemical variation along FAMOUS - Leg 37 drill site flowline; data from Flower et al. (1977a), Byerly and Wright (1978), Langmuir et al. (1977), Bryan and Moore (1977), Schilling et al. (1977), and O'Nions and Pankhurst (1977) are averaged for individual glass groups or aphyric/sparsely-phyric members of low pressure fractionation systems; element and oxide values are adjusted to equivalent values of Mg/(Mg + Fe^{+2}) (.65) to minimize discrepancies resulting from shallow level fractionation.

al., 1975), and we believe that the magnetic inclination patterns are critical to any interpretation of crustal structure. Figures 9a - 9d show syntheses of eruptive chronology at Leg 37 drill sites based on lithology, whole rock and glass compositions, and stable magnetic inclinations. Eruptive units have been defined in these sequences by assuming that the products of a single eruption have: a) Uniform glass compositions; b) whole-rock compositional variability consistent with effects of post-eruptive fractionation; c) uniform magnetic polarity and stable inclination; d) lithology consistent with the compositional variation and e) no major sedimentary or clastic intercalations.

Based on these criteria 25 eruptive units are identified in Hole 332A (330.5 meters of basement penetration), 45 at 332B in Hole 332B (583 meters), 2 at Site 334 (56 meters) and 1 at Site 335 (108 meters). In addition to single eruptive units, larger units representing apparent cycles of eruptive activity can be recognized (Figures 9a - 9d). In most cases these reflect the appearance of magma from a new or dormant system as interpreted from glass compositional data.

In Holes 332A and 332B cycle boundaries defined from glass compositions (shown as solid horizontal lines in Figure 9a and 9b) coincide with sharp changes in stable magnetic inclinations, alteration intensity, and, in some cases, magnetic polarity. Thus, crustal construction at Site 332 was strongly episodic with short-lived eruptive cycles (perhaps a few tens to a few hundreds of years) separated by lengthy periods of volcanic quiescence. Site 332B lavas were probably derived from at least four separate magma systems, and flows from these systems appear to be rhythmically interlayered in the lower parts of the hole (Figure 9b). The chemical and lithologic complexity in the lower part of Hole 332B is matched by greater complexity in the magnetic record, reflected in variations in stable inclinations and in several polarity reversals. The reversals do not indicate intrusions because this entire part of the sequence is pillowed. There is no evidence for mixing of the different magma sources prior to eruption, and it seems most likely that each magma type was derived from near-synchronous

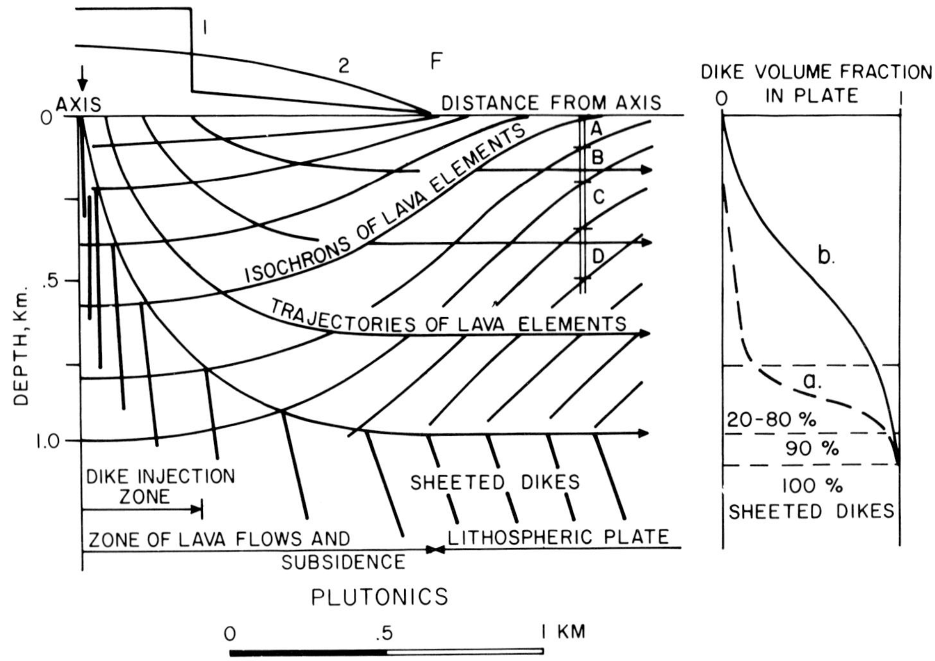

Figure 8. Schematic burial model for crust accretion, after Palmason and Saemundsson (1974); the dike-fraction encountered in drilled sections will depend on the width of the intrusive zone;
1 - distribution of intrusive dike activity (eruptive feeder conduits);
2 - distribution of lava eruption and outflow extent, resulting dike distribution curve of type a; if diking occurs outside of these limits a dike distribution will approach a curve of type b.
A, B, C, and D represent possible isochron-bounded basalt units that might be encountered in a drill hole.

systems developed independently from separate parent batches.

The lithologic and chemical variations with time shown by magmas from a single system should give at least partial information on the history of crystallization, phenocryst redistribution, and draining by eruption of a single reservoir or series of reservoirs. No gradational lithologic pattern is evident at Site 332, although complex and rapid interbedding of olivine- and plagioclase-phyric and also coarsely- and sparsely-phyric basalts may occur in any one eruptive cycle. This implies that shallow-level phenocryst-liquid fractionation did not produce a single chemical zonation of magma in a single chamber. The compositions of eruptive units probably also reflect phenocryst redistribution due to gravity settling and to flow, and perhaps also periodic magma influx from deeper-seated sources resulting in mixing of primitive and evolved magmas. Neither of these processes, however, can be invoked to explain chemical differences between magmas from different systems (Flower et al., 1977b; Byerly and Wright, 1978).

In summary, the crustal stratigraphy at Leg 37 drill sites, particularly Site 332, indicates a pattern of episodic eruption. Close coincidence of lithologic, chemical, and magnetic boundaries indicates that this stratigraphy was developed during crustal construction in the median rift. At Site 332, the early eruptive episodes were relatively frequent and accompanied by considerable deformation; later episodes reflect longer quiescent intervals between eruptions and more stable tectonic conditions.

Synthesis of Ocean Ridge Processes

Any viable ridge construction model must reconcile the evidence deduced from drilling for episodic eruption, interlayering of magmas from different systems, and substantial tectonic rotation during early stages of crustal construction with the apparent lack of tilting and the complex chemical zoning of basalts on the modern median valley floor.

Following Flower et al. (1977b), we believe that the chemical variation of FAMOUS and Leg 37 basalts indicates the co-existence of several independent shallow-level fractionation systems beneath the ridge axis. Dynamic stability of such systems will depend largely on the heat flux and the spreading rate (Sleep, 1975). If approximate isostatic balance is maintained these factors will in turn determine the overall supply of magma, the extent of fractionation, and the degree of magma entrapment at depth.

In reconstructing dynamic processes at the

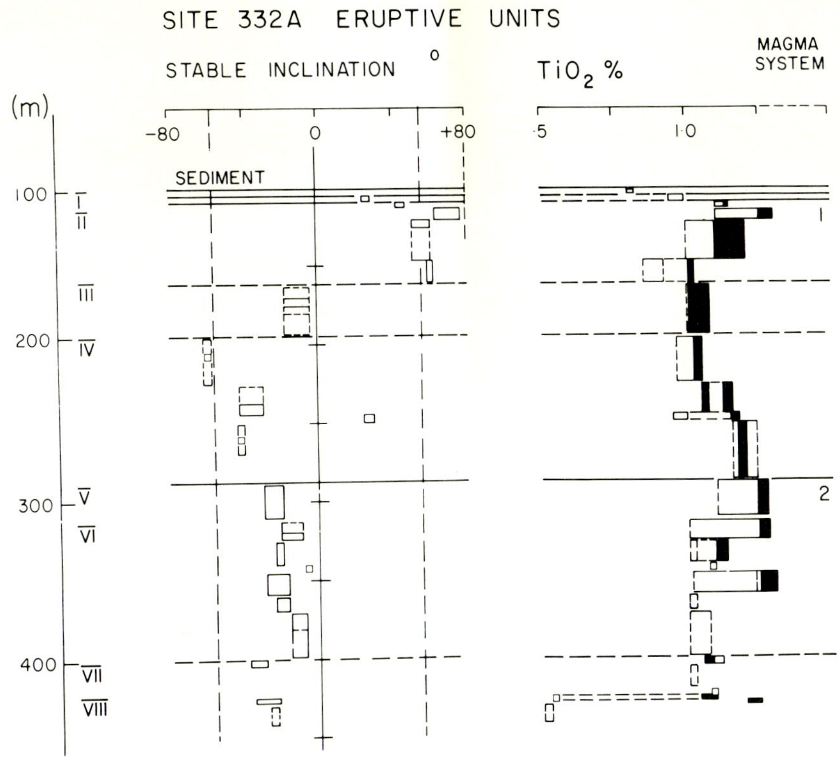

Figure 9a. Schematic downhole sections showing eruptive units inferred from variations in drill core of lithology, NRM stable inclination and chemical affinity of glass and whole-rock compositions; data sources are Hall and Ryall (1977), Flower et al. (1977a) and Byerly and Wright (1978). The range of stable inclination and TiO_2 data are shown for each eruptive unit and have been extrapolated to probable unit stratigraphic limits. Glass TiO_2 compositions are shown by solid areas; the discrepancy between glass and whole-rock compositions (unshaded areas enclosed by vertical dashed lines) is a measure of whole-rock phyric character, as TiO_2 is not incorporated by phenocryst phases. Major lithologic units and magnetic polarity changes are indicated; 'magma systems' include eruptions inferred to be cogenetic in terms of shallow-level fractionation processes from least squares analysis of glass and whole-rock compositions.

Mid-Atlantic Ridge it is useful to compare and contrast this type of active zone with other volcanic regimes. To make such comparisons, we have estimated magma supply rates for intra-plate volcanoes (Kilauea and Mauna Loa), the 'diffuse' spreading axis of Iceland, and the Mid-Atlantic Ridge (Table 4). The estimates show that supply rates from the mantle to the dilating Mid-Atlantic and Icelandic regions (for approximately equivalent conduit network areas) are 1 to 2 orders of magnitude lower than for non-dilating intra-plate systems, whereas the proportions of non-eruptive magma trapped at depth are considerably higher in the dilating regimes. Although magma mixing has been demonstrated for the Kilauea east rift (Wright and Fiske, 1971) where two systems appear to coalesce periodically, it is not a dominant process and it would seem to be of even less significance in a dilating environment. Large, void-like magma chambers are not in evidence at Kilauea (Fiske and Jackson, 1972) and it seems that magma from adjacent conduit systems of Mauna Loa and Kilauea are chemically independent (Wright, personal communication). There is seismic evidence for extensive rift-controlled magma transfer both in Kilauea and the Icelandic rift zone (Björnsson et al., 1978), similar to that we have inferred for the FAMOUS rift zone (Figure 3), suggesting that this feature is common to all three environments.

In the absence of seismic evidence for linear chambers spanning the width of the median valley (Fowler, 1976; Fowler and Matthews, 1974), and considering the chemical evidence for widespread shallow level fractionation and phenocryst accumulation (Flower et al., 1977b; Byerly and Wright, 1978), we conclude that in the FAMOUS region magma ascends along linear cracks corresponding to the axis of dilation, and is temporarily stored in a complex of dikes of extensive linear but restricted lateral extent.

The surface pattern of the median valley suggests that eruptions from multiple systems occur largely in the axial zone rather than at intermediate and flank localities (Figure 3). For the case of a narrow axial zone of eruption the compositional zoning in the basalts may either reflect short-term secular variations in

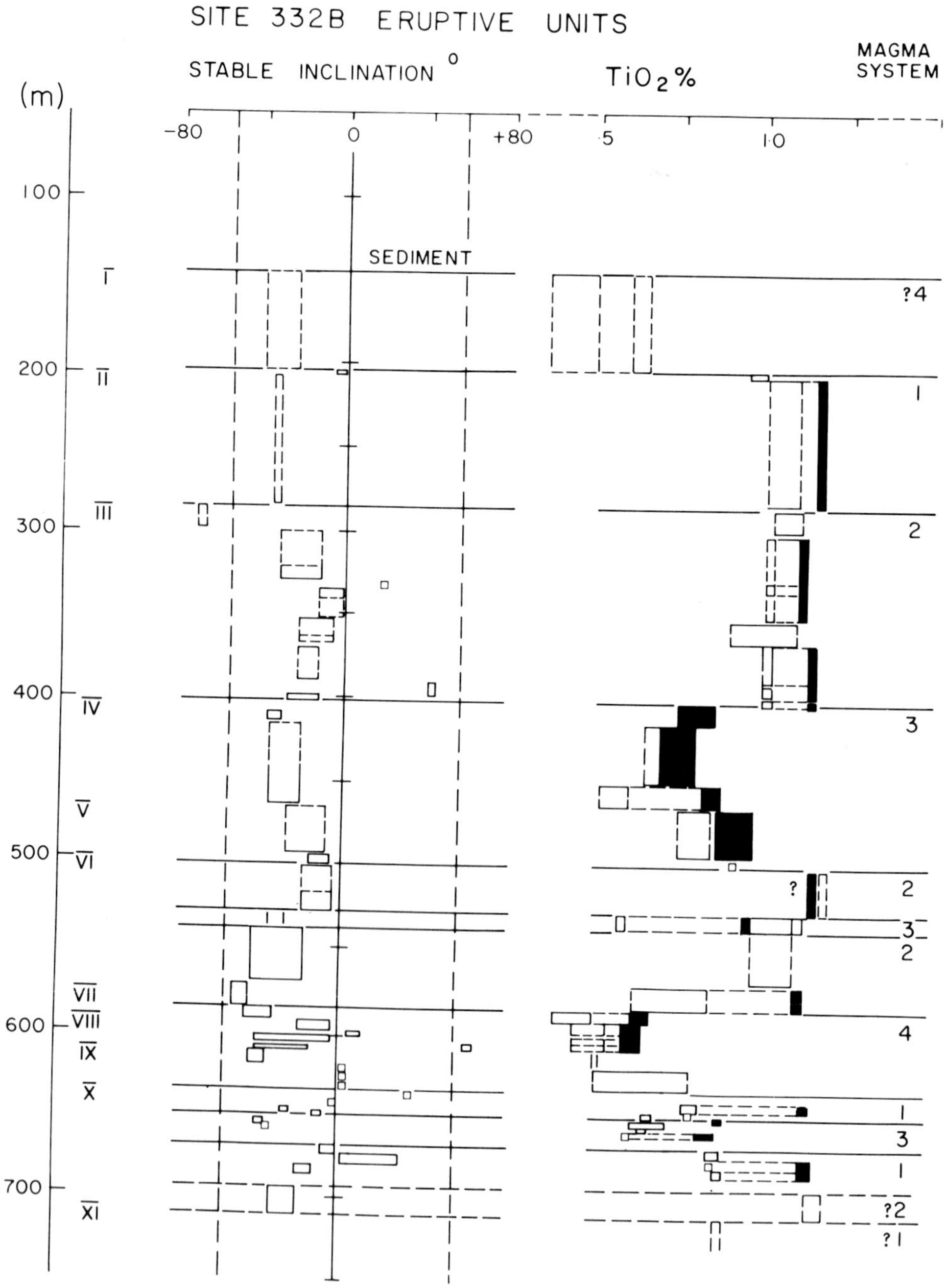

Figure 9b. Schematic downhole sections showing eruptive units inferred from variations in drill core of lithology, NRM stable inclination and chemical affinity of glass and whole-rock compositions; data sources are Hall and Ryall (1977), Flower et al. (1977a) and Byerly and Wright (1978). The range of stable inclination and TiO_2 data are shown for each eruptive unit and have been extrapolated to probable unit stratigraphic limits. Glass TiO_2 compositions are shown by solid areas; the discrepancy between glass and whole-rock compositions (unshaded areas enclosed by vertical dashed lines) is a measure of whole-rock phyric character, as TiO_2 is not incorporated by phenocryst phases. Major lithologic units and magnetic polarity changes are indicated; 'magma systems' include eruptions inferred to be cogenetic in terms of shallow-level fractionation processes from least squares analysis of glass and whole-rock compositions.

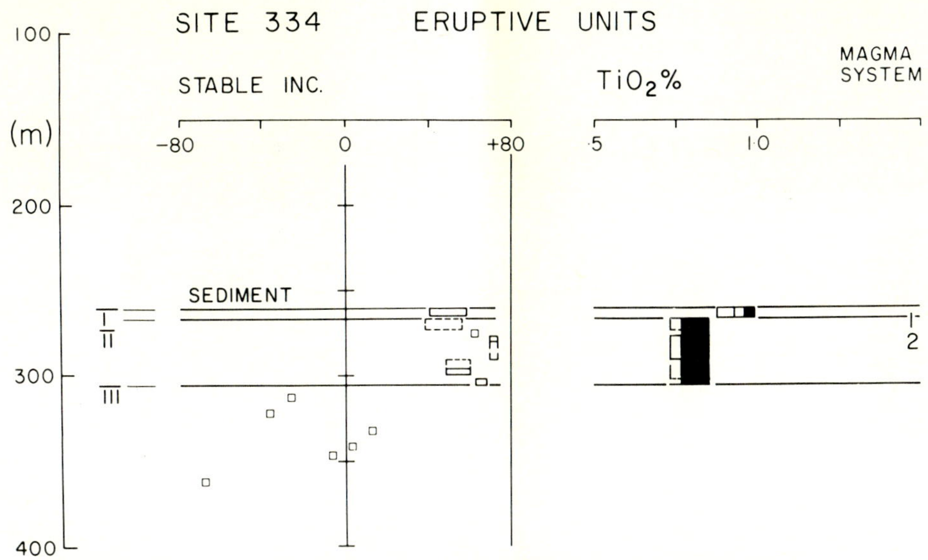

Figure 9c. Schematic downhole sections showing eruptive units inferred from variations in drill core of lithology, NRM stable inclination and chemical affinity of glass and whole-rock compositions; data sources are Hall and Ryall (1977), Flower et al. (1977a) and Byerly and Wright (1978). The range of stable inclination and TiO_2 are shown for each eruptive unit and have been extrapolated to probable unit stratigraphic limits. Glass TiO_2 compositions are shown by solid areas; the discrepancy between glass and whole-rock compositions (unshaded areas enclosed by vertical dashed lines) is a measure of whole-rock phyric character, as TiO_2 is not incorporated by phenocryst phases. Major lithologic units and magnetic polarity changes are indicated; 'magma systems' include eruptions inferred to be cogenetic in terms of shallow-level fractionation processes from least squares analysis of glass and whole-rock compositions.

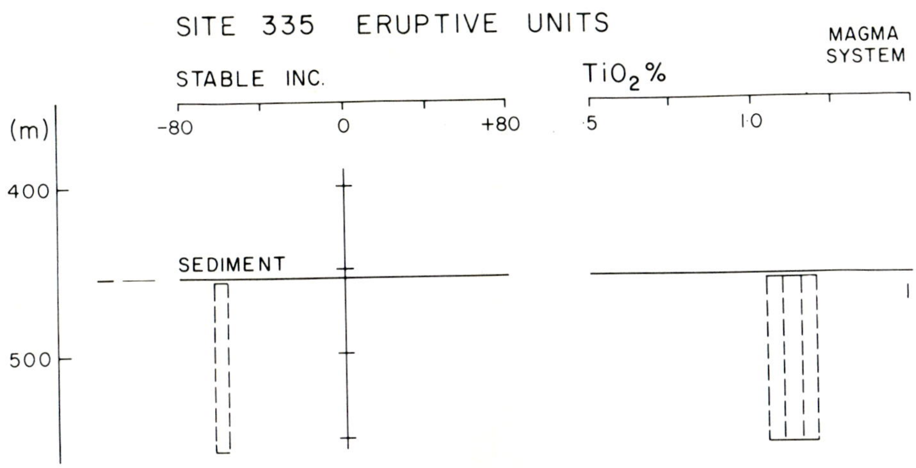

Figure 9d. Schematic downhole sections showing eruptive units inferred from variations in drill core of lithology, NRM stable inclination and chemical affinity of glass and whole-rock compositions; data sources are Hall and Ryall (1977), Flower et al. (1977a) and Byerly and Wright (1978). The range of stable inclination and TiO_2 are shown for each eruptive unit and have been extrapolated to probable unit stratigraphic limits. Glass TiO_2 compositions are shown by solid areas; the discrepancy between glass and whole-rock compositions (unshaded areas enclosed by vertical dashed lines) is a measure of whole-rock phyric character, as TiO_2 is not incorporated by phenocryst phases. Major lithologic units and magnetic polarity changes are indicated; 'magma systems' include eruptions inferred to be cogenetic in terms of shallow-level fractionation processes from least squares analysis of glass and whole-rock compositions.

Table 4

Comparative estimates for magma supply; Kilauea - Mauna Loa (Hawaii), Iceland and the Mid-Atlantic Ridge.

	Mauna Loa and Kilauea	Iceland*	M. A. R. (37°N). *	
to surface (lavas)	9×10^7 (1)	2.8×10^5 (3) 3.3×10^5 (5)	4×10^4 (4)	m^3/yr.
from mantle (lavas and intrusives)	11×10^7 (1) 13×10^7 (2)	4.4×10^5 (4)	2×10^5 (4)	m^3/yr.
eruptive layer thickness:		4000	2000	m
eruptives plus intrusives:	9000	11000	5000	m

* per 2 Km. spreading axis; half-spreading rate 1 cm/yr. (approx. equivalent conduit supply areas);
1. from Swanson (1972), estimated from eruptions since 1952 on a vesicule-free basis;
2. from estimated volume of lava pile, and age (ca. 5 my.), not allowing for vesicles, etc.;
3. from Palmason and Saemundsson (1974);
4. based on spreading rate, thickness of eruptive and intrusive layers and isostatic compensation;
5. from Jakobsson (1972): estimate for postglacial eruptions.

the magmatic regime or constitute an effect of the physical mechanism of spreading and crustal accretion. The stratigraphic record at DSDP Site 332 strongly supports the latter interpretation, and implies that the surface compositional zonation is the result of progressive burial as crustal accretion proceeds. The identification of discrete eruptive sequences in the drill core at Site 332 (Figures 9a and 9b) and the correlation of surface glass compositional groups with the topography of the median valley (Figure 3) suggest that each sequence represents a single volcano or chain of volcanoes at the ocean floor. There is some evidence (Flower et al., 1977b) that individual eruptive sequences or volcanoes show an internal progression with time from early fractionated to later more primitive compositions from a given parental magma batch. Thus, there are reasonable grounds to speculate that for continuous crustal dilation (evidenced by rifting of even the youngest volcanoes; Atwater, 1978) each eruptive sequence may form a lenticular edifice whose tilting and partial burial by succeeding eruptions would result in surface exposure of only the more fractionated lava products. The net effort of this would be to hide from view the more primitive and more tectonically deformed basalts as the accreting eruptive pile migrates laterally, giving rise to surface compositional zonation and allowing exposure only of basalts with atypically low tectonic inclinations. For any given section of oceanic crust we predict 20 to 40 such eruptive sequences contributing to layer 2. The role of an oscillating eruptive axis is hard to evaluate, but, if relevant, it would be a further important factor determining patterns of compositional zonation and magnetic inclinations.

In conclusion, we emphasize that drilling results from Leg 37 and surveys of the FAMOUS median valley form an excellent perspective for interpreting Mid-Atlantic Ridge dynamics. In particular we believe that the deep crustal structure revealed at Site 332 resulted from repeated burial of eruptive sequences in the median rift, and that similar processes are occurring in the modern active zone.

References

Ade-Hall, J. M. and others, Sources of magnetic anomalies on the Mid-Atlantic Ridge, Nature, 255, 389-390, 1975.

Atwater, T., Emplacement models of ocean crust: constraints from FAMOUS area data (abstract), Program Second Maurice Ewing Memorial Symposium, p. 2, 1978.

Aumento, F., W. G. Melson and others, Initial Reports of the Deep Sea Drilling Project 37, U. S. Government Printing Office, Washington D. C., 1977.

Ballard, R. D. and Tj. H. Van Andel, Morphology and tectonics of the inner rift valley at lat. 36° 50'N. on the Mid-Atlantic Ridge, Geol. Soc. Amer. Bull., 88, 507-530, 1977.

Björnsson, A., K. Saemundsson, P. Einarsson, E. Tryggvason, and K. Gronvold, Current rifting episode in northern Iceland, Nature, 266, 318-322, 1978.

Bryan, W. B., Regional variation and petrogenesis of basalt glasses from the FAMOUS area Mid-Atlantic Ridge, J. Petrol., (in press).

Bryan, W. B. and J. G. Moore, Compositional variation of young basalts in the Mid-Atlantic Ridge rift valley near 36° 49'N. Geol. Soc. Amer. Bull., 88, 556-570, 1977.

Bryan, W. B. and G. Thompson, Basalts from DSDP Leg 37 and the FAMOUS area: compositional and petrogenetic comparisons, Can. J. Earth Sci., 14, 875-885, 1977.

Byerly, G. and T. L. Wright, Origin of major-element chemical trends in DSDP Leg 37 basalts, Mid-Atlantic Ridge, J. Volcanol. Geotherm. Res., 3, 229-279, 1978.

Fiske, R. S. and E. D. Jackson, Orientation and growth of Hawaiian volcanic rifts: the effect of regional structure and gravitational stress, Proc. Roy. Soc. Lond., A 329, 299-326, 1972.

Flower, M. F. J., P. T. Robinson, W. Ohnmacht and H. -U. Schmincke, Petrology and geochemistry of igneous rocks: DSDP Leg 37, in: Initial Reports of the Deep Sea Drilling Project, 37, U. S. Government Printing Office, Washington D. C., 658-680, 1977a.

Flower, M. F. J., P. T. Robinson, W. Ohnmacht and H. -U. Schmincke, Magma fractionation systems beneath the Mid-Atlantic Ridge at 36 - 37°N., Contr. Mineral. Petrol., 64, 167-195, 1977b.

Fowler, C. M. R., Crustal structure of the Mid-Atlantic Ridge crest at 37°N. Geophys. J. R. Astr. Soc., 47, 459-492, 1976.

Fowler, C. M. R. and D. H. Matthews, Seismic refraction experiment using ocean bottom seismographs and sonobuoys in the FAMOUS area, Nature, 249, 752-754, 1974.

Gibson, I. L. and J. D. Piper, Structure of the Icelandic basalt plateau and the process of drift, Phil. Trans. Roy. Soc. Lond., A 271, 141-150, 1972.

Hall, J. M., Major problems regarding the magnetization of oceanic crustal layer 2, J. Geophys. Res., 81, 4223-4230, 1976.

Hall, J. M., The magnetic properties and magnetization of oceanic basalts and the implications for the history and structure of Oceanic Layer 2 (abstract), Program Second Maurice Ewing Memorial Symposium, 18-19.

Hall, J. M. and P. J. C. Ryall, Palaeomagnetism of basement rocks Leg 37, in: Initial Reports of the Deep Sea Drilling Project, 37, U. S. Government Printing Office, Washington D. C., 425-556, 1977.

Hekinian, R. and M. Hoffert, Rate of palagonitization and manganese coating on basaltic rocks from the rift valley in the Atlantic Ocean near 36° 50'N., Marine Geol., 19, 91-109, 1975.

Hekinian, R., J. G. Moore, and W. B. Bryan, Volcanic rocks and processes of the Mid-Atlantic Ridge rift valley near 36° 49'N., Contr. Mineral. Petrol., 58, 83-110, 1976.

Jakobsson, S. P., Chemistry and distribution pattern of Recent basaltic rocks in Iceland, Lithos, 5, 365-386, 1972.

Kidd, R. G. W., The nature and shape of the sources of marine magnetic anomalies, Earth Planet. Sci. Lett., 33, 310-320, 1977a.

Kidd, R. G. W., A model for the process of formation of the upper oceanic crust, Geophys. J. R. Astr. Soc., 50, 149-183, 1977b.

Langmuir, C. H., J. F. Bender, A. E. Bence, G. N. Hanson and S. R. Taylor, Petrogenesis of basalts from the FAMOUS area: Mid-Atlantic Ridge, Earth Planet. Sci. Lett., 36, 133-156, 1977.

Macdonald, K. C., Near-bottom magnetic anomalies asymmetric spreading, oblique spreading and tectonics of the Mid-Atlantic Ridge near 37° N., Geol. Soc. Amer. Bull., 88, 541-555, 1977.

Macdonald, K. C. and T. Atwater, Evolution of rifted ocean ridges, Earth Plant. Sci. Lett., 39, 319-327, 1978.

Macdonald, K. C. and B. P. Luyendyk, Deep-tow studies of the structure of the Mid-Atlantic Ridge crust near lat. 37° N., Geol. Soc. Amer. Bull., 88, 621-636, 1977.

O'Nions, R. K. and R. J. Pankhurst, Sr-isotope and rare earth element geochemistry of DSDP Leg 37 basalts, Initial Reports of the Deep Sea Drilling Project, 37, U. S. Government Printing Office, Washington D. C., 599-602, 1977.

Palmason, G. and K. Saemundsson, Iceland in relation to the Mid-Atlantic Ridge, Ann. Review of Earth Planet. Sci., 2, 22-50, 1974.

Ramberg, I. B., D. F. Gray and R. G. Reynolds, Tectonic evolution of the FAMOUS area of the Mid-Atlantic Ridge, lat. 35° 50' to 37° 20', Geol. Soc. Amer. Bull., 88, 609-620, 1977.

Schilling, J. -G., Azores mantle blob: rare earth evidence, Earth Planet. Sci. Lett., 25, 103-115, 1975.

Schilling, J. -G., R. Kingsley and M. Bergeron, Rare earth abundances in DSDP sites 332, 334 and 335 and inferences on the Azores mantle blob activity with time, Initial Reports of the Deep Sea Drilling Project, 37, U. S. Government Printing Office, Washington D. C., 591-597, 1977.

Sleep, N. H., Formation of ocean crust: some thermal constraints, J. Geophys. Res., 80, 4037-4042, 1975.

Storzer, M. D. and M. Selo, Ages par la méthode de traces du fission de basaltes préleves dans la vallée axial de la dorsale médian Atlantique aux environs de 37° N., Acad. Sci. Comptes Rendus, 279D, 1649-1651, 1974.

Swanson, D. A., Magma supply rates of Kilauea volcano 1952-1971, Science, 175, 169-170, 1972.

Walker, G. P. L., Intrusive sheet swarms and the identity of layer 3 in Iceland, J. Geol. Soc. Lond., 131, 143-161, 1975a.

Walker, G. P. L., The structure of Iceland, in: Geodynamics of Iceland and the North Atlantic area, ed. L. Kristjansson; D. Reidel publ., 177-188, 1975b.

Watkins, N. and G. P. L. Walker, Magnetostratigraphy of eastern Iceland, Amer. J. Sci., 277, 513-584, 1977.

White, W. M. and W. B. Bryan, Sr-isotope, K, Rb, Cs, Sr, Ba and rare earth geochemistry of basalts from the FAMOUS area, Geol. Soc. Amer. Bull., 88, 571-576, 1977.

White, W. M., J. -G. Schilling and S. R. Hart, Evidence for the Azores mantle plume from a strontium isotope geochemistry of the central North Atlantic, Nature, 263, 659-663, 1976.

Wright, T. L., and P. L. Doherty, A linear programming and least squares computer method for solving petrologic mixing problems. Geol. Soc. Amer. Bull., 81, 1995-2008, 1970.

Wright, T. L. and R. S. Fiske, Origin of the differentiated and hybrid lavas of Kilauea volcano, Hawaii, J. Petrol., 12, 1-65, 1971.

CONTROLS ON THE MAJOR AND MINOR ELEMENT CHEMISTRY OF MID-OCEAN RIDGE BASALTS AND GLASSES

A. E. Bence, D. M. Baylis, J. F. Bender and T. L. Grove
Department of Earth and Space Sciences
State University of New York
Stony Brook, New York 11794

Abstract. We have modelled the processes of low- and high-pressure crystal fractionation, magma mixing, and variable degree of partial melting to assess their relative importance in generating the major element spectrum observed for ∿ 1300 crystalline mid-ocean ridge basalts and ∿ 1100 basaltic glasses erupted on the ocean floor. Our analysis of these data indicates that the chemistry of crystalline basalts is controlled largely by low-pressure crystal fractionation involving olivine, plagioclase or both. Basaltic glass compositions reflect (a) initial magma compositional differences and (b) crystal-melt-controls during low-pressure crystallization. Clinopyroxene fractionation can account for the reduced CaO/Al_2O_3 ratios observed in the basaltic glasses. However, this occurs at low pressure only after significant (in excess of 30%) crystal fractionation of primitive magma compositions and must involve the prior removal of olivine and plagioclase. High-pressure fractionation of clinopyroxene is consistent with experimental and petrographic observations. Clinopyroxene is the liquidus phase for primitive MORB compositions at pressures exceeding 10 kb under anhydrous conditions. Magma mixing in an open chamber can enrich the incompatible components to levels above that expected in a simple low-pressure fractionation model without drastically lowering $Mg/(Mg + Fe^{2+})$. Multiple cycles of magma mixing accompanied by extensive fractional crystallization are required to generate the observed variation of TiO_2 in the basaltic glasses. Variable degrees of partial melting of a lherzolite source can generate the range of TiO_2 concentrations in basalt glasses having high and restricted $Mg/(Mg + Fe^{2+})$. The entire range of TiO_2 and FeO* concentrations and $Mg/(Mg + Fe^{2+})$ ratios cannot be generated by variable degrees of partial melting of a chemically homogeneous source.

Introduction

Subalkaline (tholeiitic) basalts erupted at diverging plate boundaries are the dominant lithology of the upper part of the oceanic crust (layers 2 and 2A). Alkaline varieties appear to be volumetrically insignificant and are apparently restricted to oceanic islands, seamounts, and topographic highs. It has been recognized that the minor and trace element characteristics of these two basalt types require that they must be derivatives of chemically distinct mantle source regions (see, for example, Engel *et al.*, 1965; Gast, 1965, 1968; Kay *et al.*, 1970). These discussions, based largely on studies of samples dredged from ridges, rises, fracture zones, and seamounts, emphasized the pronounced chemical differences between the two groups and the distinct but generally homogeneous character of the subalkaline mid-ocean ridge basalts (MORB's). More recently, because ocean floor coverage has increased dramatically through recoveries by the Deep Sea Drilling Project (DSDP), the International Program for Ocean Drilling (IPOD), submersible operations (e.g., Project FAMOUS) and detailed dredging surveys along ridges (e.g., the Reykjanes Ridge), significant second order chemical variations in MORB's have been noted. These variations are observed in basalts recovered from different ridge systems, from different sites along the same ridge system (e.g., Schilling, 1973) and even from the same submersible (e.g., Project FAMOUS), dredge (Miyashiro *et al.*, 1969) or drill site (e.g., Site 332A of DSDP Leg 37).

Fresh subalkaline MOR basalts and basaltic glasses exhibit a spectrum of major element compositions which are characterized by the presence of normative hypersthene and by the depletion of large-ion-lithophile (LIL) elements (Th, U, K, Rb, Zr, Hf, REE, Y, NB, Ti, etc.). Basalts with normative olivine dominate (olivine tholeiites), but quartz-normative compositions (quartz tholeiites) occur as well. The subalkaline basalts are subdivided on the basis of their abundances of the LIL elements, particularly K and the light rare earth elements (LREE), into high- and low-K tholeiites (Green, 1971). The low-K basalts, which are by far the most abundant, usually contain less than 0.1-0.3% K_2O and have LREE relative abundances that range from highly depleted $[(La/Sm)_N \sim 0.4]$ to slightly enriched $[(La/Sm)_N \sim 2]$. Heavy rare-earth patterns are flat at about 12 times chondrites on the average but range from about 8-20 x cc. The high-K tholeiites, which appear to be restricted to topographic highs, have moderate light rare-earth enrichment (La/Sm ∿ 4-7) and higher K contents (0.3-1 wt.%). These tholeiites are transitional between low-K tholeiites and alkali basalts (Kay *et al.*, 1970). Hawaiian (Schilling and Winchester, 1969) and Icelandic (Schilling, 1973) tholeiites are of this type. Deep Sea Drilling Project sites 2-10 (Frey *et al.*, 1974), 15-151 (Bence *et al.*, 1975), and 26-250A, 251A (Fleet *et al.*, 1976) recovered high-K tholeiitic basalts.

Three factors play major roles in generating the compositional variations observed in fresh basalts erupted at the mid-ocean ridges. These factors are: (1) source region chemical and mineralogical heterogeneities, (2) postmagma generation fractionation effects at low and high pressures, and (3) variable percentages of partial melting and variations in melting mechanism. All of these factors interact and, because the effects of the three mechanisms can be similar and the observed chemical variations are small, it is difficult, in some cases, to identify the extent to which each has contributed to the spectrum of compositions observed in spatially related suites such as those recovered by Project FAMOUS. Further, processes such as magma (e.g., O'Hara, 1977; Donaldson *et al.*, 1977) and source (e.g., Schilling, 1973) mixing may be superimposed, further obscuring the cause of the variations.

This paper examines the major and minor element chemistries of crystalline basalts and basaltic glasses recovered by DSDP, IPOD, submersible, and dredging operations from all the oceans to assess the relative importance of the factors that contribute to their chemical diversity. To do this, we have amassed, from the literature and from our own published and unpublished work, a library of major element analyses of crystalline basalts (now in excess of 1300 analyses) and basaltic glasses (∿ 1100 analyses). The majority of our basaltic glass data (965 analyses) are from the work of Melson *et al.* (1976). The remainder were obtained at Stony Brook. Both data sets were collected on the glass phase of sparsely phyric rinds from pillows or flow contacts. The glass data can provide information on the initial magma composition and, if petrographic data are available, on the low-pressure phase relationships and on the liquid line of descent of the magma.

Our fractional crystallization and partial melting calculations rely heavily on the experimental work of Roeder and Emslie (1970); Roeder (1974); Bender *et al.* (1978); and Fisk (1978).

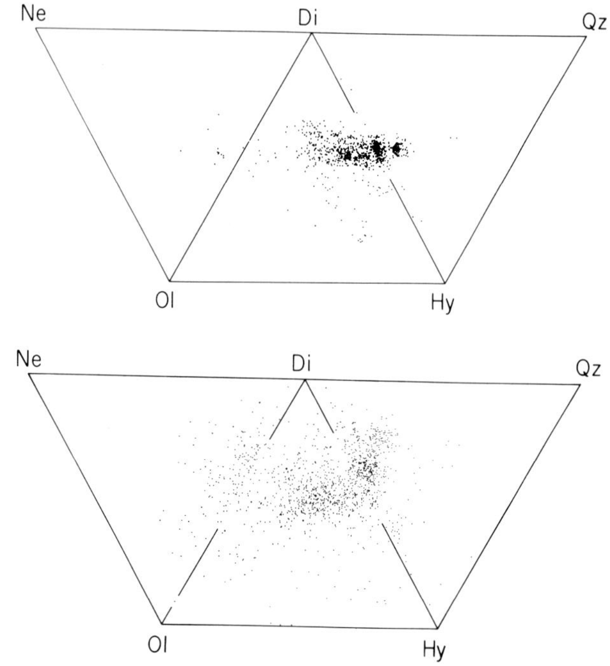

Figure 1a. Normative basalt compositions projected from plagioclase. Top: basaltic glasses (∿ 1100 analyses). Lower crystalline basalts (∿ 1300 analyses).

Compositional Variations in Mid-Ocean Ridge Basalts and Basaltic Glasses

To aid in the identification of processes that control the direction and magnitude of changes in MORB chemistry, we separate crystalline basalts from basaltic glasses in the diagrams presented below. Rocks are separated from glasses because crystal fractionation (removal or accumulation of phases) at the emplacement site is an important factor in the chemical variation of the crystalline basalts and its effects are readily distinguished. The glass compositions, on the other hand, reflect the low-pressure liquid-line of descent and reveal differences in initial magma composition. Petrographic examination (see, for example, Shido *et al.*, 1971), confirmed by experimental melting studies (Bence and Hibberson, 1975; Kushiro, 1973; Bender *et al.*, 1978; Fisk, 1978), reveals that olivine or plagioclase, or both (± spinel), are the important low- pressure liquidus phases in mid-ocean ridge tholeiites. Removal or accumulation of these phases through crystal fractionation can generate marked compositional changes. For example, nickel concentrations and the ratio $Mg/(Mg + Fe^{2+})$ (Mg#) in basalts are extremely sensitive to olivine fractionation whereas Eu and Al_2O_3 abundances are plagioclase sensitive. The fractionation effects of both phases can thus be recognized.

The normative mineralogies of the crystalline basalts and basaltic glasses, projected on the Di-Ol-Hyp plane (Fig. 1a) of the Di-Ol-Qz-Ne tetrahedron, illustrate the extreme compositional diversity of the crystalline basalts relative to the glasses. This variation is due to the combined effects of crystal fractionation and chemical weathering on the crystalline basalts. Glass chemistry is more tightly restricted, and chemical variations reflect the liquid lines of descent. For example, a plot of normative olivine-plagioclase-pyroxene reveals a linear trend that defines the low-pressure olivine-plagioclase cotectic and olivine-plagioclase-clinopyroxene multiple saturation (Fig. 1b), and indicates that multiphase fractionation at low pressures is an important process that controls basalts glass chemistry.

To aid in the identification of crystal-melt controls on mid-ocean ridge basalt chemistry, a projection scheme that is relevant to this suite of samples is desirable. The phases pertinent to the crystallization of oceanic basalts are olivine, plagioclase, high calcium clinopyroxene, spinel and silica, and a projection utilizing these five phases has been calculated. Such a projection can be useful in several ways: (1) it may reveal the phase responsible for chemical variation between a suite of samples; (2) it may be used to obtain the proportions of phases removed during a fractionation or melting sequence (all subprojections must be considered, of course); and (3) it may reveal the parental magma in a suite of samples and distinguish multiple parent magma types.

The recalculation uses molar units to calculate the proportions of olivine $(Mg,Fe)_2SiO_4$, plagioclase $Ca_{1-x}Na_xAl_{2-x}Si_{2+x}O_8$, high-Ca clinopyroxene $Ca(Mg,Fe)Si_2O_6$, silica SiO_2 and spinel $(Mg,Fe)_2TiO_4$, $(Mg,Fe)Cr_2O_4$. The scheme is useful for samples with low Fe_2O_3 and P_2O_5 and suppresses Fe/(Fe + Mg) variation, which can be considered by utilizing other projections.

Since TiO_2 and Cr_2O_3 contents are low, the recalculated spinel component is discarded and the analyses are projected into a pseudoquaternary system; olivine (Ol)-plagioclase (Plag)-clinopyroxene (Cpx)-silica (Qz). Oceanic basalts contain substantial proportions of all four components, so no simplified projection that resembles a ternary liquidus diagram can be used. Inspection of the ternary subprojections

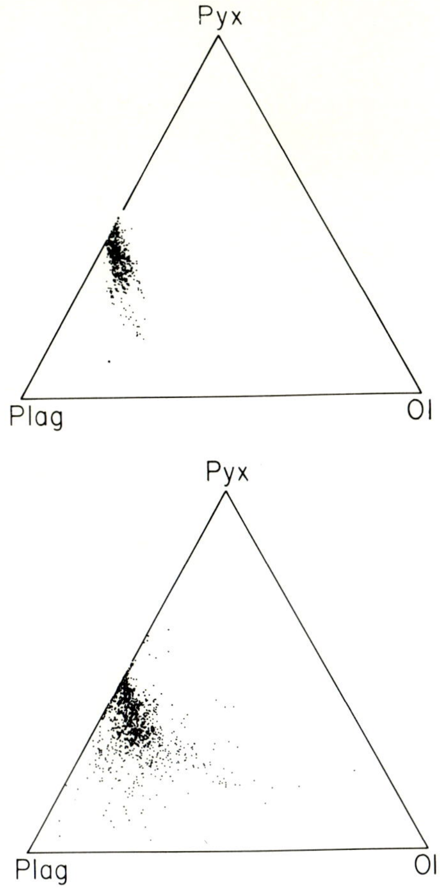

Figure 1b. MORB chemistry. Normative proportions of pyroxene, plagioclase, and olivine. Top: basaltic glasses. Bottom: crystalline basalts. Glasses plot in the vicinity of the inferred plagioclase-olivine cotectic.

in the pseudoquarternary reveals trends of significant silica enrichment and olivine-plagioclase control in molar proportions 1:2 for the basaltic glasses (Fig. 2a). The Ol-Plag-Cpx subprojection (Fig. 2b) shows a clustering of points that represent the projection, from silica, of the low-pressure olivine-plagioclase cotectic and olivine-plagioclase-clinopyroxene multiple saturation. Preliminary information from experimental studies in basaltic systems indicates that residual liquids in equilibrium with olivine, plagioclase, and clinopyroxene have compositions with a range of silica contents and Fe/(Fe + Mg) ratios but with only minor variations in the relative proportions of olivine, plagioclase, and clinopyroxene. The glass compositions plotted in Fig. 2 may define the projection of a cigar-shaped volume containing liquids that are multiply saturated with the three phases.

The position of the olivine-plagioclase cotectic defined by the glasses and the experiments is in the approximate molar proportions of one part olivine to two parts plagioclase. These proportions are used in the low-pressure modelling calculations. Phenocrysts coexisting with the natural glass compositions confirm the location of the experimentally determined cotectic (Fig. 2a). Glasses that contain olivine crystals lie in the olivine primary phase volume, glasses with

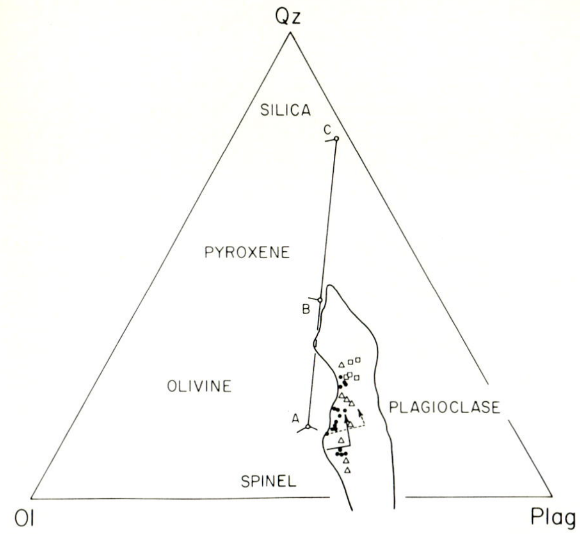

Figure 2a. Subprojection olivine-quartz-plagioclase in olivine-quartz-clinopyroxene-plagioclase pseudoquaternary. See text for details of the projection. Points A (olivine-plagioclase-spinel), B (olivine-plagioclase-pyroxene), and C (plagioclase-pyroxene-silica) from Walker *et al.* (1973) for lunar compositions with $Mg/(Mg + Fe^{2+}) = 0.7$. Field of basaltic glasses is outlined. Solid circles: glasses coexisting with olivine (± spinel) + plagioclase + clinopyroxene. Solid line is the equilibrium liquid line of descent for Project FAMOUS basalt 527-1-1 from Bender *et al.* (1978). Dashed line is the calculated liquid line of descent for 30% crystal fractionation.

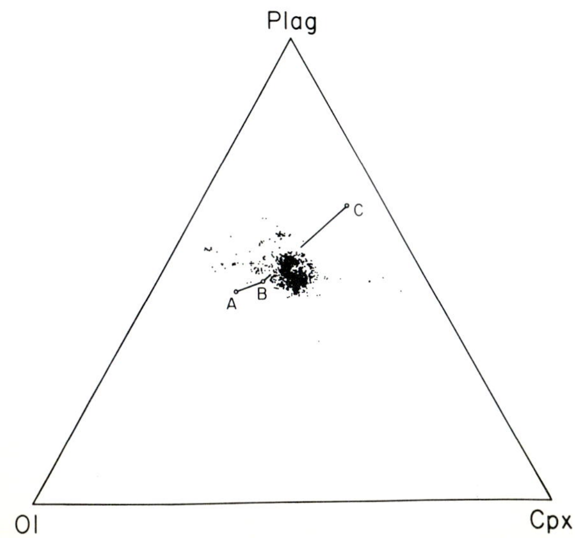

Figure 2b. Subprojection olivine-clinopyroxene-plagioclase. The composition of 1100 basalt glasses are plotted along with points A, B, and C. The dense cluster of points represents liquids saturated with olivine-plagioclase (± spinel) or olivine-plagioclase-clinopyroxene (± spinel).

olivine and plagioclase lie closest to the experimentally determined cotectic and glasses containing olivine, plagioclase and clinopyroxene lie at the upper extension of the cotectic.

Oxide-oxide and oxide-Mg/(Mg + Fe^{2+}) variation diagrams (Figs. 3-7) which serve to document the chemical diversity of the mid-ocean ridge basalts are useful in identifying the most primitive (i.e., least fractionated) magmas and the effects of low- and high-pressure fractionation and partial melting. Used in conjunction with the ternary subprojections described above, they may place constraints on MORB petrogenesis.

Primitive Magmas

The recognition of primitive (i.e., direct partial melts of the upper mantle) magmas is of critical importance to MORB petrogenesis because it is these liquids that can provide chemical and mineralogical information on their sources. Two chemical parameters, Mg/(Mg + Fe^{2+}) and Ni, are indicators of primitiveness. Mineral/melt distribution coefficients for FeO and MgO (Roeder and Emslie, 1970; Roeder, 1974; Mysen, 1975; Bender *el al.*, 1978, Fisk, 1978) and Ni (Hakli and Wright, 1967; Henderson and Dale, 1969; Leeman, 1973; Bougault and Hekinian, 1974; Sato, 1977) require that magmas generated by 20-30% partial melting of a peridotitic mantle source dominated by Fo_{90} olivine have Mg/(Mg + Fe^{2+}) ratios of $\sim$ 0.74 (see, for example, Green *et al.*, 1974; Hanson and Langmuir, 1978; Bender *et al.*, 1978) and Ni contents of $\sim$ 250-350 ppm. Only rarely are such high Ni contents observed in a noncumulate basalt (see, for example, Frey *et al.*, 1974), and noncumulate basalts do not have such high Mg/ (Mg + Fe^{2+}) ratios. The highest observed ratio is $\sim$ 0.75 which was obtained for a melt inclusion in spinel (Donaldson *et al.* 1977). Since the Mg/(Mg + Fe^{2+}) ratio is extremely sensitive to olivine fractionation but relatively insensitive to the percentage of partial melting so long as the liquid is multiply saturated with ferromagnesian phases at the source (Hanson and Languir, 1978; Bender *et al.*, 1978), it might be concluded that even the most primitive basalts as defined by the above criteria have experienced minor olivine fractionation. However, the effects of olivine fractionation can be recognized and compensated for. Liquids that are derived from source regions where they are multiply saturated will have Mg/(Mg + Fe^{2+}) $\sim$ 0.72-0.75 (atomic). From Figures 3-8 we see that liquids having these Mg values will also have the following chemistry: $TiO_2 \simeq$ 0.5-0.7 wt.%, $Al_2O_3 \simeq$ 16-17.5 wt.%, FeO* $\simeq$ 8 wt.%, and CaO $\simeq$ 13 wt.%. Parent magmas of the most primitive mid-ocean ridge basalts will have compositions in these ranges.

Crystal Fractionation at Low Pressures

Experimental and petrographic studies of mid-ocean ridge basalts confirm that the majority have olivine (± spinel) or plagioclase as liquidus phases at low pressures. Melting studies on Project FAMOUS basalt glass 527-1-1 (Bender *et al.*, 1978) indicate that the crystallization sequence for this relatively primitive composition is olivine → olivine + plagioclase → olivine + plagioclase + clinopyroxene in the pressure interval 0 $\sim$ 10 kb. Clinopyroxene crystallizes late in this composition at low P; consequently, it is unlikely to be a significant fractionating phase at low pressures.

Textural arguments indicate that Cr-spinel was the liquidus or co-liquidus (with olivine) phase in the natural sample (527-1-1) and, consequently, would be potentially an important fractionating phase. Spinel did not crystallize in the Bender *et al.* experiments, because their molybdenum capsules imposed an $\underline{f}O_2$ on the basalt that was lower than that under eruptive conditions (quartz-fayalite-magnetite buffer). Thus, the effects of spinel fractionation are not available from the experimental data on 527-1-1. If spinel fractionation is important in mid-ocean ridge basalt petrogenesis it would have a marked effect on the Cr content of the basalts. Chrome abundances in MORB's are rarely reported in the literature and in the infrequent cases that they are, the data are insufficiently precise. In subsequent modelling of basalt evolution through crystal fractionation, we have ignored the effects of spinel.

The effects of olivine and plagioclase fractionation at low pressures on the chemistry of MORB's may be recognized on Figures 2-7. In particular, the variations of Al_2O_3 with MgO and Mg/(Mg + Fe^{2+}) (Figs. 3a, b, and 4a, b) are particularly sensitive to olivine and plagioclase fractionation.

The compositional spectrum of the crystalline basalts (Figs. 3b and 4b) has a maximum concentration in the range; MgO = 7-9 wt.%, Al_2O_3 = 14-17 wt.%. Extensions of the fields towards higher concentrations of Al_2O_3 and MgO reflect the accumulation of plagioclase and olivine, respectively.

The scatter of the crystalline basalt compositions contrasts markedly with the highly restricted glass field (Figs. 3a and 4a). This restriction is a consequence of crystal-liquid controls on the composition of the *glass*. Petrographic observations on the glass samples analyzed at Stony Brook and the results of low-pressure experimental studies suggest that the region of the glass field where MgO $\gtrsim$ 7 wt.% and $Al_2O_3 \gtrsim$ 14.5 wt.% (Fig. 3a) defines the approximate position of the olivine-plagioclase cotectic. Many of the glasses with lower MgO and Al_2O_3 appear to be multiply saturated with olivine, plagioclase and clinopyroxene. Few of the crystalline basalts fall in this field.

The calculated low-pressure fractionation liquid line of descent for Project FAMOUS basalt 527-1-1 follows the inferred olivine-plagioclase cotectic (Fig. 2a). This calculation uses the equilibrium mineral/melt partition coefficients measured by Bender *et al.* (1978) and assumes ideal fractional crystallizations (i.e., crystals are removed from reactions with liquid). The process is carried out in two weight percent increments. Olivine is the liquidus phase for this composition; however, the calculation assumes that plagioclase commenced to crystallize when normative proportions of olivine:plagioclase reached 1:2 (the molar proportions defined by the position of the olivine-plagioclase cotectic for these compositions) (Fig. 2a). Olivine and plagioclase were subsequently extracted in these proportions. The calculated liquid compositions after 6% and 30% crystallization are reported in Table 1, Columns 2 and 3. The equilibrium liquid line of descent for basalt 527-1-1 (Bender *et al.*, 1978) is also shown on Figure 2a.

The width of the glass field (Fig. 2a) may reflect either a pressure effect on the olivine-plagioclase cotectic or different parental liquids. The extent of silica enrichment cannot be a consequence of the fractionation of a common parent magma but must reflect multiple parent liquids.

After approximately 50 wt.% crystallization, the calculated residual liquid composition reaches the range where clinopyroxene is observed coexisting with olivine and plagioclase in the natural samples (Fig. 2) and in the equilibrium experiments conducted on FAMOUS basalt 527-1-1 (Bender *et al.*, 1978). The majority of crystalline basalt compositions (ex-

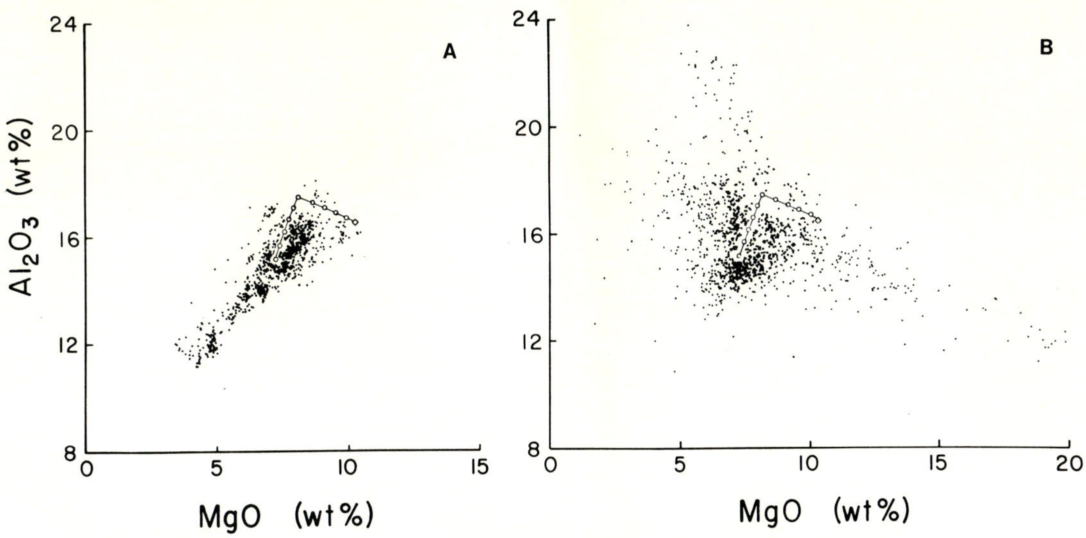

Figure 3. Al_2O_3 *vs* MgO variation diagram for basaltic glasses (a) and crystalline basalts (b). Calculated crystal fractionation liquid line of descent from FAMOUS basalt 527-1-1 composition for 30% crystallization (see text).

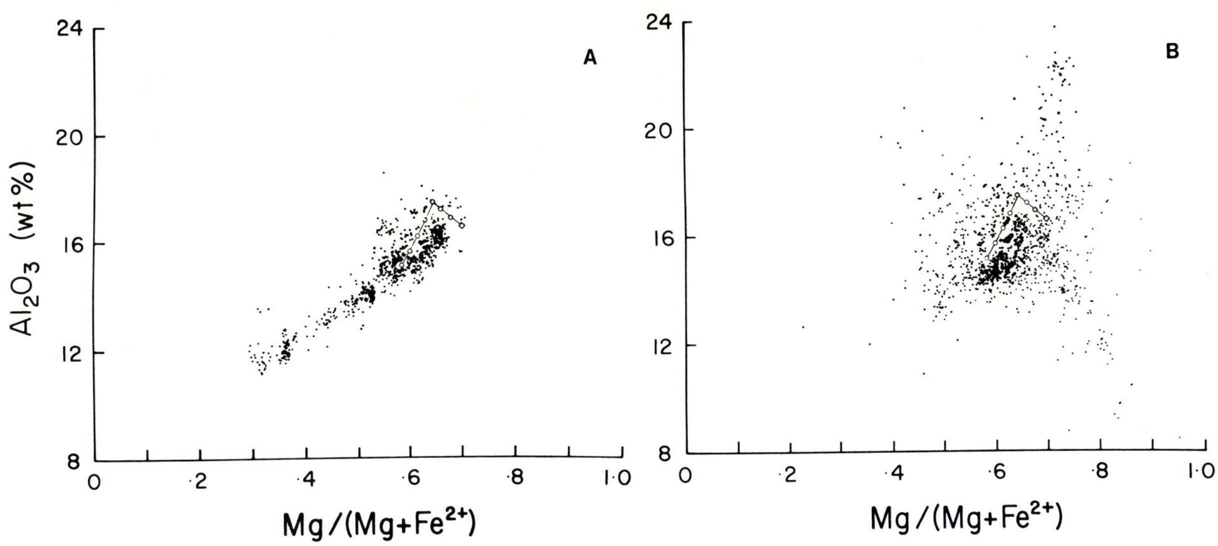

Figure 4. Al_2O_3 *vs* Mg/(Mg + Fe^{2+}). (a) Basaltic glasses. (b) Crystalline basalts. Fractionation path as for Fig. 3.

cepting those resulting from plagioclase or olivine accumulation) fall in the vicinity of the path defined by this trend (Figs. 3b, 4b, 5b, 6b, 7b). Glass compositions show much greater extremes but with more restricted ranges reflecting the degree to which they have crystallized (Figs. 3-7).

Several important observations can be made from the variation diagrams:

First, the experimental results confirm that the majority of the basaltic glass compositions reflect crystal-liquid control. Many of those with MgO and Al_2O_3 in the ranges 7-9 wt.% and 14-17.5 wt.%, respectively, fall in the vicinity of the low-pressure olivine-plagioclase cotectic. The most primitive of these glasses (Mg/Mg + Fe^{2+}) $\gtrsim$ 0.65) are characterized by Al_2O_3 $\gtrsim$ 16 wt.%, MgO $\gtrsim$ 8 wt.%, FeO* $\lesssim$ 10 wt.%, TiO_2 $\lesssim$ 1.0 wt.%. Glasses lying along the olivine-plagioclase cotectic (Fig. 2) or those more evolved compositions that appear to be multiply saturated with olivine + plagioclase + clinopyroxene have adjusted to low-pressure conditions and cannot be primary liquids.

Second, the evidence for cotectic control on the compositions of the crystalline basalts is largely obscured by the effects of plagioclase and olivine accumulation.

Third, the variation of FeO* with MgO in the glasses (Fig. 5a) is consistent with a low-pressure olivine + plagioclase fractionation path. The crystalline basalts (Fig. 5b) have three trends consistent with olivine accumulation (high MgO), olivine (+ plagioclase) fractionation (high Fe*), and plagioclase accumulation (low FeO* and MgO).

Fourth, the TiO_2-MgO and TiO_2-Mg/(Mg + Fe^{2+}) variations in the basaltic glasses (Figs. 6a, 7a) are consistent with

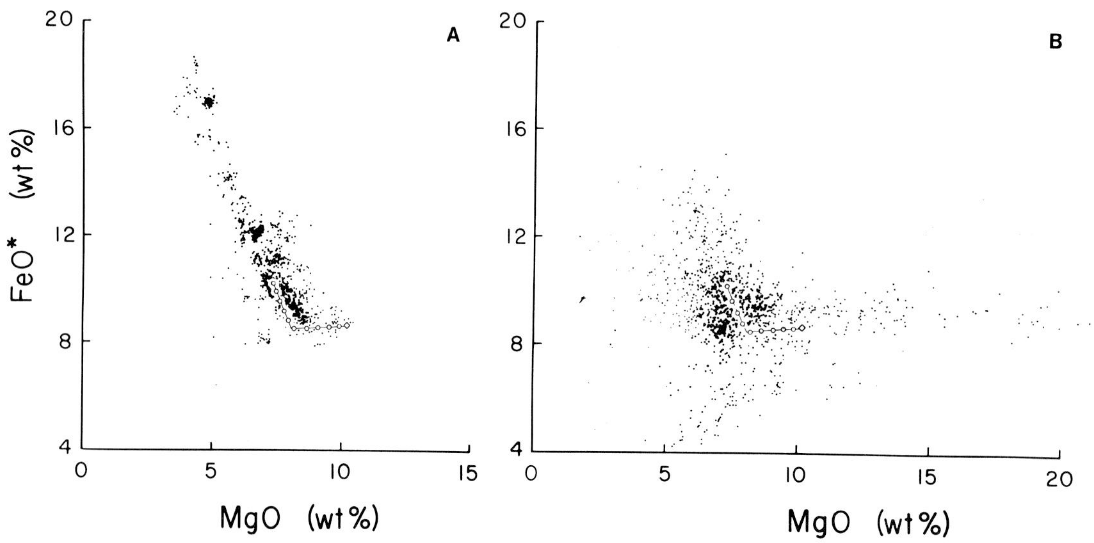

Figure 5. FeO* *vs* MgO. (a) Basaltic glasses. (b) Crystalline basalts. Fractionation path as for Fig. 3.

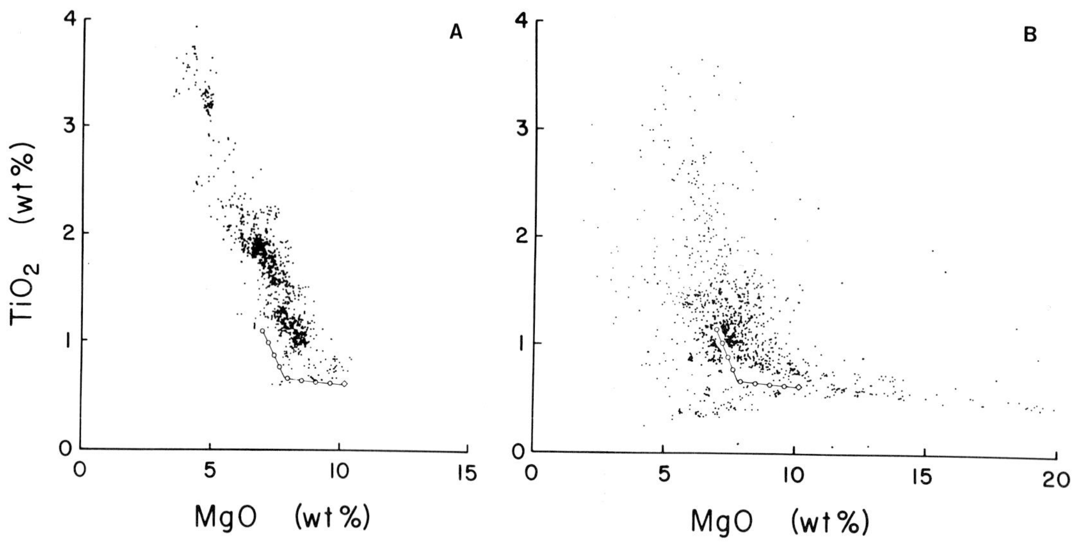

Figure 6. TiO_2 *vs* MgO. (a) basaltic glasses. (b) crystalline basalts. Fractionation path as for Fig. 3.

cotectic (olivine-plagioclase) and peritectic (olivine-plagioclase-clinopyroxene) controls. Variations of TiO_2 in the crystalline basalts (Figs. 6b, 7b) also reflect these controls (as well as the effects of olivine and plagioclase accumulation). The nearly six-fold change of TiO_2 in the noncumulate crystalline basalts cannot be due entirely to evolution of a single parent magma through simple crystal fractionation. Multiple liquids having varying TiO_2 contents appear necessary. These liquids may be derivatives of variable sources, the same source through variable degrees of partial melting, or they may arise through mixing of magmas in an open magma chamber (O'Hara, 1977; Bryan *et al.*, 1978).

We may conclude from these observations that low-pressure fractionation involving plagioclase and olivine is the most important control on MORB major- and minor-element chemistry. However, a single low FeO*-low TiO_2 magma cannot give rise to the entire spectrum of noncumulate crystalline basalt compositions through simple low-pressure crystal fractionation.

Fractionation at High Pressures

Least-squares approaches to modelling compositional variations in spatially related suites of mid-ocean ridge basalts seem to require significant amounts of clinopyroxene fractionation (see, for example, Bryan and Moore, 1977; Bryan and Thompson, 1977; Bryan *et al.*, 1978). However, experimental studies of mid-ocean ridge basalt compositions (e.g., Kushiro, 1973; Bender *et al.*, 1978; Fisk, 1978) indicate that under low-pressure anhydrous conditions, clinopyroxene

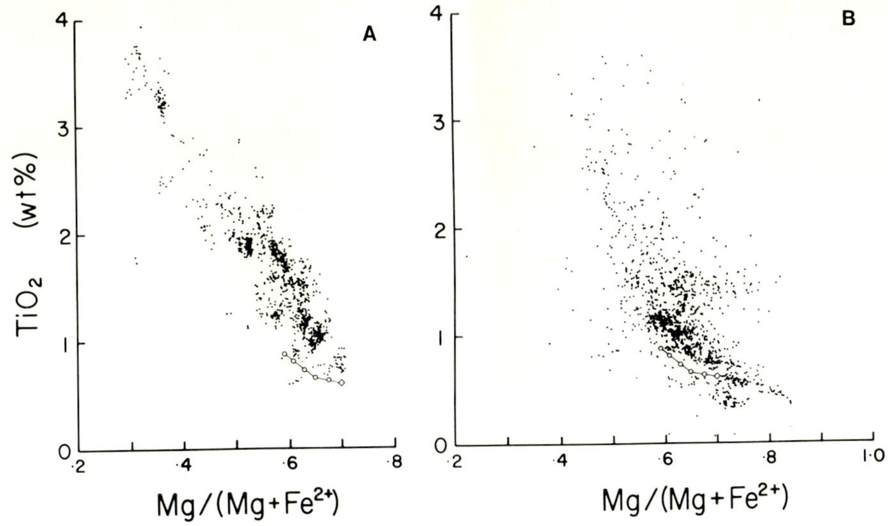

Figure 7. TiO_2 *vs* $Mg/(Mg + Fe^{2+})$. (a) basaltic glasses. (b) crystalline basalts. Fractionation path as for Fig. 3.

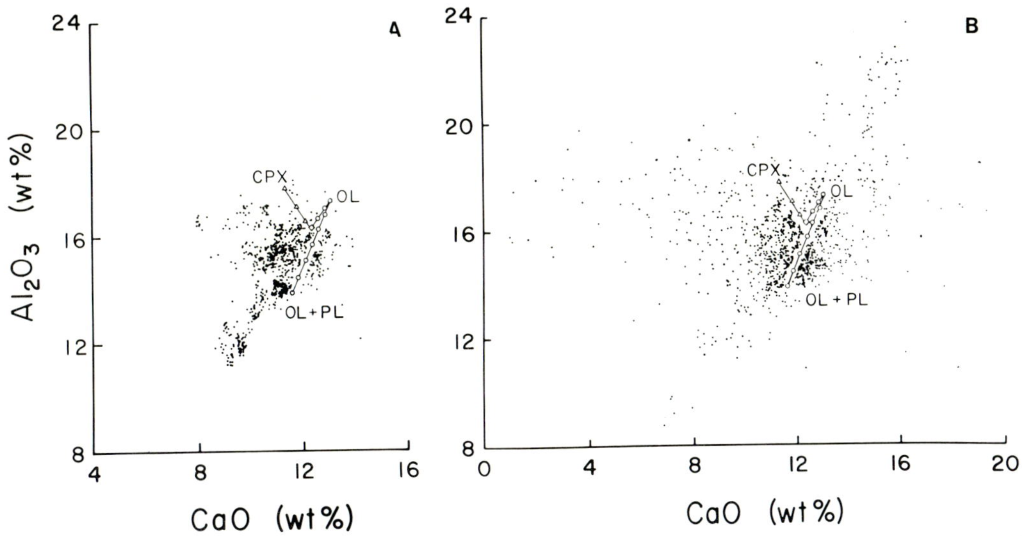

Figure 8. Al_2O_3 *vs* CaO. (a) basaltic glasses. (b) crystalline basalts. Low-pressure (olivine → olivine + plagioclase) fractionation path and high-pressure (clinopyroxene) fractionation path are shown.

follows olivine and plagioclase in the crystallization sequence. Furthermore, clinopyroxene appearance in the equilibrium experiments occurs only after considerable crystallization and consequent chemical evolution of the melt. A somewhat earlier appearance of clinopyroxene might be expected under conditions of ideal fractional crystallization, but again, extensive crystallization (and chemical evolution of the melt) prior to clinopyroxene first appearance is to be expected. It was on these grounds that Bender *et al.* (1978) concluded that low-pressure clinopyroxene fractionation was unlikely to be an important mechanism for MORB petrogenesis.

Under high-pressure ($\gtrsim$ 10 kb) anhydrous conditions, however, calcic clinopyroxene is the liquidus phase for primitive MORB compositions (see, for example, Kushiro, 1973; Bence and Hibberson, 1975; Bender *et al.*, 1978). A small amount of water ($\leq$ 0.5 wt.%) reduces the point of multiple saturation to pressures as low as $\sim$ 3 kb (Bence and Hibberson, 1975). From these results it may be concluded that clinopyroxene fractionation under moderate to high pressures is a viable mechanism for the evolution of MORB's. Petrographic observations of MORB's in the last few years have revealed small percentages of clinopyroxene phenocrysts/xenocrysts whose compositions are readily distinguishable from groundmass clinopyroxenes (Schweitzer *et al.*, 1978).

We are, therefore, concerned with the extent to which the major and minor elemental spectrum of MORB's may be a consequence of clinopyroxene fractionation under moderate pressure. However, the combined effects of olivine and plagioclase fractionation on the major elements, already recognized as an important process in MORB petrogenesis, obscure the effects of clinopyroxene fractionation. These effects can be separated by examining the variation of CaO and Al_2O_3.

Figure 8 illustrates the effects of low-pressure (olivine and plagioclase) and high-pressure (clinopyroxene) fractionation on MORB chemistry. Clinopyroxene fractionation was modelled using the high-pressure mineral/melt distribution coefficients determined by Bender *et al.* (1978). These coefficients vary markedly with composition (and temperature) and are constrained by clinopyroxene stoichiometry (Hanson and Langmuir, 1978); consequently, they must be continuously changed in a fractionation calculation.

As on the previous variation diagrams, much of the variation in basaltic glass chemistry can be attributed to cotectic control. The crystalline basalt field is dominated by the effects of both plagioclase and olivine accumulation. However, a portion of the spectrum observed on both diagrams is consistent with a small amount of clinopyroxene fractionation.

The two fractionation paths are also readily distinguished on a CaO/Al_2O_3 (weight) vs $Mg/(Mg + Fe^{2+})$ (atomic) diagram for the basaltic glasses (Fig. 9). Clinopyroxene fractionation results in lower CaO/Al_2O_3 and $Mg/(Mg + Fe^{2+})$ ratios while CaO/Al_2O_3 increase with olivine + plagioclase fractionation. Plagioclase accumulation accompanied by olivine fractionation could account for a trend having the appearance of clinopyroxene fractionation in the crystalline basalts; however, it cannot explain the ratios of $CaO/Al_2O_3 \lesssim 0.75$ in the glasses.

Consequently, it might be concluded that a significant percentage of the more evolved basaltic glasses have undergone clinopyroxene fractionation. Selected trace element data on the glasses (e.g., Sc, V, Cr, La) are needed to substantiate this conclusion.

Magma Mixing

Some of the compositional variations observed in mid-ocean ridge basalts may be a consequence of magma mixing, accompanied by crystal fractionation, in shallow magma chambers. Bryan and Moore (1977) and Bryan *et al.* (1978) suggest that fractionation in an open-system magma chamber (O'Hara, 1977) into which successive batches of magma are injected can generate the compositional relationship in basalts from the median valley of the mid-Atlantic Ridge at the Project FAMOUS study area. They suggest that this process can cause enrichment of the incompatible components above the levels that can be explained by simple crystal fractionation. Rhodes *et al.* (1978), utilizing mineralogical, geochemical, and experimental data for DSDP Leg 45 and 46 basalts, conclude that magma mixing occurred in the evolution of these basalts and that it is likely to be a fundamental process in MORB petrogenesis.

We have modelled magma mixing and crystal fractionation for the major elements varying the extent of fractionation, the compositions of the successive magma pulses, and the mixing proportions (assuming complete mixing). These calculations indicate that some of the observed enrichment of TiO_2 and FeO* in the Project FAMOUS basaltic glasses *could be* a consequence of magma mixing in an open-system magma chamber. However, such a process is not *required* to explain the major element data, and the REE data (Langmuir *et al.*, 1977) indicate that other processes must be operative to generate the observed LIL element distributions in Project FAMOUS basalts.

Calculated magma compositions for a two-stage mixing model involving low-pressure crystal fractionation prior to and following mixing are presented in Table 1. FAMOUS basalt 527-1-1 is used as the starting magma composition for these calculations. Following 30% crystal fractionation (olivine, then olivine + plagioclase) the residual liquid is mixed with a second batch of 527-1-1 composition and subsequently fractionated again (30% crystallization). The fractionation and mixing paths for TiO_2-Mg# in this process are shown on Figures 10a and b. The directions of TiO_2 and FeO* enrichments on these diagrams are consistent with but not restricted to open-magma chamber mixing and crystal fractionation. By varying the percentages of crystal fractionation, the proportions of the residual and primary magmas mixed, and the number of new magma pulses injected into the chamber, elements not incorporated in olivine or plagioclase are increasingly concentrated in the melt. At the same time, the $Mg/(Mg + Fe^{2+})$ ratio of the melt will not be lowered as rapidly as in a simple fractionation model. Although such a process is capable of generating part of the compositional

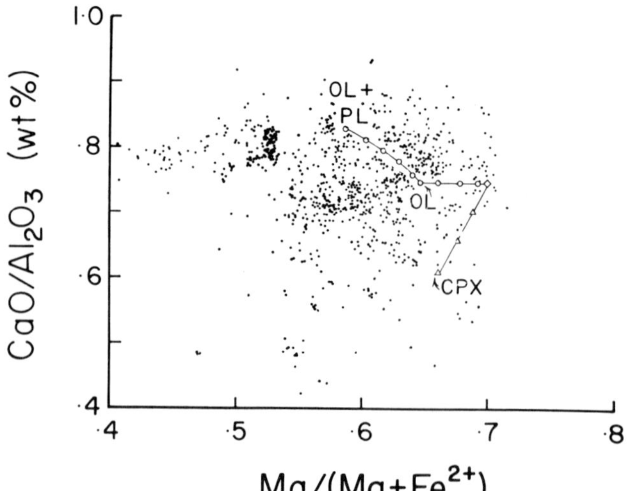

Figure 9. CaO/Al_2O_3 *vs* $Mg/(Mg + Fe^{2+})$ for basaltic glasses showing low-pressure and high-pressure (clinopyroxene) fractionation paths (see text).

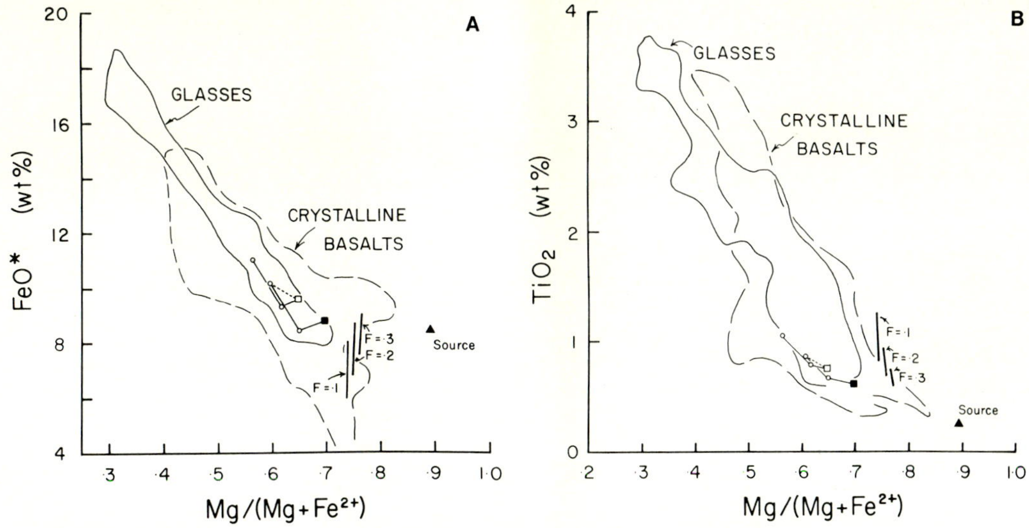

Figure 10. Plots of FeO* *vs* $Mg/(Mg + Fe^{2+})$ (a) and TiO_2 *vs* $Mg/(Mg + Fe^{2+})$ (b) showing fields for the basaltic glasses and the crystalline basalts. Magma mixing and fractionation path shown with 527-1-1 starting composition (solid square), low-pressure fractionation path (30% crystallization), magma mixing (dashed line) to new composition (open square) and second low-pressure fractionation path (30% crystallization) (see text). Melt compositions for 10, 20, and 30% partial melting of an assumed mantle source (solid triangle). Melt fields reflect variable residual mineralogy.

spectrum of MOR basalts and basaltic glasses, confirmation that it is an important process in the petrogenesis of a suite of related basalts would require assessment of the mineralogical and trace element evidence.

Partial Melting Controls on Magma Composition

The percentages of partial melting, the mechanism of partial melting, and the residual mineralogy have major controls on the trace and major element abundances in basaltic melts. Hanson and Langmuir (1978) apply trace element modelling to major elements and Bender *et al.* (1978) use this approach to show how the fraction of partial melting and the residual mineralogy can generate part of the compositional spectrum observed in the Project FAMOUS basaltic glasses. They conclude that the spectrum of TiO_2 abundances observed for glasses with restricted ratios of $Mg/(Mg + Fe^{2+})$ could be a consequence of variable degrees of partial melting of a lherzolite source containing 0.25 wt.% TiO_2.

This treatment assumes that the major elements may be modelled in a manner similar to trace elements but with the additional rigid constraint of mineral stoichiometry. The Hanson-Langmuir treatment of FeO and MgO suggests that their approach is valid for these components; however, it is much less likely to be the case for Al_2O_3, CaO, Na_2O, TiO_2, etc.

We have assumed that the approach is valid for TiO_2, which behaves largely as an incompatible element for large percentages of partial melting, and have calculated liquid compositions from the partial melting equation (Schilling, 1966; Shaw, 1970). Melt compositions for different degrees of partial melting of an assumed mantle source (Bender *et al.*, 1978) with the residual mineralogy varying between $Ol_{55\text{-}80}$: $Opx_{20\text{-}35}$:$Cpx_{0\text{-}20}$ have been calculated. Only those calculations which meet the stoichiometric restrictions of the residual minerals are assumed to be valid. The effect of variable percentages (10-30%) of partial melting on FeO*, TiO_2, and $Mg/(Mg + Fe^{2+})$ are summarized in Figures 10a and b. These results show that the variation of FeO* and TiO_2 in MORB's with relatively restricted $Mg/(Mg + Fe^{2+})$ ratios could be a consequence of variable degrees of partial melting followed by some olivine fractionation from the melt. Magmas with evolved $Mg/(Mg + Fe^{2+})$ ratios ($\lesssim 0.65$) and high TiO_2 cannot be generated solely through partial melting of a source region which contains olivine, orthopyroxene, and clinopyroxene as residual phases.

Conclusions

The following conclusions arise from this analysis of crystalline mid-ocean ridge basalt and basalt glass compositions and our modelling of petrogenetic processes.

1. The chemistry of the crystalline basalts is controlled largely by low-pressure crystal fractionation involving olivine and plagioclase.
2. Basalt glass compositions reflect (a) initial magma compositional differences and (b) crystal-melt-controls during low-pressure crystallization.
3. Clinopyroxene fractionation can account for the reduced CaO/Al_2O_3 ratios observed in the basaltic glasses. Such a process can occur at low pressures only after significant (in excess of 30%) crystal fractionation of a primitive magma composition and involves the prior removal of olivine and plagioclase. High-pressure fractionation of clinopyroxene is consistent with the experimental and petrographic observations. Clinopyroxene is the liquidus phase for primitive MORB compositions at pressures exceeding 10 kb under anhydrous conditions.
4. Magma mixing in an open chamber can enrich the incompatible components to levels above those expected in a simple low-pressure fractionation model without drastically lowering $Mg/(Mg + Fe^{2+})$.

Table 1. Calculated Melt Compositions for Low-Pressure Fractionation and Magma Mixing in an Open Magma Chamber.

	1	2	3	4	5	6
SiO_2	49.1	49.7	50.6	49.9	50.1	51.1
Al_2O_3	16.5	17.5	14.4	15.4	15.8	12.8
TiO_2	0.62	0.66	0.87	0.74	0.76	1.05
FeO*	8.78	8.57	10.5	9.63	9.55	11.7
MnO	0.15	0.16	0.21	0.18	0.18	0.25
MgO	10.3	8.00	7.94	9.12	8.37	7.52
CaO	12.4	13.2	12.9	12.6	12.9	12.7
Na_2O	1.92	2.04	2.37	2.14	2.18	2.61
K_2O	0.07	0.08	0.10	0.09	0.09	0.12
Cr_2O_3 +	0.06	0.06	0.09	0.07	0.08	0.10
P_2O_5	0.06	0.06	0.09	0.07	0.08	0.10
$Mg/(Mg + Fe^{2+})$*	0.699	0.649	0.600	0.652	0.634	0.559

1. Staring composition FAMOUS basalt 527-1-1 (normalized to 100%).
2. Cotectic composition following 6% olivine fractionation.
3. Residual liquid after 30% crystallization. Olivine and plagioclase fractionated in the molar proportion 1:2.
4. New magma composition from 50:50 mix of residual melt (column 3) and original 527-1-1 magma (column 1).
5. Cotectic composition for new magma (following 4% olivine fractionation).
6. Residual liquid after 30% crystallization. Olivine and plagioclase fractionated in molar proportions 1:2.

\+ Effect of spinel fractionation not calculated.

* Calculated assuming $Fe^{3+}/(Fe^{2+} + Fe^{3+}) = 0.1$ (atomic).

5. Variable degrees of partial melting can generate the range of TiO_2 concentrations in basaltic glasses having high and restricted $Mg/(Mg + Fe^{2+})$. Residual mineralogical proportions can significantly affect the concentration of TiO_2 in the melt during the partial melting process.

Acknowledgments. We acknowledge the assistance of Fred Allen in programming the fractionation and mixing models. This research was supported by NSF Grant OCE7622193A01 (Submarine Geology and Geophysics).

References

Bence, A. E., and W. Hibberson, Petrogenesis of Caribbean deep sea basalts: Experimental investigations, Paper presented at International Conference on the Nature of the Oceanic Crust, AGU, La Jolla, California, December, 1975.

Bence, A. E., J. J. Papike, and R. A. Ayuso, Petrology of submarine basalts from the Central Caribbean: DSDP Leg 15, *Jour. Geophys. Res., 80*, 4775-4804, 1975.

Bender, J. F., F. N. Hodges, and A. E. Bence, Petrogenesis of basalts from the Project FAMOUS area: Experimental study from 0 to 15 kb, *Earth and Planet. Sci. Lett.*, 1978 (in press).

Bougault, H. and R. Hekinian, Rift Valley in the Atlantic Ocean near 36°N: Petrology and Geochemistry of basaltic rocks, *Earth and Planet. Sci. Lett., 24*, 249-261, 1974.

Bryan, W. B., and J. G. Moore, Compositional variations of young basalts in the mid-Atlantic ridge rift valley near 36°49'N, *Geol. Soc. Amer. Bull., 88*, 556-570, 1977.

Bryan, W. B., and G. Thompson, Basalts from DSDP Leg 37 and the FAMOUS area: Compositional and petrogenetic comparisons, *Can. J. Earth Sci., 14*, 875-885, 1977.

Bryan, W. B., G. Thompson, and P. M. Michael, Compositional variation in a steady-state zoned magma chamber: mid-Atlantic ridge at 36°50'N, *Tectonophysics*, 1978 (in press).

Donaldson, C. H., and R. W. Brown, Refractory megacrysts and magnesium rich melt inclusions within spinel in oceanic tholeiites. Indicators of magma mixing and parental magma composition, *Earth and Planet. Sci. Lett., 37*, 81-89, 1977.

Engel, A. E. J., C. Engel, and R. G. Havens, Chemical characteristics of oceanic basalts and the upper mantle, *Geol. Soc. Amer. Bull., 76*, 719-734, 1965.

Fisk, M. R., Melting relations and crystal chemistry of Reykjanes ridge basalts, Ph.D. thesis, University of Rhode Island.

Fleet, A. J., P. Henderson, and D. R. C. Kempe, Rare earth element and related chemistry of some drilled southern Indian Ocean basalts and volcanogenic sediments, *J. Geophys. Res., 81*, 4257-4268, 1976.

Frey, F. A., W. B. Bryan, and G. Thompson, Atlantic Ocean floor: geochemistry and petrology of basalts from Legs 2 and 3 of the Deep Sea Drilling Project, *J. Geophys. Res., 79*, 5507-5527, 1974.

Gast, P. W., Terrestrial ratio of potassium to rubidium and the composition of the Earth's mantle, *Science, 147*, 858-860, 1965.

Gast, P. W., Trace element fractionation and the origin of tholeiitic and alkaline magma types, *Geochim. Cosmochim. Acta, 32*, 1057-1086, 1968.

Green, D. H., Composition of basaltic magmas as indicators of conditions of origin: Application to oceanic volcanism, *Phil. Trans. Roy. Soc. London, Ser. A, 268*, 707-725, 1971.

Green, D. H., A. D. Edgar, P. Beasley, E. Kiss, and N. G. Ware, Upper mantle source for some hawaiites, mugearites and benmoreites, *Contrib. Mineral. Petrol., 48*, 33-43, 1974.

Häkli, T. A., and T. L. Wright, The fractionation of nickel between olivine and augite as a geothermometer, *Geochim. Cosmochim. Acta, 31*, 877-884, 1967.

Hanson, G. N., and C. H. Langmuir, Use of distribution coefficients for major elements in mantle-melt systems, *Geochim. Cosmochim. Acta, 42*, 725-741, 1978.

Henderson, P., The partitioning of selected transition elements ions between olivine and groundmass of oceanic basalts, *Chem. Geol., 5*, 267-274, 1969/1970.

Kay, R., N. J. Hubbard, and P. W. Gast, Chemical characteristics and origin of oceanic ridge volcanic rocks, *J. Geophys. Res., 75*, 1585-1614, 1970.

Kushiro, I., Origin of some magmas in oceanic and circum-oceanic regions, *Tectonophysics, 17*, 211-222, 1973.

Langmuir, C. H., J. F. Bender, A. E. Bence, G. N. Hanson, and S. R. Taylor, Petrogenesis of basalts from the FAMOUS area, mid-Atlantic ridge, *Earth and Planet. Sci. Lett, 36*, 133-156, 1977.

Leeman, W. P., Partitioning of Ni and Ca between olivine and basaltic liquid: an experimental study, *EOS, 54*, 1222, 1973.

Melson, W. G., T. L. Vallier, T. L. Wright, G. Byerly, and J.

Nelen, Chemical diversity of abyssal volcanic glasses erupted along Pacific, Atlantic, and Indian Ocean seafloor spreading centers, in *The Geophysics of the Pacific Ocean Basin and Its Margin*, G. H. Sutton, M. H. Mangani and R. Moberly, editors, *Am. Geophys. Union Monograph 19*, 351-367, 1976.

Miyashiro, A., F. Shido, and M. Ewing, Diversity and origin of abyssal tholeiite from the mid-Atlantic ridge near 24° and 30°N, *Contrib. Mineral. Petrol.*, *23*, 38-52, 1969.

Mysen, B., Partitioning of iron and magnesium between crystals and partial melts in peridotite upper mantle, *Contrib. Mineral. Petrol.*, *52*, 69-76, 1975.

O'Hara, M. J., Geochemical evolution during fractional crystallization of a periodically refilled magma chamber, *Nature*, *266*, 503-507, 1977.

Rhodes, J. M., M. A. Dungan, D. P. Blanchard, and P. E. Long, Magma mixing at mid-ocean ridges: evidence from basalts drilled near 22°N on the mid-Atlantic Ridge, *Tectonophysics*, 1978 (in press).

Roeder, P. L., Activity of iron and olivine solubility in basaltic liquids, *Earth and Planet. Sci. Lett.*, *23*, 397-410, 1974.

Roeder, P. L., and R. F. Emslie, Olivine-liquid equilibrium, *Contrib. Min. and Petrol.*, *29*, 275-289, 1970.

Sato, H. Nickel content of basaltic magmas: identification of primary magmas and a measure of the degree of olivine fractionation, *Lithos*, *10*, 113-120, 1977.

Schilling, J.-G., Iceland mantle plume: Geochemical study of Reykjanes Ridge, *Nature, London*, *242*, 565-571, 1973.

Schilling, J.-G., Rare earth fractionation in Hawaiian volcanic rocks, Ph.D. thesis, Massachusetts Institute of Technology, 1966.

Schilling, J.-G., and J. W. Winchester, Rare earths in Hawaiian basalts, *Science*, *153*, 867-869, 1966.

Schweitzer, E. L., J. J. Papike, and A. E. Bence, Statistical analysis of clinopyroxenes from deep sea basalts, *Amer. Mineral.*, 1978 (in press).

Shaw, D. M., Trace element fractionation during anatexis, *Geochim. Cosmochim. Acta*, *34*, 237-243, 1970.

Shido, F., A. Miyashiro, and M. Ewing, Crystallization of abyssal tholeiites, *Contrib. Mineral. Petrol*, *31*, 251-266, 1971.

Walker, D., J. Longhi, T. L. Grove, E. Stolper, and J. F. Hays, Experimental petrology and origin of rocks from Decartes Highlands, *Proc. Lunar Sci. Conf. 4th*, p. 1013-1032, 1973.

ISOTOPE GEOCHEMICAL STUDIES OF NORTH ATLANTIC OCEAN BASALTS AND THEIR IMPLICATIONS FOR MANTLE EVOLUTION.

R. K. O'Nions, N. M. Evensen, S. R. Carter, and P. J. Hamilton

Lamont-Doherty Geological Observatory of Columbia University, Palisades, New York 10964

Abstract. The available Pb-, Nd-and Sr-isotope determinations on North Atlantic Ocean basalts are reviewed. It is shown that the sub-oceanic mantle which has supplied these basalts has been previously depleted in Rb relative to Sr and Nd relative to Sm, and by inference large-ion lithophile elements in general. Basaltic volcanism does not appear to have sampled mantle which is undifferentiated or primordial. The Pb-, Nd- and Sr-isotope data are compatible with continuous models of mantle differentiation, although a simplistic interpretation of the Pb-isotope data alone suggests a single mantle differentiation event about 1.6 Gy ago. The Sr-, Nd- and Pb-isotope data can be used together with trace element abundance data to test simple two-component mixing models. Such models do not satisfy the available data from either Iceland and the Reykjanes Ridge nor the Azores and more complex models involving additional components are required. The degree of heterogeneity is such that the system probably cannot be adequately modelled with a few discrete components.

Introduction

In the last few decades, petrologists and geochemists have become increasingly concerned with the geochemistry of oceanic basalts in attempts to determine the nature and chemistry of sub-oceanic mantle. Classic studies on the origin of oceanic basalts have postulated their derivation from peridotitic sources (e.g., Bowen, 1928) or alternatively from basaltic sources (e.g., Daly, 1944). Models for the generation of oceanic basalts by partial fusion of peridotite were placed beyond dispute by the results of experimental petrology in the 1960's. Similarly, the relationship of alkali and tholeiite basalt types to primary magma compositions was largely resolved (e.g., Yoder and Tilley, 1962; Green and Ringwood, 1967; O'Hara, 1965, 1968).

The study of radiogenic isotopes in oceanic basalts has proceeded in parallel with these petrological studies and become increasingly important. In particular, isotopic studies have helped clarify petrogenetic relationships between consanguineous oceanic volcanics and the nature and temporal aspects of sub-oceanic mantle differentiation. At an early stage in these isotopic studies Gast (1960) was able to demonstrate that the Earth has a lower Rb/Sr ratio than chondritic meteorites because of the paucity of ^{87}Sr in the Earth. Subsequently, Tatsumoto et al. (1965) for example, noted that basalts dredged from mid-ocean ridges posess insufficient ^{87}Rb to support their ^{87}Sr contents. This observation implied that the mantle source regions of mid-ocean ridge basalts (MORB) had been depleted in Rb relative to Sr by differentiation episodes which occurred prior to the genesis of MORB. This important observation was amplified by Gast's (1968) demonstrations that MORB source regions are depleted in other large ionic-radius lithopile elements in addition to Rb, such as Ba and the rare-earth elements. In the last few years, the measurement of Nd-isotope compositions in oceanic basalts has provided a major stimulus to the understanding and interpretation of mantle differentiation. The combination of Nd-and Sr-isotope data now provide a geochemical framework within which depletion and enrichment of mantle source regions can be both identified and quantified. From the standpoint of the Rb-Sr, Sm-Nd and U-Pb systematics of oceanic volcanics the N. Atlantic is the most thoroughly studied part of the ocean basins. Considerable attention has been devoted to the petrogenetic relationship between magmas erupted in Iceland and on the Reykjanes Ridge as a result of Schilling's (1973) important demonstration that lithopile element abundances vary systematically along the length of the Reykjanes Ridge. The body of isotopic data now accrued on drilled, dredged and subaerial basalts from the N. Atlantic permits an evaluation of the differentiation of their mantle source regions.

Basaltic volcanism has occurred in the N. Atlantic during the last 50-60 My. In principle, some of the chemical characteristics of the mantle sources which have supplied basalts over this time

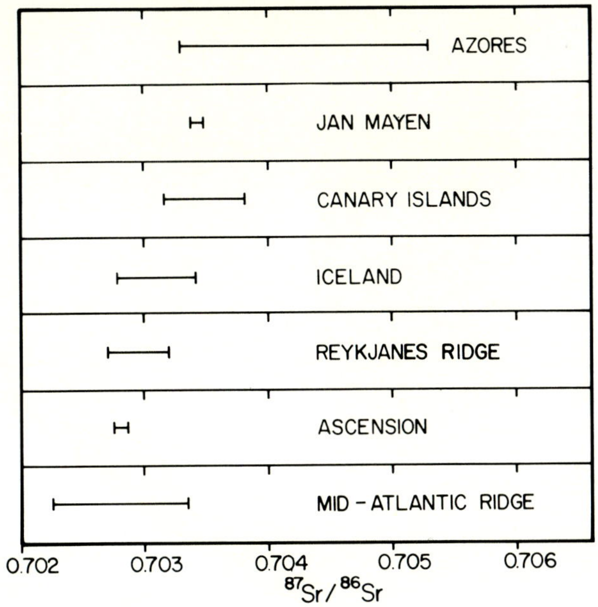

Fig. 1. Ranges of $^{87}Sr/^{86}Sr$ ratios for submarine ocean ridge and ocean island basalts in the North Atlantic. Only data with ± 0.0001 (2σ) precision, or better have been utilized. Data from Hart *et al.*, 1973; Hart 1976; White *et al.*, 1976; O'Nions and Grönvold, 1973; O'Nions and Pankhurst, 1973, 1974; O'Nions *et al.*, 1973, 1976; Sun and Jahn, 1975).

interval can be determined from isotope studies. The earliest Tertiary volcanics (outcropping in Baffin Island, E. and W. Greenland, Faroes and the British Isles) were erupted in a continental environment during the early stages of continental breakup. However, some of these basalts were contaminated during their passage through the continental crust (Carter *et al.*,1978a) which has obscured the chemistry of their mantle sources.

In the present article the Sr,- Nd- and Pb-isotope compositions of basalts erupted in the N. Atlantic ocean basin are reviewed, and their relevance to the nature and evolution of their source regions is discussed. Particular consideration is given to their relevance to continuous episodic models of mantle differentiation.

Review of Sr, Nd and Pb- Isotope Geochemistry

Sr - Isotopes. The earliest Sr - isotope measurements of N. Atlantic basalts were reported by Tatsumoto *et al.*, (1965) for mid-ocean ridge dredge basalts and Hedge and Walthall (1963) and Gast *et al.* (1964) for ocean island basalts. These measurements have now been superceded by a large number of high-precision Sr-isotope data.

The ranges of $^{87}Sr/^{86}Sr$ ratios for dredged basalts from the mid-Atlantic ridge (including the Reykjanes Ridge) are compared with basalts from Jan Mayen, Iceland, Ascension, Canaries and the Azores in Fig. 1. In general, the MORB have lower $^{87}Sr/^{86}Sr$ ratios (0.7023-0.7032) than the oceanic island basalts, although there is a significant overlap with the ranges reported for Iceland and Ascension. Of particular interest has been the demonstration of both local and regional variations in $^{87}Sr/^{86}Sr$ ratios. For example, Hart *et al.* (1973) and O'Nions and Pankhurst (1974) demonstrated significant differences in the $^{87}Sr/^{86}Sr$ ratios of basalts erupted along the Reykjanes Ridge, which more or less, parallels the variation in La/Sm ratios reported by Schilling (1973). Internally, Iceland and the Azores contain basalts with a considerable range of $^{87}Sr/^{86}Sr$ ratios. The data of O'Nions *et al.* (1976) demonstrated that significant $^{87}Sr/^{86}Sr$ variations exist in the Recent Icelandic basalts occurring in the active spreading zones and the Snaefellsnes Peninsula of western Iceland and encompass a large proportion of the variations found along the Reykjanes Ridge. Indeed, basalts of the Theistereykir region of N. Iceland, erupted over a few tens of km^2, have a similar range of $^{87}Sr/^{86}Sr$ and Ce/Yb ratios to that of the Reykjanes Ridge basalts (O'Nions *et al.*, 1976). In addition to the resolvable range of $^{87}Sr/^{86}Sr$ ratios in Recent Icelandic basalts, O'Nions and Pankhurst (1973) suggested that there was a secular decrease in the $^{87}Sr/^{86}Sr$ ratios from basalts erupted in Iceland from 16 My ago to the present day. It is possible that this apparent secular variation is an artefact resulting from inadequate sampling, although insufficient new data have been obtained to further evaluate this possibility.

In addition to the variations in $^{87}Sr/^{86}Sr$ along the Reykjanes Ridge substantial variations are well documented farther south along the mid-Atlantic Ridge between 50°N and 30°N (White *et al.*, 1975, 1976; White and Schilling, 1978), with maximum values of $^{87}Sr/^{86}Sr$ occurring at about 45°N and between 35° and 40°N with the lowest ratios recorded from 50°N.

Sr- isotope data for Tertiary basalts erupted during the initial opening phase of the N. Atlantic have not been plotted in Fig. 1. It has already been indicated that some of these have suffered significant contamination by continental crust (Carter *et al.*, 1978a). However, uncontaminated basalts (ca.60 My old) from Baffin Bay, E. and W. Greenland and the Isle of Mull in N.W. Scotland have $^{87}Sr/^{86}Sr$ ratios ranging from 0.7028 to 0.7040 (Carter *et al.*, 1979).

Nd and Sr Isotopes. $^{143}Nd/^{144}Nd$ ratios have now been measured in a number of Recent N. Atlantic basalts, including samples from Iceland, the Reykjanes Ridge, and the equatorial mid-Atlantic Ridge. The most useful of these data are those for which $^{87}Sr/^{86}Sr$ ratios have been determined on the same samples (O'Nions *et al.*, 1977; Zindler and Hart, 1978). The $^{143}Nd/^{144}Nd$ and the $^{87}Sr/^{86}Sr$ ratios of such samples are plotted in Fig. 2 and compared with the overall trend defined by oceanic basalts. The inverse correlation between $^{143}Nd/^{144}Nd$ and $^{87}Sr/^{86}Sr$, defined by the oceanic data and the N. Atlantic basalts themselves, was first recognized by DePaolo and Wasserburg (1976b) and O'Nions _et al._, (1977). This covariance between $^{143}Nd/^{144}Nd$ and $^{87}Sr/^{86}Sr$ in oceanic basalts attests to a coherence in the fractionation of Rb/Sr and Sm/Nd during sub-oceanic mantle differentiation, and leads to an improved estimate (Fig. 2) of the Rb/Sr ratio of the bulk Earth (DePaolo and Wasserburg, 1976b; O'Nions _et al._, 1977). Those parts of the sub-oceanic mantle with lower $^{87}Sr/^{86}Sr$ and higher $^{143}Nd/^{144}Nd$ ratios than the bulk Earth (Fig. 2) have been depleted in Rb relative to Sr and light REE relative to heavy REE, resulting in a reduced Rb/Sr ratio and increased Sm/Nd ratio (see Table 1 for bulk Earth values.

The inverse correlation between $^{143}Nd/^{144}Nd$ and $^{87}Sr/^{86}Sr$, illustrated in Fig. 2, has been defined only by data obtained on unaltered oceanic basalts. Island arc and altered ocean floor basalts are compared with the trend of unaltered oceanic basalts in Fig. 3, illustrating the divergence from this trend caused by the addition of seawater Sr ($^{87}Sr/^{86}Sr \simeq 0.709$).

The Sr- isotope composition of seawater has been estimated by a number of workers (see Faure and Powell, 1972) and in addition it's Nd - isotope composition has been estimated from the $^{143}Nd/^{144}Nd$ ratios of Mn- nodules, Fe - rich sediments and fish debris (O'Nions _et al._, 1978a; DePaolo and Wasserburg, 1977).

Changes in the $^{143}Nd/^{144}Nd$ ratios of ocean floor basalt induced by interactions with seawater have not been detected so far. This is presumably because of the exceeding low concentration of Nd in seawater.

Pb- Isotopes. Pb- isotope compositions of reliable accuracy have been published for N. Atlan-

Table 1. Bulk Earth Parameters

	I (ppm)	II (ppm)
K	200	460
Rb	0.67	1.3.
Sr	21.0	48.0
U	0.020	0.046
Th	0.08	0.18
Pb°	0.10	0.22
Sm	0.32	0.74
Nd	0.97	2.2
Rb/Sr	0.031	0.031
K/U	10^4	10^4
K/Rb	300	300
Sm/Nd	0.32	0.32

Column 1: Abundances assuming that Earth has the same relative abundances of Ca, U, Th, Sm, and Nd as carbonaceous chrondites, Rb/Sr = 0.031, K/U = 10^4 and $^{238}U/^{204}Pb$ = 10.4 by weight. Column 11: Abundances assuming that heat lost from the Earth equals heat generated and that the Th/U, Sm/U, K/U, Rb/Sr, and $^{238}/^{204}Pb$ ratios for the Earth are the same as Column I. Mean global heat flow from Williams and Von Herzen (1975), after O'Nions _et al._, (1978c).

tic basalts from Iceland (Sun and Jahn, 1975), the Reykjanes Ridge (Sun _et al._, 1975), the Canaries and the Azores (Sun, 1973) and the mid-Atlantic ridge (Tatsumoto, 1978). These data are plotted on a $^{207}Pb/^{204}Pb$-$^{206}Pb/^{204}Pb$ diagram in Fig. 4, referenced with a 4.55 Gy isochron and growth curves for $^{238}U/^{204}Pb(\mu)$ = 8.0 and 8.5. The N. Atlantic basalts form a linear array with

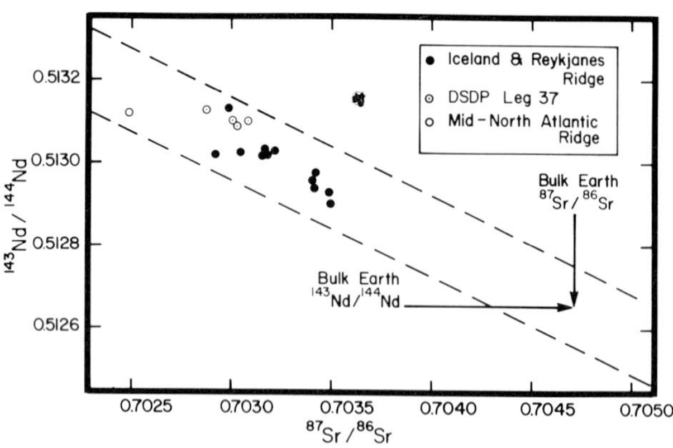

Fig. 2. $^{143}Nd/^{144}Nd$ versus $^{87}Sr/^{86}Sr$ ratios for N. Atlantic basalts from O'Nions _et al._, 1977 and Zindler & Hart (1978). The data are compared with the overall oceanic trend and the corresponding bulk Earth composition reported by O'Nions _et al._, (1978b).

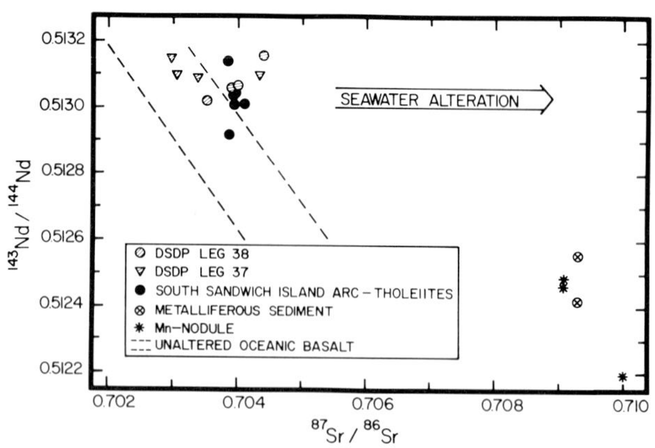

Fig. 3. Comparison of $^{143}Nd/^{144}Nd$ and $^{87}Sr/^{86}Sr$ ratios in island arc and altered ocean floor basalts with those in Mn-nodules and metalliferous sediments, illustrating the effects of seawater interaction on the ocean floor and at subduction zones. Data are from (DePaolo and Wasserburg, 1977; Hawkesworth _et al._, 1977; O'Nions _et al._, 1978b).

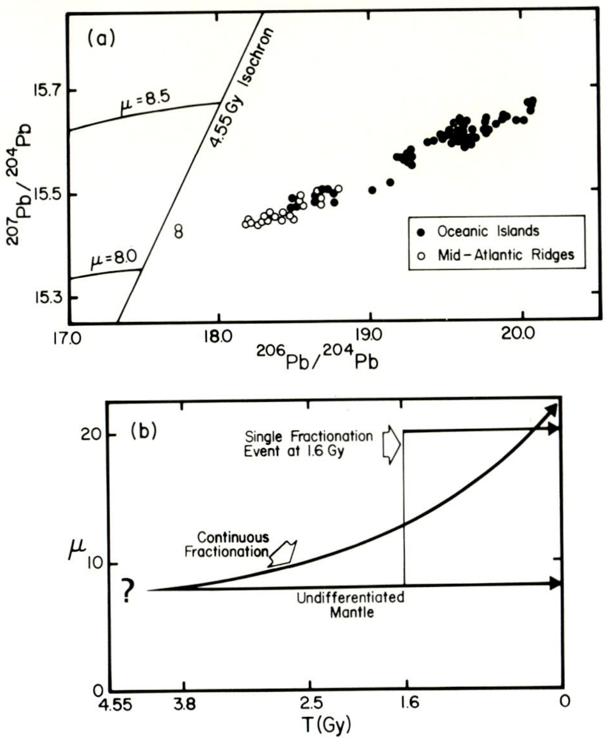

Fig. 4. (a) $^{207}Pb/^{204}Pb$ versus $^{206}Pb/^{204}Pb$ ratios for N. Atlantic basalts from Iceland, the Canaries, the Azores, mid-Atlantic ridge and the Reykjanes Ridge. Data from Tatsumoto (1978), Sun et al., 1975, Sun and Jahn (1975) and Sun (1973). For reference a 4.55 Gy isochron and μ ($^{238}U/^{204}Pb$) growth lines of 8.0 and 8.5 have been plotted.

(b) μ ($^{238}U/^{204}Pb$) versus T(Gy) diagram illustrating models capable of producing the Pb- isotope range observed in these N. Atlantic basalts. The μ value of the early Earth (before ca., 4 Gy) is uncertain because of the amount of lead (if any) incorporated into the Earth's core. Fractionation models are shown relative to an undifferentiated mantle μ value of 8.0. Two-stage model: single mantle-wide fractionation event at 1.6 Gy. Source regions of the most radiogenic volcanics would have acquired μ values as high as 20. Multi-stage model: continuous fractionation of μ from 3.8 Gy (approximate time of commencement of continental crust stabilization). Curve indicates the maximum increase of μ with time to produce the most radiogenic volcanics. From 3.8 Gy to 2.5 Gy μ could only have increased by ca., 20%.

the oceanic island basalts generally containing more radiogenic lead than the submarine ridge basalts. The true significance of such linear arrays has been debated for many years. The central problem has been to distinguish between (1) a simplistic two-stage episodic mantle evolution or (2) a more complex multi-stage continuous mantle evolution. This question will be discussed in more detail in a subsequent section of this paper.

Differentiation and Evolution of Mantle Magma Sources

The most important recent advances in the use of isotope geochemical data from oceanic basalts are based upon the improved estimates of the Sm/Nd and Rb/Sr ratios in the bulk Earth (≡ undifferentiated mantle). Estimates of the elemental abundances are also much improved (O'Nions et al., 1978c). The current best estimates and sources of these critical bulk Earth parameters are reported in Table 1. It should be noted at this juncture that it is assumed that the isotopic compositions of the erupted basalts are equal to those of their source, which was in isotopic equilibrium at the time of melting, and have not been modified during their ascent to the surface. The possibility of local disequilibrium during melting has been extensively discussed recently (e.g., O'Nions and Pankhurst, 1974; Hoffman and Hart 1978). For the situation where the mantle is partially molten for a long period prior to the extraction of melt, Hoffman and Hart (1978) have shown from kinetic considerations that the degree of isotopic disequilibrium should be negligible. Furthermore, disequilibrium melting models were found to be incompatible with the isotopic and trace element variability of oceanic basalts by O'Nions et al.(1976) and White et al. (1976).

Previously O'Nions et al. (1978b) and Carter et al. (1978b) have compared the single-stage time-integrated Rb/Sr and Sm/Nd ratios of basalt-source regions, as computed from their measured $^{87}Sr/^{86}Sr$ and $^{143}Nd/^{144}Nd$ ratios, with the bulk Earth Rb/Sr and Sm/Nd ratios. The fractional differences have been expressed in terms of ΔSR and ΔND parameters. In Fig. 5 ΔND and ΔSR values are compared for both the Recent N. Atlantic basalts (see Fig. 2) and uncontaminated, Tertiary basalts (ca60 My) from Baffin Bay, E. and W. Greenland and N.W. Scotland. Without exception these basalts have negative ΔSR and positive ΔND values. Thus, compared with undifferentiated mantle of the bulk Earth composition, the sources of all these N. Atlantic basalts have been depleted in Rb relative to Sr and Nd relative to Sm (equivalent to light REE depletion). This includes basalts erupted at so-called hot-spot sites as well as in submarine ridge environments. It seems inescapable that during the 60 My evolution of the North Atlantic (from the time of continental break-up to the present day) only depleted mantle sources have been sampled. Mantle sources which could be considered in anyway undifferentiated, primary, pristine or even enriched were not sampled by basaltic volcanism and probably do not exist in the N. Atlantic upper mantle in any significant quantity. It must be noted that the mantle sources of basalts erupted in Iceland and other so called 'hot-spots' are less unique amongst oceanic basalts than has sometimes been implied and furthermore have not been isolated from the mantle differentiation processes involved in the generation of continental crust. The overall depletion of the upper mantle caused by the removal of lithopile elements to the continental crust throughout the last 3.8 Gy of Earth history is clearly evident in the source regions of all N. Atlantic basalts discussed here.

Having compared the N. Atlantic upper mantle within the bulk Earth framework, it is germane to discuss the timing of the mantle differentiation events responsible for the observed depletion. It is now abundantly clear that this problem is closely tied to the generation of the continental crust. (e.g., O'Nions and Pankhurst 1978; O'Nions et al., 1978, a, b). Although comparatively little is known in detail about the rate of crustal growth through time, it is well-established that the earliest preserved granite crust dates from about 3.8 Gy ago. A substantial proportion of the continental crust (approximately half) had been produced by 2.5 Gy ago and the remainder has been added in a quasi-continuous manner (cf. Moorbath, 1975; O'Nions and Pankhurst, 1978;

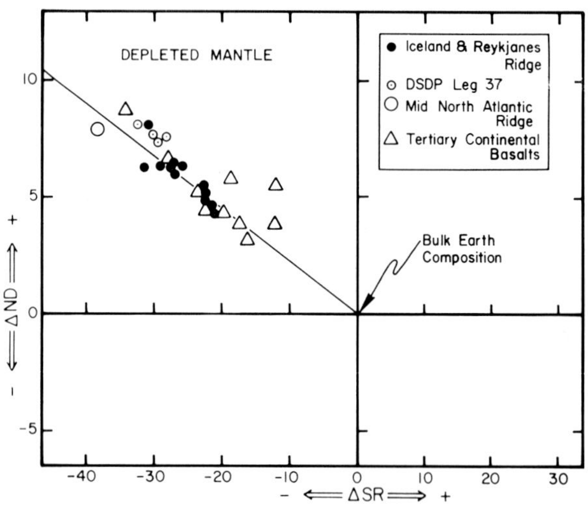

Fig. 5. ΔND versus ΔSR for Recent and Tertiary N. Atlantic basalts. Δ parameters are defined as the percentage deviation of the single-stage (SS) parent-daughter ratios from the bulk Earth (BE) composition. For example:

$$\Delta ND = \frac{{}^{147}Sm/{}^{144}Nd\ SS - {}^{147}Sm/{}^{144}Nd\ BE}{{}^{147}Sm/{}^{144}Nd\ BE} X\ 10^2$$

where

$${}^{147}Sm/{}^{144}Nd\ SS = \frac{{}^{143}Nd/{}^{144}Nd\ measured - 0.50682}{((Exp\ \lambda^{147}\ .\ T) - 1)}$$

and T = 4.55 Gy

All samples plot in the -ΔSR/+ΔND quadrant indicating that their mantle source regions were depleted in Rb and light REE relative to the bulk Earth composition.

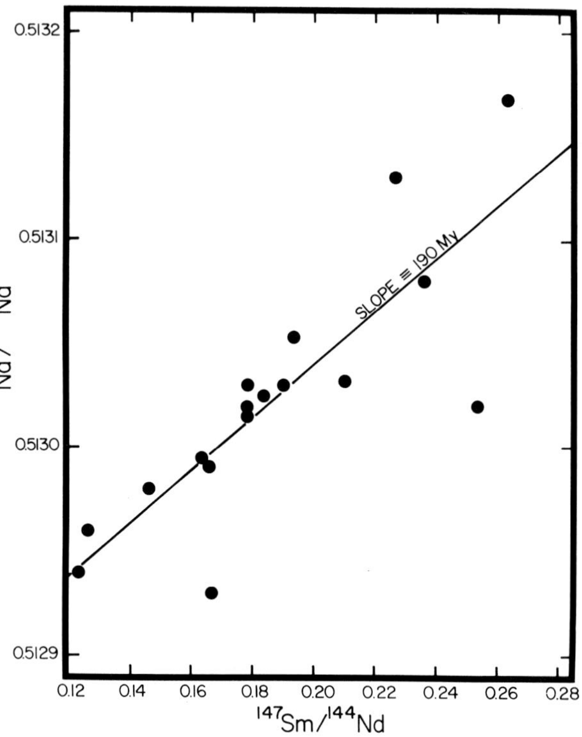

Fig. 6. Sm-Nd evolution diagram for basalts from Iceland and the Reykjanes Ridge. For amounts of melting in excess of 2%, the $^{147}Sm/^{144}Nd$ values will be within 20% of their source values. The best-fit line corresponds to an age of ca. 200 My., and the initial ratio ∿ 0.5128, see text for a discussion of the significance of the observed correlation. Data from O'Nions et al., (1977) and Zindler and Hart (1978).

McCulloch and Wasserburg, 1978). The implication of these observations is that Rb/Sr, Sm/Nd and U/Pb must have been fractionated in the mantle prior to 3 Gy ago and chemical heterogeneities must have existed (see O'Nions and Pankhurst, 1978; Sun and Nesbitt, 1977). Recognition of mantle heterogeneities dating from the earliest Archaean may be hindered by two factors: 1) Those parts of the mantle containing evidence of early mantle differentiation may not be sampled by contempory volcanism. 2) The long-lived heterogeneities may have been eradicated by convective motions resulting in mantle mixing.

The $^{147}Sm/^{144}Nd$ and $^{143}Nd/^{144}Nd$ ratios of N. Atlantic basalts are plotted on a Sm-Nd evolution diagram in Fig. 6. The slope of the best-fit line to the data plotted in Fig. 6 corresponds to an age of 190±70 My. This age is only geologically meaningful however, if the Sm/Nd ratios of the basalts are equal to those of their source regions, and all source regions had identical $^{143}Nd/^{144}Nd$ ratios 190 My ago. A similar analysis of the available Rb-Sr data for these and other samples yields apparent times of a mantle fractionation event ranging from 200 My to 500 My if assumptions analogous to those outlined above are adopted (O'Nions et al., 1976; Langmuir et al., 1978; White and Schilling, 1978).

The linear array of Pb- isotope compositions shown in Fig. 4 can also be used to estimate the timing of mantle differentiation. This array can be modelled by (i) a two-stage episodic model to imply a major fractionation event 1.6 Gy ago or alternatively (ii) a multi-stage model. In the latter case the linear array could merely result from mixing of two distinct end-member compositions as suggested by Sun et al. (1975) and Tatsumoto (1978). The μ - time relationships for a two-stage episodic model interpretation is compared with that of a continuous fractionation model in Fig. 4. Both models are capable of producing the range of Pb- isotope composition observed in the N. Atlantic basalts, but have very different implications for the evolution of the mantle.

The Sm-Nd, Rb-Sr and Pb isotope data considered above pose the following questions:

1) If the Sm-Nd data reflect a unique fractionation event in the mantle approximately 200 My ago, was the mantle undifferentiated until that time?

2) Can the two stage model interpretation of the Pb- isotope data (1.6 Gy fractionation event) be reconciled with the Sm-Nd and Rb-Sr systematics?

The answer to the first question is straightforward. The initial $^{143}Nd/^{144}Nd$ ratio of the mantle-isochron interpretation in Fig. 6 is 0.5128 whereas undifferentiated mantle is predicted to have a lower $^{143}Nd/^{144}Nd$ ratio (0.5124) 200 My ago. Therefore, the time integrated Sm/Nd ratio of these source regions between 4.55 Gy and 200 My ago was greater than the bulk Earth value, indicating the light REE depletion had already occurred during this time interval. With respect to the second question, regardless of whether the Pb-data are interpreted using a two-stage or a multi-stage model, it is evident that some fractionation event(s) affected the U-Pb system 10^9 y or more ago. Hence, both the Sm-Nd and U-Pb systems suggest the occurrence of early ($>10^9$y) mantle fractionation events, but the Sm-Nd system indicates more recent fractionations also. The data presented for the N. Atlantic basalts in this paper are compatible with the following statements:

1) U/Pb, Sm/Nd and Rb/Sr have been fractionated in the mantle since at least the beginnings of continental growth 3.8 Gy ago.

2) The present heterogeneiities in U/Pb, Rb/Sr and Sm/Nd ratios are mainly the result of more recent fractionations which occurred on a time scale closer to 10^8y than 10^9y.

3) The fractionations of Sm/Nd and Rb/Sr in the mantle during the last 5 x 10^8y or so are responsible for a considerable amount of the present variability of $^{143}Nd/^{144}Nd$ and $^{87}Sr/^{86}Sr$ in the upper mantle.

4) The variability of Pb- isotope (particularly $^{207}Pb/^{204}Pb$) compositions reflect still earlier differentiation events occurring on a time scale of $\geq 10^9$y ago. At this time ^{235}U was relatively more abundant than at present. More recent ($\sim 10^8$y) fractionations of U/Pb have produced comparatively little variability in $^{207}Pb/^{204}Pb$ because of the low abundance of ^{235}U.

All three of these parent-daughter systems appear to be compatible with continuous or multi-stage fractionation models. The tendency towards continuous eradication or 'smoothing-out' of chemical heterogeneiities produced in the earlier epochs of mantle evolution attest to the efficiency of mantle convection processes in removing them - at least within the upper mantle.

Dynamical Processes in the Mantle

Isotopic variations in oceanic basalts can be used to impose some constraints on dynamical processes in the mantle. In favorable circumstances they can be used in conjunction with concentration data to test models of source and magma mixing, and partial melting. Several attempts have been made to test source and magma mixing models for Iceland and Reykjanes Ridge basalts (Sun et al., 1975; O'Nions et al., 1976; Langmuir et al., 1978). Mixing equations have been derived by Lancelot and Allegre (1974) and Vollmer (1976) and provide the rationale for each of these attempts.

Sun et al., (1975) used Pb- isotope and concentration data for Reykjanes Ridge basalts, and demonstrated that linear plots of $^{207}Pb/^{204}Pb$ and $^{206}Pb/^{204}Pb$ versus $1/Pb$ were consistent with binary mixing models. This result implied that melting and crystallization differentiation processes had not modified the Pb- concentrations sufficiently to mask the correlations. From this observation one might expect that similar correlations should exist between $^{87}Sr/^{86}Sr$ versus $1/Sr$ and $^{143}Nd/^{144}Nd$ versus $1/Nd$. O'Nions et al., (1976) showed that a plot of $^{87}Sr/^{86}Sr$ versus $1/Sr$ was not linear and did not lend support to this model. These Sr-data together with the Nd- isotope and concentration data of O'Nions et al., (1977) are compared in Fig. 7. From this diagram it is apparent that a linear correlation does not exist between $^{143}Nd/^{144}Nd$ and $1/Nd$ either, but $^{87}Sr/^{86}Sr$ does correlate linearly with Sr. Thus, these data taken together provide little support for a simple binary mixing model. If the isotopic variations in Icelandic and Reykjanes Ridge basalts are indeed the result of mixing of two end members then the combined effect of melting and crystallization have fractionated at least Rb/Sr and Sm/Nd in such a way as to destroy consistent correlations.

Despite these inconsistencies there are reported correlations between $^{87}Sr/^{86}Sr$, $^{143}Nd/^{144}Nd$, $^{147}Sm/^{144}Nd$, $^{87}Sr/^{86}Sr$ and Ce_N/Yb_N in the N. Atlantic basalts. The significance of these first two correlations has been discussed above. O'Nions et al. (1976) attributed the third correlation to the generation of heterogeneous mantle several hundred million years ago and Langmuir et al. (1978) in a more detailed discussion have shown that mixing models involving three or more sources can also accomodate the data. Further efforts to quantify the actual number of discrete sources beneath Iceland and the Reykjanes Ridge would now appear to be a very difficult exercise. Similarly an extension of this approach to identify a particular mantle component to be included in a dynamical model of mass transport (such as a mantle plume) would be unproductive, since it is now clear that whatever form the mantle source supplying basalts to Iceland has, it must be heterogeneous.

Marked correlations between trace element abundance ratios such as Ba/Sr, Rb/Sr, and $^{87}Sr/^{86}Sr$ have been documented for the Mid-Atlantic Ridge basalts between 30°N and 50°N (White et al., 1975; White and Schilling, 1978) and models involving more than two components must be appealed to. At the present time a unique solution to the data available does not exist and it is almost a matter of personal taste whether one attributes the variations to a number of mutually intermixing mantle plumes, or simply to mantle heterogeneities without any implications concerning the physical nature of the mass transport processes.

Depletion and Enrichment Processes

Earth differentiation has proceeded with the progressive removal of large ion lithopile elements from the upper mantle to the continental crust. The production of partial melts in the mantle and their physical removal from the residium is the prime mechanism of element transport and fractionation. The transport of large ion lithopile elements from the mantle to the continental crust must have been dominated by such processes.

It is now well established that the single-stage time integrated Sm/Nd and Rb/Sr ratios of the N. Atlantic mantle sources are respectively higher and lower than those of the bulk Earth. The actual measured Sm/Nd and Rb/Sr ratios of some ridge basalts also bear this relationship to the bulk Earth composition. However other basalts, particularly ocean island basalts, have relatively high Rb/Sr ratios and are enriched in the light REE (e.g., Schilling, 1973; O'Nions et al., 1976). The relationship of the measured parent-daughter ratios of N. Atlantic basalts to the bulk Earth composition is illustrated by δ'ND and δ'SR parameters in Fig. 8. The δ' parameter is the fractional deviation of the measured parent-daughter ratio in a particular basalt from that of the bulk Earth. Also plotted on this diagram are the ΔND and ΔSR values for the N. Atlantic from Fig. 5. Small degrees of melting (2-5%) are not expected to produce more than a 20% change in δ'ND, thus if the source regions shown are applicable, then the generation of some basalts appear to require even greater fractionations of Sm/Nd. The mechanism whereby this occurs is presently obscure, but it is not the result of simple par-

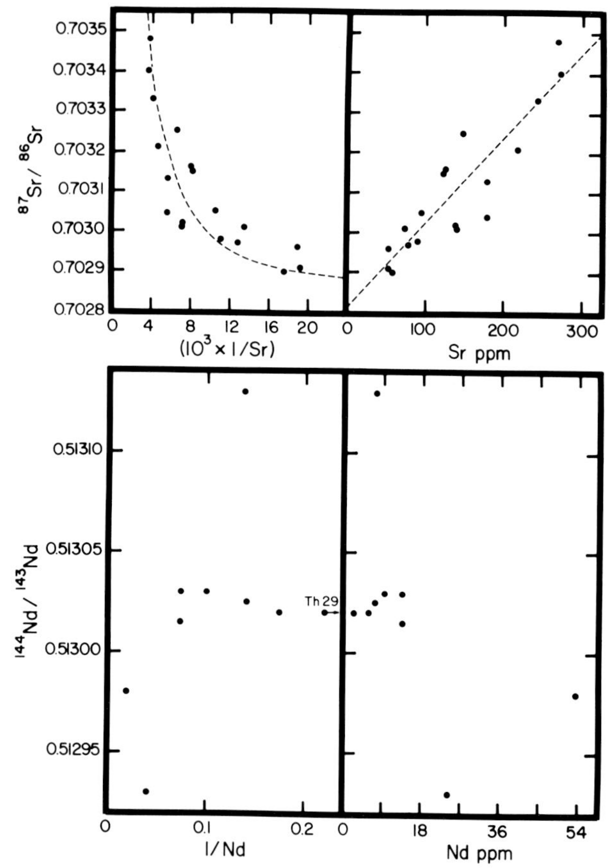

Fig. 7. Plots of $^{87}Sr/^{86}Sr$ versus Sr and $^{1}/Sr$ and $^{143}Nd/^{144}Nd$ versus Nd and $^{1}/Nd$ for basalts from Iceland and the Reykjanes Ridge. For a simple two component mixing model $^{87}Sr/^{86}Sr$ and $^{143}Nd/^{144}Nd$ are predicted to correlate with $^{1}/Sr$ and $^{1}/Nd$ respectively. These data are inconsistent with such a simple model. Data from O'Nions et al., 1976, 1977.

tial melting and must include more complex processes of enrichment. In view of the correlation evident in Fig. 6 and the correlation of Ce_N/Yb_N with $^{87}Sr/^{86}Sr$ (O'Nions et al., 1976) it seems that the enrichment may pre-date the contemporary melting event by several hundred millions of years. If such fractionations of REE commonly occur in the upper mantle then the isotope systematics of Recent basalts require that they are short-lived heterogeneiities

Concluding Remarks

Studies of oceanic basalts have proved to be one of the most successful ways of studying the composition and differentiation history of the Earth's mantle. Sufficient Nd and Sr isotope measurements exist on a world wide basis to define the basic parameters (Sm/Nd and Rb/Sr) of the bulk Earth compostion and thus provide a framework within which mantle evolution can be examined.

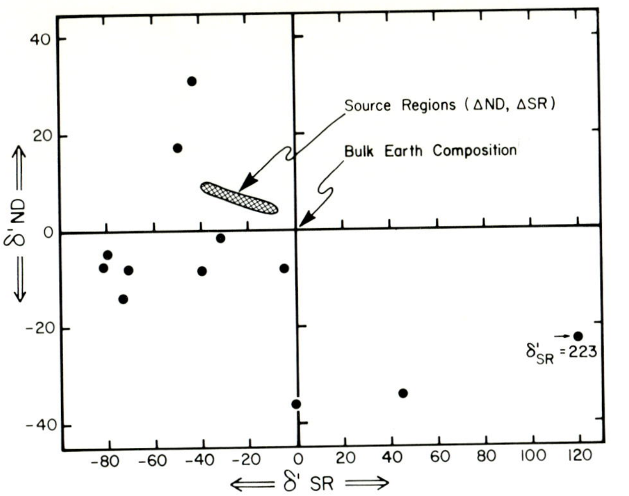

Fig. 8. Plot of δ'ND and δ'SR values for N. Atlantic basalts compared with the range of their ΔND and ΔSR values.

$$\delta'ND = \frac{(^{147}Sm/^{144}Nd)_{meas} - (^{147}Sm/^{144}Nd)_{BE}}{(^{147}Sm/^{144}Nd)_{BE}} .10^2$$

and δ'SR is defined in an analogous manner. The source regions of the basalts have time-integrated Sm/Nd and Rb/Sr ratios indicated by the plotted ΔND and ΔSR values. Partial melting of such sources produces changes in δ'ND of ca., 20% for 2% melting. Some basalts with δ'ND around -40% could not have been generated by such simple melting processes and some re-enrichment of the source or more complex melting and fractionation process is indicated.

The N. Atlantic is a well studied ocean basin, particularly in the vicinity of Iceland and the Azores. The conclusions drawn from these isotopic studies of N. Atlantic basalts may be summarised as follows:

1) Uncontaminated basalts erupted in the N. Atlantic over the last 60 My have been generated from depleted mantle. This includes early Tertiary basalts erupted in continental environments at the time of continental break-up (now preserved in Baffin Island, E. and W. Greenland and N.W. Scotland), recent ocean floor basalts and oceanic island basalts including Iceland and the Azores.

2) Single-stage time-integrated Rb/Sr and Sm/Nd as computed from the initial $^{87}Sr/^{86}Sr$ and $^{143}Nd/^{144}Nd$ ratios, are respectively less than and greater than the bulk Earth values. This requires a depletion of Rb relative to Sr, and light rare-earths relative to heavy rare-earths in their source regions.

3) The most depleted mantle in a time-integrated sense has been sampled by submarine mid-Atlantic Ridge basalts and by some Iceland and Ascension basalts, yet insignificant volumes of undepleted mantle have been sampled.

4) The timing of mantle fractionation events can be estimated from the Rb-Sr, Sm-Nd and Pb-isotope systematics of the erupted basalts. If the Rb-Sr and Sm-Nd data are interpreted as mantle isochrons they suggest fractionation events some 200 - 500 My ago. However, the Sm-Nd data indicate that the source region must already have been depleted at that time. The simplest interpretation of the Pb - isotope data yields a conflicting result and requires a mantle wide fractionation event 1.6 Gy ago. However, the isotope data are also compatible with multi-stage fractionation models.

7) Correlations between $^{87}Sr/^{86}Sr$ and Sr, $^{143}Nd/^{144}Nd$ and Nd, indicate that simple two component mixing models cannot explain the isotopic variations in Iceland and along the Reykjanes Ridge, however, the few available Pb- isotope data are compatible with such a model. More complex (3 or more) source mixing models may accomodate the data, but attempts to quantify the actual number become exceedingly difficult.

The North Atlantic has become a testing ground for petrogenetic hypotheses and models of mantle evolution and Earth differentiation. Opinions have differed considerably amongst the reseachers involved over the years, but a measure of agreement and consensus is beginning to emerge. The debates have been vigorous and furthermore added a stimulus to geochemical and petrological studies in general.

Acknowledgements. This work was supported by a grant from the National Science Foundation (EAR-75-20891).

References

Bowen, N.L., The Evolution of the Igneous Rocks. Princeton University Press, Princeton, 1928.

Brooks, C., Hart, S.R., Hoffman, A. and James, D.E. Rb-Sr mantle isochrons from oceanic regions. Earth Planet. Sci. Lett. 32, 51-61, 1976.

Carter, S.R., Evenson, N.M., Hamilton, P.J., and O'Nions, R.K. Nd- and Sr- isotope evidence for crustal contamination of continental volcanics. Science, 202, 743-747, 1978.

Carter, S.R., Evensen, N.M., Hamilton, P.J. and O'Nions, R.K. Continental volcanics derived from enriched and depleted source regions: Nd- and Sr- isotope evidence. Earth Planet. Sci. Lett. 37, 401-408, 1978b.

Carter, S.R., Evensen, N.M., Hamilton, P.J. and O'Nions, R.K. Nd- and Sr- isotope study of basalt magma sources during opening of the North Atlantic. In preparation, 1979.

Cumming, G.L. and Richards, J.R. Ore-lead isotope ratios in a continuously changing Earth. Earth Planet. Sci. Lett. 28, 155-171, 1975.

Daly, R.A. Volcanism and petrogenesis as illustrated in the Hawaiian Islands. Geol. Soc. Amer. Bull. 55, 1391-1396, 1944.

DePaolo, D.J. and G.J. Wasserburg, Nd-Isotope

Variations and Petrogenetic Models. Geophys. Res. Letts., 3, 249-252, 1976a.

DePaolo, D.J. and G.J. Wasserburg, Inferences about Magma Sources and Mantle Structure from Variations of $^{143}Nd/^{144}Nd$. Geophys. Res. Lett., 3, 743-746, 1976b.

DePaolo, D.J. and Wasserburg, G.J. The sources of Island arcs as indicated by Nd and Sr isotopic studies. Geophys. Res. Lett. 4, 465-468, 1977.

Faure, G. and Powell, J.L. Strontium Isotope Geology. Springer-Verlag, 1972.

Gale, N.H. and Musset, A.E. Episodic uranium-lead models and the interpretation of variations in the isotopic composition of lead in rocks. Rev. Geophys. Space Phys. 11, 37-86, 1973.

Gast, P.W., Limitations on the Composition of the Upper Mantle. J. Geophys. Res., 65, 1287-1293, 1960.

Gast, P.W. Trace element fractionation and the origin of tholeiitic and alkaline magma types. Geochim. Cosmochim. Acta, 32, 1057-1086, 1968.

Gast, P.W., Tilton, G.R., Hedge, C.E. Isotopic composition of lead and strontium from Ascension and Gough Islands. Science, 145, 1181-1185 (1964).

Green, D.H. and Ringwood, A.E. The genesis of basaltic magmas. Contrib. Mineral. Petrol., 15 103-190 (1967).

Hart, S.R. LIL-Element Geochemistry, Leg 34 basalts. Initial Rep. DSDP Vol. XXXIV, 283-288 (1976).

Hart, S.R., Schilling, J.-G., and Powell, J.L. Basalts from Iceland and along the Reykjanes Ridge: Sr- Isotope geochemistry. Nature Phys. Sci., 268, 707-725, 1973.

Hawkesworth, C.J., O'Nions, R.K., Pankhurst, R.J., Hamilton, P.J. and Evensen, N.M. A geochemical study of island-arc and back-arc tholeiites from the Scotia Sea. Earth Planet. Sci. Lett. 36, 253-262 (1977).

Hedge, C.E. and Walthall, F.G. Radiogenic strontium-87 as an index of geological processes. Science, 140, 1214-1217 (1963).

Hoffman, A.W. and Hart, S.R. An assessment of local and regional isotopic equilibrium in the mantle. Earth Planet. Sci. Lett. 38, 44-62, 1978.

Lancelot, J.R. and Allegre, C.J. Origin of carbonatitic magma in the light of the Pb-U-Th isotope system. Earth Planet. Sci. Lett., 22, 233-238, 1974.

Langmuir, G.H., Vocke, R.D., Jr., Hanson, G.N. and Hart, S.R. A general mixing equation with applications to Icelandic basalts. Earth Planet. Sci. Lett., 37, 380-392, 1978.

McCulloch, M.T. and Wasserburg, G.J., Sm-Nd and Rb-Sr chronology of Continental Crust formation. Science, 200, 1003-1011, 1978.

Moorbath, S. Evolution of Precambrian crust from strontium isotopic evidence. Nature, 254, 395-398, 1975.

Moorbath, S. and Bell, J.D. Strontium isotope abundance studies and Rubidium strontium age determinations on Tertiary igneous rocks from the Isle of Skye, North-West Scotland. Journal Petrol., 6, 37-66, 1965.

O'Hara, M.J. Primary magmas and the origin of basalts. Scot. J. Geol. 1, 19-40 (1965).

O'Hara, M.J. The bearing of phase equilbria studies in synthetic and natural systems on the origin and evolution of basic and ultra-basic rocks. Earth Sci. Rev., 4, 69-133, 1968.

O'Nions, R.K. and Grönvold, K. Petrogenetic relationships of acid and basic rocks in Iceland: Sr- isotopes and rare-earth elements in late and postglacial volcanics. Earth Planet. Sci. Lett., 19, 397-409 (1973).

O'Nions, R.K. and Pankhurst, R.J. Secular variations in the Sr- isotope compostions of Icelandic volcanic rocks. Earth Planet. Sci. Lett., 21, 13-21, 1973.

O'Nions, R.K. and Pankhurst, R.J. Petrogenetic significance of isotope and trace element variations in volcanic rocks from the mid-Atlantic. J. Petrol., 15, 603-634, 1974.

O'Nions, R.K. and Pankhurst, R.J. Early Archaen rocks and geochemical evolution of the Earth's crust. Earth Planet. Sci. Lett., 38, 211-236, 1978.

O'Nions, R.K., P.J. Hamilton and N.M. Evensen, Variations in $^{143}Nd/^{144}Nd$ and $^{87}Sr/^{86}Sr$ Ratios in Oceanic Basalts. Earth Planet. Sci. Lett. 34, 13-22, 1977.

O'Nions, R.K., Pankhurst, R.J. and Grönvold, K. Nature and development of basalt magma sources beneath Iceland and the Reykjanes Ridge. J. Petrol., 17, 315-338, 1976.

O'Nions, R.K., S.R. Carter, R.S. Cohen, N.M. Evensen and P.J. Hamilton. Pb, Nd and Sr-isotopes in oceanic ferromanganese deposits and ocean floor basalts. Nature, 273, 435-438, 1978a.

O'Nions, R.K., N.M. Evensen, P.J. Hamilton and S.R. Carter. Melting of the Mantle Past and Present: Isotope and Trace Element Evidence. Phil. Trans. Roy. Soc. Lond. A., 258, 546-559, 1978b.

O'Nions, R.K., Carter, S.R., Evensen, N.M. and Hamilton, P.J. Upper Mantle Geochemistry - In The Sea, VII, in press 1978c.

O'Nions, R.K., Pankhurst, R.J., Fridleifsson, I.B. and Jakobsson, S.P. Strontium isotopes and rare-earth elements in basalts from the Heimaey and Surtsey volcanic eruptions. Nature, 243, 213-214 (1973).

Schilling, J.-G. Iceland mantle plume: geochemical study of the Reykjanes Ridge. Nature, 242, 565-571, 1973.

Sun, S.S. Lead Isotope Studies of Young Volcanic Rocks from Oceanic Islands, Mid-Ocean Ridges and Island Arcs. Unpublished Ph. D. Thesis, Columbia University, 1973.

Sun, S.S. and Jahn, B. Lead and strontium isotopes in post-glacial basalts from Iceland.

Nature, 255, 527-530, 1975.

Sun, S.S. and Nesbitt, R.W. Chemical heterogeneity of the Archaean Mantle Composition of the Earth and Mantle Evolution. Earth Planet. Sci. Lett., 35, 429-448, 1977.

Sun, S.S., Tatsumoto, M. and Schilling, J.-G. Mantle-plume mixing along the Reykjanes Ridge Axis: Lead isotopic evidence. Science, 190, 143-147, 1975.

Tatsumoto, M. Isotopic composition of lead in oceanic basalt and its implication to mantle evolution. Earth Planet. Sci. Lett. 38, 63-88, 1978.

Tatsumoto, M., Hedge, C.E. and Engel, A.E.J. Potassium, rubidium, strontium, thorium, uranium and the ratio strontium-87 to strontium-86 in oceanic tholeiite basalts. Science, 150, 886-888, 1965.

Vollmer, R. Rb-Sr and U-Th-Pb systematics of alkaline rocks: the alkaline rocks from Italy. Geochim. Cosmochim. Acta., 40, 283-295, 1976.

White, W.M. and Schilling, J.G. The nature and origin of geochemical variation in Mid-Atlantic Ridge basalts from central N. Atlantic. Geochimica. Cosmochim. Acta., 42, 1501-1517 (1978).

White, W.M., Hart, S.R. and Schilling, J.- G. Geochemistry of the Azores and the Mid-Atlantic Ridge: 29°N to 60°N. Carng. Inst. Yearb. 75, 224-234, (1975).

White, W.M., Schilling, J.-G. and Hart, S.R. Evidence for the Azores mantle plume from strontium isotope geochemistry of the central N. Atlantic. Nature, 263, 659-663 (1976).

Williams, D.L. and Von Herzen, R.P. Heat loss from the Earth: New estimate. Geology, 2, 327-328 (1974).

Yoder, H.S., Jr., and Tilley, C.E. Origin of basalt magmas: an experimental study of natural and synthetic rock systems. J. Petrol. 3, 34-532, 1962.

Zindler, A. and Hart, S.R. Nd- and Sr- isotope compositions and REE abundances in Reykjanes Peninsula basalts: Evidence for mantle heterogeneity. EOS. 59, 410 (1978).

ALTERATION, FRACTIONAL CRYSTALLIZATION, PARTIAL MELTING, MANTLE PROPERTIES FROM TRACE ELEMENTS IN BASALTS RECOVERED IN THE NORTH ATLANTIC

Henri Bougault

Centre Oceanologique de Bretagne, B.P. 337, 29273 Brest Cedex, France

Jean-Louis Joron and Michel Treuil

Institut de Phusique du Globe, 4, place Jussieu, 75230 Paris cedex 05, et Laboratorie Pierre Sue, Centre d'Etudes Nucleaires, B.P. n 2, 91190 Gifsur-Yvette, France

Abstract. Alkali metals prove to be very sensitive to alteration process even if this process is moderate. It is very difficult to use these elements to trace magmatic processes since their concentration may vary within an homogeneous "unaltered" unit.

From all the basaltic units investigated, a classification of "low partition coefficient" elements is presented (Sc, Ti, V, Y, Zr, Nb, La-Tb, Hf, Ta, Th).

The incompatible elements ratios (Y/Tb, Zr/Hf, Nb/Ta) are constant in the samples studied and are nearly equal to the same ratios in chondrites. This is consistent with the existence of an homogeneous primordial mantle at the scale of the North Atlantic, compatible with a chondritic composition of the earth.

More than two distinct mantle sourced derived from the primordial mantle (invoking the possibility of mixing of sources), are needed to explain the data. Moderate "incompatible" or hygromagmaphile" element/highly "hygromagmaphile" element ratios are function of a partial melting.

Partial melting (variable), magma chamber dynamics (possibility of mixing) and only one or two eruptions is a cycle frequently repeated : variability of partial melting and limited sampling of liquids derived from the same initial liquid explain the variable rare-earth patterns encountered in one single hole and makes it difficult to determine the proportion of minerals crystallizing.

† Contribution n° du Département Scientifique, Centre Océanologique de Bretagne.

Introduction

Several sites have been drilled into the oceanic crust in the North Atlantic up to and including Leg 53 of the Glomar Challenger (DSDP-IPOD program) : Leg 37, sites 332 to 335 at 36° N along a track perpendicular to the ridge ; Leg 45 sites 395 and 396 at 22° N almost symmetric with respect to the ridge in crust about 7 and 10 m.y. old ; Leg 46, site 396 again ; Leg 49, three sites 407, 408, 409 at 63° N, along a track perpendicular to Reykjanes Ridge, one site 410 at 45° N and three sites 411, 412 and 413 at about 36° N near the FAMOUS area ; the megalec (Legs 51, 52, 53) at 25° N, at 110 m.y. old sites 417 and 418. In addition the FAMOUS operation at 36° N provided an intensive sampling both by dredging and by submersible in the Rift valley and adjacent fracture zones.

During the drilling program, we have operated the CNEXO X-ray fluorescence van on board the Glomar Challenger to obtain on board major elements analysis (and some trace elements during Legs 45 and 46).These shipboard chemical data have been useful in defining the different basalt units and for selection of samples for shore-based studies (Bougault, 1977). Our contribution to the shore based studies has been the systematic study of trace elements both through X ray fluorescence spectrometry and neutron activation analysis in samples which have been selected on board for shipboard analysis (XRF : V, Cr, Ni, Co, Rb, Sr, Y, Zr, Nb ; NAA : Sc, Ni, Co, Zr, La, Eu, Tb, Hf, Ta, Th).

Leg 53 marks the end of first phase of the IPOD program as far as drilling the crust in the Atlantic Ocean is concerned and we feel it is important to draw out major conclusions which now

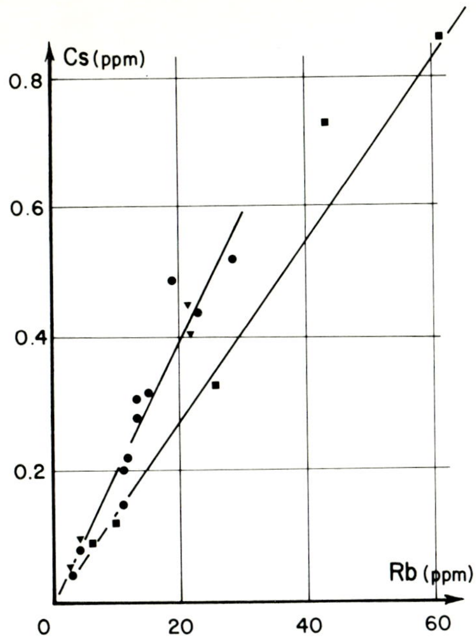

Fig. 1 : Cs versus Rb in Hole 417 A. The black dots represent the data below core 30 and the squares represent the data above core 30. The triangles represent altered samples in Hole 417 D.

can be made. Discussions related to alteration, magmatic units, fractional crystallization, magmatic cycles, partial melting, heterogeneity or homogeneity of the mantle or upper mantle can be now made as a consequence of the data which have been made available through the DSDP program and FAMOUS operation. We have analysed more than 400 samples ; results (sample by sample) are given (or will be) in DSDP Initial Reports, in Arcyana (1977) and in Bougault *et al.* (in press) and are not duplicated in this paper ; only tables for specific samples or group of samples necessary for the discussion are reported.

Alteration process. Alkali metals and strontium

Basalt alteration is discussed with respect to the behavior of alkali metals (K_2O, Rb, Cs) and Sr. Two examples are considered, sites 417 and 418 (Leg 51) on one hand, and site 395 (Leg 45) on the other. Basalts of sites 417 and 418 have experienced a high degree of alteration which resulted in extensive change of major element composition. Site 395 basalts experienced only slight alteration and can be considered to be fresh samples on the basis of their low water contents.

At site 417, the variation of alteration effects are observed at two scales, which may be related to water circulation.

- Within a single core (e.g. core 35 of hole 417 A), Rb varies from 12 to 29 ppm, Cs from 0.22 to 0.56 ppm and K_2O from 0.98 to 2.04 %. This is probably due to local cracks or fissures through which water circulates resulting in the formation of alteration minerals (Joron *et al.*, in preparation).

- Major interstice differences are observed between hole 417 A and hole 417 D (and 418 A) which are about 400 meters apart (Joron *et al.*, in preparation). Hole 417 A is highly altered ; the volcanic pile has been divided into several zones according to the occurrence of different minerals : montmorillonite and celadonite occur in all zones ; chlorite, scolecite and analcite occur in different zones (Legs 51, 52 ; scientific party ; in preparation). Potassium varies from $\sim$ 6 % K_2O in very highly altered zones (top) to classical values in oceanic tholeiites (< 0.2 %) at the bottom. Thus a general decrease of alteration is observed from the top to the bottom. At the opposite, in hole 417 D, some 400 meters from hole 417 A, strikingly fresh basalts were recovered including fresh glass.

This difference in alteration between 417 A and 417 D (110 m.y. old) only 400 meters apart is explained by a difference in topography between both holes. The 417 A altered site is a topographic high which remained unburied approximately 20 m.y. after eruption (lack of cretaceous sediments) (Leg 51, 52 ; scientific party, in preparation). According to Donnelly (1978) "we might regard the hill as a sort of upside down drain, which, lacking an impermeable sedimentary cap, becomes the principal conduit for convective effluent water".

If this interpretation is correct, it means that much of the alteration process (at least the type of alteration observed) takes place shortly after eruption. When the basalts are covered by a sufficient sedimentary thickness they are protected from further sea water circulation.

A good correlation is found between Cs and Rb (fig. 1) (K versus Rb is also similar). But the dispersion of points is far greater than the analytical precision (K_2O $\pm\sim$0.02 %, Rb $\pm$ 2 ppm, Cs $\pm$ 0.02 ppm). The abundances of these elements

TABLE 1. Comparison of uptake of alkali metals from sea-water and residence times of alkali metals in the ocean. Sea-water concentrations and residence times are from Goldberg (1965).

	K	Rb	Cs
Uptake Altered sample/ sea water proportional to	44.7	242	1 120
Residence time	$1.1\ 10^7$	$2.7\ 10^5$	$4\ 10^4$

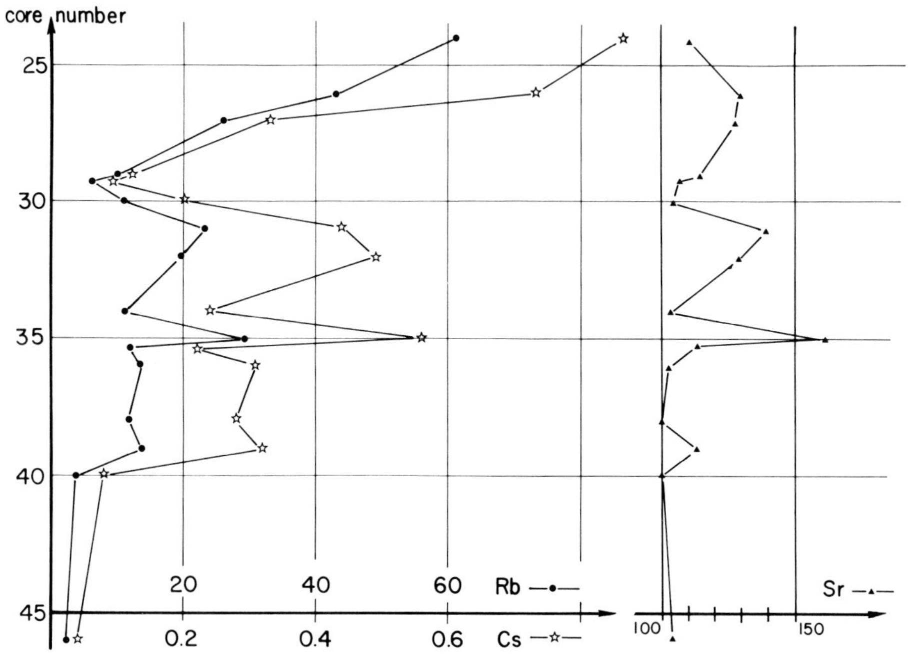

Fig. 2 : Hole 417 A : Rb (ppm), Cs (ppm), Sr (ppm) variation down the hole. Core 25 is about 270 m, core 45 is about 405 m.

correlates positively with the abundances of alteration minerals (unaltered to highly altered samples) but the elemental ratios, at least in first approximation, may be a function of the proportion of the different kinds of minerals present in the different zones. This could account for the dispersion of points observed in Figure 1. If sea water circulation is responsible for alteration, the uptake by rocks of alkali metals from sea water can be defined by the ratio of the concentration of these metals in altered rocks (in hole 417 A, initial concentration in fresh rocks can be neglected compared to concentration due to alteration) to sea water concentration. Thus, the figures mentioned Table 1, which represent the ratio of concentrations of one representative sample to sea water concentrations (for K, Rb and Cs), are proportional to the uptake of these metals. Cs uptake is higher than Rb uptake which in turn is higher than K uptake. We observe that the uptake of investigated elements varies in the opposite way compared to residence times (table 1). These observations are in agreement with a sea water origin of alkali metals

TABLE 2. Hole 395 A, aphyric unit A 3, 200 m thick. Note the possible variation in alkali metal concentrations compared to other elements.

	Ni	Ti	Tb	La	Ta	K_2O	Rb	Cs
33-2 127-129	114	10 200	0.85	4.3	0.22	0.24	2.9	0.02
35-1 53-54	113	10 260	0.91	4.0	0.23	0.18		0.03
37-1 136-141	121.	10 200	0.85	4.4	0.22	0.29	4.7	0.28
39-1 102-107	128	10 320	0.85	4.0	0.22	0.28	3.1	0.17
41-1 142-144	121	10 440	0.93	4.0	0.24	0.27	4.4	0.21
42-1 86-89	115	10 600	0.92	4.0	0.23	0.16	< 1.5	< 0.02
45-1 124-127	111	10 660	0.90	4.2	0.24	0.18	2.8	< 0.02
47-1 63-67	119	10 020	0.87	4.0	0.22	0.21	< 1.5	< 0.02
48-1 135-140	120	10 260	0.87	4.1	0.23	0.25	3.6	0.081
49-2 13-17	118	10 320	0.90	4.1	0.23	0.17	2.3	< 0.02
51-2 41-44	119	10 320	0.89	3.9	0.225	0.23	1.5	0.02
52-1 48-53	118	10 440	0.90	4.0	0.22	0.18	2.2	0.03
53-1 7-11	119	10 320	0.88	3.9	0.23	0.18	2.5	0.02
54-1 48-51	111	10 320	0.90	4.0	0.23		< 1.5	0.015
55-1 49-54	116	10 380	0.91	4.1	0.24	0.17	5.4	0.20

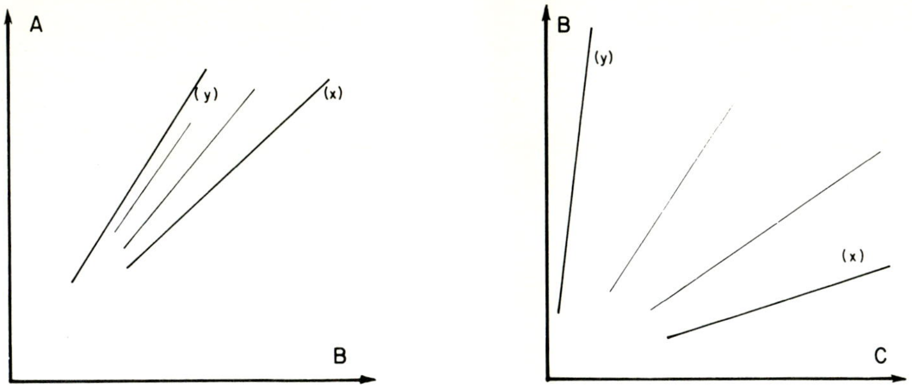

Fig. 3 : Hole 395 A ; variation of Cs and K_2O within three different basaltic units.

in altered basalts and if so, alteration of the oceanic crust could play a role in the budget of these elements in sea water.

The variation of Rb, Cs and Sr with depth in hole 417 A is given in figure 2. As previously mentioned a general decrease in concentration of Rb and Cs is observed. This agrees with the general decrease of alteration with depth (core numbers). The variation of Sr in alteration is very slight compared to the variation of alkali metals. In core 35 for instance the Rb concentration is 25 times higher than in unaltered samples while for Sr the enrichment ratio is 1.5. If we take into account that Sr occurs largely in carbonate phases, leaching moderately altered samples with acid solutions may enable us to measure reasonable values for $^{87}Sr/^{86}Sr$ (if no isotopical equilibration).

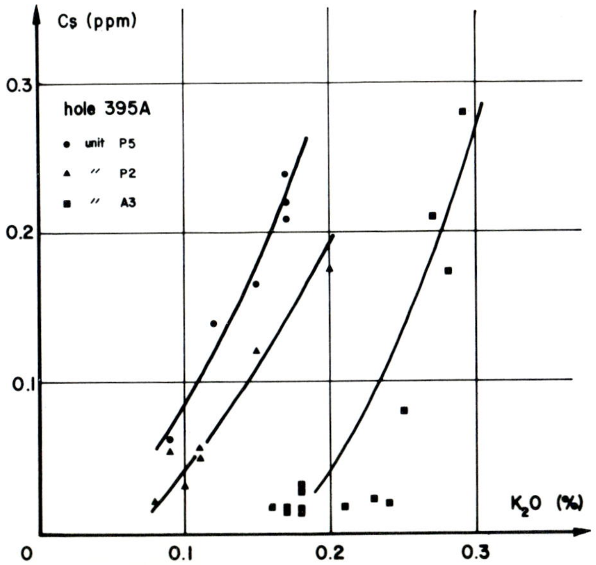

Fig. 4 : Way used to classify the elements according to their partition coefficients (see text). A versus C would show a similar figure.

In hole 395 A (leg 45, 22° N, 7 m.y. old) most samples selected for shipboard analysis can be considered to be fresh (H_2O^-, loss on ignition). In this hole, within a single homogeneous unit 200 meters thick we observe the variation of alkali metals due to moderate alteration of samples which otherwise are considered fresh. Table 2 shows the values obtained for a high partition coefficient element (bulk partition coefficient) Ni, and four low partition coefficient elements Ti, Tb, La, Ta together with K_2O, Rb and Cs. The Ni, Ti, Tb, La and Ta concentrations do not vary throughout this unit. The major elements and some other trace elements investigated confirm this homogeneity. Only the alkali metals (and Sb) show variation in the unit (Bougault et al., in press) with the greatest variation exhibited by Cs. In addition, with the exception of K_2O in 55 ≠ 1 49-54, K_2O, Rb and Cs covary. Figure 3 shows the variation of Cs versus K_2O for three different units of hole 395 A of Leg 45.

These observations indicate that the use of alkali metal concentrations in oceanic basalts as petrogenetic indicators should be made with extreme caution. Alteration processes can produce extensive variation in the concentration of these elements, even when the samples appear from other criteria to be fresh.

Partition coefficients and physico-chemical properties of trace elements

Through the behavior of trace elements, it has been attempted for several years to model the different parameters or processed involved in magma genesis. This modelling requires the knowledge of the mineral/melt partition coefficients for the trace elements and phases involved in the process to be modelled. Rare earths, K, Rb, Sr, Ba and the first transition series elements have been the object of numerous investigations of this type. This section is concerned with elements which behave incompatible or low bulk partition coefficient (< 1) or are "hygromagmaphile" (Treuil, 1973). These elements are presented in Table 3.

From the data obtained in the North Atlantic we will classify these elements according to their bulk partition coefficients and then discuss this classification with respect to their physico-chemical properties.

This classification utilizes the relative behaviors of the trace elements in fractionation processes. Considering three elements, A, B, C : A is plotted versus B and B versus C for different series of samples from series x to y (Fig. 4). If we know from previous investigation that the bulk partition coefficient for C (D^C) is lower than the bulk partition coefficient for A (D^A), the partition coefficient of B (D^B) can be classified with respect to A and C by comparing the position of the different series of samples (from x to y) and the observed range of slopes. In the given example, D^B lies between D^A and D^C.

Figures 5 and 6 show such a comparison for elements belonging to the third transition series : Tb, Hf, and Ta. Hf and Ta are fractionated to a large degree, whereas the fractionation of Tb and Hf is less. We know from other investigations (comparing Ta to La) that D^{Ta} is lower than D^{Tb}. From the position of each series of samples in each diagram, it is deduced that D^{Hf} lies between D^{Tb} and D^{Ta}. Following the same method of comparison, it can also be deduced that D^{Th} is lower than D^{Ta} (Fig. 7). In this way, a classification based on rock/melt partition coefficients is obtained. This classification is as follows :

$$D^{Sc} > D^{V} > D^{Y} \simeq D^{Tb} > D^{Ti} > D^{Zr} \simeq D^{Hf} > D^{Nb} \simeq D^{Ta} \simeq D^{La} > D^{Th}$$

Is this classification compatible with the properties of elements ? To account for the "incompatible" or "hygromagmaphile" character of these elements, two processes at least have to be considered :

i) Cation substitutions in the structure of a mineral. No stabilization through crystal field (Burns, 1970 ; Curtis, 1964) needs to be considered since all cations investigated have a rare gas configuration. The ease with which such cations enter a crystal structure depends mainly upon their

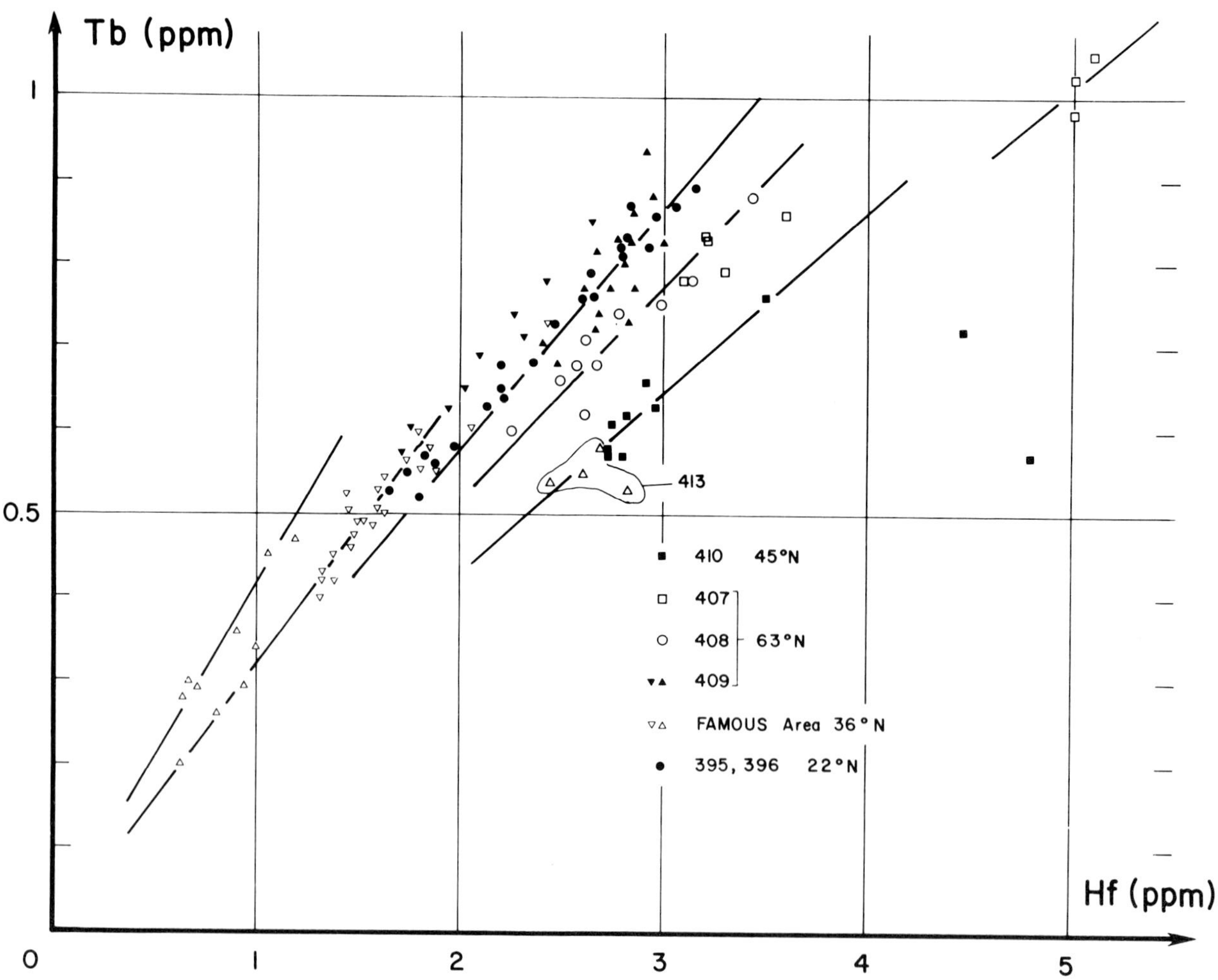

Fig. 5 : North Atlantic : Tb versus Hf.

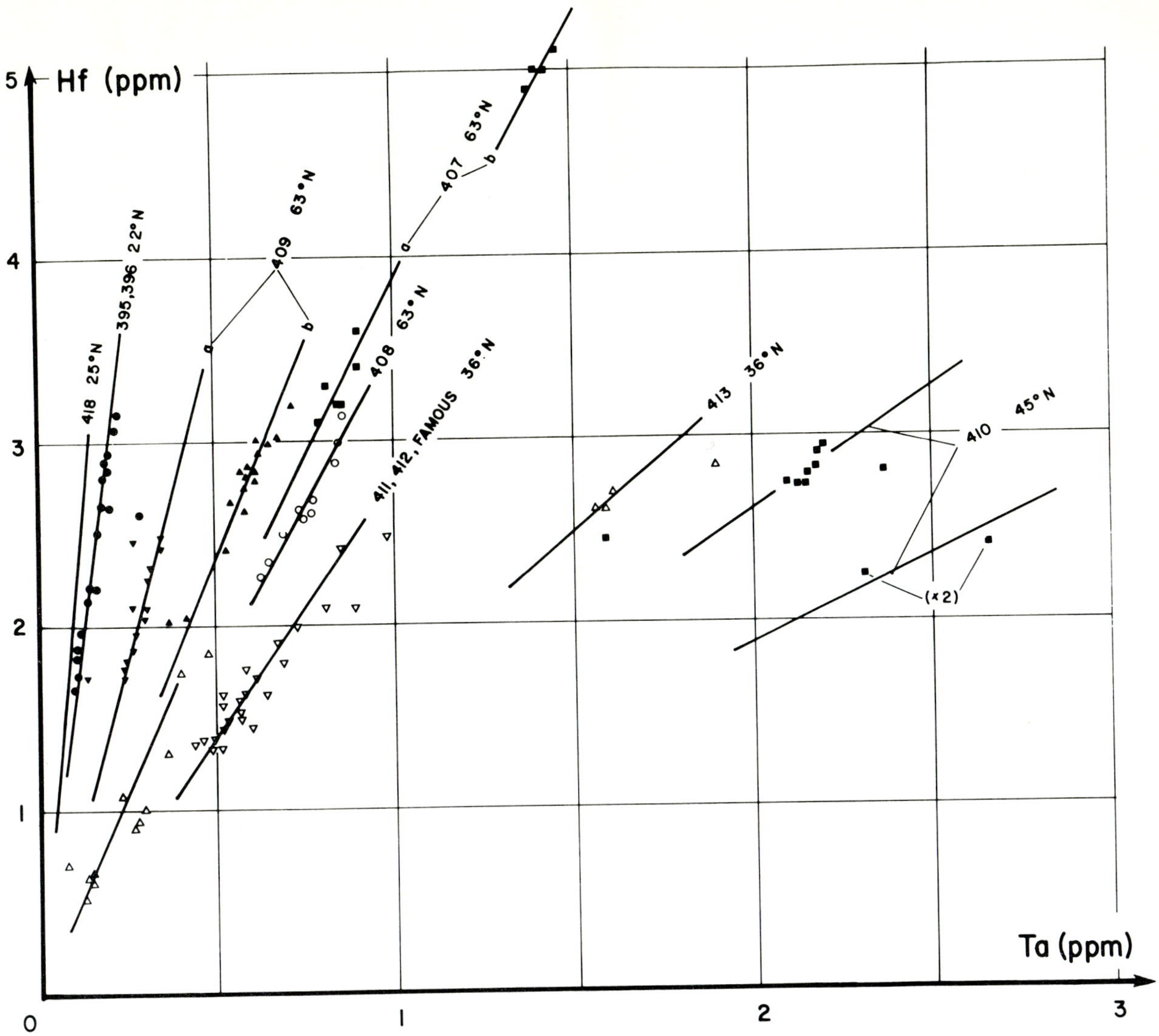

Fig. 6 : North Atlantic : Hf versus Ta.

ionic radii and their charges ; this corresponds to the "incompatible" character of the elements with respect to the host structure.

ii) Complex formation in the liquid (Treuil, 1973). This also depends upon charges and sizes of cations, but in a different way, for instance through the ratio n/r (charge/ionic radius) ; the higher is this ratio, the greater is the stability of a complex. The possibility of complex formation accounts for the "hygromagmaphile" character of these elements.

Table 3 shows for the elements of concern, the charge most common in nature, the ionic radius and charge/ionic radius ratio.

If we consider only the size of the ions, we note that the ionic radius decreases from Y to Zr to Nb (0.92, 0.79, 0.69 Å) and thus it is not possible to explain why the partition coefficients decrease from Y to Zr to Nb. But if we consider charges (3, 4 and 5 respectively, incompatibility increasing with respect to crystal structure) and even more n/r (3.26, 5.06, 7.25 ; possibility to form complexes in the liquid increasing) it is possible to explain why the partition coefficients decrease from Y to Zr to Nb.

Considering elements belonging to groupe IIIA, IV A, V A, of the second and third transition series, we observe that for each of the element pairs, Y-Tb, Zr-Hf and Nb-Ta, the two ions have the same charge and the same ionic radius (as a consequence the same n/r ratio). Consequently it can be concluded that each element in the pair processes the same incompatibility with respect to a crystal structure and the same possibility of forming complexes in the melt. The elements in each pair do not fractionate with respect to the other since they have equal or nearly equal partition coefficients.

D^{La} is close to D^{Ta} or D^{Nb} in spite of their differing charges, ionic radii and n/r ratios. La is more incompatible based on ionic radius, less incompatible based on charge and less hygromagmaphile based on n/r than Ta (see Table 3). These different effects compensate to produce similar partition coefficients for both elements. Despite the similarity in D^{La} and D^{Ta} (or D^{Nb}) it is important to keep in mind the differences (radius, charge, n/r) which exist between La and Ta when compared to Nb-Ta similarity. This point will be discussed later.

In the group IV A elements (Ti^{4+}, Zr^{4+}, Hf^{4+}, Th^{4+}) (table 3) the low value of D^{Th} is a consequence of its large ionic radius (1.02 Å) as compared to Zr^{4+} or Hf^{4+} radii (0.79 and 0.78 Å). A similar observation can be made comparing Zr^{4+} - Hf^{4+} (0.79 and 0.78 Å) to Ti^{4+} (0.68 Å). Within a column the classification of the elements according to their partition coefficients is dominated by the incompatible character rather than by the hygromagmaphile character of each element. Similar deductions can be made considering Sc^{3+}, Y^{3+} and the heavy rare earths (Tb^{3+} for instance).

From the results for the group III A (Sc, Y, heavy rare earths) and group IV A (Ti, Zr, Hf and Th) elements the following rule can be formulated :

(a) within a group (column) the higher is the ionic radius, the lower is the partition coefficient. Considering the second and third transition series (Y, Tb,Nb) and (heavy rare earths, Hf, Ta), it can be deduced :

(b) the higher is the possibility to form complexes in the liquid (n/r), the lower is the partition coefficient.

D^V is an exception to the rules : it satisfies rule (a) but does not satisfy rule (b), since its partition coefficient is classified between Sc and Ti. A possible explanation of this anomaly is that V may have a lower state of oxidation. If so, n/r would be lower and V would not be anymore in a rare gas configuration. Consequently it would be subject to some crystal field stabilization in octaedral structures.

The present study confirms known properties of these elements. It confirms the geochemical similarity of Y and heavy rare earths, and the incompatible behavior of Ti, see for example Cann (1970), Bougault (1977) and Bougault *et al.* (b) (in press).

Additional information is also available. These include the non-fractionation of Zr/Hf and Nb/Ta. Knowing D^{La} and D^{Tb} it is possible to place limits

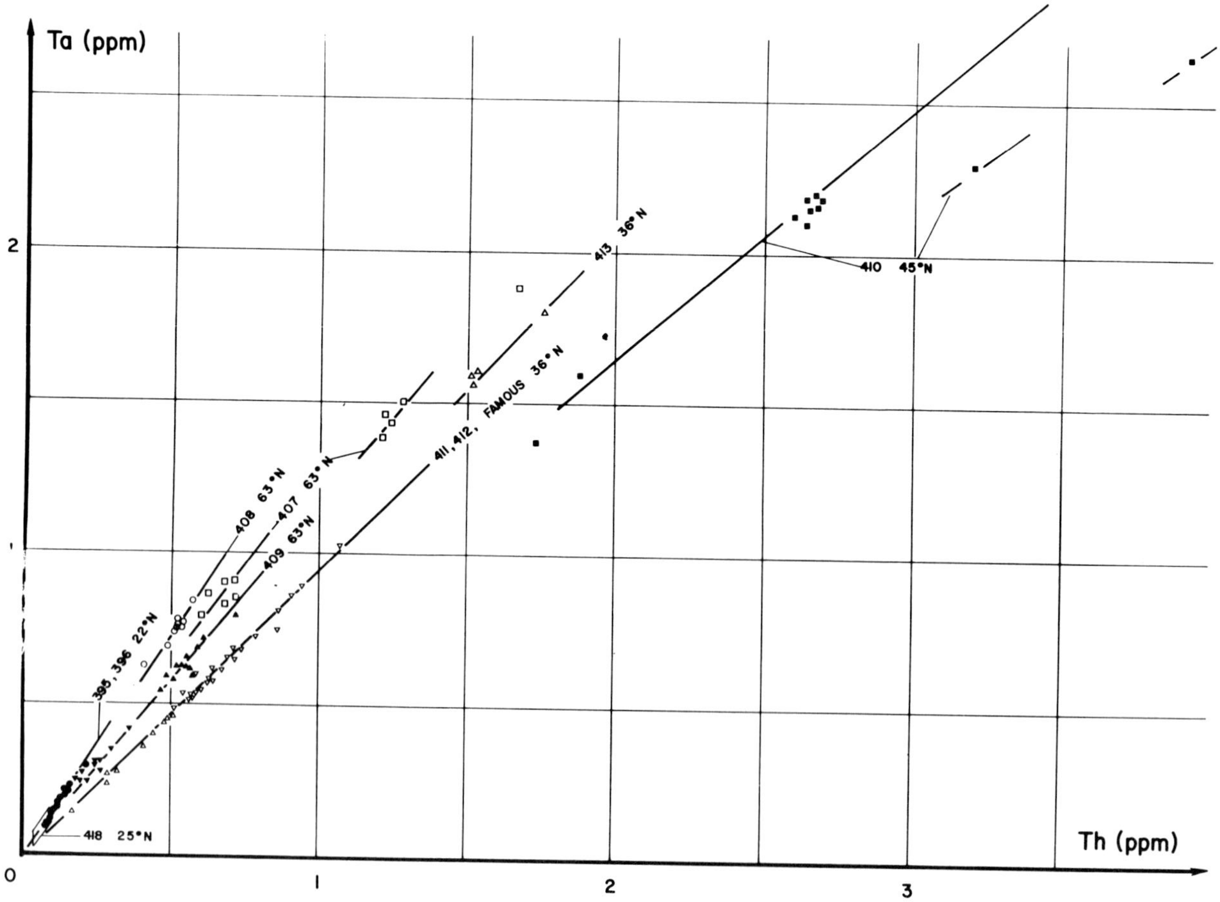

Fig. 7 : North Atlantic : Ta versus Th.

TABLE 3. "Low partition coefficient" elements investigated : ionic radii are from Goldschmidt (1926). > (or <) means that partition coefficients of the elements in left side (or up) is classified higher (or lower) than the element in right side (or down). ≈ means equal or similar, partition coefficients.

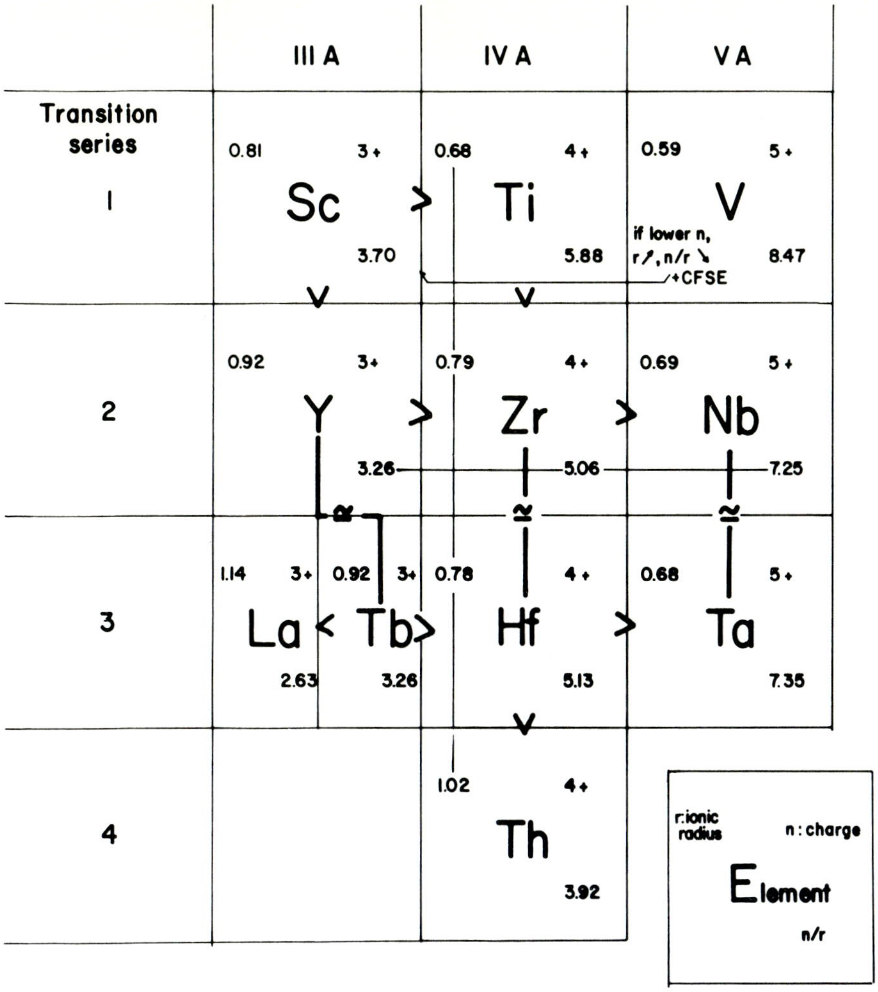

on D's for other elements and classify them accordingly.

Mantle homogeneity or heterogeneity : Y/Tb, Zr/Hf, Nb/Ta, La and Th. A comparison with chondrites and lunar samples

In the previous section, it has been shown that within each element pairs Y-Tb, Zr-Hf, Nb-Ta, fractionation of one element with respect to the other is not observed. In fact, two conditions are required to maintain Y/Tb, Zr/Hf, Nb/Ta ratios constant or nearly constant (fig. 8, 9 and 10) for all studied samples, although the absolute concentrations vary over a wide range (for instance, Nb varies between 2 to 90 ppm). These conditions are : (1) little or no fractionation of the elements in the pair has occurred, and (2) the ratios in the source regions prior to magma generations were constant and equal to that in the derivative basalts.

Because the samples studied were recovered in different parts of the North Atlantic(latitudes from 22° N to 63° N ; crustal ages from 7 m.y. at 22° N to 110 m.y. at 25° N), the point (2) can be interpreted in terms of homogeneity of the mantle at the scale of the North Atlantic. These couples of elements give us the opportunity to compare the primordial mantle with chondrites. Table 4 shows the values obtained for the ratios of these elements both in the North Atlantic samples and in chondrites. The Y/Tb ratio in chondrites is from Frey *et al.* (1968) and the range of ratios for Zr/Hf is obtained from Ehman and Rebagay (1970) and Ehman and Chi (1974). Ganapathy *et al.* (1976) give lower range of values for Zr/Hf in chondrites. Ehman *et al.* (1975) mention the difficulty in obtaining precise ratios of Zr/Hf in chondrites because of the small size of samples available and because of low abundance levels and error limits ($\pm$ 10 %).

Very little data are available for Nb and Ta. The values of 14 and 17 are from Graham and Mason (1972) and Vinogradov (1962).

The limits of the ratios reported in Table 4 for the North Atlantic samples include both analytical errors and possible fractionations within element pairs. The limits for Nb/Ta correspond very closely to analytical error limits whereas the limits for Y/Tb and Zr/Hf are approximately twice that of the analytical error limits. This suggests that despite the range of variation of concentrations (shown in Table 4 by the factor of variation), Y/Tb and Zr/Hf may fractionate a little more easily than Nb/Ta. This can be attributed to some slight differences between partition coefficients and to the orders of magnitude of the partition coefficients (Nb and Ta being the lowest).

Taking into account the chondritic data, analytical precisions, and comparisons of data from different laboratories, the ratios of Nb/Ta, Zr/Hf and Y/Tb attributed to a primordial mantle favor a chondritic composition for the earth.

Ehman *et al.* (1975) found Zr/Hf ratio to be 39 $\pm$ 1 in the Palisades Sill (New Jersey), 39 in two chilled margins and 47.3 $\pm$ 0.8 in the Littletown metamorphic series. These figures are in good agreement with our own data for oceanic rocks but contrast with the range of values that these authors have found for lunar rocks. Wanke *et al.* (1975) similarly found variations in Zr/Hf for lunar rocks as well as for Nb/Ta using the data of Willis *et al.* (1972). These authors attempt to explain the possible fractionation of Zr-Hf (and Nb-Ta) invoking possible different oxydation states of these elements. They state that achondrites which have Zr/Hf ratio higher than chondrites may reflect an even greater fractionation. This model is supported by oxygen fugacities calculated from Eu^{2+}/Eu^{3+} which decreases from achondrites to lunar ferro-basalts to terrestrial rocks. If such an interpretation is correct, *i.e.* the possibility

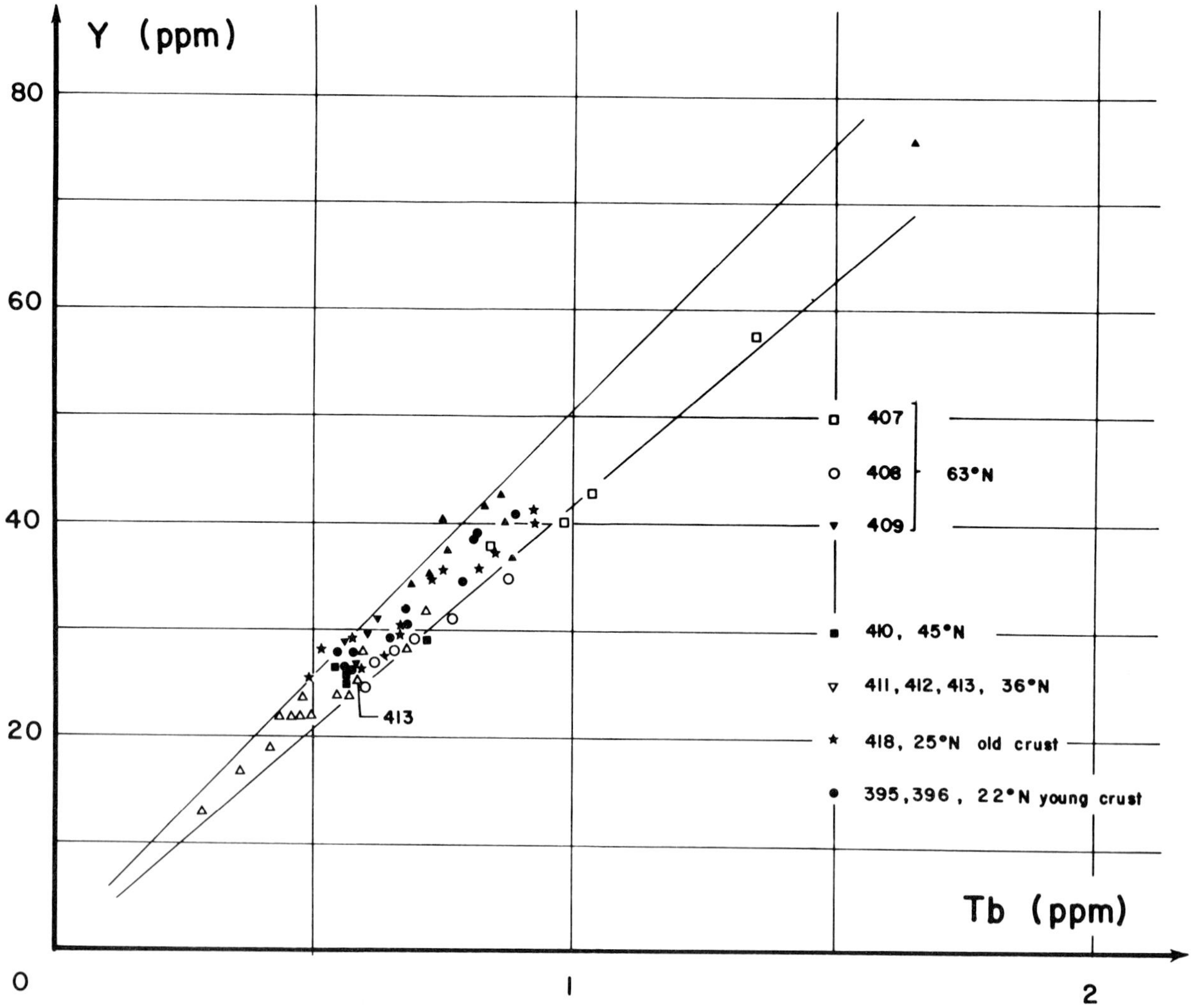

Fig. 8 : North Atlantic : Y versus Tb.

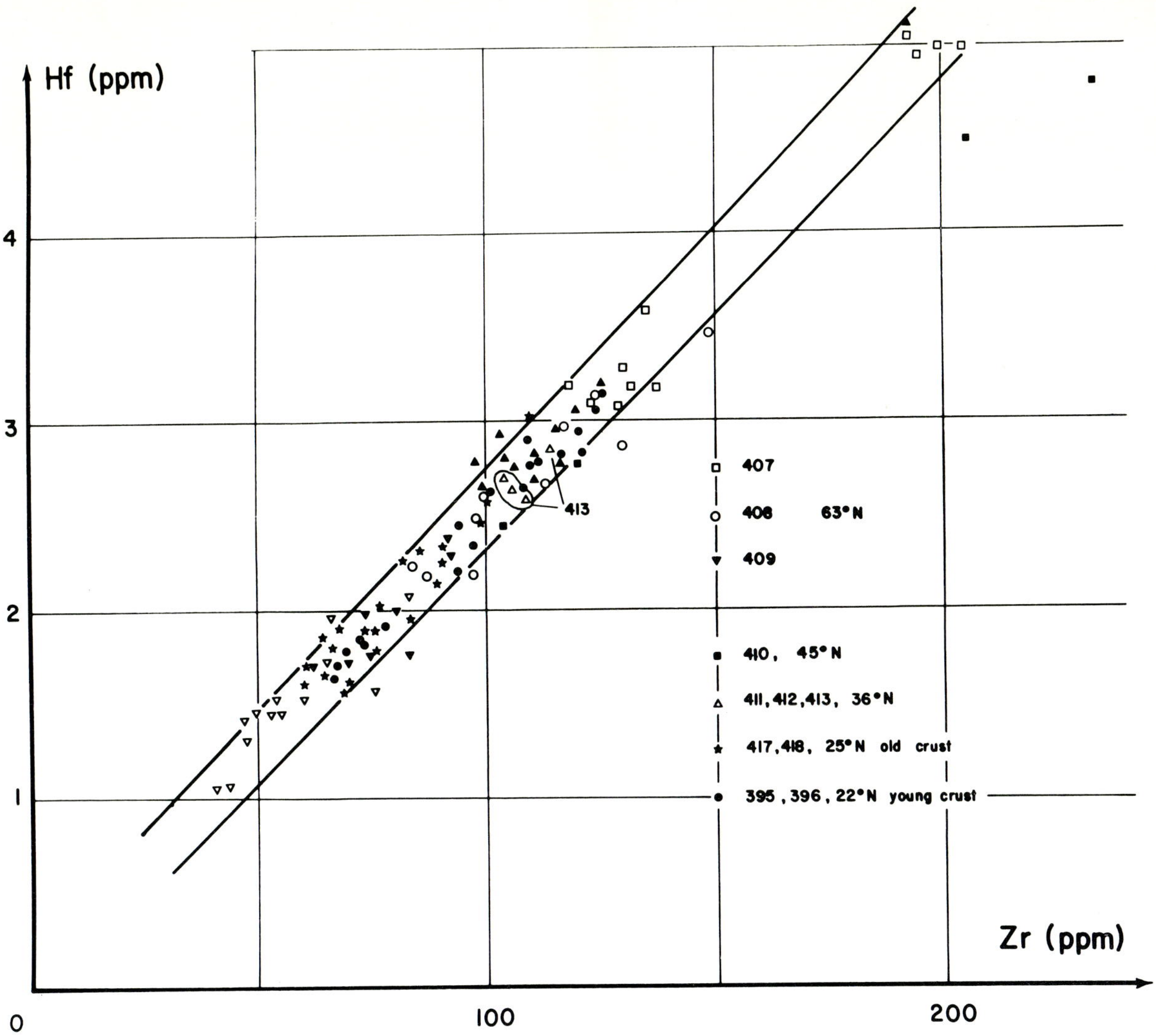

Fig. 9 : North Atlantic : Hf versus Zr.

of Zr-Hf or Nb-Ta fractionating by changing oxygen fugacity, we would conclude that relatively stable conditions exist in the terrestrial mantle.

A plot of La versus Ta for the North Atlantic samples defines two different lines. One of them includes samples recovered at 36° N, 45° N and 63° N which have large variations in absolute concentration. The second line includes samples recovered at 22° N (young crust) and 25° N (old crust). The similarity of D^{La} and D^{Ta} was deduced from the first line on the basis of a constant ratio for a large variation of absolute concentrations. We also mentioned previously that this similarity can be explained by a compensation of the incompatible and hygromagmaphile characters of La and Ta. However, because of differences between ionic radii and charges it is more likely that a fractionation may occur between La and Ta rather than between Nb and Ta. Large variations in the percentage of partial melting (see below) do not affect the La/Ta ratio. For this reason we interpret the two different La/Ta ratios as characteristic features of the mantle sources related to the two groups of samples. The highest La/Ta ratio corresponds to a characteristic "depleted" oceanic mantle (22° N and 25° N) and the lowest ratio corresponds to a less depleted mantle (36° N, 45° N, 63° N).

If a "primordial mantle" is confirmed through the constant values of Y/Tb, Zr/Hf and Nb/Ta ratios, then it may be deduced that one mantle type may derive from the other type through a mantle differentiation process. The variation of Ta/Th ratio (fig. 7), two elements which have very low but slightly different partition coefficients would reflect different degrees of this differentiation. The jump of La/Ta ratio would correspond to a major change of the solid (mineralogical for instance) through mantle differentiation.

Multiple source model, mixing of sources, mixing of magmas, partial melting

Langmuir *et al.* (1978) propose a method for testing different kinds of mixing in which an hyperbolic function represents the behavior of two elements or of ratios of two couples of elements.. We postulate that the ratio of two hygromagmaphile elements is not (or little) modified through fractional crystallization process. Then, the ratios of two elements have to be interpreted in terms of partial melting, mantle sources, mixing of magmas and (or) mixing of sources. Figure 13 represents the plot of Hf/Ta(deduced from fig. 6) versus Ta/Th (deduced from Fig. 7). If the studied samples were derived from two end members, the points defined by the various basaltic units should lie on an hyperbolic line. This is obviously not the case.

Partial melting and magma mixing. The variation of the ratios of two elements with percentage of partial melting is a question of the magnitude of the difference between their bulk D's which in turn are dependent upon the residual mineralogy. If both partition coefficients are low ($\sim$ 0.01 or lower ; ex.: Ta, Th) the ratio varies little, when the partial melting degree is sufficiently high. If the D's have different orders of magnitude (0.1 and 0.01 ; Tb/La, Ti/Th, Hf/Ta ...) the elements fractionate. We can test such possibilities of variations in the data of Leg 49 site 409 and from the FAMOUS area.

Site 409 (Leg 49): Unit b cannot be derived from unit a through fractional crystallization process. The similar Ta/Th and La/Ta ratios (fig. 7 and 11, and Table 5) and the increase of hygromagmaphile element concentrations from unit a to b (table 5) are consistent with fractional crystallization. However, at the same time Ni (and Cr) should decrease while the opposite is observed (Table 5). Partial melting can generate the observed elemental relationships. For similar Ta/Th or La/Ta ratios we observe different Ti/Th, Tb/La ratios (Table 5) and a different Hf/Ta ratio (fig. 6). In addition, one unit in hole 407 also has the same Ta/Th ratio and a different Hf/Ta ratio (Fig. 13). Rare earth patterns for these three basaltic units (Wood *et al.*, in press) are presented figure 14. These patterns, for a single Ta/Th ratio, vary in the same way as the Hf/Ta ratio.

FAMOUS area : The three samples reported in Table 5 may be considered as unfractionated on the basis of their MgO ($Mg^{2+}/Mg^{2+} + Fe^{2+}$) and Ni values. Each of them belongs to one of the three basaltic units of the FAMOUS area (Fig. 13). For similar Ta/Th ratios different concentrations in Ta, Th and other hygromagmaphile elements are

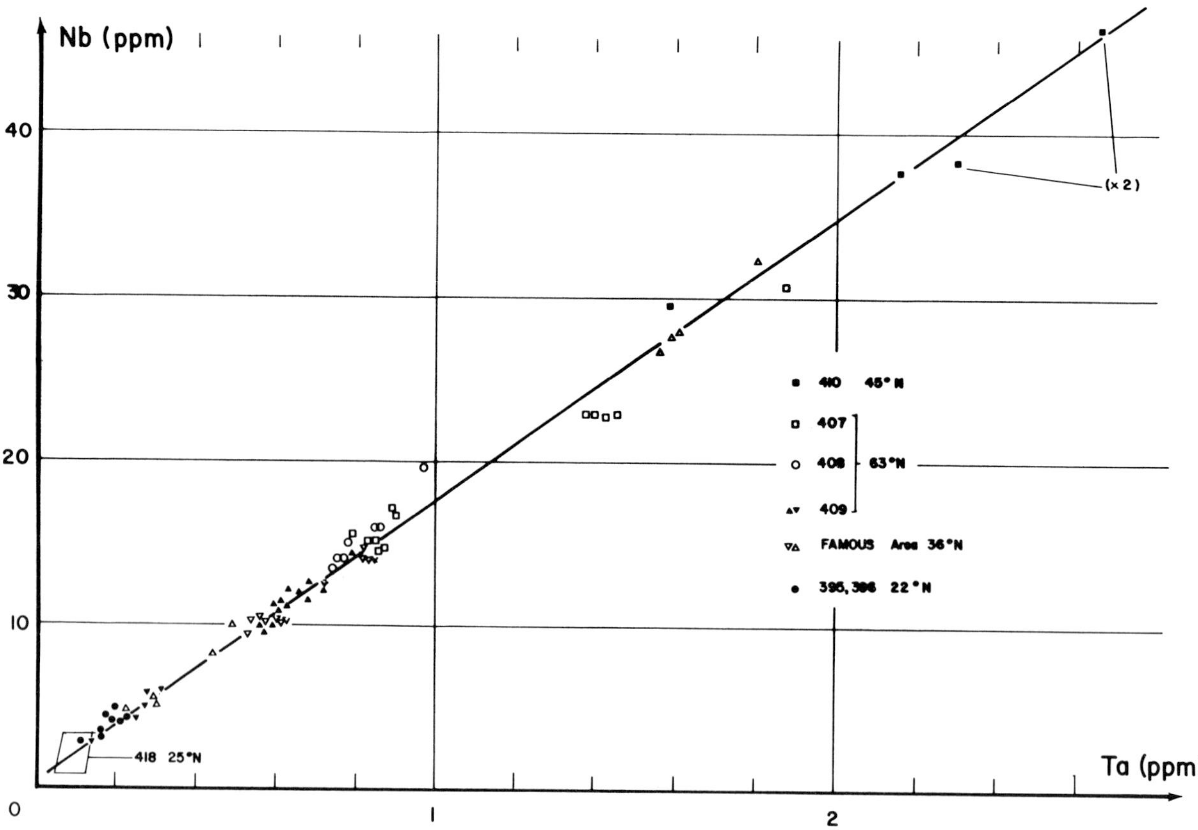

Fig.10 : North Atlantic : Nb versus Ta : (x 2) for the two right samples means that both Nb and Ta concentrations are twice higher.

TABLE 4. Y/Tb, Zr/Hf, Nb/Ta in North Atlantic basaltic samples and in chondrites.

	Y/Tb	Zr/Hf	Nb/Ta
North Atlantic ratio	42.5±5	39±5	17.5±2
Factor of variation	4.25	5	50
Chondrites ratio	41.7	30-40	15-17

observed. These features are believed to be due to different degrees of partial melting. Different Hf/Ta ratios (fig. 13), different Ti/Th and Tb/La ratios (Table 5) are observed together with different rare earth patterns (fig. 15).

The hole 409 example, no more than the FAMOUS example, is not compatible with magma mixing. Rhodes *et al.* (in press) interprete the different Ti concentrations observed for a single Mg/Fe ratio as the result of mixing a new unfractionated magma with a magma which has already undergone some fractional crystallization. The different Ti (and other low partition coefficient element) concentrations related to unfractionated magmas (FAMOUS example) cannot be interpreted in the same way. This observation does not mean that magma mixing does not exist but we would not consider this process as a rule as suggested by Rhodes *et al.*

Mixing of mantle sources or evolution of separate sources. The variation of Hf/Ta ratio (fig. 13) or of rare earth patterns (fig. 14, 15) with partial melting makes it difficult to recognize mixing of sources on Hf/Ta (or Sm/La) versus Ta/Th (or isotopic ratios) plots. Data for a large number of samples are required to see and recognize through the effects of partial melting variations in Hf/Ta representative of the source or as a consequence of the mixing of sources (Schilling, 1973). Langmuir *et al.* (1978) suggest that if mixing can be diagnosed from ratios of elements it should be confirmed by other elements or ratios of elements. On figure 13, an hyperbolic line could fit the points representing the different groups. If it is assumed that the scatter within groups is due to variable degrees of partial melting it could be concluded that

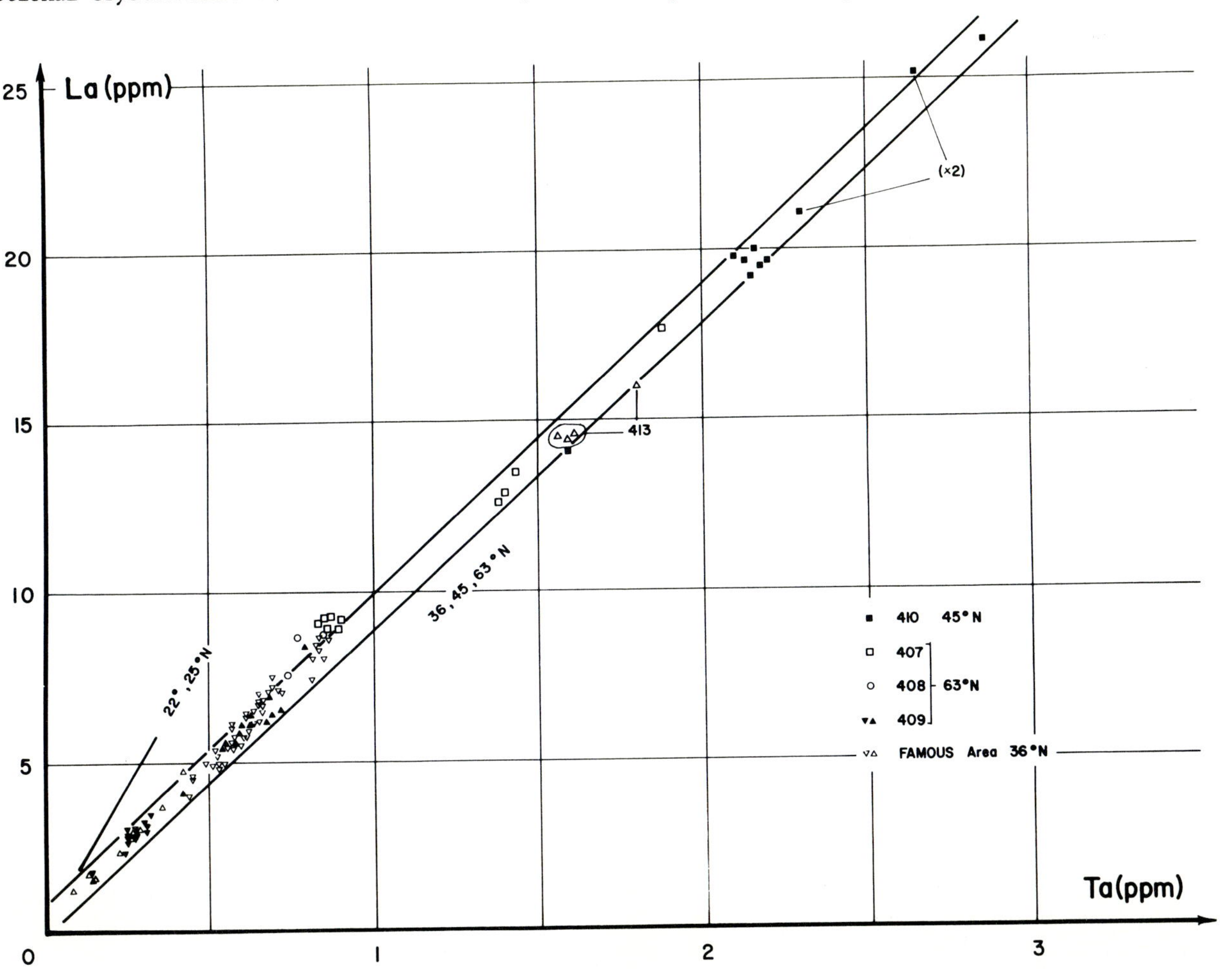

Fig.11 : North Atlantic : 36° N, 45° N, 63° N : La versus Ta (x2) for the two right samples means that both La and Ta concentrations are twice higher.

mixing of two different mantle sources is possible. But, considering another ratio, La/Ta (Fig.11, 12), only two values have been found (Fig. 13) making impossible an hyperbolic fit of La/Ta versus Ta/Th. Consequently, it is necessary to consider more than two sources. Langmuir *et al.* (1978) using other elements and other samples conclude that several models are possible. Mixing of sources is one of them, but even then more than two mantle sources are required. If several mantle sources are required, it is necessary to account for a multiple source model (O'Nions *et al.*, 1976). Each source would be derived from a primordial mantle which is evidenced by Y/Tb, Zr/Hf, Nb/Ta.

Co, Ni, Cr : fractional crystallization process

Because of their ionic configuration (presence of d electrons), Co^{++}, Ni^{++} and Cr^{+++} ions are subject to crystal field stabilization in octaedral structures (Burns, 1970 ; Curtis, 1964). The partition coefficients found agree with this observation (Bougault and Hékinian, 1974 ; Dale and Henderson, 1972). Crystal field stabilization has the effect of giving these elements a quite different behavior than low partition coefficient elements.

From theoretical considerations Co, Ni and Cr (to a lesser extent) should vary within a narrow range in the liquid phase as partial melting proceeds. This is independant of the chosen model and the type of initial solid material, except chromium which may depend upon spinel content (Bougault *et al.* (a), in press). Olivine being the major constituant of the solid and melting in low proportion relative to clinopyroxene for instance acts as a buffer for Co and Ni during the melting process. In addition, the results found for Ni, Co, Cr in the undifferentiated samples mentioned above fit with the theoretical calculation obtained from concentrations observed in peridotites and currently accepted bulk partition coefficients (Bougault *et al.*, in press (a)). Thus, variations in the concentrations of these three elements in basalts are largely due to fractional crystallization.

Because these three elements are enriched in minerals like olivine, clinopyroxene and spinel, it is important to note that basalt samples used for interpretation of the Ni, Co and Cr abundances should be free of these minerals. Glasses, aphyric basalts and plagioclase phyric basalts are appropriate. Plagioclase is a diluant of Co, Ni and Cr.

In the samples that we have analysed Co varies within a narrow range 40-50 ppm, Ni between 30 to 250 ppm and Cr between 30 to 600 ppm. This is observed for most of the oceanic tholeiites with the condition of selecting samples as mentioned above. Two ways of interpretation of these data can be given according that oxide and sulphide phases are involved or not in the process of fractionation.

i) The large variation of Ni reflects crystallization of olivine (Kd = 12) and or clinopyroxene (Kd = 4.4). However the narrow range of variation of Co (40-50 ppm) precludes that more than a few percents of olivine or (olivine + spinel) have crystallized because the Co partition coefficients in these minerals are 3 and 5 respectively. More likely the bulk partition coefficient of Co is probably close to one in order to explain its narrow range of concentration. This implies that a low Kd mineral with respect to Co has also crystallized, *i.e.* plagioclase.

Accepting the Rayleigh distribution during crystallization, the relationship between Cr and Ni in the liquid is :

$$\log [Cr] = \frac{D^{Cr} - 1}{D^{Ni} - 1} \log [Ni] + K$$

D^{Cr} and D^{Ni} being respectively the bulk partition coefficients of Cr and Ni ; K is a constant depending upon partition coefficients and initial concentrations of Cr and Ni. Data for glasses, aphyric samples or plagioclase phyric samples are

TABLE 5. Concentrations of some elements in the two units of Hole 409 and three samples in the Famous area

	MgO	Mg/(Mg+ Fe^{2+})	Ni	Ti	Tb	Th	La	Ta	Ti/Th	Tb/La	La/Ta	Ta/Th
409												
Unit a	∿ 8	-	80	7500	0.6	0.25	3.1	0.27	$3\ 10^4$	.19	11.5	1.08
Unit b	∿ 7.5	-	110	10200	0.8	0.53	6.0	0.63	$1.9\ 10^4$	.13	9.5	1.19
Famous area												
411-1-1 78-82	11.5	0.70	265	4184	.44	.28	2.4	.23	$1.49\ 10^4$	.18	10.4	0.82
Cyp 31-35	10.6	0.72	253	5100	.42	.58	5.4	.52	$.88\ 10^4$	.08	10.4	0.90
413 3-1 44-48	10.6	0.68	240	8573	.58	1.53	14.6	1.61	$.56\ 10^4$	.04	9.1	1.05

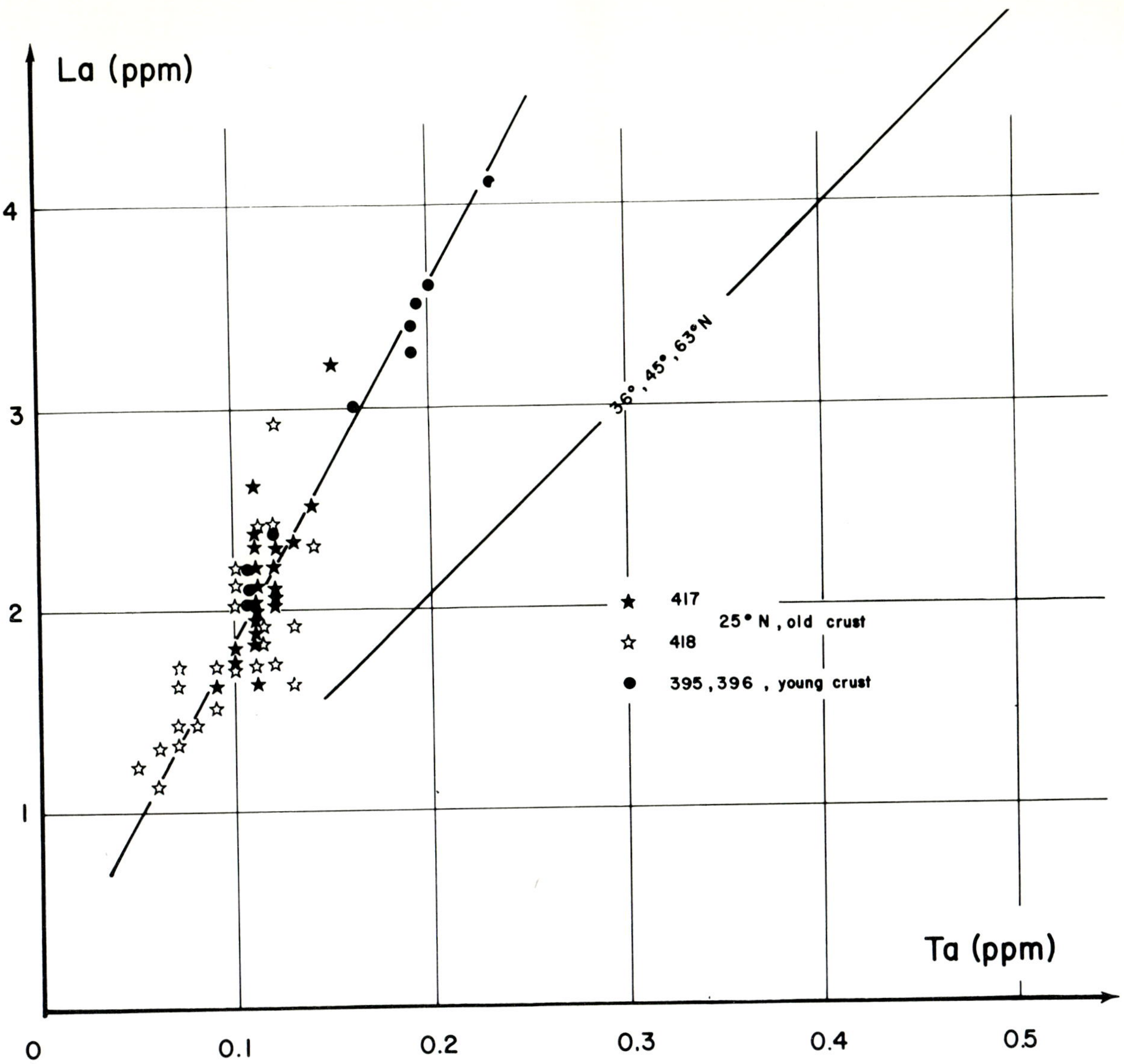

Fig.12 : North Atlantic : 22° N (young crust) and 25° N (old crust) La versus Ta.

plotted in figure 16. The scatter may be partly due to the presence of olivine, spinel or clinopyroxene phenocrysts in some of these samples. The upper part of the field (Ni : 250 ppm, Cr : 600 ppm) correspond to the most primitive liquids. The field cannot be a consequence of the fractionation of olivine alone ; Ni would be depleted extensively whereas Cr would be unchanged in the liquid (Kd^{Cr-}_{ol} is around 1). Olivine + spinel is possible in the early stages but we have seen above that this possibility is also limited by the Co range of variation. The scattering in the upper part of the field (early stages of crystallization) is probably due to different proportions of olivine (+ small amount of spinel) and plagioclase removal. The lower part of the field tends to be narrower and the slope

$$\frac{D^{Cr} - 1}{D^{Ni} - 1}$$

is compatible with plagioclase-clinopyroxene fractionation with very little or no olivine. The FAMOUS data, which fall along the average line which could be defined for the overall field is not a straight line (as suggested by the relationship from the Rayleigh law) ; the variation of slope needs a variation in bulk partition coefficients attributed to variation of partition coefficients themselves and the proportions of minerals as crystallization proceeds. Precise calculations of the amount of minerals crystallizing at different steps is difficult for several reasons. The basalts investigated to not derive from the same initial liquid produced by the same partial melting, even in a single hole. Ti, and other hygromagmaphile

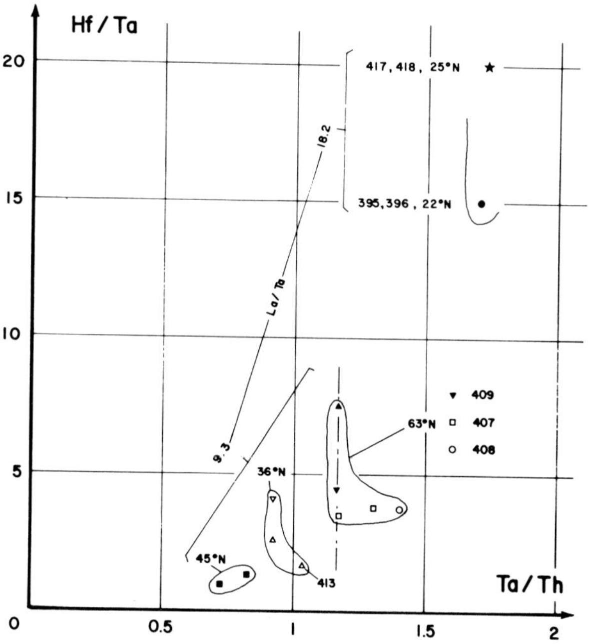

Fig.13 : North Atlantic : Hf/Ta versus Ta/Th.

elements do not increase regularly as Cr and Ni decrease (Bougault, 1977). Major element relationships show the same results : Ti does not fit (Bryan and Thompson, in press). Magma mixing (Rhodes *et al.*, in press) is an additional possibility.

ii) Oxides and sulphides even if in low proportions are an additional difficulty in estimating, through trace elements, the nature and the proportions of crystallizing minerals. If opaque minerals occur as crystallizing phases, they would probably deplete the liquid in Co and Ni and in addition in Cu and Zn as well. Such a fractionation does not fit the observed behavior of Co, Cu and Zn since these elements are not depleted. However, if sulfides are considered as an immiscible liquid, sulphides could be a buffer in respect of these

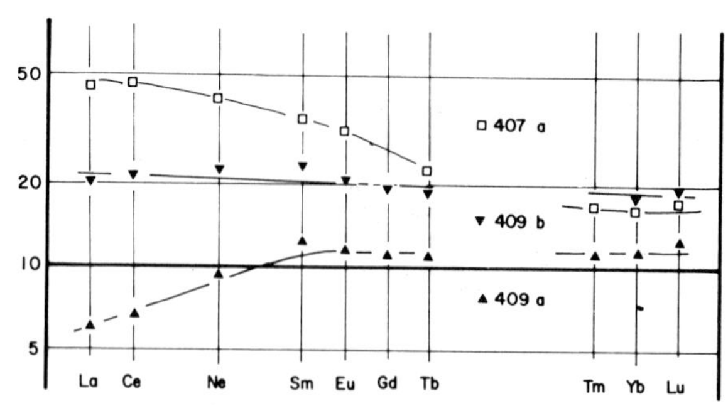

Fig.14 : Variation of rare earth patterns in Hole 409 (from Wood *et al.*, in press). 407 a pattern is also shown : both units of 409 and 407 a have similar Ta/Th and La/Ta ratios.

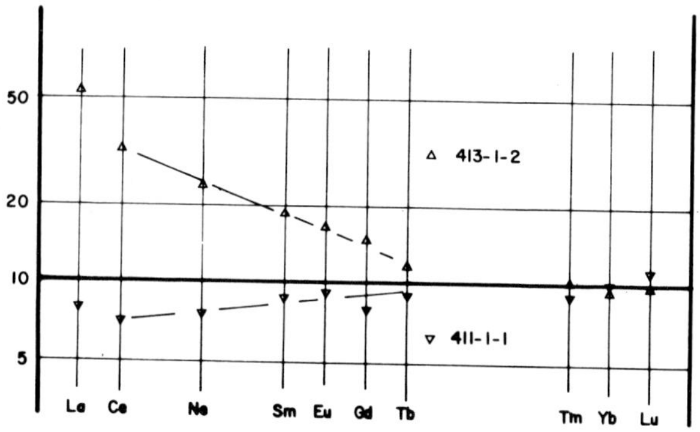

Fig.15 : Variation of rare earth pattern in the Famous area (from Wood *et al.*, in press). Both samples have similar Ta/Th and La/Ta ratios.

elements except for elements whose bulk D relative to crystallizing phases is higher than D relative to the immiscible liquid (Cr and Ni possibly). This late interpretation is only tentative as liquid (sulphides) - liquid (silicate) partition coefficients are not known for the investigated elements in these types of rocks.

Before drilling into the oceanic crust started, the hypothesis to find in a single hole different units deriving one from each other from the same initial liquid through fractional crystallization process seemed reasonable. This hypothesis now appears to be wrong. No more than one or two

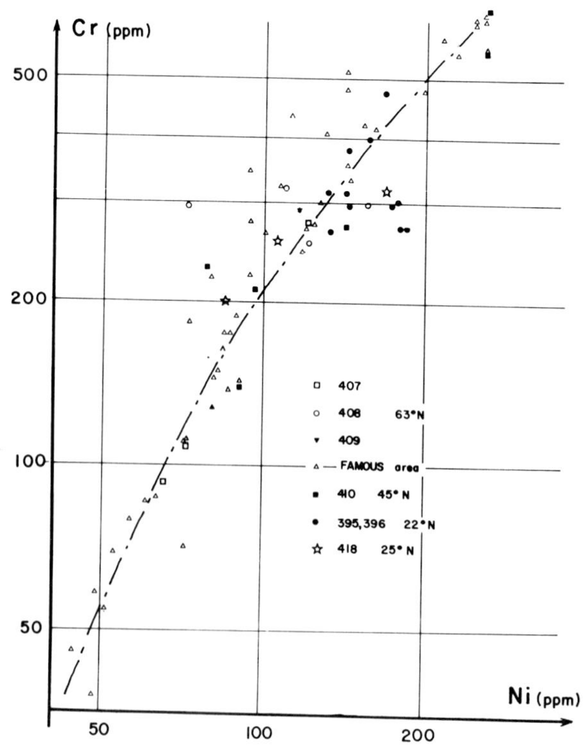

Fig. 16 : North Atlantic : log Cr versus log Ni.

TABLE 6. Oceanic basalt genesis : summary of the behavior of trace elements.

Basalts	Alteration →	Cs, Rb, K
↑		
Magmatic Chamber liquids	Crystallization →	Cr, Ni, Co
↑	Partial Melting →	different Ti/Th, Hf/Ta, Tb/La for a single Ta/Th
Mantle Sources	Mantle differentiation →	Ta/Th , Nb/Th
↑	22°,25°N ≠ 36°,45°,63°N →	La/Ta
Primordial, homogeneous Mantle	→	Nb/Ta, Zr/Hf, Y/Tb

eruptions occur involving the same initial liquid. Partial melting, formation of a magma chamber (possibility of mixing), crystallization and one or two eruptions seem to be a cycle that is frequently repeated. This corresponds to the notion of magma cycles proposed by Flower *et al.* (1977). This is why, accepting the hypothesis of dynamic melting (Langmuir *et al.*, 1977) or melting a solid which is already a residue (Bougault *et al.*, in press), different percentages and mechanisms of melting can explain the different rare earth patterns found within a single hole.

Conclusion

The major features resulting from this study are summarized in table 6. These include :

Y/Tb, Zr/Hf and Nb/Ta ratios are constant in basalts from the North Atlantic (22° N to 63° N) and do not vary with age. The values of these ratios, which are close to chondrite values, are in agreement with an homogeneous primordial mantle and with a chondritic composition of the earth.

Highly "incompatible" or "hygromagmaphile" element ratios (La/Ta, Ta/Th) require several mantle sources derived from the initial primordial mantle.

The variability of partial melting is responsible for the variation of moderately "incompatible" to highly "incompatible" element ratios.

High partition coefficient elements bring some constraints about the proportions of minerals which crystallize; it is not possible to deduce quantitatively these proportions of minerals since very few basaltic units, even in a single hole, derive from the same initial liquid (variability of partial melting).

The variation of alkali metals cannot be used to trace magmatic processes because of their extreme sensitivity to alteration.

References

Arcyana, Rocks collected by bathyscaph and diving saucer in the Famous area of the mid-Atlantic Rift Valley : petrological diversity and structural setting, Deep-Sea Research, 24, 565-589, 1977.

Bougault, H., and R. Hékinian, Rift valley in the Atlantic ocean near 36° 50' N : petrology and geochemistry of basaltic rocks, Earth Planet. Sci. Lett., 24, 249-261, 1974.

Bougault, H., Major elements : analytical chemistry on board and preliminary results, in Initial Reports of the Deep-Sea Drilling Project, vol. 37, 1977, Wahsington (U.S. Govt. Printing Office).

Bougault, H., First transition series elements : fractional crystallization and partial melting, Leg 37 DSDP, in Initial Reports of the Deep-Sea Drilling Project, vol. 37, 1977, Washington (U.S. Government Printing Office).

Bougault, H., Cambon, P., Corre, O., Joron, J.L., Treuil, M. (a), Evidence for variability of magmatic processes and upper mantle heterogeneity in the axial region of the Mid-Atlantic Ridge near 22° N and 36° N, Tectonophysics, in press.

Bougault, H., Treuil, M., and Joron, J.L. (b), Trace elements from 22° N and 36° N in the Atlantic ocean : fractional crystallization, partial melting and heterogeneity of the upper mantle, Leg 45, DSDP, in Initial Reports of the Deep-Sea Drilling Project, vol. 45, in press, Washington (U.S. Government Printing Office).

Bryan, W.B., Thompson, G., Compositional variation in a steady-state zoned magma chamber : Mid-Atlantic ridge at 36° 50' N, Tectonophysics, in press.

Burns, G., Mineralogical application of the crystal field theory, Cambridge Earth Sci. Series, 1970, Cambridge Univ. Press, 224 p.
Cann, J.R., Rb, Sr, Zr and Nb in some ocean floor basaltic rocks. Earth and Planet. Sci. Lett., 19 , 10, 7-11.
Curtis, G.D., Application of the crystal field theory : the inclusion of trace transition elements in mineral during magmatic differenciation : Geochim. Cosmochim. Acta, 1964, v. 28, p. 389-403.
Dale, S.M., and P. Henderson, The partition of transition elements in phenocrysts bearing basalts and the implications about melt structure, 24th IGC Section 10, 1974.
Donnelly, T.W., Low temperature alteration of the oceanic crust : an exceptionally well developped case from DSDP Megaleg. 2nd Ewing symposium, Lamont-Doherty Geological Laboratory, 1978. See also this volume.
Ehman, W.D. and Rebagay, Zr and Hf in meteorites by activation analysis, Geochim. Cosmochim. Acta, vol. 34, pp. 649-658, 1970.
Ehman, W.D. and L.L.Chyi, Zr and Hf in meteorites, Earth and Planet. Sci. Lett., 21, 230-234, 1974.
Ehman, W.D., L.L. Chyi, A.N. Garg, B.R. Hawke, M. S. Ma, M.D. Miller, W.D. James and R.A. Pacer, Chemical studies of the lunar regolith with emphasis on Zirconium and Hafnium, Proc. Lunar Sci. Conf. 6th, p. 1351-1361, 1975.
Flower, M.J.F., P.T. Robinson, H.U. Schminke and W. Ohnmacht, Deep-sea drilling project, Leg 37 : geochemical evidence for processes of crustal construction, Bull. Soc. Géol. France, 7, XVIII, n° 4, p. 819-823, 1976.
Flower, M.F.J., P.T. Robinson, H.U. Schminke and W. Ohnmacht, Petrology and geochemistry of igneous rocks : DSDP Leg 37, in Initial Reports of the Deep-Sea Drilling Project, vol. 37, Washington (U.S. Government Printing Office), 1977.
Frey, F.A., M.A. Haskin, J. Ann Poetz, and L.A. Haskin, Rare earth abundances in some basic rocks, Journal of Geophysical Research, vol. 73, 18, 6085-6098, 1968.
Ganapathy, R., G.M. Papia and L. Grossman, The abundance of Zr and Hf in solar system, Earth and Planet. Sci. Lett., 29, 302-308, 1976.
Goldberg, E.D., Minor elements in sea water, in Chemical Oceanography, Ed. by J.P. Riley and G. Skirrow, vol. 1, 163-196, Academic Press, 1965.
Goldschmidt, V.M., Probleme and Methoden der Geochemie, Gerlands Beitr. Geophys., 15, 38-50, 1926.
Graham, A.L. and B. Mason, Nb in meteorites, Geochim. Cosmochim. Acta, 36, 917-922, 1972.
Joron, J.L., C. Bollinger, J.P. Quisefit, H. Bougault and M. Treuil, Trace elements in basalts at 25° N, old crust, in the Atlantic ocean : alteration, mantle and magmatic processes, in Initial Reports of the Deep-Sea Drilling Project, Washington (U.S. Government Printing Office) (in preparation).
Langmuir, C.H., R.D. Vocke, G.N. Hanson and S.R. Hart, A general mixing equation with application to icelandic basalts, Earth and Planet. Sci. Lett. 37, 380-392, 1978.
O'Nions, R.K., R.J. Pankhurst and K. Gronvold, Nature and development of basalt magma sources beneath Iceland and the Reykjanes Ridge, J. Petrol., 17, 315-338, 1976.
Rhodes, J.M., M.A. Dungan, D.P. Balnchard and P.E. Long, Magma mixing at mid-ocean ridges : evidence from basalts drilled near 22° N on the Atlantic ridge, Tectonophysics, in press.
Schilling, J.G., Iceland mantle plume : geochemical study of Reykjanes Ridge, Nature, 242, 565-571, 1973.
Treuil, M., Critères pétrologiques géochimiques et structuraux de la genèse et de la différenciation des magmas basaltiques : exemple de l' Afar, Thèse, Université d'Orléans, 1973.
Vinogradov, A.P., Atomic abundance of the chemical elements in the Sun and in stony meteorites, Geokhimiya, 4, 291-295, 1962.
Wanke, H., H. Palme, H. Baddenhausen, G. Drelbus, E. Jagoutz, H. Kruse, C. Palme, B. Spettel, F. Teschke, R. Thacker, New data on the chemistry of lunar samples : primary matter in the lunar highlands and the bulk composition of the moon, Proc. Lunar Sci. Conf. 6th, 1313-1340, 1975.
Willis, J.P., A.J. Erlank, J.J. Gurney, R.A. Theil, and L.H. Ahkens, Major, minor and trace element data for some Apollo 11, 12, 14 and 15 samples, Proc. Lunar Sci. Conf. 3rd, 1269-1273, 1972.
Wood, D.A., J. Varet, H. Bougault, O. Corre, J.L. Joron, M. Treuil, H. Bizouard, M.J. Norry, C.J. Hawkesworth, and J.C. Roddick, The petrology, geochemistry and mineralogy of North Atlantic basalts : a discussion based on IPOD Leg 49, in Initial Reports of the Deep-Sea Drilling Project, vol. 49, Washington (U.S. Government Printing Office), in press.

VERY-LOW-TEMPERATURE HYDROTHERMAL ALTERATION OF THE OCEANIC CRUST AND THE PROBLEM OF FLUXES OF POTASSIUM AND MAGNESIUM

Thomas W. Donnelly

Department of Geological Sciences State University of New York Binghamton, New York

Geoffrey Thompson

Woods Hole Oceanographic Institution Woods Hole, Massachusetts

Paul T. Robinson

Department of Earth Sciences University of California Riverside, California

Abstract. A 200 m section of drilled basalt in the Cretaceous basement of the western North Atlantic (hole 417A, Deep Sea Drilling Project) recovered the most altered basalt found to date in the world ocean. The very high uptake of potassium (690 g/cm^2 for the 200 m thickness) requires the passage of very large volumes of sea water through the rock. The mineralogy of the rocks (k-feldspar, celadonite, smectite, zeolites) suggests very low temperatures, and we conclude that the water of alteration was the effluent warm water discharging from a cell of convecting water in the cooling oceanic crust. The character of the rocks and the absence of much of the earlier sediment support this idea.

The principal implication of this discovery is that only a small fraction of oceanic floor altering in this way could account for the excess of potassium inferred to be entering the oceans over that accounted for in oceanic sedimentation. A further implication is that the direction of both potassium and magnesium fluxes is opposite to that inferred for higher temperatures of alteration.

Introduction

Extensive alteration of the oceanic crust has been recognized for several years (e.g., R. Hart, 1970; Matthews, 1971), but nearly all of the examples have been of dredged basalts, metabasalts, and gabbros. Attempts to relate mineralogical and chemical character of these samples to the problem of elemental fluxes (Wolery and Sleep, 1976; Humphris and Thompson, 1978a) suffer in two ways from this limitation: 1) rocks available for dredging are those which occur on fault scarps, and a special structure such as this might imply a special process, such as hydrothermal alteration concentrated along the faults; and 2) rocks which are dredged regardless of their location are those which have been exposed directly to sea water, and their alteration might be typical of only a few meters of uppermost oceanic crust.

Numerous observations in recent years have led to the conclusion that water circulates through oceanic crust and provides an important means of heat transfer. The observations include the discrepancy between observed and calculated heat flows in younger crust (Wolery and Sleep, 1976), the direct observation of hot springs on the ocean floor in very young crust (Edmond, et al., 1977), and several instances of anomalously high and low heat flow values in younger oceanic crust (Anderson, et al., 1978). Experimental studies of basalt-sea water interaction (e.g., Mottl, 1976) and observations of Icelandic hydrothermal wells (Tomasson and Kristmannsdottir, 1972) illustrate the importance of higher temperature alteration both in speed of reaction and amount of elemental exchanges, but tell us relatively little about lower temperature processes. Finally, considerations of elemental fluxes into and out of the oceanic reservoir (e.g., Maynard, 1976; Drever, 1974), emphasize the problem of imbalances of certain elements, such as potassium and magnesium, and imply that the oceanic crust might be extensively involved as a sink or source for several chemical species. Thus, an example of severely altered oceanic crust for which extensive warm water circulation can be inferred is of special interest, and the calculated elemental fluxes are of potentially great significance.

A site drilled during Leg 51, Deep Sea Drilling Project, yielded the first appreciable recovery of significantly altered oceanic crust, and as such deserves careful consideration for its implications for oceanic crustal alteration. Hole 417A was located on crust about 110 million years of age, and is approximately equidistant from Puerto Rico and Bermuda. It penetrated 200 meters of crust, with about two thirds recovery, and showed extensive alteration even in the deepest core recovered. Hole 417D, which was about 450 meters away, penetrated an equivalent basalt section, but with two significant differences: the basalt was generally very fresh and the top of the basement was reached about 140 meters deeper with the entire sedimentary section present, whereas about 15-20 million years of sediment was missing from site 417A. We concluded (Donnelly, et al., in press) that the altered basalts from hole 417A represented extensive alteration which resulted from long-term circulation of warm (about 30°C) water through the crust, that this circulation was thermally driven and resulted ultimately from crustal cooling, and that the contrast in alteration between the two holes resulted ultimately from the difference in topography. Hole 417A was located on a topographic mound that remained free of sediment and became an upside-down drain for effluent convecting warm water, while hole 417D was located in a topographic depression and was quickly covered by sediment which retarded extensive water circulation.

Character of Site 417 Basalts

The basalts at hole 417A are brownish to brown-gray in color, especially in the more altered portions, while the basalts at hole 417D are gray in color. The mineral content and chemical composition at hole 417D vary little with depth. Fresh glass is common in this hole, and although there are scattered chemical and mineralogical signs of mild alteration (higher K_2O, oxidation of iron, occurrence of veins with calcite, smectite, celadonite), the phenocrysts are commonly unaltered and the basalts as a whole would be characterized as very fresh. The basalts at hole 417A are sufficiently altered that inferences about its original composition would be very difficult to draw if this hole had been the only one drilled. However, the phenocryst content (the phenocrysts are extensively pseudomorphed but can, in most cases, be identified) is fairly uniform and similar to that at hole 417D, leading to the view that the original compositions were probably very similar. Also, one lithologic unit, a massive, ophitic basalt in the lower part of the hole, and portions of thick pillowed flows above this unit are fairly fresh, based on K_2O content and on oxidation state of the iron. These are similar to the fresher basalts at hole 417D, leading again to the conclusion that the original basalt at hole 417A was probably essentially the same as that at 417D. Minor element distributions (especially Zr and Y) also support the conclusion that these basalts were originally essentially the same composition.

In detail the two sections are not correlative. At hole 417A there is a high content of hyaloclastite and a distinctive coarse-grained, ophitic, massive basalt occurs near the base of the drilled section. At hole 417D there is far less hyaloclastite, and a finer-grained massive basalt appears at a different horizon. We conclude that the two sections cannot be demonstrated to be precisely correlative sections which were separated by post-extrusive faulting, and we suggest that the shallower hole 417A represents an original accumulation of flows and hyaloclastites which formed a topographic high on the sea floor. Readers familiar with the results of drilling at these two holes will recall that a fault has been proposed (largely on paleomagnetic grounds) separating the two holes. The existence or non-existence of this fault does not affect the present discussion. We are concerned here with two drilled basement sites which were originally essentially identical in chemical composition and mineralogical character but which are not correlative on a layer-by-layer basis. For a reason which is not entirely clear but which may be the result of original constructional morphology, one of the occurrences (417A) existed as a topographic high shortly after eruption and the other (417D) occupied a much lower position.

The mineralogy of the basalts at hole 417A is consistent with a very low temperature alteration, and indeed consists dominantly of minerals characteristic of authigenesis in oceanic sediments.[1] A more extended and detailed account of the mineralogy of both the altered and fresh basalts of these two holes (as well as of a third hole, 418A, located about 6.5 km from 417A) will be found in the Initial Reports of the Deep Sea Drilling Project (Humphris, et al., in press; Juteau, et al., in press; Scheidegger and Stakes, in press; Rusinov, et al., in press; Pertsev and Rusinov, in press; Mevel, in press; Pritchard, in press). Basically all of the basalts at these sites are plagioclase-phyric with a widespread development of olivine and clinopyroxene phenocrysts. Although the mineralogical evidences of alteration are far more prominent at hole 417A, the differences between 417A and 417D are largely of degree and not of kind. Thus the basalts at 417A are highly oxidized with extensive replacement of plagioclase and groundmass by K-feldspar. Calcite and clay minerals (smectites and a green, micaceous mineral) are ubiquitous at 417A but more restricted at 417D. At both sites we find a dioctahedral smectite which is more or less interlayered with mica layers. At 417D a trioctahedral saponitic smectite occurs at lower levels. The micaceous phase is K- and Fe-rich with variable Mg. It is transitional in composition and x-ray character between a interlayer smectite and a mica (celadonite). Generally the micaceous phase is more abundant at 417A.

At hole 417A zeolites are conspicuous and include analcime, natrolite, phillipsite, and chabazite. The zeolites tend to occur somewhat deeper than the K-feldspar-rich upper samples, but the overlap is broad and no genetic implication is suggested by this poor zonation. At holes 417D (and 418A) zeolites are found in more limited quantities.

[1]J.R. Cann (this volume) names the mineral assemblage found at site 417A the "brownstone facies" and identifies it as a very-low-temperature facies transitional to weathering.

Pyrite is scarce but widespread at hole 417D. In hole 417A hydrated iron oxides appear to take its place, suggesting, in parallel with the bulk chemistry (below), a far more advanced degree of alteration at 417A. Such mineralogical differences as exist between the two holes would appear to suggest that at 417A there was a long exposure to oxidizing solutions, while 417D was exposed more briefly to reducing solutions.

The dominant mineralogy (K-feldspar, celadonite, analcime, smectite, calcite) is characteristic of very low temperatures of alteration. Oxygen isotopic evidence (Muehlenbachs, in press) indicates alteration at temperatures of about 30°C. The extent of alteration is not simply related to original textural and structural variations in the basalt. Thus, the abundance of hyaloclastite and the thinness of the pillow basalt units in the higher portions of the hole may explain the more intense alteration here. Conversely, two lithic units near the bottom of the hole have minimal fracturing and minimal alteration. However, beneath these the lowest unit at Hole 417A is a highly altered basalt breccia.

Chemistry of the Altered Basalts of 417A

Samples were selected so as to be as representative as possible of the basalt types. Chemical analyses were by wet (rapid) techniques; and the methods and results are given and the method discussed in more detail in Donnelly, et al., (in press).

The potassium values of the shallower, more altered samples, are the highest ever recorded for oceanic basalts, and in one case exceed 9 weight percent K_2O on a water-free basis. The highest K_2O values occur in the most severely altered basalt ("red-brown") and in basalt breccia fragments in dominantly hyaloclastic units. The lowest K_2O values are from the centers of the larger pillows in the lower units and from the massive, ophitic basalt near the bottom of the cored section. For budgetary calculations (tables 1 and 2), we have assumed an original K_2O value of 0.11, based on the eleven "freshest" 417D basalts analyzed. The excess potassium at 417A is found to be 690 g/cm^2 K for the 200 meter column drilled.

The relationship of potassium with depth for Holes 417A and 417D is shown in Figure 1. The clustering of low values for 417D, with a scattering of higher values, is broadly related to the degree

Table 1 Budgetary Calculations

	Composition - water free basis				Weight - 200 m column				
	417A hyalo-clastite	417A average basalt	417A hole average	417D average freshest samples	417A	417D-equiv. assume constant aluminum	change	417D-equiv. assume constant titanium	change
SiO_2	53.6	49.6	49.9	49.4	24979	26088	-1120 (4%)	24701	+278 (1%)
TiO_2	1.13	1.53	1.50	1.50	752.9	795.2	-42.3 (5%)	752.9	0
Al_2O_3	18.1	17.9	17.9	17.0	8979	8979	0	8508	+471 (6%)
Fe_2O_3	10.9	6.87	71.6	1.55	3583	817 (1634)		774 (1547)	
FeO	0.40	3.62	3.39	8.09	1700	4275 (3540)		4048 (3352)	
Fe_2O_3*	11.4	10.9	10.9	10.5	5472	5568	-96 (2%)	5472	+199 (4%)
MnO	0.18	0.15	0.15	0.16	76.9	87.1	-10.2 (12%)	82.5	-5.6 (7%)
MgO	5.80	5.41	5.44	6.14	2723	3246	-523 (16%)	3073	-350 (11%)
CaO	3.68	10.8	10.3	13.55	5139	7162	-2023 (28%)	6781	-1642 (24%)
Na_2O	1.70	2.25	2.21	2.40	1108	1269	-161 (13%)	1202	-94 (8%)
K_2O	4.36	1.59	1.79	0.12	895.3	65.4	+829.9	61.9	+833.4
P_2O_5	0.04	0.27	0.25	0.13	125	70.8	+54.2	67.0	+58
			calculated original density			2.73		2.51	
			calculated original porosity			.06		.13	

Table 1 gives average water-free compositions for hyaloclastite, basalt, and the total for 417A; and the sum of the weights of each oxide for 417A, assuming a 200 meter column 1 cm^2 in cross section. These are calculated on a core-by-core basis, using average densities and compositions for each core. The average composition of the eleven freshest 417D samples is also given. Assuming that the 417A rocks were originally the same in composition as 417D, the calculated weight of each oxide prior to alteration for 417A is calculated for two cases; assumed constant aluminum during the alteration, and assumed constant titanium. The original calculated density and porosity (the latter assuming an original rock density of 2.9 g / cm^3) are also given.

of fracturing and is typical of other drilled sites. The relative potassium values for the less altered hole 417D are more erratic than for 417A; consequently, the estimate of average potassium contents core-by-core is less meaningful. We find an average composition of about 0.28 percent K_2O for Site 417D, but with a very high standard deviation, the excess potassium calculated from this value on the order of 33 g/cm^2 K for a 200 meter column. This value may be typical of older "fresh" basalts, but quantitative data to support this speculation are still largely lacking.

Figure 2 shows the relationship between K_2O and H_2O. The samples from 417D show the typical rise of K_2O with absorbed

TABLE 2. Budgetary Calculations

	Case 1 assume constant aluminum gained	Case 1 lost	Case 2 assume constant titanium gained	Case 2 lost
	17.62 K^+	0.14 Mn^{++}	17.69 K^+	0.08 Mn^{++}
	0.76 P^{5+}	12.97 Mg^{++}	0.82 P^{5+}	8.73 Mg^{++}
	82.14 HCO_3^-	36.07 Ca^{++}	55.91 HCO_3^-	29.28 Ca^{++}
		5.20 Na^+		1.52 Na^+
		82.14 CO_3^{2-}		55.91 CO_3^{2-}
		41.07 H_2O		27.96 H_2O
Oxygen uptake				
a) original Fe_2O_3 = 1.50	8.06 O_2		10.04 O_2	
b) original Fe_2O_3 = 3.0	5.50 O_2		7.62 O_2	
Equivalent precipitated				
excess $CaCO_3$		24 meters		20 meters
volumes of void fillings		1570		689 volumes

Table 2 assumes that Si, Fe total, Al, and Ti are all conserved during alteration within the errors inherent in this calculation. It converts the gains and losses in the 200 meter column of rock into total moles. The electrical balance is maintained by consuming H^+ in the form of HCO_3^- and converting this to CO_3^{2-}. The evolved oxygen is assumed to be H_2O. The oxygen consumed to produce ferric iron is calculated for two cases: original Fe_2O_3= 1.50 and Fe_2O_3 = 3.0. The excess Ca and CO_3 are calculated as $CaCO_3$, which is converted to an equivalent column, 1 cm^2 in cross section. Finally, the amount of water that must be moved through the rock is calculated as a multiple of the void volumes for the column, using the calculated original porosities. This assumes that K^+ is the limiting species and that it is completely removed from seawater.

water seen in most oceanic basalt occurrences. The 417A samples, however, show a tendency towards migh higher values of K_2O at given water values, especially in the brecciated basalts and "red-brown" basalt margins. This figure suggests that these rocks have absorbed considerable quantities of water but might have been altered under different conditions, leading to a relatively higher "efficiency" of potassium absorption relative to water absorption.

As Figure 3 shows, calcium is depleted as strikingly as potassium is enriched, with an inverse relationship between these elements. The shallower "fresh"-looking basalts have lost approximately two-thirds of their original calcium, and the most severely altered rocks about nine-tenths. The removal of magnesium parallels that of calcium (Figure 4), but the amounts removed are far less. In other environments considered to represent higher temperature alteration, magnesium is enriched rather than depleted (Humphris and Thompson, 1978a). In the present case, there is little doubt that depletion is the general rule. The flux of this element depends critically on the conditions of the alteration process (Thompson and Humphris, 1977). Hyaloclastites are not so depleted in magnesium as the altered basalts, but in no case are they enriched in magnesium. Although fewer analyses of sodium were available for this study, this element appears to be slightly depleted (Figure 5). Fresher samples have abundant plagioclase, but more altered rocks have

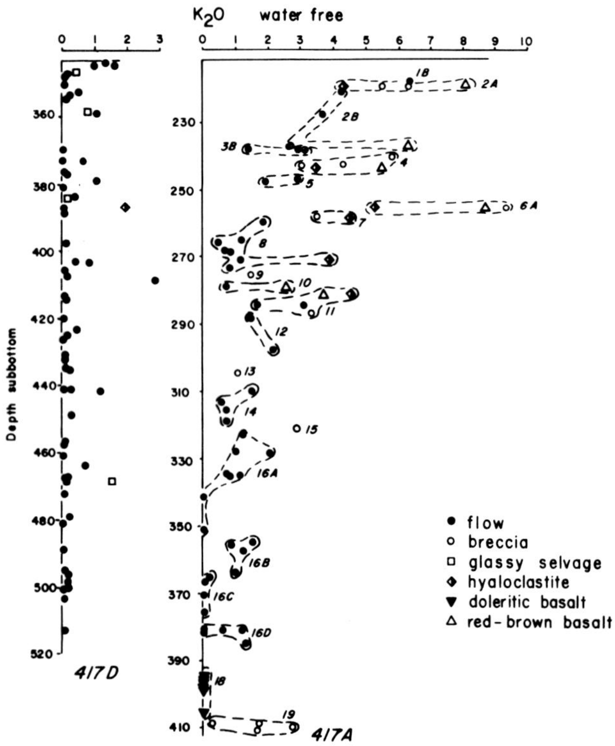

Figure 1. Diagram relating K_2O, calculated on a water-free basis, to depth, for Holes 417A and 417D. The tops of the basalt have been matched in position. Various symbols indicate lithologic types. Lithologic units are circled where more than one sample is given, and identified by italicized number. "Doleritic basalt" refers to massive basalt of unit 18.

K-feldspar and zeolites (variably Na-, Ca-, or K-rich) replacing the plagioclase.

Several minor elements were analyzed during this study. However, the relatively large changes in major elements are paralleled in only a few of the minor element species. Thus, boron, barium, and lithium (Donnelly, et al., in press), and rubidium and cesium (Joron, et al., in press) are all increased in the altered rocks with distribution patterns similar to that of potassium. Transition metals (nickel, vanadium, copper, etc., Donnelly, et al., in press) are relatively unchanged. Elements which are regarded as the most stable during low-temperature alteration processes (zirconium, yttrium, titanium, Donnelly, et al., in press) are somewhat depleted in hyaloclastites, with yttrium virtually absent. Possibly some of the smectite component of the hyaloclastite represents aqueous transport and precipitation of most of the constituents of basalts as dissolved species, with these three elements having been conspicuously untransported.

Budget of Alteration

The relatively high recovery of rock at Hole 417A permits an estimate of the total flux of major elements for the 200 meter column of rock drilled here. The calculation (Tables 1 and 2) was made by averaging the composition of the basalt core-by-core, and calculating the weight of each oxide in each core using measured rock densities. Hyaloclastite was found (six samples analyzed) to have a consistent chemical composition throughout, and its density was taken as 1.5 g/cm^3. The fraction of basalt and hyaloclastite was measured from shipboard photographs, and unrecovered intervals were assigned the values of adjacent recovered intervals. The composition of the presumed original fresh basalt was taken by averaging the freshest samples from Site 417D. Of

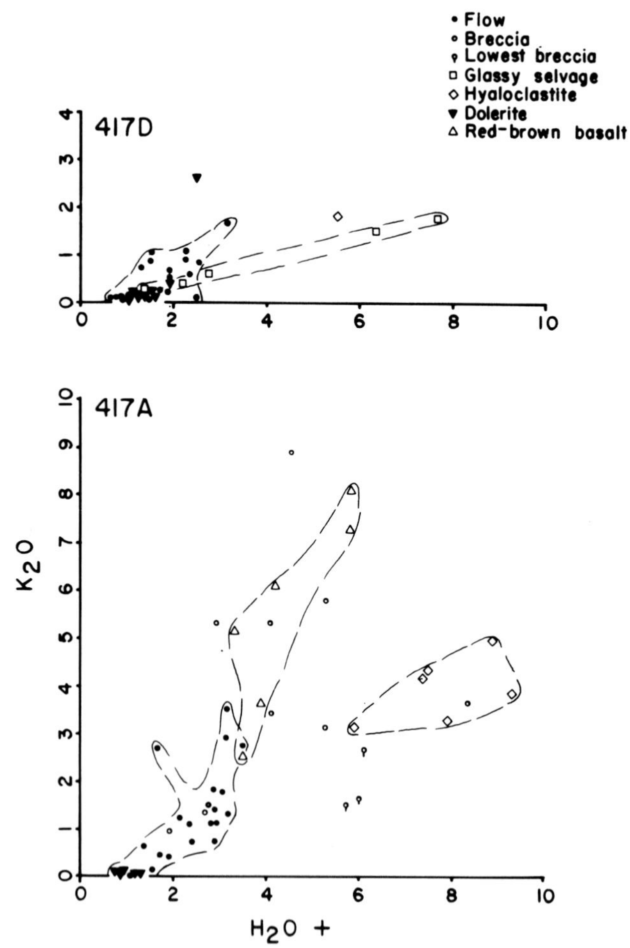

Figure 2. Diagram relating K_2O, calculated on a water-free basis, and H_2O^+ for Holes 417A and 417D. Symbols for various lithologies indicated. "Lowest breccia" refers to unit 19, "Dolerite" refers to unit 18, Hole 417A.

course, mild alteration has occurred at that site, and a minor degree of potassium absorption for the hole has been allowed for, as noted above. The formation density and porosity of the original basalt at hole 417A is not known, but can be estimated if we assume that some chemical element has been unchanged during alteration, and we know the weight percentage of this element at Hole 417D. As Table 1 shows, if we assume constant aluminum, then the basalt at Hole 417A must have had a porosity of 6 percent, if the rock density was originally 2.9 g/cm^3. There is some reason to believe that aluminum might not be the best normalizing element; the abundance of phyric plagioclase casts some doubt on the constancy of this element, and the amount of Al relative to Ti, Fe, and Si is slightly higher at Hole 417D, even though the latter three elements are fairly constant with respect to each other. A second calculation assuming that titanium, instead of aluminum, is constant gives an original porosity of 13 percent, which seems to be more in accord with measured and inferred values for shallow portions of the crust. The relative constancy of TiO_2 in the Hole 417D basalts supports normalization against this element.

A further problem in calculating the chemical exchange is that the original oxidation state of the iron is not known. Calculations have been made using differing values of original ferric iron,

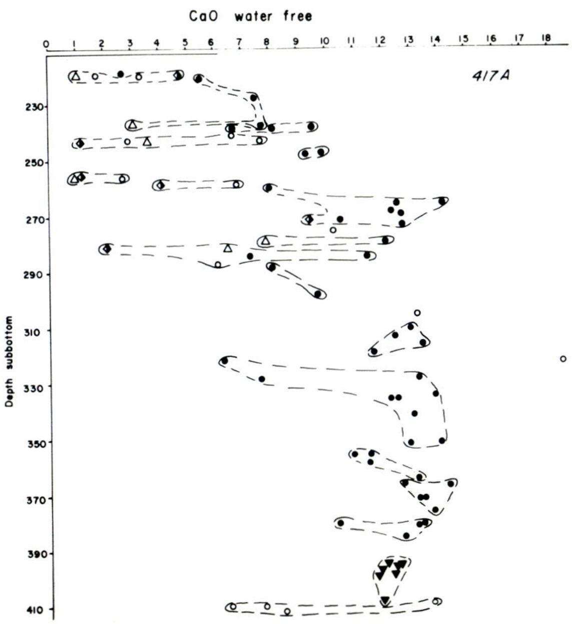

Figure 3. Diagram relating CaO, calculated on a water-free basis, to depth, Hole 417A. Sumbols as in Figure 1.

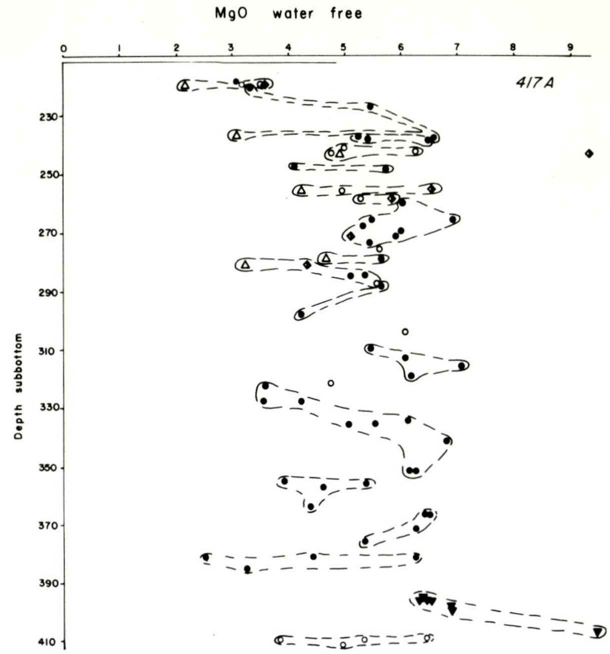

Figure 4. Diagram relating MgO, calculated on a water-free basis, to depth, Hole 417A. Symbols as in Figure 1.

and their effect on the total chemical change is fairly minor.

As Figure 6 shows, the loss of calcium, magnesium, and sodium is not balanced by the gain of potassium. The only ion available to maintain electrostatic constancy is hydrogen, which we assume must come from the breakdown of bicarbonate ion, producing hydrogen ion for alteration and creating carbonate ion. The balance of chemical species gained and lost under differing assumptions is shown in Table 2. We are assuming that the relative gains and losses of total iron, silicon, aluminum, and titanium are sufficiently small that they are not included in the balance. Potassium is the principal metallic ion gained from seawater, but hydrogen is apparently the major consumed species. Calcium is the principal effluent metallic ion, and carbonate the principal ionic species. There is a slight uptake of phosphorus and a loss of manganese shown by these equations, but neither element changes very much. The alteration process, then, takes up potassium, phosphorus, and bicarbonate, and releases calcium, manganese, sodium, carbonate, and oxygen (probably as water). The minor phosphorus (as well as barium, rubidium, and cesium) could be obtained from small amounts of trapped calcareous ooze, but the very large amounts of potassium have

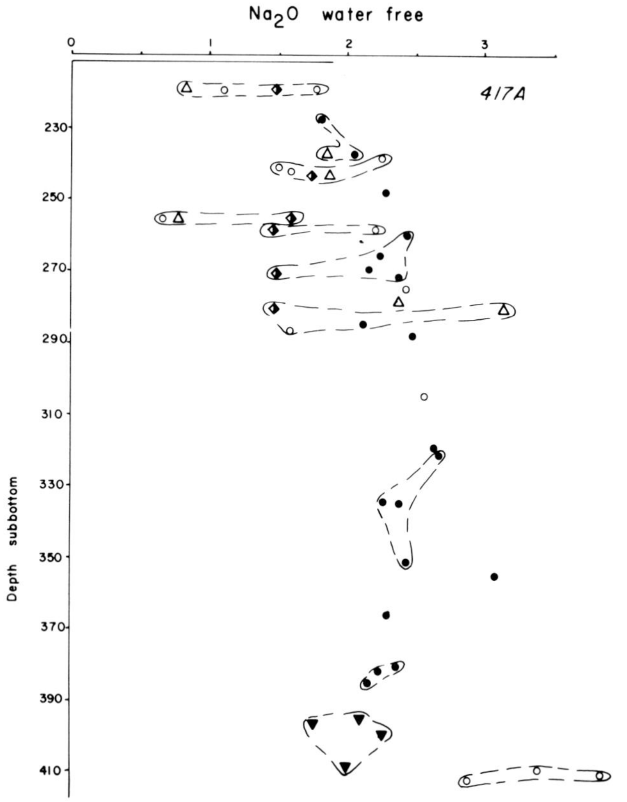

Figure 5. Diagram relating Na_2O, calculated on a water-free basis, to depth, Hole 417A. Symbols as in Figure 1.

to come ultimately from seawater.

The hydrogen ion taken up during alteration combines with oxygen in the silicate minerals to form water, as shown in Table 2. Further oxygen consumed during oxidation of the iron must come from dissolved oxygen in seawater. If we assume that the oxygen content of the seawater is about 4 ml/l, then the molecular ratio K^+ / O_2 is about 56, and as Table 2 shows, oxygen might be a limiting species, requiring on the order of twenty times as much water for alteration as does potassium. Bicarbonate poses a parallel problem; it is normally about one-fifth the molecular abundance of potassium in seawater, and a similarly large amount of water might be required to furnish the necessary bicarbonate. However, in the case of oxygen we know neither the oxygen content of the water, nor, more important, whether the oxidation was completed during the process of chemical exchange. In fact, the higher contents of ferric iron in the otherwise fresh (low K) units of hole 417A as well as the whole of 417D, especially when compared to such well studied younger basalts as holes 332A and 332B (Robinson, et al., 1977), suggest that oxidation may continue long after the other alteration has been completed. In the case of bicarbonate, we do not know the HCO_3^- content of the circulating water; if it passed initially downward through sediment before entering the basement, it might have acquired considerable initial alkalinity. Thus, in assuming that potassium rather than oxygen or bicarbonate is the limiting species, we are proposing a conservative limiting case. Doubtless more water moved through the system than these estimates suggest.

For the two cases (titanium constant or aluminum constant), we calculate slightly differing volumes of the amount of seawater required to alter the basalt, assuming that the limiting element (potassium) is completely removed. Because the two cases lead to differing original porosities, we find two quite different values for the ratio between volume of water required and volume of voids, i.e. number of times the voids must be filled. In the aluminum-constant case, to supply the necessary potassium we find an original porosity of 6 percent, a required volume of 1.76 m^3 per cm^2 column 200 meters long, and 1570 volumes of the voids. For constant titanium, these figures are 13 percent, 1.66 m^3, and 642 void fillings.

The discharge of calcium and carbonate ion supersaturates sea water for calcium carbonate. The total excess calcite corresponds to the equivalent of about 24 meters of $CaCO_3$ in the constant-aluminum case, and 20 meters in the constant-titanium case. Much of this excess calcite is now seen as the ubiquitous void and fracture fillings, but some discharged into seawater. Seawater is generally close to

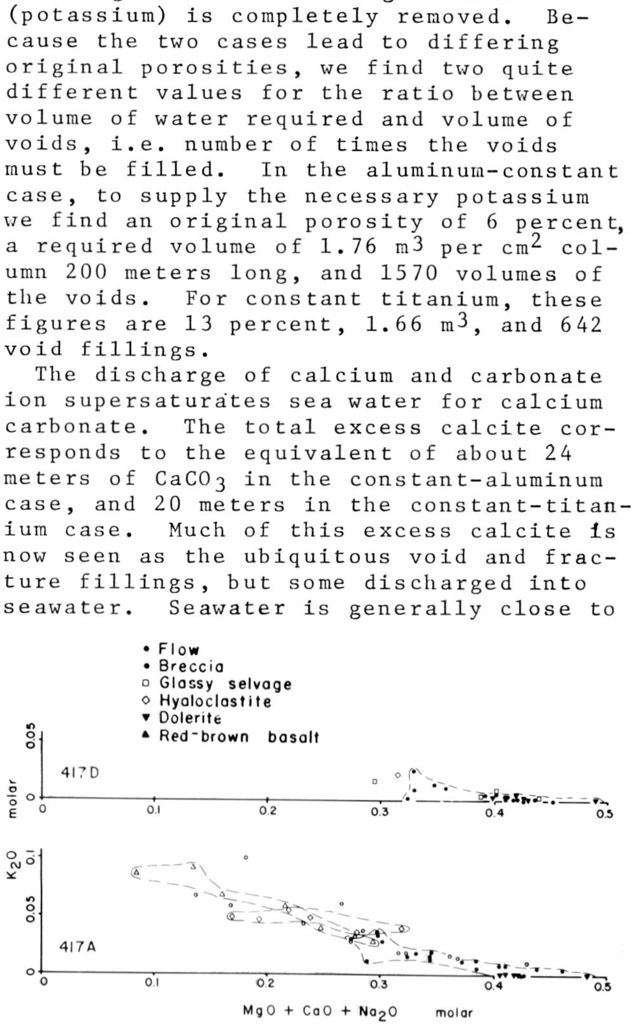

Figure 6. Diagram relating K_2O to $MgO + CaO + Na_2O$ for Holes 417A and 417D. Units are moles per 100 grams. Symbols as for Figure 2.

saturation at the depths of most mid-ocean ridges; the effect of discharge of Ca^{2+} and CO_3^{2-} ions into seawater might be to supersaturate it, or raise it to saturation level if it is slightly undersatured. A common observation, exemplified clearly at hole 417D, is that the basal meter or so of sediment has nearly perfect preservation of the calcareous fossils, and is overlain directly by sediment with no preserved carbonate. We might speculate that the interval of preservation corresponds to an important discharge of the products of basement alteration, even though the vent from which the effluents issue may be several hundred meters away. The recovery of this calcareous sediment could mistakenly be interpreted as accumulation above the lysocline.

The recognition of extensive alteration by warm water to depths of at least 200 m sub-basement requires a mechanism to drive the large volumes of water and thus rules out simple weathering. We propose that the water is the effluent portion of a convection cell driven by

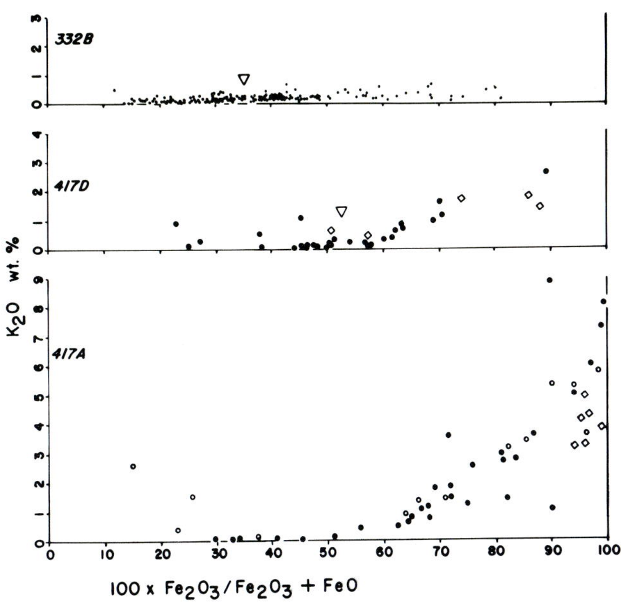

Figure 7. Diagram relating K_2O to iron oxidation, plotted as ratio of ferric to total iron, for basalts of holes 332B (data from Aumento, Melson, et al., 1977), 417D, and 417A. Symbols for latter two holes as in figure 1, except that symbols for dolerite and "red brown" basalt have been merged with flow symbol, for 417A and glassy selvage has been merged with hyaloclastite for 417D. Inverted triangle shows median sample.

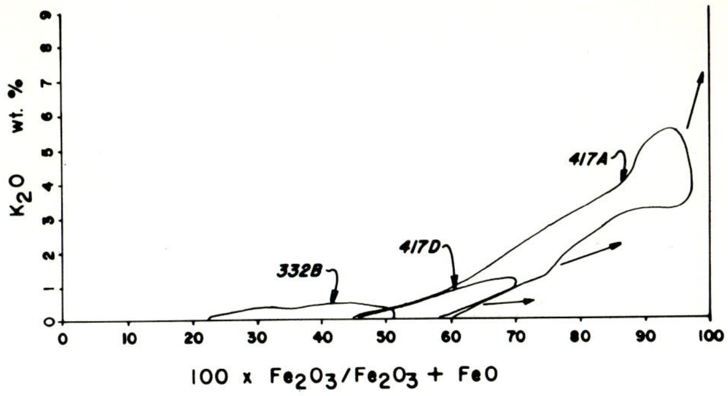

Figure 8. Summary diagram relating K_2O to iron oxidation for sites 332B, 417D, and 417A. Circled fields enclose central two-thirds of samples. Arrows show trends towards more oxidized samples. Data from figure 7.

the cooling of basalt. The inferred contrast in original topography between 417A and 417D completes the model: the elevated basalt area remains free of sediment cover for several (up to about 20) millions of years and persists as a porous, upside-down drain. The "recharge" area (using a hydrologic analogy) is the lower, sediment-covered area. The relative areas of "recharge" areas affected by downward moving of cold water to "drains" with upward moving warm water is probably high, and is equivalent to the probability of drilling blindly into a "drain" as well as to the ratio of the amounts of water affecting the basement rocks in the two areas. Thus, the geometry of the convective cells results from early-formed topographic singularities.

The minimum heat necessary to drive the convective water can be calculated by assuming a temperature rise of 30 degrees celsius (coresponding to oxygen isotopic measurements). The heat required for the cases of constant aluminum and constant titanium are nearly the same, because the volumes of water moved are nearly the same. They correspond to about 5×10^7 calories for a 1 cm^2 cross section column 200 meters long. The original heat in a column of basalt, assuming 100 cal/g for latent heat of cooling and 0.3 cal/g°C for specific heat of basalt cooling from 1100°C shows that the 200 meter column could not in itself supply more than half the necessary heat. We could heat no more than 0.8 m^3 of water with this basalt, but we need to move through at least 1.7-1.8 m^3. Thus, the heat necessary must come in part from depths greater than the observed alteration, i.e. the cooling crust in general.

Ageing vs Mesothermal Alteration of the Oceanic Crust

The recovery of 500+ meter drilled sections of oceanic crust of approximately similar composition during Legs 37 and 51 through 53 of the Deep Sea Drilling Project affords us an invaluable opportunity to add the dimension of time to the alteration process. As figure 7 shows, the relatively fresh basalts of the upper 200 m of hole 417D are similar to the basalts of Site 332B (age approximately 3 million years) except that the older basalts are more oxidized and tend towards higher K_2O values at a given level of oxidation. Figure 8 summarizes the groupings of basalts (circles enclose about two-thirds of the available analyses and arrows indicate trends towards more oxidized samples). If we assume that the basalts of 417D represent what those of 332B might look like after the passage of about 107 million years, then we note the relatively greater importance of oxidation than of potassium uptake with ageing.

Petrographic study of the 332B basalts show two stages of alteration. The first stage alters olivine and glass but produces a little chemical change, other than hydration of the glass. A later but more restricted alteration occurs along fractures and produces yellow to red oxidized basalt containing some high-K smectites. The limit of this alteration would presumably be met when production of hydrated phases causes the tiny fractures to close, terminating water circulation. The more advanced alteration at hole 417D indicates that the basalts at hole 332B have not reached this stage, or, more probably, that some minor tectonic activity re-opened fractures.

As shown also in figure 2, the uptake of potassium in 417A contrasts with that at 417D in more than degree: at equivalent water contents and levels of iron oxidation, the 417A basalts have taken up more potassium than those at 417D. Samples from the two holes shown in figure 8 have different trends, and we suggest that the difference represents the nature of the early alteration process. The warm temperatures may be responsible for the difference in potassium uptake. Interestingly, there are several samples from 417D with more than 2 percent K_2O, which is a very unusual value in even the more altered pillow basalts. Thus, the very slight temperature differences between dredged pillow basalts at 0^oC and ageing crust with a probable temperature slightly higher than zero may account for this mild potassium uptake in older crust. The assessment of potassium uptake or release by the crust will depend critically on an extensive knowledge of temperature distribution and water access during the ageing process.

Evaluation of Elemental Fluxes - the Problem of Dredged vs Drilled Basalts

Previous studies of dredged pillow basalts (e.g., S. Hart, 1969; R. Hart, 1970; Thompson, 1973) clearly indicated that basaltic rock exposed to seawater undergoes reaction even at ambient bottom water temperatures. Chemical changes included uptake from seawater by the rock of H_2O, K, P, Pb, Cs, B, Li and occasionally Ba and Sr. Elements leached from the rock included Si, Ca and Mg. Ferrous iron was oxidized but elements such as total Fe, Mn and Na show less obvious effects; Ti and Al, and trace elements such as Zr and Y, were apparently unaffected.

The rates of reaction during submarine weathering of the rocks were apparently slow. R. Hart (1970) calculated rates from plotting the composition of oceanic basalts reported in the literature against their calculated age. In 1973 Hart again calculated rates of exchange using the relationship of change in density with change in composition of weathered basalts, and the rate of change in seismic velocity seen in layer 2 across ocean basins. Thompson (1973) calculated rates of exchange from detailed studies and mass balance in individual dredged pillows or cooling units from DSDP cores. In a more systematic and detailed study of dredged pillow basalts recovered on a transect from the median valley of the Mid-Atlantic Ridge at 23^oN out on to the western flank and representing an age range of 0 to 57 million years, Thompson and coworkers (Thompson and Rivers, 1976; Rivers, 1976; Thompson, et al., in preparation) calculated rates of alteration at low temperature. Thompson and Humphris (1977) summarized these and previous calculations by Hart, and they are shown here in Table 3. Trace element uptakes (Rb, Li, B, Cu and Zn) are of the order 1 to 2 x $10^{-12} g/cm^3 yr$. These exchanges were considered to represent a maximum rate for low temperature weathering, the exposed basalts representing a large water/rock ratio. It was indicated that these fluxes operating over an 85 million year period, equivalent to 200 x 10^6 km^2 of sea floor, would have to be representative of the upper 0.6km of oceanic basement, or even greater depth,

Table 3 Rates of exchange between dredged basalt pillows and seawater at low temperatures.

Units = 10^{-9} g / cm^3 yr

	Range	Average
H_2O	+ 1.0 to + 6.1	+ 3.1
K_2O	+ 0.2 to + 1.2	+ 0.66
SiO_2	- 2.2 to - 11.5	- 5.3
CaO	- 2.0 to - 4.6	- 2.7
MgO	- 0.4 to - 3.8	- 1.6
Na_2O	- 0.1 to - 0.3	- 0.2

+ taken up by rock from seawater

- lost from rock to seawater

Rates calculated assuming constant volume; rates do not significantly change if calculated assuming constant Ti or Al. Data from R. Hart (1970, 1973) Thompson (1973) and Thompson, et al (in preparation)

to account for the uptake of the annual river input of K_2O (Maynard, 1976). Such effects were apparently not seen at depth in the oceanic crust when deeper DSDP drilling cores were recovered, e.g. sites 319, 320, 321, of Leg 34 in the Nazca plate (Thompson, et al., 1976), site 332, Leg 37 in the Atlantic (Aumento, Melson, et al., 1977) and even site 417D or 418A, Leg 51-53 (Staudigel et al., in press, a,b). In dredged pillow basalts, original K_2O contents of about 0.08 weight percent were increased to 0.2 to 0.4 weight percent (except palagonitized margins which reached as high as 1.0 weight percent K_2O). The DSDP cores typically showed K_2O increasing to 0.15-0.18 weight percent although K_2O ranged higher in small alteration veins or palagonitized regions. The DSDP basement lithic units also showed less Si, Mg and Ca loss, with the Ca commonly being reprecipitated as carbonate in the basaltic pile.

At hole 417A the direction of chemical fluxes, e.g., uptake of K, H_2O, P, Ba, Li, B, Rb, Cs (Donnelly, et al., in press, Joron, et al., in press) and removal of Ca, Mg, and some Na, as well as oxidation of iron, is in the same direction as for dredged basalts, but reaches far higher values. Assuming an available time period of 20 million years before sedimentation sealed the system to water flow, and taking the estimated 820 g/cm^2 per 200 meter column as the added K_2O we find that the flux is about 2 x 10^{-9} g/cm^3 year, which is beyond the range shown in table 3.

Before consideration of the potential role of such fluxes in affecting seawater composition and elemental budgets in the marine environment, it is pertinent to realize that other and different reactions of seawater and basalt occur. From experimental studies (e.g. Mottl, 1976; Bischoff and Dickson, 1975), observations in natural seawater hydrothermal systems, such as at Reykjanes in Iceland (Tomasson and Kristmannsdottir, 1972) and from studies of hydrothermally altered basalts from the ocean floor (Humphris and Thompson, 1978a, b), it is clear that high temperature ($\geq 200^oC$) reactions of seawater and basalt result in rapid chemical exchanges. In general at higher temperatures Mg is taken up by the rock from seawater and also probably Na, whereas Si, Ca and K and probably Fe and Mn are leached from the rock. Of the trace elements B, Li, Cu, Ni, Zn, Ba, and Sr are mobilized and sometimes removed: elements such as Zr, Y and the rare earths are apparently largely unaffected.

In Table 4 we show the major chemical fluxes that may result from high temperature ($\geq 200^oC$) basalt seawater interaction as calculated by a number of workers from different observations. The magnitude and effect of such fluxes depends of course on the amount of oceanic crust undergoing such reaction with seawater. If one-fifth of the annually produced new oceanic crust (8.8 x 10^{15}g/yr, density 2.9g/cm^3) undergoes reaction and is hydrothermally altered, this could result in fluxes of Mg uptake from seawater up to 348 x 10^{12}g/yr magnesium, shown in parentheses in Table 4. These values compare with an estimated input of Mg into the oceans by the rivers of 130 x 10^{12}g/yr (Garrels and Mackenzie, 1971). Drever (1974) has indicated that a sink of 69 x 10^{12}g of Mg per year is all that is required - clearly quite compatible with the hydrothermal estimated sink. At these high temperatures potassium is lost, but K_2O input from hydrothermal alteration seems relatively small compared to other fluxes. Humphris and Thompson (1978a) noted that hydrothermally altered basalts recently exposed on the seafloor did not greatly differ in their K_2O contents from their fresh precursors, possibly due to uptake of K_2O from seawater at low temperatures when reexposed.

The net effect of basalt-seawater interactions on oceanic budgets must consider cold weathering, warm alteration

Table 4 Comparison of estimates of major element fluxes during hydrothermal ($\geq 200^{\circ}C$) alteration of basalts. Units = g / cm^3

[1]from oceanic metabasalts	[2]from experimental studies	[3]from heat flow estimates
MgO - 1.8 to + 11.4 (-54.5 to + 345 x 10^{12} g / yr	+ 4.12 (125 x 10^{12} g / yr)	+ 5.74 to + 11.5 (174 to 348 x 10^{12} g/yr)
Na_2O - 2.6 to + 3.1	0 to + 2.81	
SiO_2 + 3 to - 24		- 0.99 to - 2.96
CaO - 0.01 to - 17.5	- 3.39 to - 9.31	- 2.15 to - 4.44
K_2O - 0.46 to + 0.59	- 0.35	

\+ taken up by rock from seawater
\- lost from rock to seawater

[1]from Humphris and Thompson (1978a)

[2]from Mottl (1976)

[3]from Wolery and Sleep (1976)

involving very large volumes of water, and high-temperature ($\geq 200^{\circ}C$) hydrothermal alteration. R. Hart (1973) considered that weathering was the major sink for K_2O, and hydrothermal alteration the major sink for Mg and Na, and he calculated a net effect without consideration of the opposite effects of these two processes. Wolery and Sleep (1976) considered that the fluxes calculated for low temperature weathering of dredged material were indicative of only the upper few tens of meters of oceanic crust and thus not important: hydrothermal alteration, because of the large volumes of seawater that must be pumped through the oceanic crust near the spreading centers to cool the lithosphere, was for them the major process governing chemical exchange between the oceanic basement and seawater.

However, if site 417A is at all representative of a few percent of the upper part of layer 2 of the ocean crust, warm water alteration may be also be important process. In this case the net Mg sink would be lessened or reversed, and the K_2O sink increased. Reactions similar to those that occur at site 417A must be restricted to oceanic crust younger than 15 million years, where sufficient heat is still available in the cooling lithosphere to drive the water. At temperatures greater than about $100^{\circ}C$ the chemical nature of the alteration process would change.

Conclusions

The basalts of hole 417A exhibit a degree of alteration hitherto unseen in oceanic rocks. If such sites comprise even a percent or two of the oceanic crust, then the elemental fluxes seen here would be of first order importance. Although none of the other dozen or so deeply drilled holes (that is, holes drilled many tens of meters into the oceanic crust) revealed such altered basalts, several earlier holes with penetration of a meter or so recovered suspiciously highly potassic samples. These sites include sites 137 and 138 (eastern Atlantic), 192 (northern Pacific), and 235 (western Indian Ocean), all of which recovered basalts with K_2O values of 2 percent or higher.

There is little doubt that the forced passage of warm to hot water through a cooling oceanic crust could result in immense elemental fluxes. At high temperatures Mg is taken up and K released, and at lower temperatures the reverse is observed; thus, it is conceivable that these effects might cancel each other. Much depends on the length of time that water is exposed to rock at various temperatures; under some conditions highly undersaturated hot water could be discharged directly into seawater with little other than a local thermal effect. Or water with high K and with low Mg might lose its K and recover Mg if allowed to interact with wall rock at lower temperatures prior to discharge. Further recovery of dredged rocks will do little to settle the controversy, because of the problems listed above. Only extensive crustal drilling to relatively shallow depths (one hundred meters should be fully adequate to assess the type and extent of alteration) will provide the answers needed. Drilling of older, sediment covered crust provides an opportunity absent in younger crust: the opportunity to drill topographic highs which cannot be penetrated in the absence of a sediment cover. It is not possible *a priori* to assign relative importances to the two temperature ranges, but the recovery of one well documented, clear example and four additional possible examples with very limited penetration and recovery emphasize the importance of the warm water process.

Acknowledgements. Discussions with several 'Megaleg' co-workers have been most valuable, especially Matthew Salisbury, James Lawrence, and Karlis Muehlenbachs. Mike Mottl and Bill Bryan critically reviewed a related paper. Chemical analyses were performed by Max Budd, Department of Geological Sciences, State University of New York, Binghamton, and D. C. Bankston and M. Sulanowska at the Woods Hole Oceanographic Institution. NSF Grant GA-22971 is gratefully acknowledged. We are especially grateful to John Tarney for a thoughtful review.

References

Anderson, R.N., M.A. Hobart, and M.G. Langseth, Convective heat transfer in the oceanic crust and sediment on the flanks of midocean ridges in the Indian Ocean (abstract, *Eos Trans., AGU, 59*, 384, 1978.

Aumento, F., and W.G. Melson, et al., eds., *Initial Reports of the Deep Sea Drilling Project, Volume 37*. Washington, U.S. Government Printing Office, 1977.

Bischoff, J., and F.W. Dickson, Seawater-basalt interaction at 200°C and 500 bars: implications as to the origin of sea floor heavy metal deposits and regulation of seawater chemistry, *Earth Planet. Sci. Letters, 25*, 385, 1975.

Donnelly, T.W., G. Thompson, and M. Salisbury, The chemistry of altered basalts at Site 417A, DSDP Leg 51, in *Initial Reports of the Deep Sea Drilling Project, Volume 51-53*, Washington, U.S. Government Printing Office, in press.

Drever, J.I., The magnesium problem, in E. Goldberg, ed., *The Sea*, volume 5, New York, Wiley, 337, 1974.

Edmond, J.M., L.I. Gordon, and J.B. Corliss, Chemistry of hot springs on the Galapagos Ridge axis (abstract), *Eos Trans., AGU, 58*, 1176, 1977.

Garrels, R.M., and F.T. Mackenzie, Evolution of Sedimentary Rocks, New York, W.W. Norton, 1971

Hart, R., Chemical exchange between sea water and deep ocean basalts, *Earth and Planet. Sci. Letters, 9*, 269, 1970.

Hart, R., A model for chemical exchange in the basalt -sea water system of oceanic layer II, *Can. Jour. Earth Sci., 10*, 799, 1973.

Hart, S., K, Rb, Cs contents and K/Rb, K/Cs ratios of fresh and altered submarine basalts, *Earth and Planet. Sci. Letters, 6*, 295, 1969.

Humphris, S., and G. Thompson, Hydrothermal alteration of oceanic basalts by seawater, *Geochim. Cosmochim. Acta, 42*, 107, 1978a.

Humphris, S., and G. Thompson, Trace element mobility during hydrothermal alteration of oceanic basalts, *Geochim. Acta, 42*, 127, 1978b.

Humphris, S., R.N. Thompson, and G.F. Marriner, The mineralogy and geochemistry of basalt weathering, holes 417A and 418A, in *Initial Reports of the Deep Sea Drilling Project, Volume 51-53*, Washington, U.S. Government Printing Office, in press.

Joron, J.L., C. Bollinger, J.P. Quisefit, H. Bougault, and M. Treuil, Trace elements in basalts at 25° N, old crust, in the Atlantic Ocean: alteration, mantle, and magmatic processes, in *Initial Reports of the Deep Sea Drilling Project, Volume 51-53*, Washington, U.S. Government Printing Office, in press.

Juteau, T., Y. Noack, H. Whitechurch, Mineralogy and geochemistry of alteration products in holes 417A and 417D basement samples (D.S.D.P., Leg 51), in *Initial Reports of the Deep Sea Drilling Project, Volume 51-53*, Washington, U.S. Government Printing Office, in press.

Matthews, B.M., Altered basalts from Swallow Bank, an abyssal hill in the N.E. Atlantic and from a nearby seamount, *Roy. Soc. London, Philos. Trans. A, 268*, 551, 1971.

Maynard, J.B., The long-term buffering of the oceans, Geochim. Cosmochim. Acta, 40, 1523, 1976.

Mevel, C., Mineralogy and chemistry of a secondary phases in low temperature altered basalts from D.S.D.P. legs 51, 52, and 53, in Initial Reports of the Deep Sea Drilling Project, Volume 51-53, Washington, U.S. Government Printing Office, in press.

Mottl, M., Chemical exchange between seawater and basalt during hydrothermal alteration of the oceanic crust, Ph.D. dissertation, Harvard University, 1976.

Muehlenbachs, K., The alteration and aging of the basaltic layer of the sea floor: oxygen isotopic evidence from DSDP/IPOD legs 51, 52, and 53, in Initial Reports of Deep Sea Drilling Project, Volume 51-53, Washington, U.S. Government Printing Office, in press.

Pertsev, N.N. and V.L. Rusinov, Mineral assemblages and alteration process in basalts at Sites 417 and 418, in Initial Reports of the Deep Sea Drilling Project, Volume 51-53, Washington, U.S. Government Printing Office, in press.

Pritchard, R.G., The alteration of basalts from D.S.D.P. Legs 51, 52, and 53 Megaleg), holes 417A and 418A, in Initial Reports of the Deep Sea Drilling Project, Volume 51-53, Washington, U.S. Government Printing Office, in press.

Rivers, M., The chemical effects of low-temperature alteration of sea-floor basalt, undergraduate thesis, Harvard University, 1976.

Robinson, P.T., M.F.J. Flower, H.-U. Schmincke, and W. Ohnmacht, Low temperature alteration of oceanic basalts, DSDP Leg 37, in, Aumento, F. and W.G. Melson, et al., Initial Reports of the Deep Sea Drilling Project, Volume 37, Washington, U.S. Government Printing Office, 775, 1977.

Rusinov, V.L., B.P. Gradusov, I.P. Laputina, G.N. Muravitskaja, and B.B. Zvjagin, Clay minerals in basalts from Sites 417 and 418, in Initial Reports of the Deep Sea Drilling Project, Volume 51-53, Washington, U.S. Government Printing Office, in press.

Scheidegger, K.F., and D.S. Stakes, X-ray diffraction and chemical study of secondary minerals from DSDP Leg 51 sites 417A and 417D, in Initial Reports of the Deep Sea Drilling Project, Volume 51-53, Washington, U.S. Government Printing Office, in press.

Staudigel, H., W.B. Bryan, and G. Thompson, Chemical variation in glass-whole rock pairs from individual cooling units in holes 417D and 418A, in, Initial Reports of the Deep Sea Drilling Project, Volume 51-53, Washington, U.S. Government Printing Office, in press, a.

Stuadigel, H., F.A. Frey, and S.R. Hart, Rare earth element and Sr isotopes in basalts and glass from sites 417D and 418A: fresh rock composition and alteration effects, in Initial Reports of the Deep Sea Drilling Project, Volume 51-53, Washington, U.S. Government Printing Office, in press, b.

Thompson, G., A geochemical study of the low temperature interaction of seawater and oceanic igneous rock (abstract), Eos Trans., AGU, 54, 1015, 1973.

Thompson, G., W.B. Bryan, F.A. Frey, J.S. Dickey, and J. Suen, Petrology and chemistry of basalts from DSDP Leg 34, Nazca Plate, in, Hart, S.R. and S.R. Yeats, et al., Initial Reports of the Deep Sea Drilling Project, Volume 34, Washington, U.S. Government Printing Office, 215, 1976.

Thompson, G. and S. Humphris, Seawater-rock interactions in the oceanic basement, in, Pacquet and Y. Tardy, eds., Proc. 2nd International Symposium on Water-Rock Interaction, vol. III, Univ. Louis Pasteur, Strasbourg, 3, 1977.

Thompson, G. and M. Rivers, The low-temperature weathering of oceanic basalts by seawater: 1, major element fluxes (abstract), Geol. Soc. Amer. Abstracts, 8, no. 6, 1137, 1976.

Thompson, G., M. Rivers, S. Heinrichs, and D.C. Bankston, the low-temperature weathering of oceanic basalts by seawater, in preparation.

Tomasson, J., and H. Kristmansdottir, High temperature alteration minerals and thermal brines, Reykjanes, Iceland, Contr. Mineral. and Petrol., 36, 123, 1972.

Wolery, T.J., and N.H. Sleep, Hydrothemal circulation and geochemical flux at mid-ocean ridges, Jour. Geology, 84, 249, 1976.

Woods Hole Contribution No. 4191.

RIDGE CREST - HYDROTHERMAL METAMORPHISM AT THE GALAPAGOS SPREADING CENTER AND REVERSE WEATHERING

John M. Edmond

Department of Earth and Planetary Sciences, Massachusetts Institute of Technology, Cambridge, Massachusetts

John B. Corliss and Louis I. Gordon

School of Oceanography, Oregon State University, Corvallis, Oregon 97331

Introduction

Weathering on the continents takes place in a flow system. Dilute rain and ground waters, of quite constant composition at a given location are renewed continuously at the exposed mineral surfaces. This reaction medium is relatively acidic and poor in cations the average ratio of cation equivalents to protons being of order 10^3. The net result, for silicate weathering, is a replacement of the cations in the alumino silicate lattices by protons. Resulting structural accomodations lead to release of silica along with the cations and the production of stripped or degraded phases, clays. If primary weathering is regarded as the titration of quenched, high temperature, disequilibrium igneous rock by acidic volcanic volatiles then the continents are inevitably over-titrated, acting as a sink for protons.

Oceanic weathering is a distinctly different process. The environment is static: the cation to proton ratio is of order 10^7: low-temperature reactions between basalts and sea-water therefore involve changes in the cation ratios of the two phases with comparatively minor net exchange of protons. The trend is towards equilibrium rather than reduction of the weathering substrate to a refractory residue.

The primary source of protons for continental weathering reactions is the acid volatiles released by volcanoes. Of these the strong acids hydrochloric, sulphuric, nitric and hydrofloric are neutralized rapidly by the rocks. Of the weak acids only carbonic is important, boric being present in minor quantities. The centrality of carbonic acid lies in its being not only weak but also volatile as CO_2: it is the only acidic species which can be regenerated - i.e. neutralized, and then recycled through the system. Mass balance calculations indicate that the rate of weathering on the continents depends critically on this recycled acid (e.g. Garrels and MacKenzie, 1971). If there is nowhere a back-titration of the products of continental weathering then the process will run down rapidly as the available protons are fixed in clays.

An appreciation of this problem by Sillen (1961) led him to hypothesize that the degraded products of continental weathering are in fact reconstituted in the ocean by re-reaction with seawater. That is to say the reaction in a proton-rich, cation poor environment is reversed by interaction with a cation rich, proton poor solution - seawater. The observed load of degraded and partially degraded alumino-silicates carried by rivers to the ocean is ample to allow this process (MacKenzie and Garrels, 1966). In addition, by its very nature the overall cycle acts to stabilize the major ion chemistry of the ocean since it is closed.

The profundity and generality of this reverse weathering hypothesis and its implications for the stability of seawater composition over geologic time have led to its gradual acceptance and elaboration since the original proposal in 1959. This despite the almost complete lack of supporting field or laboratory evidence for the occurrence of the proposed reactions under oceanic conditions (e.g. Drever, 1974 for summary).

Over the past decade our understanding of the marine geochemical cycle has been radically transformed by the realization that crustal formation in the intrusion zone at the ridge axes is accompanied by pervasive hydrothermal activity (Elder, 1965, Deffeyes, 1970, Talwani et al., 1971). The generally observed negative anomaly in conductive heat

TABLE 1. Contributions of the various proton carriers to the reverse weathering process

Species	eq/kcal	annual flux x 10^{12}	% of total (10^{13} eq/yr)
A_t	-30	1.5	15
CO_2	15	0.75	7.5
Ca	38-80	1.9-4	19-40
H_2S	0-40	0-2	0-20
Mn	2-7.6	0.1-0.38	1-3.8
Fe	0-5	0-0.25	0-2.5
Total			42.5-89

loss from young crust (Anderson et al., 1977), the seismic properties of the upper crust (Fox et al., 1973) and the isotope equilibrium temperatures in dredged samples (Muehlenbachs, 1976) are together compelling evidence for the global significance of this process. The drastic chemical transformations observed in laboratory studies of seawater-basalt interactions at high temperatures and sea floor pressures indicate major associated effects on the chemistry of the circulating waters (Bischoff and Dickson, 1975, Hajash, 1975, Seyfried and Bishop, 1977, Mottl and Holland, 1978). The discovery and sampling of extensive fields of hot springs on the Galapagos Spreading Center (Sclater and Klitgord, 1973, Lupton et al., 1977, Corliss et al., 1978) provided the first direct evidence for the existence of the process. The chemical data confirm the experimental work at least in general: coupled with the global flux estimates derived from the He-3, heat relationship (Jenkins, et al., 1978) these data provide strong evidence that ridge crest hydrothermal metamorphism exerts a pronounced influence on the major and minor element chemistry of the ocean. The possibility therefore exists of an alternative to the Sillen hypothesis at least as general and perhaps more amenable to direct investigation. In this paper therefore the existing chemical data from the hot springs will be discussed in relation to the reverse weathering problem.

Requirements

For ridge-crest hydrothermal metamorphism to be the mechanism of reverse weathering the seawater-basalt interactions must result in a net production of protons i.e. the charge balance must be conserved through hydrogen ion rather than any other cation. If not then a budget problem is created at the expense of one resolved. This consideration constrains quite sharply the kind of processes which can act as effective back-titration reactions.

They must produce protons or 'proton carriers' e.g. CO_2 itself, calcium through the subsequent reaction:

$$Ca^{2+} + HCO_3^- = CaCO_3 + H^+$$

reduced iron or manganese via the oxidation reactions:

$$2Fe^{2+} + 1/2\ O_2 + 3H_2O = 2Fe\ O(OH) + 4H^+$$

$$Mn^{2+} + 1/2\ O_2 + H_2O = MnO_2 + 2H^+$$

or H_2S:

$$H_2S + 2O_2 = SO_4^= + 2H^+$$

The required proton flux can be estimated from the published mass balance schemes developed to 'confirm' the Sillen hypothesis (MacKenzie and Garrels, 1966). The 'primary' bicarbonate, that not derived from weathering of carbonate rocks amounts to about 35% of the total annual river flux of approximately 3×10^{13} eq/yr i.e. the hydrothermal solutions must provide a net proton flux of 10^{13} eq/yr.

The Galapagos Data

1. Temperature. The ^3He data (Jenkins et al., 1978) give a ratio to heat of 22×10^{-18} moles/calorie. Assuming this ratio is general for all hot springs and accepting the average annual flux of ^3He to the ocean as 4 ± 1 atoms/cm^2/sec (Craig et al., 1975) then the associated heat flux is $4.9 \pm 1.2 \times 10^{19}$ cal/yr in remarkable agreement with the geophysical estimates (Williams and Von Herzen, 1974, Woolery and Sleep, 1976). If the average exit temperature of the hot springs is 15°C (the highest sampled at the Galapagos site) then a volume of seawater equivalent to the whole ocean must pass through the ridge crest every 360,000 years. This exit temperature cannot however be representative of that in the reaction zone given the extensive chemical modifications observed. It is almost certainly an artifact of dilution by 'ground waters' of approximately ambient composition of the primary end member solution as it ascends through the pillow pile at the ridge crest (Corliss et al., 1978). If it is assumed that equilibrium is reached at some high temperature in the reaction zone, that no subsequent reactions take place during ascent to the surface and that dilution is indeed with waters of ambient temperature and composition then various solution geothermometers can be used to estimate the conditions in the equilibrium regime.

The silica data from all the hot springs sampled define a single linear relationship with temperature (Fig. 1). Due to various

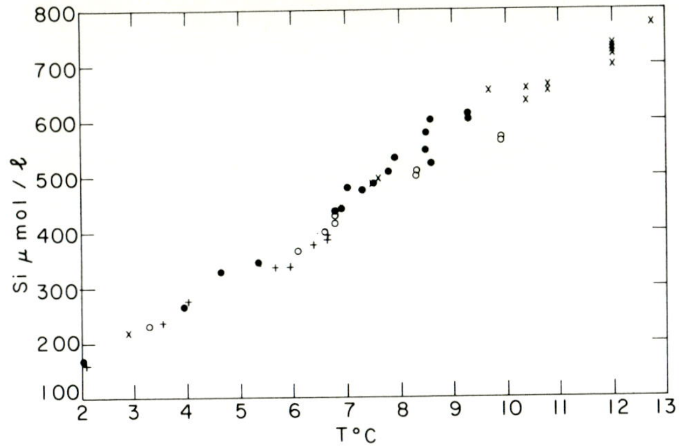

Fig. 1. Silica versus temperature. The four vent fields are identified by symbol: Clambake •; Dandelions +; Oyster Beds O; Garden of Eden X. The scatter in the data reflects the uncertainty in the temperature which, due to entrainment of ambient water at variable rates during sampling, can be as much as ±0.5°C for some samples.

sampling and instrumental difficulties the uncertainty in the temperature measurements ranges up to about ±0.5°C. This appears to be the major cause of scatter in the Si-T relationship in that plots of all other parameters against silica are much tighter than against temperature (Corliss et al., 1978).

In many continental hot spring systems the subsurface reservoir temperature can be estimated with high accuracy from the silica data by assuming equilibrium saturation with respect to quartz at the end-member, no requilibration during transport from the reaction zone to the surface and no mixing with extraneous waters (Fournier and Rowe, 1966). Corrections are made for adiabatic cooling and for compositional effects associated with flash boiling. The relationship given by Truesdell (1975) is:

$$T = 1533.5/(6.989 - \log Si) - 273.15$$

where T is the reservoir temperature in degrees Centigrade and the silica concentration is in micromoles per liter. The general success of this approach, confirmed by down hole temperature measurements validates these assumptions.

In the Galapagos system dilution with local 'ground water' apparently does occur. Hence the mixing assumption must be relaxed to include entrainment of water putatively of the temperature and composition of the ambient seawater. One therefore wishes to determine the point of intersection of the extrapolation of the measured Si-T relation with the quartz solubility curve. The Si-T equation is

$$Si = a + b\Delta T$$

where a is the ambient value (∿165 μmol/l), b is the Si gradient in μmol/ℓ/°C and ΔT is the temperature increment. Substituting in the Truesdell equation:

$$6.989T - (T \pm 273.15)\log(a + b\Delta T) + 375.6 = 0$$

Solving for T (ΔT = T - 2.05) gives the extrapolated quartz saturation temperature at 285°C. A more elaborate treatment of the problem, following Walther and Helgeson (1977), does not seem warranted by the data. Secondary quartz is a common high temperature reaction product in sea floor basalts (see e.g. Humphris and Thompson, 1978; Coleman, 1977; Cann, 1969). The amounts of silica involved in solution reactions is probably a small proportion (<1%) of the total available in the rock. Equilibrium control by a volumetrically minor phase is therefore quite reasonable.

It can be objected that during mixing cristobalite or another soluble silica phase may precipitate as is observed in some continental systems: alternatively the waters in fact may be dissolving volcanic glass of high but indeterminate solubility. In either case, the 'quartz saturation temperature' is merely an artifact of the final observed Si-T dilution line. It is important therefore to establish other measures of the end-member temperature. Based on the experimental work the most obvious is the 'zero magnesium temperature'. This parameter is derived from the observation of rapid uptake of the seawater magnesium by basalt at elevated temperatures to form a sepiolite or chlorite (Bischoff and Dickson, 1975). In systems with are not rock limited the magnesium concentration is reduced effectively to zero in a few hours. Given the observed gradients in 'soluble' elements, e.g. Li and Ba, such is certainly the

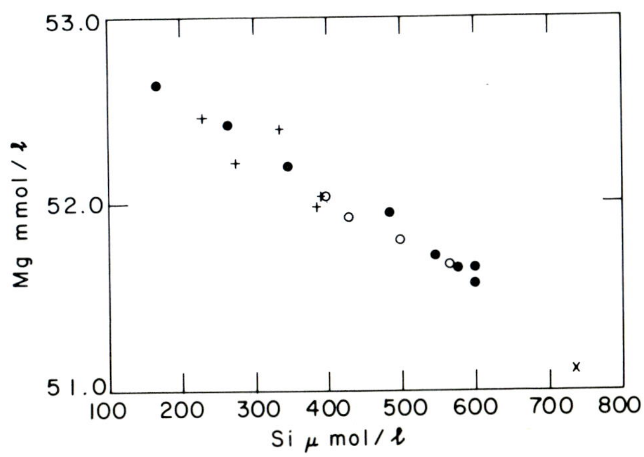

Fig. 2. Magnesium versus silica. Data are plotted versus silica since the uncertainty in the temperature introduces a spurious scatter.

case at the Galapagos. The data for Mg fall on a uniform decreasing trend with temperature (Fig. 2) which extrapolates to give zero concentration at 295°C.

Recent data for sulphate; to be published elsewhere, also extrapolates to zero at around 300°C. Application of the empirical Na-K-Ca geothermometer of Fournier and Truesdell (1973) is beset with complications. The sodium variation in the hot springs is imperfectly known being a few tenths of a percent at most at the highest temperatures sampled. More fundamentally the success of the scheme apparently is based on cation exchange equilibria with feldspars and hence of dubious application to tholeiitic systems. Olafsson and Riley (1978) found for the Reykjanes brines that the computed temperature based on the compositional data gave values about 60°C below those measured. Application of the geothermometer to the experimental data of Bischoff and Dickson (1976) and Mottl (1976) failed to give any systematic relation - a calibration - with the reported laboratory temperatures. With the present data (Fig. 3,4) the geothermometer gave a maximum temperature for the case of invariant sodium, the value being 180°C.

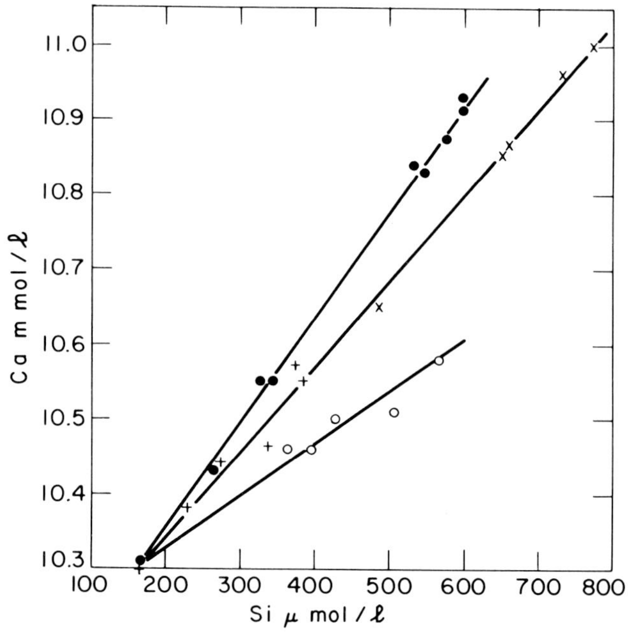

Fig. 4. Calcium versus silica. Regression lines for the various fields are drawn. Garden of Eden and Dandelions display the same relationship within analytical accuracy.

From this discussion it would appear that the water leaves a reaction zone depleted in magnesium (and sulphate) and at equilibrium with quartz with a temperature of between 275 and 300°C. During ascent to the surface vents major entrainment of 'ground water' of approximately ambient composition occurs the exit temperature being determined by the extent of this latter process. Further, during actual sampling of the hot springs further admixture of ambient waters may occur. Hence the measured data fall on dilution lines whose slopes are characteristic for each element and, in general, for each field.

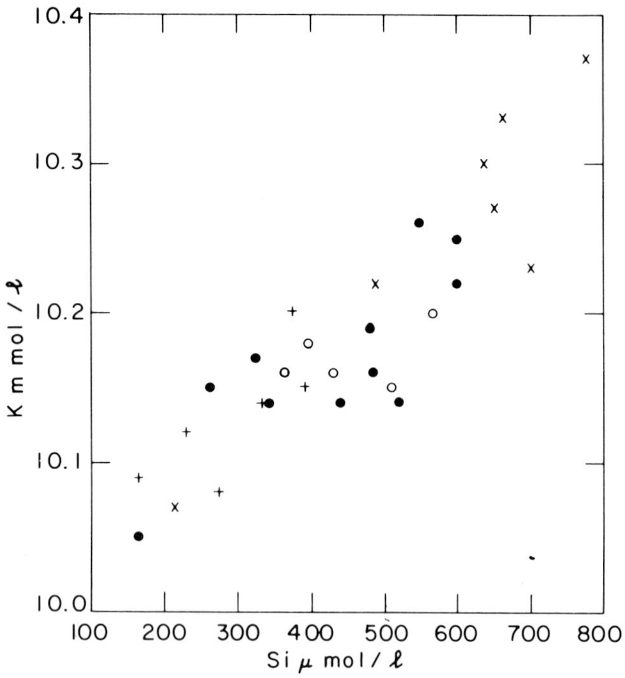

Fig. 3. Potassium versus silica. Each data point is the average of between two and six replicate measurements.

The Carbonate System: A_t, ΣCO_2, Ca

For the ridge crests to be a significant factor in the global weathering balance the flux of protons and proton carriers into the water column must be on the order 10^{13} eq/yr. Accepting the global hydrothermal heat flux as 5×10^{19} cal/yr gives a net concentration gradient with hot spring temperature of 200 μeq/ℓ/°C. Of course this assumes constant relative composition gradients in all hot springs, a condition not fulfilled even between the four vent-fields sampled to date! However the calculation is important in placing the ridge crest processes in the overall context of the aqueous geochemical cycle.

The alkalinity decreases with increasing temperature (Fig. 5). The average slope is approximately -30 μeq/ℓ/°C. Two groups of points fall markedly off the trend: those for several but not all of the Dandelion vents are high and apparently real: those for the Oyster Bed samples collected on Dive 726 are systematically low by as much as 100 μeq/ℓ. This may be caused by oxidation of reduced

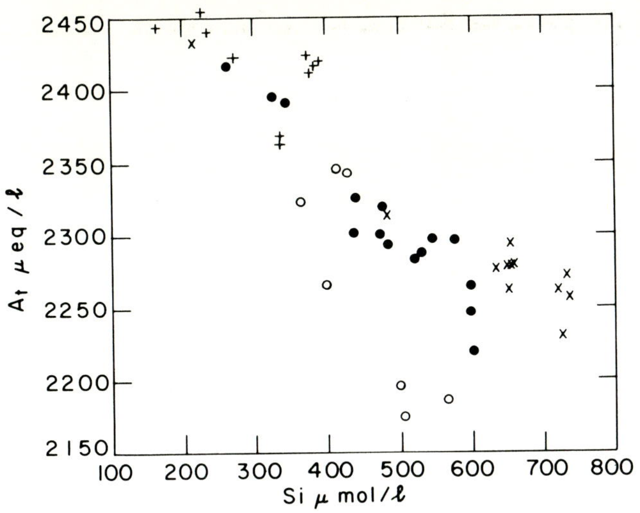

Fig. 5. Alkalinity versus silica.

sulphur compounds before analysis since these samples, unlike the others, had to be stored for a day before processing. The data from the Garden of Eden fall slightly above the main trend. In the absence of samples in the intermediate range this is assumed to be a real effect. The scatter in the data overall is appreciably greater than the analytical uncertainty (∿0.25%, most samples were run in duplicate): the same is true for H_2S suggesting that there is a range of products formed during oxidative reaction of the sulphide in the sub-surface mixing zone producing variable additions of protons. This has been shown to be the case in fresh water systems where thiosulphate, polysulphide, and native sulphur can amount to over half the non-sulphate sulphur (Boulegue, 1977).

The ΣCO_2 content of the hot spring waters (Fig. 6) increases with temperature. However, there is considerable scatter in the data much of it probably analytical since the measurements were made at sea using a simple gas chromatograph. The Garden of Eden samples fall low at approximately 5 µmol/ℓ/°C and the Dandelions high at about 30. Clambake and Oyster Beds scatter about a slope of 15. Extrapolation of this trend to the reaction zone temperature gives a CO_2 excess of 4200 µmol/ℓ: a similar calculation for He-3 gives 6.2×10^{-12}moles/ℓ. As pointed out by Corliss et al. (this volume) this indicates a very high extraction efficiency for helium but a much lower one for CO_2. Clearly much more precise data will be required to resolve this question.

Calcium increases with temperature in all the vent fields (Fig. 4). However there are variations in the slope between the different fields. Clambake is highest at 80 µmol/ℓ/°C, Dandelion and Garden of Eden are intermediate at 64 µmol/ℓ/°C and Oyster Beds is markedly lower at 38 µmol/ℓ/°C. Other measured species e.g. Li, Ba, Mn, Fe also show significant variations between fields. However the systematics are not the same for each element. Because of this a simple aging effect can not be the only factor in determining the spread in the trends.

The effective flux of protons carried by the alkalinity - CO_2 - calcium system is obtained from the sum of the slopes for Ca and CO_2 minus that for alkalinity. On the basis of the data for the latter two species it is possible to assign only crude averages to the trends: these are taken as 15 µmol/ℓ/°C and -30 µeq/ℓ/°C respectively. The total flux hence varies between 125 and 83 µeq/ℓ/°C of proton equivalents. The extreme ranges of the data give 156 and 64 µeq/ℓ/°C. It is clear that around 50% of the primary flux of bicarbonate from continental weathering of non-carbonate rocks is back-titrated by ridge crest hydrothermal solutions. Calcium is the responsible proton carrier for about half of this.

The annual consumption of alkalinity is 1.5×10^{12} eq/yr, about 5% of the river flux or 14% of that not balanced by formation of oceanic carbonates (MacKenzie and Garrels, 1966). The CO_2 flux is very approximately 0.75×10^{12} moles/yr equivalent to about 7% of the annual sedimentation rate as carbonate and about 4.5% of the annual consumption of CO_2 by continental weathering. For comparison the flux in the photosynthesis respiration cycle is estimated at 4.7×10^{15} moles/yr and that from anthropogenic fossil CO_2 release at 4×10^{14}moles/yr.

The addition of calcium is between 1.9 and 4×10^{12} moles/yr, 15 and 30% of the river

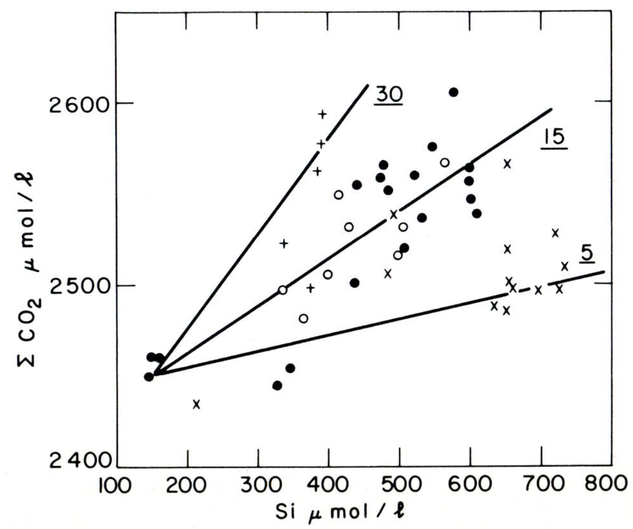

Fig. 6. Total CO_2 versus silica. Lines of gradient 30, 15 and 5 µmoles CO_2/ℓ/°C are indicated.

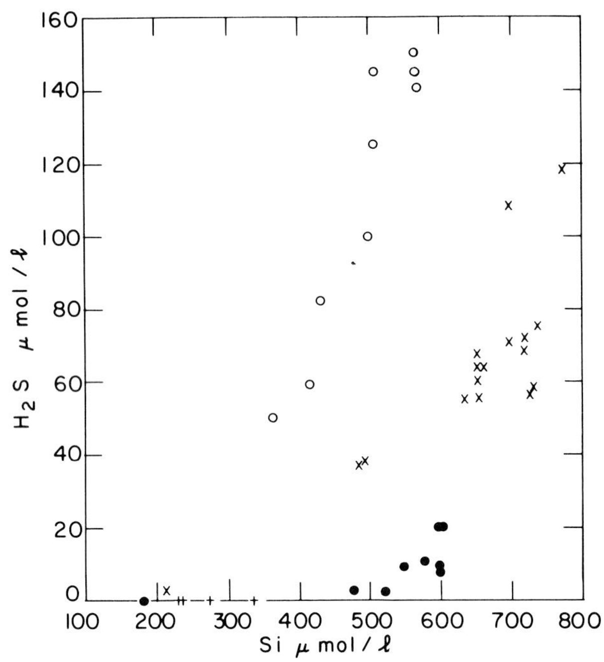

Fig. 7. H_2S versus silica.

flux. Holland (1978) estimates that about 75% of the latter is derived from the solution of sedimentary carbonate rocks. Hence the ridge crests are a source of 'primary' igneous calcium comparable to that from the continents. Approximately one third of the carbon required to precipitate this calcium as carbonate is derived from the ridge crest. The remainder, an average of 2 x 10^{12} moles/yr, must be obtained from metamorphic processes and primary outgasing in non-ridge crest environments. In a recent compilation Holland (1978) estimated this combined flux at 7.5±2.5 x 10^{12} moles/yr. While the necessary amount of CO_2 is within the quoted uncertainties the general arguments supporting the value itself are very extended and will require a great deal of information before they can be considered substantiated.

Reduced Species: H_2S, Mn^{2+}, Fe^{2+}

The H_2S levels vary widely between vent areas but more or less linearly within a given field (Fig. 7). A crude average is about 7.5 μmol/ℓ/°C equivalent to 15 μmol H^+/ℓ/°C. Manganese shows the same separation between fields as calcium (Fig. 8). The Mn/Ca ratio is about 0.04 resulting in an average flux of less than 3 μmol Mn/ℓ/°C equivalent to 6 μmol H^+/ℓ/°C. The iron trends vary greatly between vent fields (Fig. 9). In Clambake the values are very high and after a sharp increase at low temperature rise linearly with temperature (and H_2S!) with a gradient of 2.5 μmol/ℓ/°C. In the other fields the highest values are less than 1 μmol/ℓ. These trends all show a maximum generally at the higher end of the temperature range sampled. There is no obvious explanation for this behavior. Examination of particles filtered *in situ* by SEM and EDAX indicates that Fe-S bearing phases are rare (Corliss, personal communication). Taking, for purposes of illustration the average gradient as 0.5 μmol/ℓ/°C as ferrous iron the equivalent proton flux on oxidation is 1 μmol/ℓ/°C.

Since these reduced species show widely varying behavior between vent fields the calculated average fluxes are of course subject to very wide uncertainty. The contributions to the proton balance of the figures presented here is roughly 10% of the total. The oxygen 'demand' is about 16.6 μmol/ℓ/°C of which 90% is from H_2S oxidation. This corresponds to an ocean wide consumption rate of 8.3 x 10^{11} moles/yr or less than 0.1% of the marine photosynthesis-respiration cycle and about 10% of the total weathering cycle.

The Mn-Fe ratio varies from 1.2 in Clambake to between 10 and 100 in the other fields. The spread in the manganese temperature slopes may be related to ageing effects in the hot spring systems since it is similar to that

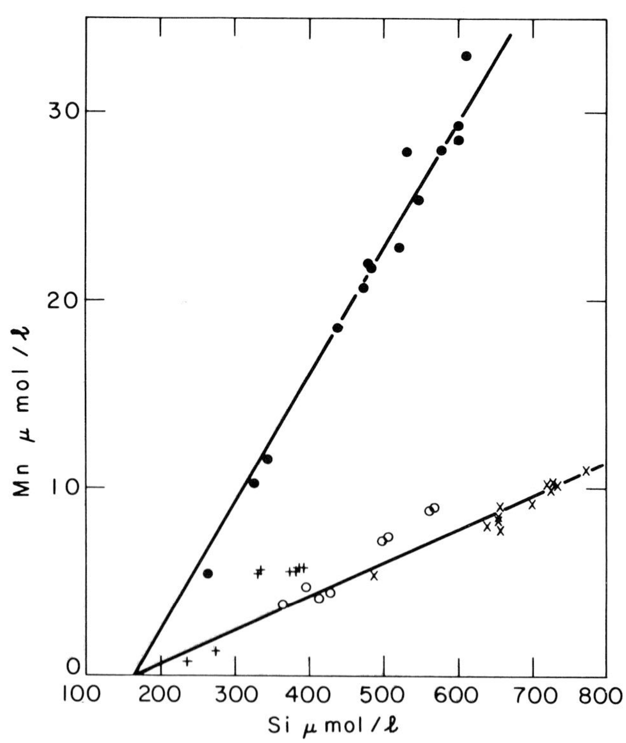

Fig. 8. Manganese versus silica. Regression lines for Clambake and Garden of Eden are drawn.

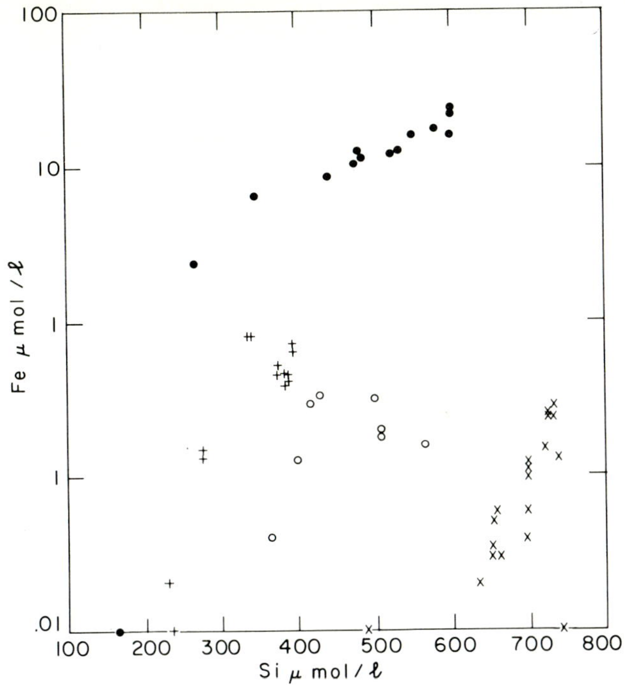

Fig. 9. Iron versus silica. Notice the logarythmic concentration scale for iron.

observed for barium (Corliss et al., 1978). However the ratio of the slopes is about 30, much lower than the average tholeiite values of about 330, suggesting either that manganese is not as labile as barium in the primary water - rock interaction or that it undergoes secondary removal processes.

Conclusions

The contributions of the various proton carriers to the reverse weathering process are summarized in Table 1. The calculated extreme range is between about 40 and 90% with the main contribution (∿50%) coming from calcium. Alkalinity consumption is next in importance although the oxidation of H_2S is comparable in two of the vent fields. The CO_2 flux is highly uncertain but appears to be larger than for manganese or iron.

The charge balance in the hydrothermal solutions is maintained by a decrease in sulphate, an increase in potassium, and positive and negative variations in chloride and sodium (calculated). The elegance of the Sillen hypothesis where cation removal is balanced by immediately concomitant proton production is therefore lost. The ridge crest contribution to reverse weathering - the back titration of the bicarbonate fixed during continental weathering - is decoupled from the cycles of the major ions and to a large extent independent of them.

Clearly the discussion presented above involves extrapolations of very large magnitude. However there are internal regularities in the system which make these justifiable. Of the major ions i.e. species strongly subject to charge balance constraints, magnesium and sulphate are quantitatively removed; potassium release is strictly rock limited and therefore relatively constant; calcium is modulated by uptake or release of sodium and chloride. The severity of these constraints can only be established through the discovery and exploration of additional hot spring areas over the range of ridge crest environments.

References

Anderson, R.N., M.G. Langseth, and J.G. Sclater, The mechanism of heat transfer through the the floor of the Indian Ocean, J. Geophys. Res., 82, 3391-3409, 1977.

Bischoff, J.L. and F.W. Dickson, Seawater - basalt interactions at 200°C and 500 bars: implications for the origin of sea-floor heavy metal deposits and regulation of seawater chemistry, Earth Planet. Sci. Lett. 25, 385-397, 1975.

Cann, J.R., Spilites from the Carlsberg Ridge, Indian Ocean, J. Petrol., 10, 1-19, 1969.

Coleman, R.G., Ophiolites, Springer-Verlag, NY, Ch. 2-3, 1977.

Corliss, J.B., J. Dymond, L.I. Gordon, J.M. Edmond, R.P. von Herzen, R.D. Ballard, K. Green, D. Williams, A. Bainbridge, K. Crane, Tj. H. van Andel, Submarine thermal springs on the Galapagos Rift, Science, (in press).

Craig, H., W.B. Clarke and M.A. Beg, Excess ^3He in deep water on the East Pacific Rise, Earth Planet. Sci. Lett., 26, 125-132, 1975.

Deffeyes, K.S., The axial valley: a steady-state feature of the terrain in The Megatectonics of Continents and Oceans, Ch. 9, H. Johnson and B.L. Smith, Eds., Rutgers University Press, New Brunswick, NJ, 1970.

Drever, J.I., The magnesium problem in The Sea, Vol. 5, Ch. 10, E.D. Goldberg, ed. Wiley-Interscience, NY, 1974.

Elder, J.W., Physical processes in geothermal areas in Terrestrial Heat Flow, Geophys. Monogr. Ser., Vol. 8, W.H.K. Rec, ed., AGU Washington, DC, 1965.

Fournier, R.O. and J.J. Rowe, Estimation of underground temperatures from the silica content of water from hot springs and wet-steam wells, Am. J. Sci. 264, 685-697, 1966.

Fournier, R.O., and A.H. Truesdell, An empirical Na-K-Ca geothermometer for natural waters, Geochim. Cosmochim. Acta, 37, 1255-1275, 1973.

Fox, P.J., E. Schreiber, and J.J. Peterson, The geology of the oceanic crust: compressional wave velocities of oceanic rocks, J. Geophys. Res., 78, 5755-5772, 1973.

Garrels, R.M., and F.T. MacKenzie, Evolution of Sedimentary Rocks, Norton, NY, 1971.

Hajash, A., Hydrothermal processes along mid-ocean ridges: an experimental investigation, Contrib. Miner. Pet., 53, 205-226, 1975.

Holland, H.D., The Chemistry of the Atmosphere and Oceans, Wiley, Interscience, NY, 1978.

Humphris, S.E. and G. Thompson, Hydrothermal alteration of oceanic basalts by seawater, Geochim. Cosmochim. Acta, 42, 107-125, 1978.

Jenkins, W.J., J.M. Edmond, and J.B. Corliss, Excess ^{3}He and ^{4}He in Galapagos submarine hydrothermal waters, Nature, 272, 156-158, 1978.

Lupton, J.E., R.F. Weiss, and H. Craig, Mantle helium in hydrothermal plumes in the Galapagos Rift, Nature, 266, 603-604, 1977.

MacKenzie, F.T. and R.M. Garrels, Chemical mass balance between rivers and oceans, Am. J. Sci. 264, 507-525, 1966.

Mottl, M.J., Chemical exchange between seawater and basalt during hydrothermal alteration of the oceanic crust, Ph.D. Thesis, Harvard University, 188 pp., 1972.

Mottl, M.J., and H.D. Holland, Chemical exchange during hydrothermal alteration of basalt by seawater - I. Experimental results for major and minor components of seawater, Geochim. Cosmochim. Acta, 42, 1103-1116, 1978.

Muehlenbachs, K., Oxygen isotope geochemistry of DSDP Leg 37 rocks, Initial Rep. DSDP 37, 1976.

Olafsson, J. and J.P. Riley, Geochemical studies on the thermal brine from Reykjanes (Iceland) Chem. Geol. 21, 219-237, 1978.

Sclater, J.G., and K.D. Klitgord, A detailed heat flow, topographic and magnetic survey across the Galapagos spreading center at 86°W, J. Geophys. Res., 78, 6951-6975, 1973.

Seyfried, W. and J.L. Bischoff, Hydrothermal transport of heavy metals by seawater: the role of seawater/basalt ratio, Earth Planet. Sci. Lett. 34, 71-77, 1977.

Sillen, L.G., The physical chemistry of seawater in Oceanography, Mary Sears ed. AAAS Washington, DC, 1961.

Talwani, M., C.C. Windisch, and M.G. Langseth, Reykjanes Ridge crest: a detailed geophysical study, J. Geophys. Res., 76, 473-577, 1971.

Truesdell, A.H., Geochemical techniques in exploration in Proc. 2nd UN Symposium on the Development and Use of Geothermal Resources, 1, 53-58, 1975.

Walther, J.V., and H.C. Helgeson, Calculation of the thermodynamic properties of aqueous silica and the solubility of quartz and its polymorphs at high pressures and temperatures, Am. J. Sci. 277, 1315-1351, 1977.

Williams, D.L., and R.P. von Herzen, Heat loss from the earth: New estimate, Geology, 2, 327-328, 1974.

Woolery, T.J., and N.H. Sleep, Hydrothermal circulation and geochemical flux at mid-ocean ridges, J. Geol., 84, 249-275.

SOME IMPLICATIONS OF HEAT/MASS RATIOS IN GALAPAGOS RIFT HYDROTHERMAL FLUIDS FOR MODELS OF SEAWATER-ROCK INTERACTION AND THE FORMATION OF OCEANIC CRUST

John B. Corliss and Louis I. Gordon

School of Oceanography, Oregon State University

John M. Edmond

Department of Earth and Planetary Sciences, Massachusetts Institute of Technology

High heat flow was recognized very early as one of the unique features of mid-ocean ridges (Hess, 1962). As the plate tectonic model for mid-ocean ridges became accepted, quantitative models of their thermal structure were developed (Turcotte and Oxburgh, 1967; McKenzie, 1967) which gave rough agreement with this observed trend of increasing heat flow approaching the ridge axis. Although these models predict high heat flux at the spreading axis, detailed surveys of conductive heat flow on mid-ocean ridges show considerable scatter in measured values and mean values lower than those predicted by the models (Talwani et al., 1971; Lister, 1972). These authors suggested that convective circulation in sea water through the newly formed crust was responsible, and Lister (1972, 1974, 1977) has pursued this model in some detail.

Another early discovery regarding mid-ocean ridges was the discovery, by Bostrom and Peterson (1966) that they appear to be the locus of Fe and Mn enrichments in metalliferrous deep sea sediments. They attributed this to volcanic emanations from the mantle accompanying submarine eruptions at the ridge axis. Corliss (1971) proposed that the metal enrichments in these sediments resulted from the hydrothermal convection of sea water through newly formed basaltic crust.

In early 1977, the first in situ observations of these mid-ocean rift thermal springs were made from the deep submersible ALVIN on the Galapagos Rift. In the initial report of the results of this study (Corliss et al., 1978), it is shown that the ratio of the concentration of various elements in the fluids to the temperature, and thus the "heat content", or enthalpy, of the fluids show consistent patterns in several vent areas. These data, combined with estimates of the initial composition of the magma can be used to examine heat and mass fluxes of the hydrothermal systems.

The Chemistry of the Vent Waters

Samples of submarine hydrothermal fluids were collected by ALVIN from four vent areas on the axis of the Galapagos Rift (Figure 1). Eighty-eight 8.8 liter samples of vent waters and ambient waters were collected, and have been extensively analyzed both at sea and in shore based labs.

The properties of the water samples can be considered the result of two mixing processes. The first is *subsurface mixing*, in which an initial end member fluid, which interacts at high temperatures with the rocks at depth, mixes, as it rises through the crust to the sea floor, with a descending flow of cold sea water. The second is *surface mixing*, in which this water, as it flows from fissures in the rocks, mixes with ambient bottom water during sampling (Corliss et al., 1978). Since the ultimate end members are the same, this can be considered to be *one continuous mixing process*. In this model, the temperatures of the waters sampled, or more precisely, their enthalpy, will fall on a linear mixing line between the enthalpy of the high temperature end-member interacting with rocks and ambient seawater. The same is true of all conservative components of the fluid,. i.e. elements which are not dissolved by or precipitated from the fluids during mixing and cooling. Table 1 lists several elements which have been measured in the vent waters and which appear to behave conservatively in these systems. Figures 2 a, b and c are plots of Si, He^3 and Ba in the samples.

Since silica is supersaturated during ascent and mixing, this assumption may be questioned for this component. In fact, this is the basic assumption used for application of the silica thermometer to hydrothermal systems, and research on land geothermal systems indicates that the ascending supersat-

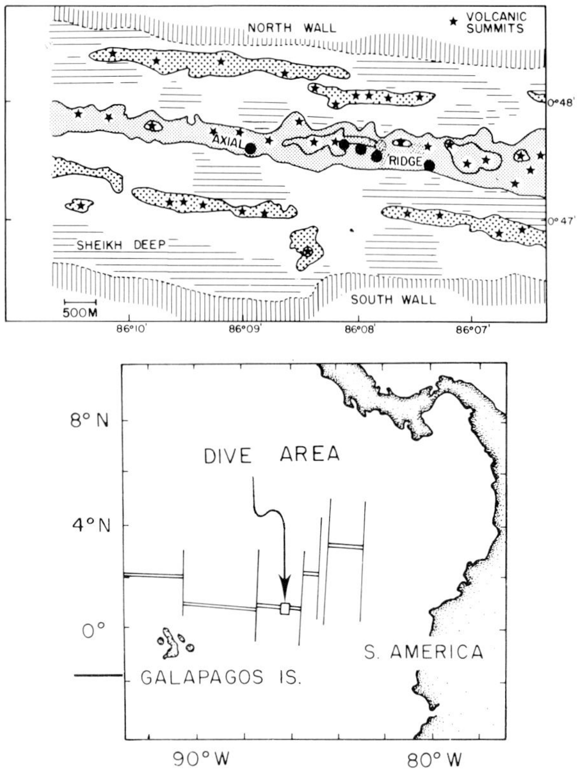

Fig. 1. A map of the thermal spring area on the Galapagos Rift. The active vents were located along the axial ridge, a topographical high in the center of symmetry of the rift. The active thermal springs are named, from west to east, the Clambake, the Dandelions, the Oyster Bed, and the Garden of Eden.

urated solutions do not dissolve or precipitate silica if the flow rate is rapid (Fournier and Truesdale, 1974). We measured vertical flow rates of 2 to 10 cm/sec immediately above vent opening where the water was rising and mixing with ambient water. At this rate, the fluids would rise 1 km in 3 to 14 hours. This flow rate is for the diluted fluid reaching the surface. Assuming that the vertical heat flux is constant, the vertical mass flux of high temperature fluid is considerably lower, approximately inversely proportional to temperature, but the vertical velocity could be as high or higher if the cross sectional area of the fissure system through which the plume rises at depth was smaller. Given the same permeability the higher density contrasts at high temperatures should lead to greater bouyant forces and resultant vertical velocities. The remaining elements in Table 1 are undersaturated in the reducing H_2S-bearing seawater solutions and thus the assumption of conservative behavior seems justified. Transition metals which form stable sulfides (Fe, Cu, Ni, Zn, Cd, etc.) precipitate from these fluids at low temperatures and thus are not conservative (Edmond et al., in prep.).

The slope of the mixing lines in Figure 2 provides the most important input to this discussion of thermal models of the hydrothermal heat flux: the ratio of element flux to heat flux in the hydrothermal fluids. These values, from Corliss et al., (1978) or unpublished data, are listed in Column 2 of Table 1. This ratio for Si is consistent for all vent areas. The same is essentially true for ^{3}He, although the ratio is slightly offset to higher temperatures in the Garden of Eden (Jenkins, et. al., 1978). For Ba and Mn, three vent areas, the Dandelions, Oyster Bed and Garden of Eden fall on one mixing line while the Clambake falls on a separate line with higher Ba contents. For Li, the Clambake, Dandelions and Garden of Eden fall on one line, while the Oyster Bed has lower Li. The Sm and Eu gradients are from Clambake samples analyzed by Conard and Corliss (unpublished data). The K gradient is small compared to the ambient value, but does reveal a significant flux from the rocks.

Composition of the Erupted Magmas

The other important inputs to this discussion are estimates of the composition of the Galapagos rift erupted magmas with regard to the elements listed in Table 1. The estimates used in this paper for the Galapagos lavas are listed in Column 4. For some components, the abundances measured in the rocks are not likely to have changed during eruption and cooling, but for He, CO_2, K, Li, we must use care in selecting data, since the gases can be lost during cooling, and the alkalis removed by hydrothermal leaching from non-glassy slowly cooled rocks, or added from sea water at low temperature.

It has been shown that the rapidly quenched glassy portions of erupted mid-ocean basalts (MORB) retain gases dissolved in the melt (e.g., Funkhouser, et al., 1968). This has lead to the use of glassy pillow margins for a number of studies of the gases and other volatile phases dissolved in oceanic tholeiite lavas (e.g., Dymond and Hogan, 1978; Craig and Lupton, 1976; Moore et al., 1977; Delaney et al., 1978). The major uncertainty in using the abundance of gases (e.g. He, CO_2) in glassy margins as representative of the initial magma concentrations results from the strong gradients in gas concentration across the margin resulting from degassing from the interiors of the rocks during slow cooling.

Dymond and Hogan (1978) present evidence to support the conclusions that (1) the relatively uniform values for He^3/He^4 found in oceanic rocks imply a constant proportion of primordial and radiogenic He in the upper mantle and, (2) loss of He during quenching leads to large variations in He concentration in glassy rocks. Since these isotopes are derived from completely different sources, these observations imply that the total initial He in MORB magmas is uniform, and that the maximum observed values in glassy rocks are the best approximations of this initial value, regardless of their geographic location. One value used here for He^4 is the average of analyses of glass from four very fresh young basalts collected by ALVIN on the Galapagos rift in the area of hydrothermal vents (J. Dymond, unpublished data). The He is determined by the consistent $^3He/^4He$ ratio reported from the Galapagos vent waters by Jenkins et al. (1978). The four values are tightly clustered ranging from 4.20-4.84 x 10^{-6} cc/gm (STP) with a mean of 4.51 $\pm$ .30 cc/gm (STP). This uniformity is somewhat surprising given the above discussion of degassing and large concentration gradients within pillows.

The other value used, 20 μcc/gm, which is the highest reported for He in MORB, also provided by Dymond (unpublished data), is from Cayman Trough basalts collected by R. Ballard. These rocks were erupted at a depth of 4000 meters, compared to 2500 meters, for the Galapagos Rift. The ambient pressure difference is presumably responsible for retention of more gas in the Cayman rocks.

For CO_2, no measurements are available for the Galapagos Rift rocks, but two recent studies have provided considerable insight into the distribution of CO_2 in mid-ocean ridge basalt magmas. Moore et al. (1977) found that vesicles in the glass fraction of mid-ocean ridge basalts are ~95% CO_2, and estimated the CO_2 content of these rocks by measuring the volume percent vesicles. Using values for the solubility of CO_2 and assuming that saturation of the melt was indicated by the presence of the vesicles, they arrived at an estimate of 800-1600 ppm CO_2 in the melt. Delaney et al. (1978) measured CO_2 directly, obtaining values from 800 to 2300 in glassy pillow basalt margins. In the absence of data on the Galapagos Rift rocks, we shall use these two extremes in our calculations.

The use of analyses of fresh glassy samples also minimizes the problem of low temperature alteration and high temperature hydrothermal leaching. The Li data in Engel et al., (1965) show the effect of rock alteration. In eight rocks with Fe_2O_3/FeO ratios less than .40, and H_2O^+ less than .70%, the Li content averages 4.7 $\pm$ 1.3 ppm. In two altered samples with Fe_2O_3/FeO of .72 and .57, and H_2O^+ of 1.50 and 1.06%, the Li content is 27 and 21 ppm respectively, presumably reflecting addition of Li during low temperature oxidative interaction with sea water. There remains the possibility that the low Li samples are holocrystalline rocks from which Li segregated into a residual component of the melt and extracted by high temperature interaction with sea water with minimal alteration of the rock (Corliss, 1971). Until these analyses can be completed on the Galapagos Rift glasses, we must use the mean of Engel et. al.'s values on unaltered rocks (5 ppm).

The Ba concentration used (12 ppm) is the average of analyses by isotope dilution given by Hart (1976). It is important to note that White and Bryan (1977) have reported on fresh, glassy

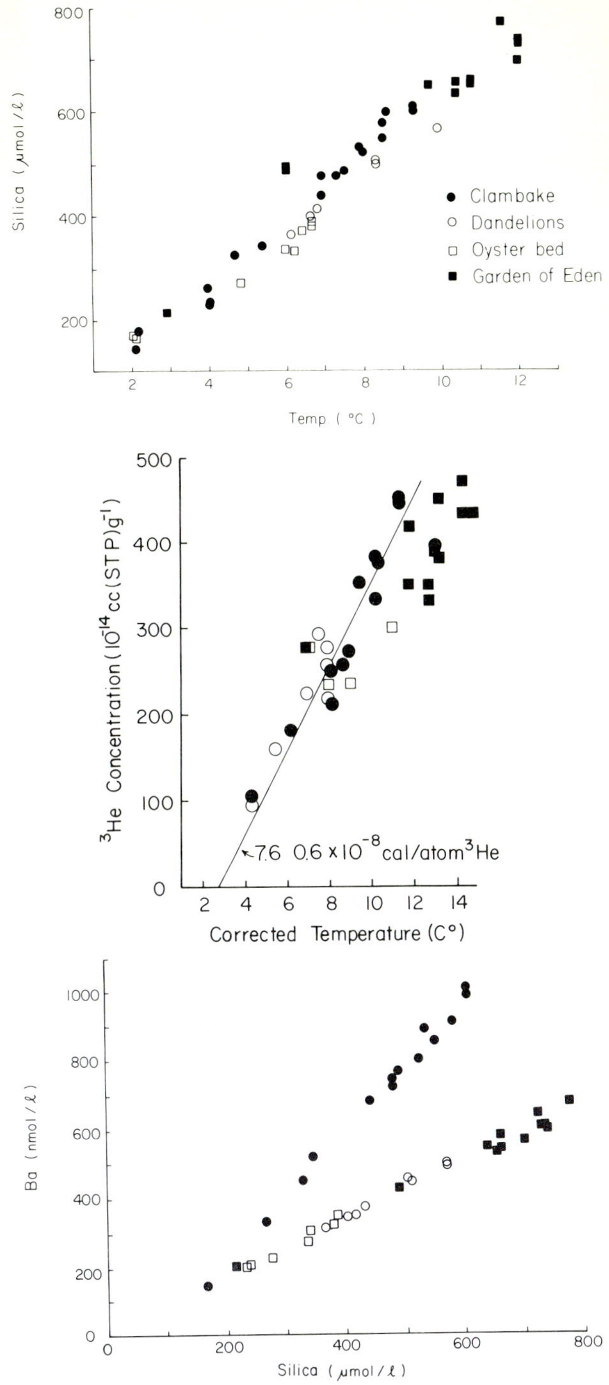

Fig. 2. Plots showing mixing lines for Si, He^3 and Ba in the thermal vent waters. The surface mixing of ambient seawater and the hydrothermal waters is shown here. Note that Ba behaves differently in different vent areas. Presumably, this mixing continues in the sub-surface and these lines can be extrapolated to the high temperature fluid at depth. 2a Si vs. Temperature 2b He^3 vs. Temperature 2c Ba vs. Silica

TABLE 1. Input Parameters and Results of Heat/Mass Budget Calculations

	(1) Ambient Water µmol/l	(2) Mixing Slope µmol/l°C	(3) R Heat/Mass Ratio cal/µmole	(4) Rock Concentration ppm*	(5) C_r Rock Concentration µmol/kg	(6) Q_e $=RFC_r$ cal/kg (F = 1)	(7) F' Apparent Extraction Efficiency	(8) D_e Depth of Extraction (Km) $D_e = D_rF'$ D_r = 3.23 km
He	1.8×10^{-3}	1.96×10^{-3}	5.10×10^5	4.51×10^{-6}	.20	1.0×10^5	3.9	12.6
He	1.8×10^{-3}	1.96×10^{-3}	5.10×10^5	20×10^{-6}	.89	4.5×10^5	.88	2.84
CO_2	2.4	25	40	800	1.8×10^4	7.2×10^5	.55	1.78
CO_2	2.4	25	40	2300	5.2×10^4	21×10^5	.19	.61
Li	26	1.5	670	5	1300	8.7×10^5	.46	1.49
Li	26	2.6	380	5	1300	4.9×10^5	.81	2.62
Ba	0.13	.049	20400	12	87	17×10^5	.23	.74
Ba	0.13	.130	7700	12	87	6.7×10^5	.60	1.94
Mn	1.0	1.5	670	1350	2.46×10^4	165×10^5	.024	-
Mn	1.0	3.9	256	1350	2.46×10^4	63×10^5	.063	-
K	10	.040	25000	400	1.02×10^4	2.6×10^8	1.6×10^{-3}	-
Sm	3×10^{-6}	3.5×10^{-7}	2.9×10^9	3.7	24.6	7.0×10^{10}	6×10^{-6}	-
Eu	8×10^{-7}	3.3×10^{-8}	3.0×10^8	1.2	7.9	2.4×10^9	2×10^{-4}	-
SiO_2	165	57	18	50%	8.3×10^6	1.5×10^8	3×10^{-3}	-

*Rock data for He are in cc/gm (STP), for Si in %

Note: Some elements appear twice to reflect a range of values estimated for rock concentration, C_r (He, CO_2), or for heat/mass ratio, R, from different vent areas (Li, Ba, Mn).

basalts from the Famous area on the mid-Atlantic ridge and find a range of values from 30 to 91 ppm. Ba covaries somewhat with the rare earths, alkalies and other alkaline earths, suggesting that this is a primary fractionation not related to alteration. Since it appears from the Galapagos thermal water data that Ba is readily leached from the rocks, some of the low values for Ba previously reported may be data on non-glassy rocks which have been leached at high temperatures by sea water. In the absence of Ba data for the Galapagos rock, we use Hart's average in our calculations.

For the remaining elements (Si, Mn, K, Sm, Eu), the data are from our analysis of the Galapagos Rift glass (Corliss, 1978). The rocks are predominately high in Fe, Ti, and the rare earths. The data used here are from averages of the non-fractionated samples.

A Conceptual Model for the Galapagos Rift Thermal Springs

Lister (1977, 1978) has proposed that seawater convection through oceanic crust occurs in two stages. The first, the active stage, occurs as water penetrates into newly erupted rock by a process of cooling, contraction, fracturing, and penetration of water resulting in increased cooling, etc., a feedback process which Lister (1974) has exhaustively examined. In the active stage, the water penetration into the rock by fracturing is rapid (~10 m/year), the water temperatures reach 200-350°C and the heated water ascends to the rock surface in a few hours. The system has a lifetime on the order of hundreds of years,and since the penetration rates are much greater than the spreading rate, the phenomena must be episodic. According to Lister, this initial stage cools the crust to some depth where active penetration is halted by the inability of the rock to maintain open fractures at high temperatures and high overburden pressure, or alternatively, by encountering ultramatic rock which undergoes a volume increase during alteration (serpentinization) which offsets the volume decrease due to thermal contraction.

The second, or passive stage of hydrothermal circulation, occurs when this permeable, water-saturated crust is heated from below by conduction. This passive system continues for as long as thermal gradients and permeability are sufficient, and in fact may continue at some level throughout the life of the oceanic crust. The pattern of heat flow observed and the large area of hydrothermal mounds south of the Galapagos Rift (Williams et al., 1974; Corliss et al., 1978 are presumably produced early in this passive stage.

This model seems to fit very closely the conditions we find at the Rift axis springs. The active thermal springs which we located and sampled at the spreading axis are all within 100 meters of the center of topographical symmetry, along the axial ridge (Figure 1). The SiO_2 thermometer suggests temperatures at ~300°C, the magnesium mixing line extrapolates to zero at about this same temperature, and the observed flow rates suggest ascent rates of hours. Preliminary work by Turekian (pers comm) on the Pb^{210} content of clams living in the thermal vent areas suggests that the age of the clams and, in turn, the vents is tens of years or less. The occurrence of dead vent areas with Mn stained rocks and slowly dissolving clam shells reveals the episodic nature of the process.

Ophiolite complexes provide a very useful source of information to allow us to see beneath the sea floor, since they record the tectonic, magnetic and hydrothermal processes which form the upper oceanic crust. Figure 3 is a steady-state cross section of the upper oceanic crust, including convecting seawater, at the spreading

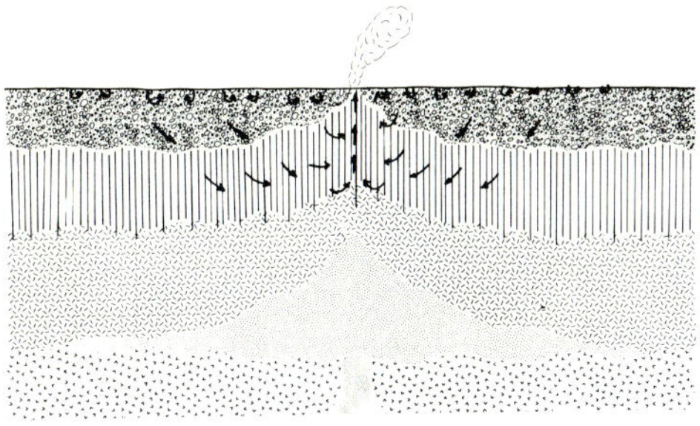

Fig. 3. A steady-state model of the formation of oceanic crust (modified from Dewey and Kidd, 1977). This figure reveals several aspects of the formation of the crust which are relevant to this discussion:

(1) The width of the dike intrusion zone is ~50 meters whereas the extrusive rocks can flow to a distance 2 to 4 km from the rift axis leading to an accumulation of up to 1 km of extrusive rocks. The cooling of these surface flows can provide ~25% of the missing conductive heat flux on the Galapagos Rift.

(2) The dike complex is formed by rapid quenching between intrusions resulting from seawater circulation.

(3) The isotropic gabbro is cut by the dikes, which are quenched against it, suggesting that this gabbro is plated onto the upper surface of the magma chamber as a result of the removal of heat by seawater.

(4) The "magma chamber", the stippled zone underlying the isotropic gabbro, is inferred from Sleep's (1975) thermal model which does not take into account hydrothermal heat loss at the axis. It may consist of a "mush", a self supporting framework of 70% crystals and 30% residual liquid. (Sleep, 1975) If all the remaining 75% of the missing conductive heat at the Galapagos Rift is extracted at the axial vents we sampled, the cooling of the crust should extend to depths greater than the depth of the magma chamber.

axis, modified from Dewey and Kidd (1977), with input from Kidd (1977), Parmentier and Spooner (1978), Coleman (1977) and our observations on the Galapagos Rift.

One important implication of these observations relates to the effects of hydrothermal cooling on the dike complex and underlying cumulate complex. The dike injection zone is restricted to a width of about 50 meters (Kidd, 1977), has a vertical thickness of up to 1.5 km., and is composed of individual dikes from 10 cm. to 5 m. thick. (Coleman, 1977, P. 55-56). The active springs we located on the Galapagos Rift are restricted to a narrow band along the axial ridge. Thus it appears that the initial "active" cooling stage occurs at the point of dike formation.

The dikes are intruded into the previously cooled, water saturated, permeable dike complex. Cooling of each newly intruded dike results from the penetration of water into vertical fractures which form as the dike is cooled and subsequent convection of the heat to the surface. Dredged holocrystalline rocks with parallel margins which show effects of sea water leaching (Corliss, 1971) are presumably formed here.

The cooling effect of the hydrothermal circulation must also extend down into the magma chamber. This assertation is supported by the statement of Coleman (1977, P. 54) that "Nearly all exposures studied (by Coleman) and cited in the literature provide clear evidence that the diabase dikes are chilled against the gabbro and that they also pinch out downward into it". As stated by Dewey and Kidd (1977), "the fact that the dikes...are "rootless" requires that they were fed from a magma chamber and that the gabbro they cut was plated onto the roof of that chamber." Thus, the thickness of the dike complex and the solidification of the roof of the magma chamber are related to the rate and extent of heat extraction by the rift axis thermal springs.

Lister (1974) suggested that the existence of a layer of "sheeted dikes" up to 1 km thick requires cooling by convective circulation of sea water, for if they were cooled only by conduction to the surface, the individual dikes would not cool enough to solidify between eruptions. Lister (1977) subsequently concluded on the basis of a model by Sleep (1975) that water percolation probably does not play a major role in the formation of the sheeted dikes. Sleep modeled the formation of oceanic crust assuming only conductive cooling and derived the temperature distribution within the crust at various spreading rates. The 1185°C isotherm, which is approximately the liquidus of the magma and which Sleep suggests is the top of the "magma chamber", appears at increasing depths at decreasing spreading rates: approximately 700 m at 5 cm/year, 1100 m at 3 cm/year and ~3.5 km for 1 cm/year. The solidus of these liquids, if we assume they are equivalent to the Hawaiian tholeiites of the Alae lava lake, is 980°C (Peck, et al., 1966). The depth to this solidus in Sleep's model is approximately 400 m, 850 m and 1800 m for the same spreading rates. Presumably, the temperature required to quench dike margins is nearer the solidus temperature and from observations of dikes cutting gabbro, they can extend below the sheeted dikes. Studies of the grain size changes along dike margins in ophiolites at various depths into the dike complex may allow more insight into the temperature contrasts between the magma and previously formed dikes.

The presence of hydrothermal vents at the Rift axis clearly shows that convective cooling of the dikes occurs. Nothing in the thermal models eliminates the possibility that the dike complex is cooled to low temperatures (e.g. 300°) at the spreading axis, and that the isotropic gabbro plates onto the base of the dike complex as a result of continuing convective heat transfer. Cooling of the gabbro could continue some distance from the actual site of dike injection. The vertical chimney of rising water cooling the youngest dikes could draw water horizontally across the upper portions of the isotropic gabbro which conducts heat from the crystallizing magma below. It should be possible to incorporate this convective cooling into a thermal model such as Sleep's to examine its effects quantitatively; this is beyond the scope of the present discussion.

Extracted Heat and Available Heat

We shall proceed now to consider the heat/mass relationships in the hydrothermal fluids in the light of this knowledge of the geologic setting of the convective systems. At this point it is useful to state explicitly the assumptions of the model encountered to this point:

1. That the sampled fluids represent mixtures of ambient sea water and fluids produced by the interaction of sea water with rocks at high temperature.
2. That during the mixing of these end-members several components of the fluid behave conservatively, i.e. are not precipitated or dissolved.
3. That enthalpy is conserved during mixing, i.e. heat is not gained by, or lost from, the fluid by conduction during mixing and ascent.
4. That the heat/mass ratios of the sampled fluids are representative of the fluids produced by these systems.
5. That the concentration of various components in the initial melt can be estimated by analysis of the quenched erupted liquid.
6. That some of these components of the initial basaltic magma are quantitatively extracted from the rock into the hydrothermal fluid.

The assumption that the sampled fluids are representative is supported by the fact that four vent areas were sampled which are presumably in different stages of their life cycle. The possibility exists that radical changes in fluid composition during the evolution of the systems may occur requiring modification of the generalizations made here.

The assumption that some components are quantitatively extracted from the rocks by sea water interactions seems well founded. The rapid and complete loss of gases during cooling of pillow interiors is shown by the common observation that they are completely depleted in rare gases, relative to the glassy margins. This indicates that the cooling and crystallization of the magma results in rejection of the gases into residual phases where they can be dissolved and transported by sea water. For some other elements, it is clear that only a very small fraction of the element present is extracted.

If we make the assumption, for the present, that the extraction is complete for all elements (F = fraction extracted = 1), we can multiply the heat extracted in calories per μmole of fluid component (R) by the μmoles of each component present in a kilogram of the magma (C_r) and arrive at a figure for the heat extracted per kilogram of rock indicated for each element.

$$Q_e = RFC_r \tag{1}$$

Interesting features of these figures, given in column 6 of Table 1, are:

1. The model gives values for extracted heat which are constant within a factor of 20 for components whose concentration in the rock varies over a factor of 10^5 (He, CO_2, Li and Ba).
2. The Mn value is somewhat higher than these elements, and K, Si and the rare earths have very high apparent basalt heat contents.

This last observation is clearly due to the fact that these elements are not removed quantitatively from the basalt, so that a mass of basalt greater than a kilogram provided the component and thus could provide significantly more heat per unit mass of element.

Keeping in mind the uncertainties in the data, it is interesting to consider how much heat would reside initially in a kilogram of magma which solidifies and cools to form part of the crust, and to compare the value with these calculated values of extracted heat.

If we assume that the convective system cools crustal rocks initially emplaced as liquids, the heat extracted from a given mass of rock is the heat required to cool that mass of liquid from some initial temperature to the liquidus, crystallize it, and then cool it from the solidus to its final temperature, or

$$Qr = (\rho C\Delta T + L)\frac{1}{\rho} \tag{2}$$

ρC = volume heat capacity of the basalt solid and liquid, (cal/cm^3 °C)

L = the latent heat of crystallization of basalt, (cal/cm^3)

ΔT = the temperature change, from the initial temperature of the erupted magma to the final temperature after cooling. (°C)

TABLE 2. Heat extractable from basaltic magma by hydrothermal cooling
$Qr = (\rho c\Delta T + L)\ 1/\rho$; ρc = .91 cal/cm^3°C, L = 246 cal/cm^3 (Sleep, 1975); ρ = 2.9 gm/cm^3

ΔT	Q_r (cal/kg)
1200	4.61×10^5
1100	4.30×10^5
1000	3.99×10^5
900	3.67×10^5
800	3.36×10^5

Sleep (1974, 1975) has considered vertical temperature distribution at the spreading axis within the rising solid mantle diapir (from 100-33 km), the zone of partial melting (33-5 km) and the crust (5-0 km). Assuming cooling along the solid adiabat is ~1°C/km, the initial temperature difference in the crust from this effect is negligible for our purposes. Sleep assumes an initial erupted magma temperature at the surface of 1180°C.

The final cooled temperature is less readily estimated. If the initial estimates of the deep end-member temperatures in the Galapagos vent areas of 300° are correct, and the mixing model is real, the surface temperatures are 20°C and the average temperature in the column is 160°C. Presumably when the "active" stage ends, the temperature will be somewhat lower. Table 2 lists values of Qr for a range of reasonable temperatures. If both the heat and the fluid components leached from the rocks were quantitatively extracted from the same mass of rock, the apparent quantity of heat extracted, along with the components of the magma which are quantitatively removed by the sea water, (Column 6, Table 1) would be identical to the amount of available heat (Table 2). Given the uncertainties of the model, the fact that estimates of the available heat (Q_r) fall in the middle of the range of apparent extracted heat contents (Q_e) for the readily extracted elements is encouraging.

It is useful to categorize the possible causes for discrepencies between these values; this is done in Table 3. The only component of the fluid for which Q_e is clearly less than Q_r is He if we use the data from the Galapagos Rift rocks. Loss of heat from the rising plume by conduction to the rock and subsequent horizontal conduction within the rock mass is not likely to be significant. The primary reason is that the area away from the vertically rising warm plume is presumably saturated with flowing cold water which forms the descending portion of the convection cell and recycles the heat into the system. The possibility that R is too low because of gain of He from previously cooled rocks is also very unlikely; rapid degassing during initial cooling is clearly demonstrated by the absence of He in pillow interiors compared to quenched pillow margins. It is more likely, in spite of the consistent values from four different rocks, that He is lost from these rocks, and

TABLE 3. Possible Causes of Discrepancies Between Extracted Heat (Q_e) and Available Heat (Q_r). $Q_e = RFC_r$ and $Q_r = (\rho c\Delta T + L)\frac{1}{\rho}$ (See Text)

$Q_e > Q_r$	
R too high	(1) loss of fluid component during ascent of fluids, e.g. precipitation. (2) conduction of heat into the site of water rock interaction
F too high	(1) F<1, conponent not quantitatively extracted from rock
C_r too high	(1) The measured concentration in the rock is <u>greater than</u> that in the magma cooled by water. Not likely for abundances measured in quenched glass; if the data is from non-glassy rocks the possibility of seawater contamination is present. e.g. low temperature alteration of analyzed rock
$Q_e < Q_r$	
R too low	(1) gain of fluid component during ascent, e.g. leaching at low temperatures without heat transfer (2) loss of heat to rocks by horizontal conduction away from plume
F too low	(1) F>1, requires migration of a component within the rock mass to the site of water-rock interaction.
C_r too low	(1) measured concentration in the rock is <u>less than</u> that in the magma cooled by water; for gases measured in glassy rocks; loss during quenching. For other component measured in holocrystalline rocks; loss during high temperature thermal leaching.

that the true value of C_r for He is closer to the Cayman Trough data. One remaining explanation, considered in a later section, is the migration of He from within the magma chamber up to the site of water rock interaction.

Extraction Efficiency

For the remaining elements, the estimated Q_e is either comparable to Q_r or somewhat larger (CO_2, Li, Ba, Mn) or dramatically higher (K, Sm, Eu, SiO_2). It is clear that for the latter elements the assumption of complete extraction does not apply. These elements are <u>not</u> quantitatively removed from the basalt, so that the water must interact with a mass of basalt greater than one kilogram to extract the mass of the component present in one kilogram of rock. Since heat is extracted more efficiently, apparent extractable heat per kilogram of rock (Q_e) is greater than that actually available from a kilogram of rock (Q_r). If we assume a value of $Q_r = 4 \times 10^5$, and that $Q_e = Q_r$ for F = 1, we can calculate a value for apparent extraction efficiency, F´, from the relation $F' = Q_r/Q_e$. These values are tabulated in column 7 of Table 1. These numbers suggest that a very small fraction of the rare earths and silica are extracted from the rocks, somewhat more potassium, and from 2.5 to 6.0% of the manganese. The holocrystalline rocks studied by Corliss (1971) had lost ~10% of their manganese compared to the glassy pillows they were associated with, which is in reasonable agreement, but were depleted by ~20% in Sm, which does not agree with the smaller extracted fraction suggested by the thermal spring data. It is possible that the rare earth elements and for that matter, K and SiO_2, may precipitate through low temperature interaction with rocks during ascent of the fluids, which would have the same effect on Q_e, but this is clearly an <u>ad hoc</u> argument.

The calculated values of Q_e for CO_2, Li, Ba and Mn are not well enough known, given the data we have on rock abundance, to allow us to attribute the variations to specific processes cited in Table 3, but it is useful to discuss these processes. The sequence Mn, Ba, Li would appear to be a sequence of increasing extraction efficiency with the present data suggesting that the removal of Ba and Li is as efficient as the extraction of CO_2. Since CO_2 should be quantitatively extracted, it would appear that the values of C_r for Li and Ba must be too low, that the rocks in which this data was measured may have been holocrystalline rocks from which Li and Ba had been previously extracted. For elements for which $Q_e > Q_r$ and F = 1, e.g. CO_2 with the present data, one interesting possibility is the conduction of heat to the site of water rock interaction.

The Heat Budget and Depth of Cooling of Hydrothermal Systems

In the model of Figure 3, from Dewey and Kidd (1977), we can see that the cooling and crystallization of the "isotropic" gabbro which is plating onto the upper surface of the magma chamber could contribute significantly to the heat which drives the seawater convection. This heat is conducted through the rock layer which separates the upper margin of the magma chamber from the circulating water. It is important, in this context, to distinguish between depth of water penetration, depth of cooling and depth of extraction. Mobility of some components of the melt, e.g., He, might allow extraction from depths greater than the actual depth of water penetration. Similarly, conduction of heat allows hydrothermal cooling to penetrate further than the water actually penetrates. We can pursue this possibility by examining the amount of heat required to cool the rock at the rate new magma is supplied.

One way to calculate the depth of hydrothermal cooling at the spreading axis is to estimate the heat loss due to seawater convection by comparing the measured conductive heat flux with the models of total heat input, and then calculate the depth of rock which must be cooled to provide this heat flux. Given the half spreading rate, u (3.5 cm/yr for the Galapagos Rift), the heat required to cool a given mass of rock, Q_r (in cal/kg, Table 2), the rock density, ρ(2.9 gm/cm^3), and the depth of cooling D_r(km) the rate of heat removal by hydrothermal convection is, in dimensionless form,

$$q_r = 2\rho u Q_r D_r \quad (3)$$

(For q_r in cal/cm sec, and other terms in the units given, the right side is multiplied by 3.17×10^{-6} to convert years to seconds and km to cm.) For $\Delta T = 1000°C$, $Q_r = 3.99 \times 10^5$ cal/kg (Table 2), the relationship between convective heat loss and depth of heat extraction is $q_r = 25.7\ D_r$.

Von Herzen, Green and Williams (in Corliss et al., 1978) have estimated the hydrothermal convective heat loss at the Galapagos Rift from extensive heat flow data on crust from 100,000 to 1,000,000 years old. The relationship $h=12t^{-1/2}$ gives the theoretical variation of conductive heat flux with crustal age where h is in heat flow units (10^{-6} cal/cm^2 sec), and t is the age of crust in m.y. (Lister, 1977). This equation, integrated over the last 1 m.y., gives a mean heat loss of 24 H.F.U. The mean measured conductive heat loss for this interval measured by Von Herzen et al., is 8 H.F.U. This indicates a missing heat flux of 16 H.F.U., or, including both sides of the spreading axis, a value of 112 cal/cm sec for the hydrothermal heat flux per unit length of ridge crest. If this heat is vented entirely during the initial cooling of the crust, it would require extraction of the heat from the upper 4.36 km of the crust ($\Delta T = 1000°$ and including the latent heat).

Consideration of figure 3 reveals an important modification which must be made to these calculations. As Kidd (1977) and Dewey and Kidd (1977) have pointed out, the width of dike injection may be narrow (Kidd estimates ~50 meters) whereas the width of extrusion is a few kilometers. This results in the accretion of ~1 km of surface flows by the time the crust has spread to the edge of the zone of extrusive accretion (Kidd, 1977). These extrusive flows are cooled by seawater convection, but the time lapse between eruptions is presumably long enough to allow each flow to be cooled separately. This circulation produces the typical chemical effects seen in young extrusive rocks, e.g. deposition of yellow and orange staining which is observed photographically on exposed rocks and also on dredged rocks. This cooling of extrusive rocks also contributes all of their content of gases (CO_2, He, etc.) and heat to the global hydrothermal flux, and may be important in providing a flux of elements such as Fe and Mn to the deep sea environment. The cooling of these flows will also contribute to the missing 112 cal/cm sec of heat flux at the mid-ocean ridge, and may be the source of small off-axis thermal anomalies observed during the Deep Tow surveys of the Galapagos Rift. The heat flux (q_r) contributed by the cooling of these extrusives can be calculated from equations (2) and (3), assuming T = 1150°C, since the final temperature of these surface rocks is lower than the mean for the dike complex and upper isotropic gabbro. This results in a flux of 29 cal/cm sec from this source leaving 83 cal/cm sec to be accounted for by axial vents. This modifies the depths of extraction of heat for the axial vents to D_r = 3.23 km.

We can substite Q_e, the apparent heat extracted for each component, into equation (3)

$$qe = 2\ uQ_eD_e$$

D_e is the depth from which the given component must be extracted to give the observed heat/mass ratios in the fluids, providing the actual heat supplied from the rocks is Q_r (3.99 cal/kg for $\Delta T = 1000°$), the heat is extracted from 3.23 km of rock, and the heat flux, q_r = 83 cal/cm sec.

To clarify the meaning of this relationship, we substitute from (1):

$$q_e = 2\rho uRFC_rD_e \quad (5)$$

and see that $2\rho uFC_rD_e$ is the rate of supply, by the processes that generate new crust, of a given rock component to that portion of the crust (depth = D_e) from which that component is extracted by seawater. Thus, q_e is the hydrothermal heat flux deduced from this mass flux plus the heat/mass ratio for the components. If we assume a heat flux of 83 cal/cm sec ($q_e = q_r$) and D_r = 3.23 km, we can calculate the apparent depth of extraction for each component, which is simply, from (3) and (4),

$$D_e = D_rQ_r/Q_e = D_rF' \quad (6)$$

These values, for elements which are strongly extracted from the rocks, are listed in Table 1, column 8.

Depths of Penetration and Extraction

The values for D_e indicate that He is extracted from 3 to 12 km of the crust, CO_2 from ~1 to 2 km of crust, while the heat is extracted from ~3.2 km of crust, the depth of cooling. A somewhat ad hoc explanation can be proposed for this:

(1) The CO_2 is initially dissolved in the magma, remains uniformly distributed due to low diffusion

rates, is fractionated into residual phases during crystallization of the magma and quantitatively removed from the rock during penetration of the water into the rock. Thus the depth from which CO_2 is extracted is the depth of actual initial water penetration at the spreading axis.

(2) The conduction of heat upward from the roof of the magma chamber where the isotropic gabbro is plating out, to the circulating seawater, combined with convection within the magma, which delivers heat to the roof of the magma chamber, allows extraction of heat from the upper 3.2 km of crust. It is worth noting that this depth assumes cooling through an average ΔT of 1000°C. This conductive cooling beyond the depth of penetration is likely to become less effective at somewhat higher final temperatures. Thus the "final" temperature in the isotropic gabbro (below 1 km) at the cessation of the "active" rift axis convection may be somewhat lower, and more rock must be cooled to provide the same convective flux. If the "final" mean temperature below 1 km was 500°C, the total depth of cooling would be 4.26 km.

(3) The He contained in the magma may be extracted from a greater depth than the water penetrates, which requires transfer of He from the convecting melt to the site of water rock interaction, or enrichment of the rocks interacting with the water by prior fractional crystallization. There is a limit to the depth of extraction of He imposed by the nature of the formation of the crust. Since the crust is produced by partial melting and is underlain by residual harzburgite depleted by the melting process, the maximum depth of extraction is equal to the thickness of the crust. This means that the maximum value for De for He must be ~5 km. Given these constraints, the minimum mean concentration of He in crustal rocks is ~7.9×10^{-6} cc/gm STP.

As Craig and Lupton (1976) have pointed out, diffusive transfer of He should lead to fractionation and thus changes in the He^3/He^4 ratio. Their model dealt with molecular diffusion through glasses at low temperature and may not be quantitatively relevant to this discussion. The He^3/He^4 ratio in the vent water (10.8×10^{-6}) (Jenkins et al. 1978) is slightly lower than the value for "high He^3" from glassy rocks reported by Craig and Lupton ($11.9 - 13.2 \times 10^{-6}$). If the Galapagos magmas have He^3/He^4 ratios in this range, then the role of extensive diffusion which fractionates isotopes in delivering He to the convecting seawater is limited. Crystallization of the magmas during plating of the isotropic gabbro to the roof of the magma chamber would fractionate He into intergranular boundaries, where it could migrate without significant isotopic fractionation down the temperature gradient, i.e., up to the penetrating water. The same effect would result if the basalts interacting with the water were enriched in He, and other gases for that matter, by fractional crystallization or vertical migration of gas bubbles (vesicles) in the melt. Thus the depth of extraction of He and other gases, like the depth of cooling, might be somewhat greater than the depth of penetration of the convecting seawater.

Summary and Conclusions

We have arrived at a model which encompasses all of the observed variations in Q_e/Q_r, i.e. the ratio of "extracted" heat to available heat in the rocks of the Galapagos spreading center. Limited extraction from the rocks ($F \ll 1$) accounts for those components whose concentration in the vent water is relatively low ($Q_e \gg Q_r$, e.g. SiO_2, Sm, Eu). For some elements with Q_e approaching Q_r (Mn, Ba, Li) we assume F is approaching 1 and that the depth of extraction is equal to the depth of penetration. For those components which we presume are quantitatively extracted (F=1), the variations in Q_e/Q_r are accounted for by assuming that heat is conducted to the site of water-rock interaction ($Q_e > Q_r$, e.g. CO_2) or that the component can migrate from the magma to the site of water-rock interaction ($Q_e < Q_r$ e.g. He).

In spite of some ad hoc reasoning, the model gains support from some independent observations on the expected behavior of the chemical components considered and from studies of exposed ophiolite complexes. The values of F´ derived are not unreasonable from our knowledge both of the chemistry of the elements and observations of fresh and altered crustal rocks.

For the values of D_e to be applicable to the vents we sampled, we must assume that all of the missing heat is extracted by these rift axis systems, which in turn requires cooling of the large fraction (~3.2 to ·4.2 km) of the oceanic crust at the axis. The evidence from ophiolites suggests that cooling penetrates into the magma chamber at the site of dike injection. This seems to conflict with recent seismic studies of the rift axis on the East Pacific Rise at 9°N which suggest that a low velocity zone interpreted to be a magma chamber persists for some time. (Orcut, et al., 1975, Rosendahl, 1976, Rosendahl, et al., 1976) The resolution of this conflict is not clear. The spreading rate is 6 cm/yr in the region studied compared to 3.5 cm/year for the Galapagos Rift. Sleep (1975) would suggest a narrower magma chamber on a slower spreading rift. The crustal model proposed by Rosendahl et al. does include a layer overlying the magma chamber analogous to the plated isotropic gabbro. If off-axis hydrothermal activity removes heat by convection, the heat supplied by axial water circulation (qr) is less, and the required depth of heat extraction (D_r) is reduced. Some undetected convective heat loss may occur at the sediment mounds, 10 km south of the axis. (Williams, et al., 1974, Corliss, et al., 1978)

The results inferred from the model calculations are strongly dependent on the assumptions of the model, which have been explicitly stated. These assumptions are subject to testing by further work currently under way, including further analyses of rocks and additional sampling of hydrothermal vents. The measured heat/mass ratios in the fluids, the inferred heat/mass ratios in the basaltic magmas with which they interact, theoretical models regarding water-hot rock interactions and observations from ophiolite complexes can be combined to examine several aspects of the thermal and chemical behavior of hydrothermal systems at mid-ocean ridges. The following general statements can be made:

(1) If Kidd (1978) is correct, convective cooling of the dike complex and at least the upper part of the gabbro takes place in the immediate vicinity of the rift axis (within ~100 meters). In fact, the existence of the dike complex may be dependent on hydrothermal cooling.

(2) The heat/mass ratios calculated for elements which are expected to be quantitatively or nearly quantitatively extracted from the rocks show close agreement with the heat/mass ratios inferred for the magma.

(3) The calculated extraction efficiency of the system for elements not quantitatively extracted also agrees with their expected behavior during water-rock interaction.

(4) The ophiolite model implies that ~25% of the missing conductive heat flux predicted from heat flow data on the Galapagos Rift is released by convective cooling of the extrusive pillow lavas and sheet flows over the width of the axial valley.

(5) If we assume that all of the remaining missing conductive heat flux is removed from the crust by the rift axis hydrothermal systems, the crust will be cooled to a depth of 3.2 to 4.2 km at the rift axis.

(6) If the inferred magma compositions are correct, they imply that the depth of water penetration, the depth of heat extraction and the depth from which various components of the magma are removed can be different, and additional data on magma composition, and heat/mass ratios in the hydrothermal fluids, should allow us to characterize these variables with increasing accuracy.

References

Bostrom, K., and M.N.A. Peterson, Precipitates from hydrothermal exhalations of the Pacific Rise, Econ. Geol., 61, 1258-1265, 1966.

Bromley, L.A., A.E. Diamond, E. Salami, and D.G. Wilkins, Heat capacities and enthalpies of sea salt solutions to 200°C, Jour. Chem. Eng. Data, 15, 246-253, 1970.

Coleman, R.G., Ophiolites, Springer Verlag, New York, 1977.

Corliss, J.B., Elemental distribution patterns in rocks collected by DSRV ALVIN on the Galapagos Rift, (to be submitted to Journ. Geophys. Res.)

Corliss, J.B., J. Dymond, L.I. Gordon, J.M. Edmond, R.P. von Herzen, R.D. Ballard, K. Green, D. Williams, A. Bainbridge, K. Crane, Tj.H. van Andel, Exploration of submarine thermal springs on the Galapagos Rift. (submitted to Science), 1978.

Corliss, J.B., The origin of metal-bearing submarine hydrothermal solutions, J. Geophys. Res., 76, 8128-8138, 1971.

Craig, H. and J.E. Lupton, Primordial neon, helium and hydrogen in oceanic basalts, Earth and Planet Sci. Lett. 31, 369-385, 1976.

Davis, E.E. and C.R.B. Lister, Fundamentals of ridge crest topography, Earth Planet Sci. Lett., 21, 405-413, 1975.

Delaney, J.R., D.W. Muenow, and D.G. Graham, Distribution of volatiles in the glassy rims of submarine pillow basalts, Geochim Cosmochim. Acta., in press, 1978.

Dewey, J.F. and W.S.F. Kidd, Geometry of Plate Accretion, Geol. Soc. Amer. Bull. 88,960-963,1977.

Dymond, J. and L. Hogan, Factors controlling the noble gas abundance patterbs if deep-sea basalts, Earth Planet Sci. Lett., 38, 117-128, 1978.

Engel, A.E.J., C.G. Engel and R.G. Havens, Chemical characteristics of oceanic basalts and the upper mantle, Bull. Geol. Soc. Amer., 76, 719-734, 1965

Fournier, R.O. and A. M. Truesdell, Geochemical indicators of subsurface temperature - part 2. Estimation of temperature and fraction of hot water mixed with cold water, Journ. Res. U.S. Geol. Surv., 2, 263-270, 1974.

Funkhouser, J.G., D.E. Fisher and E. Bonatti. Excess argon in deep-sea rocks, Earth and Planet Sci. Lett., 5, 95, 1968.

Hart, S.R. LIL element geochemistry leg 34 basalts, in Yeats, R. 5, S. R. Hart, et al. Inital Reports of the deep sea drilling project, Washington, D. C., U.S. Govt. Printing Office, 34, 83-110, 1976.

Hess, H.H., History of ocean basins in Petrologic Studies: A Volume to Honor A.F. Buddington, edited by A.E.J. Engel et al., p. 599-620. Geological Society of America, New York 1962.

Jenkins, W.J., J.M. Edmond, and J.B. Corliss, Excess He^3 and He^4 in Galapagos submarine hydrothermal vents, Nature, 272, 156, 1978.

Kidd, R.G.W., A model for the process of formation of the upper oceanic crust, Geophys. J.R. Astron. Soc., 50, 149-183, 1977.

Lister, C.R.B., On the thermal balance of a mid-ocean ridge, Geophys. J. Roy. Astron. Soc., 26, 515-535, 1972.

Lister, C.R.B., On the penetration of water into hot rock, Geophys. J. Roy. Astron. Soc., 39, 465-509, 1974.

Lister, C.R.B., Qualitative models of spreading center processes, including hydrothermal penetration, Tectonophysics, 37, 203-218,1977.

Lister, C.R.B., "Active" and "passive" hydrothermal systems in the oceanic crust: predicted physical conditions, in Benthic Boundary Layer Volume, Joint Oceanographic Assembly, Edinburgh, Scotland (1976), in press, 1978.

Lubimova, E.A., and V.N. Nikitina, On heat flow singularities over mid-ocean ridges, J. Geophys. Res., 80, 232-243, 1975.

McKenzie, D.P., Some remarks on heat flow and gravity anomalies, J. Geophys. Res. 72, 6261-6273, 1967.

Moore, J.G., J.N. Batcheler and C.G. Cunningham, CO_2-filled vesicles in mid-ocean basalt, J. Vol. and Geotherm. Res., 2, 309-327, 1977.

Oldenburg, D.W., A physical model for the creation of the lithosphere, Geophys. J. Roy. Astron. Soc., 43, 425-451, 1975.

Orcutt, J., B. Kennet, L. Dorman, and W. Prothero, A low velocity zone underlying a fast spreading rise crest, Nature, 256, 475-476, 1975.

Parker, R.L., and D.W. Oldenburg, Thermal model of ocean ridges, Nature Phys. Sci., 242, 137-139, 1973.

Parmentier, E.M. and E.T.C. Spooner, A theoretical study of hydrothermal convection and the origin of the ophiolitic sulfide ore deposits in Cyprus, Earth and Planet Sci. Lett., 40, 33-40, 1978.

Peck, D.L., T.L. Wright and J.G. Moore, Crystallization of tholeiite basalt in Alae Lava Lake, Hawaii, Bull. Volcanol., 29, 629-656, 1966.

Rosendahl, B.R., Evolution of oceanic crust 2. constraints, implications and inferences, J. Geophys. Res., 81, 5305-5314, 1976.

Rosendahl, B.R., R.W. Raitt, L.M. Dorman, L.D. Bibee, D.M. Hussong and G.H. Sutton, Evolution of oceanic crust,1, a physical model of the East Pacific Rise crest derived from seismic-retraction data, J. Geophys. Res., 81, 5294-5304, 1976.

Sleep, Norman H., Secgregation of magma from a mostly crystalline mush, Geol. Soc. Amer. Bull., 85, 1225-1232, 1974.

Sleep, Norman H., Formation of oceanic crust: some thermal constraints, J. Geophys. Res., 80, 4037-4042, 1975.

Talwani, M., C.C. Windisch, and M.G. Langseth, Reykjanes ridge crest: A detailed geophysical study, J. Geophys. Res., 76, 473-517, 1971.

Turcotte, D.L., and E.R. Oxburgh, Finite amplitude convective cells and continental drift, J. Fluid Mech., 28, 29-42, 1967.

White, W.M. and W.B. Bryan, Sr isotope, K, Rb. Cs, Sr, Ba and rare earth geochemistry of basalts from the FAMOUS area, Geol. Soc. Amer. Bull. 88, 571-576, 1977.

Williams, D.L., R.P. von Herzen, J.G. Sclater, and R.N. Anderson, The Galapagos spreading centre: Lithospheric cooling and hydrothermal circulation, Geophys. J. Roy. Astron. Soc., 38, 587-608, 1974.

CHEMISTRY OF RIDGE-CREST SEDIMENTS FROM THE NORTH ATLANTIC OCEAN

Martha R. Scott, Patricia F. Salter, and Leo A. Barnard

Department of Oceanography, Texas A&M University, College Station, Texas 77843

Abstract. The influence of ridge crest hydrothermal activity on sediment chemistry has been compared at several sites on the Mid-Atlantic Ridge. For this study gravity cores and surface sediment grab samples from the FAMOUS Area at 37°N were analyzed for metal content and found to be enriched in metals with respect to average North Atlantic pelagic sediment: Fe(x3), Mn(x2), Cu(x4), Ni(x2), Co(x3), Cr(x3) and Zn(x2). Enrichment factors of sediment from the TAG hydrothermal field are similar: Fe(x1.5), Mn(x1.5), Cu(x3), Ni(x4), Co(x5) and Cr(x4). Neither area shows the extreme enrichments reported for East Pacific Rise sediments (Bostrom and Peterson,1969). All seven of the cores from the FAMOUS Area showed an upper red brown layer up to 50 cm thick with pronounced metal enrichment, and a lower section that was grey, had less metal enrichment, and had intermittent thin, dark Mn-rich layers with higher metal content. Only two of the cores showed slight Mn depletion in the lower section with respect to average North Atlantic Pelagic clay.

Sediment accumulation rates were determined by the ^{230}Th excess method for four cores from the FAMOUS Area. The analyses led to the discovery of anomalously high U concentrations and ^{234}U/^{238}U activity ratios in these sediments. The distribution of the anomalous values suggests post-depositional injection of hydrothermal fluids from below. A maximum value of 45 ppm U (carbonate-free basis) was observed. The high U concentrations are associated with anomalously low ^{230}Th/^{234}U ratios, thus making the use of the ^{230}Th excess method extremely difficult. Best estimates of accumulation rates from radiometric, paleontological and spreading rate data yield rates of .96, .94, .50 and .35 cm/10^3 y for the cores. These rates are minimum estimates, and are considerably slower than the ^{14}C rate of 2.9 cm/10^3 y measured by Nozaki et al. (1978) for the FAMOUS Area, and 1.8 cm/10^3 y for the TAG area sediment (Scott et al., 1978). Metal accumulation rates for Fe and Mn in the TAG area, the FAMOUS Area and 45°N (Cronan, 1972) are much slower than for East Pacific Rise sediments. The same relationship holds for trace metals with the exception of TAG area sediments which appear to be accumulating Ni, Co, Cu and Cr at rates approaching East Pacific Rise values.

A scanning electron microscope and EDAX survey of ferromangenese micronodules in one of the FAMOUS Area cores showed significant variation in the nature of these concretions with depth. The core top micronodules were well formed and contained significant amounts of Fe, Mn and Cu. Below the color change the micronodules appear to have had much of their ferromanganese cement dissolved in reducing conditions. Pyrite concretions are also found below the color change, commonly inside foram tests.

The distribution of metals in FAMOUS Area sediments appears to result from three processes: metal enrichment by hydrothermal fluids delivering dissolved material which precipitates from the overlying sea water; diagenetic remobilization of metals from a reduced layer probably formed by anaerobic metabolism of organic matter; and post-depositional injection of hydrothermal fluids at the base of the sediment column.

Introduction

Metal enrichment of ridge crest sediments has been one of the many clues to resolving the processes that occur during formation of new oceanic crust. The work of Bostrom (Bostrom and Peterson, 1966, 1969; Bostrom et al., 1969; Bostrom and Valdes, 1969) and others (Bender et al., 1971; Arrhenius and Bonatti, 1965; Heath and Dymond, 1977; Cronan, 1972) have shown that general enrichments of sediment in Fe, Mn, Co, Ni, Cu, Hg, As and other trace metals can be widely observed at spreading centers, and in basal sediments to either side of spreading centers (von der Borch and Rex, 1970; von der Borch et al., 1971; Cronan et al., 1972; Cronan and Garrett, 1973; Horowitz and Cronan, 1976; Dymond et al., 1976; Bostrom et al., 1973; Dymond et al., 1977). Detailed chemical studies of the East Pacific Rise sediments (Dasch et al., 1971; Bender et al., 1971; Dymond et al., 1973; Corliss et al., 1978) and of the Mid-Atlantic Ridge (Cronan, 1972; Horowitz and Cronan, 1976; Bostrom et al., 1969; M. Scott et al., 1974, 1978) have led to the conclusion that large amounts of sea water circulate down fissures in the newly formed crust, become heated and laden with materials dissolved from the basalt, and exit through hydrothermal vents at the sea floor, adding the newly

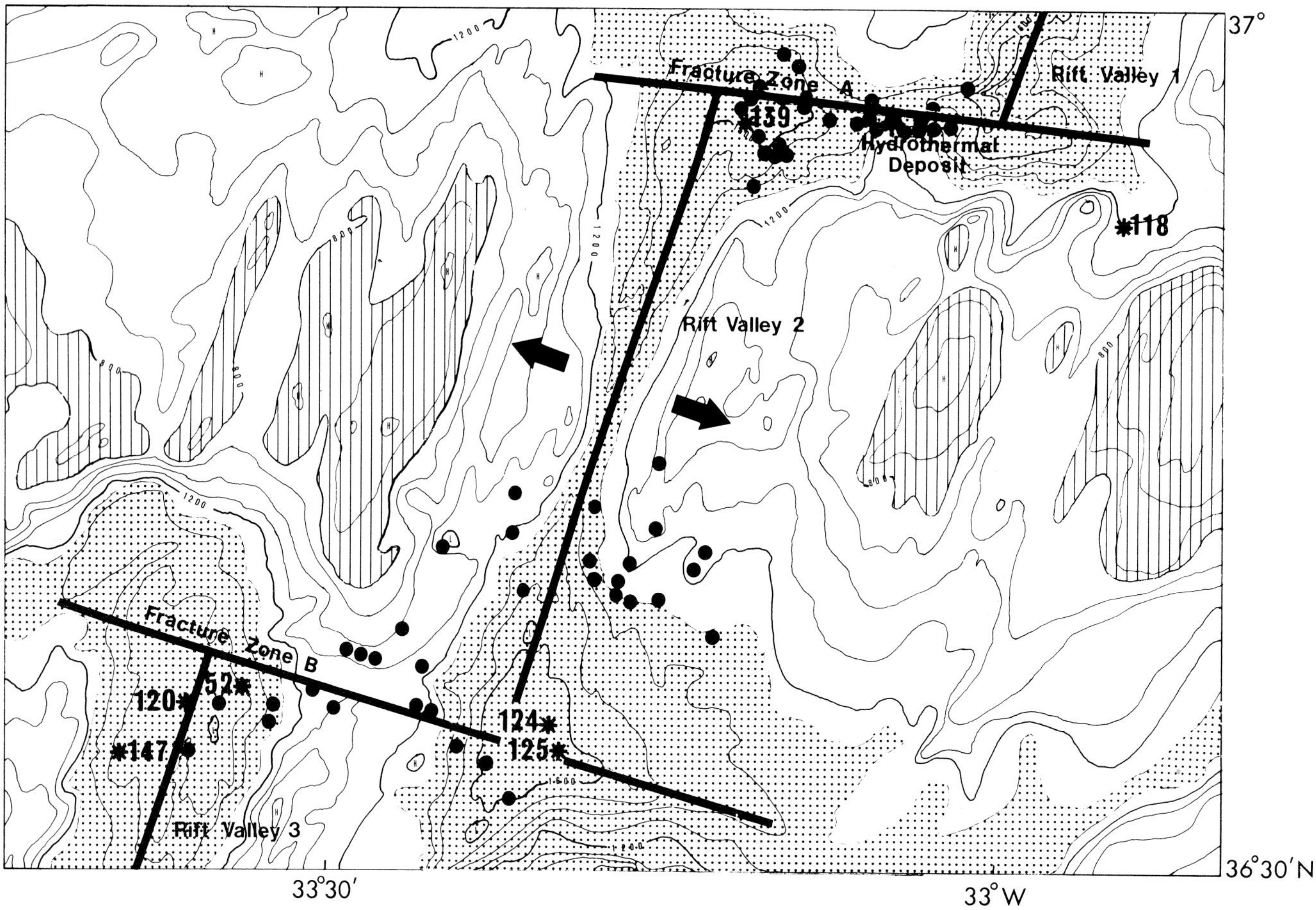

Figure 1. Location of gravity cores taken in the FAMOUS Area. Asterisks represent gravity cores; circles locate earthquake epicenters (from Reid and MacDonald, 1973; Spindel et al., 1974; Francis et al., 1977). The triangle marks hydrothermal deposits. The inner valley floor is stippled, and the outer walls of the rift valley are marked with vertical stripes.

acquired metals to the marine system. This general concept of the hydrothermal processes associated with active spreading centers is supported by geophysical studies and models for heat flow in these systems. Wolery and Sleep (1976), Lister (1972, 1974), Williams and von Herzen (1974), Spooner and Fyfe (1973) and others have concluded that the circulation of sea water through newly generated basaltic crust is the process that controls the heat flow distribution at active ocean ridges.

The effect of this hydrothermal system on the chemistry of the ocean may be enormous. Wolery and Sleep (1976) estimate that the entire ocean may be cycled through the active ridge crest system in only 5-11 million years. One effect is the addition of Fe and Mn, much of which precipitates shortly after entering the ocean. The patchy and localized distribution of the large Fe and Mn oxide deposits discovered so far (M. Scott et al., 1974; ARCYANA, 1975; Moore and Vogt, 1976; Corliss et al., 1978) indicate that such concentrated deposits form only near the vents themselves. The more general case is Fe and Mn enrichments of the sediments accumulating in the ridge crest area.

This paper examines in detail a series of sediment cores and surface sediment samples from the FAMOUS Area at 37°N, and compares these samples with sediments from the TAG hydrothermal field (R. Scott et al., 1974) and from other parts of the ridge crest in the North Atlantic Ocean.

The ridge crest sediments are enriched in a number of trace constituents such as Co, Ni, Cu, Cd, Zn and others which cannot clearly be shown to come from the hydrothermal process itself, but may be derived from sea water by the precipitating Fe and Mn oxides (Dasch et al., 1971; Dymond et al., 1973). This seems all the more probable because of the low trace metal content of the oxide deposits near vents (M. Scott et al., 1974: Moore and Vogt, 1976; Hoffert et al., 1978) compared to the metal enriched sediments. The Fe and Mn added to the sediment probably traveled some distance through sea water in particulate form.

Geologic Setting

The area of the Mid-Atlantic Ridge selected for study by the FAMOUS project is located southwest of

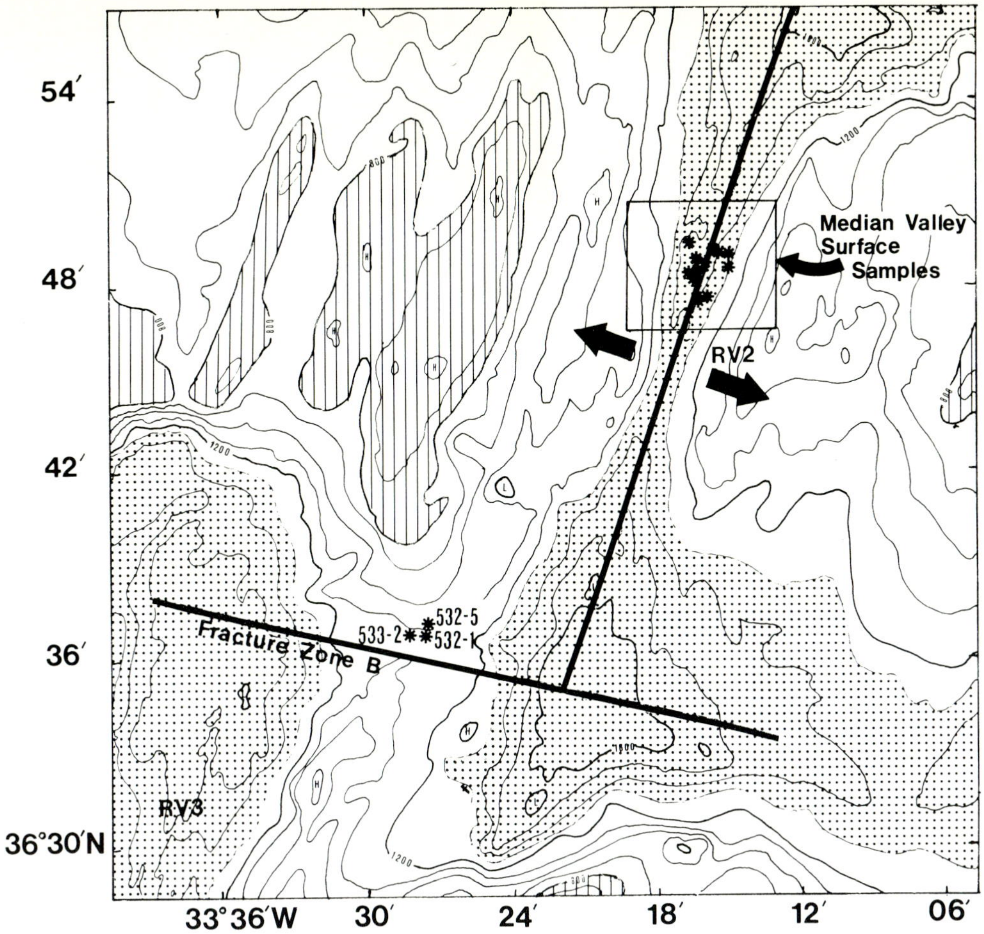

Figure 2. Location of surface sediment samples taken in the FAMOUS Area. Asterisks represent the sample locations. The area within the box includes surface samples taken near Mount Pluto. The inner valley floor is stippled and the outer walls of the rift valley are marked with vertical stripes.

the Azores between 36°N and 37°N. Detailed investigations have been made of geological and geophysical processes involved in the formation of new oceanic crust at this active spreading center. The intensive studies resulted in the collection of magnetic, gravity, seismic refraction, earthquake, heat flow and bottom current data for the area. This research provides the best available background information for studies of the geochemical processes taking place at the ridge crest. This study of sediment chemistry has been conducted on seven gravity cores taken by the R/V KNORR and fifteen surface grab samples taken by the ALVIN as part of the FAMOUS project.

The detailed bathymetry and location of the FAMOUS sediment samples are shown in Figures 1 and 2. The rift valley in the study area is approximately 45 km long and 30 km wide. The valley is offset at each end by east-west trending transform faults. The northern fault area is designated as Fracture Zone A (FZA) and the southern fault area as Fracture Zone B (FZB). The three sections of the rift valley separated by these fractures zones are designated from north to south as Rift Valley 1, 2 and 3 (RV1, RV2, RV3), as shown in Figure 1.

Most of the surface samples used in this study are located between Mount Venus and Mount Pluto, which appear to be the sites of the most recent volcanic activity (Ballard and Van Andel, 1977). No temperature anomalies or hydrothermal deposits were found in the ridge valley (Fehn *et al.*, 1977; Ballard *et al.*, 1975; ARCYANA, 1975). But periodic Hg anomalies observed by Carr *et al.* (1975) may be related to volcanic activity. The talus deposits along the scarps bordering the valley floor may mask the actual hydrothermal deposits (ARCYANA, 1975). The faults forming the valley wall exhibit a high level of earthquake activity averaging 10 to 30 events per day (Spindel *et al.*, 1974; Reid and MacDonald, 1973; Francis *et al.*, 1977).

Sediment thickness in the FAMOUS area rift valley is highly erratic but generally increases with distance from the valley axis (MacDonald and Luyendyk, 1977). The inner valley floor is covered with a thin veneer of sediment (ARCYANA, 1975; Luyendyk and MacDonald, 1977), although sediment cover up to 1-2 m thick has been observed in some areas of the

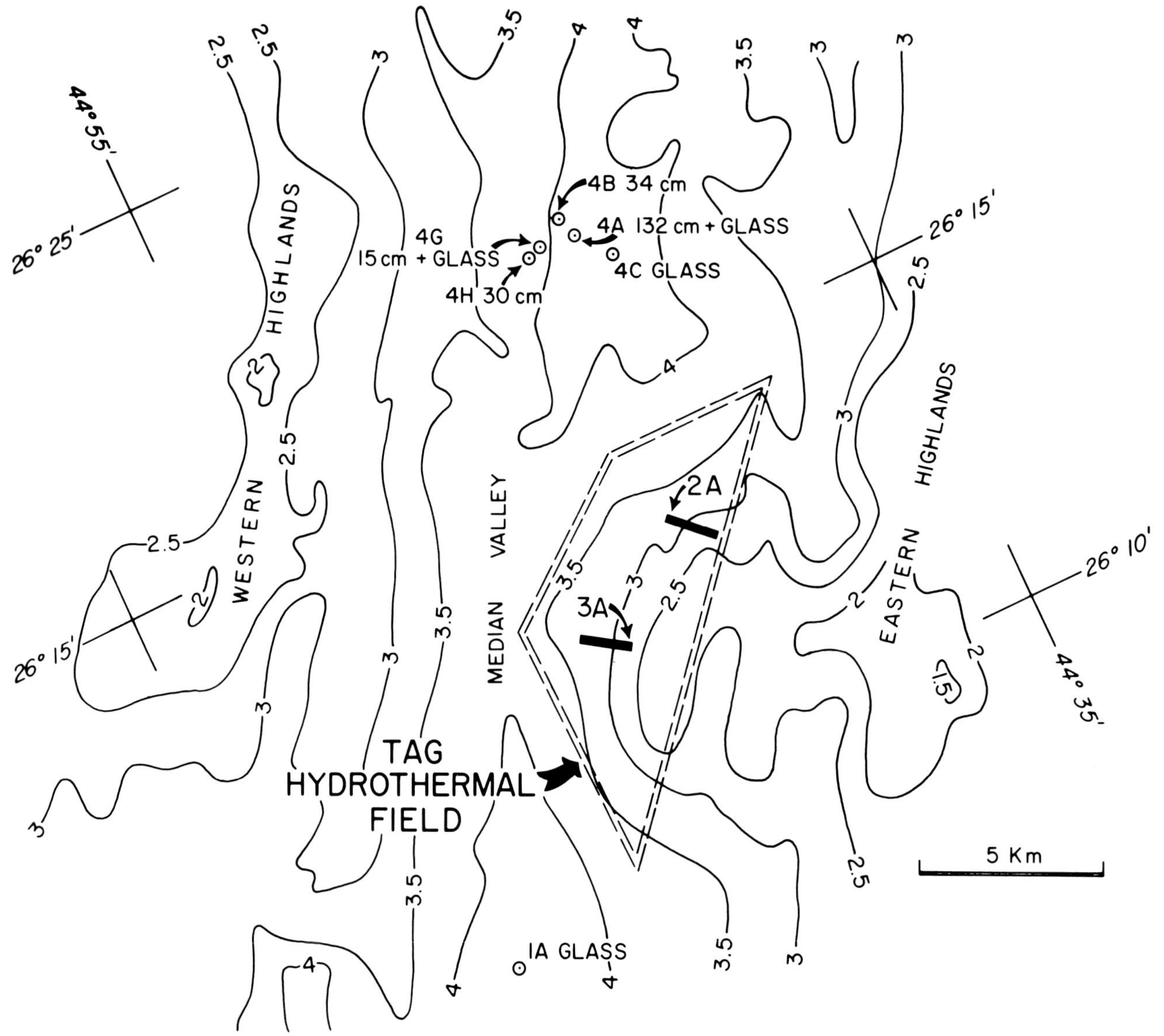

Figure 3. Detailed map showing TAG Hydrothermal Field (R. Scott et al., 1974) and location of sediment cores and dredge samples. Total length of core recovered is noted by sample locations.

inner valley floor (Needham and Francheteau, 1977). The sediment outside the inner valley floor accumulates in ponds, usually in faulted depressions. The sediment distribution is dominated by downslope movement which is probably the result of fairly fast bottom water currents (up to 20 cm/sec) (Keller, et al., 1975) and high degree of tectonic activity present in the area.

Fracture Zone A transects the Rift Valley. The intersections between the fracture zone and the separated rift valley segments are marked by circular depressions separated by a low sill. Gravity core 139 was taken in the western depression and Core 118 was taken on a terrace area just south of the eastern depression (Fig. 1). Most of the terrain in the fracture zone is covered by sediment with rock outcrops usually present at the summit of fracture blocks (ARCYANA, 1975). The fracture zone is the site of considerable tectonic activity indicated by abundant earthquake activity (Fig. 1) (Reid and MacDonald, 1973; Spindel et al., 1974; Francis et al., 1977). Heat flow measurements along FZA show a bimodal distribution, with values of about 0.6 H.F.U. to the west and values averaging 2.2 to the east (Williams et al., 1977). Two hydrothermal vents were observed by the ARCYANA (1975) in the central portion of the fracture zone (Fig. 1). The vents are located near the summits of two hills capped by black Mn encrustations from 10 to 50 cm thick. The deposits show the extreme fractionation of Fe and Mn typical of hydrothermal deposits (Fe/Mn ranges from 0.2 to 237) (ARCYANA, 1975), but no

Table 1. Chemical Composition of Sediments from the FAMOUS Area

Sample	H_2O %	$CaCO_3$ %	Fe %	Mn %	Al %	Cu ppm	Ni ppm	Co ppm	Cr ppm	Zn ppm
Core 52										
0-2 cm	44.4	84.1	2.22	.083	0.67	55.1	21.7	14.4	57.9	27.6
7-9	46.2	84.8	2.37	.090	0.69	70.7	21.2	15.7	38.7	30.4
13-15	45.5	84.0	2.27	.091	0.79	63.1	20.1	16.1	16.3	28.2
22-24	46.6	84.6	2.13	.096	0.83	70.1	20.1	16.8	16.3	32.1
30-32	42.7	83.2	1.23	.097	1.07	38.0	24.6	18.1	19.4	27.3
44-46	44.8	84.1	1.04	.088	1.11	23.3	24.5	18.0	20.4	29.9
50-52	45.8	81.8	1.01	.071	1.11	39.7	33.5	17.6	26.5	36.9
63-65	49.8	80.6	1.10	.053	1.34	52.6	19.0	18.1	31.1	37.6
75-77	45.8	80.0	1.35	.028	1.82	32.5	23.4	17.2	26.0	41.8
80-82	46.1	70.1	1.61	.029	1.79	38.9	26.8	15.6	27.2	44.5
86-88	54.4	74.2	2.26	.030	1.46	64.6	45.8	41.9	21.4	51.4
Avg.	46.6	81.0	1.69	.069	1.05	49.9	25.5	19.0	27.4	35.2
Avg. (top 32 cm)	45.1	84.1	2.04	.091	0.81	59.4	21.5	16.2	29.7	29.1
Core 118										
0-2 cm	42.5	87.5	1.58	.098	0.71	40.2	34.6	16.8	104.9	30.5
8-10	43.6	86.4	1.82	.110	0.67	45.3	19.0	20.6	21.9	31.7
13-15	42.2	88.5	1.57	.076	0.73	41.9	18.5	16.2	15.3	27.9
24-26	38.0	89.4	0.91	.059	0.79	22.8	19.0	17.5	13.2	21.8
30-32	39.3	89.7	0.73	.115	0.79	24.4	25.7	23.4	12.7	22.9
45-47	41.7	42.4	0.87	.049	0.51	16.8	19.0	16.4	13.7	20.5
Avg.	41.2	89.0	1.24	.084	0.70	31.9	22.6	18.5	30.3	25.9
Avg. (top 15 cm)	42.8	87.5	1.66	.095	0.70	42.5	24.0	17.9	48.4	30.0
Core 124										
0-2 cm	48.8	84.8	2.89	.109	0.75	83.7	27.9	16.8	72.2	35.5
8-10	47.4	82.9	2.82	.105	0.77	72.2	19.0	16.8	16.3	32.6
12-14	48.6	84.0	2.75	.099	0.71	65.4	24.6	16.0	26.5	32.9
29-31	45.3	82.2	1.79	.387	1.11	58.3	35.7	19.8	24.4	32.4
39-41	44.4	81.0	1.29	.176	1.24	30.6	29.0	23.2	23.9	31.9
41-42	46.8	83.5	1.27	.392	1.23	82.6	44.4	43.4	26.5	29.7
45-47	46.4	79.5	1.53	.049	1.24	43.0	17.9	9.8	20.4	29.0
62-64	50.7	80.6	0.99	.050	1.30	44.5	25.7	11.4	25.7	35.2
90-92	50.4	79.4	0.99	.064	1.30	45.6	36.8	22.9	43.8	45.3
Avg.	47.6	82.0	1.81	.159	1.07	58.4	29.0	20.0	31.0	33.8
Avg. (top 31 cm)	47.5	83.5	2.56	.175	0.84	69.9	26.8	17.4	34.8	33.4
Core 125										
0-2 cm	48.8	87.4	2.05	.099	1.13	62.8	31.9	24.6	27.8	45.2
5-7	49.1	80.5	1.09	.058	1.27	43.3	34.0	24.6	29.2	47.8
9-11	51.3	78.3	1.08	.038	1.31	56.6	32.9	19.3	22.4	45.2
25-27	48.3	80.7	1.00	.041	1.42	44.6	35.7	20.1	23.4	40.7

Table 1. Continued

Sample	H_2O %	$CaCO_3$ %	Fe %	Mn %	Al %	Cu ppm	Ni ppm	Co ppm	Cr ppm	Zn ppm
Core 125 (Continued)										
35-37 cm	50.9	79.4	0.98	.037	1.42	42.7	37.9	22.6	20.9	41.4
40-45	49.4	81.7	1.05	.041	1.44	40.8	29.0	18.4	32.5	38.5
Avg.	49.6	81.3	1.21	.052	1.33	48.5	33.6	21.6	26.0	43.1
Core 120										
0-2 cm	46.2	83.1	2.39	.091	0.73	56.6	20.1	15.5	19.4	28.5
12-14	46.8	85.7	2.44	.095	0.71	61.7	32.4	16.0	48.9	28.4
25-27	43.3	84.2	1.87	.097	0.95	51.5	19.0	15.9	15.8	26.9
45-47	40.7	82.2	1.08	1.36	1.19	56.6	71.4	51.1	37.7	77.1
48-50	45.2	78.3	1.56	.441	1.19	32.5	24.6	15.3	19.4	28.9
62-64	49.3	80.6	1.06	.038	1.22	25.9	24.5	10.6	26.9	33.6
80-82	44.9	80.6	0.98	.043	1.07	38.7	32.4	19.6	27.0	33.7
93-95	47.7	82.2	0.99	.037	1.19	43.1	24.6	14.9	19.4	35.8
110-112	41.9	81.6	0.74	.060	1.23	29.2	40.2	17.4	21.4	37.3
Avg.	45.1	82.0	1.45	.251	1.05	44.0	32.1	19.6	26.2	36.7
Avg. (top 27 cm)	45.4	84.3	2.23	.094	0.80	56.6	23.8	15.8	28.0	27.9
Core 139										
0-2 cm	48.4	81.2	2.56	.110	0.83	60.6	19.0	16.5	16.3	30.4
13-15	46.5	81.8	2.49	.113	0.83	67.5	24.6	17.6	34.6	30.7
30-32	46.6	81.2	2.78	.166	0.87	84.0	21.2	19.0	18.3	32.7
45-47 cm	41.4	81.4	1.53	.122	1.26	45.0	35.7	20.6	54.0	29.6
58-60	46.7	80.5	1.78	.062	1.23	41.2	22.3	12.9	20.4	40.9
60-62	48.1	80.9	1.60	.057	1.26	29.7	25.7	13.6	21.4	28.8
88-90	43.4	79.8	0.96	.050	1.19	35.7	20.1	15.4	19.4	29.1
112-114	37.4	81.6	1.03	.046	1.15	35.6	22.3	16.8	19.9	35.9
125-127	42.9	82.4	0.89	.041	1.11	41.7	24.6	17.8	19.4	37.1
Avg.	44.6	81.2	1.74	.085	1.08	49.0	23.9	16.7	24.8	32.8
Avg. (top 32 cm)	47.2	81.4	2.61	.130	0.84	70.7	21.6	17.7	23.1	31.3
Core 147										
0-2 cm	44.8	85.0	2.25	.086	0.69	53.2	19.5	14.6	21.4	32.2
10-12	44.8	84.8	2.37	.098	0.79	60.1	19.0	16.9	15.3	31.0
35-37	42.0	85.2	1.12	.103	1.11	40.8	23.4	19.7	66.2	31.0
47-49	43.8	82.6	1.02	.407	1.19	54.8	52.5	46.2	28.5	35.3
52-54	43.2	81.9	1.05	.173	1.23	34.6	27.9	13.7	30.6	32.0
64-66	47.0	81.7	0.81	.030	1.23	46.6	31.2	14.2	23.4	39.0
80-82	44.8	83.7	0.81	.032	1.03	30.2	24.6	15.2	32.6	35.3
95-97	46.5	82.8	0.81	.039	1.07	32.8	29.0	18.4	17.3	33.2
115-117	45.8	81.5	0.74	.053	1.31	36.4	23.5	12.4	54.0	38.7
Avg.	44.7	83.2	1.22	.113	1.07	43.3	27.8	19.0	32.1	34.2
Avg. (top 37 cm)	43.9	85.0	1.91	.096	0.86	51.4	20.6	17.1	34.3	31.4

Table 1. Continued

Sample	H_2O %	$CaCO_3$ %	Fe %	Mn %	Al %	Cu ppm	Ni ppm	Co ppm	Cr ppm	Zn ppm
Core Avg. Carbonate free			9.0	.70	6.2	278	161	114	158	201
Surface Core Avg. Carbonate free			10.7	.6	5.6	292	141	118	198	172
Avg. N. Atlantic Pelagic Clay*			5.02	.40	9.04	115	79	39	80	129

*Data taken from TUREKIAN and IMBRIE, 1966; CRONAN, 1972; BOSTROM and VALDES, 1969; HOROWITZ and CRONAN, 1976; VAN DER WEIJDEN et al., 1970; EL WAKEEL and RILEY, 1961; BRUTY et al., 1972.

Table 1. Chemical Composition of Sediment Cores from the FAMOUS Area.

temperature anomalies were noted in this area (Fehn *et al.*, 1977).

Fracture Zone B is located at the south end of RV2 and can be divided into two sections, the western depression (> 2700 m depth) and the eastern depression (> 2900 m depth) (Fig. 1). The depressions are formed by the intersections of the fracture zones with the two sections of the rift valley it offsets. The depressions are separated by a sill about 800 m high and 9 km wide on which three of the surface grab sample sites were located (Fig. 2). The fracture zone has heavy sediment cover with local exposures of weathered basalt talus and greenstones (Ballard *et al.*, 1975). The average heat flow in the western depression is 5 H.F.U. (Williams *et al.*, 1977) compared to a world average of 1.5 H.F.U. Three of the gravity cores analyzed in this study were taken from the high heat flow area (Fig. 1). The remaining two gravity core locations are in the eastern depression of FZB (Fig. 2).

A number of other studies have dealt with the chemistry of Mid-Atlantic Ridge sediments (Bostrom *et al.*, 1969; Cronan, 1972; Horowitz, 1974; Horowitz and Cronan, 1976) but only the TAG hydrothermal field has had reported hydrothermal oxide deposits as well as metal enriched sediments (R. Scott *et al.*, 1974; M. Scott *et al.*, 1974; R. Scott *et al.*, 1976; Rona *et al.*, 1976; M. Scott *et al.*, 1978). Figure 3 is a sediment sample location map of the TAG area (M. Scott *et al.*, 1978). The hydrothermal field itself, shown in Figure 3, is a 100 km square area on the eastern valley wall of the median valley covered with thick, rapidly deposited MnO_2 coatings (R. Scott *et al.*, 1974; M. Scott *et al.*, 1974). The deposit has apparently formed at a vent site at the base of the valley wall covered by talus. Sediment thickness in the area is quite variable. Samples 4A, 4B and 4H represent cores of ponded carbonate sediment collected from the median valley. A study of bottom photographs of this section of the median valley shows ripple-marked sand and other evidence of the strong bottom currents responsible for sweeping sediment off some portions of the valley floor into sediment ponds of uneven distribution (Temple *et al.*, 1978). Betzer *et al.* (1974) have reported that iron and manganese rich particles are far more common over the hydrothermal site than to either side of the ridge crest. The TAG area has more numerous fault blocks than the FAMOUS area, and lacks the terraces between the inner and outer valley wall (Temple *et al.*, 1978). Otherwise the two areas are similar.

Chemical Analyses

Sediment samples from the FAMOUS Area were analyzed for Fe, Mn, Al, Cu, Ni, Co, Cr and Zn by atomic absorption spectrophotometry (Salter, 1977). The reference standards used include CRPG-BR, CSJ-JB-1, USGS-BCR-1, and matrix standards made by adding Ca to salt solutions of the elements of interest. Relative precisions for the elements were Fe ± 4%, Mn ± 4, Al ± 3%, Cu ± 5%, Ni ± 8%, Co ± 5%, Cr ± 10% and Zn ± 5%. The data are listed in Table 1. Uranium and thorium isotopes were analyzed by alpha spectrometry using a procedure similar to that of Ku (1965). Table 2 includes the isotopic data on the samples; the errors listed represent 1σ from the counting statistics. The $CaCO_3$ content of the sediments was determined by an EDTA titration of the acid soluble Ca, modified from the procedure of Turekian (1956).

Results

The most conspicuous chemical trends in the FAMOUS Area Sediments were evident on visual inspection of

Table 2. Chemical Composition of Surface Sediments from the FAMOUS Area

Sample	H_2O %	$CaCO_3$ %	Fe %	Mn %	Al %	Cu ppm	Ni ppm	Co ppm	Cr ppm	Zn ppm
525-1	45.4	75.6	2.70	.110	1.78	58.4	35.7	21.9	46.8	34.2
525-3	41.8	76.6	2.51	.125	1.50	61.2	27.8	23.2	47.8	36.1
526-4	29.5	82.3	2.01	.106	0.95	50.4	23.4	22.7	32.6	29.9
527-4	51.8	81.8	2.06	.113	0.95	50.2	23.4	21.6	33.6	30.3
528-1	46.8	82.3	2.06	.121	0.95	62.0	23.4	23.4	26.5	35.5
529-4	40.2	74.8	2.06	.109	1.50	54.8	37.9	23.0	52.9	32.8
529-5	45.4	78.0	2.34	.146	1.32	52.9	38.0	29.4	46.3	32.4
530-1	43.2	80.3	2.24	.110	1.18	60.0	25.6	19.2	33.5	30.1
530-4	64.2	56.4	3.54	.085	3.44	65.6	70.6	35.3	305.0	57.4
532-1	50.1	87.6	1.52	.108	0.53	52.6	25.0	23.3	--	30.2
532-5	56.2	85.2	1.45	.104	0.69	53.1	24.5	20.3	15.7	32.4
534-1	47.9	81.8	2.07	.116	0.94	61.8	22.2	23.2	27.4	36.7
534-2	48.2	83.2	1.96	.115	0.91	52.2	25.6	20.2	65.1	28.8
533-2	44.4	81.8	1.84	.109	0.87	50.0	19.0	20.8	63.2	28.2
535-1	58.5	60.1	3.16	.125	2.69	83.5	46.9	28.4	79.4	60.2
Sample Avg. Carbonate free			10.7	.60	5.6	292	141	118	198	172
Avg. N. Atlantic Pelagic Clay*			5.02	.40	9.04	115	79	39	80	129

*Data taken from TUREKIAN and IMBRIE, 1966; CRONAN, 1972; BOSTROM and VALDES, 1969; HOROWITZ and CRONAN, 1976; VAN DER WEIJDEN et al., 1970; EL WAKEEL and RILEY, 1961; BRUTZ et al., 1972.

Table 2. Chemical Composition of Surface Samples from the FAMOUS Area.

the cores. All of the surface grab samples and the upper parts of the gravity cores were markedly enriched in Fe and Mn, and had a reddish-orange color which faded into a light tan or grey with depth in the core. Intermittent dark, Mn-rich bands appeared in several of the cores. The high concentration of biogenic $CaCO_3$ (Table 1), common in ridge crest cores, obscures variations in metal concentrations. This effect has been typically corrected for by ratioing the metal concentrations to aluminum in a variety of ways. In this paper a direct metal to aluminum ratio is used. Aluminum is relatively immobile during weathering, so that it is considered a reliable indicator of detrital material. It does not appear to be a significant component of hydrothermal solutions formed by sea water leaching of basalts (Hajash, 1975; Bischoff and Dickson, 1975; Seyfried and Bischoff, 1977). Figures 4a-4g show variations with depth in the seven gravity cores from the FAMOUS Area discussed here. Figure 5 is a similar diagram for Core 4A from the TAG area (M. Scott et al., 1978). Included on the diagrams are the metal to aluminum ratios for North Atlantic average pelagic clay (Turekian and Imbrie, 1966; Cronan, 1972; Bostrom and Valdes, 1969; Horowitz and Cronan, 1976; Van der Weijden et al., 1970; El Wakeel and Riley, 1961; Bruty et al., 1972). The Fe and Mn enrichments in the upper parts of the cores (Table 1, Figs. 4a-4g) decrease below the color change to values near the average for North Atlantic Pelagic clays. The trace metals Cu, Ni, Co, Cr and Zn are also enriched in these sediments in the upper Fe and Mn enriched layers. In general the trace metal enrichment is lower below the color change than above it. There appears to be an increase in some of the trace metals toward the bottom of Cores 139, 118, 52 and 124 (Figs. 4a, 4b, 4d and 4f). This feature is shown by the uranium profiles for the four cores on which uranium analyses were made (Figs. 4a, 4b, 4d and 4e). In addition to enrichment at depth the uranium concentrations showed enrichment toward the tops of these same cores relative to average pelagic clay.

Metal enrichment of the FAMOUS Area sediments varied in both space and time. In addition to the major color change boundary in these sediments, there are abundant dark layers in some of the cores. Analyses of several of these layers showed them to be highly enriched in manganese and other metals. Cores 120, 124 and 147 (Figs. 4b, 4d and 4g) from Fracture Zone B show several dark layers in each core, but the layers were not noted in Cores 52 and 125 from the same area (Figs. 4a and 4e). These Mn rich bands may represent periodic, localized pulses of intense

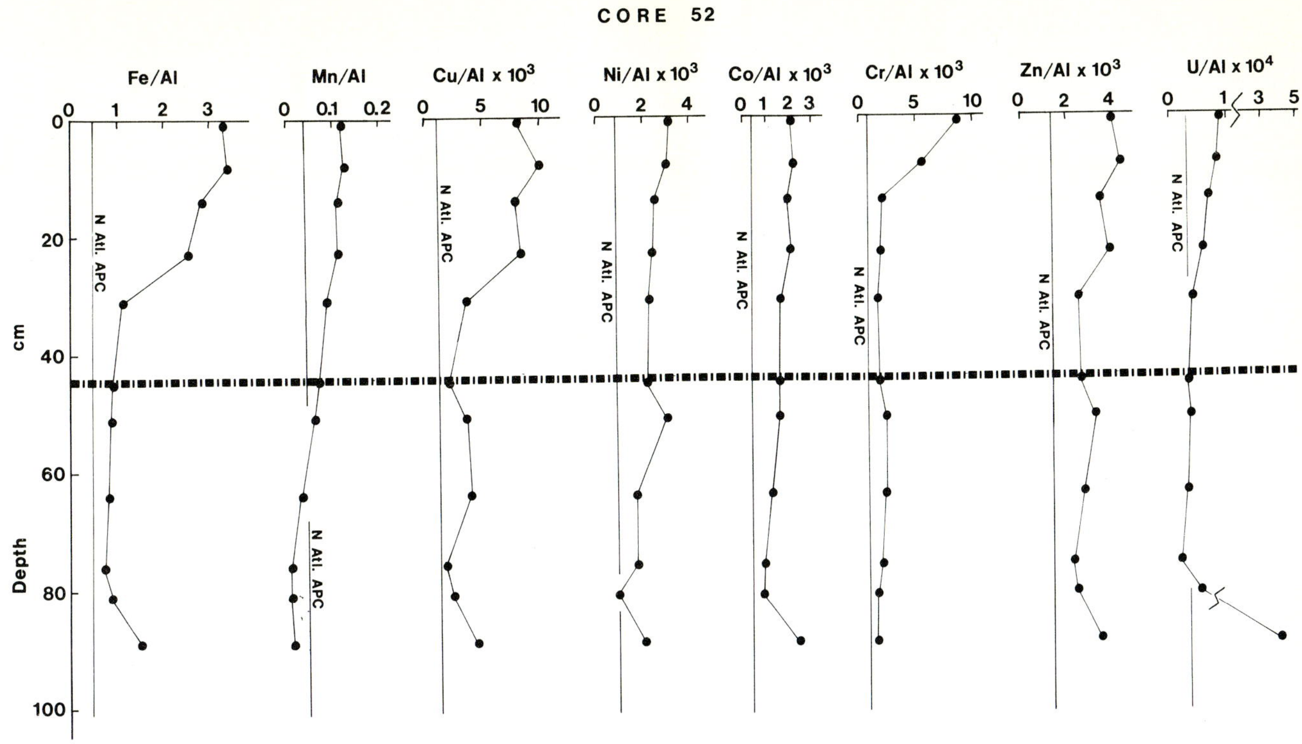

Figure 4a. Core 52 Vertical distribution of metals in FAMOUS Area cores plotted as metal to Al ratios. The average North Atlantic pelagic clay ratios are shown by vertical lines labeled N. Atl APC (Table 1). The heavy horizontal line with vertical stripes represents the location of the color change in the cores. The thinner horizontal lines show the positions of dark Mn-rich bands.

hydrothermal activity in the area. The samples analyzed (Figs. 4b, 4d and 4g) show increased Co, Ni and Cu. These features will be discussed below.

Uranium and Th isotope measurements were made on four of the cores (Table 3) so that sediment and metal accumulation rates could be determined for the area. Cores 52, 124 and 125 were found to have anoma lously high concentrations of U near the bottom of the cores. All of the cores also show anomalously high $^{234}U/^{238}U$ activity ratios in some samples (Table 3). These unusual U concentrations and isotope activity ratios are related to the hydrothermal activity in the FAMOUS Area, and make it necessary to use caution in interpreting ^{230}Th excess data for sediment accumulation rates.

The plots of exponential decrease of the ^{230}Th excess with depth in the sediment are shown in Fig. 6 for Cores 52, 118, 124 and 125. The sedimentation rates given with the plots were derived from the slopes of the lines through the data points. The semi-log plot of ^{230}Th versus depth for Core 52 is linear and the slope gives a sedimentation rate of 0.98 cm/10^3 y. Two data points with anomalously high U were omitted from the plot (75-77 cm and 86-88 cm, Table 3). Points at 30-32 and 50-52 cm fall off the line because of anomalous U activity ratios (Table 3).

The ^{230}Th excess versus depth plot is linear for Core 118 despite some anomalous $^{234}U/^{238}$ ratios in the data. The slope of the line gives a sedimentation rate of 0.35 cm/10^3 y. This rate is much lower than determined for the other cores, but 118 is located on the slope of a scarp rather than in a depression as the other cores are, and might be expected to have a lower accumulation rate. The high $^{234}U/^{238}U$ ratios for the two bottom points result in an overcorrection for supported ^{230}Th contributing to the lower apparent rate (Table 3).

Efforts to arrive at a sedimentation rate for Core 124 from the isotopic data lead to a value of 0.35 cm/10^3 y (solid line, Fig. 6), but anomalous Th and U contents make this value suspect. Moreover, for a total core length of 120 cm, this sedimentation rate results in an age for the bottom of the core of 343,000 years. Paleontological data (S. Gartner, personal communication) for this core indicates that the sediments are late Pleistocene - early Holocene (less than 100,000 years old). The age of the crust in the inner valley floor determined by paleomagnetic data and spreading rates is about 100,000 ± 30,000 years (Needham and Franchetau, 1974; Hekinian and Hoffert, 1973; MacDonald, 1977). It is not known whether the core reached basement, so a minimum sedimentation rate of 0.94 cm/10^3 y is calculated for Core 124 from paleontological data and core length (dashed line, Fig. 6).

Core 125 is located close to 124 and is expected to have a similar sedimentation rate. The isotopic rate of 0.19 cm/10^3 y is unrealistically low for the same reasons discussed for Core 124, and a more re-

Table 3. Uranium and Thorium Isotopic Analysis of the FAMOUS Area Sediments

Sample	$CaCO_3$ %	U ppm	Th ppm	Th/U	$^{234}U/^{238}U$	$^{230}Th/^{232}Th$	$^{230}Th/^{234}U$	$^{230}Th/^{238}U$	^{230}Th dpm/g	^{234}U dpm/g	$^{230}Th_{ex}$ dpm/g
Core 52											
0-2 cm	84.1	.59±.07	3.77± .12	6.41± .79	1.20±.19	8.03± .25	14.30±1.64	17.13±2.11	7.45±.14	.52±.06	6.93±.15
7-9	84.8	.57±.01	1.35± .06	2.37± .11	1.05±.03	20.91± .84	15.69± .11	16.45± .54	6.95±.16	.44±.01	6.51±.16
13-15	84.0	.56±.02	3.30± .07	5.85± .34	1.17±.07	8.98± .18	14.94± .66	17.48± .82	7.29±.08	.49±.02	6.80±.08
22-24	84.6	.49±.01	2.06± .10	4.23± .19	1.01±.03	16.26± .66	22.66± .87	22.85± .88	8.22±.22	.36±.01	7.86±.22
30-32	83.2	.46±.03	3.30± .09	7.25± .66	1.44±.14	6.97± .18	11.64± .96	16.74±1.31	5.66±.08	.49±.03	5.17±.09
44-66	84.1	.37±.01	1.98± .06	5.35± .22	0.96±.02	9.92± .27	18.33± .44	17.62± .46	4.83±.09	.26±.00	4.57±.09
50-52	81.8	.41±.04	3.21± .12	7.94± .84	1.66±.21	5.91± .22	9.34± .80	15.49±1.63	4.66±.11	.50±.04	4.16±.11
63-65	80.6	.43±.01	2.58± .28	6.01± .67	1.00±.02	8.16± .72	16.30±1.22	16.30±1.22	5.19±.38	.32±.01	4.87±.38
75-77	80.0	.50±.01	18.09±1.15	35.81±2.38	1.02±.02	3.47± .11	40.44±2.61	41.29±2.68	15.43±.91	.38±.01	15.05±.91
80-82	70.1	.92±.09	4.47± .20	5.13± .55	1.21±.14	5.92± .25	7.85± .74	9.52± .97	6.51±.18	.83±.07	5.68±.20
86-88	74.2	6.71±.07	2.44± .10	0.36± .02	1.13±.01	7.64± .28	0.82± .08	0.92± .02	4.60±.12	5.60±.06	-1.00±.13
Core 118											
0-2 cm	87.5	.56±.02	4.67± .11	8.40± .34	1.17±.04	4.99± .11	11.89± .36	13.95± .45	5.73±.08	.48±.01	5.25±.08
8-10	86.4	.44±.01	1.24± .04	2.79± .11	0.97±.03	16.94± .53	16.16± .58	15.74± .55	5.18±.09	.32±.01	4.86±.09
13-15	88.5	.49±.02	2.24± .04	4.57± .20	1.32±.08	8.92± .15	10.23± .45	13.55± .68	4.92±.04	.48±.02	4.44±.05
24-26	89.4	.28±.01	1.06± .12	3.74± .43	1.03±.02	7.78± .83	9.38± .67	9.68± .69	2.03±.14	.22±.00	1.81±.14
30-32	89.7	.47±.01	2.81± .05	5.95± .16	1.42±.05	3.34± .06	4.63± .11	6.60± .20	2.31±.03	.50±.01	1.81±.03
45-47	92.4	.48±.02	1.02± .02	2.11± .10	1.80±.07	5.94± .15	2.29± .07	4.15± .13	1.49±.02	.65±.02	0.85±.03
Core 124											
0-2 cm	84.8	----	2.48± .04	----	----	7.87± .14	----	----	4.79±.04	----	----
8-10	82.9	.50±.01	2.93± .18	5.86± .38	1.12±.03	18.49± .91	32.17±1.59	36.00±1.77	13.32±.54	.41±.01	12.91±.54
12-14	84.0	.43±.02	1.23± .03	2.86± .15	1.26±.08	29.79± .70	22.31±1.06	28.26±1.52	9.01±.07	.40±.02	8.61±.08
29-31	82.2	.27±.01	1.83± .09	6.81± .42	1.06±.04	13.99± .62	29.67±1.65	31.61±1.83	6.29±.18	.21±.01	6.07±.18
39-41	81.0	.36±.01	1.97± .06	5.47± .04	1.64±.08	13.32± .39	14.71± .03	24.16± .04	6.45±.09	.44±.01	6.01±.09
45-47	79.5	----	1.80± .14	----	----	12.87± .96	----	----	5.71±.19	----	----
62-64	80.6	2.92±.06	2.82± .06	0.97± .03	1.10±.02	7.08± .15	2.07± .05	2.28± .06	4.92±.06	2.34±.05	2.55±.08
90-92	79.4	1.63±.05	2.96± .06	1.81± .07	1.04±.04	8.63± .16	4.98± .15	5.20± .16	6.28±.06	1.26±.04	5.02±.07
Core 125											
0-2 cm	87.4	.56±.02	2.87±.06	5.13± .21	1.12±.04	8.39± .17	12.77± .38	14.31± .44	5.93±.06	.46±.01	5.46±.06
5-7	80.5	.45±.02	3.45±.06	7.63± .36	0.97±.05	5.69± .10	14.82± .16	14.42± .62	4.83±.04	.33±.01	4.50±.05
9-11	78.3	1.92±.05	3.45±.17	1.80± .10	1.05±.03	8.72± .36	4.97± .21	5.22± .23	7.40±.25	1.49±.03	5.91±.26
25-27	80.7	2.03±.04	1.80±.05	0.53± .03	1.07±.02	8.77± .23	2.41± .06	2.59± .06	3.88±.05	1.61±.03	2.27±.06
35-37	79.4	9.28±.04	0.22±.11	0.02± .02	1.10±.02	17.92±8.97	0.12± .16	0.14± .16	0.97±.15	7.58±.29	6.61±.32
45-47	81.7	1.40±.03	1.88±.04	1.35± .04	1.32±.03	10.44± .22	3.53± .08	4.67± .11	4.84±.05	1.37±.03	3.48±.05

Table 3. Uranium and Thorium Isotopic Analysis of FAMOUS Area Sediments.

liable minimum rate of 0.50 cm/10^3 y is calculated (dashed line) based on core length and paleontological and paleomagnetic evidence.

All of the sedimentation rates determined for the FAMOUS cores above are slower than the ^{14}C rate of 2.9 cm/10^3 y determined for a core taken by the ALVIN in the inner valley near the center of Rift Valley 2 (Nozaki et al., 1977). The discrepancy is not surprising considering the uneveness of the sediment cover discussed above and the nature of the U data from the hydrothermally altered sediments.

Ku et al. (1972) have measured sedimentation rates for the Mid-Atlantic Ridge at 42°N and 20°N by ^{230}Th excess, ^{231}Pa excess and paleontological methods, and M. Scott et al. (1978) have measured the sedimentation rate for Core 4A from the TAG area at 26°N. A comparison of the sedimentation rates along the ridge is given in Table 4. Cores located in the median valley are marked with an asterisk in this table. Median valley cores analyzed in this work show lower sediment accumulation rates than those reported by Nozaki et al. (1977) and M. Scott et al. (1978). As discussed above, these should be considered minimum rates. The carbonate free sedimentation rates determined for FAMOUS median valley cores in this study range from .19 to .04 cm/10^3 years, values close to those reported by Goldberg et al. (1964) for 24°N on the ridge. However, Ku et al. (1968) have suggested the latter rates are too low because of the assumption of constant $^{230}Th/^{232}Th$ with depth in the cores. The sedimentation rates listed in Table 4 are variable from place to place, reflecting the uneven sediment cover, high bottom currents and tendency toward ponding in depressions. The rates of detrital material accumulating in the median valley appear to be lower than in the abyssal plains at the same latitude. The ^{230}Th excess method must be used with caution in these sediments because of anomalous U concentrations and isotope ratios.

Discussion

Metal to aluminum ratios for FAMOUS Area sediments are higher than the same ratios for average North Atlantic pelagic clay by factors that vary from element to element. The average enrichment factors are Fe(x3), Mn(x2), Cu(x4), Ni(x2), Co(x3), Cr(x3) and Zn(x2). These factors are similar to the ones reported for sediment from the TAG Hydrothermal

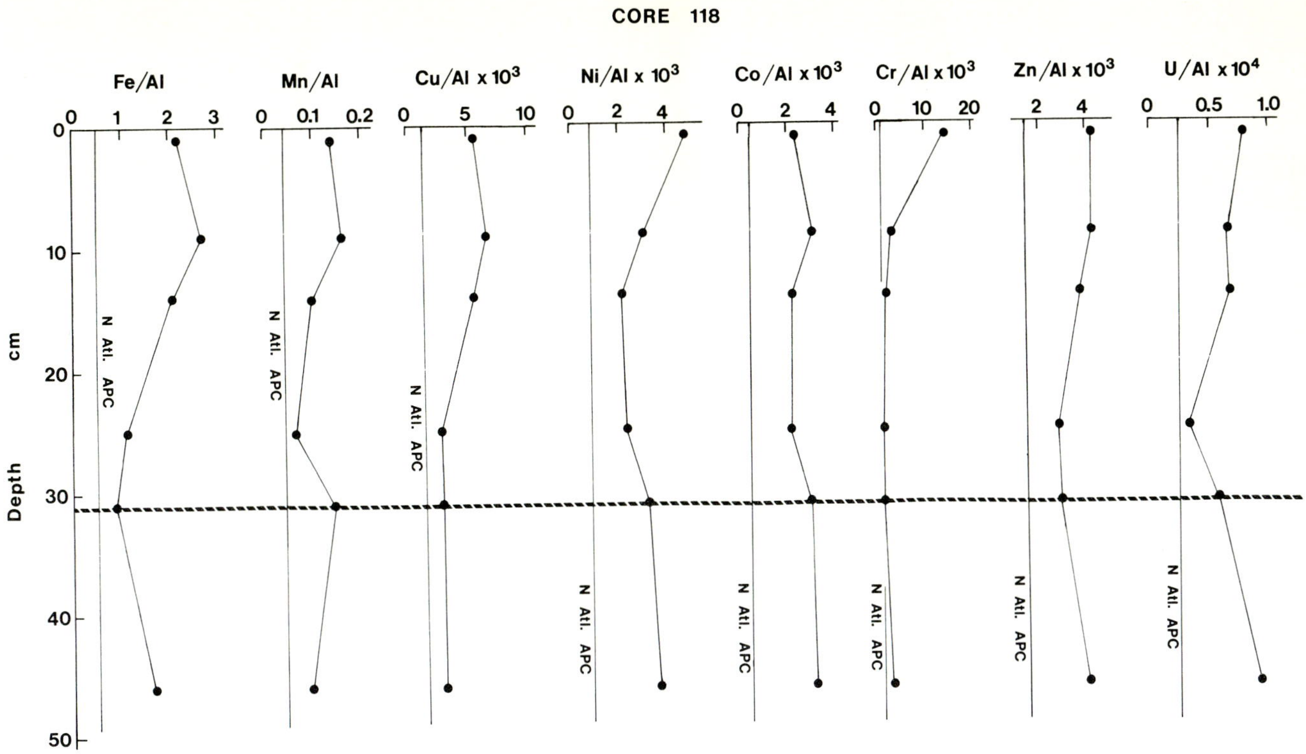

4b. Core 118

Field (M. Scott et al., 1978): Fe(x1.5), Mn(x1.5), Cu(x3), Ni(x4), Co(x5) and Cr(x4). Neither area has the extremes of metal enrichment reported for the East Pacific Rise by Bostrom and Peterson (1969) and others.

The surface samples taken near Mount Pluto in the inner rift valley were analyzed to locate possible hydrothermal vents by using metal distribution patterns. However, there were neither enough samples nor the correct spacing to serve that purpose. A transect from the west wall across the top of Mount Pluto to the east wall of the inner rift valley (Fig. 7) does show a slight increase in metal concentrations from WSW to ESE. There is also a decrease in the Fe/Mn ratio in the sediments from west to east across the rift valley. The ratio varies from 42 near the west wall to 17 near the east wall. This distribution suggests the possible location of a hydrothermal vent somewhere along the scarp that marks the west wall, because the Mn would be expected to remain in solution longer than the Fe (Krauskopf, 1957). A description of a factor analysis study of the FAMOUS Area surface and gravity core samples will be presented elsewhere (Salter, 1978, in prep.)

Table 5 gives the chemical composition and metal to Al ratios for sediments from a variety of areas influenced by ridge crest hydrothermal activity. The ratios for FAMOUS Area sediments are similar to those in sediments from other areas along the Mid-Atlantic Ridge at 45°N (Horowitz and Cronan, 1977), at 26°N (M. Scott et al., 1978) and in basal sediments of DSDP cores in the Atlantic (Dymond et al., 1973; Horowitz and Cronan, 1976). A comparison of the profiles of metal to Al ratios for the FAMOUS cores (Fig. 4) and TAG Core 4A (Fig. 5) further illustrates this general similarity. However, the TAG sediment did not show a color change with depth. The decrease in metals at 30 cm coincides with a layer of basaltic glass fragments. All of these Mid-Atlantic Ridge sediments are much less metal rich than sediments from the East Pacific Rise, particularly in the case of Mn. The Fe/Mn ratios of FAMOUS sediments average about 15, compared to a ratio of about 3 for the East Pacific Rise sediments. The differences observed in the sediments from these two active spreading centers may be related to water circulation differences in the hydrothermal systems of fast and slow spreading ridges (R. Scott et al., 1978). However, the general metal enrichment trends appear to be similar for all these active ridge sediments.

Metal accumulation rates for the FAMOUS Area sediments have been calculated from the metal analyses (Table 1), the sediment accumulation rates (Fig. 6) and the dry sediment density calculated by the method of Lyle and Dymond (1976). These data are listed in Table 6. The accumulation rates of hydrothermally derived metals in the FAMOUS Area are minimum rates only because of the uncertainties in the sedimentation rates discussed above. The metal accumulation rates for other areas on the Mid-Atlantic Ridge and for East Pacific Rise sediments (Table 6) show the same trends observed in the metal to Al ratios. The data for Mid-Atlantic Ridge sediments in Table 6 have been corrected for the accumulation of metal

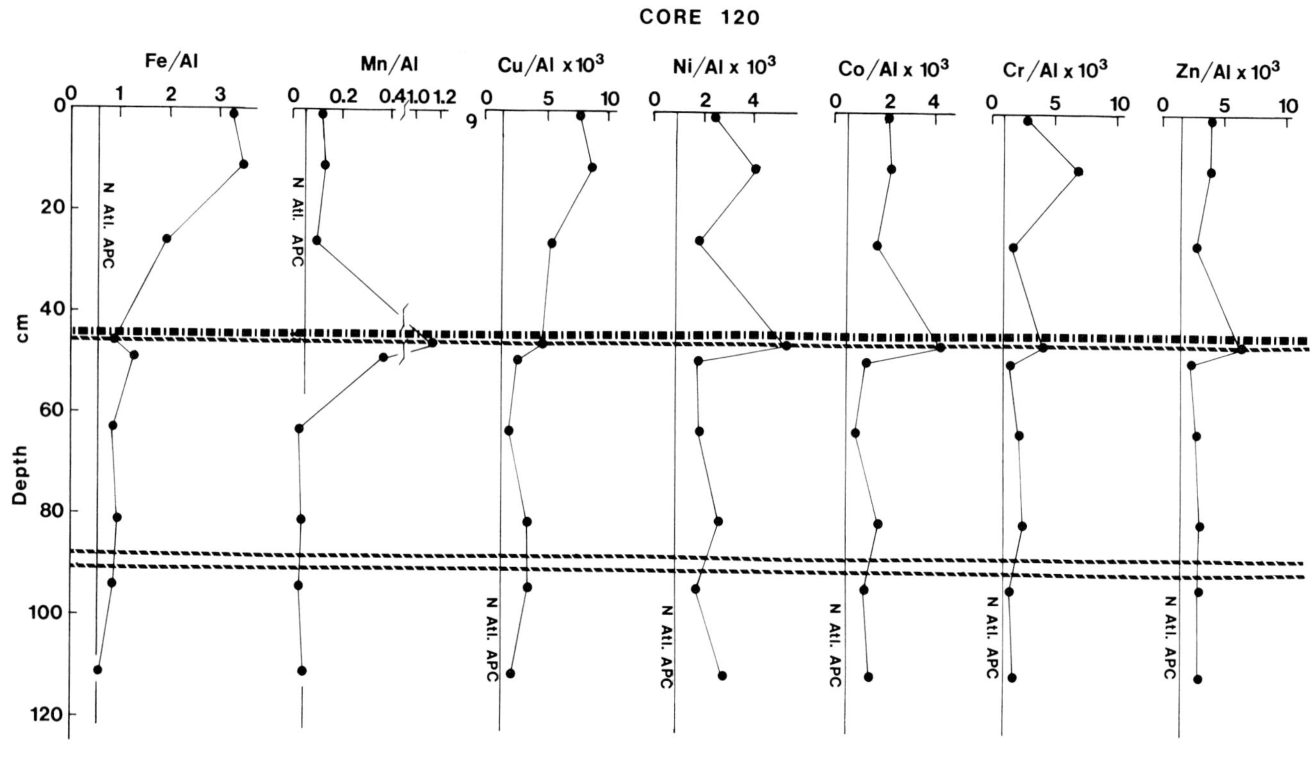

4c. Core 120

in detrital components by using the metal to Al ratios of average North Atlantic pelagic clay. The non-detrital accumulation rates are given in parentheses. The TAG area and 45°N sediments are accumulating hydrothermal Fe and Mn at rates much faster than indicated for the FAMOUS Area, and slower than rates for the East Pacific Rise. The low FAMOUS Area rates may be an artifact of low sediment accumulation rates based on anomalous radiometric data, as mentioned. Non-detrital accumulation rates for Ni, Co, Cu and Cr in FAMOUS Area sediments are similarly low. TAG area sediments are accumulating these metals at a conspicuously higher rate, equivalent to or higher than the East Pacific Rise rates.

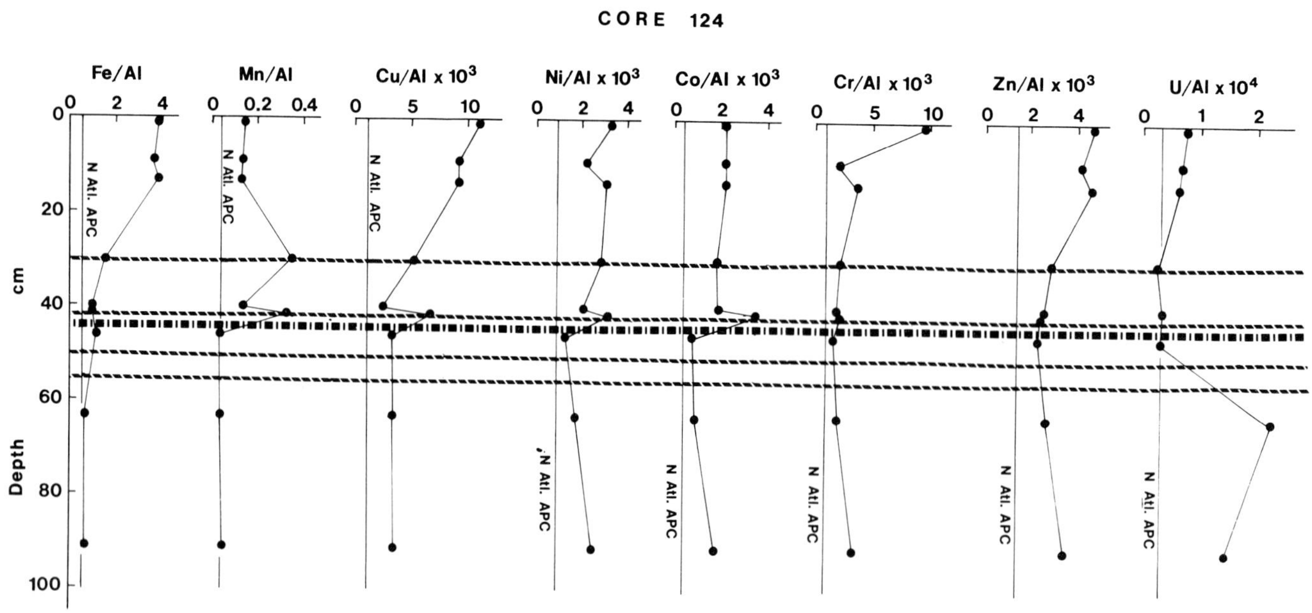

4d. Core 124

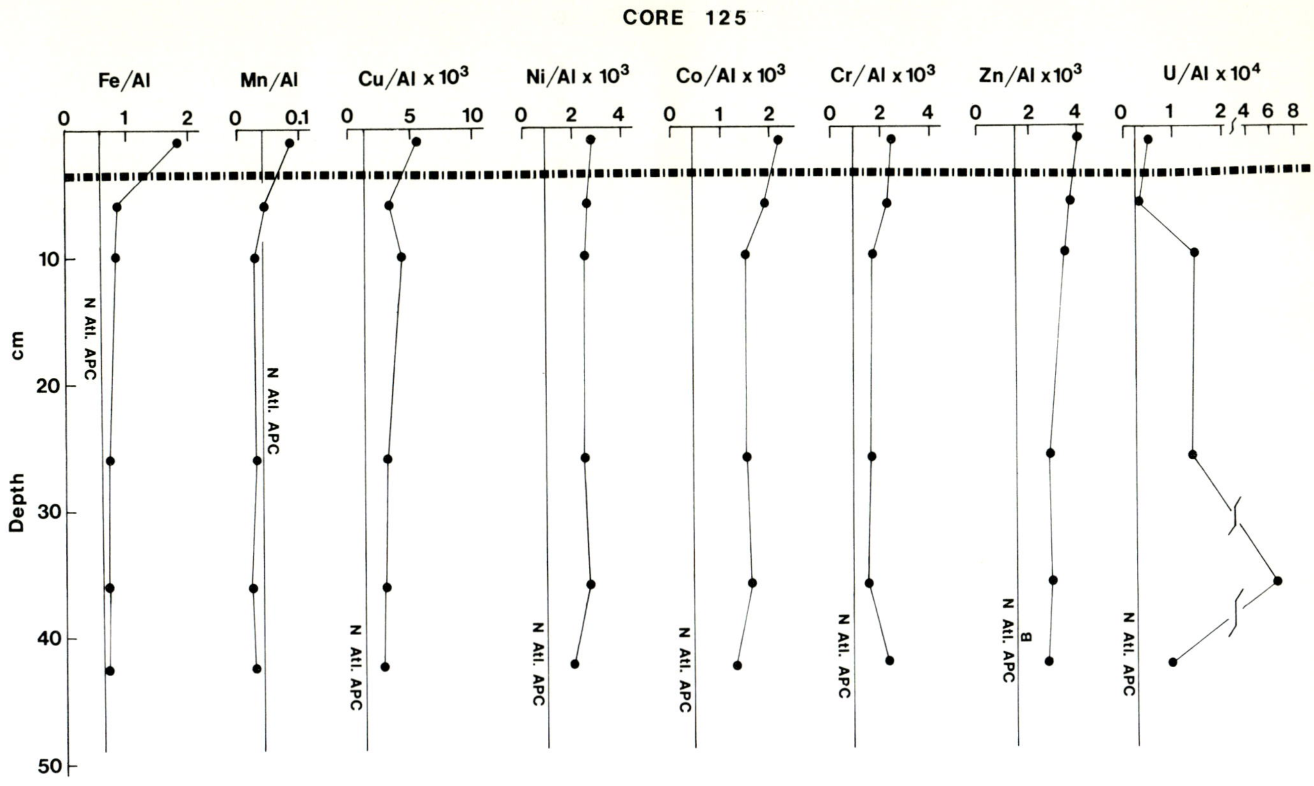

4e. Core 125

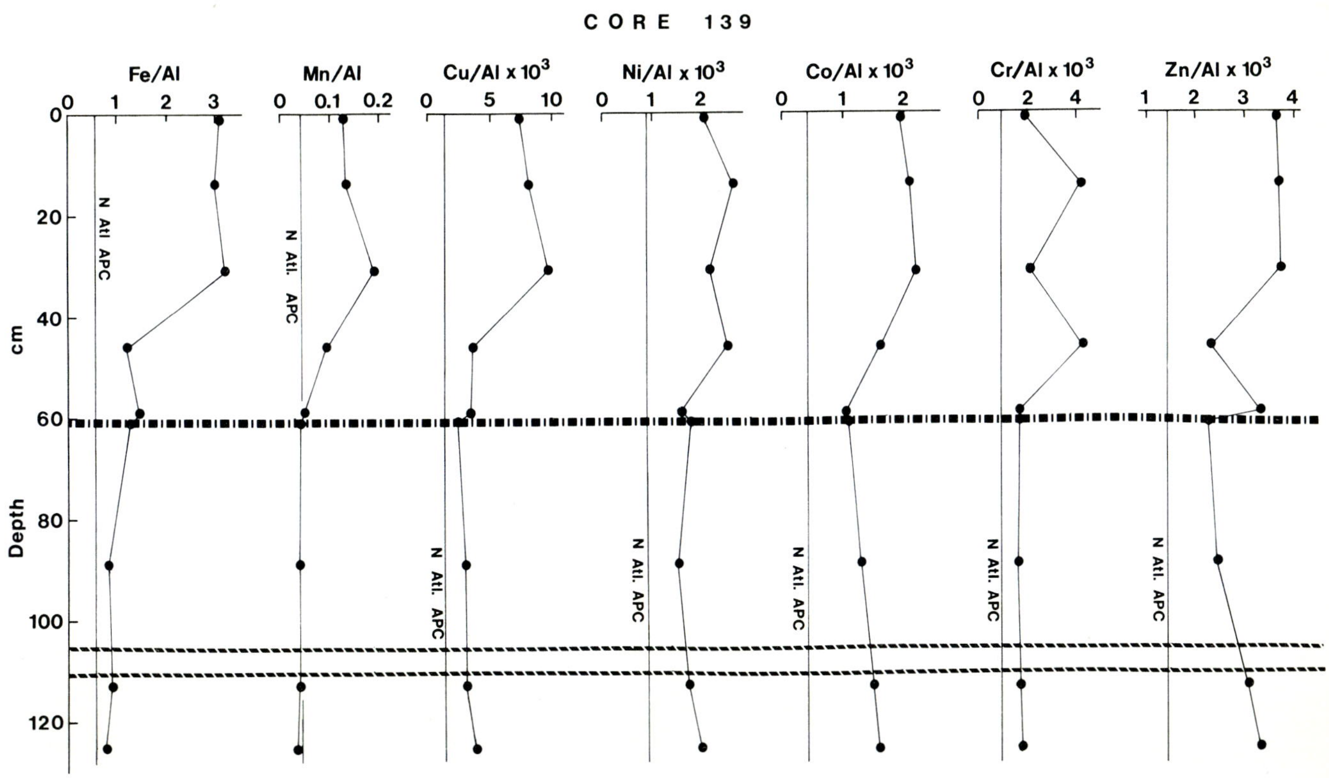

4f. Core 139

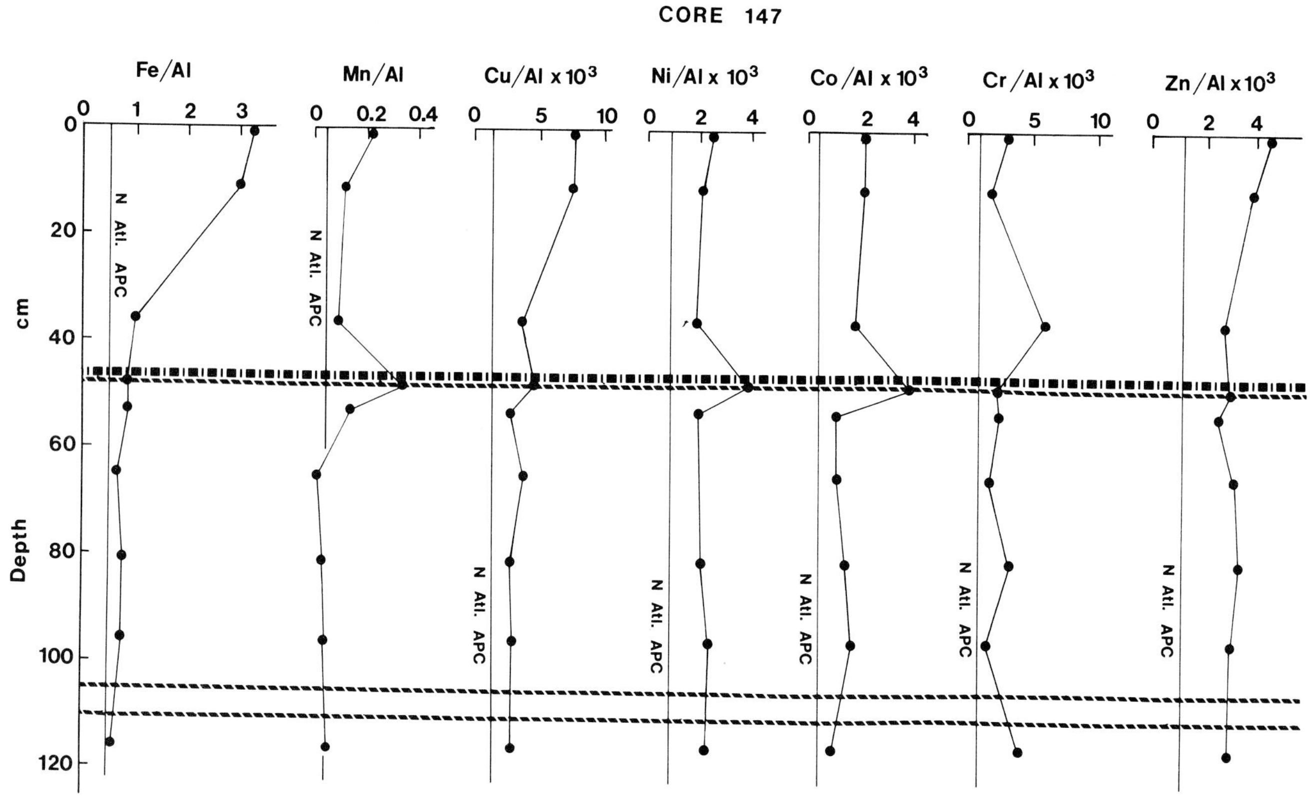

4g. Core 147

Only one core from the TAG area was analyzed, so it is not clear whether this is a widespread phenomenon in that locale. The 45°N sediments show trace metal accumulation rates intermediate to those for TAG and FAMOUS sediments.

There are few geophysical or geological differences among the three areas along the Mid-Atlantic Ridge. The TAG area appears to have a more highly fractured basaltic crust than does the FAMOUS Area. The less extensive fracture system in the FAMOUS Area could restrict the surface expression of sea water-basalt interaction or cause the expression to be more localized. A more thorough survey of other ridge crest areas is needed before we can assess chemical trends in hydrothermal systems along ridge crests.

Examination of the chemical and isotopic analyses of FAMOUS Area sediments suggests that several processes have operated on these sediments to produce the observed chemistry. The overall pattern of metal enrichment suggests that Fe and Mn were added to the water column by hydrothermal fluids venting at the sea floor. The other metals may have been added by the same process or extracted from sea water by the precipitating Fe and Mn oxides, which would then carry them to the sediment.

The presence of a marked color change strongly suggests that the sediments have become reducing at depth. This change would cause remobilization and upward diffusion of Fe, Mn and other metals, and their subsequent reprecipitation in the oxidizing upper layers of the sediment. Ku and Bischoff (1970) (reported in Ku *et al.*, 1972) have described brown oxidizing sediments overlying greyish sediments in sediments collected from the Mid-Atlantic Ridge at 40°N. The pore water Mn concentration was as high as 3 ppm in the greyish sediments. Turekian and Bertine (1972) have suggested that organic-rich sediments in sediment ponds along ridge crests may cause enrichment of sediments in Mo and U. This process could also remobilize sedimentary Mn and Fe. However, note that nearly all of the samples of FAMOUS sediments analyzed had Fe/Al and Mn/Al ratios equal to or higher than the value for average North Atlantic pelagic clay (Fig. 4). The overall metal enrichment at the surface is not generated soley by oxidation-reduction reactions, and such processes operating in the lower layers have not depleted Fe and Mn values below average pelagic values.

Comparison of the metal profiles for the FAMOUS sediments with metal profiles for sediments known to have undergone reduction and remobilization of metals (Bonatti *et al.*, 1971), indicates that simple diagenetic remobilization of metals alone cannot account for the distribution of metals in the FAMOUS sediments (Figs. 4a-4g). Co and Ni do not generally exhibit a depletion with depth in these sediments, although they would be expected to follow Mn to some extent during remobilization. Cores 52, 124 and 139 even show an increase in Co and Ni at depth. The Fe and Cu profiles shown by Bonatti *et al.* (1971) indicate little change in the Fe and Cu concentration

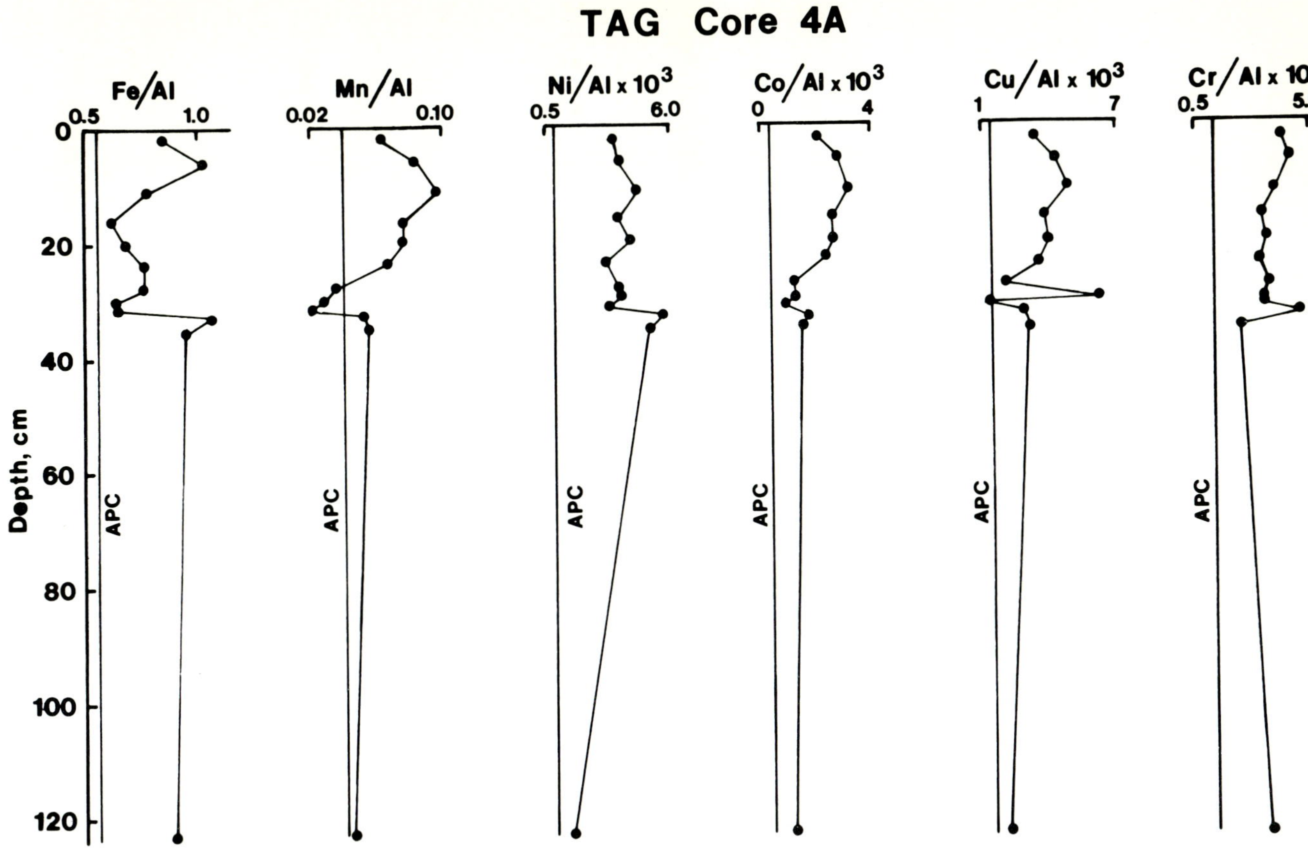

Figure 5. Vertical distribution of metals in Core 4A from the TAG area (M. Scott et al., 1978) shown as metal to Al ratios. The vertical lines marked APC represent metal to Al ratios for average North Atlantic pelagic clay.

across the redox boundary, but the FAMOUS Area sediments show an increase in these elements above the color change. Furthermore, in the core analyzed by Bonatti et al. (1971) Cr increased sharply below the redox boundary, but the FAMOUS sediments show higher Cr in the upper layers and little variation in the concentration below the color change. It is possible to build at least a circumstantial case for a temporal increase in hydrothermal input of Fe, Mn and other metals during the deposition of the upper layers of the sediment. A diagenetic remobilization of the metals superimposed on this effect could produce the observed distribution.

Post-depositional injection of hydrothermal fluids at the base of the sediment column has also been operating in the FAMOUS Area. A well-documented case of this phenomenon in the Galapagos Rift has been described by Corliss et al. (1978). Evidence for this process in the FAMOUS Area is most easily seen in the U data for the cores (Table 3; Figs. 4a, 4b, 4d and 4e). The U concentrations at the bottoms of Cores 52 and 125 (Figs. 4a and 4e) are an order of magnitude greater than in the rest of the cores (6.7 and 9.3 ppm). Core 124 also exhibits high U concentrations. The $^{230}Th/^{234}U$ activity ratios associated with the anomalously high U concentrations in these cores should still be considerably out of equilibrium (much greater than 1) because these sediments are less than 100,000 years old. However, the measured $^{230}Th/^{234}U$ activity ratios range from 0.12 to 3.53 (Table 3). These anomalously low $^{230}Th/^{234}U$ activity ratios without an associated anomalously high $^{234}U/^{238}U$ ratio suggest that the U at this depth in the core is not contemporaneous with the time of sediment deposition. That is, there has been post-depositional injection of U-rich hydrothermal solutions at depth in these sediments.

The possibility exists that the U enrichment is the result of the diagenetic process described by Turekian and Bertine (1971). This seems unlikely, because the source of the U would have to be sea water, and the U enrichment should begin closest to the source in the vicinity of the redox boundary (color change) and work its way downward. Instead the major enrichment in the FAMOUS cores shows up at the base of the cores and decreases upward. There is also a slight enrichment at the tops of the cores, which probably represents scavenging of U out of sea water by the precipitating Fe hydroxides.

Uranium enrichment of ridge crest sediments has been previously reported by a number of authors (Veeh and Bostrom, 1971; Scott et al., 1972; Piper et al., 1975; Fisher and Bostrom, 1969; Rydell and Bonatti, 1973). Injection of U-rich hydrothermal

Table 4. Radiometric Sediment Accumulation Rates on the Northern Mid-Atlantic Ridge

Core and Location		Reference	Sedimentation rate, cm/10^3 y	
			Bulk	Carbonate Free
Near 42°N				
CH 82-24, Ridge flank, on foothill		1	2.7	0.95
CH 82-26, Ridge flank on plain among low hills		1	2.1	0.53
CH 82-30(1), Ridge flank on small rise		1	1.5	0.27
CH 82-41, Foothills west of rift valley		1	1.3	0.31
Near 37°N, FAMOUS Area				
*Core 52	Intersection of FZB and RV3 (Fig. 1)	2	0.98	0.19
*Core 124	Intersection of FZB and RV2 (Fig. 1)	2	0.94	0.17
*Core 125	Intersection of FZB and RV2 (Fig. 1)	2	0.50	0.10
*Core 118	Near FZA on scarp (Fig. 1)	2	0.35	0.04
*Core 527-3	Floor of inner valley, near center of RV2	3	2.9	0.78
TAG Area 26°N				
*Core 4A	Median valley floor	4	1.8	0.74
24°N				
ZEP 12	Ridge flank	5	-	0.45
*ZEP 13	Ridge crest valleys	5	∿0.4	∿0.2
*ZEP 15	Ridge crest valleys	5	∿0.22	∿0.05
*ZEP 18	Ridge crest valleys	5	∿0.25	∿0.11
ZEP 22	Ridge flank	5	0.25	-
20°N				
AII 42-13	Ridge flank	1	0.5	0.17
*AII 42-17	Ridge crest; hills adjacent to median valley	1	0.73	0.13
AII 42-33	Ridge flank	1	0.47	0.31
AII 42-41	Abyssal hill west of ridge	1	1.3	1.26

*Median Valley Cores

References: (1) KU et al., 1972; (2) This work; (3) NOZAKI, et al., 1977; (4) M. SCOTT et al., 1978; (5) GOLDBERG et al., 1964.

Table 4. Radiometric Sediment Accumulation rates on the Northern Mid-Atlantic Ridge.

fluids at the base of the sediment column has been suggested by Rydell *et al*. (1974). Clogging of hydrothermal vents by sediments and continued emission of fluids from the vents appears to be the most probable mechanism, and fits the description of the well-documented Galapagos Rifts sediment mounds (Corliss *et al*., 1978). Cores 52 and 124 (Figs. 4a and 4e) show increases in Fe, Cu, Ni, Co, Cr and Zn at the base; these enrichments may also be related to injection of slightly reducing, possibly acidic metal-rich hydrothermal fluids from below. Core 52 is located in the high heat flow

Table 5. Comparison of the Composition of Metalliferous Sediments Found Associated with Mid-Ocean Ridges*

		Al %	Fe %	Mn %	Cu ppm	Ni ppm	Co ppm	Cr ppm	Zn ppm	Fe/Al	Mn/Al	Cu/Al x 10^3	Ni/Al x 10^3	Co/Al x 10^3	Cr/Al x 10^3	Zn/Al x 10^3	Fe/Mn
East Pacific Rise	(1)																
Ridge Crest		1.1	14.8	8.1	709	782	-	-	293	13.4	7.4	64.4	71.1	-	-	26.6	1.8
Ridge Flank		1.0	4.7	1.6	500	144	-	-	133	4.7	1.6	50.0	14.4	-	-	13.3	2.9
East Pacific Rise	(2)																
Ridge Crest		0.5	18.0	6.0	730	430	105	55	380	36.0	12.0	146.0	86.0	21.0	11.0	76.0	3.0
Ridge Flank		4.7	10.5	3.0	960	675	230	32	290	2.2	0.6	20.6	14.5	4.9	6.9	6.2	3.5
East Pacific Rise - Avg.	(2)	3.5	12.3	3.9	870	615	210	36	315	3.5	1.1	24.9	17.6	6.0	1.0	9.0	3.2
Bauer Deep	(3)	1.4	18.2	5.7	1100	950	90	-	600	13.0	4.1	78.6	67.8	6.4	-	42.8	3.2
	(4)	3.2	14.5	5.3	806	1129	226	75	-	4.5	1.6	24.8	34.7	7.0	2.3	-	2.8
		4.4	6.5	3.4	782	870	317	209	-	1.5	0.8	17.6	19.6	7.1	4.7	-	1.9
NE Pacific Delwood Seamount	(5)	1.0	28.5	2.0	10	54	5	12	450	28.5	2.0	1.0	5.4	0.5	1.2	45.0	14.2
SE Pacific Amph D-2 Seamount	(6)	0.5	32.5	1.9	74	400	35	-	-	65.0	3.9	14.8	80.0	7.0	-	-	16.8
Lau Basin Rise	(7)	5.8	11.7	1.6	295	144	135	70	178	2.0	0.3	5.1	2.5	2.3	1.2	3.1	7.3
DSDP, Pacific	(8)	2.4	23.6	6.7	1070	630	86	16	600	9.8	2.8	44.6	26.2	3.6	0.7	25.0	3.5
DSDP, Atlantic	(9)	3.4	6.6	0.2	85	83	-	118	148	1.9	0.05	2.5	2.4	-	3.4	4.3	36.9
Mid-Atlantic Ridge 45°N	(10)	5.3	9.6	0.5	399	214	-	128	228	1.8	0.1	7.5	4.0	-	2.4	4.3	18.5
Mid-Atlantic Ridge 26°N TAG Core 4A - Avg.	(11)	6.2	4.9	0.3	214	235	120	204	-	0.8	0.1	3.5	3.9	2.0	3.3	-	14.3
Mid-Atlantic Ridge 36°N	(12)																
FAMOUS-Core Avg.		6.2	9.0	0.7	278	161	114	158	201	1.4	0.11	4.5	2.6	1.8	2.5	3.2	12.7
FAMOUS-Surface Sample Avg.		5.6	10.7	0.6	292	141	118	198	172	1.9	0.11	5.2	2.5	2.1	3.5	3.1	17.8
FAMOUS-Surface Core Avg.		4.8	15.0	0.6	392	147	108	234	240	3.1	0.13	8.2	3.0	2.2	4.4	4.1	25.0
Avg. Pacific Pelagic Clay	(13)	8.3	5.1	0.9	400	300	110	78	-	0.61	0.11	4.8	3.6	1.3	0.9	-	5.4
Avg. Atlantic Pelagic Clay	(14)	9.0	5.0	0.4	115	79	39	80	129	0.56	0.04	1.3	0.9	0.4	0.9	1.4	12.6

(1) DYMOND and VEEH, 1975; (2) BOSTROM and PETERSON, 1969; (3) SAYLES and BISCHOFF, 1973; (4) MCMURTRY and BURNETT, 1975; (5) PIPER et al., 1975; (6) BONATTI and JOENSUU, 1966; (7) BERTINE, 1974; (8) DYMOND et al., 1973; (9) HOROWITZ and CRONAN, 1976; (10) HOROWITZ and CRONAN, 1977; (11) M. SCOTT et al., 1978 (12) This paper; (13) TUREKIAN and IMBRIE, 1966; LANDERGREN, 1964 from BISCHOFF and ROSENBAUM, 1977; DYMOND et al., 1973; (14) See TABLE 1.

* **All data on a calcium carbonate free basis.**

Table 5. Comparison of the Composition of Metalliferous Sediments Found Associated with Mid-Ocean Ridges.

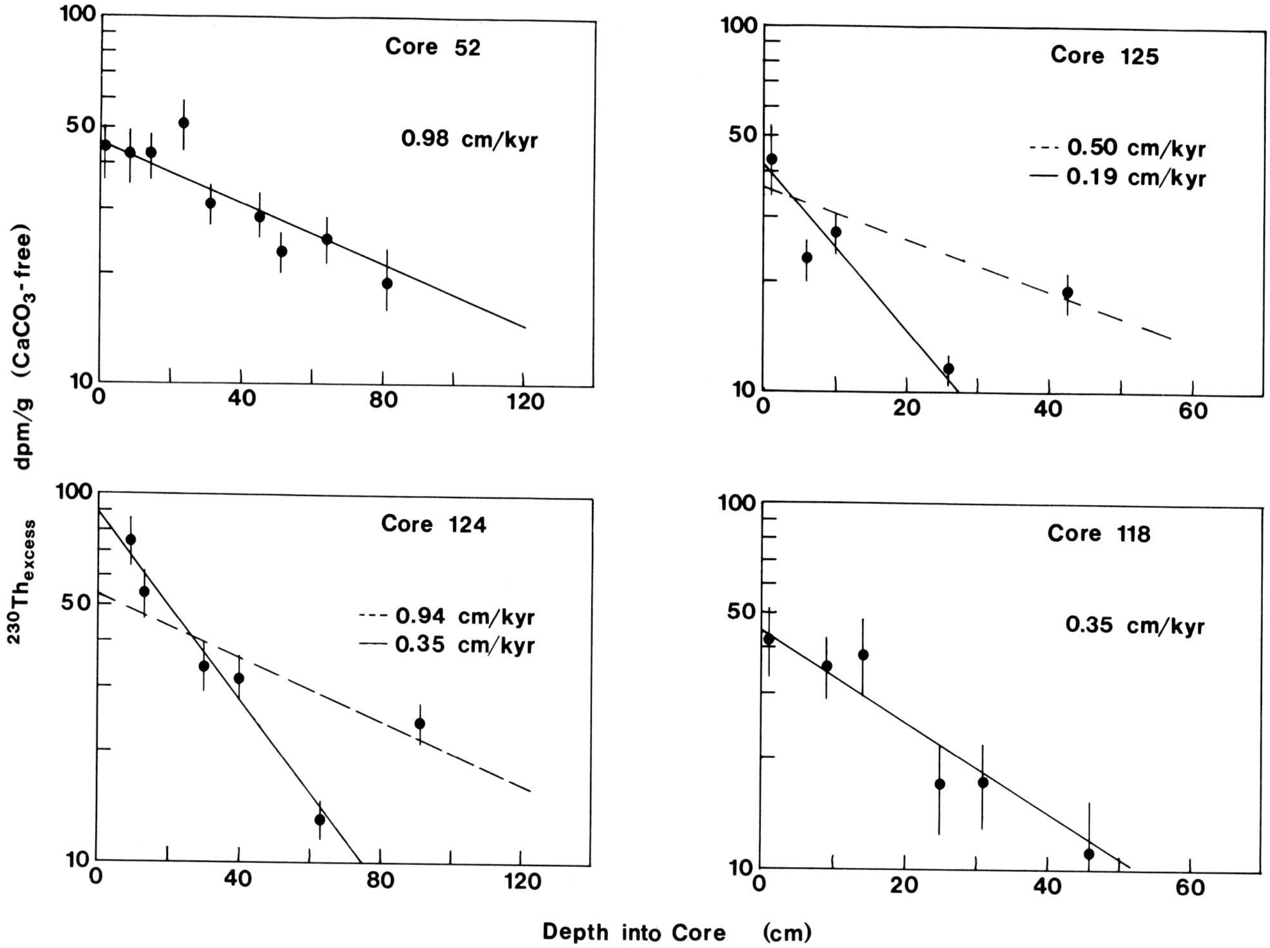

Figure 6. Sedimentation rates for gravity cores in the FAMOUS Area. ^{230}Th excess on a carbonate free basis is plotted as a function of depth into the core. The indicated sedimentation rates are based on solid lines best fitting the ^{230}Th excess data. The dashed lines for Cores 124 and 125 are sedimentation rates based on paleontological and geophysical data.

area of Fracture Zone B; all of the gravity cores are located in or near the fracture zones.

Uranium activity ratios in the FAMOUS Area cores are anomalous in several of the samples, with $^{234}U/^{238}U$ ratios well above the sea water value of 1.15 ± .02 (Table 3). The relationship between uranium enrichment and anomalous ratios is not a simple one. Figure 8 shows that the samples with the highest activity ratios do not occur in the layers most enriched in uranium. Unusual uranium concentrations and isotope ratios were noted by Ku et al. (1972) in Mid-Atlantic Ridge sediments, by Rydell and others (1974) in an East Pacific Rise core, and by Rydell and Bonatti (1973) in marine metalliferous deposits. Rydell and Bonatti (1973) have developed a mixing model for sea water uranium and hydrothermal uranium to explain the distribution of isotope ratios and concentrations of U in metalliferous deposits. Their model is based in part on a plot of $^{234}U/^{238}U$ ratios versus U/Al ratios similar to the one shown for FAMOUS sediments (Fig. 9). Samples with very high ratios have low U concentrations; samples with high concentrations have ratios close to the sea water value of 1.15. The explanation suggested by Rydell and Bonatti (1973) is that sea water uranium is adsorbed by Fe and Mn particles forming in the water column, thus diluting the high isotope ratios typical of the hydrothermally derived U. But in FAMOUS Area sediments most of the uranium enrichment appears to come from below rather than from particle scavenging of the overlying seawater. If such injections of uranium are occurring at the base of the sediments, they are not associated with uniformly high isotope ratios as suggested by Rydell and Bonatti (1973).

The anomalous U activity ratios in Core 124 at 41 cm (Fig. 4d) and in Core 118 at 31 cm (Fig. 4b) are associated with Mn rich bands. These bands may represent pulses of hydrothermal contributions to the sediment that are localized enough to retain a hydrothermal fingerprint in their $^{234}U/^{238}U$ activity ratios. Other samples with anomalous ratios are not associated with Mn rich layers. Both Mn and ^{234}U may be redistributed by hydrothermal fluids injected into

Table 6. Metal Accumulation Rates in Ridge Crest Sediments

	Fe	Mn	Al	Ni	Co	Cu	Cr	$CaCO_3$ Free Sedimentation Rate
	mg/cm²/10³ y			μg/cm²/10³ y				mm/10³ y
Mid-Atlantic Ridge								
45°N[1]	34(25)*	3 (2)	15	62(49)	36(29)	78(59)	28(15)	4
37°N[2]								
FAMOUS Area								
Core 52	11(7)	0.3(0.1)	7	16(10)	13(9)	31(23)	17(11)	1.9
Core 118	3(2)	0.2(0.1)	1	5(3)	4(3)	7(5)	6(5)	0.4
Core 124	10(7)	0.9(0.6)	6	16(11)	11(8)	32(24)	17(12)	1.7
Core 125	3(1)	0.2(0)	4	10(6)	6(5)	14(9)	8(4)	0.9
26°N[3]								
TAG Area								
Core 4A	50(15)	4 (1.2)	62	235(180)	120(93)	214(136)	204(150)	6.8
Average North Atlantic Pelagic Clay[4]	9	0.7	16	14	7	20	14	2.5
East Pacific Rise[5]								
a.	82	28	0.61	160	--	--	--	0.5
b.	58	18	3	153	47	254	21	5.3

* Numbers in parentheses are non-detrital accumulation rates determined by subtracting a detrital metal component for North Atlantic pelagic clay based on the detrital metal to Al ratio.

1. Rates given in BERTINE 1974) calculated from data of CRONAN (1972) and TUREKIAN and IMBRIE (1966) with sedimentation rates from ERICSON *et al.* (1961) and TUREKIAN and STRIVER (1964).
2. This paper.
3. M. SCOTT *et al.* (1978).
4. See Table 1.
5. a. DYMOND and VEEH (1975)
   b. BERTINE (1974) calculated from data of BENDER *et al.* (1971); BOSTROM and PETERSON (1969); DASCH *et al.* (1971).

Table 6. Metal Accumulation Rates in Ridge Crest Sediments.

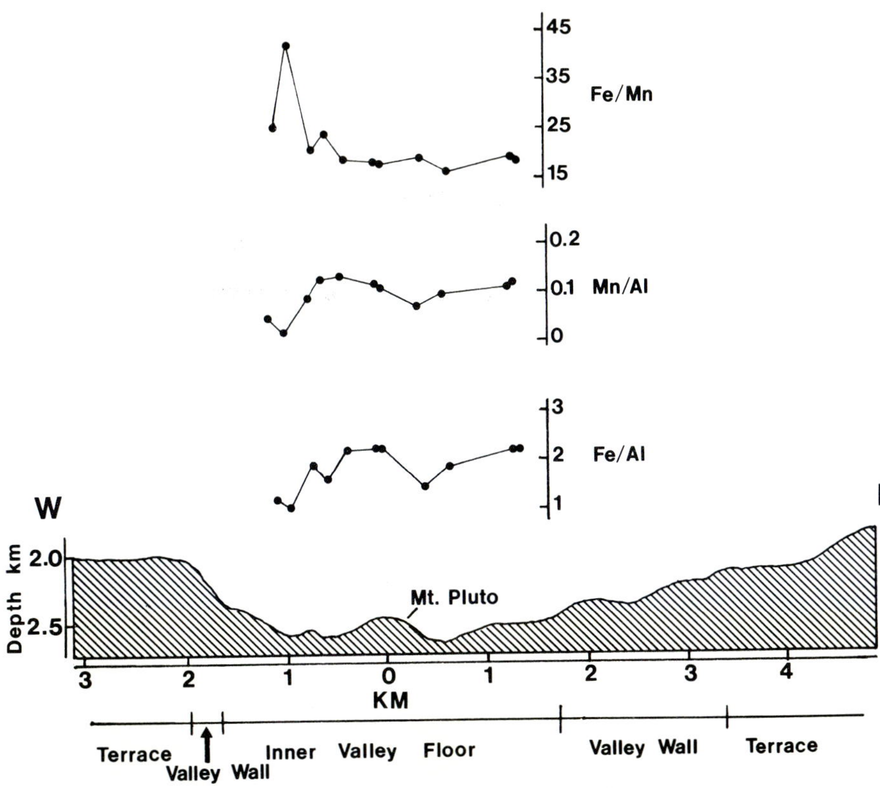

Figure 7. Surface Metal Distribution Across Mount Pluto, Inner Valley Floor. Vertical exaggeration: 1.4x.

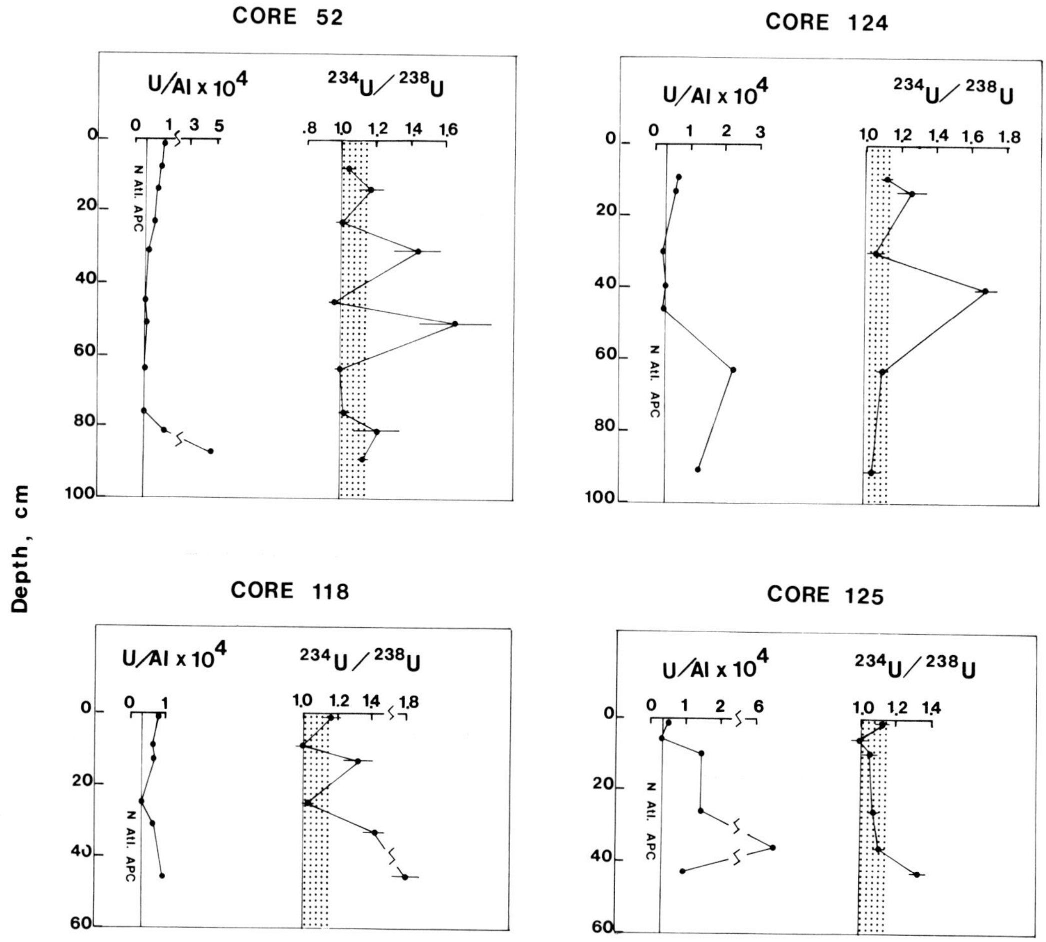

Figure 8. Vertical distribution of uranium and $^{234}U/^{238}U$ activity ratios in FAMOUS sediments. The average North Atlantic pelagic clay U/Al ratio is shown on the graphs as a vertical line. The stippled area covers the expected sea water range of activity ratios.

the core, and circulation patterns of such fluids through the sediments may be quite complex (Corliss et al. 1978).

As pointed out by Rydell and Bonatti (1973) U is much less mobile under reducing conditions than oxidizing conditions. Thus a hydrothermal solution carrying dissolved U would have to be only mildly reducing at most, or contain high concentrations of a ligand such as $CO_3^=$ to keep U in solution.

Uranium anomalies in these sediments may result from remobilization of uranium in the sediments by circulating fluids rather than by interactions between the hydrothermal fluids and basalts. It should be noted that the Th concentrations in some of the samples are unusually low or high (Table 3). Similar discrepancies have been noted by Ku et al. (1972) and Rydell et al. (1974), but the anomalies are not consistent and do not suggest large scale Th remobilization.

An additional line of evidence concerning the chemical evolution of the FAMOUS sediments is a study of the ferromanganese micronodules found in the sediment. Similar concretions have been reported in other metalliferous sediments (Bischoff and Rosenbauer, 1977; Heath and Dymond, 1977; Cronan, 1972). Micronodules were found at all levels in Core 52, but the morphology and chemistry of these microconcretions varied with depth in the sediment. The sediment samples were sieved through several nylon screens (100 to 400 mesh) and the dark micronodules were hand picked under a binocular microscope. The particles were studied with a JOEL JSM-U3 Scanning Electron Microscope and subjected to X-ray analysis with an EDAX energy dispersive spectrometer and multichannel analyzer.

Figure 10a is a typical micronodule from the 0-2 cm level. It is well formed and smooth, with little evidence of the nucleus around which it is growing. Fe, Mn and Cu are strongly represented on the EDAX record. Micronodules from deeper in the core below the color change are much different in appearance. Although still black to brown in color, these concretions appear to have had much of their ferromanganese cement dissolved in reducing conditions.

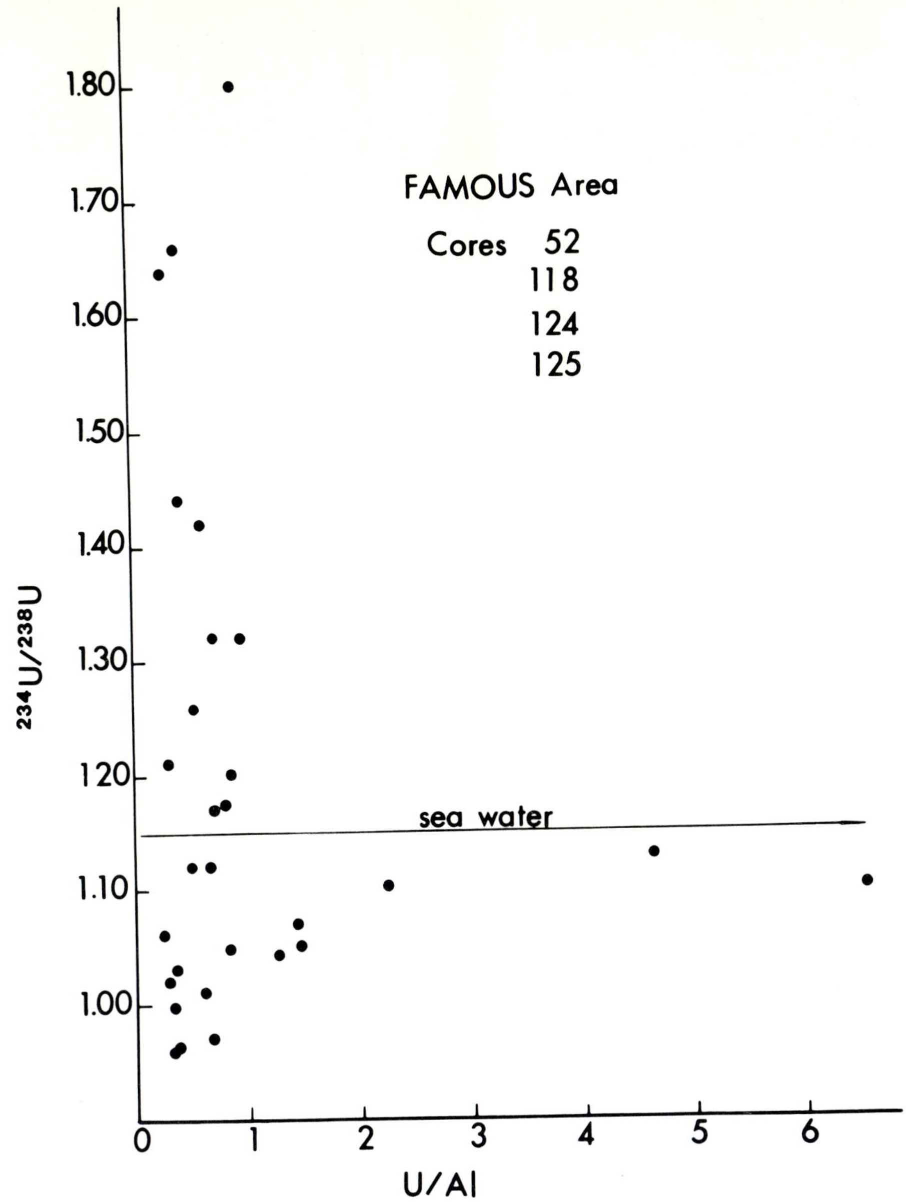

Figure 9. Plot of uranium isotope ratios versus uranium to aluminum ratios for FAMOUS Area sediments.

Figure 10b, a photo of a micronodule from the 84-86 cm level of Core 52, illustrates this phenomenon. Note that the biogenic debris around which the micronodules nucleate and grow is conspicuous in this sample. The EDAX scan shows Mn, a trace of Fe, and Cu. Strong fractionation of Fe and Mn is typical for micronodules in the lower part of the core. The change from well preserved, rounded grains to highly eroded grains is progressively more exaggerated with depth below the color change.

Figure 11 is a picture of pyrite concretion from the 96-98 cm layer of Core 52. Similar pyrite concretions occur throughout the sediment below the color change, commonly as fillings in foraminifera tests.

The changes in micronodule morphology and chemistry with depth in Core 52 mirror the changes in sediment chemistry. Oxidized upper layers contain well formed microconcretions. Sediments below the color change, considered to be a redox boundary, contain micronodules that are severely eroded by chemical solution of their ferromanganese cement. The pore fluids and local environments in the sediment become reducing enough so that pyrite concretions form during $SO_4^=$ reduction. Exaggeration of the amount of deterioration of the micronodules toward the base

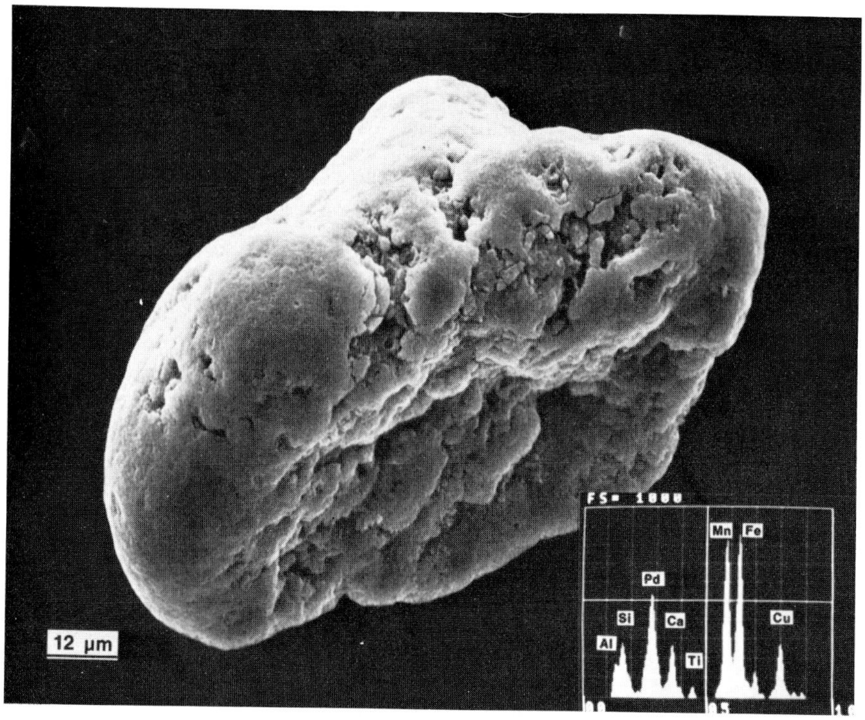

Figure 10a. Ferromanganese micronodule from the 0-2 cm layer of Core 52. The palladium peaks on the EDAX scans represent metallic overcoatings necessary for SEM inspection.

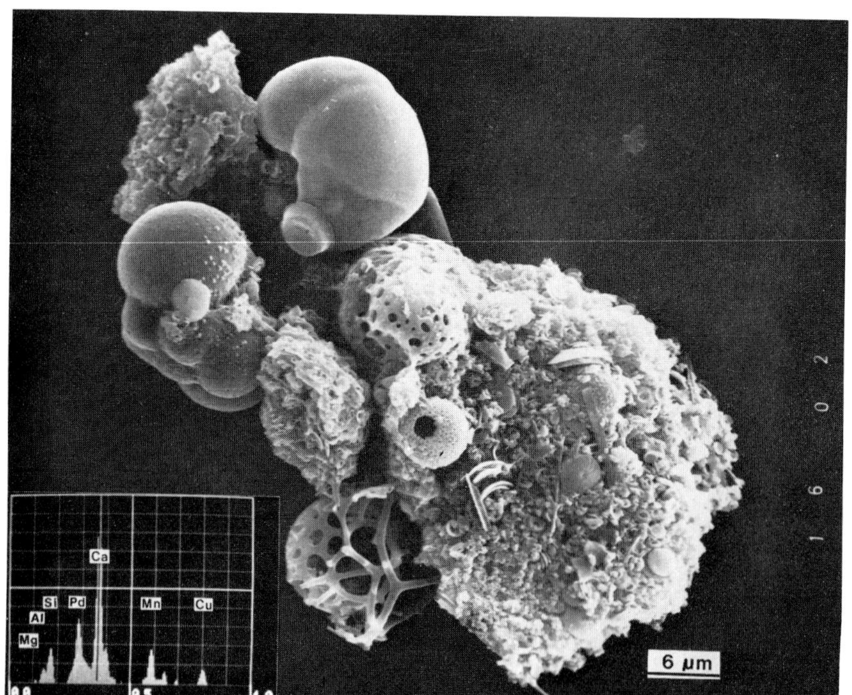

Figure 10b. Ferromanganese micronodule from the 84-86 level of Core 52. Despite the well preserved biological debris included in the concretion, the color was black under reflected light.

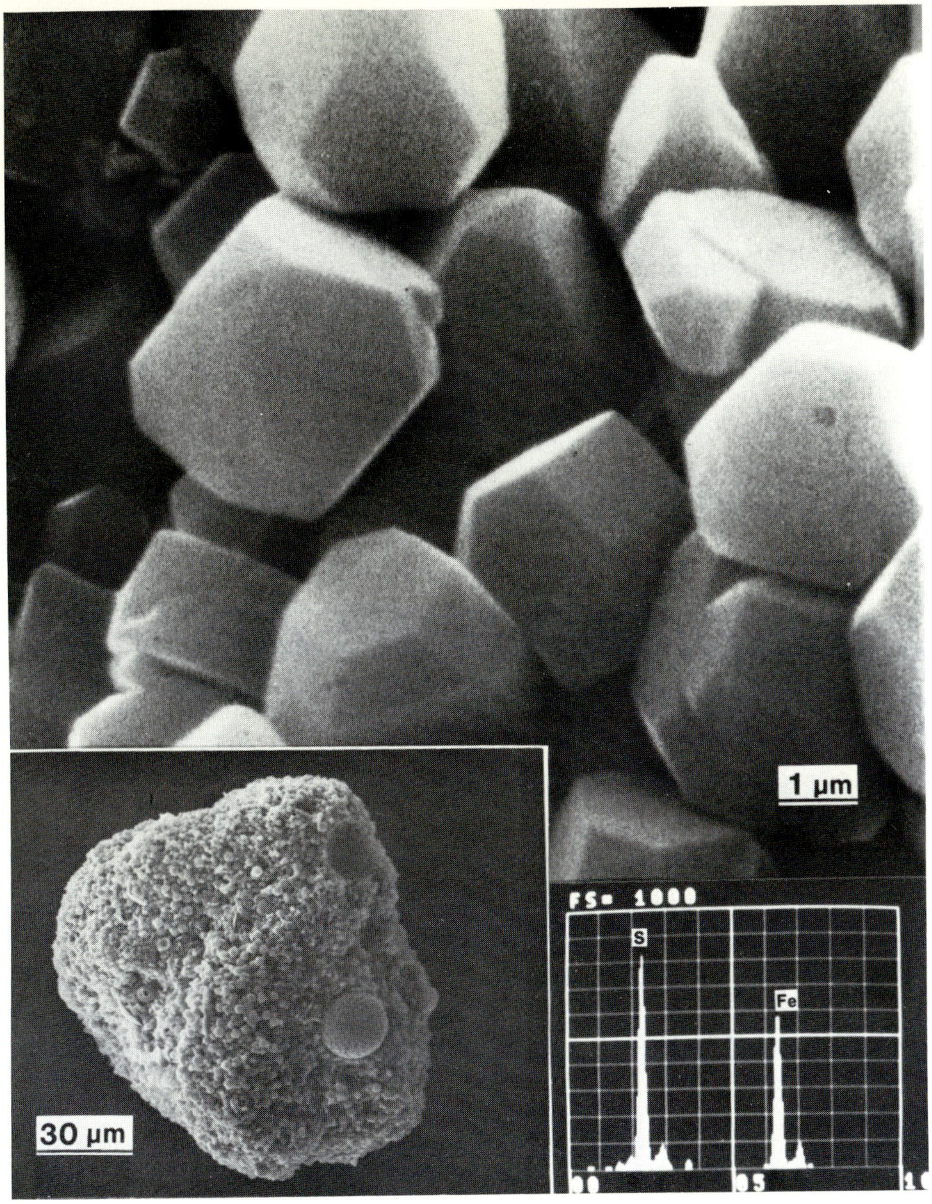

Figure 11. Pyrite concretion from the 96-98 cm level of Core 52. The spherical object is biogenic $CaCO_3$.

of the core suggests that reducing hydrothermal fluids may be in part responsible for dissolving the Fe and Mn.

Conclusions

Mid-Atlantic Ridge crest sediments in the FAMOUS Area show patterns of metal enrichment similar to those for sediments elsewhere on this same active ridge and to sediments on the East Pacific Rise. The major differences observed in ridge crest sediments include the much greater metal to Al ratios and metal accumulation rates for East Pacific Rise sediments, and the lower Fe/Mn for the East Pacific Rise sediments. These differences may be related to the differences in hydrothermal circulation in fast and slow spreading active ridge areas.

FAMOUS Area sediments have undergone several processes affecting the metal concentrations and distributions with depth. In addition to the general enrichment in Fe, Mn and trace metals above average North Atlantic pelagic clay values, the cores exhibit a marked increase in these metals above a conspicuous color change present in all of the cores. The color change is strongly suggestive of a redox boundary formed by anaerobic metabolism of organic matter at depth in the cores. The Fe and Mn distributions are consistent with such a model, but these metals are only rarely depleted below average pelagic values in the lower parts of the sediment. Moreover, the distribution of trace metals with respect to the color change boundary and the presence of numerous Mn-rich bands in several of the cores suggest that fluctuations in the hydrothermal input of metals may be partly responsible for the observed depth profiles of the metals.

Anomalously high concentrations and $^{234}U/^{238}U$ activity ratios exist in all of the cores analyzed for U and Th isotopes. The U enrichment is most pronounced at the base of the cores, and is paralleled by slight enrichments in Cr, Zn, Co and Ni. These trends suggest the influx of hydrothermal fluids at the base of the core. Such fluids would probably be reducing and would contribute to redistribution of metals in the sediment column.

Ferromanganese micronodules have been recovered from all levels of FAMOUS Core 52. The micronodules are well preserved in the core top and are increasingly eroded at depth, probably by solution of Fe and Mn under reducing conditions. Pore fluids and local microenvironments have attained low enough Eh for the formation of pyrite aggregates and framboids below the color change.

The distribution of metals in FAMOUS Area sediments is considered to result from three superimposed processes: metal enrichment by hydrothermal solutions delivering dissolved material which subsequently precipitates from the overlying sea water; diagenetic remobilization of metals from a reduced layer formed by anaerobic metabolism of organic matter; and post-depositional injection of hydrothermal fluids at the base of the sediment column.

Acknowledgements: This work has been supported by NSF Grants ID075-00250 and OCE 77-11372.

References

ARCYANA, Transform fault and rift valley from bathyscape and diving saucer, Science, 190, 108, 1975.

Arrhenius, G.O.S., and E. Bonatti, Neptunism and volcanism in the ocean, in Progress in Oceanography, (ed. M. Sears), 3, 7, Pergammon, 1965.

Ballard, R.D., W.B. Bryan, J.R. Heirtzler, G. Keller, J.G. Moore, and Tj. van Andel, Manned submersible observations in the FAMOUS Area: Mid-Atlantic Ridge, Science, 190, 103, 1975.

Ballard, R.D., and Tj. van Andel, Morphology and tectonics of the inner rift valley at lat 36°50'N on the Mid-Atlantic Ridge, Bull. Geol. Soc. Am., 88, 507, 1977.

Bender, M.L., W.S. Broecker, V. Gornitz, V. Middel, R. Kay, S. Sun, and P. Biscaye, Geochemistry of three cores from the East Pacific Rise, Earth Planet. Sci. Lett., 12, 425, 1971.

Bertine, K.K., Origin of Lau Basin Rise sediments, Geochim. Cosmochim Acta, 38, 629, 1974.

Betzer, P.R., W.G. Bolger, B.A. McGregor, P.A. Rona, The Mid-Atlantic Ridge and its effects on the composition of particulate matter in the deep ocean, Trans. Amer. Geophys. Union, 55, 293, 1974.

Bischoff, J.L., and F.W. Dickson, Sea water-basalt interaction at 200°C, 500 bars: implications for orgin of sea floor heavy metal deposits and regulation of sea water chemistry, Earth Planet. Sci. Lett., 25, 385, 1975.

Bischoff, J.L., and R.J. Rosenbauer, Recent metalliferous sediment in the North Pacific manganese nodule area, Earth Planet Sci. Lett., 33, 379, 1977.

Bonatti, E., D.E. Fisher, O. Joensuu, and H.S. Rydell, Post-depositional mobility of some transition elements, P, U, and Th in deep sea sediments, Geochim. Cosmochim.Acta, 35, 189, 1971.

Bostrom, K., and M.N.A. Peterson, Precipitates from hydrothermal exhalations on the East Pacific Rise, Econ. Geol., 61, 1258, 1966.

Bostrom, K., and M.N.A. Peterson, The orgin of Al-poor ferromanganoan sediments in areas of high heat flow on the East Pacific Rise, Mar. Geol, 7, 427, 1969.

Bostrom, K., M.N.A. Peterson, O. Joensuu, and D. Fisher, Aluminum poor ferromanganoan sediments on active ocean ridges, J. Geophys. Res., 74, 3261, 1969.

Bostrom, K., and S. Valdes, Arsenic in the Ocean Floor, Lithos., 2, 351, 1969.

Bostrom, K., and D. Fisher, Volcanogenic uranium, vanadium, and iron in the Indian Ocean sediments, Earth Planet. Sci. Lett., 11, 95, 1971.

Bostrom, K., T. Kraemer, and S. Gartner, Provenance and accumulation rates of opaline silica, Al, Ti, Fe, Mn, Cu, Ni, and Co in Pacific pelagic sediments, Chem. Geol., 11, 123, 1973.

Bruty, O., R. Chester, L. G. Royle, and H. Elderfield, Distribution of zinc in North Atlantic deep sea sediments, Nature Phys. Sci., 237, 86, 1972.

Carr, R.A., M.M. Jones, T.B. Warner, C.H. Cheek, and E.R. Russ, Variation in time of mercury anomalies at the Mid-Atlantic Ridge, Nature, 258, 588, 1975.

Corliss, J.B., M. Lyle, D. Dymond, and K. Crane, The chemistry of hydrothermal mounds near the Galapagos Rift, Earth Planet. Sci. Lett., 40, 12, 1978.

Craig, H., W.B. Clarke, M.A. Beg, Excess ^{3}He in deep water on the East Pacific Rise, Earth Planet. Sci. Lett., 26, 125, 1975.

Cronan, D.S., The Mid-Atlantic Ridge near 45°N: XVII. Al, As, Hg, and Mn in ferrugenous sediments from the median valley, Canad. J. Earth Sci., 9, 319, 1972.

Cronan, D.S., G.R. Heath, Tj. van Andel, R.H. Bennett, R.S. Yeats, M.G. Denkelman, D. Bukry, S. Charleston, A. Kaneps, and K. Rodolfo, Iron rich basal sediments from the Eastern Equatorial Pacific, Leg 16 Deep Sea Drilling Project, Science, 175, 61, 1972.

Cronan, D.S., and D.E. Garrett, Distribution of elements in metalliferous Pacific sediments collected during the Deep Sea Drilling Project, Nature Phys. Sci., 242, 88, 1973.

Dasch, E.J., J.R. Dymond, and G.R. Heath, Isotopic analysis of metalliferous sediment from the East Pacific Rise, Earth Planet. Sci. Lett., 13, 175, 1971.

Dymond J., J.B. Corliss, G.R. Heath, G.W. Field, E.J. Dasch, and H.H. Veeh, Origin of metalliferous sediments from the Pacific Ocean, Bull. Geol. Soc. Am., 84, 3355, 1973.

Dymond, J., J.B. Corliss, and R. Stillinger, Chemical composition and metal accumulation rates of metalliferous sediments from sites 319, 320, and 321, Deep Sea Drilling Project, Leg 34, Initial Reports Deep Sea Drilling Project, 34, 575, 1976.

Dymond, J., J.B. Corliss, and G.R. Heath, History of metalliferous sedimentation at Deep Sea Drilling site 319 in the Southeastern Pacific, Geochim. Cosmoschim Acta, 41, 741, 1977.

El Wakeel, S.K., and J.P. Riley, Chemical and Mineralogical studies of deep sea sediments, Geochim. Cosmochim Acta, 25, 110, 1961.

Fehn, V., M.D. Siegel, G.R. Robinson, H.D. Holland, D.L. Williams, A.J. Erickson, and K.E. Green, Deep water temperatures in the FAMOUS Area, Bull. Geol. Soc. Am., 88, 488, 1977.

Fisher, D.E., and K. Bostrom, Uranium-rich sediments on the East Pacific Rise, Nature, 224, 64, 1969.

Francis, T., I. Porter, and J. McGrath, Ocean bottom seismograph observations on the Mid-Atlantic Ridge near lat 37°N, Bull. Geol. Soc. Am., 88, 644, 1977.

Goldberg, E.D., M. Koide, J.J. Griffin, and M.N.A. Peterson, A geochronological and sedimentary profile across the North Atlantic Ocean, in Isotopic and Cosmic Chemistry, (eds. H. Craig, S. Miller, and G. Wasserburg), 211, North Holland, 1964.

Hajash, A.J., Hydrothermal processes along mid-ocean ridges: an experimental investigation, Contrib. Mineral. Petrol., 53, 205, 1975.

Heath, G.R., and J. Dymond, Genesis and transformation of metalliferous sediments from the East Pacific Rise, Bauer Deep, and Central Basin, Northwest Nazca Plate, Bull, Geol. Soc. Am., 88, 723, 1977.

Hekinian, R., and M. Hoffert, Rate of palagonitization and manganese coating on basaltic rocks from the rift valley in the Atlantic Ocean near 36°50'N, Mar. Geol., 19, 91, 1975.

Hoffert, M., A. Perseil, R. Hekinian, P. Choukroune, H.D. Needham, J. Franchetau and S. LePichon, Hydrothermal deposits sampled by diving saucer in Transform Fault "A" near 37°N on the Mid-Atlantic Ridge, FAMOUS Area, Oceanol. Acta, Vol. 1, 73, 1978.

Horowitz, A., Geochemistry of sediments from the North Reykjanes Ridge and the Iceland Faroes Ridge, Mar. Geol., 17, 103, 1974.

Horowitz, A., and D.S. Cronan, The geochemistry of basal sediments from the North Atlantic Ocean, Mar. Geol, 20, 205, 1976.

Horowitz, A. and D.S. Cronan, More ferrugenous sediments from the Mid-Atlantic Ridge at 45°N, (in press).

Keller, G.H., S.H. Anderson, and J.W. Lavelle, Near-bottom currents in the Mid-Atlantic Ridge rift valley, Canad. J. Earth Sci., 12, 703, 1975.

Krauskopf, K.B., Separation of manganese from iron in sedimentary processes, Geochim. Cosmochim Acta, 12, 61, 1957.

Ku, T.L., Uranium series disequilibrium in deep sea sediments, Ph.D. Thesis, Columbia University.

Ku, T.L., W.S. Broecker, and N. Opdyke, Comparison of sedimentation rates measured by paleomagnetic and the ionium methods of age determination, Earth Planet. Sci. Lett., 4, 1, 1968.

Ku, T.L., J.L. Bischoff, and H. Boersma, Age studies of Mid-Atlantic Ridge sediments near 42°N and 20°N, Deep Sea Res., 19, 233, 1972.

Lister, C.R.B., On the thermal balance of mid-ocean ridges, Geophys. J. Royal Astron. Soc., 26, 515, 1972.

Lister, C.R.B., Water percolation in the oceanic crust?, Trans. Amer. Geophys. Union, 55, 740, 1974.

Luyendyk, B.P., and K.C. MacDonald, Physiography and structure of the inner floor of the FAMOUS rift valley: observations with a deep-towed instrument package, Bull. Geol. Soc. Am., 88, 648, 1977.

Lyle, M.W. and J. Dymond, Metal accumulation rates in the Southeast Pacific - errors introduced from assumed bulk densities, Earth Planet. Sci. Lett., 30, 164, 1976.

MacDonald, K.C., Near bottom magnetic anomalies, asymmetric spreading, oblique spreading, and tectonics of the Mid-Atlantic Ridge near lat 37°N, Bull. Geol. Soc. Am., 88, 541, 1977.

MacDonald, K.C. and B.P. Luyendyk, Deep-tow studies of the structure of the Mid-Atlantic Ridge crest near lat 37°N, Bull. Geol. Soc. Am., 88, 621, 1977.

Moore, W.S. and P.G. Vogt, Hydrothermal manganese crusts from two sites near the Galapagos spreading axis, Earth Planet. Sci. Lett., 29, 349, 1976.

Needham, H.D., and J. Francheteau, Some characteristics of the rift valley in the Atlantic Ocean near 36°48'N, Earth Planet. Sci. Lett., 22, 29, 1974.

Nozaki, Y., J.K. Cochran, K.K. Turekian, and G. Keller, Radiocarbon and ^{210}Pb distribution in submersible-taken deep sea cores from Project FAMOUS, Earth Planet. Sci. Lett., 34, 167, 1977.

Piper, D.Z., H.H. Veeh, W.G. Bertrand, and R.L. Chase, An iron-rich deposit from the Northeast Pacific, Earth Planet. Sci. Lett., 26, 114, 1975.

Reid, I., and K.C. MacDonald, Microearthquake study of the Mid-Atlantic Ridge near 37°N using sonobuoys, Nature, 246, 88, 1973.

Rona, P.A., G.B. Bassinger, R.N. Harbeson, R.B. Scott, and A.J. Nalwalk, Tectonic fabric and hydrothermal activity of the Mid-Atlantic Ridge crest at 26°N, Bull. Geol. Soc. Am., 87, 661, 1976.

Rydell, H.S., and E. Bonatti, Uranium in submarine metalliferous deposits, Geochim. Cosmochim Acta, 37, 2557, 1973.

Rydell, H.S., T. Kraemer, K. Bostrom, and O. Joensuu, Post-depositional injections of uranium-rich solutions into East Pacific Rise sediments, Mar. Geol., 17, 151, 1974.

Salter, P.F., The geochemistry of metal enriched sediments from the FAMOUS Area, Mid-Atlantic Ridge (37°N lat), (in prep.), 1978.

Scott, M.R., J.K. Osmond, and J.K. Cochran, Sedimentation rates and sediment chemistry in the South Indian Basin, Antartic Oceanology II.: The Australian-New Zealand Sector - Antartic Research Series, (ed. D.E. Hayes), 19, 317, Amer. Geophys. Union., 1972.

Scott, M.R., R.B. Scott, P.A. Rona, L.W. Butler, and A.J. Nalwalk, Rapidly accumulating manganese deposit from the median valley of the Mid-Atlantic Ridge, Geophys. Res. Lett., 1, 355, 1974.

Scott, M.R., R.B. Scott, J.W. Morse, P.R. Betzer, L. W. Butler, and P.A. Rona, Metal enriched sediments from the TAG Hydrothermal Field, Nature, 276, 811, 1978.

Scott, R.B., P.A. Rona, B.A. McGregor, and M.R. Scott, The TAG Hydrothermal Field, Nature, 251, 301, 1974.

Scott, R.B., J. Malpas, P.A. Rona, and G. Udintsev, Duration of hydrothermal activity at an oceanic spreading center, Mid-Atlantic Ridge (lat 26°N), Geol., 4, 233, 1976.

Scott, R.B., D.G. Temple, and P.R. Peron, Nature of hydrothermal exchange between oceanic crust and sea water at 26°N lat, Mid-Atlantic Ridge, (in press), 1978.

Seyfried, W., and J.L. Bischoff, Hydrothermal transport of heavy metals by sea water: the role of the sea water/basalt ratio, Earth Planet. Sci. Lett., 34, 71, 1977.

Spindel, R.C., K.C. MacDonald, S.B. Davis, J.D. Phillips, and R.P. Porter, Microearthquake survey of the median valley of the Mid-Atlantic Ridge at 36°30'N, Nature 248, 577, 1974.

Spooner, E.T.C., and W.S. Fyfe, Sub-sea floor metamorphism, heat, and mass transfer, Contrib. Mineral. Petrol., 42, 287, 1973.

Temple, D.G., R.B. Scott, and P.A. Rona, Geology of a submarine hydrothermal field, Mid-Atlantic Ridge, 26°N lat, J. Geophys. Res., (submitted), 1978

Turekian, K.K., Rapid technique for determination of carbonate content of deep sea cores, Bull. Am. Assoc. Petrol. Geol., 40, 2507, 1956.

Turekian, K.K., and J. Imbrie, The distribution of trace elements in deep sea sediments of the Atlantic Ocean, Earth Planet. Sci. Lett., 1, 161, 1966.

Turekian, K.K., and K.K. Bertine, Deposition of Mo and U along the major ocean ridge systems, Nature, 299, 250, 1971.

Van der Weijden, C.H., R.D. Schuiling, and H.A. Das, Some geochemical characteristics of sediments from the North Atlantic Ocean, Mar. Geol., 9, 81, 1970.

Veeh, H.H., and K. Bostrom, Anomalous $^{234}U/^{238}U$ on the East Pacific Rise, Earth Planet. Sci. Lett., 10, 372, 1971.

Von der Borch, C.C., and R.W. Rex, Amorphous iron oxide precipitates cored during Leg 5, Deep Sea Drilling Project, Initial Reports Deep Sea Drilling Project, 5, 541, 1970.

Von der Borch, C.C., W.D. Nesteroff, and J.S. Galehouse, Iron-rich sediments cored during Leg 8 of the Deep Sea Drilling Project, Initial Reports Deep Sea Drilling Project, 8, 829, 1971.

Williams, D.L., and R.P. Von Herzen, Heat loss from the earth: new estimate, Geol., 2, 327, 1974.

Williams, D.L., R.P. Von Herzen, K.E. Green, M.A. Hobart, and T.C. Lee, A geothermal study fo the Mid-Atlantic Ridge near 37°N, Bull. Geol. Soc. Am., 88, 531, 1977.

Wolery, T.J., and N.H. Sleep, Hydrothermal circulation and geochemical flux at mid-ocean ridges, J. Geol., 84, 249, 1976.

OPHIOLITIC ROCKS AND EVIDENCE FOR HYDROTHERMAL CONVECTION OF SEA WATER WITHIN OCEANIC CRUST (Abstract)

E.T.C. Spooner

Department of Geology, University of Toronto, Toronto, Ontario M5S 1A1, Canada.

A significant amount of evidence in support of the general hypothesis of sea water circulation within oceanic crust has now been obtained from examination of ophiolitic complexes. Although it is uncertain how representative of ocean crust ophiolitic sequences actually are, information from the Troodos Massif, Cyprus, in particular, indicates that formation of cupriferous pyrite ore deposits, which are economically exploitable on land, was an integral part of the convective process. It is possible, therefore, that comparable deposits might be present in oceanic crust. Observations on Cyprus permit the geometry of circulation to be deduced. In this particular case, it appears that flow approximated to a series of axially symmetric cells which contained central plumes of hot ascending fluid the positions of which were fixed through the time with respect to enclosing rock (Spooner, 1977). Studies of hydrothermal convection in a permeable medium with an open top and a cylindrical shape using finite difference approximations suggest that this mode of circulation is possible at low to moderate values of the Rayleigh number (Parmentier and Spooner, 1978).

The clearest evidence for water/rock interaction in ophiolitic rocks is provided by the occurrence of non-isochemical, hydrothermal metamorphism of ocean-floor origin. This has been documented from ophiolitic rocks in Cyprus (Gass and Smewing, 1973; Spooner et al., 1977), Newfoundland (Williams and Malpas, 1972; Coish, 1977), E. Liguria, Italy (Spooner and Fyfe, 1973; Spooner et al, 1977) and S. Chile (Stern et al, 1976). Zeolite to amphibolite facies metamorphic assemblages, which usually pseudomorph original igneous textures, have been observed in metabasic pillow lavas, sheeted dykes, the upper parts of layered plutonic sequences, and associated lithologies. Usage of isotopes (H/D; $^{13}C/^{12}C$; $^{18}O/^{16}O$; $^{87}Sr/^{86}Sr$) as geochemical tracers has confirmed that the interacting fluid was of sea water origin (Spooner et al., 1974; Heaton and Sheppard, 1977; Spooner, 1977):-

e.g. 1 Zeolite to amphibolite facies metabasic rocks from the Troodos Massif, Cyprus are variably enriched in ^{87}Sr relative to fresh rocks (Spooner et al., 1977). Initial $^{87}Sr/^{86}Sr$ ratios as high as 0.70760 ± 0.00003, for an interstitial zeolite sample, and 0.7069, for a metabasic rock, have been recorded. These compare with low values obtained for fresh gabbroic rocks of 0.70338 ± 0.00010 to 0.70365 ± 0.00005. Upper Cretaceous sea water, with a ratio of ∿ 0.7076, was the only possible contaminant (Spooner et al., 1977).

e.g. 2 Heaton and Sheppard (1977) have shown that the isotopic composition of hydrogen in water in equilibrium with chlorite and amphibole samples from metadolerite dykes and metagabbros from Cyprus was indistinguishable from that of sea water.

Whereas hydrothermal metamorphism occurred during in-flow and through-flow of sea water it appears that, in Cyprus, sulphide ore deposits formed at the positions of discharge. In general terms these deposits consist of a lens of massive ore underlain by an approximately pipe-shaped stockwork. The massive ore consists of a porous, crudely colloform textured mass of fine-grained pyrite and chalcopyrite with accessory quantities of marcasite, sphalerite and galena. The voids in the massive ore are frequently partially filled with sooty fine-grained framboidal pyrite. The lenses occur intercalated within the pillow lava sequence and their attitudes are conformable with the local stratigraphy. (Description summarized from Hutchinson and Searle, 1971; Constantinou and Govett, 1973). A large lens, for example that at Skouriotissa, had a maximum thickness of ∿ 50m, and was approximately elliptical in plan (long axis = ∿ 500 m; short axis = ∿ 350 m) (Constantinou and Govett, 1973). It contained pre-production ore reserves of ∿6 million tonnes at 2.3% Cu (Bear, 1963). An important point is that the evidence suggests that this massive ore actually formed <u>on</u> the ocean floor -- it does not replace pre-existing rock, can show alteration near the top to a goethitic material (ochre), which has been interpreted as a submarine weathering product of ore (Constantinou and Govett, 1972), and is either overlain by unmineralized deep-water marine sediments itself, or by unmineralized pillow lavas which are overlain by pelagic sediments (Robertson, 1975).

Stockworks, which occur below massive ore and which were clearly feeder zones, consist of highly altered and mineralized material of basaltic origin. Mineralization occurs as veins, veinlets and disseminated impregnations of sulphides. The alteration silicate mineralogy of the reconstituted pillow lava consists of quartz-chlorite-illite-sphene. Uneconomic alteration pipes characterized by pyritisation penetrate 1-2 km down into the ophiolitic sequence. Stockworks are normally elliptical in cross-sectional shape. A good idea of dimensions is provided by the Limni deposit in western Cyprus with a long axis of ∿ 800 m and a short axis of ∿ 400 m (Trennery and Pocock, 1972; Adamides, 1975). It contained pre-production reserves of ∿ 4 million tonnes at 1.37% Cu (Bear, 1963) and has proved to be economically exploitable for ∿ 200 m below the original ocean floor.

About 20 deposits have been worked in Cyprus. They are normally a few million tonnes in size (e.g. ave. of eleven listed deposits = 3.6 million tonnes) but can be as large as 15 million tonnes (e.g. Mavrovouni). Copper grades are extremely variable but are typically between 0.8 and 2 wt%. Along the 80 km long northern pillow lava outcrop in Cyprus 15 deposits occur with an imperfectly regular distribution characterized by a mean half-spacing of 2.6 km ± 1.4 km (1σ)(Spooner, 1977).

Examination of fluid inclusions and the hydrogen, oxygen and strontium isotopic composition of ore deposit material from Cyprus has shown that the sulphides were precipitated from a hydrothermal fluid of sea water origin (Spooner and Bray, 1977; Heaton and Sheppard, 1977; Chapman and Spooner, 1977):-

e.g. 1 Fluid inclusion studies on ore material from Cyprus have shown that samples of the ore forming fluid trapped in small cavities in quartz intergrown with sulphides have a freezing point indistinguishable from that of sea water ($-1.9^{\circ}C \pm 0.4^{\circ}C$; 205 measurements; Spooner and Bray, 1977). Since the freezing point of a solution is a reflection of its salinity this information indicates a close bulk compositional similarity between the hydrothermal fluid, which was at temperatures between ∿ $300^{\circ}C$ and ∿ $350^{\circ}C$ (Spooner and Bray, 1977), and sea water.

e.g. 2 Mineralized material from 4 deposits in Cyprus is significantly enriched in ^{87}Sr (Chapman and Spooner, 1977). Initial $^{87}Sr/^{86}Sr$ ratios as high as upper Cretaceous sea water have been recorded (e.g. 0.7075 ± 0.0002).

e.g. 3 The hydrogen isotope composition of the hydrothermal fluid has been shown to be close to that of sea water (Heaton and Sheppard, 1977).

Hence, this evidence completes successful identification of the complete cycle of sea water circulation. The following points can now be made about the nature of the convective process which occurred in the ophiolitic rocks of the Troodos Massif, Cyprus (summarized from Spooner, 1977 and, particularly, from Parmentier and Spooner, 1978):-

(a) The flow geometry appears to have consisted of axially symmetric cells containing a central plume of hot ascending fluid which was positionally fixed through time with respect to enclosing rock. This is, perhaps, a rather unexpected result.

(b) Strong hydrothermal metamorphism does not significantly affect the middle and lower parts of the gabbroic sequence on Cyprus. This suggests that strong sea water penetration and circulation occurred in a permeable layer ∿ 2-3 km thick.

(c) The mean half-spacing between ore deposits along the northern pillow lava outcrop in Cyprus is 2.6 ± 1.4 km (1σ). This suggests that the original rising plumes were separated by a distance comparable to the thickness of the permable layer -- an observation consistent with a convective model.

(d) In Cyprus, ore deposits are frequently cross-cut by unmineralized dykes and/or overlain by unmineralized pillow lavas. This suggests that formation of the ore deposits occurred within the volcanically active zone at a spreading ridge. For a half-width of ∿ 6 km and a spreading rate of ∿ 5 cm/yr the ore deposits may have formed in a time on the order of 10^5 years. The mean internal upward flow rate may have been ∿ 10^{-6} g $cm^{-2}s^{-1}$ (Spooner, 1977) -- a flux which has been shown to be reasonable by later finite difference calculations (Parmentier and Spooner, 1978).

(e) Homogenisation temperatures of fluid inclusions in three mineralized stockworks in Cyprus indicate that the temperature of the plume of rising hydrothermal fluid was $300^{\circ}C$ - $350^{\circ}C$ (Spooner and Bray, 1977). This range is in good agreement with a value of ∿ $300^{\circ}C$ estimated for the temperature of last equilibrium with quartz of end-member hydrothermal fluid from the Galapagos Ridge axis (Edmond, 1978), and is consistent with a value of ∿ $285^{\circ}C$ (max.) estimated by Crane and Normark(1977) for hydrothermal activity on the East Pacific Rise at 21°N.

(f) Parmentier and Spooner (1978) have modelled hydrothermal circulation relevant to the origin of the Cyprus deposits by finite difference approximations, and have found that diffuse plumes of rising fluid contain hot core zones which are comparable in size to stockworks in Cyprus. Such hot core zones remain approximately isothermal to within ∿ 300 m of the ocean floor, where they cool abruptly to ocean bottom water temperatures. In the model, conductive cooling alone was considered. In the real situation cooling by mixing with cool sea water in a near surface zone is probably more important (e.g. Edmond, 1978).

A larger scale implication of the possible presence of sulphide deposits in oceanic crust, which could be comparable in size and density of occurrence to those in Cyprus, is suggested by the isotopic composition of contained sulphur. It appears to indicate that the ore sulphide is largely of reduced sea water sulphate origin (Bachinski, 1977; Spooner, 1977; Heaton and Sheppard, 1977).

Hence, large quantities of sulphide, in association with hydroxyl and chloride, could be added to oceanic crust. Release of water from the descending slab at subduction zones may cause wet melting of mentle material, and could produce volatile rich siliceous magmas enriched in chlorine and sulphur. Such a simple model could explain the amounts of magmatically released water, chlorine and sulphur required for the formation of porphyry Cu ± Mo ± Au deposits. Water is needed as the transport medium, chloride for metal complexing and sulphur for fixing the metals as solid phases.

Finally, it should be re-emphasized that it is uncertain how representative of oceanic crust the Troodos Complex, in particular, and ophiolitic sequences, in general, might be. The general nature of the findings discussed above is, therefore, likewise subject to uncertainty.

Acknowledgements. I am very grateful to Dr. Manik Talwani for inviting me to attend. The Second Maurice Ewing Memorial Symposium, and would also like to thank Drs. R.N. Anderson and S.R. Hart for helpfully reviewing this abstract.

References

Adamides, N.Y., Geological history of the Limni concession Cyprus, in the light of the plate tectonics hypothesis, Trans. Inst. Min. Metall. Sect.B: Appl. Earth Sci., 84, 17-23, 1975.

Bachinski, D.J., Sulfur isotopic composition of ophiolitic cupriferous iron sulfide deposits, Notre Dame Bay, Newfoundland, Econ. Geol., 72, 243-257, 1977.

Bear, L.M., The mineral resources and mining industry of Cyprus, Bull. Geol. Surv. Dep. Cyprus 1, 208 pp, 1963.

Chapman, H.J. and Spooner, E.T.C., ^{87}Sr enrichment of ophiolitic sulphide deposits in Cyprus confirms ore formation by circulating sea water, Earth Planet. Sci. Lett., 35, 71-78, 1977

Coish, R.A., Ocean floor metamorphism in the Betts Cove ophiolite, Newfoundland, Contrib. Mineral. Petrol., 60, 255-270, 1977.

Constantinou, G. and Govett, G.J.S., Genesis of sulphide deposits ochre and umber of Cyprus, Trans. Inst. Min. Metall., Sect. B: Appl. Earth Sci., 81, 32-46, 1972.

Constantinou, G., and Govett, G.J.S., Geology, geochemistry and genesis of Cyprus sulfide deposits, Econ. Geol. 68, 843-858, 1973.

Crane, K., and Normark, W.R., Hydrothermal activity and crustal structure of the East Pacific Rise at 21 $^{\circ}$ N, J. Geophys. Res., 82, 5336-5348, 1977.

Edmond, J.M., Chemistry of the hot springs on the Galapagos ridge axis (abstract), The Second Maurice Ewing Memorial Symposium (abstracts), 13.1978

Gass, I.G., and Smewing, J.D., Intrusion, extrusion and metamorphism at constructive margins: evidence from the Troodos Massif, Cyprus, Nature, 242, 26-29, 1973.

Heaton, T.H.E. and Sheppard, S.M.F., Hydrogen oxygen isotope evidence for sea-water-hydrothermal alteration and ore deposition, Troodos complex, Cyprus, Geol. Soc. Lond. Spec. Publ., 7,42-57, 1977.

Hutchinson, R.W., and Searle, D.L., Stratabound pyrite deposits in Cyprus and relations to other sulphide ores., Soc. Mining Geol. Japan, Spec. Issue, 3, 198-205, 1971.

Parmentier, E.M., and Spooner, E.T.C., A theoretical study of hydrothermal convection and the origin of the ophiolitic sulphide ore deposits of Cyprus, Earth Planet. Sci. Lett., 40, 33-44 1978

Robertson, A.H.F., Cyprus umbers: basalt -- sediment relationships on a Mesozoic ocean ridge, J. Geol. Soc. London, 131, 511-531, 1975 .

Spooner, E.T.C., Hydrodynamic model for the origin of the ophiolitic cupriferous pyrite ore deposits of Cyprus, Geol. Soc. Lond. Spec. Publ., 7, 58-71, 1977.

Spooner, E.T.C., Beckinsale, R.D., Fyfe, W.S., and Smewing, J.D., ^{18}O enriched ophiolitic metabasic rocks from E. Liguria (Italy), Pindos (Greece) and Troodos (Cyprus), Contrib. Mineral. Petrol 47, 41-62, 1974 .

Spooner, E.T.C., Beckinsale, R.D., England, P.C., and Senior, A., Hydration, ^{18}O enrichment and oxidation during ocean floor hydrothermal metamorphism of ophiolitic metabasic rocks from E. Liguria, Italy, Geochim. Cosmochim. Acta, 41, 857-871, 1977.

Spooner, E.T.C., and Bray, C.J., Hydrothermal fluids of sea water salinity in ophiolitic sulphide ore deposits in Cyprus, Nature, 266, 808-812, 1977.

Spooner, E.TC., CHapman, J.J., and Smewing, J.D., Stontium isotopic contamination and oxidation during ocean floor hydrothermal metamorphism of the ophiolitic rocks of the Troodos Massif, Cyprus Geochim. Cosmochim, Acta, 41, 873-890, 1977.

Spooner, E.T.C., and Fyfe, W.S., Sub-sea-floor metamorphism, heat and mass transfer, Contrib. Mineral. Petrol. 42, 287-304, 1973.

Stern, C., De Wit, M.J., and Lawrence, J.R., Igneous and metamorphic processes associated with the formation of Chilean ophiolites and their implication for ocean floor metamorphism, seismic layering and magnetism, J. Geophys. Res, 81, 4370-4380, 1976.

Trennery, T.O., and Pocock, B.G. Mining and milling operations at Limni mine, Cyprus, Trans. Inst.Min. Metall., Sect. A: Min. Ind. 81, 1-12, 1972.

Williams, H., and Malpas, J., Sheeted dikes and brecciated dike rocks within transported igneous complexes, Bay of Islands, Western Newfoundland, Can. J. Earth Sci., 9, 1216-1229, 1972